W9-AMC-872

ECOLOGY
Theories and Applications

Third Edition

Peter Stiling

University of South Florida

Prentice Hall
Upper Saddle River, New Jersey 07458

Library of Congress Cataloging-in-Publication Data

Stiling, Peter D.
Ecology: theories and applications / Peter D. Stiling—3rd ed.
p. cm.
Includes bibliographical references and index.
ISBN 0-13-915653-4
1. Ecology. I. Title.
QH541.S674 1999
577--dc21 98-4743
CIP

Senior Editor: Teresa Ryu
Executive Editor: Sheri L. Snavely
Editor-in-Chief: Paul F. Corey
Editorial Director: Tim Bozik
Assistant Vice President of Production and Manufacturing: David W. Riccardi
Executive Managing Editor: Kathleen Schiaparelli
Project Management: J. Carey Publishing Service
Senior Marketing Manager: Jennifer Welchans
Manufacturing Manager: Trudy Pisciotti
Manufacturing Buyer: Benjamin Smith
Director of Creative Services: Paula Maylahn
Assistant Creative Director: Amy Rosen
Art Director: Heather Scott
Art Manager: Gus Vibal
Art Editor: Karen Branson
Text Designer: Geri Davis, Davis Group, Inc.
Cover Designer: Bruce Kenselaar
Cover Illustration: © RimPacific
Photo Researcher: Beaura Ringrose
Editorial Assistant: Lisa Tarabokjia
Art Studio: Kandis Elliot
Text Composition: Preparé/Emilcomp

Printed in the United States of America

10 9 8 7 6 5 4 3 2 1

ISBN 0-13-915653-4

All noncredited photographs were supplied
by the author with permission.

Prentice-Hall International (UK) Limited, *London*
Prentice-Hall of Australia Pty. Limited, *Sydney*
Prentice-Hall Canada Inc., *Toronto*
Prentice-Hall Hispanoamericana, S.A., *Mexico*
Prentice-Hall of India Private Limited, *New Delhi*
Prentice-Hall of Japan, Inc., *Tokyo*
Simon & Schuster Asia Pte. Ltd., *Singapore*

About the Author

The photograph shows the author with daughters Zoe and Leah at the coast near his field sites.

Peter Stiling is an associate professor of biology at the University of South Florida, Tampa. He has taught classes in ecology, environmental science, and community ecology and in 1995 received a teaching award in recognition of classroom excellence in these areas. Dr. Stiling obtained his Ph.D. from University College, Cardiff, Wales, and completed postdoctoral research at Florida State University. It was while teaching ecology at the University of West Indies, Trinidad, that the idea for this book was conceived. Dr. Stiling's research interests include plant-insect relationships, parasite-host relationships, biological control, restoration ecology, and effects of elevated CO_2 on native communities. He has published many scientific papers in journals such as *Ecology*, *Oikos*, and *Oecologia* and is currently an editor for *Oecologia*. His field research has been supported by the National Science Foundation, U.S. Fish and Wildlife Service, and the Nature Conservancy.

Brief Contents

Contents

Preface

To most people today, ecology is associated with the broad problems of the human environment, especially pollution. To others, ecology is synonymous with conservation and saving the whales or the forests. Most researchers, however, know that ecology does not simply deal with pollution or conservation, it is related to the environment much as physics is to engineering. Ecology provides the scientific framework upon which conservation programs or pollution-monitoring schemes can be set up. Other key beliefs about ecology are that a "balance of nature exists" and that species have a reason for their existence (Westerby 1997). However, researchers are aware that change and fluctuation are pervasive in nature and that there are no moral imperatives to species. This can create tension among students seeking moral enlightenment from an ecology class as well factual knowledge because many professors' life philosophy is to teach the integrity of evidence and rigorous logic, not a life philosophy of ecology as a quasi religion with moral positions. Perhaps we could use different words for the academic and life philosophy aspects of ecology? *Biophilia* has been suggested to describe the love of organisms (Kellert and Wilson 1993).

The scientific framework of ecology is often erected not from studies of rare, exotic animals or studies of oil spills but from studies of invertebrates such as insects and small, relatively unappealing plants because these can best be manipulated experimentally to test ecological theories. For example, it is rather difficult to experiment with blue whales or Florida panthers, of which there are relatively few. This book is thus peppered with studies of the teeming hordes of common life. The biosphere is composed of living tissue and covers the surface of the globe in a layer proportionately thinner than the skin of an apple. If all the material were evenly distributed, it would be less than 1 centimeter (cm) thick and would weigh about 3.6 kilograms per meter squared (kg m^{-2}) (Anderson 1981), but the Earth's biomass is not evenly distributed. It shows great variation, from about 45 kg m^{-2} in tropical forests to 0.003 kg m^{-2} in the oceans. Ecology is concerned with explaining this variation; it asks why plants and animals are found in certain areas and what controls their numbers.

The biggest change in the third edition has been the integration of applied ecology into the text as a whole. The impetus for this was twofold. First, I learned that the applied ecology section, coming as it did at the end of the second edition, would not be reached by the professors or students in some courses. Second, I wanted to make clear that every facet of ecology has real-world applications. The answer was to include applied material in every chapter in the form of boxes. Thus, global warming is discussed in Chapter 7 on abiotic factors, acid rain in Chapter 22 on nutrient cycles, and optimal yield in Chapter 10 on predation.

Another major change has been to expand the coverage on behavioral and ecosystems ecology. For behavioral ecology, I have written another chapter on life history theory (Chap. 5), which includes sections on mating systems, sex ratios, habitat selection, dispersal, and age structure. For ecosystems ecology I have beefed up coverage on energy flow (Chap. 21) and carved out a separate chapter for nutrient cycles (Chap. 22).

The core of the book still remains the large sections on population ecology and community ecology (Sections 4 and 5). In population ecology I have doubled the coverage on abiotic factors to include new discussions of such factors as nutrients, soil, light, wind, pH, and salinity. This is based partly on my own experience of how important such environmental factors are to the distribution and abundance of species and partly in response to reviewers' suggestions.

The chapters on life tables, mutualism, competition, predation, herbivory, and parasitism have all been throughly modernized. The last chapter in this section, Chapter 13, comparing the strength of mortalities, has been thoroughly stripped down, expanded, and updated to include all the latest work in this area. This, to me, remains one of the most exciting areas of ecology.

Finally, the community ecology section (Section 5) has seen huge expansions in coverage. This is partly because there is much current interest in species richness and biodiversity and partly because there is currently no satisfactory textbook in this area. I have given species richness, diversity, stability, succession, and island biogeography their own chapters, each thoroughly up to date. Thus, the stability chapter now discusses nonequilibrium theories, and the diversity chapter touches on cluster analysis and ordination.

Besides improving the book scientifically, I have also made some pedagogical changes. I have provided expanded section headings that try to provide the take-home message of the next few paragraphs. And, most importantly, I have taken greater pains to try and explain the many equations in the book—why we need them and how they work. I know from teaching my own students that this is vital.

An additional new feature is vignettes of active ecologists. This is designed to humanize ecology, to show students how people from all walks of life contribute to ecological theory. It also conveys some of the trials and tribulations of ecological research and gives a good impression of how and why people started their ecological careers and, indeed, continue them. I am grateful to all the ecologists who participated in this idea.

Finally, I am grateful to the following users for their critiques of the second edition: Karen Olmstead, University of South Dakota; Craig Benkman, New Mexico State; Bruce Grant, Widener University; Alan Stiven, University of North Carolina; Kevin Dixon, Arizona State University West; Donald Batzer, University of Georgia; Peter Meserve, Northern Illinois University; Kate Lajtha, University of Oregon; Sally Holbrook, University of California–Berkeley; Bob Holt, University of Kansas, Natural History Museum; Cliff Amundsen, University of Tennessee; John Abern, University of New Hampshire; and, especially, Andrea Lloyd of Middlebury College. The third edition was further refined after reviews by William Teska, Furman University; Gerardo Camilo, St. Louis University; Jonathan Newman, Southern Illinois University; Kathy Williams, San Diego State University; Tim Mousseau, University of South Carolina; Mitchell Cruzan and Cliff Amundsen, University of Tennessee; Chris Migliaccio, Miami-Dade Community College; Frank Romano, Jacksonville State University; and Gleen Turnipseed, Arkansas Tech University. To all these individuals I am grateful. Please let me know if additional errors of fact or emission remain. I am particularly grateful for the efforts of my new editor, Teresa Ryu, for gathering these reviews and critiques. Once again, Jacqui Stiling deserves special credit for helping me in all phases of this book.

I believe this third edition is broad in its coverage yet hopefully, retains the brevity of style that made the first two editions successful.

To Don and Dan

SECTION

1

Introduction

Life is not distributed evenly on Earth, but rather in patches. Ecology seeks to explain this phenomenon. Here, in the Okavango delta, Botswana, papyrus reedbeds and other vegetation are distributed throughout the region. Islands may begin as termite mounds. Abiotic variables, such as water levels, may affect the distribution of plants as may herbivores. *(Gregory G. Dimijian, Photo Researchers, Inc. 7L5327.)*

There is a widespread belief that people of preindustrial civilizations did far less damage to their environment than do their modern industrial counterparts. This belief supposes that hunter-gatherers lived in harmony with nature, practicing a conservation ethic and somehow avoiding short-sighted, destructive exploitation. They did not. For example, on every oceanic island for which we have adequate knowledge, the first arrival of humans was quickly followed by extermination of all or most large animals (Diamond 1986a). Easter Island, home of the famous monolithic statues, was once covered with palms, trees, and shrubs. Polynesians reached the island around 400 A.D. By 1500, seven thousand people lived there; they had deforested the island so completely that its tree species are now extinct. The deforestation had serious implications: no logs were available to be made into canoes, so offshore fishing was curtailed, and the huge statues could not be erected without log levers. Once the population exceeded the carrying capacity of the island, warfare was rampant, as were chronic cannibalism and slavery. Spear points were manufactured in enormous quantities, and people reverted to living in caves for defense.

The scale on which destruction occurred was not limited to small islands. In the deserts of the U.S. Southwest stand huge empty communal houses or pueblos, relics of the Anasazi, one of the most advanced pre-Columbian civilizations in North America. When construction began, the cliffs were covered with pinyon-juniper woodland. Collection for firewood and construction denuded the area completely to a radius of 40 to 70 kilometers (km) by the time the site was abandoned.

Easter Island, off the coast of Chile, is one the most isolated islands on Earth. At one time, this island was covered with palms, trees, and shrubs. Deforested by Polynesian colonists by 1500 A.D., the lack of trees on this island is testament to the destructive capabilities of humans. (*Anna E. Zuckerman, Tom Stack & Associates 1690-4-4.*)

Now the effects of humans have begun to change the entire globe (Vitousek et al. 1997b; Matson et al. 1997; Noble and Dirzo 1997). Between one-third and one-half of the land surface has been transformed by human action. Acid rain is carried from one country to another. Carbon dioxide pumped out by the industrial centers of developed nations has increased the atmospheric CO_2 levels worldwide, from the poles to the equator. More nitrogen is fixed by humanity than by all natural sources combined. Pesticides, powerful human-made poisons, have been detected in human breast milk and in the tissues of penguins—both of whom were completely unintended targets. About one-quarter of the bird species on earth have been driven to extinction. Now, more than ever, there is a strong impetus to understand how natural systems work, how humans change those systems, and how in the future we can reverse these changes.

Ecologists are among the best-equipped scientists to study natural systems. Before 1960 ecologists were few in number, and their activities were dominated by taxonomy, natural history, and speculation about observed patterns. Their equipment included sweep nets, quadrats, and specimen jars. Since that time, ecologists have become active in investigating environmental change on regional and global scales. They have embraced reductionist analyses and experimentation and have adapted concepts and methods derived from agriculture, physiology, biochemistry, genetics, physics, chemistry, and mathematics (Grime 1993). Their equipment now includes portable computers, satellite-generated images, and sophisticated chemical auto-analyzers. The challenge for ecologists is to come together and agree on solutions to the world's ills. The alternative, as Stanford University ecologist Hal Mooney has been quoted as saying (Baskin 1994), is "Frank Sinatra" science where ecologists always try to do things "my way."

CHAPTER

1

Why and How to Study Ecology

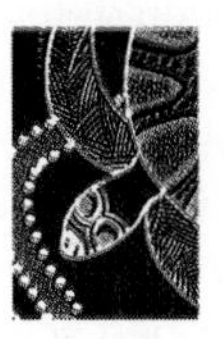

Ecology is the study of interactions between organisms and between organisms and their environments. The development of ecology as a field has paralleled, and some would argue has been spurred by, the increasing awareness of the environmental impacts of human activities. These impacts range from worldwide problems like global warming, anthropogenic nitrogen deposition, and biodiversity depletion to local or regional problems caused by specific human actions. One well-studied example of the unforeseen regional ecological problems associated with developmental impacts is provided by the Aswan Dam in Egypt. For example, in 1970, after eleven years and an expenditure of $1 billion, construction of the Aswan High Dam, depicted in Photo 1.1, was completed. Located in southern Egypt on the world's longest river, the Nile (Fig. 1.1), it is the largest dam of its kind in the world. It contains more than four times the capacity of Lake Mead, the reservoir behind Hoover Dam, which is the largest dam in the United States. The Aswan High Dam was projected to provide several years of irrigation reserve, to add 526,000 hectares (ha) (1.3 million acres) to the arable lands of Egypt, to produce 10 billion kilowatts of electric power annually, and to protect the

■ **PHOTO 1.1** Aswan High Dam, Egypt. (*Bruce Watkins, Animals Animals/Earth Scenes IND 051WAB00101.*)

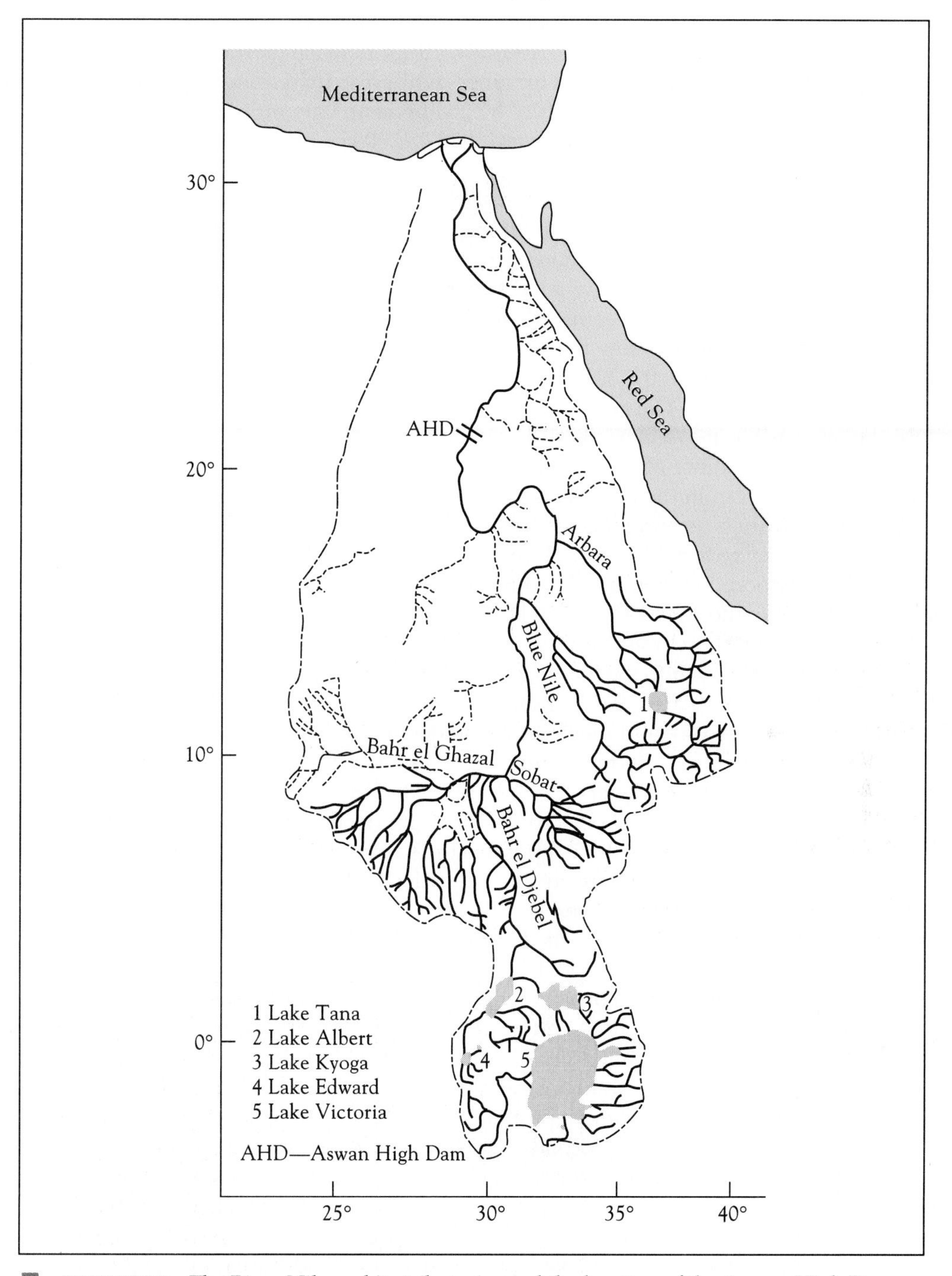

FIGURE 1.1 The River Nile and its tributaries and the location of the Aswan High Dam.

country from catastrophic floods. Today, the dam does produce more than 50 percent of Egypt's electrical power. The reservoir of water saved the rice and cotton crops during the droughts of 1972 and 1973. It has facilitated the cultivation of two or three crops annually rather than one (Azim Abul-Atta 1978), thus increasing productivity by 20 percent for some crops and by 50 percent for others while also increasing governmental and national annual income from agriculture by 200 percent. One million hectares of additional farmland can be irrigated year round, and 380,000 ha of desert are being irrigated for the first time (Moore 1985).

However, it has been argued that in many ways the dam stands as a monument to ecological ignorance. First, loss of water to seepage through bedrock meant that the dam was still not full by 1988, though it had been predicted to fill by 1970. This problem may have been compounded by twenty years of below-average rainfall in the area (Wright 1988). Until unusually heavy rains fell in August 1988, the volume of water in the impoundment had fallen so low that serious conservation measures were about to be implemented, and the electric turbines were about to stop, which would have caused widespread power cuts. The volume of the reservoir had dropped to around 3 billion cubic meters (m^3), out of a capacity of 110 billion m^3.

Second, the incidence of schistosomiasis (a debilitating parasitic disease) caused by the tropical flatworm in the area was argued to have increased from 47 to 80 percent because the parasite's secondary hosts, snails, reproduce year round in the reservoir and thus are no longer reduced by drought (van der Schalie 1972). This problem may since have abated (Moore 1985) and, according to some sources, may never have been as severe as was once thought (Walton 1981). The variety of schistosomiasis now prevalent, however, is a much more severe one.

Third, reduced flow into the Mediterranean reduced phytoplankton blooms and fish harvest in the discharge area. Sardine catches alone dropped from 15,000 tons annually to 500 tons (George 1972), and yields from the new fishing areas behind the dam have been low. Fourth, reduced silt deposition along the floodplain (Petts 1985) has increased the need for commercial fertilizers to the tune of $100 million annually. The new fertilizer plants use much of the hydroelectric power from the dam. Fifth, most farmers over-water their land, and as the water evaporates the salt is left behind. In 1986 almost half the irrigated area in Egypt was affected by salt (Kishk 1986).

It is interesting to note, however, that many of the adverse opinions about the dam are based on "Western" standards; the Egyptians themselves have chosen to look more positively upon the dam (Fahim 1981). Nevertheless, it is clear that a deeper ecological knowledge would have helped predict what would happen in nature.

Theoretical ecology has, more and more, been winning its spurs in the real world (Moffat 1994). Roy Anderson and Robert May of Oxford University in England modeled the spread of disease in nonhuman hosts. Following the explosion of the AIDS epidemic these two scientists tried to model its spread. They were among the first to show that a relatively few promiscuous individuals could be disapproportionately responsible for the rapid spread of AIDs. Population biologist Russell Lande of the University of Oregon in Eugene modeled how big the size of the home ranges of the spotted owl ought to be in the old-growth forests of the Pacific Northwest to prevent extinction. The models showed that the proposed set-asides of the Forest Service and Bureau of Land Management were too small. After these agencies were forced by the federal courts to consider these findings, new, larger patches of forest were set aside.

Ultimately, ecologists will be instrumental in determining every phase of the services that nature provides to the world. A recent paper in the journal *Nature* (Constanza et al. 1997) made the first such attempt (see Applied Ecology: The Value of the World's Natural Services).

1.1 Ecologists investigate a wide variety of questions, from detailed studies of physiology to continental or global-scale studies of ecological communities.

The type of questions that scientists ask are those that define specific scientific disciplines. Ecologists investigate a wide variety of phenomena occurring on a range of spatial and temporal scales. What types of ecological questions should we be asking? Slobodkin (1986a) has suggested that the "big questions" in ecology and evolution (what is life, how do higher taxa evolve, what determines the number of species in one locality?) may be too large to be amenable to theoretical formulation. He maintains that the most useful approach is to confine ourselves to more specific questions, such as, "How does meat dissolve in the stomach of a kite?" He suggests that in such minimal systems it may be possible to see meaning and derivations more clearly and to examine the criteria for theoretical quality. Colwell (1984), Slobodkin (1986b), Bartholomew (1986), and others have all urged a return to focusing on organisms and real problems rather than on the external criteria of philosophers and mathematicians. Those "pesky biological details" matter a great deal; they often violate the assumptions of mathematical theory, and theorists who ignore them risk wasting their efforts.

There have been several recent surveys of the memberships of ecological societies to try and determine what ecologists think are important issues (Table 1.1). It is hard to be precise in comparing these surveys because each survey contained a somewhat different list of concepts that members were asked to rank. Nevertheless, it is clear that most ecologists seem to believe community-oriented concepts like succession and ecosystems are more important than population-oriented concepts like competition and plant-animal interactions. They also believe both theoretical ecology, life history theory, conservation biology, and ecosystem fragility (applied ecology) are important. While this may be what people think, what do they actually do? Stiling (1994) answered this question by surveying published papers in three mainstream ecological journals, *Ecology*, *Oikos*, and *Oecologia* (Table 1.2). There is some disparity between what ecologists think and what they do because population-oriented concepts and autecology (the ecology of single species) are actually studied more than community ecology and synecology (the ecology of communities) and applied ecology. Perhaps this is because the former are easier to study despite the interest of the latter. Or perhaps the contributors to the journals *Ecology*, *Oikos*, and *Oecologia* do not represent mainstream views. Stiling (unpublished) surveyed ecological articles from twenty-six other journals over roughly the same time period (Table 1.2). The concordance of rank of concept between the two sets of journals (how they co-varied together) was suprisingly high, suggesting that most ecologists do study population-oriented phenomena, regardless of taxonomic discipline or area of interest. However, bear in mind that most ecologists publishing in these journals were considered scholars or researchers, not "practical" or "applied ecologists." The authors of the twenty-six ecological journals surveyed were primarily in education (78.4%) or research (13.6%), and only 8 percent were practicing conservation biologists, agriculturalists, foresters, or fisheries people.

The Value of the World's Natural Services

APPLIED ECOLOGY

Ecologists and economists have recently joined forces to estimate the value of the services that the world's ecosystems and natural capital provide (Constanza et al. 1997). There are many direct goods (such as food) and indirect services (such as waste assimilation) that ecological systems provide (Table 1) on a renewable basis. This excludes nonrenewable fuels and minerals. Many ecosystems provide more than one service. For example, swamps are important in disturbance regulation, water supply, waste treatment, and habitat refugia (Photo 1). In turn, most services are provided by more than one ecosystem, for example, many systems are involved in nutrient recycling.

Constanza and colleagues estimated that the world's ecosystems provide at least U.S.$33 trillion worth of services annually (Table 2). This staggering figure is more than the total global gross national product, which is around U.S.$19 trillion per year. The majority of the services are currently outside the market system and include gas regulation, waste treatment, and nutrient cycling. Nutrient cycling, at U.S.$17 trillion, was by far the most expensive service performed, but even if we eliminate this the total annual value would still be a whopping U.S.$16 trillion.

About 63 percent of the estimated value is contributed by marine systems, with most of this coming from coastal systems (U.S.$10.6 trillion per year). About 37 percent of the estimated value came from terrestrial systems, mainly from forests (U.S.$4.7 trillion) and wetlands (U.S.$4.9 trillion).

If ecosystem services were actually paid for, as they should be, the global price system would be very different from what it is today. The price of commodities would skyrocket. However, because ecosystem services are largely outside the market they are usually ignored or grossly undervalued.

TABLE 1 Partial list of the world's ecosystem services.

SERVICE	EXAMPLE
Indirect	
Atmospheric gas regulation	CO_2, O_3, SO_x levels
Climate regulation	CO_2, NO_x, methane, CFC levels
Disturbance regulation	Storm protection, flood control, drought
Waste treatment	Pollution control
Soil erosion control	Loss of topsoil, siltation of lakes
Nutrient recycling	N, P, C, and other cycles
Direct	
Water supply	Irrigation, water for industry
Pollination	Plant pollination
Biological control	Population regulation
Refugia	Habitat for species
Food production	Crop, livestock
Raw materials	Fuels, timber
Genetic resources	Medicines, genes for plant resistance
Recreation	Eco-tourism
Cultural	Aesthetic value

■ PHOTO 1 What price is this swamp at Skuppernog Creek, Kettle Morraine State Forest, Wisconsin? Useless land or valuable asset? A recent evaluation by Robert Constanza and his colleagues (1997) valued swamps such as these at $19,580 per acre because of their value in regulating water supply and in waste treatment and habitat refugia. (*Terry Donnelly, Tom Stack WISP 40.67-X2.*)

TABLE 2 Valuation of the world's ecosystem services and natural capital (U.S. trillion $ per year). (After Constanza et al. 1997)

BIOME	TOTAL GLOBAL VALUE	TOTAL VALUE (PER HA)	MAIN SERVICES PROVIDED
Open ocean	8,381	252	Nutrient cycling
Estuaries	4,100	22,832	Nutrient cycling
Seagrass/algal beds	3,801	19,004	Nutrient cycling
Coral reefs	375	6,075	Recreational/disturbance regulation
Coastal shelf	4,283	1,610	Nutrient cycling
Tropical forest	3,813	2,007	Nutrient cycling/raw materials
Temperate forest	894	302	Climate regulation/waste treatment
Grasslands	906	232	Waste treatment/food production
Tidal marsh	1,648	9,990	Waste treatment/disturbance regulation
Swamps	3,231	19,580	Water supply/disturbance regulation
Lakes, rivers	1,700	8,498	Water regulation
Desert	0	0	
Tundra	0	0	
Ice/rock	0	0	
Cropland	128	92	Food production
Urban	0	0	
Total	33,268		

TABLE 1.1 Recent surveys of members of ecological societies: Cherrett (1989), British Ecological Society; Travis (1989), Ecological Society of America; Sugden (1994), readers of the magazine *Trends in Ecology and Evolution.*

RANK	CHERRETT 1989 (N = 37)	TRAVIS 1989 (N = 40)	SUGDEN 1994 (N = 33)
Top			
1	The ecosystem	Community ecology	Conservation biology
2	Succession	Ecosystems studies	Animal ecology
3	Energy flow	Animal population ecology	Life history theory
4	Conservation of resources	Plant population ecology	Behavioral ecology
5	Competition	Plant-animal interactions	Microevolution
6	Niche	Theoretical ecology	Biogeography
7	Materials cycling	Plant physiological ecology	Macroevolution
8	The community	Phytosociology	?
9	Life history strategies	Conservation biology	?
10	Ecosystem fragility	Ecosystem theory	?
Bottom			
1	The diversity/stability hypothesis	Landscape architecture	Microbial ecology
2	Socioecology	Plant systematics	Oceanography
3	Optimal foraging	Regional planning	?

Note: Respondents checked off what they thought were important topics from lists of various concepts (*n* values). Lists contained somewhat different items in each survey.

There is no doubt that for just about any ecological question more data are needed to definitively answer it. Simon (1986) and others have noted this lack of information, and they use it to say, for instance, that the data are insufficient to show that deforestation is causing a great reduction in the numbers of plant and animal species on earth. To take another example discussed in a paper by Pilgrim and Western (1986), Parker and Douglas-Hamilton have independently used the same data to discuss elephant conservation and the ivory trade. Parker concludes that a substantial portion of harvested tusks comes from natural mortality and that hunting for profit is a serious threat in only a few areas. He further claims that increases in hunting deaths are due to the elephants' competition with Africa's rapidly expanding human population. Douglas-Hamilton disagrees with these contentions on every point. He claims that Parker considerably exaggerates the number of deaths due to natural mortality, that elephants are overhunted in all but the most inaccessible areas, and that the high price of ivory is to blame. When such disparate positions can be reached from identical information, more data are needed. The issue is especially relevant given a 1997 international ruling to permit Botswana, Namibia, and Zimbabwe to sell ivory to Japan (Bagla 1997). Critics worry that poachers will take elephants in other nations, and in Asia too, because of the difficulty of identifyng an ivory source.

TABLE 1.2 Top ten and bottom three concepts studied by ecologists in 3,108 papers published between 1987 and 1991 in the journals *Ecology, Oikos,* and *Oecologia.* (After Stiling 1994)

CONCEPT RANK	FROM ECOLOGY, OIKOS, OECOLOGIA	FROM 26 "SUBJECT-ORIENTED" JOURNALS
Top		
1	Ecological adaptation, physiological ecology	Life history strategies
2	Life history strategies	Ecological adaptation
3	Plant-herbivore	Habitat selection, spatial variation
4	Competition and coexistence	Optimal foraging
5	Habitat selection, spatial variation	Predator-prey
6	Predator-prey	Mating behavior
7	Nutrient cycling	Competition and coexistence
8	Population regulation	Dispersal, migration
9	Optimal foraging	Plant-herbivore
10	Stability, disturbance	Stability, disturbance
Bottom		
1	Restoration	Restoration
2	Pest control	Pest control
3	Pollution	Pollution

Note: The data for the right-hand column came from 2,289 ecological papers from one year over the same period in the journals *American Naturalist, American Midland Naturalist, Trends in Ecology and Evolution, Journal of Animal Ecology, Journal of Ecology, Biological Conservation, Conservation Biology, Journal of Wildlife Management, Wildlife Society Bulletin, Journal of Applied Ecology, Journal of Biogeography, Biotropica, Marine Biology, Ecological Entomology, Journal of Fish Biology, Annals of the American Entomological Society, Auk, Condor, Herpetologica, Copeia, Journal of Mammalogy, American Journal of Botany, Animal Behavior, Behavioral Ecology and Sociobiology, Evolutionary Ecology* and *Evolution.* The total number of different concepts scored was 29 in both sets of journals.

The preservation of biological diversity has been justified based on assessments of the ecological and economic value of diversity as well as on moral and ethical grounds.

During the latter half of the 1980s, the reduction of the Earth's biological diversity emerged as a critical issue and was perceived as a matter of public policy (U.S. Congress 1987). A major concern was that loss of plant and animal resources would impair future development of important products and processes in agriculture, medicine, and industry. For example, *Zea diploperennis*, an ancient wild relative of corn, could be worth billions of dollars to corn growers around the world because of its resistance to seven major diseases plaguing domesticated corn. Two species of wild green tomatoes discovered in an isolated area of the Peruvian highlands in the early 1960s have contributed genes for a marked increase in fruit pigmentation and soluble-solids content, currently worth $5 million per year to the tomato-processing industry (U.S. Congress 1987). Loss of tropical forests could mean the loss of billions of dollars in potential plant-derived pharmaceutical products. About 25 percent of the prescription drugs in the United States are derived

■ **PHOTO 1.2** The nine-banded armadillo. This is the only animal other than humans to contract leprosy and is, therefore, valuable to medical researchers. *(Photo © copyright by Florida Game and Fresh Water Fish Commission, A.V. Department, negative no. 4365.)*

from plants, and as long ago as 1980 their total market value was $8 billion per year. On a smaller scale, individual species often thought worthless can actually be very valuable for research purposes. Armadillos, for example, are the only known species, other than humans, that can be used in research on leprosy (Photo 1.2). Desert pupfishes, found in the U.S. Southwest, tolerate salinity twice that of seawater and are valuable models for research on human kidney diseases. The technology does not exist to re-create ecosystems or even individual species. Once a species or a system is gone, it is lost forever.

More than this, humans benefit not just from individual species but from whole ecosystems too. Forests soak up carbon dioxide, maintain soil fertility, and retain water, preventing floods. The loss of biodiversity could disrupt an ecosystem's ability to carry out such functions. This consideration bears on the debate about the level of diversity necessary to carry out the world's ecosystem functions. MacArthur (1955) and Elton (1958) had proposed a linear relation between diversity (here meant as numbers of species) and stability (here meant as community persistence in an unchanged state) (Fig. 1.2). In 1981, Stanford ecologists Paul and Anne Ehrlich (1981) proposed the "rivet hypothesis." The diversity of life, said the Ehrlichs, is like the rivets on an airplane, with each species playing a small but critical role in keeping the plane airborne. The loss of each rivet weakens the plane and causes it to lose a little airworthiness. The loss of a few rivets could probably be tolerated, however, while the loss of more rivets would prove critical. Another alternative idea is the so-called redundancy hypothesis (or passenger hypothesis), proposed a decade later by Australian ecologist Brian Walker (1992) (see Ecology in Practice: Brian Walker). According to this hypothesis, most species are like passengers on a plane—they take up space but do not add to the airworthiness. This is caused by the activity of just a few species, perhaps the "pilot," "co-pilot," "stewards," and so on (Fig. 1.2). To round out the alternatives, Lawton (1994) included the possibility of a null model, that is, there is no relationship between community richness and ecosystem function.

As is usual, the "truth" may vary according to the system under study (Johnson et al. 1996). Although the null model and the linear relation models are not widely discussed, the rivets and redundancy hypotheses are still widely debated. Although there may be many rivets, the loss of some of them probably does not significantly reduce airworthiness. American forest communities are still functional despite the loss of the passenger pigeon and the Carolina parakeet. However, there comes a critical point where the loss of more rivets may cause the plane to crash. Thus, a compromise between the rivets and

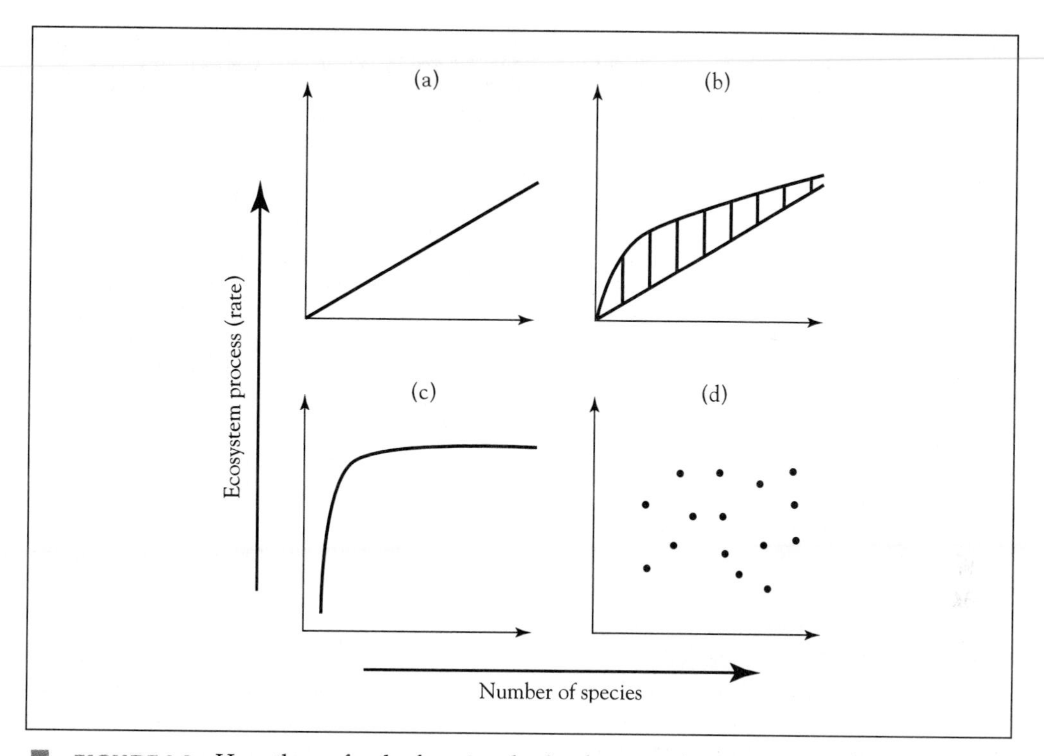

■ **FIGURE 1.2** Hypotheses for the functional role of species diversity in ecosystems: (**a**) Diversity-stability hypothesis. (**b**) Rivet hypothesis. (**c**) Redundancy hypothesis. (**d**) Idiosyncratic hypothesis. (*After Johnson et al. 1996.*)

redundancy hypotheses seems likely to explain ecosystem function in nature. On a larger scale, it is noteworthy that the productivity of temperate forests of the Northern hemisphere is virtually identical despite huge differences in tree species richness: East Asia has 876 species, North America 158, Europe 106. The conclusion is that average annual forest productivity tops out in the range of ten to forty species, despite the existence of many more (Baskin 1994). But the presence of more species may ensure a supply of "backups" should some of the most productive species fail through insect attack or disease—consider, for example, the demise of the American chestnut. The diversity-ecosystem efficiency connection is discussed in more detail in the Applied Ecology box in Chapter 15.

Finally, good arguments can be made against ecological mismanagement and the loss of biotic diversity on moral and ethical grounds. We simply have no right to destroy species and the environment around us. Philosophers like Tom Regan argue that animals should be treated with respect because they have a life of their own and therefore have value apart from anyone else's interests (Gunn 1990). Other philosophers like Christopher Stone, a law professor at the University of Southern California, have argued that entities such as trees, or even natural features such as lakes, could be given legal standing just as corporations, by a legal fiction, are treated as persons for certain purposes. Corporations can sue and be sued; the "interests" of corporations are represented legally by actual persons.

ECOLOGY IN PRACTICE

Brian Walker, (CSIRO), Australia

Academic History

Born and raised in Zimbabwe, I developed an early interest in the ecology of wildlife ecosystems. I obtained a B.Sc. (Agriculture) degree majoring in Rangelands Science from the University of Natal in South Africa in 1961 and after that spent three years as an agricultural officer back in the tribal lands in Zimbabwe. Realizing I needed a better understanding of ecological processes I returned to university, this time to the university of Saskatchewan in Canada, and obtained my higher degrees in Ecology. My thesis research was on the plant ecology of Saskatchewan's wetlands. Back in Zimbabwe, after a year as a rangelands research officer I lectured in ecology at the University of Zimbabwe (then Rhodesia) and co-directed the M.Sc. course in Tropical Resource Ecology. After six years of research on savanna ecology, I moved to South Africa to a chair in the Department of Botany at the University of the Witwatersrand, where I directed the Centre for Resource Ecology and continued my work in savannas. In 1985 I moved to Australia to take up my present position as chief of the Division of Wildlife and Ecology in the Commonwealth Scientific and Industrial Research Organization (CSIRO).

Hurdles and Difficulties

Much of my life has been spent working with managers of wildlife areas and ranches, and one of the greatest difficulties I have experienced is getting managers to appreciate that disturbance and change are natural and necessary characteristics of ecosystems. A misinterpretation of the meaning of stability often leads many managers to try to prevent change by keeping the area under their control constant in terms of animal population numbers and attempting to make the area as spatially homogeneous as possible, for example, by trying to get grazing pressures evenly distributed through the use of water, nutrient supplements, and fences. In fact, of course, maintenance of ecological resilience depends on disturbance and change, and the prodigious efforts of managers to prevent change so often leads to local loss of species and unpleasant surprises in the dynamics of the ecosystems they are trying to conserve.

Underlying much of the misunderstanding about ecological resilience is a more basic lack of understanding about the relationships between ecosystem function and ecosystem structure; more specifically, ecosystem species composition. Scientists too have been guilty of misinterpretation through misunderstanding, either deliberately (because of adherence to a ruling paradigm) or through superficial reading. As an example, I published a paper on biodiversity and ecological redundancy in 1992, which elicited over a hundred responses. Many were negative, some of them quite vehemently so, accusing me of saying that we can afford to lose some species when there are several species within a functional group. Most of these scientists were engaged in the debate on how to reverse the rapid rate of biodiversity loss, and their immediate opposition to my paper was triggered by the word *redundant*. In hindsight, I should have used some other word even though, in a strict scientific sense, it was the correct word to use. A careful reading of the paper shows quite clearly that no suggestion of allowing species to go extinct is made. I wrote a follow-up paper in the same journal in 1995, using different words in the title ("Conserving biological diversity through ecosystem resilience"). It clearly explained the same hypothesis that had been put forward in the first paper, using some examples. It was a better paper overall, and though it had a fairly good response there was nothing like the number of replies generated by the first one. There is still frequent reference to the first paper, more often than not misquoting or misinterpreting it because of the word *redundancy*, and I suspect that many of those who quote it do so on the basis of what they have read about it in a paper by someone else. This is unfortunately an all too common practice in writing scientific papers. The example highlights two important lessons: be sure to refer to the original work before interpreting and quoting other scientists, and be very careful in the use of emotive words, especially in the title of a paper!

A pervasive hurdle in today's research climate is the control of research funding by bureaucracies that are driven by a philosophy of immediate economic relevance. The distinction between pure and applied research is false. There is just as much good and difficult science to do in applied problems as there is in so-called pure research. But economic rationalism does not support research that is not directly linked to economic benefit, and in ecology it is becoming progressively more difficult to obtain support for research into such topics as complex adaptive systems or for large manipulative experiments to understand how populations of different kinds of species go extinct. However, having been on both sides of the argument for funding I can understand how ecologists have often mishandled their demands for more funds to do basic science. In many complex resource use issues ecological science is only one (often quite small) part of what is needed for a solution.

From my perspective, as a scientist in a strategic research organization, I see two major areas of development in ecology over the next couple of decades. The first is the science of integration. How to bring together the expertise of different fields in biology, ecology, and the physical sciences, together with economics and sociology, to address the huge number of regional scale problems related to wise or best use of natural resources—the issue of sustainable use. It does not mean tagging on a bit of economics to an ecological study, or vice versa. It calls for a new approach to studying the combined, interacting biophysical, economic, and social system, as one system. The second area that needs much more effort is the functional role of biodiversity. We need a much better understanding of the reciprocal effects of species on ecosystem function (primary production, nutrient cycling, hydrological budgets) and of ecosystem function on the species composition of ecosystems. To what extent are ecosystems self-organizing systems? What is an appropriate measure of the functional diversity of organisms in an ecosystem? How do we devise experimental protocols for manipulating ecosystems that inherently cannot be replicated?

Influential People

I have been less influenced by famous scientists from the early stages of ecological science than by colleagues and excellent scientists with whom I've been fortunate enough to work. Of course, I have been influenced by the classic papers of famous contemporary ecologists, but they will in any case become known to those who read this book.

Can animals (or plants) "count" in their own right? Consider a tragic accident in 1974 in which a schooner sank off the eastern coast of the United States (Johnson 1990). The captain, his wife, their eighty-pound Labrador retriever, and an injured crewman occupied a lifeboat to which two youths, aged nineteen and twenty, as well as a forty-seven-year-old Navy veteran, tied themselves with ropes while floating in the freezing waters. The captain refused repeated requests by the swimmers that he throw the dog overboard to make room for (some of) those in the water. He later explained that he could not bear to do it. After nine hours, the youths perished from exposure. All the occupants of the lifeboat, dog included, were subsequently rescued. After an initial investigation, the Coast Guard recommended that no criminal action be brought. However, in May 1975, the captain was indicted in a federal court for manslaughter for refusing to eject his dog in lieu of (some of) the swimmers who died. Can a dog count morally, count directly, for its own sake, rather than because of some human's interest in it? Can wildlife count for more than humans in some cases?

1.2 Ecology, like all scientific disciplines, employs the scientific method to acheive advances in understanding.

The scientific method is a formalized system of observation.

Can science in general and the study of ecology in particular help overcome the loss of biotic diversity and all the problems it entails? Yes it can. Although acid rain, toxic wastes, air and water pollution, and nuclear radiation are often seen by the public as direct results of scientific "progress"—and science is seen not as the "hero" but as the "goat"—one must bear in mind that these problems are the results of the misuse of scientific knowledge. Solutions to ecological problems can also be found through scientific methods.

The scientific method is at the heart of the acquisition of scientific knowledge. There are two possible processes involved here, induction and deduction. In induction, the first step is sound observation repeated to determine the frequency of an event and verified and confirmed by independent observers. After observations, the next step is the construction of a hypothesis to explain the observed events. In deduction, the scientist first makes general assumptions using logic and existing theory. Good hypotheses carried out inductively or deductively should be tested by further observations or, better still, by experiments. Mere correlation of one variable with another is not sufficient, as variation may be caused by a third, independent variable. For example, insect numbers may be correlated to plant quality (nitrogen levels) but be actually changed by parasitism levels which co-vary with nitrogen levels by chance alone (Stiling, Brodbeck, & Strong 1982). If a hypothesis stands up to repeated testing, it may reach the status of a theory or law. Some philosophers, like Popper (1972a), have emphasized the important point that science progresses not by trying to confirm theories but by attempting to falsify them. It is usually possible to find at least some confirmatory evidence for any hypothesis; one piece of negative data, on the other hand, refutes the hypothesis absolutely. Unfortunately, manuscripts that report the nonoccurrence of something are often distrusted, and as a result it is often easier to get confirmatory papers accepted than to publish "negative" results. Mahoney (1977) submitted two sets of contrived research papers differing only in the results they reported. He found that papers reporting "negative" results were less likely to be accepted for publication and less likely to be rated methodologically sound than were those reporting confirmatory results. In a recent review of the ecological literature, Csada, James, and Espie (1996) showed that only 8.6 percent of papers published in biological journals contained nonsignificant results.

The scientific method, of course, seems rather formal and tedious to the layperson and sometimes even to the scientist. Roughgarden (1983) and Quinn and Dunham (1983) argue that sometimes hypotheses can be developed less formally by the type of commonsense logic that we use in everyday life, which usually involves the research for confirmatory evidence. Simberloff (1983) has replied that, although this type of approach is most seductive, it is likely to be wrong. "Common sense" leads millions to conclude the existence of a deity and millions more to deny such existence. As Strong (1983) noted, common sense led many to believe the world was flat. Nowadays, we have books with such titles as *The Common Sense of Drinking* (Peabody 1931) and even *Common-sense Sui-*

cide (Portwood 1978). At the very least, formalization should help us become more efficient in using our research time. Besides as Simberloff points out, even perception and the search for confirmatory evidence itself imply the tacit mental construction of some part of a hypothesis.

In practice, scientific advances probably occur by several routes. Isaac Newton claimed that scientists work from the particular to the general, first observing phenomena and only later deriving generalizations from them (induction). Popper argued that imagination comes first, then hypotheses are tested by experimentation (deduction). Both avenues have undoubtedly proved valuable in research. Popperian science has its drawbacks, of course. Popper (1972b) has denied that Darwin's theory of evolution is a valid scientific theory; it seems untestable as a whole and is best formulated from many separate lines of evidence (see Section 2). The same arguments apply to the science of astronomy.

Creationists have exploited this view in an attempt to refute the entire theory of evolution. For this reason, some philosophers (for example, Suppe 1977) and ecologists (Dunbar 1980) have abandoned Popper's type of approach. Fagerström (1987) has even made the controversial suggestion that it would be better to erect a theory and throw out the data if they are not in agreement. He argues that this approach is not as absurd as it at first sounds. For example, ecological data are not always "hard facts"; they are produced, digested, and interpreted with the aid of theories and apparatuses and by people who are constrained by prejudices and previous experiences. Data are, in actuality, theory-laden. Ecological theories are judged more often by their simplicity, beauty, and intuitive appeal than by strict agreement with data. Even the simplest ecological statement assumes more about nature than can be concluded from observations alone. There can never be a complete match between theories and data no matter how much empirical evidence is gathered. Murray (1986) suggests that, in fact, most ecologists interpret their data by formulating the question, "How can I explain my data in terms of theory X," a procedure that often leads to a rash of "me too" papers in the wake of the publication of a new theory and that can slow scientific progress by leading to the uncritical acceptance of any plausible hypothesis. It would be preferable simply to ask, "How can I explain my data?"

Dyson (1988) has suggested that among the scientists in all fields there are "diversifiers," who are content to explore the details of phenomena, and "unifiers," who strive to unearth general principles. He believes that biology is the natural realm of diversifiers, much as physics is the realm of unifiers. As a result, biology lacks general themes. Dyson suggests that physicists "have a driving passion to find general principles," but biologists "are happy if they leave the universe a little more complicated than they found it."

It has been further argued that biologists think inductively and that physicists and chemists think deductively. For example, from a series of observations a physicist might predict the next element in the periodic table. Biologists group facts together and draw conclusions from them but are loathe to make subsequent predictions. Einstein and Infeld (1938) illustrated this by discussing the motion of a cart. If we give the cart a push it will travel a certain distance. If we reduce friction by oiling the wheels and smoothing the road it will travel farther. If we remove all sources of friction, the cart will never stop. We can never do this experiment; we can only think it. Murray (1992) has argued that biologists would spend their time performing experiments to determine the effects of the slope of the road, the number of wheels on the cart, and other factors on the motion of the cart, while a physicist would be the one to make the ultimate jump to a law of inertia, much as Newton did in 1729: "every body continues in its state of rest, or of uniform motion in a straight line, unless it is compelled to change that state by forces impressed upon it."

However, I believe that many biologists, especially ecologists, *have* been making predictions, and testing them, especially over the past ten to twenty years. In applied ecology especially, politicians and policymakers are constantly forcing uncomfortable ecologists to put numbers and estimates to rates of environmental degradation. Very often, general hypotheses are tested in ecology. Strong (1980) has suggested that the set numbers of particles in matter make explanations simpler for physicists than biologists, where the number of particles, or species, is very much higher.

1.3 Experiments, studies that involve the deliberate manipulation of a factor, provide the most rigorous test of hypotheses.

Ecological hypotheses are perhaps best tested by experiments. Hairston (1989) has noted that the percentage of field studies involving experiments has risen from less than 5 percent in the 1950s to around 10 percent in the 1970s and over 30 percent in 1987. Experiments can be classified in several ways. Diamond (1986b) distinguishes three main types: laboratory experiments, field experiments, and natural experiments. In practice, these types form a continuum. Diamond further divides the natural experiments into two categories: Natural trajectory experiments are comparisons of an ecosystem or species before and after some dramatic perturbation like a storm, a volcanic eruption, or the introduction of another species. Natural snapshot experiments compare natural areas that differ from one another in only one or two characteristics, for example, the presence or absence of certain predators. Such differences have often been maintained throughout recent history. Strengths and weaknesses of these different types of experiments are outlined in Table 1.3. For example, the spatial scale of laboratory experiments is likely to be limited to the size of a constant-temperature laboratory room, around 0.01 ha, and that of field experiments to usually less than 1 ha. Natural experiments, however, may be virtually unlimited in scale and often use large islands or continents.

Laboratory experiments can regulate more exactly all abiotic factors, from light, temperature, and moisture to available nutrients. They are valuable in investigations of the effects of these factors. The biotic community represented in a laboratory experiment, however, is likely to be limited at best. Laboratory experiments are best used to study the physiological responses of individual plants or animals rather than the population dynamics of reproducing populations.

Field experiments are conducted outdoors and have the advantage of operating on natural rather than synthetic communities. The most commonly used manipulations include local elimination of a species, local introduction of a species, and erection of a fence or cage. Darwin used a field experiment to demonstrate that either mowing or the introduction of grazing animals increases plant species diversity on a lawn (by preventing some species from outcompeting others). Field experiments commonly manipulate systems through the use of phenomena (like cages or fences) unlikely to be generated by nature itself.

TABLE 1.3 The strengths and weaknesses of different types of experiments in ecology. (After Diamond 1986a)

	LABORATORY EXPERIMENT	FIELD EXPERIMENT	NATURAL TRAJECTORY EXPERIMENT	NATURAL SNAPSHOT EXPERIMENT
1. Regulation of independent variables	Highest	Medium/low	None	None
2. Site matching	Highest	Medium	Medium/low	Lowest
3. Ability to follow trajectory	Yes	Yes	Yes	No
4. Maximum temporal scale	Lowest	Lowest	Highest	Highest
5. Maximum spatial scale	Lowest	Low	Highest	Highest
6. Scope (range of manipulations)	Lowest	Medium/low	Medium/high	Highest
7. Realism	None/low	High	Highest	Highest
8. Generality	None	Low	High	High

Natural experiments are usually the sole technique for following the trajectory of a perturbation beyond a few decades. Simberloff (1976a) was able to examine defaunation and recolonization on mangrove islands for several years, but only on the island of Krakatau has the process been followed in the long term, for over one hundred years (see Chapter 19). One of the few exceptions is the experimental fertilization of selected experimental plots at Rothamstead Experimental Station, England, for over one hundred years, beginning in fact in 1843 (Williams 1978). The weather is frequently shown to be vital in influencing the population densities of many species, but we cannot easily manipulate the weather. Natural experiments involving drought situations or floods commonly provide some of the best types of data on this subject. Furthermore, natural experiments often have general implications for the ecological system because they sample from a wider range of natural variation among sites than do field experiments. In summary, it is apparent that there is no best type of experiment; the choice depends on what one is investigating. This point has ramifications relating to the preceding section on the scientific method—no one "right" type of experiment has inherent superiority over others.

Krebs (1988) has suggested that, of the three types of experiments, field experiments are preferable. He has detailed how laboratory experiments in the 1940s and 1950s failed to provide any useful insights into the population dynamics of rodent populations, but that field experiments, begun in the 1950s, have been successfully used to test single- and multifactor hypotheses of population regulation. That experiments must be correctly set up should be obvious. A simple example illustrates the point. Otto Korner (cited by

Sparks 1982) performed experiments in the early 1900s to see whether fish could hear. He engaged a well-known opera singer to perform before his aquariums. He watched for the signs of enthusiasm from the fish that would surely result as their piscine hearts were uplifted. None was forthcoming. Korner concluded that fish could not hear. It was left to Von Frisch to show that if fish were "given a reason" to respond to sounds, perhaps by association of sound with food, they would respond.

Experiments should be planned for greatest usefulness. For example, the simple removal of one species to examine possible competitive effects on another might document a phenomenon, say, the elevation of density of one species, and a hypothesis might then be developed that species *a* and *b* do compete, but no idea of the mechanism involved is provided. In such cases, the ability to predict the outcome of other pairwise or multispecies interactions is limited because no idea of mechanisms has been gleaned. Experiments that take the phenomenological approach and merely document a phenomenon (such as competition) are often inconclusive. Mechanistic approaches, which also attempt to explain why the experimental results are obtained (in this case, say, resource competition, allelopathic effects, or effects of shared predators), are often more useful because the level of prediction obtained is greater (Tilman 1987).

Among the most useful experiments are randomized blocks. Here the treatments, say, for example, the addition of fertilizer to plants, are randomized along an environmental gradient, say, moisture (Fig. 1.3). The Latin square design in Fig. 1.4 is generally the most powerful design because it can block along two perpendicular environmental gradients, and each treatment is represented once only in every row and column (Scheiner and Gurevitch 1993).

One problem with experiments is that they take a lot of effort. It takes time, money, and careful planning to run an experiment, and even then it is often only one factor that is being tested at a time. Low levels of replication often lead to what is known as a type I error, the declaration of an hypothesis to be false when it is actually true, for example, fertilization does not increase plant growth. On the other hand, in observational data we can get type II errors, the failure to falsify the hypothesis when it is actually false. For example, Stiling et al. 1991 detected parasitism rates of 8.33 percent and 8.24 percent on two species of planthoppers inhabiting salt marshes. Because the sample sizes were 16,044 and 7,637, respectively, this difference in parasitism rate was statistically different.

Somewhere along the line, when it came to statistical testing of results the probability value, or P-value, of something happening by chance alone became all important. Thus, if something had a P-value of < 0.1 it would have a less-than-one-in-ten chance of occurring by chance alone. A value of $P < 0.05$ became a magic number: reject the hypothesis if the probability of observing a parameter of a certain value by chance alone is less than 5 percent; do not reject if the probability is greater. However, one must consider the power of the test. In some situations, it may be acceptable to use a higher P-value, say 0.1, as a cutoff for statistical significance because the power of the test is so low (Scheiner 1993).

A recent concern over type II errors has been that in literature reviews "vote counting," or determining whether each of a series of studies does or does not demonstrate an ecological phenomenon, has been a standard procedure. The concern is that if most of the studies have low statistical power, the failure to demonstrate a phenomenon will be perpetuated. The few studies with good statistical power that demonstrate a phenomenon will be outweighed by the hordes of badly designed experiments that fail to show it. One technique for detecting the "true" strength of such seemingly infrequent

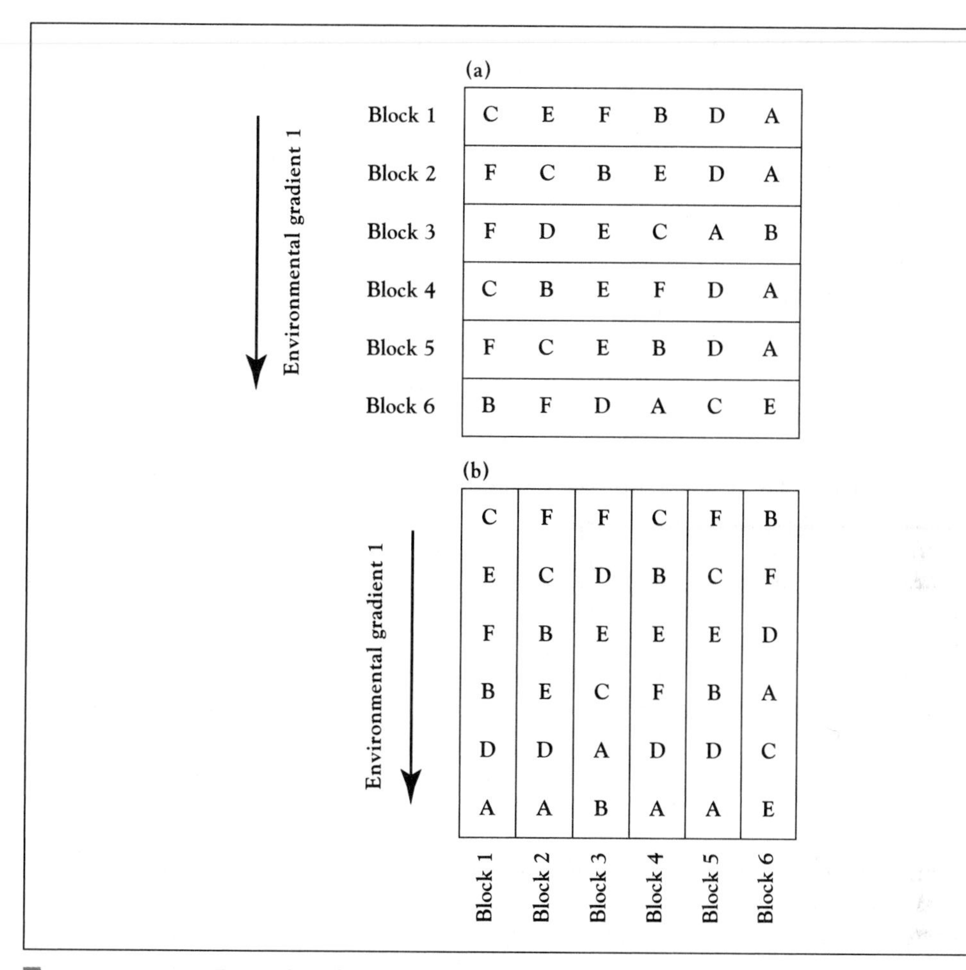

■ **FIGURE 1.3** Examples of a layout for a randomized block design to compare six levels of treatment factor (A-F) (e.g., genotypes of sunflowers): (**a**) appropriate and (**b**) erroneous layout in the presence of one environmental gradient.

processes is meta-analysis, a method for combining the results from different experiments (Gurevitch et al. 1992), which gives different studies weights based primarily on their sample size. The downside of meta-analysis is that it cannot replace multifactoral studies: for instance, it cannot disentangle the separate effects of two independent variables (e.g., age and size) by analysis of studies that each deal with either of these two variables separately (Arnquist and Wooster 1995).

A null hypothesis in ecology states that observed patterns arose solely by chance.

Often, the working hypothesis that is set up in ecological research is that of "no effect"—the hypothesis that observed ecological patterns arise by chance and not through any effect of the forces of nature. This hypothesis of "no effect" is called the null hypothesis (Connor and Simberloff 1979; Strong, Szyska, and Simberloff 1979), and there exists a wide range of statistical tests designed to challenge it (Conover 1980; Fleiss 1981; Zar 1984). Of course, this type of hypothesis is not the only one that is useful in ecology. For example, one might have

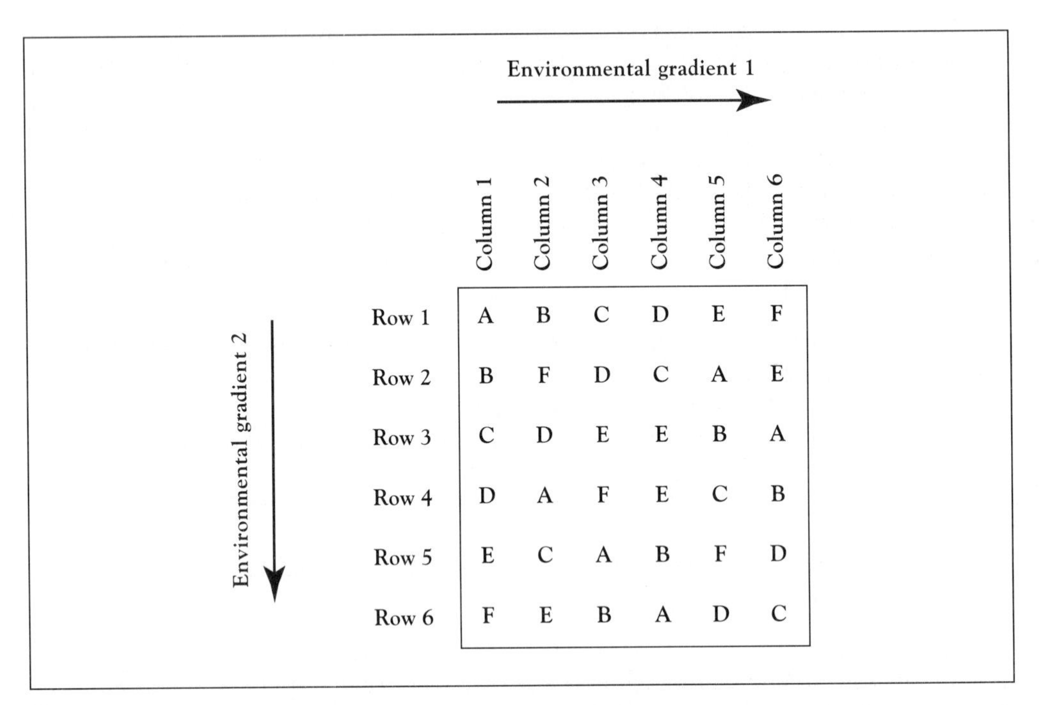

■ **FIGURE 1.4** Example of layout for a Latin square design to compare two treatments or environmental gradients on six genotypes or types of subject. The design appropriately accounts for two treatments or environmental gradients.

reason to erect the hypothesis that the disjunct distribution of two species in a particular habitat is due to competition and that they therefore live in mutually exclusive zones. However, the null hypothesis is often the most logical place to start (Simberloff 1983).

With the right type of observation, experimentation, and attempts at falsification, hypotheses can be solidified into ecological theories. J. B. S. Haldane (1963) has observed, rather cynically, that this process is normally reflected in four stages through which the hypothesis passes in the regard of the scientific community:

1. This is worthless nonsense.
2. This is an interesting, but perverse, point of view.
3. This is true but quite unimportant.
4. I have always said so.

1.4 Ecological phenomena occur on a variety of spatial and temporal scales.

Just as laboratory experiments, field experiments, and natural experiments are performed on different scales, it has become obvious that the effects of scale on ecological research and conclusions are staggering (Wiens et al. 1986; Levin 1992). What is the "right" spatial scale over which to look for a phenomenon? Ecologists must address this question in planning their experiments.

The size of the study area and the duration of an investigation can limit what one can see of an ecological system (Dayton and Tegner 1984). Wiens et al. (1986) have made the analogy that studying ecology is comparable to what it would be like to study chemistry if the chemist were only a few angstroms long and lived for only a few microseconds. If the chemist were no larger or longer lived than the molecules and processes under study, the overall course of chemical reactions would be difficult to distinguish from the random collisions of molecules. Wiens et al. emphasize that scale is, of course, a continuum, but that five major points on that continuum can be recognized:

- A space occupied by a single individual sessile organism or a space in which a mobile organism spends its entire life.
- A local patch occupied by many individuals.
- A region large enough to include many patches or populations linked by dispersal.
- A space large enough to contain a closed ecosystem (one receiving no migration)—an unlikely scenario in practice.
- A biogeographic scale large enough to encompass different habitats and climates.

Investigations on these different scales yield answers to different types of questions, and it is important to realize this point (see Table 1.4). For example, in studying the distribution and abundance of zooplankton, different phenomena cause different patterns at different scales (Fig. 1.5). For example, large-scale patterns may be caused by ocean

TABLE 1.4 The effects of scale on ecological investigations.

SCALE	PHENOMENA
1. Individual space	Physiological ecology, sociobiology, foraging ecology, reproductive biology (Section 4) (special problems arise for migratory species, for which important behavior may occur at another location entirely).
2. Local patch/ecological neighborhood	Predation, herbivory, parasitism, and pollination (Section 4).
3. Regional scale	Immigration, emigration, outbreaks, habitat preference (Section 3).
4. Closed system	Nutrient cycling, ecological energetics (Section 6).
5. Biogeographical scale	Climatic limits, evolutionary ecology (Section 2).

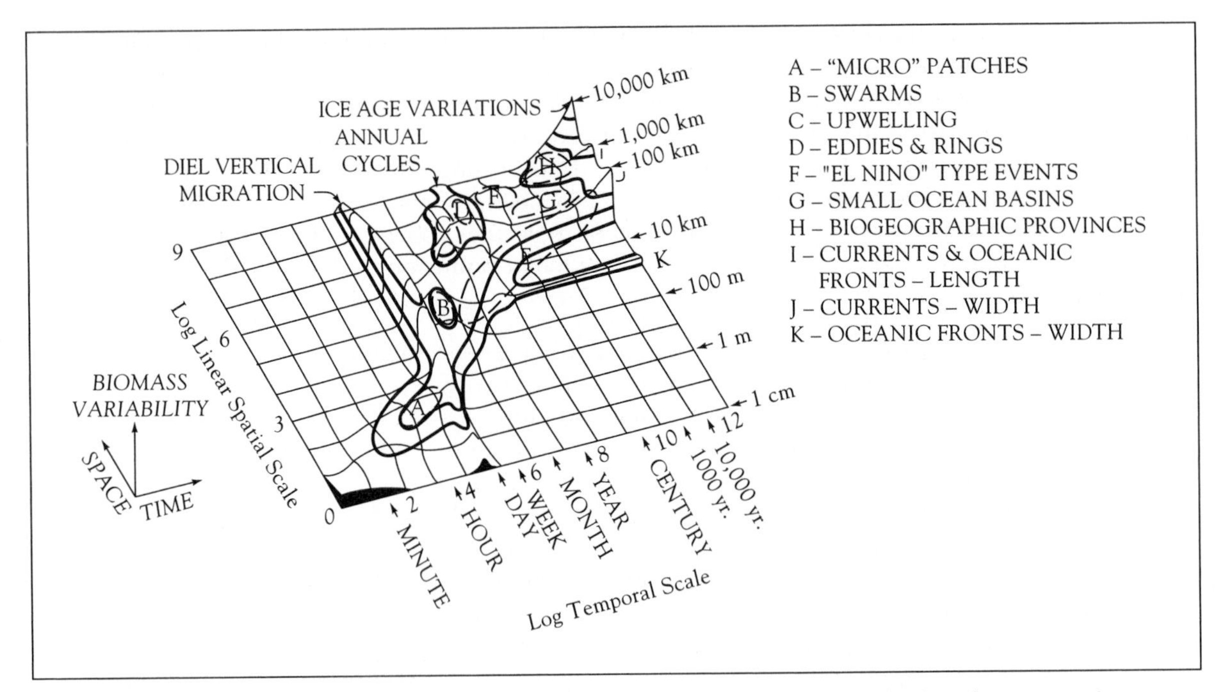

■ **FIGURE 1.5** Stommel diagram of spatial and temporal scales of zooplankton biomass variability. *(From Haury et al. 1978.)*

currents or climatic events. Small-scale patterns may be caused by local upwelling or diel vertical migration.

Addicott et al. (1987) suggest that the correct scale depends exactly on the question being asked, so there will never be a single ecological neighborhood for a given organism but rather a number of neighborhoods, each appropriate to a different process. The correct scale might depend on the ambit of the organisms involved. For example, imagine ten trees each 50 meters (m) apart (Fig. 1.6). Each tree contains caterpillars and wasps

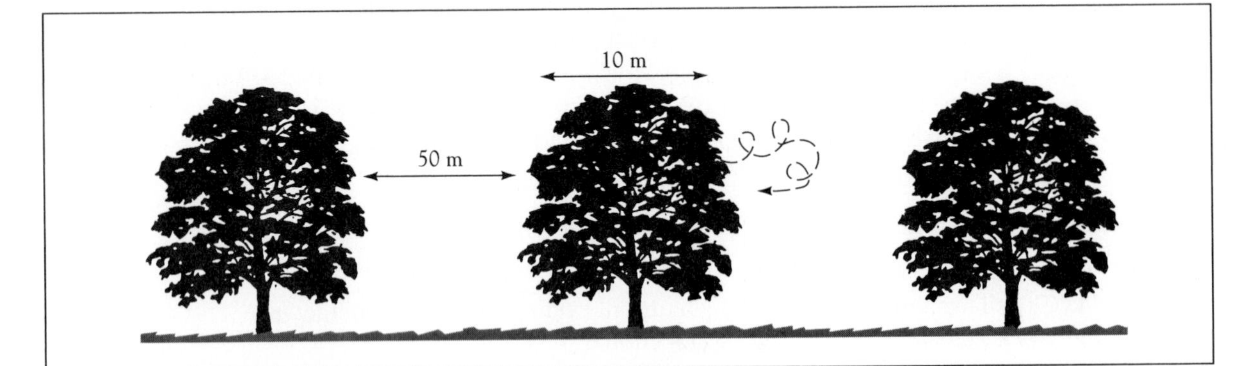

■ **FIGURE 1.6** The effect of neighborhood distance on spatial scale of investigation. If the ambit of a parasitic wasp is only 10 m then it will not be able to respond to differences in densities of caterpillars, its prey, between trees spaced 50 m apart.

that parasitize the caterpillars. In a study of how the wasps utilize the caterpillars (perhaps in a study of density dependence, see Chapter 13) one might examine all ten trees to see if the wasps aggregate on the tree with the most caterpillars. However, if the wasps can only fly a few meters, they would not be able to fly from tree to tree. They would only be able to respond to the density differences of caterpillars within a tree. The correct scale of ecological investigation might then be within trees, not between them.

Just as with spatial scale, the appropriate choice of temporal scale depends on the phenomenon and the species to be studied. One would choose a relatively short time over which to study behavioral responses, a longer one for population dynamics, and a still longer scale for studies of genetic change and evolution (Wiens et al. 1986).

Finally, the issue of scale can relate to the scale of effort involved in research. If an investigation involves relatively few observations or experiments, we have less confidence in the results than if the experiment were well replicated. If the investigation has few experimental replicates, it has low statistical power—even if treatment had a relatively dramatic effect we might not be able to detect it statistically. Conversely, if a treatment had a negligible effect but we performed scores of replicates, even a relatively unimportant effect would show up as significant—as having a meaningful effect. But is such minor consistency biologically meaningful? Sometimes we cannot perform many replicates of some experiments—for example, defaunations of whole islands, nutrient additions to large lakes—and even though such effects are large, they are not statistically significant. This is why it is often difficult to use strict statistical procedures to attach meaning to results of experiments.

SUMMARY

1. Ecology is to environmental science as physics is to engineering. A knowledge of ecology is therefore important for understanding such phenomena as conserving species, preserving biodiversity, maintaining ecosystem functions, preserving soil fertility, studying the effects of pollutants, preventing global warming, and calculating sustainable yields in fisheries, forestry, and hunting.

2. Are all species integral to maintaining life on Earth? There is still much debate as to whether species act as rivets on a plane, where the loss of each rivet weakens the aircraft, or whether species are like the people on board—only a few (the captain and other crew members) are necessary and the rest are like passengers, superfluous. The truth is probably somewhere in between, but to which end of this spectrum does it lie?

3. Progress in understanding ecological systems can be gained through observations of natural systems, through field or laboratory experimentation, through mathematical models, or from reviews and syntheses of existing studies. Great care must be exercised that appropriate sample sizes are used so that the conclusions reached are statistically valid. The effect of the scale of these investigations can also have a large influence on the conclusions.

DISCUSSION QUESTION

How can we determine when loss of biodiversity will affect ecosystem function?

SECTION

2

Evolutionary Ecology and Conservation Biology

Conservation of many endangered species, such as this Siberian tiger, may depend on a knowledge of genetics and how to overcome loss of genetic diversity, as well as how to preserve suitable habitats. A knowledge of evolutionary and behavioral ecology is therefore vital to modern ecologists. *(Tom & Pat Leeson, Photo Researchers, Inc. 2S6960.)*

If ecology is concerned with explaining the distribution and abundance of plants and animals and with the control of their numbers, then evolutionary ecology is an important part of the discipline. For example, one may argue about what controls penguin numbers in the Southern Hemisphere, but a nagging question remains—why are there no penguins in the Northern Hemisphere? The answer is not insufficient food or too many predators. Penguins simply evolved in the Southern Hemisphere and have never been able to cross the tropics to colonize northern waters. Bromeliads (members of the Bromeliadaceae, the pineapple family) are virtually all neotropical, and the Myrtaceae (*Eucalyptus* and its relatives) are restricted to Australia. The Columbidae (the doves) are virtually cosmopolitan. Flight is not the only reason—the Todidae (small kingfisher-like birds) can fly but are restricted to the Caribbean islands, whereas skinks (Scincidae) are entirely terrestrial lizards but are cosmopolitan in most tropical and temperate areas.

South America, Africa, and Australia all have similar climates, ranging from tropical to temperate, yet each is clearly characterized by its inhabitants. South America is inhabited by sloths, anteaters, armadillos, and monkeys with prehensile tails. Africa possesses a wide variety of antelopes, zebras, giraffes, lions, baboons, the okapi, and the aardvark. Australia, which has no placental mammals except bats and the introduced mouse and dingo, is home to a variety of marsupials such as kangaroos and the peculiar egg-laying mammals or monotremes, the duck-billed platypus and the echidna. A plausible explanation is that each region supports the fauna best adapted to it, but introductions have proved this explanation incorrect; rabbits introduced into Australia proliferated rapidly, and *Eucalyptus* grows well in California. The best explanation is provided by macroevolution and **continental drift.** Other, more recent geological phenomena can explain the distributions of some plants and animals. For example, during the glacial periods of the Pleistocene, lakes covered much of what is now desert in the U.S. Southwest. With the retreat of the ice, the lakes disappeared. Desert pupfishes *(Cyprinodon)* and other aquatic organisms were once widely dispersed throughout the Death Valley region but now occur only in isolated springs. Similarly, tapir fossils are found widely around the globe, and their present spotty distribution in South America and Malaysia merely represents the relic populations of once-widespread groups. A knowledge of evolution and historical ecology is clearly of paramount importance to contemporary ecology. A study of evolution and of contemporary conservation biology also necessitates a knowledge of genetics.

First, the rate of evolutionary change in a population is proportional to the amount of genetic diversity available. Reduction of genetic diversity reduces future evolutionary options. Second, heterozygosity or high genetic variation is often positively related to fitness (Allendorf and Leary 1986; Koehn, Diehl, and Scott 1988; but see also Hedrick and Miller 1992). Although many species have naturally low levels of heterozygosity and there is no set "norm," any loss of heterozygosity may indicate cause for concern. An often cited example of a correlation between low heterozygosity and low fitness is the African cheetah. Fifty-five cheetahs from several populations examined had no detectable genetic diversity at forty-seven allozyme loci and very low heterozygosity of 150 soluble proteins (O'Brien et al. 1983, 1985). These animals were so genetically uniform that skin tissue transplants were routinely accepted among all individuals—their immune systems could obviously not distinguish between themselves and other individuals. It was claimed that the animals had difficulty breeding in captivity, that the

A mother cheetah and her cubs resting at Nairobi National Park, Kenya. Is low genetic diversity to blame for low population densities, or is the problem heavy predation on juveniles? (*Rick Edwards, Animals Animals/Earth Scenes 678314-M.*)

males had sperm counts ten times lower than related cat species, and that there were higher rates of infant mortality in the wild than in captive populations. Low heterozygosity due to population bottlenecks was blamed.

However, as is often the case in ecology, there is a flip side to the story. Other scientists (Caro and Laurenson 1994) presented convincing evidence that low population densities and poor recruitment in the wild are due to heavy predation on juveniles rather than lack of genetic diversity. They have suggested that poor performance in captivity is merely due to poor husbandry and that male cheetahs in zoos, despite abnormal sperm, can fertilize females with high efficiency. The bottleneck responsible for the low heterozygosity in cheetahs may be due to events following the retreat of the last ice sheet thousands of years ago. Human assaults on the cheetahs range, and numbers have certainly not aided its recovery from the effects of this bottleneck. But it is interesting that similar effects of inbreeding resulting from recent population bottlenecks are seen in relic populations of lions in the Gir forest of western India and in the natural Ngorongoro crater in the Serengeti of Kenya.

Third, the global pool of genetic diversity represents all of the information for all of the biological processes on earth. Wilson (1985) calculated that all the bits of information encoded in this genetic library would, if translated into English text, fill all the editions of the *Encyclopedia Britannica* published since 1768. Conservation programs of rhinoceroses, grizzly bears, Siberian tigers, and many other organisms are benefiting from the input of geneticists.

CHAPTER

2

The Diversity of Life

As Grant (1977, p. 3) wrote, "The world of living organisms exhibits several general features that have always aroused feelings of wonder in mankind." First, there is a tremendous diversity of life forms—over 4,000 species of mammals, 9,040 species of birds, and about 19,000 species of fish, which together with other organisms give a total of 43,853 known vertebrates. Diversity of other groups, especially the arthropods and molluscs, is even more staggering (Table 2.1). It has been tentatively estimated that at one point or another during the history of life there have been in the neighborhood of one billion species.

Second, among these organisms there is a great structural complexity and an apparently purposeful adaptation of many characteristics to the environment. Darwin's own example was that of the woodpecker, with its chisel bill, strengthened head bones and neck muscles, extensile tongue with barbed tip, and stout tail for balance while chiseling. Another classic example is provided by the variety of feet in birds, from the perching feet of warblers, to the grasping feet of eagles, the walking feet of quail, the wading feet of herons, the swimming feet of ducks, and of course the specialized feet of woodpeckers (see Fig. 2.1). How can we explain this diversity?

2.1 Several nineteenth-century naturalists, the most famous of whom is Charles Darwin, grappled with the idea of evolutionary change.

Lamarck proposed a theory of evolution by inheritance of acquired characteristics.

A number of naturalists and philosophers, beginning with the ancient Greeks, had supposed that many forms of life evolved from each other. The first actually to formalize and publish a theory of **evolution**—"transformism" as he called it—was Jean-Baptiste

TABLE 2.1 Numbers of described species of living organisms.

KINGDOM AND MAJOR SUBDIVISION	COMMON NAME	NO. OF DESCRIBED SPECIES	TOTAL
Virus			
	Viruses	1,000 (order of magnitude only)	1,000
Monera			
Bacteria	Bacteria	3,000	
Myxoplasma	Bacteria	60	
Cyanophycota	*Blue-green algae*	1,700	4,760
Fungi			
Zygomycota	Zygomycete fungi	665	
Ascomycota (including 18,000 lichen fungi)	Cup fungi	28,650	
Basidiomycota	Basidomycete fungi	16,000	
Oomycota	Water molds	580	
Chytridiomycota	Chytrids	575	
Acrasiomycota	Cellular slime molds	13	
Myxomycota	Plasmodial slime molds	500	46,983
Algae			
Chlorophyta	Green algae	7,000	
Phaeophyta	Brown algae	1,500	
Rhodophyta	Red algae	4,000	
Chrysophyta	Chrysophyte algae	12,500	
Pyrrophyta	Dinoflagellates	1,100	
Euglenophyta	Euglenoids	800	26,900
Plantae			
Bryophyta	Mosses, liverworts, hornworts	16,600	
Psilophyta	Psilopsids	9	
Lycopodiophyta	Lycophytes	1,275	
Equisetophyta	Horsetails	15	
Filicophyta	Ferns	10,000	
Gymnosperma	Gymnosperms	529	
Dicotolydonae	Dicots	170,000	
Monocotolydonae	Monocots	50,000	248,428

(*continued*)

TABLE 2.1 (continued) Numbers of described species of living organisms.

KINGDOM AND MAJOR SUBDIVISION	COMMON NAME	NO. OF DESCRIBED SPECIES	TOTAL
Protozoa			
	Protozoans:	30,800	
	Sarcomastigophorans, ciliates, and smaller groups		30,800
Animalia			
Porifera	Sponges	5,000	
Cnidaria, Ctenophora	Jellyfish, corals, comb jellies	9,000	
Platyhelminthes	Flatworms	12,200	
Nematoda	Nematodes (roundworms)	12,000	
Annelida	Annelids (earthworms and relatives)	12,000	
Mollusca	Mollusks	50,000	
Echinodermata	Echinoderms (starfish and relatives)	6,100	
Arthropoda	Arthropods	751,000	
Insecta	Insects		
Other arthropods		123,161	
Minor inverebrate phyla		9,300	989,761
Chordata			
Tunicata	Tunicates	1,250	
Cephalochordata	Acorn worms	23	
Vertebrata	Vertebrates		
Agnatha	Lampreys and other jawless fishes	63	
Chrondrichthyes	Sharks and other cartilaginous fishes	843	
Osteichthyes	Bony fishes	18,150	
Amphibia	Amphibians	4,184	
Reptilia	Reptiles	6,300	
Aves	Birds	9,040	
Mammalia	Mammals	4,000	43,853
TOTAL, all organisms			1,392,485

Lamarck (1744–1829), though similar ideas had been thought of by others, including the Comte de Buffon (1707–1788) (Greene 1959). (See Photo 2.1.)

Whereas Buffon conceived variation as a process of degeneration or random deviation from innumerable ancestral forms, Lamarck viewed it as evolution from simple beginnings. Lamarck's (1809) work, however, was largely ignored because the mechanism he proposed to explain how evolution works was based on the inheritance of **acquired characteristics**. For example, Lamarck supposed that giraffes, in their continual struggle to reach the highest leaves on trees, stretched their necks by a few millimeters in the course of their lifetimes. This increase in neck length was passed on to their offspring, who con-

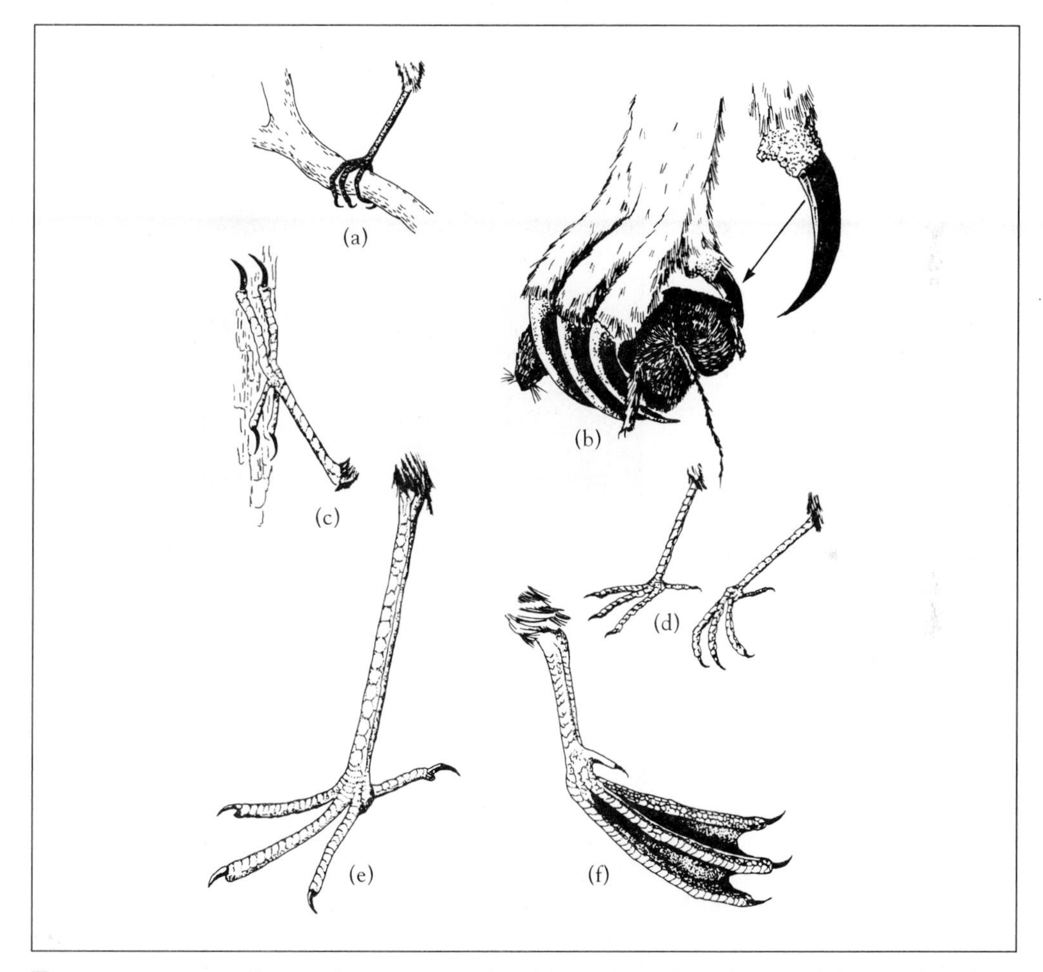

■ **FIGURE 2.1** Different adaptive types of bird feet. (**a**) Perching foot (Audubon warbler, *Dendroica audubon*). (**b**) Grasping foot with strong talons (horned owl, *Bubo virginiarius*). (**c**) Climbing foot (acorn woodpecker, *Melanerpes formicivorus*). (**d**) Walking and scratching foot (California quail, *Lophortyx californicus*). (**e**) Wading foot (green heron, *Butorides virescens*). (**f**) Swimming foot (pintail duck, *Anas acuta*). Drawings not to the same scale. (*Redrawn from Grant 1963.*)

■ **PHOTO 2.1** Jean Baptiste Lamarck, 1744-1829. *(Courtesy Department of Library Services, American Museum of Natural History, negative no. 124768.)*

tinued the process until the necks of giraffes reached their current proportions. In a similar way, Lamarck explained racial variation in skin color by assuming that the suntan developed by ancestral races was transmitted to their descendants, who were in turn a little darker than their parents. In 1988 Lamarckian evolution received a boost when John Cairns, a geneticist at Harvard University, and his colleagues, published a study of spontaneous genetic mutations in the bacterium *Escherichia coli* (Cairns, Overbaugh, and Miller 1988). They put bacteria unable to digest lactose into a petri dish with only lactose for nourishment—and found that the bacteria preferentially acquired the crucial mutations they needed to become lactose eaters. Later Barry Hall of Rochester University, New York, found a similar occurrence in other bacteria (Hall 1991) and scientists uncovered similar work by other geneticists in the 1940s and 1950s (Luria and Delbruck 1943; Ryan 1955).

However, as Lenski and Mittler (1993) have pointed out, *directed* mutation has still not been demonstrated. Rates of mutation in bacteria and yeast certainly do increase during starvation, but mutations may still be random. There are now just more chances that a mutation will be advantageous. Nevertheless, the very fact that mutation rates increase under stress is fascinating. Is this adaptative, in the sense of having evolved by natural selection for alleles that specifically promote increased genetic variation under stress? Or is the process merely an unavoidable consequence of physiological deterioration under starvation or of the induction of mechanisms to repair damage to DNA? What is needed is an assay to detect mutations—at present it's hard to prove the absence of other mutations because few, if any, survive.

After a five-year expedition in which he served as a scientific observer on the H.M.S. Beagle, Charles Darwin proposed the theory of evolution by natural selection.

Lamarck's theories were replaced by the ideas of Charles Robert Darwin (1809–1882), the founder of modern evolutionary theory (Photo 2.2). Darwin was born into a prosperous background. (His birthday, February 12, 1809, was the same as that of Abraham Lincoln.) His father was a successful doctor, and his mother and wife were both Wedgwoods, of china fame. His family's fortune enabled Darwin, educated at Edinburgh and Cambridge Universities, to accept an unpaid job as scientific observer on board H.M.S. *Beagle*, which sailed on a five-year world survey from 1831 to 1836, concentrating on South America. In some ways, Darwin was "primed" to accept the theory of gradual biological change and evolution because he had with him a copy of *Principles of Geology* (1830), newly published by Charles Lyell, who had taken the unprecedented step of describing the physical world as one that changed gradually through physical processes, not through a few catastrophic events (the view held firmly up until that time by the clergy and the masses).

During the voyage of the *Beagle* up and down both coasts of South America, Darwin was able to view diverse tropical communities, some of the richest fossil beds in the world in Patagonia, and the Galapagos Islands, six hundred miles west of Ecuador. The Galapagos contain a fauna different from that of mainland South America, exhibiting tortoises and other animals different in form on virtually every island. By the time he had finished the expedition, Darwin had amassed a wealth of data, described an astonishing array of animals, and built up a vast collection of specimens. A year after his return, Darwin read a revolutionary book on human population growth by the English

■ **PHOTO 2.2** Charles Robert Darwin, age 51, 1860. *(Courtesy Department of Library Services, American Museum of Natural History, negative no. 326668.)*

clergyman Thomas Malthus, written (anonymously) in 1798, which argued that populations had the capacity for geometric growth (that is, they could grow as rapidly as the series 1, 2, 4, 8, 16; see Fig. 6.5), yet food supply increased only arithmetically (that is, like the series 1, 2, 3, 4). Malthus proposed that—because the earth was not overrun by humans as it should be—food shortage, disease, war, or conscious control must limit population growth. Darwin quickly established that the **Malthusian theory of population** would apply to plant and animal populations. He made the logical deduction that these factors would act to the detriment of weaker, less well-adapted individuals and that only the strongest would survive.

Darwin had formulated his theory of **natural selection**, survival of the fittest: a better-adapted plant or animal would leave more offspring. In this way, giraffes born, by genetic chance, with longer necks would be better fed and able to reproduce more successfully. This trait would be passed on to their offspring, and long necks would become common. Only rarely would such long-necked mutants evolve independently; much more commonly, distinct traits, such as neck length, are inherited unchanged.

Incredibly, Darwin did not immediately publish his theory. He waited for nearly twenty years, collecting data on a wide range of organisms. Eventually, he was pushed into publication (Darwin 1859) by the arrival of a manuscript by Alfred Russel Wallace, who had independently and years later arrived at the same conclusions.

Alfred Russel Wallace arrived simultaneously at the idea of evolution by natural selection, and is credited as its codiscoverer.

Alfred Russel Wallace (1823–1913) was born into poverty and farmed out to an older brother in London at age 14, after only six years of schooling (Photo 2.3). He held down a succession of jobs until his brother died and left him some money. Wallace set sail immediately for the Amazon. A year later, a second brother joined him. Both men contracted yellow fever, and the brother died. After four years in the jungle, Wallace sailed for home with his precious collections. En route the ship caught fire and sank. After ten days in the open sea in a small boat, Wallace was saved, but four years of labor went down with the ship. Back in England, Wallace began to prepare for a second voyage, this time to the Malay archipelago as a professional collector and naturalist in the company of W. H. Bates (for whom Batesian mimicry is named). It was there, during another bout of fever, that Wallace conceived the idea of natural selection. Wallace's one major advantage over Darwin was that he was persuaded before he left on his voyages that species evolve; Darwin did not abandon his creationist beliefs until after his return. Thus, Wallace was able to gather data with an eye to his evolutionary hypothesis.

Unfortunately, Wallace's earlier papers had been ignored by the scientific community, and he was faced with the problem of lack of recognition. His solution was to send his manuscripts to Darwin, with whom he had previously corresponded. Darwin's higher standing in the scientific community made it more likely that he would be taken more seriously and that he would garner most of the credit—a phenomenon that still probably exists today. Darwin immediately sought the advice of friends (geologist Lyell and botanist Hooker) and, as a result of their suggestions, Darwin and Wallace had their theories presented jointly at a historic meeting of the Linnean Society of London on July 1, 1858. One year later, Darwin at last published his *Origin of Species*, an abbreviated version of the manuscript based on his twenty years of work.

■ **PHOTO 2.3** Alfred Russel Wallace, age 46, 1869. *(Courtesy Department of Library Services, American Museum of Natural History, negative no. 326812.)*

Although Wallace deserves full credit as a codiscoverer of the chief mechanism of evolution, Darwin's subsequent work continued to explore the same ideas and principles inherent in the original work. Furthermore, Darwin has been credited as a pioneer of the hypothetico-deductive method, that is, the testing of hypotheses through the use of observations to confirm deductions from them. In Darwin's day, most science was done by induction—conclusions were drawn from an accumulation of individual observations.

Darwin's main conclusions about the origin of species were two. First, all organisms are descended with modification from common ancestors. All the prominent scientists of the day were convinced of this point within twenty years, although the religious community of course was skeptical and still is. (For those wishing to read about the fascinating and ongoing creationism-evolution debate, an excellent entry into the literature and the rhetoric is provided by Numbers [1982], Godfrey [1983], Wilson [1983], the National Academy of Sciences [1984], and Futuyma [1995].) Second, the mechanism for evolution was natural selection. This conclusion convinced few people at the time and was not fully accepted until the late 1920s, partly because of a widespread belief in blending inheritance, in which the traits of the parents were thought to be blended in the offspring, like the colors of two paints blending to produce an intermediate color. Natural selection would not work in such a system. For example, if a long-necked giraffe mated with a short-necked giraffe, the offspring would have a neck of medium length, and the advantage of a long neck would be lost. Furthermore, the belief that environmentally induced variation could be inherited was widespread and provided an alternative to natural selection. Darwin's theory of natural selection had one serious flaw. He knew nothing of the causes of hereditary variation and could not well answer questions on that subject. The evidence of genetics and Mendel's laws of heredity were available but had passed into obscurity and were only resurrected in the early part of the twentieth century.

Gregor Mendel's classic experiments with peas provided the earliest clues about mechanisms by which traits are inherited.

Gregor Johann Mendel (1822–1884) was an Austrian who entered monastic service at Brno in 1843. As the monastery was a center of learning, Mendel was sent to Vienna University and later taught at Brno Technical School during 1856–1864, where he performed his revolutionary work on peas. He became abbot in 1868 but still carried on his hybridization work on plants, although his results were not widely disseminated.

Mendel crossed tall and dwarf strains of peas and found in the second generation a 3:1 ratio of tall to dwarf plants. He could therefore conclude that the parents differed with respect to a single gene (known today to be a single unit of DNA) that controlled size. The gene for height in pea plants existed in two different forms (or alleles), tall and short. We know today that these alleles can always be found at the same point on a chromosome, the locus of that gene. During meiosis, DNA is replicated (doubled in quantity), and the resulting two sets of chromosomes recognize each other and become precisely aligned on the meiotic spindle. During the first meiotic division, homologous chromosomes are drawn to opposite poles.

Mendel's experiments formed the basis of a genetic understanding of the acquisition of inherited traits, although much controversy would still exist for many years because his simple experiments did not work for all species. For example, the equivalent cross of tall and short humans does not yield a neat 3:1 ratio in the second generation because in more complex species features like height are often controlled by many genes, each of which alone has only a slight effect on size. Each of the individual genes controlling these so-called polygenic traits follows the principles Mendel outlined, however. Occasionally, genes that contribute to a single character or to functionally related characters show strong linkage; that is, they are located close to one another on the same chromosome and are often inherited together. Such clusters of genes are often called **supergenes**. There are only a few cases in which parental experience can influence offspring characteristics. These are usually due to the maternal origin of the cytoplasm of the egg (the sperm contributes almost nothing except DNA) and involve self-replicating bodies like mitochondria and chloroplasts.

Mendel's work showed that inheritance is generally particulate, with the exception of genes on sex-linked chromosomes. He showed that inherited factors did not contaminate one another and could be passed down from ancient ancestors in the same form. Furthermore, genes appeared exceptionally stable. As Weismann (1893) was to show, germ plasm is often entirely separate from, and immune to any influences from, the soma (the rest of the body), so environment has no influence on heredity. Interestingly, statistical evidence suggests Mendel may have "fudged" this data to make it fit more precisely to his hypothesis.

Yet some variation must occur in populations if selection is to work at all, and it is readily obvious that not all members of the same species are identical. This point is addressed in Section 3, but it is worth reconsidering here the many apparently purposeful adaptations discussed at the beginning of this section.

It is important to realize that the development of adaptations does not always proceed via natural selection for those adaptations. Feathers, for example, may have evolved as insulation for warm-blooded birds and only secondarily become useful in flight. Sometimes chance alone may influence evolution. Perhaps more commonly, the developmental biology and other morphological constraints may impose selection by coupling one trait with another. Because genes exert their effects on the **phenotype** through biochemical reactions, other reactants must be incorporated into the equation. A phenotypic effect at one locus often depends on the genotype at one or more other loci: this is the phenomenon of **epistasis**. Finally, some features may actually be neutral with respect to "fitness." For example, blue eyes and brown eyes confer equal visual ability on the majority of humans; they are merely different morphological solutions to the same problem.

2.2 Novel genotypes originate from point mutations and chromosomal rearrangements.

The variation that we recognize in our observations of the world is phenotypic variation, or variation in the appearance of organisms. Phenotypic variation arises in large part from genetic variation or variation in the genetic makeup of different organisms. In all organisms (except RNA viruses), the genetic material is DNA, deoxyribonucleic acid. In **prokaryotes** the DNA exists in the form of a single circular chromosome; in **eukaryotes** it is arranged into a set of linear chromosomes that reside in the cell nucleus. (The exact structure of the famous double-stranded DNA molecule is given in most introductory biology texts and should be familiar to biology students.) When the DNA code is copied for delivery into the gametes, mistakes are possible; these mistakes are the source of much genetic variation.

Increases in the amount of genetic variability present in the gene pool arise chiefly from **mutations** (a term coined in 1901 by Hugo Devries) during copying, of which there are two kinds: gene or point mutations (the most important in enriching the gene pool) and chromosome mutation (the most important in rearranging it). Most mutations result in a loss of fitness; the effects of mutations are at random with respect to adaptiveness (but note again Cairns 1988), and the chances are small that a random change to an already well-adapted system would result in an improvement.

Point mutations involve changes in the sequence of nucleotide bases.

A point mutation results from a misprint in DNA copying (Fig. 2.2). Most point mutations are thought to involve changes in the nucleotide bases that make up the DNA base pairs (adenine, thymine, guanine, and cytosine) at single locations only (cistrons), for example, from adenine to guanine or from adenine to thymine. Two such changes, which result in the change of the sequence GAA to CUA, combine to substitute the amino acid valine for glutamic acid, the change that causes the abnormal beta chains in sickle-cell hemoglobin. Because of the redundancy of the genetic code, about 24 percent of these codon (base triplet) substitutions do not change amino-acid sequences and thus do not alter phenotype. The genetic code is universal among prokaryotes and eukaryotes (Jukes 1983); the same triplets code for the same amino acids in all organisms.

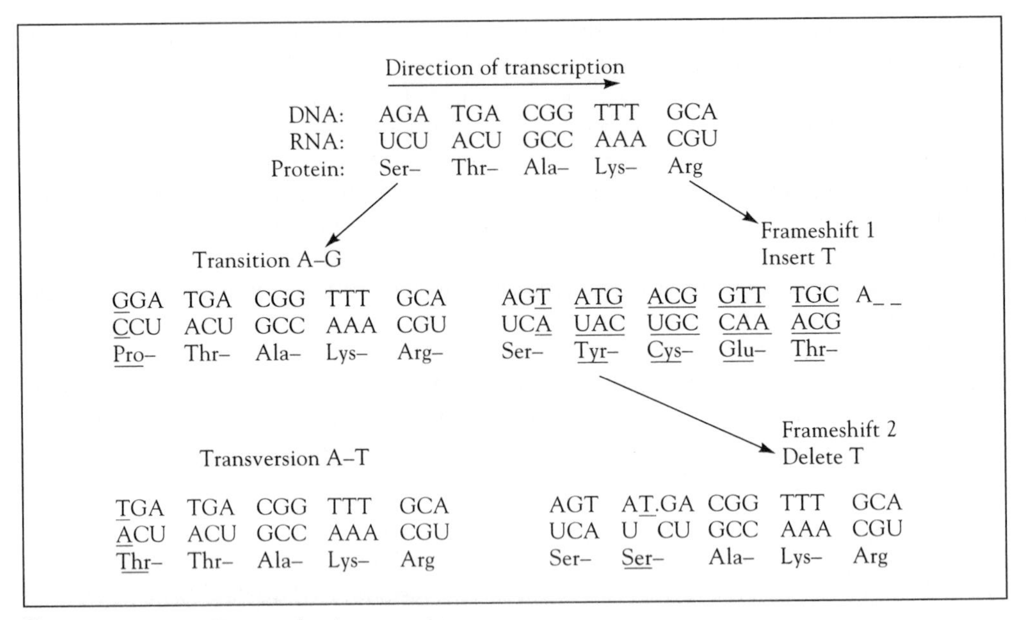

FIGURE 2.2 Types of point mutation.

More drastic changes in amino-acid sequences are caused by frameshift mutations. An addition to (or deletion from) the amino-acid triplet sequence alters the whole reading frame, leading to drastic and often fatal mutations. A second such change at another site may rectify the original reading frame so that only a short nucleotide sequence is then read in altered triplets. Genes themselves consist of many long sequences of base pairs. A gene for collagen in chickens is forty thousand base pairs (bp) (40 kilobases, kb) long. The size of the complete genome (all the genes) varies greatly among organisms, from less than four hundred bp in some viruslike particles to more than 10^{11} bp in some vascular plants. It has been estimated that the average mammalian genome is sufficient to code for more than three hundred thousand genes, that of *Drosophila*, one hundred thousand. In *Drosophila*, however, only about ten thousand genes are recognized. Much DNA therefore codes for repetitive sequences or codes for nothing; it is essentially junk DNA. In prokaryotes, there are many single copies of each DNA sequence.

Although mutations may be accelerated by man-made radiation, UV light, or substances such as colchicine, in nature such mutagens (usually in the form of weak cosmic rays) are too rare to produce many mutations. Nevertheless, it has been conservatively estimated that mutations occur in nature at the rate of one per gene locus in every one hundred thousand sex cells (Dobzhansky 1970; Neel 1983). Higher organisms contain at least ten thousand gene loci, so one out of ten individuals carries a newly mutated gene at one of its loci. Of course, most mutations are deleterious—they arise by chance, and the genes cannot know how and when it is good for them to mutate. Only one out of one thousand mutations may be beneficial, and thus one in ten thousand individuals carries a useful mutation per generation. In actuality, every individual *Drosophila* fly, corn plant, and human being seems to carry at least one abnormal, mutant allele (Spencer 1957; Morton, Crow, and Muller 1956), some originating in that organism and some inherited from ancestors.

If we estimate one hundred million individuals per generation and fifty thousand generations for the evolutionary life of a species, then five hundred million useful mutations would be expected to occur during this span. It has been estimated that only five hundred mutations may be necessary to transform one species into another, so only one in one million of the useful mutations needs to be established in a population in order to provide the genetic basis of observed rates of evolution. The chief factor limiting the supply of variability is therefore not the rate of new mutations. In fact, variability is limited mainly by gene recombination and the structural patterns of chromosomes. Even more astonishing is that some scientists have argued, with good evidence, that pure genetic drift is probably sufficient to cause directional genetic change in species with effective population sizes of under five million (Charlesworth 1984), which is probably greater than the population size of many mammals. Lande (1976) showed mathematically how even very weak selection on horses' teeth, two selective deaths out of one million individuals per generation, would be sufficient to explain the dramatic evolution of horse teeth through the ages.

Chromosome mutations introduce genotypic variability by altering the sequence of genes on a chromosome.

Chromosomal mutations do not actually add to or subtract from gene-pool variability; they merely rearrange it. Followed by natural selection, chromosomal mutations make certain adaptive gene combinations easier for the population to maintain.

Chromosomes can undergo four types of changes (deletions, duplications, inversions, and translocations) in which the order of base pairs within the gene is unaffected but the order of genes on the chromosome is altered (Fig. 2.3). A deletion is the simple loss of part of a chromosome. A deletion is usually lethal unless, as in some higher organisms, there are many duplicated genes.

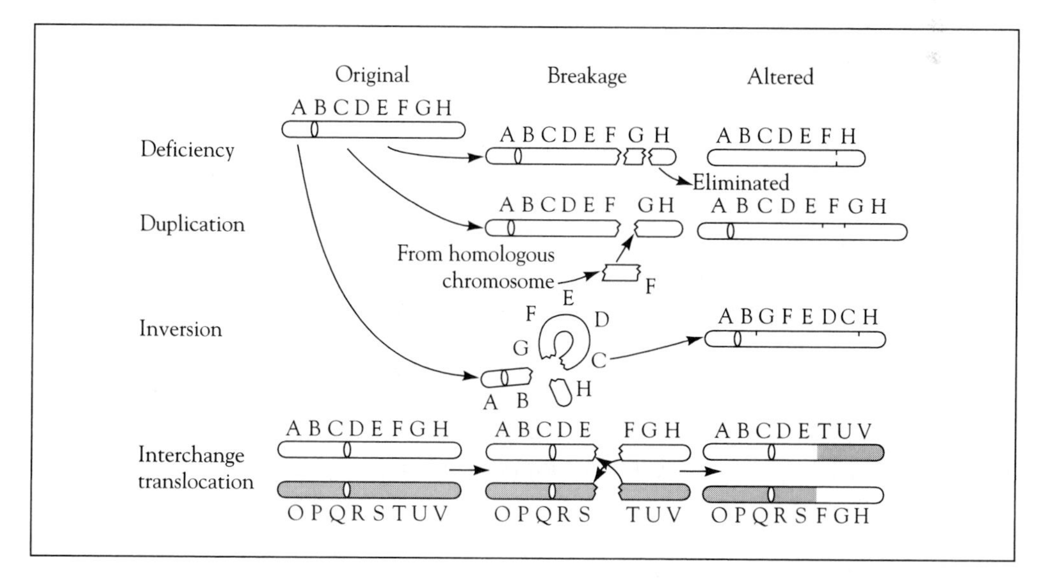

FIGURE 2.3 Chromosome breakage and reunion can give rise to four principal structural changes.

When two chromosomes are not perfectly aligned during crossing over, the result is one chromosome with a deficiency and one with a duplication of genes. The change may be advantageous in that greater amounts of enzymes may be coded for. In yeasts, for example, an increase of acid monophosphatase enables cells to exploit more efficiently low concentrations of phosphate in the medium.

An inversion occurs when a chromosome breaks in two places and then re-fuses with the segment between the breaks turned around so that the order of its genes is reversed with respect to that on the unbroken chromosome. Such breaks probably occur at prophase, during which the chromosomes are long and slender and often bent into loops. In translocations, two nonhomologous chromosomes break simultaneously and exchange segments.

Together, gene and chromosomal mutation provides most of the genetic variability in a population. However, some other mechanisms for promoting genetic diversity do exist and will be mentioned briefly. If the first meiotic division fails to occur, unreduced **diploid** ($2n$) gametes are formed. Normally, the union of a diploid gamete with a normal **haploid** (n) gamete forms a triploid ($3n$) zygote. Such individuals, if they can exist at all, are usually sterile; they cannot produce gametes with effectively balanced complements of chromosomes. However, if two unreduced ($2n$) gametes unite, the resulting tetraploid ($4n$) zygote may be fertile. Such a phenomenon can occur in some plants where, because of a doubled gene dosage, these polyploids are often bigger and more robust than diploids (Levin 1983). Tetraploidy is also found, rarely, in animals that are **parthenogenetic** (those in which females can produce daughters without fertilization by male gametes), like brine shrimp, *Artemia salina*, in which tetraploids, pentaploids, octoploids, and even decaploids have been found.

The existence of all this variation was something of a shock in the 1920s and 1930s, before which time genetic uniformity had been taken for granted, and it led to a new groundswell of opinion, led by Theodosius Dobzhansky, that populations were in fact huge collections of diverse genotypes (Futuyma 1986). How much change in the amount of DNA, in the genome size, or in chromosome structure is necessary to cause speciation? There is no simple correlation. Two species of deer in the same genus, *Muntiacus reevesi* and *M. muntjac*, even have vastly different complements of chromosomes; $2n = 46$ and $2n = 6$, respectively (White 1978)!

2.3 Because selection should act to favor the most fit genotype in each environment, it would seem that genetic diversity should decline over evolutionary time as less fit genotypes are eliminated. There are a number of processes, however, that act to maintain genetic diversity at relatively high levels.

The Hardy-Weinberg theorem describes how the frequency of alleles and genotypes should maintain equilibrium in a population not experiencing selection.

Although variation is created by mutation and chromosomal rearrangements, given simple Mendelian genetics and dominance, how is it that a dominant allele, responsible for a 3:1 numerical ratio of phenotypes, does not gradually supplant all other types of alleles, given that all alleles confer equal fitness on the organism? That is, if a gene pool in one generation consists of 70 percent A alleles and 30 percent a alleles, what stops the proportion of A alleles from increasing dramatically? What will the proportion of alleles be in the next generation? This very question was posed and independently answered by, a British mathematician, G. H. Hardy, and a German physician, W. Weinberg, in 1908. They assumed three things: (1) populations are large, which thus negates sampling errors; (2) individuals contribute equal numbers of gametes; and (3) mating is random. In short, gametes carrying different alleles combine in pairs in proportion to their respective frequencies in the gamete pool.

Assume that a diploid (normal) population polymorphic for A contains the following proportions of genotypes: 60 percent AA, 20 percent Aa, and 20 percent aa. The genotype frequencies are 0.60 AA + 0.20 Aa + 0.20 aa = 1. The frequency of the A allele (p) is obtained by adding up all of the A alleles in the population (2 from the AA genotype = 0.6 + 0.6, and one from the Aa genotype = 0.2) and dividing by twice the total frequency of genotypes (there are two alleles per genotype, so $2 \times 1 = 2$). This is done for each allele. Thus, the allele frequencies must then be

$$p = \frac{0.60 + 0.60 + 0.20}{2} = 0.70$$

$$q = \frac{0.20 + 0.20 + 0.20}{2} = 0.30$$

The paired combinations of matings are then

FEMALE GAMETES		MALE GAMETES		ZYGOTIC FREQUENCIES	
0.70 A	×	0.70 A	=	0.49 AA	
0.70 A	×	0.30 a	=	0.21 Aa	
0.30 a	×	0.70 A	=	0.21 Aa	= 0.42 Aa
0.30 a	×	0.30 a	=	0.09 aa	

The gene pool of the second generation now contains the two alleles in the following frequencies:

$$p = \frac{0.49 + 0.49 + 0.42}{2} = 0.70$$

$$q = \frac{0.42 + 0.09 + 0.09}{2} = 0.30$$

The allele frequencies are the same as they were in the first generation, which is why the proportion of alleles is unchanging, even though the frequencies of genotypes has changed.

In general, if p is the frequency of allele A and q the frequency of allele a ($q = 1 - p$), then the combinations of A and a gametes will produce zygotes in proportions given by the expansion of the binomial square $(p + q^2) = p^2 + 2pq + q^2$, that is, p^2 AA, $2pq$ Aa, and q^2 aa.) This generalization is known as the Hardy-Weinberg theorem. For three alleles whose frequencies are p, q, and r, the equilibrium frequency of genotypes is given by the trinomial square $(p + q + r)^2$. In general, for an n-ploid organism, the genotype frequencies are given by $p + q^n$.

Populations can deviate from the predictions of the Hardy-Weinberg theorem if mating is nonrandom with respect to genotype.

Of course, genotype frequencies are not always constant over time. Species change and evolve as a result of deviations from the Hardy-Weinberg assumptions. For example, not all mating is random but is often **assortative**. Bateson, Lotwick, and Scott (1980) have shown that individual Bewick's swans (*Cygnus columbianus*) can distinguish other individuals by their face markings. Swan families tend to have characteristic family markings, and young birds prefer to mate with partners whose faces are clearly different from their own. This preference reduces inbreeding, which can dramatically increase the proportions of genotypes in favor of homozygotes (Grant 1977). Furthermore, sexual selection (see Chap. 5.3) in many organisms ensures that individuals do not always contribute equal numbers of gametes. In territorial species some males contribute more genes than others; in those with harem-holding males, just one individual will fertilize the majority of the females. New genes can also often enter the population via the migration of new individuals from different areas. The result is that there are very few truly **panmictic** species, in which all the individuals from one species form a single randomly mating population. One of the few examples is the common eel, *Anguilla rostrata,* in which individuals from U.S. eastern-seaboard drainages and Europe as well migrate to one area of the ocean near Bermuda, the Sargasso Sea, to breed.

The effects of environmental variance can also cause populations to deviate from the predictions of Hardy-Weinberg.

In addition to disruption of the Hardy-Weinberg assumptions by nonrandom mating, each genotype may be phenotypically variable to some extent because its development is directly affected by the environment. Fly weight depends often on the amount of medium available to the larvae and whether or not they have been competing for it. Certain crop plants and domesticated animals perform better in particular climates and agricultural regimes. This dependence is termed environmental variance, V_E, and affects variance of phenotypes in nature, the Hardy-Weinberg theorem notwithstanding. It is in contrast to the genetic variance V_G. Total phenotypic variance V_P would thus be

$$V_P = V_G + V_E$$

Which type of variance is most important? This is the old "nature-versus-nurture" or genetics-versus-environment debate. The answer usually is "both," though a rule of thumb is 40:60 genes versus environment.

Sometimes, the reaction to a difference in environment differs among genotypes, so there is a genotype-by-environment interaction (V_{G+E}), which also contributes to the phenotypic variance. In other words, certain genotypes do better in some habitats, and other genotypes do better in others. Thus:

$$V_P = V_G + V_E + V_{G+E}$$

For example, resistance in different varieties of wheat to the wheat stem sawfly was correlated to the solidness of the straw; larvae could not move to obtain sufficient food if the pith was firm and compact (Painter 1951). However, expression of stem solidness was influenced strongly by the amount of light that the stem received when it was elongating (Platt 1941). Maddux and Cappuccino (1986) showed that differences in aphid numbers on clones of goldenrod were apparent only when the plants were well watered, and the effect was not found in drier environments. Kennedy and Barbour (1992) believe that relatively few studies clearly document the occurrence of genotype-by-environment interactions as they affect resistance in natural systems. In a study of the fifteen species of insect on coastal plants in Florida, Stiling and Rossi (1995) found environment to affect insect densities much more often than plant genotype. Moreover, only one out of fifteen possible interactions was significant.

In most populations, it is hard to know whether a gene exists for a particular trait, and it is hard to study its properties. The reason is that alternative alleles at that locus are hard to identify; studies of variation have therefore focused on distinct polymorphisms such as red eye versus white eye in *Drosophila*. In nature, however, species are seldom found with two or more discrete phenotypes that lend themselves well to Mendelian crosses (see Chap. 3 for some examples). In many of these cases, the two phenotypes or morphs were originally described as distinct species. The differential survival of these **morphs** in nature and how sources of mortality act differentially upon them fall into the realm of natural selection.

Genetic variation can also be maintained in the presence of selection in certain circumstances.

When selection acts at a locus and a homozygous genotype is most fit, the less advantageous alleles should be eliminated. This is the concept of natural selection (which will be developed further in Chap. 3). This situation contrasts with the Hardy-Weinberg theorem, which showed how two different alleles are maintained in the same population, generation after generation, if one is not favored over the other. Natural selection is obviously a common and potent force in the real world. How then does variation continue to exist in nature? Much of the effort of population geneticists has been devoted to this question. There are three common answers:

1. Selection acts on the locus so as to maintain a stable polymorphism, in which different genotypes are most fit under different situations; the "wild type" allele is not always most fit (see Chap. 3).
2. Fixation by selection is counteracted by mutation (for example, mutation may explain the reoccurrence of albinos in populations from which they are eliminated by natural selection).
3. Fixation by natural selection is counteracted by **gene flow**.

2.4 Recent technological advances have improved the ability of scientists to estimate how much genetic variation exists in nature.

Given the existence of so many mechanisms by which variation is produced, it is interesting to speculate on how much genetic variation exists in nature. For a long time, such speculation was difficult because it was impossible to tell how many loci exhibited no genetic variation—no clues were given by the phenotype. In the 1960s it was realized that most loci actually code for proteins, especially enzymes. Different forms of, say, alcohol dehydrogenase are coded for by different alleles. Two individuals with the same form of enzyme are presumably genetically identical at that locus. Therefore, invariant gene loci can be found through a search for invariant enzymes. The most common technique for distinguishing different genetic forms of enzymes (allozymes) is gel electrophoresis. In this technique, specially prepared samples of specimen tissue are placed in a porous gel, and an electrical potential is applied, causing the electrically charged enzymes to migrate through the gel along the lines of electrical force. Because slightly different forms of the same amino acid differ in charge, they migrate at different rates and separate into bands at different distances from the original specimen. These bands become visible when the gel is first flooded with a substrate on which the enzyme acts and then stained with a substance that colors the reaction products. Such enzyme differences have proved to be of genetic origin, and experienced workers have identified much genetic variation among individuals in populations whose phenotypes looked identical. For example, as Photo 2.4 shows, morphologically identical strains of bacteria have proved to be distinctly different when examined electrophoretically.

The first assays of genetic variation by protein electrophoresis were published in 1966 by Harris, who reported on variation in ten enzyme loci in humans, and by Lewontin and Hubby (1966), working on *Drosophila*. The results were something of a surprise; no one had suspected just how much variation existed in nature. In *Drosophila* the average population was polymorphic at about 30 percent of its loci, with about two-to-six alleles per polymorphic locus. An average fly was likely to be heterozygous at 12 percent of its loci, so any two flies would, on average, differ at about 25 percent of their loci. In humans, Harris's data indicated 30 percent of the loci were polymorphic and about 10 percent were heterozygous. Moreover, it has since been shown that the enzyme products of some alleles have similar electrophoretic mobility, and variation would not be distinguished by the gel technique (Coyne 1976; Coyne, Felton, and Lewontin 1978). Therefore, existing estimates of variation in nature could err on the low side.

Selander's genetic variation for a wide variety of taxa is summarized in Table 2.2. Soulé (1976) compiled the first evidence that levels of genetic variation in wildlife were related to population size (Fig. 2.4). In general, Soulé argued, vertebrates are less polymorphic than invertebrates; species that form small local populations or are inbred have reduced heterozygosity.

Frankham (1996) suggested that the hypothesis that genetic variation is related to population size leads to the following predictions:

- Genetic variation within species should be related to population size.
- Genetic variation within species should be related to island size (because population size is often related to island size).
- Genetic variation should be related to population size within taxonomic groups.
- Widespread species should have more genetic variation than restricted species (widespread species may be more abundant).
- Genetic variation in animals should be negatively correlated with body size (larger animals have longer generation times and often smaller numbers, see Chap. 6).

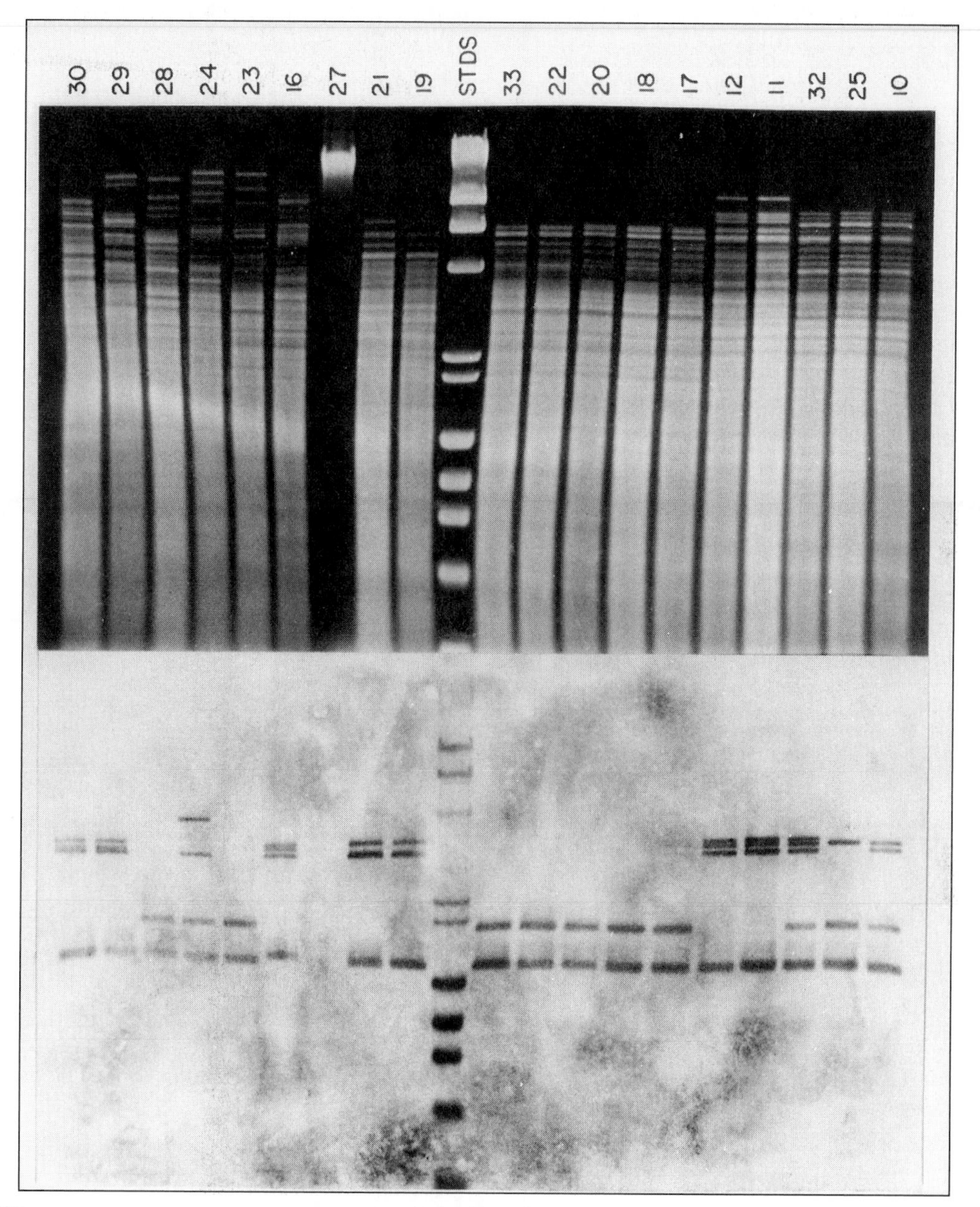

■ **PHOTO 2.4** Genetic variation within a species. The upper photo is DNA from nineteen different cultures of *Xanthomonas maltophilia* taken from a Tallahassee, Florida, hospital in 1989. The lower photo is the same DNA, subjected to an additional treatment. All the cultures in each photo were treated in the same way (according to a method that produces a DNA "fingerprint"). Despite the superficial similarity of the cultures, their DNAs are clearly different. The additional techniques used in the lower photo show that even some of those that appear identical in the upper photo (for example, 10, 25, and 32) are unique. *(Photos courtesy of R.H. Reeves, Florida State University, Tallahassee.)*

TABLE 2.2 Genetic variation at allozyme loci in animals and plants. (After Selander 1976)

	NUMBER OF SPECIES EXAMINED	AVERAGE NUMBER OF LOCI PER SPECIES	AVERAGE PROPORTION OF LOCI	
			POLYMORPHIC PER POPULATION	HETEROZYGOUS PER INDIVIDUAL
Insects				
Drosophila	28	24	0.529	0.150
Others	4	18	0.531	0.151
Haplodiploid wasps[a]	6	15	0.243	0.062
Marine invertebrates	9	26	0.587	0.147
Marine snails	5	17	0.175	0.083
Land snails	5	18	0.437	0.150
Fish	14	21	0.306	0.078
Amphibians	11	22	0.336	0.082
Reptiles	9	21	0.231	0.047
Birds	4	19	0.145	0.042
Rodents	26	26	0.202	0.054
Large mammals[b]	4	40	0.233	0.037
Plants[c]	8	8	0.464	0.170

[a] Females are diploid; males haploid.
[b] Human, chimpanzee, pigtailed macaque, and southern elephant seal.
[c] Predominantly outcrossing species.

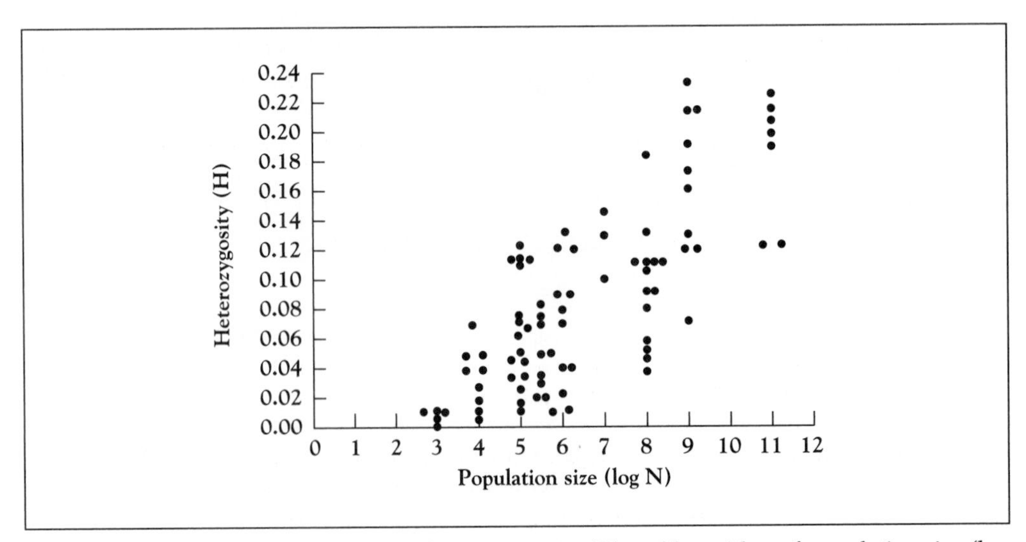

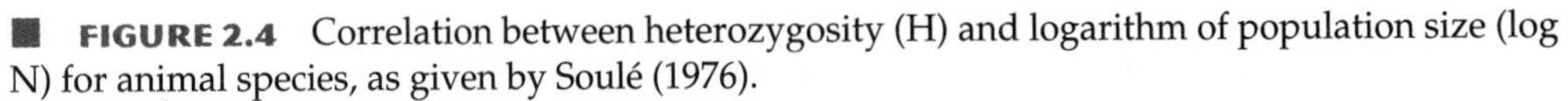
■ **FIGURE 2.4** Correlation between heterozygosity (H) and logarithm of population size (log N) for animal species, as given by Soulé (1976).

- Genetic variation should be negatively correlated with rate of chromosome evolution.
- Genetic variation across species should be related to population size.
- Vertebrates should have less genetic variation than invertebrates or plants (again due to higher numbers of plants and invertebrates).
- Island populations should have less genetic variation than mainland populations.
- Endangered species should have less genetic variation than nonendangered species (because endangered species generally have smaller population sizes).

More recent empirical observation (Nei and Graur 1984; Nero, Bieles, and Ben-Schlomo 1984; Coyne 1984; Wooten and Smith 1985; Karran 1987; Hamrick and Godt 1989) support all these hypotheses. Perhaps this is not surprising since they are all built on the same premise that population size affects genetic variation. There can be no doubt that genetic variation is related to population size, as Soulé proposed. Small population size reduces the evolutionary potential of wildlife species but at the same time predisposes them to greater chances of genetic drift.

It must always be noted, however, that some enzymes are not soluble and cannot be examined electrophoretically; whether the same degree of polymorphism exists in these is not known.

Very recently, the development of DNA technology has provided us with a means to sample genetic variation directly at the DNA level. At present, however, these methods are more expensive and generally more difficult than allozyme techniques. Therefore, there are not yet the large amounts of data on species variation that are available from allozyme studies. This situation will probably change as the technology becomes less expensive and more widely available.

Because different parts of DNA evolve at different rates, we can choose to study particular segments to answer particular questions. Some genes change very slowly and can be used to study relationships between groups of organisms that diverged from one another hundreds or even thousands of millions of years ago. Other regions of DNA change at such a rapid rate that most individuals in a population are distinct. Still other regions of DNA show intermediate levels of variability that are useful in studies of variation that and between populations of a species or of variation between closely related species.

One approach to measuring genetic variation uses enzymes to cut DNA at known locations and measures the length of the fragments that result; Variation in fragment length is thought to provide an estimate of genetic variation in that section of DNA.

The DNA-based techniques most widely used for studies of within-and-between population variation make use of the properties of enzymes derived from various species of bacteria, which use them to protect themselves from infection by viruses. The viruses are almost pure DNA, and they reproduce by inserting their own DNA into the bacteria DNA and having the host replicate it, essentially producing more viruses. The cutting (restricting) enzymes of the bacteria are very specific in the DNA sequence. They recognize and cut foreign DNA, as they attempt to cut out the viral portions. These enzymes form the background of the technology of DNA manipulation. If DNA from an individual is extracted and cut with a restriction enzyme, and the resulting fragments are separated by length in an electrophoretic gel, a pattern is obtained. Other individuals might have slight changes in their DNA, which produce additional or fewer sites that are recognized by the enzymes and a different pattern of restriction fragments is seen on the gel. Thus, by repeating this process with many individuals and different restriction enzymes, patterns of variation can be seen and analyzed to estimate the amount of variation in the DNA sequences among the individuals.

DNA sequencing, in which the exact sequence of nucleotide bases is determined for a segment of DNA, provides an even more powerful tool for measuring genetic variation.

A yet more powerful, and more expensive, method for assessing genetic variation is to sequence a portion of DNA itself. This has been made possible by the advent of the polymerase chain reaction (PCR) technique, which can be used to make millions of copies of a particular region of DNA. Because of this, it is now possible to obtain DNA sequence data from a wide variety of organisms much more quickly than was possible previously. Because PCR techniques can amplify even minute amounts of DNA, we can obtain data from very small organisms that contain little tissue. We can also use tiny samples of tissue from larger organisms without having to kill or otherwise injure them. For example, a drop of blood, a hair root, or a feather are now adequate material for DNA sequence-based work. This has obvious importance for dealing in rare and endangered species, and is much in vogue in conservation ecology. The finest scale of analysis of genetic material is known as genetic fingerprinting and was discovered by Dr. Michael Jefferies and colleagues in the 1980s at the University of Leicester in the United Kingdom. Genetic fingerprinting makes use of a common but peculiar DNA circumstance known as minisatellites. These are disposed throughout the genome and consist of randomly repeated copies of short sequence units. High levels of variation in the numbers of these repeated units are exploited in fingerprinting to identify close relatives. Since this discovery, the "genetic fingerprinting" of individuals has found a number of applications in human genetics, including forensic medicine and paternity analysis. Ecologists have also found it useful in determining kinship and relationships of family members in behavioral ecology studies. Some conservationists are curious whether the new advances in DNA technology will permit more species to be cloned (Applied Ecology: Can Cloning Help Save Endangered Species?).

The amount of genetic variation, and its geographic distribution, can provide insight into evolutionary processes.

Studies of evolutionary ecology often partition the measured genetic variation within a species into two components; variation within populations and variation among populations. Information on the amount of genetic variation within and among populations provides insight into a specie's recent evolutionary history (how long have two populations been separated?), current evolutionary processes (how much gene flow occurs among populations?), and future evolutionary challenges (does this species have enough genetic diversity to overcome disease epidemics?).

Very often, allele frequencies differ at loci more from one population to another than they do between populations. Genetic diversity in a species thus consists of within-population diversity and among-population diversity (Fig. 2.5). A simple genetic model of this diversity is

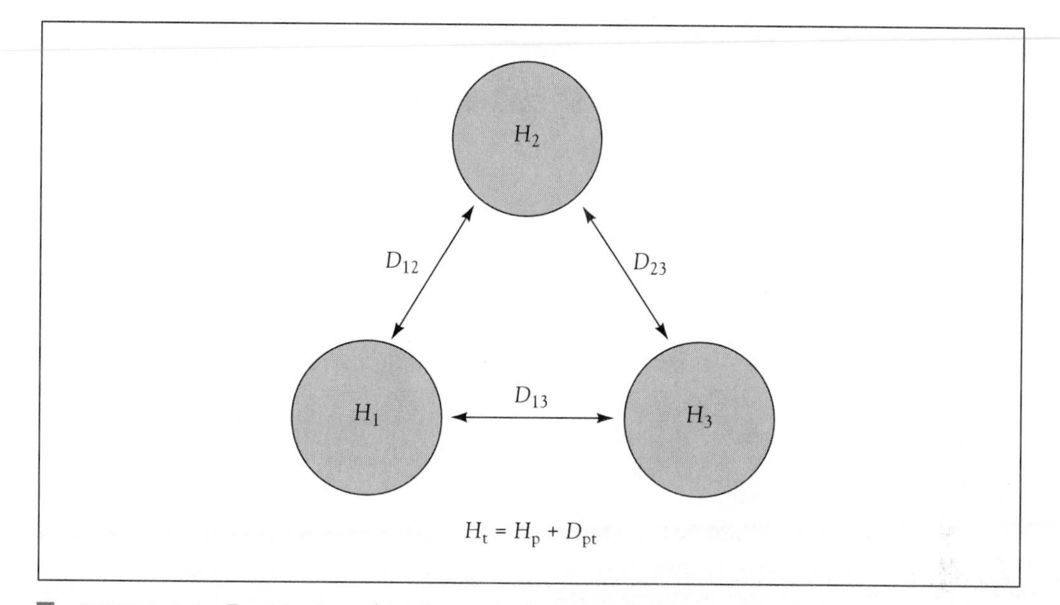

■ **FIGURE 2.5** Partitioning of total genetic diversity, H_t, into within- and among-population variation. This figure represents a species with three populations, each with some level of within-population heterozygosity (H_1, H_2, and H_3); mean heterozygosity is H_p. Among-population divergence (D_{12}, D_{23}, and D_{13}) is represented by the arrows between populations; mean divergence is D_{pt}.

$$H_t = H_p + D_{pt}$$

where $H_t =$ total genetic variation in the species, $H_p =$ average diversity within populations, and $D_{pt} =$ average diversity among population across the total species range (Nei 1975).

The point is that a species of little genetic variation may be partitioned into component parts: within- versus among-population diversity. With that approach, one can determine how variation is spatially distributed and thus define areas of particular conservation interest. For example, Stangel, Lennartz, and Smith (1992) found that the endangered red-cockaded woodpecker (*Picoides borealis*) in the southeastern United States had an overall mean allozyme heterozygosity level of 0.078 (or 7.8 percent), which is about typical for most bird species that have been sampled (Photo 2.5). Of the total genetic variation measured, 14 percent consisted of among-population differentiation and 86 percent was mean genetic diversity within populations. This may seem like most of the genetic diversity is within populations. Even though it seems low, 14 percent is relatively high for birds, and most birds have widespread dispersal, which results in high genetic exchange and little local genetic differentiation. The woodpeckers are more site specific than many birds, and consequently local populations tend to diverge genetically. Conservation programs for this species should therefore protect both components of genetic diversity, among and between populations, in order to retain the maximum amount of variation and maintain a natural population genetic structure. Studies of other organisms, like plants, bear out the generalization that within-population diversity is much greater than among-population diversity. A review of the available allozyme literature from 449 species of plants showed that 78 percent of the diversity was found within populations (Hamrick and Godt 1990).

APPLIED ECOLOGY

Can Cloning Help Save Endangered Species?

In 1997 Ian Wilmut and his colleagues at Scotland's Roslin Institute published their famous paper on the cloning of the now-famous sheep, Dolly (Photo 1), from mammary cells of an adult ewe. Conservation biologists are curious whether the same technology could be used to save endangered species (Bawa Menon, and Gorman 1997). However, there are a number of technical hurdles to overcome.

■ **PHOTO 1** Dolly, a sheep cloned by researchers in Scotland in 1997. Can conservation biologists use the same technology to save endangered species? *(Najlah Feanny, Stock Boston NJF0348C.)*

1. We would have to develop an intimate knowledge of every species' reproductive cycle. For sheep this was routine, but eggs of different species, even if they could be harvested, often require different nutritive media in laboratory cultures.
2. We would preferably have to identify surrogate females that would carry babies to term. Because it is desirable to leave natural mothers available for breeding, so closely related species would therefore have to be tried. A good candidate might be to try implanting into a horse an embryo from the endangered Przewalski's horse. Although no one knows if this would work, a zebra embryo has been grown successfully in a horse.
3. Because the collection of cells of many endangered species is fibroblast, that is, cells made during wound healing, it would be valuable to see if they could be used instead of mammary cells.
4. Cloning is currently so expensive that conservation resources might be better spent elsewhere, for example, in buying up habitat.
5. Cloning might not be able to do much to increase the genetic variability of the population. However, if it were possible to use cells in collections from long–deceased animals then these old clones would theoretically reintroduce lost genes back into the population. Furthermore, if an organism, say a giant panda, has only one natural offspring, 50 percent of its genes are lost. But if we clone the panda ten times, 95 percent of its genetic information would be maintained. This may actually allow zoos to maintain smaller stocks, for less money, yet still maintain diversity.

Variation among populations exists on much larger geographic scales as well. Commonly, the farther apart populations are, the more different they are in allele frequencies and in phenotype characteristics. A gradual change along a geographic transect is called a **cline**. Clines may extend over the whole geographic range of a species. For example, body size in white-tailed deer (*Odocoileus virginianus*) increases gradually with increasing latitude over most of North America—this phenomenon is common in mammals and birds and has been termed **Bergmann's rule**. In the clover *Trifolium repens* (Photo 2.6), the proportion of plants that produce cyanide increases in warmer locations in Europe (Fig. 2.6). This difference in frequency is caused by a balance between the

■ **PHOTO 2.5** The endangered redcockaded woodpecker about to enter its nest on a slash pine in Florida. A relatively high proportion of genetic variation is found between populations of this species, which means that in order to preserve genetic diversity, it will be necessary to conserve many areas and many populations. *(Jeff Lepore, Photo Researchers, Inc. 7X5584.)*

■ **PHOTO 2.6** White clover, *Trifolium repens*, can produce cyanide in its leaves. Freezing temperatures rupture cell membranes allowing the release of this toxin, killing the plant. For this reason, the production of cyanide increases in warmer climates where the likelihood of freezing decreases. *(E. R. Degginger, Animals Animals/Earth Scenes 559321.)*

advantage cyanogenic plants derive from being protected from herbivory and the disadvantage of being killed when frost disrupts the cell membrane, releasing toxins into the plant's tissues (Foulds and Grime 1972*a*, *b*; Jones 1973, Dirzo and Harper 1982). Cyanogenesis occurs in over eight hundred plant species spanning one hundred plant families including the rubber tree (Lieberei et al. 1989), so this is not a unique phenomenon. However, in some other species the distribution patterns are not influenced by

■ **FIGURE 2.6** Frequency of the cyanide-producing form in populations of white clover (*Trifolium repens*), represented by the black section of each circle. The cyanogenic form is more common in warmer regions. Lines are January isotherms. (*Redrawn from Daday 1954.*)

environmental gradients (Compton, Newsome, and Jones 1983) so a general explanation of the distribution of these plants is, as yet, lacking.

2.5 There are a number of processes that act to reduce genetic diversity; most of these are associated with small population sizes.

Reduced genetic diversity can arise from four factors that are all a function of population size: inbreeding, genetic drift, neighborhoods, and bottlenecks. All are of vital importance in conservation biology (Frankham 1995).

FIGURE 2.7 Inbreeding depression and hybrid vigor. The two corn plants at the left are of two inbred strains; the larger plant to their right is a hybrid of the two. All the plants to the right are successive self-fertilized generations from the hybrid. *(Drawn after a photo by Jones 1924.)*

Matings between closely related individuals can decrease the genetic variation in a population and in some cases lead to the expression of recessive deleterious alleles.

The phenomenon of inbreeding depression has long been known. **Inbreeding** (mating among close relatives) is certainly a reality in many social animals (Chap. 4). One severe form is self-fertilization or "selfing" in plants. The effects are usually deleterious, and, as is dramatically illustrated in Figure 2.7, crossbreeding can reverse those effects. Viability and especially **fecundity** (or yield of crops) decline as populations become more inbred in the laboratory (Table 2.3). Generally, the more inbred the population, the

TABLE 2.3 Inbreeding depression in rats. (After Lerner 1954)

YEAR	NONPRODUCTIVE MATINGS (PERCENT)	AVERAGE LITTER SIZE	MORTALITY FROM BIRTH TO FOUR WEEKS (PERCENT)
1887	0	7.50	3.9
1888	2.6	7.14	4.4
1889	5.6	7.71	5.0
1890	17.4	6.58	8.7
1891	50.0	4.58	36.4
1892	41.2	3.20	45.5

Note: The years 1887–1892 span about 30 generations of parent x offspring and sib matings.

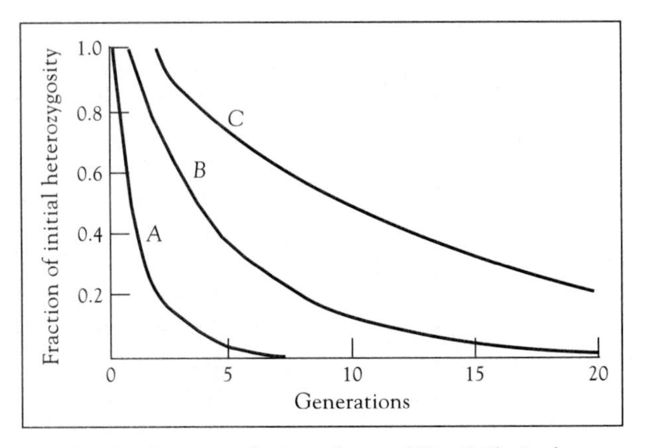

■ **FIGURE 2.8** Decrease in heterozygosity due to inbreeding. Systems of mating are exclusive self-fertilization (curve A), sib mating (curve B), and double-first cousin mating (curve C). *(Redrawn from Crow and Kimura 1970.)*

quicker the proportion of heterozygosity in the population drops (Fig. 2.8). In human populations, the consequences include higher mortality, mental retardation, albinism, and other physical abnormalities (Stern 1973). The reason that inbreeding is so disadvantageous is that the frequency of homozygotes for recessive alleles is thought to increase. Recessive alleles are more likely to be deleterious for the simple reason that dominant deleterious traits are more quickly selected out of the population, leaving behind their recessive counterparts.

Ralls and Ballou (1983) showed the effects of inbreeding on juvenile mortality in captive populations of mammals (Fig. 2.9). In both ungulates, primates, and small

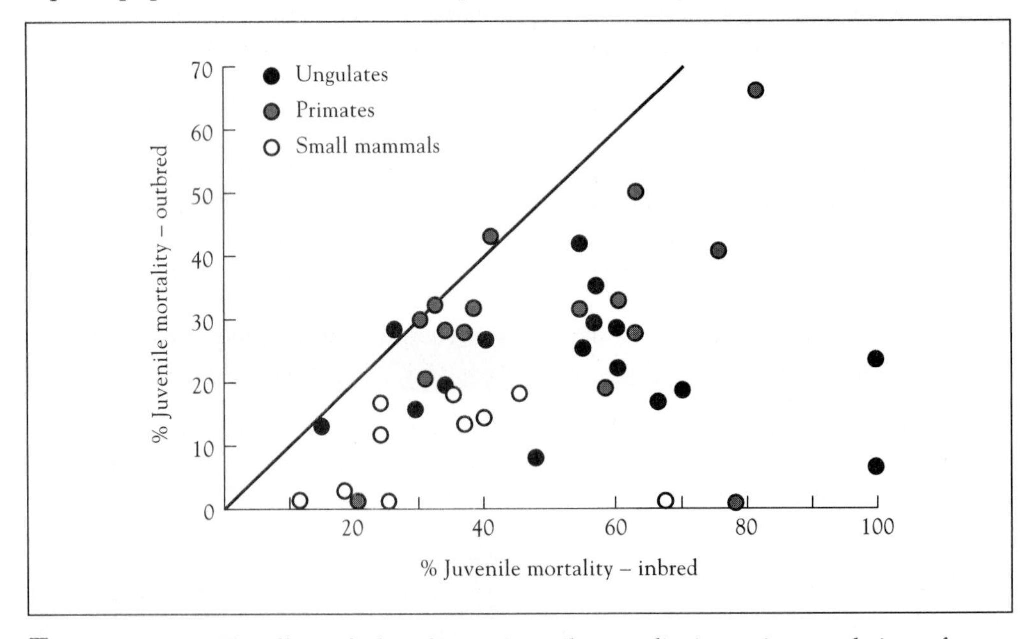

■ **FIGURE 2.9** The effects of inbreeding on juvenile mortality in captive populations of mammals. Each point compares the percentage of juvenile mortality for offspring of inbred and noninbred matings. The line indicates equal levels of mortality under the two breeding schemes. Points above the line represent higher mortality from noninbred matings; points below the line, higher mortality from inbred matings. The distance of a point from the line indicates the strength of the effect of level of inbreeding. *(From Ralls and Ballou 1983.)*

mammals there was higher mortality from inbred matings than from noninbred matings. Data from domesticated animals indicate that an expected increase of inbreeding per generation of 10 percent will result in a 5 to 10 percent decline in individual reproductive rates such as clutch size or survival rates. In aggregate, total reproductive attributes may decline by 25 percent (Frankel and Soulé 1981). Thornhill (1993) and Jimenéz et al. (1994) have gathered together much of the recent data on inbreeding and shown that there is clear evidence of inbreeding depression in the wild. One of the best examples comes from Vrijenhoek (1994) who showed that a genetically variable sexual species of fish numerically dominated a related parthenogenetic species until a drought eliminated their habitat. When the populations were subsequently reestablished, the sexual species possessed reduced genetic variation from a founding event and was consistently less abundant than the parthenogenetic species. The sexual species reestablished its numerical dominance following the deliberate addition of genetic variation via replacement of thirty sexual individuals by fish from elsewhere. Part of the mechanism here had to do with parasite load—increased genetic variation reduced parasite attack. Unfortunately, a substantial number of populations have very low levels of allozyme heterozygosity (Table 2.4) indicating high levels of inbreeding.

The effects of inbreeding are more extreme in small populations. The smaller the population N, the faster heterozygosity declines (Fig. 2.10). This result has important consequences in the real world, where plant and animal populations are constantly declining because of shrinking habitats, and conservation science has become particularly concerned with the genetics of small populations. A rule of thumb has been that a population of at least fifty individuals is necessary to prevent inbreeding for the immediate future. One species of concern is the California condor, shown in Photo 2.7. It remains to be seen whether the offspring of twenty-six captive individuals—the sole remnants of the species in 1986—can overcome the effects of inbreeding and form a fit and healthy population when released into the wild. As of 1993, forty-nine California condors had been bred in the San Diego Wild Animal Park and Los Angeles Zoo, and eight had been released back into the wild (Toone and Wallace 1994). One bird was killed by ethylene glycol, a toxin found in car antifreeze, but the other seven were doing well, feeding on provided carcasses.

Not all inbreeding is cause for alarm. Many natural populations have apparently experienced low levels of inbreeding for many generations with no ill effects. Inbreeding depression is probably more important in a species or population with historically large population sizes that now occurs in small populations. In historically small populations, such as those on islands, species have had to deal with relatively high rates of inbreeding for thousands of years.

The opposite, outbreeding, which increases the number of heterozygotes and conceals more of the recessive alleles, is termed **heterosis**. As well as inbreeding depression, we can get outbreeding depression. Often, local populations adapt to their regional environment, particularly if dispersal among populations is limited. Local adaptation, or the formation of **demes**, has been shown to occur for species of insects on long-lived plants (Karban 1989a; Mopper and Strauss 1997). Species like insects can undergo hundreds of generations on hosts like trees or clonal bushes, which live for many years. The insects can therefore remain one step ahead of the plants in the evolutionary arms race. Deme formation may also be more likely where limited gene flow occurs between demes—

TABLE 2.4 Levels of genetic variation in endangered species and populations. (After Frankham 1995)

LOW	NORMAL	HIGH
Mammals		
Cheetah	Indian rhinoceros	Speke's gazelle
Greater panda	Humpback whale	
Asiatic lion		
Black-footed ferret		
Northern hairy-nosed wombat		
Cotton-top tamarin		
Lion tamarin		
European bison		
Arabian oryx		
Ethopian wolf		
White rhinoceros		
Pere David's deer		
Black rhinoceros		
Przewalkski's horse		
Channel Island fox		
Florida panther		
Isle Royale gray wolf		
St. Lawrence beluga whale		
Sand gazelle		
Birds		
Hawaiian goose	Red-cockaded woodpecker	
Spotted owl	Palila	
Whooping crane	California condor	
Puerto Rican parrot		
Kakapo		
Fish		
Topminnow		
Insects		
Fritillary butterfly		
Plants		
Howellia aquaticus		
Pedicularis furbishiae		
Ptilmnium nodosum		
Amsincaia grandiflora		
Torrey taxifolia		
Trifolium stoloniferum		

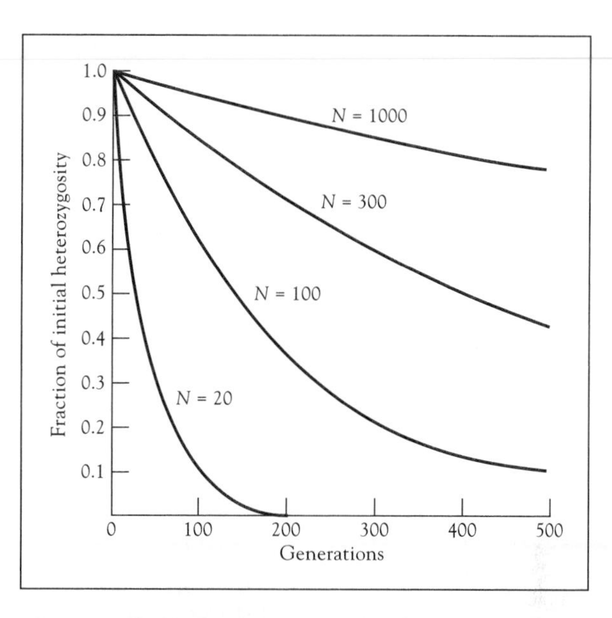

▪ **FIGURE 2.10** Decrease in heterozygosity due to finite population size. Variation is lost randomly through genetic drift. N equals population size. (*Modified from Strickberger 1986.*)

▪ **PHOTO 2.7** California condor. Because the total population of condors was only 26 birds in 1986, many conservationists warned that insufficient genetic diversity existed to ensure the continued success of the species. (*Jerry L. Ferrara, Photo Researchers, Inc. 2R7134.*)

where the insects are brachypterous or without wings. However, two recent studies have shown deme formation in mobile, winged insects, which underlies the fact that demes may form despite the presence of some gene flow. In these two cases, both insects (a leaf miner and gall-maker) were endophytic (fed inside of the host plant as opposed to exophytic, feeding externally), so had probably developed particularly close relationships with their host plants. However, outbreeding depression is commonly thought of as less important than inbreeding depression (Frankham 1995).

In small populations, genetic diversity is lost over time just by chance.

In small populations there is a good chance that some individuals, purely by chance, will fail to mate successfully. If an individual that fails to mate possesses a rare allele, that allele may not be passed on to the next generation, resulting in a loss of genetic diversity from the population. The likelihood of an allele being represented in just one or a few individuals is higher in small than in large populations, so small, isolated populations are particularly vulnerable to this type of reduction in genetic diversity. Isolated populations will lose some percentage of their original diversity over time, approximately at the rate of $1/2N$ per generation. A population of a thousand will retain 99.95 percent its genetic diversity in a generation, while a population of fifty will retain only 99.0 percent. Such losses seem insignificant but are magnified over many generations. Thus, after twenty generations a population of a thousand will still retain over 99 percent of its original variation, but the population of fifty will retain less than 82 percent. For organisms that breed annually, this could mean a substantial loss in variation over a very few years. Once again, this effect becomes more severe as the population size decreases.

Another rule of thumb is that a population size of at least five hundred is necessary to lessen the effects of genetic drift. Thus, the "50/500" rule has entered the conservation literature as a "magic" number in conservation theory (Simberloff 1988) (fifty being the critical size to prevent excess inbreeding and five hundred the critical size to prevent genetic drift [Franklin 1980]). Lacey (1987) showed that the effects of genetic drift could be countered by immigration of individuals into a population. Even relatively low rates of immigration of one immigrant every generation would be sufficient to counter genetic drift in a population of 120 individuals (Fig. 2.11). Recently, Lande (1995) argued that even five hundred might be too low a number for conservation purposes since many mildly deleterious mutations would accumulate in small populations and gradually reduce fitness. He therefore suggested five thousand as an appropriate goal for conservation biology. One of the best tests of the idea that fifty or five hundred individuals may be a minimum number for conservation purposes was provided by Berger's (1990) study of 120 bighorn sheep *(Ovis canadensis)* populations in the U.S. Southwest. The striking observation was that 100 percent of the populations with fewer than fifty individuals went extinct within fifty years, while virtually all of the populations with more than one hundred individuals persisted for this time period (Fig. 2.12). The exact causes of extinction were difficult to pinpoint and undoubtedly included disease, predation, and starvation as well as inbreeding and genetic drift.

Genetic drift has often been argued to apply more to neutral alleles (those that do not differ in their effect on survival or reproduction) than on other alleles. Whether or not many alleles conform to this assumption is a subject of considerable controversy.

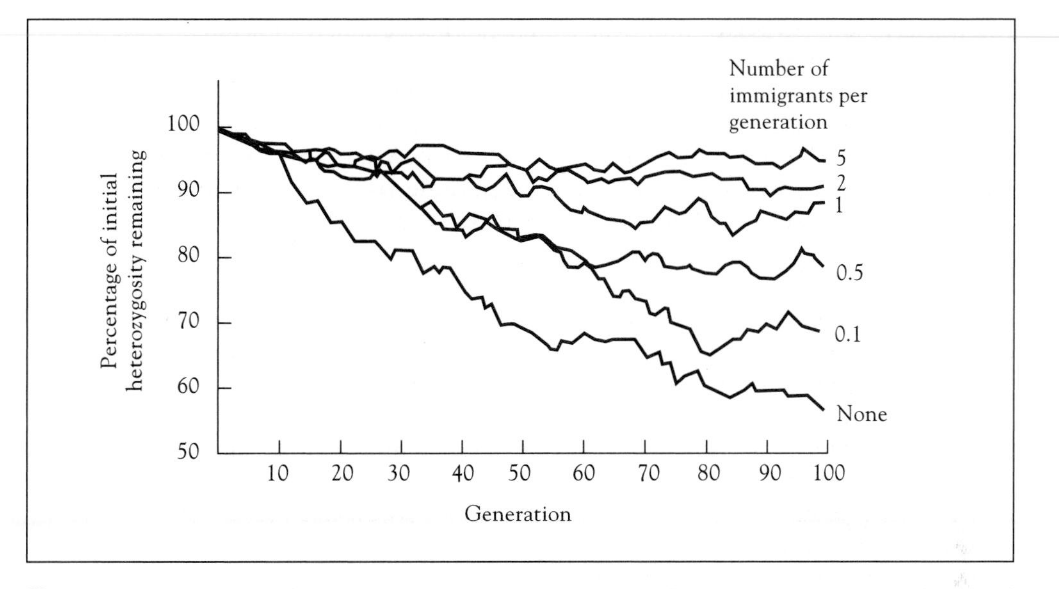

FIGURE 2.11 The effect of immigration on genetic variability in 25 simulated populations of 120 individuals each. Even low rates of immigration (one immigrant per generation) can prevent the loss of heterozygosity from genetic drift. *(After Lacey 1987.)*

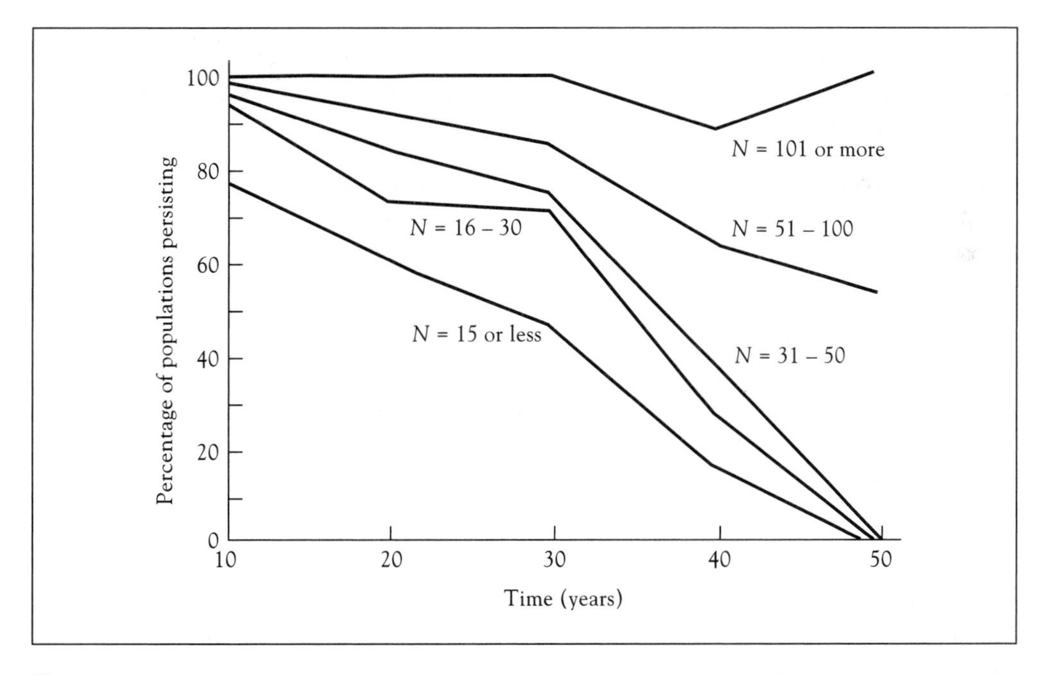

FIGURE 2.12 The relationship between the size of a population of bighorn sheep and the percentage of populations that persist over time. The numbers on the graph indicate population size (N); populations with more than 100 sheep almost all persisted beyond 50 years, while populations with fewer than 50 individuals died out within 50 year. *(After Berger 1990.)*

Kimura (1983*a* and 1983*b*) and Nei (1983) hold that much molecular variation in allozymes, DNA, and proteins is neutral and that any divergence detected among species is likely to be the result of genetic drift.

An interesting aside is that population theory predicts that loss of genetic variation by genetic drift is not as important as loss of rare alleles from the population. Rare alleles may contribute relatively little to overall genetic variation, but they may be important to a population during infrequent or periodic events such as temperature extremes or exposure to new parasites or new pathogens and can offer unique responses to future evolutionary changes. The analogy here is that global climate change may not affect populations very much, but the frequency of extreme events like hurricanes, extreme drought in the summer, or freezing temperatures in the winter could well be the straw that breaks the camel's back for many species. So it is with lack of diversity. The lack of diversity may not be as big a problem as loss of rare alleles. Rare alleles are lost much more rapidly from small populations, even though much of the overall genetic diversity is retained (Denniston 1978).

Genetic diversity will tend to decline in populations in which individuals mate primarily with close neighbors.

Even in large populations, the effective population size may actually be quite small because individuals only mate within a neighborhood. The number of individuals in a neighborhood is given by $4\pi s^2 D$, where D is the **density** of the population and s is the standard deviation of the distances between the birth sites of individuals and the birth sites of their offspring. By marking individual deer mice (*Peromyscus maniculatus*), Howard (1949) showed that at least 70 percent of males and 85 percent of females breed within 150 m of their birthplaces. Even migratory species, such as birds, usually return to the vicinity of their birthplaces to breed. Barrowclough (1980) has also shown that, for noncolonial birds such as wrens and finches, local effective population sizes range from 175 to 7,700 individuals, and Levin (1981) has argued that, despite seed and pollen dispersal, effective size of local populations for plants is often in the dozens or hundreds.

Furthermore, even within a neighborhood, some individuals do not reproduce. If only half the individuals in a population breed, half the heterozygosity is lost per generation. In a population of fifty, if only half the members breed, then the population has an effective size N_E of 25. In territorial species and those with harems, N_E is even lower because the few males that reproduce contribute disproportionately to the subsequent generations and genetic drift is inflated. If a population consists of N_m breeding males and N_f breeding females, the effective population size N_E is given by:

$$N_E = (4\ N_m\ N_f)/(N_m + N_f).$$

In a population of 500, a 50:50 sex ratio and all individuals breeding, $N_E = (4 \times 250 \times 250)/(250 + 250) = 500$. However, if 450 females bred with 50 males $N_E = 180$ or 36 percent of the actual population size. Many species have mating structures that greatly skew N_E..

A knowledge of effective population size is vital to ensure the success of many conservation projects. Notable among these are the reserves designed to protect grizzly bear populations in the contiguous forty-eight states of the United States (Photo 2.8). The grizzly bear (*Ursus arctos horribilis*) has declined in numbers from an estimated one hundred

▪ **PHOTO 2.8** Because not all individuals in a population of grizzly bears breed, effective population sizes may be much less than real population sizes. An exchange of bears between populations, or even the addition of bears from zoos, might do much to alleviate the problems of inbreeding. (*Tom & Pat Leeson, DRK Photos 221761.*)

thousand in 1800 to less than one thousand at present. The range of the species is now less than 1 percent of its historical range and is restricted to six separate populations in four states. Some computer simulations of population size suggested a target of less than a hundred individuals in a population. However, research by Fred Allendorf of the University of Montana (Allendorf 1994) indicated that populations of a hundred would likely lose heterozygosity because effective population sizes are only about 25 percent of actual population sizes. Thus, even fairly large isolated populations, such as the two hundred bears in Yellowstone National Park are vulnerable to the harmful effects of loss of genetic variation. Allendorf and his colleagues proposed that an exchange of bears between populations or even zoo collections would help tremendously in promoting genetic variation. Even an exchange of two bears between populations would greatly reduce the loss of genetic variation. Allendorf was left to conclude that viable populations could only be maintained by maintaining large reserves, by promoting artificial gene flow among reserves, and by promoting gene flow between zoos and reserves. A combination of various techniques would thus become necessary for the preservation of the grizzly bear and probably for the preservation of many large animals in the United States.

Population bottlenecks happen when population size decreases sharply for a generation or more.

A sudden reduction in population size, such as might result from a natural disaster, can reduce genetic diversity if individuals containing unique alleles perish. Effective population sizes can be further lowered if populations vary in size from generation to generation. The effective size N_E in this instance is estimated by the harmonic mean population size

$$\frac{1}{N_E} = \left(\frac{1}{t}\right) \sum_{i=1}^{t} \left(\frac{1}{N_i}\right)$$

where t is the generation.

Thus, if a population goes through five generations of size 100, 150, 25, 150, and 125 individuals, N_E is about 70 as opposed to the arithmetic mean of 110. N_E is more strongly affected by lower than by higher population sizes. The actual magnitude of the loss of genetic diversity in a bottleneck is about 50 percent when N_E is reduced to 1 but is virtually 0 percent with a large N_E of 1,000. This assumes the bottleneck is only maintained for one generation. If the bottleneck is maintained for longer and the population growth of the population afterward is low, some additional genetic variation is lost. However, a bottleneck rarely has a severe genetic or fitness consequence if population size quickly recovers in a generation or two. It is a surprising fact that the gametes of just one individual can carry, on average, 50 percent of the genetic diversity of the population.

The northern elephant seal (*Mirounga angustirostris*) went through a severe bottleneck in the 1890s when its numbers were reduced to only about twenty by hunting. Remarkably, it recovered, and more than thirty thousand are alive today. Because of the harem mating system (see Chap. 5), the effective population size must have been less than twenty. Bonnell and Selander (1974) were able to demonstrate by electrophoresis that there is no genetic variation among today's northern elephant seals in a sample of twenty-four loci, whereas in the southern elephant seal (*M. leonina*) normal genetic variation exists. Southern elephant seals were never drastically reduced in numbers. To this point, the loss of genetic variation does not seem to have adversely affected the northern elephant seals, though what might happen in the case of a disease epidemic is not certain.

Widely divergent views have been expressed about the magnitude of N_E/N in wildlife. Empirical estimates of 0.2 to 0.5 have commonly been reported (Mace and Lande 1991) while theoretical considerations suggest it should usually be greater than 0.25 (Nunney and Campbell 1993). Empirical estimates that include the effects of unequal sex-ratio, variance in family sizes, and fluctuations in population size average 0.11 (Frankham 1995), a value perhaps lower than commonly assumed. In many colonizing species a bottleneck occurs when a new habitat is colonized for the first time. This type of bottleneck is known as the **founder effect**. In dispersing species, for example, it is not unusual for only one or two members of a population to make it successfully to an island. A colony founded by a pair of diploid individuals can have at most four alleles at a locus, though there may be many more in the population from which they came. Bottlenecks and founder effects are, therefore, common occurrences for weeds and pests and should not always be regarded as universally deleterious.

The main conclusion is that, despite limitations by bottlenecks and neighborhoods, much variation exists in nature. The question of whether much of it is evolutionarily meaningful or is maintained by selection is open to debate (Lewontin 1974; Koehn, Zera, and Hall 1983). Probably the answer is that variation is neutral on prevailing genetic backgrounds in many environments but affects fitness on different genetic backgrounds in some environments. The cases in which it does are examples of natural selection operating in nature.

SUMMARY

1. The distribution patterns of plants and animals on Earth depend not only on contemporary factors such as climate, habitat quality, and the presence or absence of competitors, predators, and diseases but also on the evolutionary past. It is valuable to know where species originally evolved and whether or not they have been able to disperse to areas outside of this center of origin.

2. Variation in nature results from gene and chromosome mutations. A knowledge of how much of this variation is maintained within and between populations is vital to conservation programs. If little variation occurs within a population and large variation exists between populations, then many populations will need to be conserved to preserve this variation.

3. Population variation is reduced by inbreeding, genetic drift, neighborhoods, and bottlenecks. The effects of all these increase in severity with decreasing population size. How big do populations have to be to prevent inbreeding or genetic drift? The 50/500 rule suggests that 50 is the critical size to prevent inbreeding, and 500 is the critical size to prevent genetic drift.

4. Is it feasible for humans to move individuals from one wild population to another to simulate natural dispersal? Some models suggest that moving even one or two individuals per generation would counter genetic drift.

DISCUSSION QUESTION

Why are there so many different species of plants and animals? Why aren't there fewer or more?

CHAPTER

3

Natural Selection, Speciation, and Species Loss

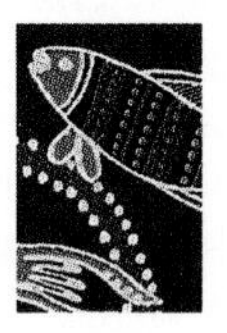

Natural selection, the backbone of Darwin's theories on evolution and the mechanism that can explain how evolution occurs, is in essence the differential survival and reproduction of some individuals in a population and the death without issue (or with fewer issue) of others. Important though random changes and genetic drift are, a more active process, **natural selection**, is best invoked to explain many natural features of evolution in the real world. The feet of different types of birds are features that fit them to their ways of life and are best explained by natural selection. Never, however, is there a predestined form to which plants or animals are shaped as they evolve (Dawkins 1989); as the environment changes so usually do the organisms that live in it—no moral or ethical forces impinge on natural selection.

3.1 Natural selection is the process whereby some types of individuals survive and reproduce more successfully than others in a given environment.

Theoretically, natural selection can operate whenever different kinds of self-reproducing entities differ in survival or reproduction. Sometimes the genes themselves have been regarded as these entities (Dawkins 1989), but more commonly individual organisms are regarded as the units (see Sober and Lewontin [1982] for a critique of the selfish gene theory). Natural selection, then, occurs when genotypes differ in fitness.

Changes in the frequency of dark-colored mutants of the peppered moth over the past century have been attributed to natural selection for cryptic coloration in a polluted environment.

One of the best-analyzed examples of natural selection of genotypes in operation is the change in color that has taken place in certain populations of the peppered moth, *Biston betularia*, in the industrial regions of Europe during the past one hundred years. Originally, moths were uniformly pale gray or whitish in color; dark-colored or melanic individuals were rare and made up less than 2 percent of the total individuals (Kettlewell 1973; Bishop and Cook 1980). Gradually, the dark-colored forms came to dominate the populations of certain areas—especially those of extreme industrialization such as the Ruhr Valley of Germany and the Midlands of England. Genetic tests showed that the dark allele was dominant and that crosses of dark with pale individuals produced typical Mendelian segregation. Pollution did not directly affect mutation rates. For example, caterpillars feeding on soot-covered leaves did not give rise to dark-colored adults. Rather, it promoted the survival of dark morphs on soot covered trees. Melanics were normally quickly eliminated in nonindustrial areas by adverse selection; birds found them conspicuous. This phenomenon, an increase in the frequency of dark-colored mutants (*carbonaria* forms) in polluted areas, is known as **industrial melanism**. A similar pattern occurred in North American forms of the peppered moth around the industrial areas of southern Michigan (Grant, Owen, and Clarke 1995). Another North American equivalent of this story is the *swettaria* form of *Biston cognataria* that showed up in industrialized areas such as Philadelphia, New Jersey, Chicago, and New York in the early 1900s. By 1961, it constituted over 90 percent of the population in parts of Michigan (Owen 1961, 1962).

The operation of natural selection on the peppered moth was illustrated by Professor H. B. D. Kettlewell of Oxford University. He argued that normal pale forms are cryptic when resting on lichen-covered trees, whereas dark forms are conspicuous. In industrialized areas, lichens are killed off, tree barks become darker, and the dark moths are the cryptic ones. Figure 3.1 illustrates the two forms of *Biston betularia*. Kettlewell suspected that birds were the selecting force, and he set out to prove it by releasing thousands of moths marked with a small spot of paint into urban and industrialized areas (Kettlewell 1955). In the nonindustrial area of Dorset he recaptured 14.6 percent of the pale morphs released but only 4.7 percent of the dark moths. In the industrial area of Birmingham, the situation was reversed; 13 percent of pale morphs but 27.5 percent of dark morphs were recaptured. Birds were clearly implicated in differential predation, eating more pale morphs in industrial habitats and more dark morphs in nonindustrial areas. As a test of his field observations, Kettlewell and companions set up blinds and watched birds voraciously gobble up moths placed on tree trunks. The action of natural selection in producing a small but highly significant step of evolution was seemingly demonstrated.

■ **FIGURE 3.1** Industrial melanism. (**a**) Typical (light-colored) form of Biston betularia. (**b**) Carbonaria (dark-colored) form of Biston betularia. *(Drawn from photo by Kettlewell 1995.)*

However, the black form has not become fixed even in the most industrial of locales, so other factors may play roles in the maintenance of melanic frequencies (Lees 1981). Furthermore, many authors have pointed out inconsistencies in the peppered moth story (summarized in Berry 1990). Kettlewell was a general medical practitioner who took up a fellowship at Oxford University at the age of fourty-five years. He was "the best naturalist and almost the worst professional scientist," making rapid diagnoses and refusing to be diverted by what he regarded as irrelevant evidence (Berry 1988). Various authors have, at one time or another argued for the following:

a. A physiological advantage of *carbonaria* even in rural areas (Lees and Creed 1975).
b. A higher dispersal rate of *carbonaria*, which influences correlations of pollution levels with melanic frequencies (Brakefield 1990).
c. A failure of adult moths to select an appropriately camouflaged background (Grant and Howlett 1988; Liebert and Brakefield 1987).
d. A direct effect of pollution on moth phenotypes. A similar decline in the *carbonaria* form has occurred in the industrial Detroit region over the past thirty years (1959–1995) despite the fact that lichen densities on trees have not changed (Grant, Owen, and Clarke 1995), although SO_2 and particulate matter have dropped dramatically.

There is clearly more to melanism than meets the eye. Neverthless, it is probably still true to say that the majority of ecologists still view color polymorphism in the peppered moth as one of the best examples of natural selection in action.

Interestingly enough, the white form of the peppered moth is making a strong comeback. In Britain, a Clean Air Act was passed in 1965. Sir Cyril Clarke has been trapping moths at his home on Merseyside, Liverpool, since 1959. Before about 1975, 90 percent of the moths were dark, but since then there has been a steep decline in *carbonaria* forms, and in 1989 only 29.6 percent of the moths caught were melanic (Fig. 3.2a) (Clarke, Clarke, and Dawkins 1990). The mean concentration of sulfur dioxide pollution fell from about 300μg m^{-3} in 1960 to less than 50g m^{-3} in 1975 and has remained fairly constant since then (Fig. 3.2b). If the spread of the *typica* (light-colored) form of the moth continues at the same speed as *carbonaria* spread in the last century, then the melanic form will again be only an occasional mutant in the Liverpool area by the year 2010. Although it appears that peppered moths may be a reasonable indicator species of high or low levels of environmental pollution, it is disconcerting that the numbers of dark morphs only decreased (1978) when pollutant levels had been at a low value for a number of years.

The case of the peppered moth notwithstanding, much evidence still shows that many alleles or genotypes vary in fitness according to their environment. There is no automatic "better" genotype (remember the genotype $\times$ environment interactions discussed in Chap. 2). Other cases of industrial melanism are known (Bishop and Cook 1981), and other examples of rapid evolutionary change have become apparent as, for example, more and more pests become resistant to insecticides and more and more bacteria become resistant to antibiotics. Antibiotics are chemicals produced by organisms, generally fungi or bacteria, that are inhibitory to the growth of other microorganisms. Although antibiotics have been known since 1928, when Sir Alexander Fleming in England isolated penicillin from the mold *Penicillium*, the industry did not undergo rapid expansion until after the Second

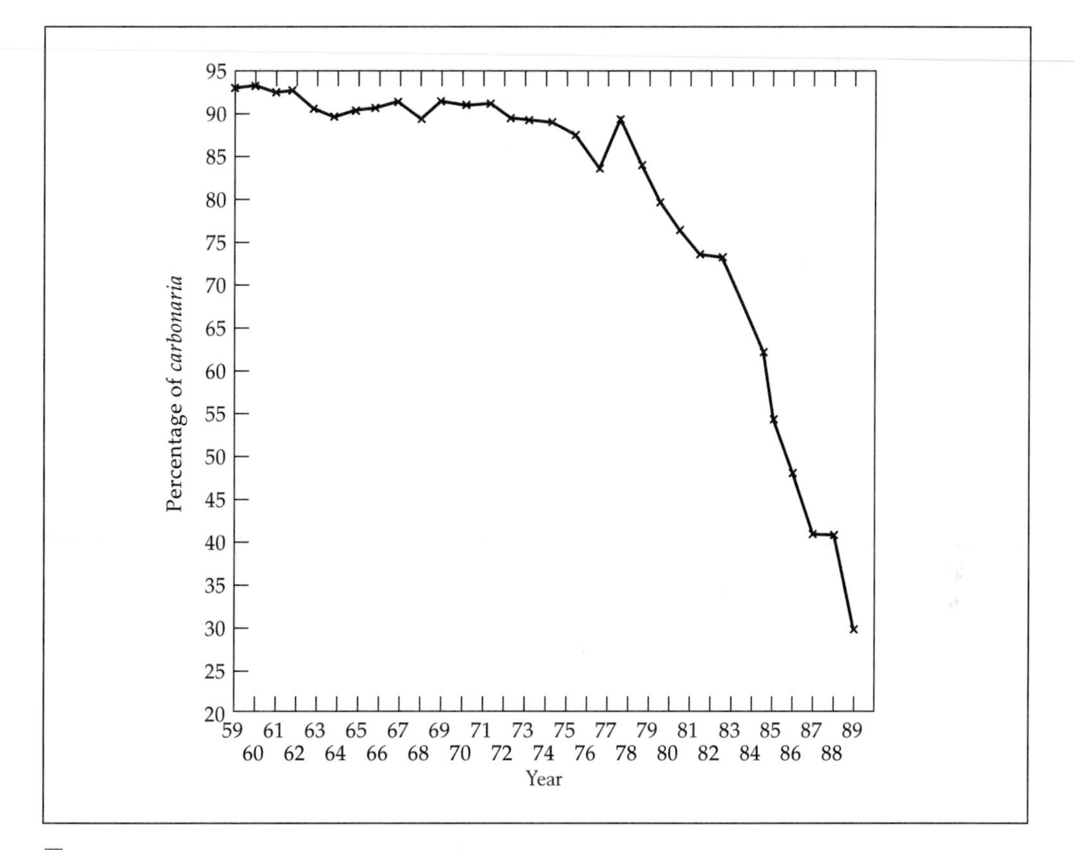

■ **FIGURE 3.2A** Decline in the proportion of form carbonaria in West Kirby, England. The total number of *B. betularia* trapped in the 31-year period was 14,882. (*After Clarke et al. 1990.*)

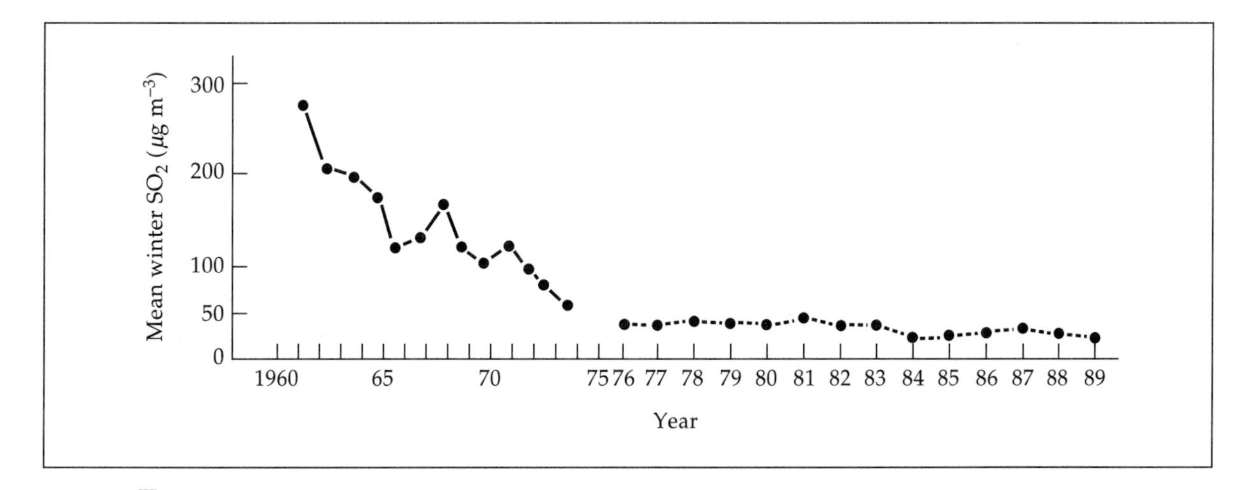

■ **FIGURE 3.2B** Sulfur dioxide concentrations in West Kirby (1976-1989) and near West Kirby (1961-1974). (*After Clarke et al. 1990.*)

World War. Today, several thousand types of antibiotics and their chemical derivatives are known, and the use of antibiotics has revolutionized the treatment of infectious disease. Indeed, antibiotics have been misused in medicine by being administered to patients with nonspecific ailments or virus-caused conditions such as the common cold or influenza. Misuse has been particularly brazen in animal husbandry, where antibiotics have been incorporated into livestock feed as "preventatives" against pathogens. As might be expected, the upshot of the continuous presence of antibiotics in the bacterial environment has been the evolution of antibiotic-resistant strains. This same thing has happened in the pest control industry. Growers were encouraged by chemical companies to use sprays on their crops as a preventative measure. Eventually, the insect pests became resistant, and more sprays of higher toxicity had to be used against them. This in turn led to the development of still more resistant types of pests, spawning the term *pesticide treadmill.*

Even such a normally harmful allele as sickle-cell hemoglobin (which causes severe anemia) can be advantageous in areas where malaria is prevalent because heterozygous individuals, who carry the allele but are not severely affected by the anemia, are more resistant to malaria than are normal individuals. This mechanism for the maintenance of a polymorphism is known as *heterozygous advantage*. In the absence of malaria, the sickle-cell trait is quickly lost. In Norway rats, the allele for resistance to the pesticide warfarin lowers the animal's ability to synthesize vitamin K. Resistant varieties are thought to be at a 54 percent disadvantage to wild types in nature (Bishop 1981), but the allele is maintained in the population by the advantage it gives individuals that encounter warfarin. Finally, trappers in Canada preferred the pelts of silver foxes over red foxes (Elton 1942). Because of this the proportion of silver foxes in the catch declined from 15 percent in 1934 to only 5 percent in 1983.

In some cases, natural selection will act to maintain multiple alleles at a given locus, giving rise to a genetically polymorphic population.

At its simplest, natural selection will tend to lead to an increase in the frequency of the allele that confers the highest fitness in a given environment. In some cases, however, selection will act to maintain a number of alleles in a population, creating a genetic polymorphism. Polymorphisms may be maintained if the relative fitness of different alleles changes on a fine spatial or temporal scale.

The unstable existence of two or more morphs in a population is called a transitional or directional **polymorphism**; one allele is replacing another in the population. In many cases, however, one allele does not completely replace the other, and the stable intermediate frequency is called a **balanced polymorphism**. An example can be seen in the land snail *Cepaea nemoralis* in which six alleles affect the color of the shell, which can be several shades of brown, pink, and yellow (Fig. 3.3). The relative abundances of the various forms differ, even between localities less than a mile apart, and fossils show that this polymorphism has persisted at least since the Pleistocene epoch (Diver 1929). This European snail is common in a variety of habitats, woods, meadows, and hedgerows, and the maintenance of the polymorphism has again been shown to be due to bird-predation pressure, this time by the song thrush, *Turdus ericetorum* (Cain and Sheppard 1954a). Thrushes hunt by sight, and shell color plays the central role in the concealing coloration of these snails. The birds break open shells on suitable stones, or anvils, providing the experimenter with an ideal opportunity to compare the proportion of vari-

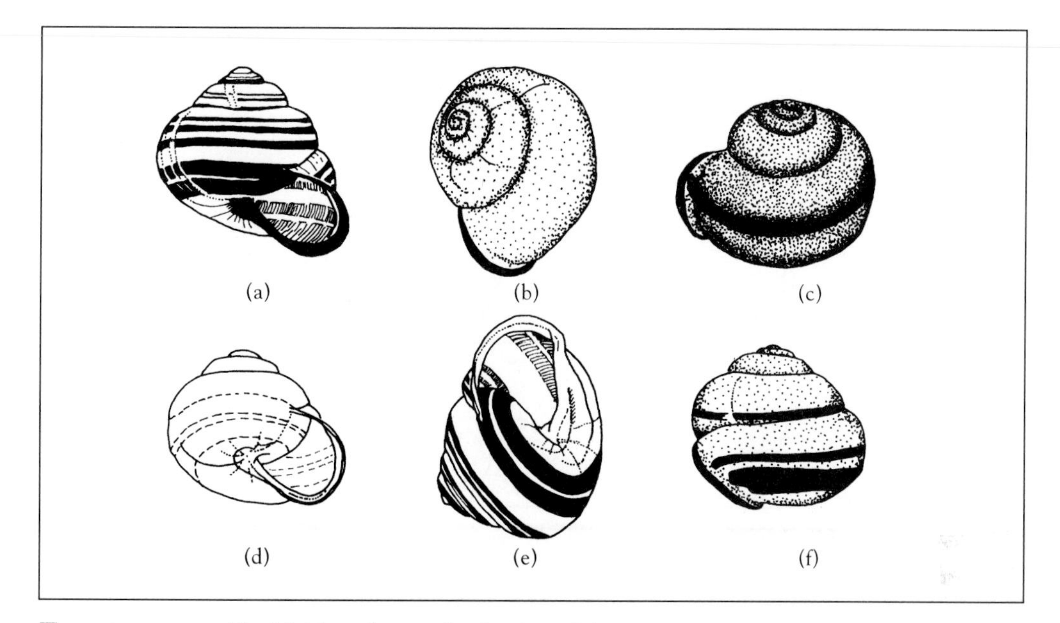

FIGURE 3.3 The highly polymorphic land snail *Cepaea nemoralis*. (**a**) A yellow shell with five bands and a dark lip at the mouth of the shell. (**b**) A pink shell with no bands and a dark lip. (**c**) A brown shell with only the central band present. (**d**) A yellow shell with the bands present but unpigmented, making them translucent (the lip is also unpigmented). (**e**) A yellow shell with pigmented bands but an unpigmented lip. (**f**) A pink shell with the first two bands missing so that it has only the central and two lower bands present.

eties in a colony with those taken from it by song thrushes. In beech woods, the leaf litter is red-brown, and the snails are brown and pink. In grassland habitats, red and brown forms are rare, and yellow forms predominate.

The story is somewhat complicated by the fact that genotypes differ in their susceptibility to extreme temperatures (Jones, Leith, and Rawlings 1977), and yellow shells may also be more common in colder areas (Cameron 1992). In addition, the snails have a complicated pattern of thin black bands. This characteristic is again linked to habitat type, as unbanded snails are cryptic in uniform habitats and banded ones are cryptic in "rough" habitats (as are zebras) (Cain and Sheppard 1954a). Interestingly, an African land snail exhibits a similar variation in shell color and banding patterns (Owen 1966), so the phenomenon is likely to be of wide ecological occurrence. Similar predator-driven polymorphisms may exist in insects too (Stiling 1980).

The peppered moth and land snail examples show, fairly conclusively, the operation of natural selection on populations and the survival of the fittest individuals. They also show that natural selection can operate in more than one way; it can drive populations toward one type of morph or act to maintain more than one. Often it is difficult to tell whether variations on a theme in nature represent truly different species or merely different morphs of the same thing. Many different morphs of species are so different that they have been categorized as different species. In general, three types of natural selection are recognized: stabilizing, in which one morph is favored; directional, in which the population is driven to exhibit a different type of morph (industrial

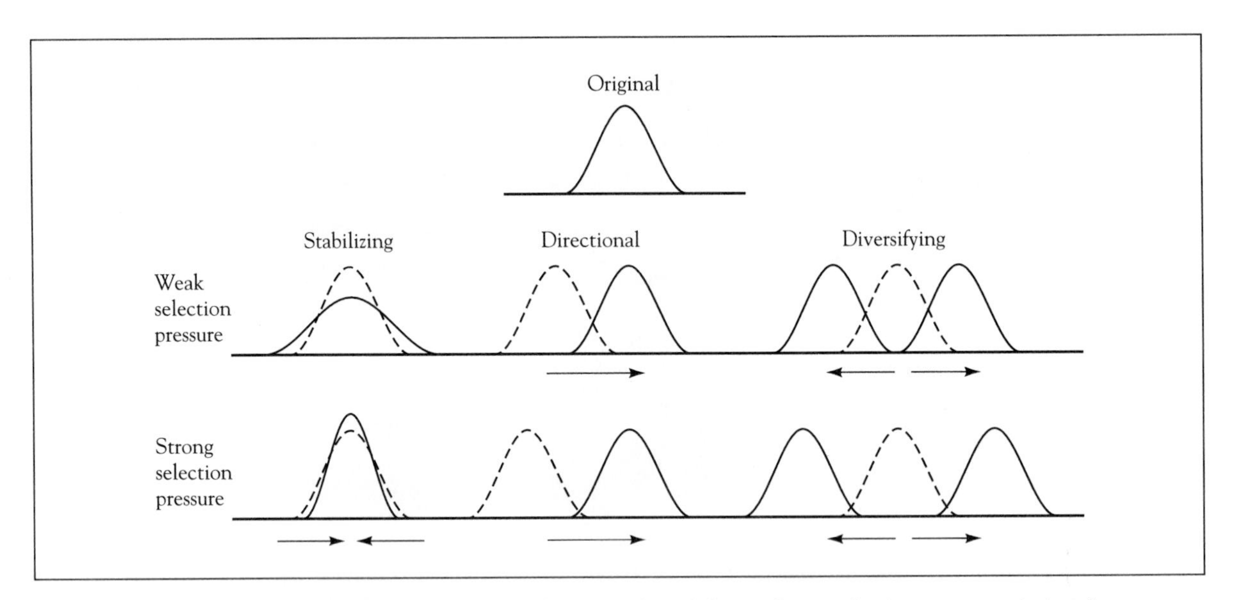

FIGURE 3.4 Effects of stabilizing, directional, and diversifying selection upon variation for a quantitative character.

melanisms); and diversifying or disruptive, in which, as in the case of the land snail, more than one morph can be favored (Fig. 3.4). Presumably, when disruptive selection is strong enough, speciation results.

3.2 Speciation is the evolutionary process by which genetically distinct groups become subdivided into separate species.

Biologists do not agree on exactly what constitutes a species, but in general a species is defined as a group of potentially interbreeding indviduals.

Many of all the smaller taxonomic groups specified, like races, subspecies, biotypes, even genera and families, are quite arbitrarily defined (Wilson and Brown 1953; Futuyma 1986), and their value is questionable. Unfortunately, estimates of the rate of evolution are often based on the origin and extinction of families or other taxa, so some knowledge of higher systematics is essential. For many biologists, however, one taxonomic category is considered real and nonarbitrary—the **species**.

In any discussion of speciation, it is valuable to have a working concept of species. Perhaps the best known is that of Mayr (1942): "groups of populations that can actually or potentially exchange genes with one another and that are reproductively isolated from other such groups."More recently, it has been noted that for many species with widely separate ranges we have no idea if the reproductive isolation is by distance only or whether there is some isolating mechanism (Donoghue 1985). Thus, our judgment of a species is largely based on a surrogate method: morphological criteria. Also bear in mind that asexually reproducing organisms are given names by taxonomists and that paleontologists recognize different temporal portions of the same lineage as distinct—

for example, *Homo erectus* and *H. sapiens*—so the above quotation is but a good working definition and is not watertight.

It is clear that the biological species concept is not universally accepted (Levin 1979; Rojas 1992). However, before we can come to the conclusion that a definition of a biological species is only of interest to academics, consider that the definition of species may be critical in conservation. Adoption of different species concepts could alter the way we define and conserve species. This could have a real effect on conservation of biodiversity. If we accept Mayr's definition of species, commonly called the **biological species concept** or BSC, then present estimates vary from about five to thirty million species in the world today (Wilson 1988). Another popular definition of species today is the **phylogenetic species concept** (PSC). This concept depends on the branching, or cladistic, relationships among species or higher taxa. The PSC definition is based on the concept of shared derived characters (Cracraft 1983). Adopting PSC would result in more recognized species than at present (McKitrick and Zink 1988). Many currently recognized species or distinct populations would be elevated to species status (Photo 3.1 and Fig. 3.5). Populations of species not presently endangered because they are widespread and abundant at some localities might attain endangered status as new species, if split under the PSC system. We should note however, that the U.S. endangered species law has provisions to recognize certain populations of vertebrates as rare even if the remainder of the species is secure (as in the cases of the Southern Bald Eagle and the Florida panther), so this change would only affect plants and invertebrates.

PHOTO 3.1 The black racer, *Coluber constrictor*, exists as many different color forms or races throughout the U.S. Shown here is the Northern black racer. Some authorities argue that each race should actually be elevated to the status of a species. (*Z. Leszczynski, Animals Animals/Earth Scenes R-6722.*)

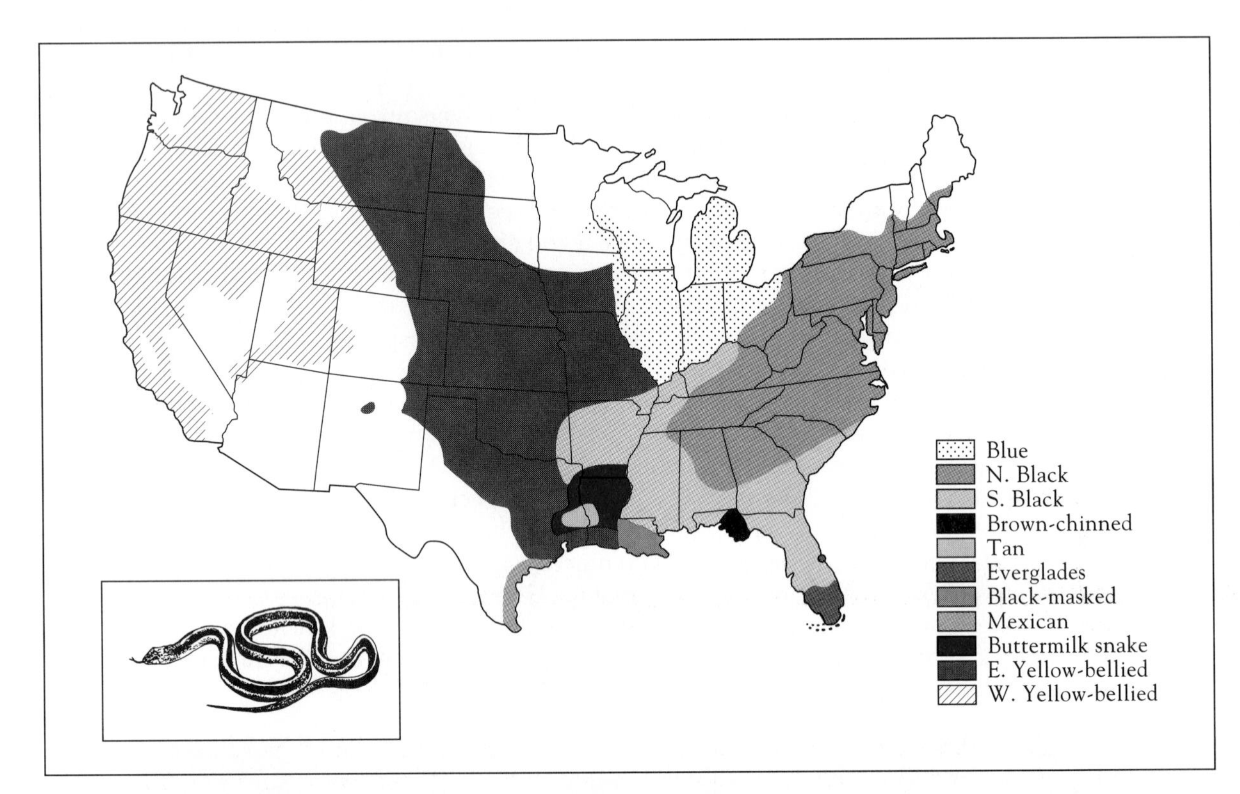

■ **FIGURE 3.5** Distribution of the various named subspecies of the racer (*Coluber constrictor*), a common snake of North America. Each of the subspecies would probably be designated a full species under the PSC approach (see text). (*Modified from Conant 1975.*)

It is quite plausible that raising the number of species that are endangered could create even more problems than it solves. There could perhaps be strong public and political backlash against their legal protection. It is also possible that restoration efforts might be hampered because scientists might be more unwilling to move individuals from populations to other populations that they are trying to recover.

Mayr's definition of a species can result in an overly optimistic view of global biodiversity. This is because a species can continue to exist even if many of its populations are destroyed. Those lost populations might represent a decline in biodiversity if they contained unique, genetic traits. A species approach could not tell us that diversity has been lost—because the species count remains unchanged. However, using a PSC approach to defining species would reduce this problem. Using the PSC approach, it is clear that we have been losing much more global diversity in the form of populations than is often recognized (Fig. 3.6).

Mayr was an ornithologist, and his definition of species seemed to fit birds very well—breeding groups are often clearly delineated (but see Applied Ecology: Hybridization and Extinction). However, especially in plants, fertile hybrids often form between species that greatly blur species distinctions. Oak trees provide a particularly good example of confusion in species definitions. Oaks often form reproductively viable hybrid populations. That is, oaks from different species interbreed and their offspring are themselves viable, capable of reproducing with other oaks. For this reason, one might

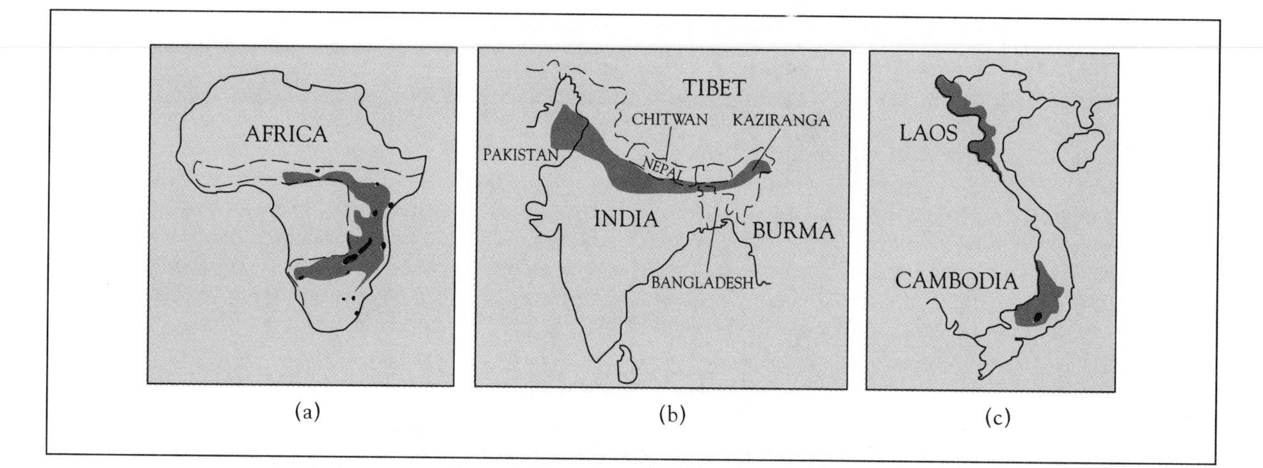

■ **FIGURE 3.6** Present and former distributions of three species of rhinoceros, showing loss of populational diversity and retreat to a few refuges. (**a**) The black rhinoceros (*Diceros bicornis*), showing historical distribution (black outline), distribution in 1900 (shaded area), and distribution in 1987 (black areas). (**b**) The greater one-horned rhinoceros (*Rhinoceros unicornis*), formerly distributed across the shaded area, is now reduced to two populations at Chitwan and Kaziranga reserves. (**c**) The Javan rhinoceros (*Rhinoceros sondaicus*), showing historical (shaded) and present (black) distributions. (*After G. K. Meffe and C. R. Carroll 1994.*)

question whether the parental species should be called "species" at all. For example, *Quercus alba* and *Q. stellata* form natural hybrids with eleven other oaks species in the eastern United States. It could be argued that, if these oaks cannot tell each other apart, why should biologists impose different species names on them? However, Davis (1983) has noted that many examples of viable hybrid formation in nature occur when historically isolated species are brought into contact through climate change, landscape transformation, or transport by humans beyond their historic boundaries. Oaks were absent from the eastern deciduous forests during most of the Pleistocene, and in many places their co-occurrence is no older than one hundred generations. There are numerous other examples of fertile hybrids forming when once-isolated species of plants or animals are rapidly thrust into contact (Arnold and Hodges 1995). In fact, extinction by hybridization with introduced species is a real problem in conservation (Applied Ecology: Hybridization and Extinction).

Calls have often been made to the U.S. Congress to revise the endangered species act to make only full species subject to protection. Protecting subspecies and populations, opponents argue, causes undue economic strain. It is interesting to note that a survey of 492 listings or proposed listings of plants and animals from 1985 to 1991 indicated that 80 percent were full species, only 18 percent were subspecies, and just 2 percent were distinct populations (Wilcove, McMillan, and Winston, 1993). At the present time, species-level taxonomy is still driving endangered species legislation.

Allopatric speciation happens when populations become isolated by a geographic barrier and diverge.

How much of a genetic difference is necessary to make a new species? Nobody knows. To paraphrase Mayr (1963), to try to determine the difference between species in terms of nucleotide pairs of DNA would be absurd, much as it would be to compare how different

APPLIED ECOLOGY

Hybridization and Extinction

Nonindigenous species can bring about a form of extinction of native flora and fauna by hybridization either through purposeful or accidental introductions by humans or by habitat modification that brings previously isolated species together (Rhymer and Simberloff 1996). For example, mallard ducks, *Anas platy rhynchos*, which are a holarctic species, have been introduced into many places such as New Zealand, Hawaii, Australia, and south Florida. The mallard has been implicated in the decline, through hybridization, of the New Zealand grey duck, *A superciliosa*; the Hawaiian duck, *A. wyvilliana*; and the Florida mottled duck, *A. fulvigula*. The Northern American ruddy duck, *Oxyura jamaicensis*, similarly threatens Europe's rarest duck, the whiteheaded duck, *O. leucoephala*, which now exists only in Spain. Even the northern spotted owl, *Strix occidentalis*, is threatened in the Pacific northwest by the recent invasion of the barred owl *S. varia*. Hybrids and fertile offspring have been found.

There are several mammalian examples as well. Feral housecats, *Felix catus*, threaten the existence of the wildcat, *F. silvestris*, in Scotland. Eighty percent of wild cat individuals studied by a number of genetic analyses had domestic cat traits (Hubbard et al. 1992), raising the question, when is an endangered species not a pure species anymore? In the United States, hybridization between the wolf, *Canis lupus*, and domestic and feral dogs, *C. familiaris*, have led to calls to delist the wolf from the endangered species list because it no longer satisfies the criteria (Hill 1993). Similarly, the Florida panther, *Felis concolor coryi*, a subspecies of the cougar, is listed as endangered by the U.S. Fish and Wildlife Service (Photo 1). Fewer than forty individuals remain in the wild, all in south Florida. Of the two largest groups, one, in the Everglades, consists exclusively of hybrids between the Florida panther and individuals of another subspecies from South America, and the other, in the Big Cypress Swamp, consists mainly of "pure" Florida panthers but with a few hybrids. Migration is believed to occur between the two groups, and, again, some question whether the Florida panther should be delisted (Fergus 1991). However, now the thinking is turning around, and the introduction of cougars from Texas has been recommended to prevent inbreeding depression. Hybridization has already occurred anyway, the feeling goes, and the Florida panther was not very different genetically from other subspecies to begin with. The whole story makes interesting reading (Maehr and Caddick 1995). In 1995, eight female Texas cougars were released in south Florida.

■ **PHOTO 1** Some scientists believe that the Florida panther should be de-listed from the endangered species list because most individuals resulted from hybridization with a subspecies from South America. Others take the completely different view that more cougars from Texas should be introduced to prevent hybrid depression. (*Thomas Kitchin, Tom Stack, TK30377D-166-991.10.*)

two books were on the basis of the number of letters they use. Species have many genes in common, much as books have words in common; it is the particular arrangement of each that is so critical.

Most evolutionists consider **allopatric speciation** to be the most likely mechanism for the evolution of a species (Mayr 1942, 1963). Allopatric speciation involves spatial separation of populations by a geographical barrier. For example, nonswimming populations separated by a river may gradually diverge because there is no gene flow between them. Alternatively, the upthrusting of mountains often divides populations into many units, among which speciation then proceeds. In an area only twenty by five miles on Hawaii, twenty six subspecies of land snail, *Achatinella mustelina*, have been recognized, each in a different valley separated from the others by mountain ridges. Some of the best-known instances of divergence among isolated populations are on islands, where a species that is rather homogeneous over its continental range may diverge spectacularly from the continental form in appearance, ecology, and behavior. Darwin's finches have speciated extensively within the Galapagos archipelago. Could habitat fragmentation from environmental development promote biodiversity by increasing allopatric speciation? Probably not; local extinctions because of resultant small population vulnerabilities are more likely.

In many situations, the ranges of species that have evolved allopatrically come to overlap, and the species become sympatric once again. The isolating mechanisms between the species, however, have usually continued to evolve separately, and as a result the males and females of the different species are likely to have become incompatible. These isolating mechanisms are a result of divergence, not a cause. Some important isolating mechanisms are outlined in Table 3.1. Among birds and fish, species-specific male coloration seems to be an important isolating mechanism, whereas in insects it is often smell (moths), sound (grasshoppers), or even the correct flight path and flash patterns of lights (Lampyridae, fireflies); in frogs and toads, chorusing is important.

TABLE 3.1 **Summary of the most important isolating mechanisms that separate species of organisms.**

A. *Prezygotic mechanisms.* Fertilization and zygote formation are prevented.

1. *Habitat.* The populations live in the same regions but occupy different habitats; e.g., different spadefoot-toad species occupy different soil types (Wasserman 1957).
2. *Seasonal or temporal.* The populations exist in the same regions but are sexually mature at different times, e.g., flowers that bloom in different months (Grant and Grant 1964) or fireflies that mate at different times of the night (Lloyd 1975).
3. *Ethological* (animal only). The populations are isolated by different and incompatible behavior before mating, e.g., courtship songs of birds or frogs, flash patterns of fireflies.
4. *Mechanical.* Cross fertlization is prevented or restricted by differences in structure of reproductive structures (e.g., genitalia in animals, flowers in plants).
5. *Physiological.* Gametes fail to survive in alien reproductive tracts.

B. *Postzygotic mechanisms.* Fertilization takes place, and hybrid zygotes are formed, but these are inviable or give rise to weak or sterile bybrids.

1. *Hybrid inviability or weakness.* Hybrids of the frogs *Rana pipiens* and *R. sylvatica* do not develop beyond the gastrula stage (Moore 1961).

Sympatric speciation, a more controversial mechanism of species formation, arises when populations diverge without becoming separated geographically.

The alternative to allopatric speciation is **sympatric speciation**, the appearance of new species in an area not geographically separated from other members of the population. Most models of sympatric speciation are highly controversial. Even if some members of a population began to inhabit slightly different parts of a population's range, gene flow between the two groups would probably still be sufficient to prevent speciation. Any slight morphological or behavioral changes in one group would be conveyed to the other. If a single mutation or chromosomal change were to confer complete reproductive isolation in one fell swoop, its bearer would be reproductively isolated. Given the unlikely scenario of the same mutation's occurring at the same time in another individual of the opposite sex, and given that these two individuals could find each other, a new species might form. Close inbreeding (self-fertilization or mating with sibs) may promote the likelihood of such events, and thus sympatric speciation has been proposed for some insect groups such as the Chalcidoidea, that is, parasitic Hymenoptera in which many individuals develop from a single egg (a process called **polyembryony**) laid in a host (Askew 1968). Indeed, because insects are themselves so speciose, comprising an estimated thirty million species (Erwin 1982), it is sometimes difficult to imagine allopatric speciation in this order all by means of geographic barriers, especially given the dispersal abilities of insects (Johnson 1969). In many groups of insects, closely related species are restricted to different host plants, and it has been argued that sympatric speciation has occurred there (Bush 1975a and 1975b; Wood and Guttman 1983; Bush 1994). However, in the few cases that have been analyzed, host preference in insects appears to be controlled by many genes, a situation that would not be favorable to a quick, one-genetic-step method of sympatric speciation (Futuyma and Peterson 1985). Furthermore, even when a species utilizes a new host plant to feed on, it often still interbreeds with individuals reared on other host plants. Thomas et al. (1987) showed that the butterfly *Euphydryas editha* has extended its range onto a new plant, *Plantago lanceolata*, within the past one hundred years but that populations still interbreed with individuals reared on the old host, *Collinsia parviflora*.

How rapidly does speciation occur? The answer, of course, differs for different organisms. J. B. S. Haldane theorized that species of vertebrates might differ at a minimum of one thousand loci and that at least three hundred thousand generations would be necessary for the formation of new species. Indeed, a great many of the populations isolated for thousands of generations by the Pleistocene glaciations did not achieve full species status. American and Eurasian sycamore trees (*Platanus* sp.) have been isolated for at least twenty million years, yet still form fertile hybrids (Rhymer and Simberloff 1996). The selective forces on these two continents have obviously not been sufficient to cause reproductive isolation between these ecologically general species. However, several genera of mammals, for example, polar bears (*Thalarctos*) and voles (*Microtus*) do appear to have originated relatively recently, in the Pleistocene (Stanley 1979). Many of the Hawaiian species of *Drosophila* flies have arisen in just a few thousand years, although their generation time is, of course, much shorter than that of mammals. Lake Nabugabo in Africa has been isolated from Lake Victoria for less than four thousand years, yet it contains five endemic species of fish (Fryer and Iles 1972). Lake Victoria itself is only 500,000 to 750,000 years old but harbors about 170 species of cichlid. Again in

Hawaii, at least five species of *Hedylepta* moth feed exclusively on bananas, which were only introduced by the Polynesians some one thousand years ago.

3.3 The rate at which species appear and become extinct can be estimated from the fossil record.

What are the rates of formation of new species and the rates of extinction of old ones? If we knew this then we could possibly use the information to decrease current rates of extinction. Usually, the best patterns issue from the fossil record. Marine invertebrates have left the best fossil records and have been the most intensely analyzed. Sepkoski (1984) documented a steady rise in the number of these families, which reached a plateau in the Ordovician, suffered a major extinction in the Permian, and have shown a steady increase in diversity ever since; at least nineteen hundred families of marine invertebrates are now recognized. Similar patterns are shown for many other taxa (Fig. 3.7). Some of

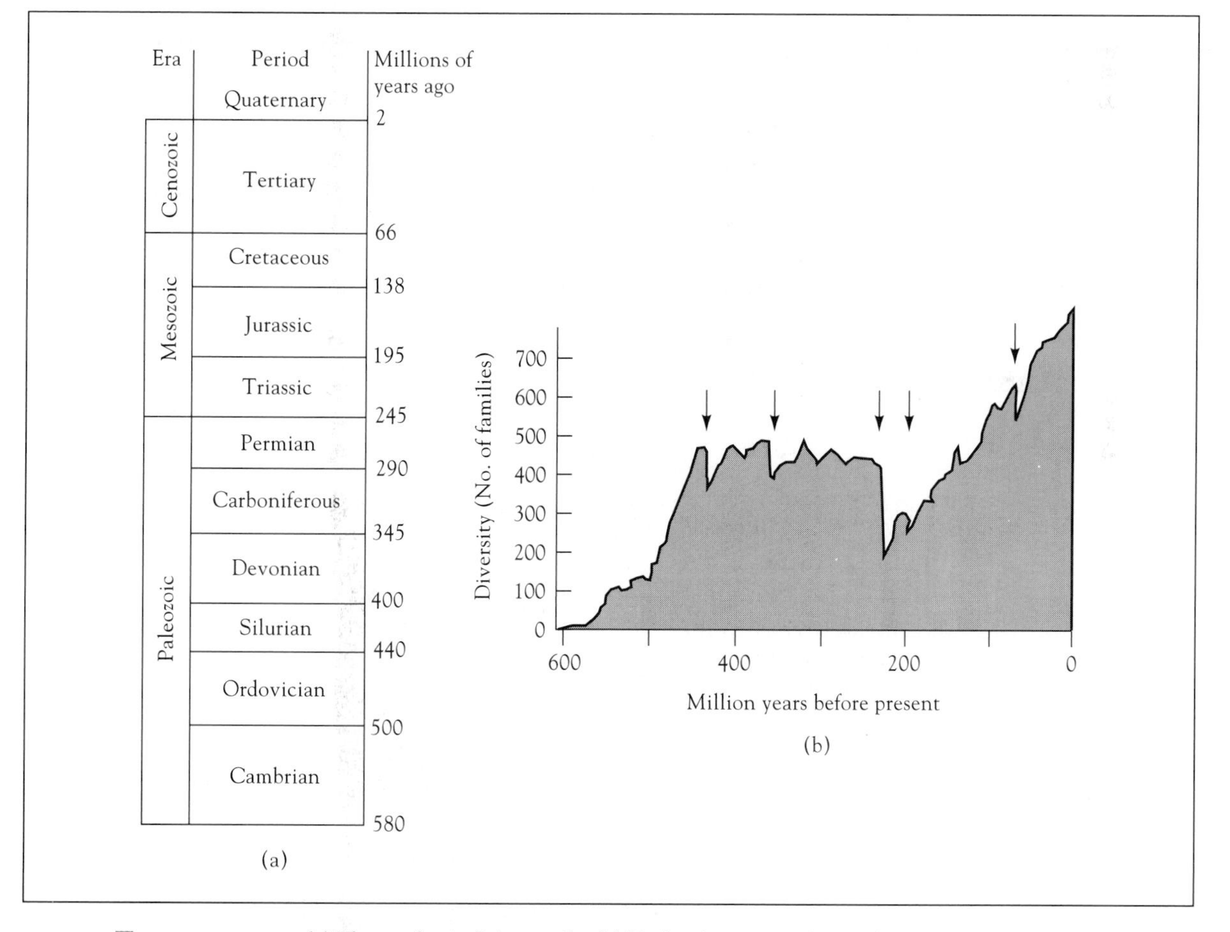

■ **FIGURE 3.7** (**a**) The geological timescale. (**b**) Extinction events in marine organisms. Arrows indicate the five major extinction events (see text). *(After Erwin et al. 1987.)*

this increase has been attributed to **provincialization**, the appearance of differentiated regional biotas, but Bambach (1983) has concluded that the increase in community diversity is a consequence of the addition of organisms with new ways of life, for example, new methods of feeding. Patterns for individual families, of course, vary. Diversity in the gastropods and bivalves is staggering and appears to be still on the increase. Other species—the so-called living fossils—coelacanths *(Latimeria chalumnae)*, horseshoe crabs *(Limulus)*, and ginkgo trees *(Ginkgo biloba)* represent the last members of once-diverse lines. There are five major mass extinction events, one in each of the Ordovician, Devonian, Permian, Triassic, and Cretaceous periods. The causes of these extinctions have been much debated. For the Permian extinction, geologically rapid changes in climate, continental drift, and volcanic activity are probably most important. A single catastrophic event, such as a meteor strike, is probably not to blame. Ordovician extinction appears correlated with a huge global glaciation (the Hirnation glaciation), and the Cretaceous extinction may have been associated with a single catastrophic event—(a meteor strike?)—although this is far from certain. The causes of the Triassic and Devonian extinctions are not well known. Besides these "major" extinctions, a series of minor extinctions can be noticed with a periodicity of twenty six to twenty eight million years. In sum, these minor extinctions account for more total extinctions than the five major extinctions combined.

The fossil records for groups other than the marine invertebrates is not so good. Tetrapods have been subject to at least six mass extinction events since their appearance in the late Devonian, and fishes have experienced eight extinctions since their origin in the Silurian. Most of these events coincide with each other and with those extinctions recorded for marine invertebrates. Again, the most significant event is the late Permian, which is the largest recorded extinction both for fishes (44 percent of families disappearing) and tetrapods (58 percent of families disappearing). The late Cretaceous extinction was far more significant for tetrapods than for other groups, with 36 of the 89 families in the fossil record disappearing at this time. These families were mainly confined to three major groups: the dinosaurs, plesiosaurs, and pterosaurs. Somewhat surprisingly, most other major vertebrate taxa were very much unaffected. Currently, correlations between the more minor extinctions events in vertebrates and the postulated periodic extinction in marine invertebrates is poor.

For plants, the fossil record does not clearly show the same sudden mass extinction effect seen in the animal record. For vascular plants, the diversity appears to have increased from the Devonian to the Permian, dropped, then risen to a plateau that was maintained until the Mesozoic. In the Upper Triassic, angiosperms diversified, a trend that has continued to the present (Niklas, Tiffney, and Knoll 1980; Niklas 1986). The only general similarity with animal extinction events is the end-Cretaceous catastrophe, which appears to have had a major influence on the structure and decomposition of terrestrial vegetation and on the survival of species. Perhaps 75 percent of species at this time became extinct.

Is there any overall pattern to the global diversity of all life through time? Does diversity fluctuate around some preordained level, or does it constantly increase until knocked back by some catastrophic natural disaster or extinction event? Biologists ask the same types of questions about what (if anything) regulates the numbers of species in communities of modern animals and plants (see Section 5), and some analogies can be drawn.

Raup et al. (1973) performed a computer simulation of changes in diversity on the assumptions that a lineage could branch, go extinct, or remain unchanged in a given time period. Extinctions and speciation events were assumed to occur randomly and to occur, on average, with the same frequency. The results mirrored many real historical patterns of diversity, suggesting that the available fossil record is largely a result of random extinctions and speciation events through time. Some particular biological phenomena were not well predicted: the long-term survival of "living fossils" was one; another was that lineages often increase and diversify in the fossil record much more rapidly than they would at random.

Often the origin of new taxa is correlated with the extinction rate (Stanley 1979). The result, of course, is that diversity remains unchanged, leading some to believe that even historically distant communities were saturated with species and that new ones could succeed only in the place of old extinct ones. Alternatively, it may be more likely that environmental changes affect both processes concurrently. Stanley (1975) assumed that, for newly arisen taxa, the number of species was able to increase in an exponential fashion following the equation $N_t = N_O e^{rt}$ (see Chap. 6) because there were no competing species to usurp the existing "niche space." In this formula, r, the rate of increase, equals $O - E$, the rate of origin or speciation minus the rate of extinction. If the time t since origin is known from the fossil record, the initial number of species is assumed to be 1, E is calculated from the average life span of fossils, and N_t is the number of species existing, then O could be solved for. Stanley concluded that speciation rates of mammals were higher than those of bivalves (Table 3.2).

Sepkoski (1978, 1979, and 1984) has made the analogy between historical diversity and another population model, the logistic (see Chap. 6). This model assumes initial rapid population growth followed by a leveling off at an asymptote as resources for growth become limiting; the result is a **sigmoid curve**. The upper asymptote is depressed if other competing species are present to lower the level of available resources still further. For marine invertebrates in the entire Paleozoic, diversity fits a logistic model. This result implies that, as diversity increases, speciation rates decline and extinction rates rise, possibly because of competition between species. However, one must

TABLE 3.2 Estimate rates of speciation O, extinction E, and increase in diversity R. (After Stanley 1975)

	T (MILLION YEARS)	N (SPECIES)	R	$\overline{R}$	E	O
				(PER MILLION YEARS)		
Bivalvia				0.07	0.17	0.24
Veneridae	120	2400	0.06			
Tellinidae	120	2700	0.07			
Mammalia				0.20	0.50	0.70
Bovidae (cattle, antelopes)	23	115	0.21			
Cervidae (deer)	23	53	0.17			
Muridae (rats, mice)	23	844	0.29			
Cercopithecidae (OW monkeys)	23	60	0.18			
Cebidae (NW monkey)	28	37	0.13			
Cricetidea (mice)	35	714	019			

remember that in many cases a group has diversified only shortly after the demise of another lineage created a potential "empty niche." Thus, the crocodilians invaded their present habitats only after the phytosaurs became extinct. The great decline of branchiopods at the end of the Permian was followed by an explosive radiation of clams. In fact, the evidence for this type of event outweighs that for competitive exclusion of one lineage by another (Raup 1984; Jablonski 1986a).

It is important to realize that extinction is the rule rather than the exception. Because the average species lives five to ten million years (myr) and the duration of the fossil record is 600 myr, then the Earth's current number of plant and animal species represents about 1 to 20 percent of species that have ever lived (Lawton and May 1995). For most species in geologic time, and even for some in historical time, very little is known about the immediate causes of extinction (Simberloff 1986b). Predation, parasitism, and even competition can have severe impacts on the populations of many species, but habitat alteration as a result of climatic change is probably the prime moving force in evolution over geologic time. Over evolutionary time, Van Valen (1973) suggested that within most taxonomic groups the probability of the extinction of a genus or family is independent of the duration of its existence. Old lineages do not die out more readily than younger ones. For example, among marine invertebrates, the average lifetime of a genus has been 11.1 myr (Raup 1978) for marine animals 4 to 5 myr (Raup 1991; Sepkoski 1992) and for mammals just 1 myr (Martin 1993). Within these classifications, the figures vary tremendously, from 78 myr for bivalves to 7.3 myr for ammonites. Again, great care must be exercised in interpreting these records. From early to late geologic times, the number of species described per genus has generally increased. Under the influence of such a trend, because extinction of a family requires extinction of all its species extinction rates of families will decline even if the probability of extinction of species is constant (Flessa and Jablonski 1985). The same is true of genera. This is known as the pull of the recent.

The assertion that the basal rates of extinction have speeded up at certain times is also a matter of some contention. The so-called periods of mass extinction at the ends of five geological eras have been argued to be simply the quantitatively extreme cases in a basal array of extinction rates (Quinn 1983). Raup and Sepkoski (1984) have suggested that mass extinctions occur regularly, with a periodicity of about 26 myr. If they are right, then past adaptations of species provide little preadaptation to extraordinary periodic conditions.

There is some evidence that the survivors of mass extinctions tended to be the more ecologically and morphologically generalized species (Fig. 3.8). Specialization can hamper adaptation to changing conditions, and overspecialization actually can work against a species. Morphologically complex and specialized species tend to occur in more specialized environments and are thought to be more susceptible to local environmental changes. Generalists tend to have a greater breadth of geographic distribution, which appears to be important in enhancing survival. The bottom line is that for many extinct lineages there is still no adequate explanation as to why extinction happened. The horse, *Equus*, evolved entirely in North America, but eight to ten thousand years ago it went extinct there. Yet, when introduced by the Spaniards in 1519, herds became established very quickly. Sometimes after extinctions similar morphological forms reappear in the fossil record, derived from new ancestors. More often, the chief impact of extinctions is

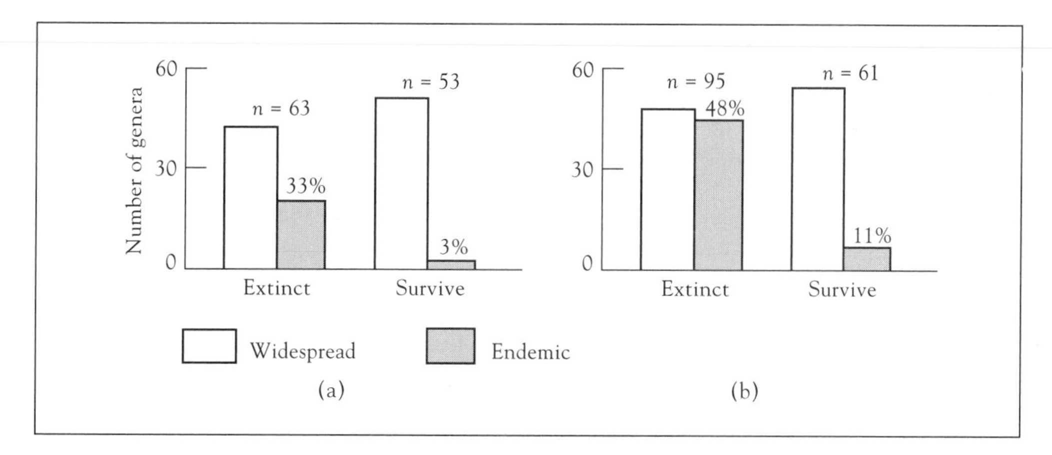

■ **FIGURE 3.8** The proportion of genera that survived the extinction event at the end of the Cretaceous period. (**a**) Bivalves. (**b**) Gastropods. Among both bivalves and gastropods, the proportion that survived was greater for geographically widespread forms than for genera with narrow geographical distributions (endemics). The histograms also show the percentage survival and extinction of bivalves and gastropods of the Gulf and Atlantic coastal plain of North America, distinguishing those restricted to the region from those that were more widespread. *(Based on Jablonski 1986b.)*

the obliteration of forms of life whose like never reappears (Strathmann 1978). In summary, it is certainly hard to say why some species die and others persist. Out of the many genera of horselike animals, why did *Equus* alone survive? Was it really structurally better adapted than the others, or was it merely lucky that its habitat persisted?

3.4 The current extinction crisis can be understood by examining the traits that increase sensitivity to extinction and by investigating the human activities that cause extinctions.

Not all species are equally likely to go extinct: Ecologists have identified several characteristics that seem to increase a species' likelihood of extinction.

Species extinction is a natural process. Fossil records show that the vast majority of species that have ever existed are now extinct, with extinct species outnumbering living species by a factor of perhaps 1,000:1. The current fear is that natural extinction rates are being increased by human activity. Therefore, we should try to determine how human activity increases the chance for species to go extinct.

Species become extinct when all individuals die without producing progeny. They disappear in a different sense when a species' lineage is transformed over evolutionary time or divides into two or more separate lineages (so-called pseudoextinction). The relative frequency of true extinction and pseudoextinction in evolutionary history is not yet known.

In order to predict what type of human causes are critical in the extinction of wild species, some knowledge of life history traits that may be correlated with high levels of extinction is desirable. At least seven life history traits have been proposed as factors affecting a species' sensitivity to extinction (Karr 1991; Lawrence 1991; Angermeier 1995):

1. *Rarity*. Generally, rare species are more prone to extinction than common ones. This may sound like a truism at best and a platitude at worst. However, it is not as intuitive as it might first be thought. For example, a very common species might be very susceptible to even the slightest change in climate whereas rare species, although they may exist at very low numbers of individuals, may be more resistant to climatic change and thus more persistent in evolutionary time as, for example, global warming proceeds. Rabinowitz (1981) has shown that "rarity" itself may depend on three factors: geographic range, habitat breadth, and local population size. A species is often termed rare if it is found only in one area, regardless of its density there. A species that is widespread but at very low density can also be regarded as rare. Conservation by habitat management is much easier, and more likely to succeed, for species of the first type than for those of the latter.
2. *Dispersal ability*. Species that are capable of migrating between fragments of habitat, such as between mainland areas and islands, may be more resistant to extinction. Even if one small population goes extinct in one area, it may be "rescued" by immigrating individuals from another population (more on this later in "Metapopulations," Chap. 13).
3. *Degree of specialization*. It is often thought that organisms that are specialists, for example, those organisms that can only feed on one type of plant, like pandas, which only feed on a single species of bamboo, are more likely to go extinct. Animals that have a broader diet may be able to switch from one food type to another in the case of habitat loss and are thus less prone to extinction. Plants that can live only in one soil type may be more prone to extinction.
4. *Population variability*. Species with relatively stable populations, that is, those that are generally maintained at some equilibrium level, may be less prone to extinction than others. For example, some species, especially those in northern taiga ecosystems, show pronounced cycles. Lemmings reach very high numbers in some years, and the population crashes in others. It is thought that these might be more likely to go extinct than others.
5. *Trophic status* (animals only). Animals occupying higher trophic levels usually have small populations. For example, birds of prey or Florida panthers are far fewer in number than their food items and, as noted earlier, rarer species may be more vulnerable to extinction.
6. *Longevity*. Species with naturally low longevity may be more likely to become extinct. Again, this is not as obvious as it first sounds. Imagine two species of birds, one of which lives for seventy or eighty years, like a parrot, and the other, which is about the same size but breeds earlier in life, for example at age 2 or 3, and only lives to the age of ten years. The parrot, with its eighty year life span, may be able to "weather the storm" of a fragmented habitat for ten years without breeding, then it can pick up and

breed again when the habitat becomes favorable. Species with naturally low longevity are not able to do this.

7. *Intrinsic rate of population increase*. Species that can reproduce and breed quickly may be more likely to recover after severe population declines, say, following the introduction of new exotic diseases, than those that cannot. Populations of species that breed only slowly may be more likely to suffer a double setback, for example, a cold winter following a summer when an exotic disease was introduced. Thus, it is thought that those organisms with a high rate of increase, especially small organisms, bacteria, insects, and small mammals, are less likely to go extinct than larger species like elephants, whales, and redwood trees. The passenger pigeon laid only one egg per year, and this probably contributed to its demise.

Humans have caused extinctions by altering habitat, directly exploiting species, and introducing new species to an area.

How long do species last on Earth? The average life span of a species in the fossil record is around four myr, which would give, at a very gross estimate, a background extinction rate of four species each year out of a total number of species of around ten million. However, it can be argued that the fossil record is heavily biased toward successful, often geographically wide-ranging species, which undoubtedly have a far longer average persistence time. If background extinction rates were ten times higher than this, extinctions among the four thousand or so living mammals today would be expected to occur at a rate of one every four hundred years, and among birds one every two hundred years. It is indisputable that the extinction rate in recent times has been far higher than this.

It is easy to suggest that humans have been the overwhelming cause of recent extinctions though it is equally easy to suggest that the calculated background rates of extinction are still an underestimate. Nevertheless, the arrival of humans on previously isolated continents, around thirteen thousand years ago in the case of Australia and eleven thousand years ago or possibly earlier for North and South America, seems to coincide with large-scale extinctions in certain taxa. Australia lost nearly all its species of very large mammals, giant snakes, and reptiles, and nearly half its large flightless birds around this time. Similarly, North America lost 73 percent and South America 80 percent of their genera of large mammals around the time of the arrival of the first humans. The probable cause is hunting, but the fact that climate changed at around this same time leaves the door open for natural changes as a contributing cause of these extinctions. However, the rates of extinctions on islands in the more recent past seem to confirm the devastating effects on humans. The Polynesians, who colonized Hawaii in the fourth and fifth centuries A.D., appear to have been responsible for exterminating around fifty of the one hundred or so species of endemic land birds in the period between their arrival and that of the Europeans in the late eighteenth century. A similar impact probably was felt in New Zealand, which was colonized some five hundred years later than Hawaii. There, an entire avian megafauna, consisting of huge land birds, was exterminated by the end of the eighteenth century. This was probably accomplished through a combination of direct hunting and large-scale habitat destruction through burning. Steadman (1995) suggests that loss of bird life in the tropical Pacific islands alone at this time may exceed two thousand species, mostly flightless rails, and this represented a 20 percent worldwide reduction in the number of species of birds.

On Madagascar, the giant elephant bird, the largest bird ever recorded; twenty species of lemur, most of them larger than any surviving species; and two giant land tortoises have become extinct within the last fifteen hundred years. In the Caribbean, at least two ground sloths became extinct before Europeans arrived at the end of the fifteenth century. Once again, climate change, particularly progressive desiccation, may have played a part as well. However, it is indisputable that recorded extinctions have increased dramatically in recent years, just as the population of humans has been seen to skyrocket (Fig. 3.9). Certain generalized patterns of extinction do emerge on examination of the data. Perhaps one of the most important of these is the preponderance of extinctions on islands versus those in continental areas. Although both the mainland and islands have similar overall numbers of recorded extinctions, there are lower numbers of species on islands than on continents, making the percentage of taxa extinct on islands greater than on continents (Table 3.3). The reason for high extinction rates on islands is perhaps because many species effectively consist of single populations on isolated islands. Adverse factors are thus likely to affect the entire species and bring about its extinction. Also, species on islands may have evolved in the absence of terrestrial predators and may often be flightless. Tameness, flightlessness, and reduced reproductive rates appear to be major contributory factors to species extinction, especially when novel predators are introduced.

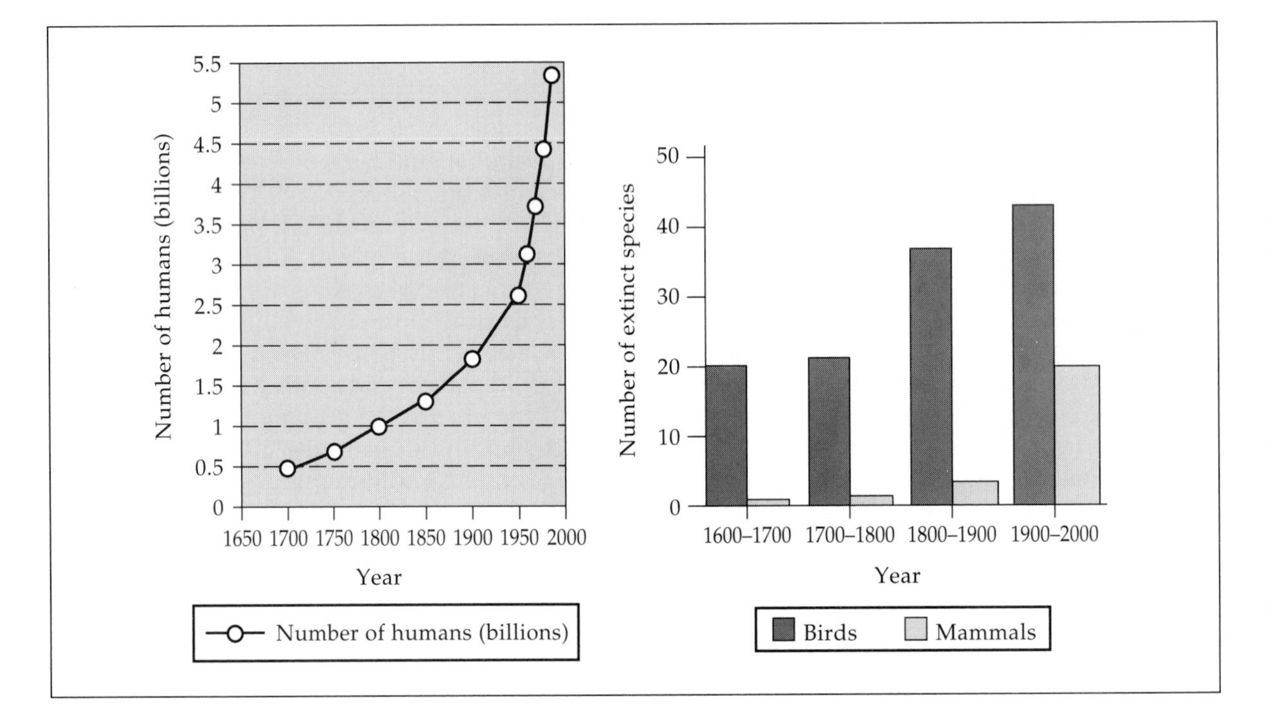

■ **FIGURE 3.9** Population growth and animal extinctions. (**a**) Geometric increase in the human population. (**b**) Increasing numbers of extinctions in birds and mammals. These figures suggest that, as human numbers increase, more and more living species are exterminated.

TABLE 3.3 Recorded extinctions, 1600 to present, on mainland areas (larger than 10^6km^2, the size of Greenland) islands and the ocean.

TAXA	MAINLAND	ISLAND	OCEAN	TOTAL	APPROXIMATE NUMBER OF SPECIES	PERCENTAGE OF TAXA EXTINCT SINCE 1600
Mammals	30	51	2	83	4,000	2.1
Birds	21	92	0	113	9,000	1.3
Reptiles	1	20	0	21	6,300	0.3
Amphibians	2	0	0	2	4,200	0.0
Fish	22	1	0	23	19,100	0.1
Invertebrates	49	48	1	98	1,000,000+	0.5
Vascular plants	380	219	0	599	250,000	0.3
Total	505	431	3	936		

When we look at the possible causes of extinction, most often no cause has been assigned (Fig. 3.10). Of those causes that have been assigned, introduced animals and direct habitat destruction by humans have been major factors involved in extinctions, being implicated in 17 percent and 16 percent of cases, respectively. These are equivalent to 39 percent and 36 percent, respectively, of the extinctions for which causes are known. Hunting and deliberate extermination also contribute significantly (23% of

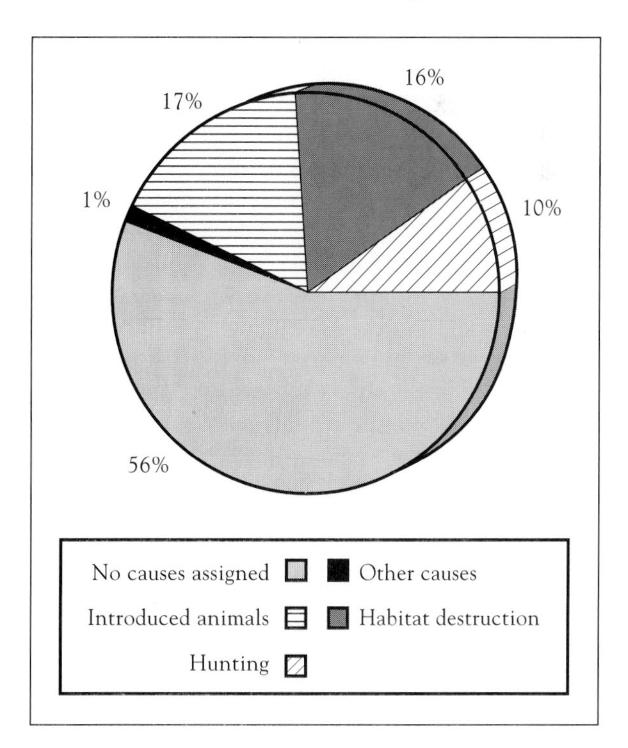

FIGURE 3.10 The causes of animal extinction, based on knowledge of 484 extinct species. *(After data from World Conservation Monitoring Centre 1992.)*

extinctions with known cause). Thus, three factors appear to be of paramount importance in causing extinction. In order of importance, they are as follows:

1. Introduced species.
2. Habitat destruction.
3. Direct exploitation.

We can break up the "introduced species" category a little further. The effects of introduced species can be assigned to competition from introduced species, predation from introduced species, or disease and parasitism from introduced species. Competition may exterminate local populations, but it has not yet been clearly shown to extirpate entire populations of rare species. The best examples of competition-caused extinction occurred long ago in evolutionary time when the land bridge between North and South America formed and species of North American megafauna (large animals often greater than 100 kilograms [kg]) migrated into South America and vice-versa. However, there is much evidence of the competitive effects of introduced species even today (see Chap. 9). For predation there have been many recorded cases of extinction. Introduced predators such as rats, cats, and mongooses have accounted for at least 112 of 258 recorded extinctions of birds on islands, or 43.4 percent (Brown 1989). Of the seventy five species of birds and mammals that have vanished during the past three hundred years or so, predation was a major factor in twenty-five. Parasitism and disease by introduced organisms is also important in causing extinctions. Avian malaria in Hawaii, facilitated by the introduction of mosquitoes, has been thought to have contributed to the demise of many local Hawaiian birds. Similarly, the American chestnut tree and European and American elm trees have been severely impacted by introduced plant diseases, though neither of these has yet become extinct.

In terms of habitat alteration, direct habitat destruction, like deforestation, is the prime cause of the extinction of species. More subtle alteration of habitat by events like modified climate due to pollution events has not yet been shown to have caused any extinctions. Direct exploitation, particularly the hunting of animals, has caused many extinctions. Stellers sea cow, a huge species of manatee-like mammal, was hunted to extinction on the Bering Straits only twenty-seven years after its discovery in 1740. The Dodo was hunted to death on Mauritius soon after it became a Dutch colony in 1644. In the United States the passenger pigeon was the most common bird in the entire bird population of North America. Unbelievably, hunting was the primary reason for its eventual extinction by 1900. The Carolina parakeet suffered the same fate (Photo 3.2). The list of species hunted to extinction goes on and on. Interestingly, in the case of the passenger pigeon, in the end there may not be enough individuals to stimulate synchronous breeding because a number of females were needed to stimulate ovarian development. This is common in some colonial species. Furthermore, some species of parasite probably followed all these species into extinction. Stork and Lyal (1993) note that two species of obligate ectoparasites, the lice *Columbicola extinctus* and *Campanulotes defectus*, were associated with the passenger pigeon.

Ecological data on species' vulnerability to extinction can be used to classify species as threatened or endangered.

Knowing why species have gone extinct helps us to prioritize the threats that are likely to deplete species numbers. This may make it easier to protect threatened species. A threatened species is one thought to be at risk of extinction in the foreseeable future.

■ **PHOTO 3.2** The Carolina parakeet, *Conuropsis carolineisis*, the only parrot native to the United States, went extinct in 1914, the last individual dying at the Cincinnati Zoo, one month later than the last Passenger pigeon. Hunting and habitat destruction were featured prominently in its demise. (*Runk/Schoenberger, Grant Heilman Photography, Inc. 3PPAC-11A.*)

The growth in public awareness of the problem of depletion and the possible extinction of species perhaps started with the publication of the so-called Red Data Books, a concept founded by Sir Peter Scott during the 1960s. This was an attempt to categorize species at risk according to the severity of the threats facing them and the estimated imminence of their extinction. More recently, the Red Data Books have been compiled on a global basis by the International Union for Conservation of Nature and Natural Resources (IUCN), now called the World Conservation Union. Red Data Books were first compiled for terrestrial vertebrates but have been expanded to include books for plants and invertebrates. The animals' Red List has been compiled every two years since 1986 by the World Conservation Monitoring Center in collaboration with the IUCN. In addition to these global books on threatened species, individual countries or areas may issue their own type of Red Data Book. This is because species may be more threatened on a national basis than they are globally. For example, the osprey may be thought of as a rare species in the United Kingdom but is common in other parts of the world, for example, coastal areas of the southeast United States. As of 1992, most bird species in the world had been comprehensively reviewed for rarity, about 50 percent of mammal species, probably less than 20 percent of reptiles, 10 percent of amphibians, and 5 percent of fish.

The main threat categories used for threatened species are as follows:

1. *Endangered*. Taxa in danger of extinction whose survival is unlikely if the causal factors continue to operate. Included are taxa whose numbers have been reduced to a critical level or whose habitats have been so drastically reduced that they are deemed to be in imminent danger of extinction.
2. *Vulnerable*. Taxa believed likely to move into the endangered category in the near future if the causal factors continue operating. Included here are taxa that are decreasing because of overexploitation, extensive destruction of habitat, or other environmental disturbance.
3. *Rare*. Taxa with small world populations that are not at present endangered or vulnerable but are at risk. Taxa in this category usually have very localized distributions, for example, populations that are wholly maintained on small oceanic islands.
4. *Indeterminate*. Taxa are known to be endangered, vulnerable, or rare but where there is not enough information to say which of the three categories is appropriate.

Species currently classified as threatened or endangered are largely at risk because of human activities.

Most of the causal factors currently threatening species are anthropogenic in nature. These factors include:

1. Habitat loss or modification. Causes include pastoral development, cultivation and settlement, forestry, and pollution.
2. Overexploitation for commercial or subsistence reasons, including hunting for meat, fur, or for the pet trade.
3. Accidental or deliberate introduction of exotic species that may compete with, prey on, or hybridize with native species.
4. Disturbance, persecution, and uprooting, including deliberate eradication of species considered to be pests. This is perhaps most serious for predatory species such as wolves or tigers.
5. Incidental take, particularly the drowning of aquatic reptiles and mammals in fishing nets.
6. Disease, both exotic and endemic, exacerbated by the presence of large numbers of domestic livestock or introduced species.

In many of these cases, individual species are faced by several of these threats simultaneously. Some understanding of the relative importance of these different threat types, as measured by frequency of occurrence, has been estimated from an examination of threats facing the terrestrial mammals of Australia and the Americas and those facing the birds of the world. For mammals, these threats are summarized in Figure 3.11.

Of the 119 species of mammals from these continents considered threatened, 75 percent are threatened by more than one factor and, of these, twenty-seven face four or more threats. The major threat, which affects 76 percent of species, is habitat loss and modification. This has a variety of causes, of which the most frequent is cultivation and settlement. Overexploitation affects half the species, the most significant cause being hunting

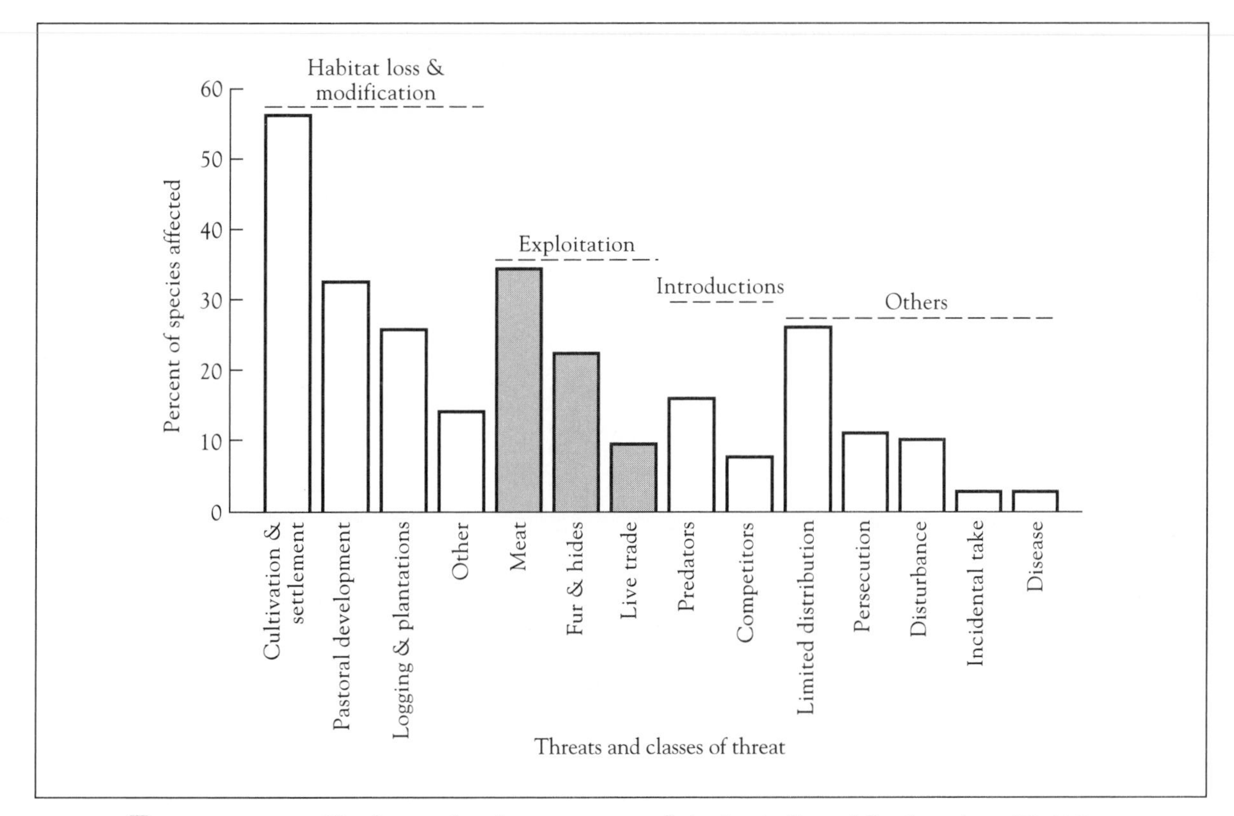

■ **FIGURE 3.11** The factors that threaten mammals in Australia and the Americas. (*World Conservation Monitoring Centre 1992.*)

for meat. Introduced predators and competitors affect 18 percent of the threatened species. The most serious other factor is limited distribution, which affects one-quarter of species. An interesting point here is that the factors that threaten endangered species rank differently in importance to those that are known to actually to have caused extinctions, where introduced species rank highest.

Overexploitation affects some animals more than others. Many fur-bearing animals, including chinchilla, giant otter, many species of cats, and some species of monkeys, have declined to very low population sizes because their pelts are prized. Valuable timber species, including populations of the West Indies' mahogany in the Bahamas and the caoba (mahogany) of Ecuador, have been severely depleted, and the Lebanese cedar has been reduced to a few scattered remnants of forest. Overexploitation is a more selective threat to species survival than is habitat loss and primarily threatens vertebrates and certain taxa of plants and insects. More specifically, carnivores, ungulates, primates, sea turtles, showy tropical birds, and timber species have been overharvested. Many species of butterflies and orchids have been overharvested for commercial interests too, and rare plants have been threatened by collectors. Predator control efforts have also significantly reduced the population sizes of many vertebrates, including sea lions, birds of prey, foxes, wolves, various large cats, and bears.

Threatened species are not evenly distributed among taxonomic groups.

In all, some 835 animal species were listed as endangered in the 1990 IUCN Red List, or much less than 0.1 percent of the world's estimated total of over 1.5 million described animal species. The class with the greatest number of endangered species is fish. Some orders seem to have a disproportionately high number of endangered or threatened species. For example, two out of two elephant species are threatened; manatees and dugongs have four out of four species threatened; and primates, carnivorous cats, and antelopes are also highly threatened with 53 percent, 32 percent, and 31 percent, respectively, of their constituent species listed. Although these latter three orders combined only contain some 20.6 percent of the world's mammals species, they account for just under half of the listed threatened species and just over half of the endangered species. Vertebrates are probably more vulnerable to extinction than invertebrates because they are much larger and require more resources and larger ranges. For example, the Florida panther population is now extremely low but may have only been about sixteen hundred animals at its maximum, before humans appeared. On the other hand, many invertebrates have extremely small ranges, but populations may still run in the millions. Nevertheless, there probably are many endangered invertebrates, like insects; it's just that the other orders like fish, and especially birds and mammals, are better monitored. On a worldwide basis, mammals are officially acknowledged as the most endangered taxa on earth, followed by fish, birds, reptiles, amphibians, and invertebrates (Fig. 3.12).

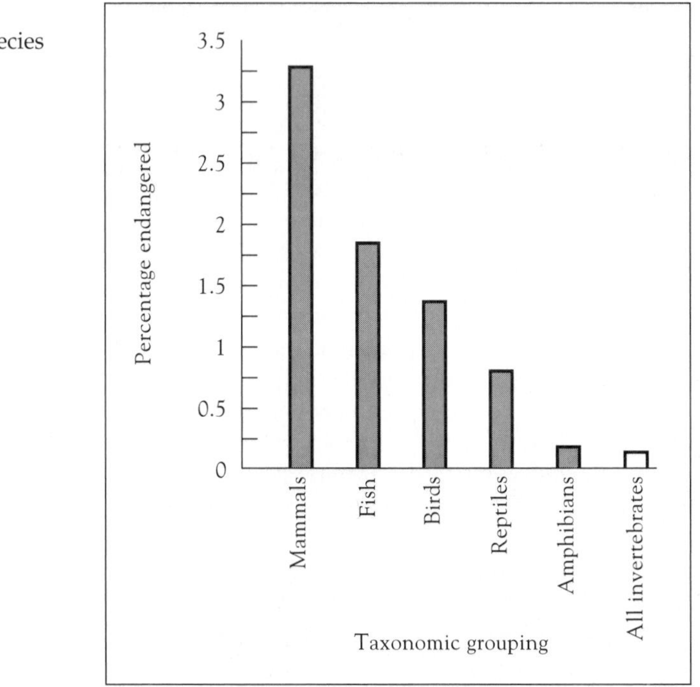

FIGURE 3.12
Percentage of known species classed as endangered.

Threatened species are not evenly distributed among geographical areas.

We can also break down the data to show which geopolitical areas contain the most threatened species (Table 3.4). The majority of threatened mammals occur in mainly tropical countries, with the highest numbers recorded from Madagascar (53), Indonesia (49), China, (40), and Brazil (40). Such countries may have large numbers of threatened animals simply because they have more animals in general; therefore, if the same percentage were threatened in each country, a country having a larger number of animals to begin with would have a higher number threatened. We can obtain some idea of what a country might be expected to have in terms of numbers of threatened species by drawing a graph of that country's area against the number of threatened animals. This is done in Figure 3.13. Madagascar and Indonesia, in particular, have more threatened mammals in relation to country area than would be predicted statistically (points above the line), whereas the United States, for example, has fewer. The patterns for birds are similar to those for mammals, the majority of endangered species being concentrated in Southeast Asia, Mexico, and South America. In comparison, Europe, Africa, Canada, the Middle East, and the Arabian peninsula have relatively few globally threatened bird or mammal species. It may come as a surprise to find that the United States heads the list in terms of the numbers of threatened reptiles, amphibians, and fishes (perhaps because other countries monitor them less well?).

We can also break the data down into the habitat types that contain the most numbers of threatened species. This type of data is available for the birds of the world and the mammals of Australia and the Americas (Fig. 3.14). The habitat type in which the largest number of threatened species, both bird and mammal, occur is clearly tropical forests. In general, the world's threatened bird species occupy the range of habitat types similar to the threatened mammals. However, there are a couple of noticeable differences. Most important of these are that there are far more threatened birds than mammals on oceanic islands. In fact, the data shown in Figure 3.14 does not include mammals

TABLE 3.4 The "top-ten" list of countries for different types of endangered species.

MAMMALS		BIRDS		REPTILES		AMPHIBIANS		FISHES	
COUNTRY	TOTAL	COUNTRY	TOTAL	COUNTRY	TOTAL	COUNTRY	TOTAL	COUNTRY	TOTAL
Madagascar	53	Indonesia	135	USA	25	USA	22	USA	164
Indonesia	49	Brazil	123	India	17	Italy	7	Mexico	98
Brazil	40	China	83	Mexico	16	Mexico	4	Indonesia	29
China	40	India	72	Bangladesh	14	Australia	3	South Africa	28
India	39	Colombia	69	Indonesia	13	India	3	Philippines	21
Australia	38	Peru	65	Malaysia	12	New Zealand	3	Australia	16
Zaire	31	Ecuador	64	Brazil	11	Seychelles	3	Canada	15
Tanzania	30	Argentina	53	Colombia	10	Spain	3	Thailand	13
Peru	29	USA	43	Madagascar	10	Yugoslavia	2	Sri Lanka	12
Vietnam	28							Cameroon	11

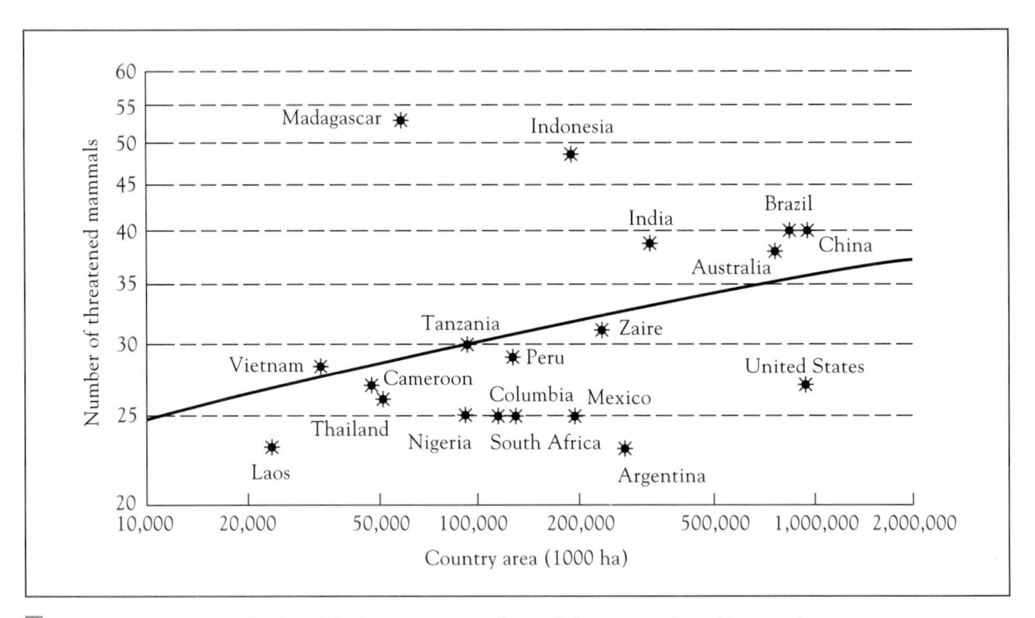

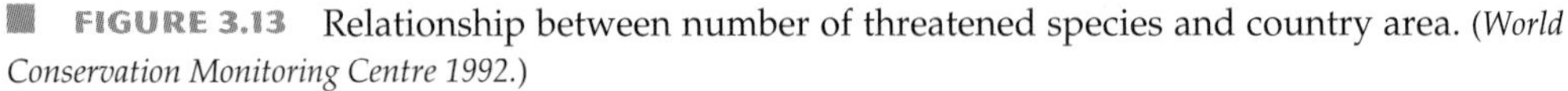

■ **FIGURE 3.13** Relationship between number of threatened species and country area. (*World Conservation Monitoring Centre 1992.*)

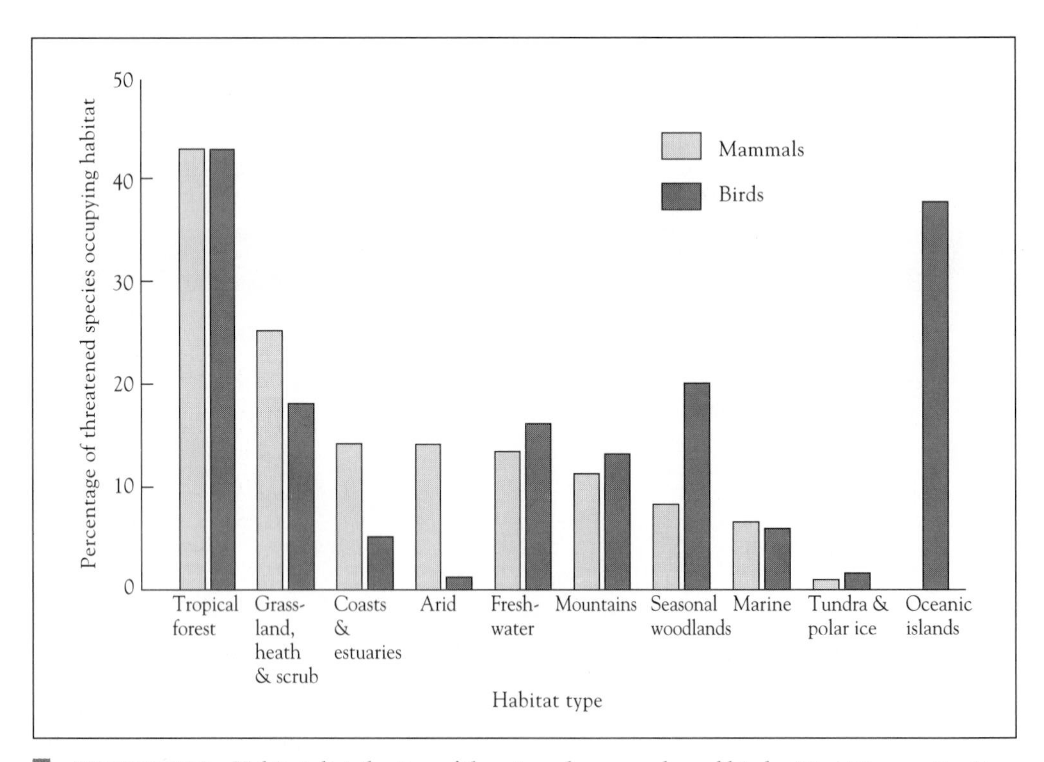

■ **FIGURE 3.14** Habitat distribution of threatened mammals and birds. (*World Conservation Monitoring Centre 1992.*)

on oceanic islands, but it is known that there are few mammals on such islands, and therefore the number of threatened species on such habitats is likely to be small.

Oceanic islands are also known to have very many species of threatened plants. For example, Hawaii has 108 endemic taxa that have already gone extinct, 138 of which are endangered, 37 vulnerable, 126 rare, and 9 indeterminant. Hawaii has one of the most distinctive and most threatened floras in the world. Other oceanic islands show similar patterns. St. Helena, in the Atlantic Ocean, has 46 endemic species that are threatened. Bermuda has 14 of 15 endemic species threatened.

In the United States, Dobson et al. (1997) used geographic distribution data for endangered species in the United States to locate "hot spots" of threatened biodiversity. The hot spots for plants, birds, fish, and molluscs rarely overlapped, so it would be hard to use just one taxa to act as a surrogate or indication of rarity for other taxa. However, endangered plant species maximized the incidental protection of all other species groups. This is because more land is needed to preserve endangered plants than for any other taxa, so it is expected that land management for endangered plants maximizes protection of other taxa. When area is controlled for, birds and reptiles actually provide the best indicators for any particular area. The amount of land needed to protect currently endangered and threatened species in the United States was quite small (Fig. 3.15). This of course does not represent the best way to protect biodiversity—merely endangered

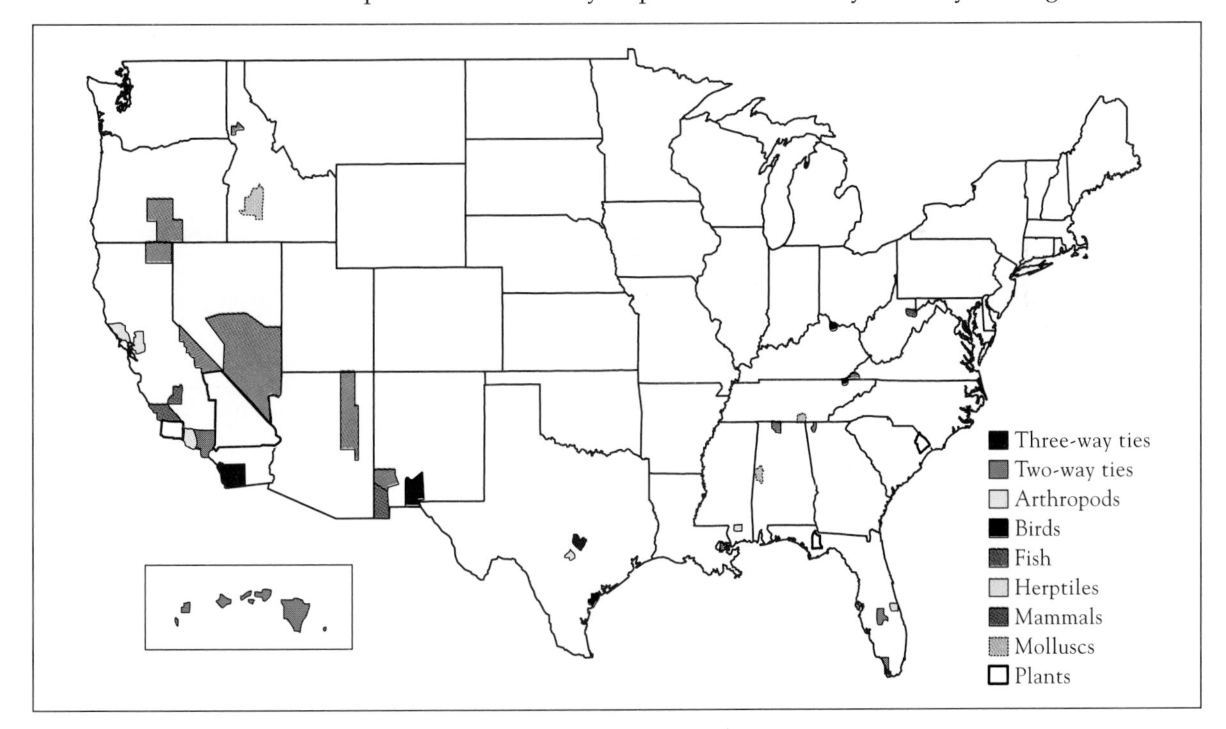

■ **FIGURE 3.15** Complementary set of counties that contains 50% of the listed species for each taxonomic group. The analysis identified two counties that contain large numbers of endangered species from three groups and nine counties that contain large numbers of species from two groups (Hawaii not to scale). The two counties that are hot spots for three groups are San Diego, California (fish, mammals, and plants), and Santa Cruz, California (arthropods, herptiles, and plants). (*After Dobson et al. 1997.*)

species. Half of the currently listed plant species were found in just thirteen counties (the highest thirteen ranked), totaling 1.33 percent of the U.S. land mass. The equivalent figures for other groups are given in Table 3.5a.

The greatest numbers of endangered species occurred in Hawaii, southern California, southern Appalachia, and the southeastern coastal states. Forty-eight percent of plants and 40 percent of endangered arthropods are restricted to a single county. Of course, for some species a whole county might not be needed to ensure its survival; for others, like grizzly bears, one county may not be sufficient. Furthermore, some counties, like San Diego, are much bigger than others. San Diego county is also larger than the state of Rhode Island. The average number of counties in which a listed plant was found was 3.9. Comparable data for other taxa are given in Table 3.5b.

TABLE 3.5A Endangerment of species in the United States data. (From Dobson et al. 1997)

TAXA	NUMBER OF COUNTIES NEEDED FOR 50% OF ENDANGERED SPECIES	% AREA OF U.S. LAND MASS NEEDED TO PROTECT 50% OF THE ENDANGERED SPECIES
Plants	13	1.3
Molluscs	6	0.14
Arthropods	9	0.46
Fish	14	2.04
Reptiles	7	0.34
Birds	4	0.28
Mammals	7	0.40

TABLE 3.5B The restriction of endangered species to single counties in the United States and the average number of counties a listed species occurs in. (From Dobson et al. 1997)

TAXA	% OF SPECIES RESTRICTED TO SINGLE COUNTIES	AVERAGE NUMBER OF COUNTIES IN WHICH A LISTED SPECIES IS FOUND
Plants	48%	3.9
Arthropods	40%	4.4
Birds	36%	62.7
Mammals	26%	32.9
Fish	31%	8
Herptiles *(reptiles and amphibians)*	14%	18.8
Snails	57%	2.1
Clams	3%	12.1

Note: Mean values for birds are inflated by the inclusion of peregrine falcons and bald eagles, which occur in many counties. Without these the figures are 37% and 31.7%.

Finally, Dobson and colleagues examined associations between the density of endangered species in each state and the intensity of human economic and agricultural activities, climate, topology, and vegetative cover. They found that two variables, agricultural output and temperature or rainfall, explained 80 percent of the variance in density of endangered species. Agricultural output was the key variable for plants, mammals, birds, and reptiles.

Kerr and Currie (1995) did a similar analysis on a countrywide scale using ninety countries and six indices of human activities (Table 3.6). Twenty-eight to 50 percent of the variation in the proportion of threatened bird and mammal species in each country was attributable to these variables, with human population density most important for birds (Fig. 3.16a) and per capita GNP for mammals (Fig. 3.16b). In addition, mammalian population density was found to strongly correlate with the extent of protected area per country. There were, in fact, a variety of relationships between indices of extinction risk and socioeconomic factors (Table 3.7), and Kerr and Currie used a technique called path analysis to test the relative importance of which way, directly or indirectly, these variables worked (Figs. 3.17 a, b). The path analysis diagrams for both threatened birds and mammals were quite similar. One major difference was that hunting mammals for food is important, while it is not for birds. Thus, the numbers of threatened mammals are more tightly linked to GNP.

TABLE 3.6 Factors hypothesized to influence susceptibility to extinction among birds and mammals. (After Kerr and Currie 1975)

FACTOR	RATIONALE
Human population	High populations lead to habitat loss that increases numbers of threatened species and to reduced population densities.
Per-capita GNP	Subsistence hunting and uncontrolled sport hunting and poaching are more extensive in poorer countries, and damaging land-use practices are less likely to be regulated.
Extent of protected area	Protected areas may be established to safeguard threatened species, and they may permit higher population densities.
Total cropland	Loss of habitat attributable to agricultural activity increases numbers of threatened species and reduces population densities.
Birth rates	High birth rates are related to poor socioeconomic conditions.
Per capita industrial CO_2 emissions	Greater industrialization may result in higher pollution levels.

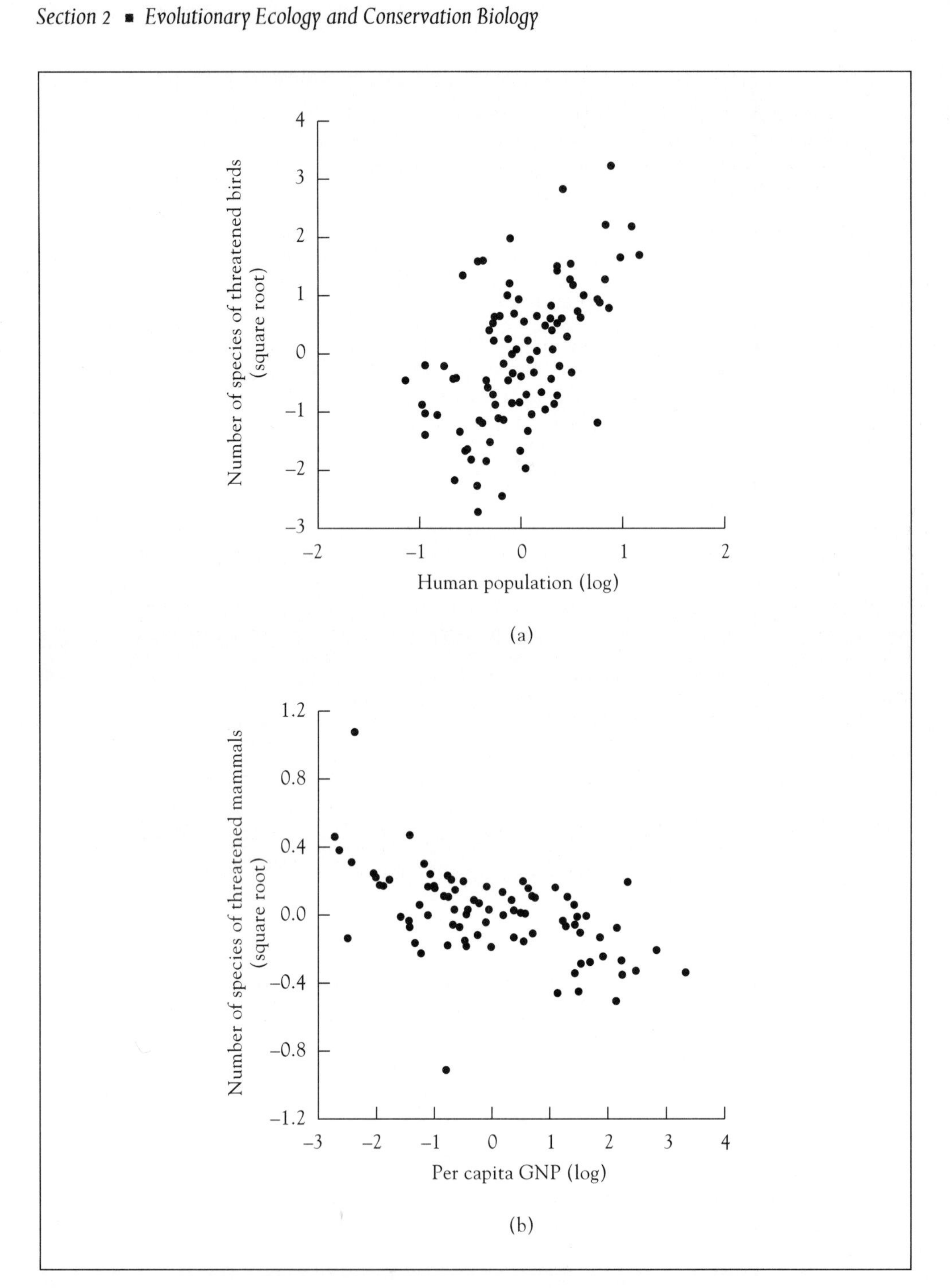

■ **FIGURE 3.16** (**a**) Partial plot of numbers of threatened bird species per country against human population per country. The effects of land area and species richness per country have been statistically controlled. (**b**) Similar plot for mammals. *(After Kerr and Currie 1995.)*

TABLE 3.7 Simple Pearson correlations between indices of extinction risk, the major nonanthropogenic covariates of extinction risk, and several socioeconomic measures of anthropogenic activity. The higher the number, the tighter the relationship between the variables. Negative values indicate negative relationships. Log or square root indicates how the data was transformed prior to analysis. (After Kerr and Currie 1995)

	INDICES OF SUSCEPTIBILITY TO EXTINCTION			NON ANTHROPOGENIC COVARIATES		
FACTOR	THREATENED BIRD SPECIES (SQUARE ROOT; N = 95)	THREATENED MAMMAL SPECIES (SQUARE ROOT; N = 82)	MAMMALIAN POPULATION DENSITY (LOG; N = 78)	LAND AREA (LOG; N = 78 TO 95)	TOTAL BIRD SPECIES RICHNESS (SQUARE ROOT; N = 95)	TOTAL MAMMAL SPECIES RICHNESS (SQUARE ROOT; N = 82)
Land area (log)	0.56	0.60	NS	—	0.59	0.64
Body size (log)	—	—	−0.68	NS	—	—
Human population (log)	0.71	0.43	−0.45	0.70	0.47	0.52
Per-capita GNP (log)	NS	−0.50	0.68	NS	−0.38	−0.31
Extent of protected area (log)	NS	NS	0.75	NS	NS	0.60
Total cropland (log)	0.63	0.47	−0.25	0.78	0.46	0.54
Birth rates (log)	−0.32	0.46	−0.60	NS	0.36	0.30
Per-capita industrial CO_2 emissions (log)	0.27	−0.32	NS	NS	NS	NS
Total bird species richness (square (root)	0.72	—	—			
Total mammal species richness (square root)	—	0.77	—			

Note: NS = no significant relationship.

SUMMARY

1. Some of the best-analyzed examples of natural selection involve industrial melanism—a change in color that takes place in areas that have been heavily polluted. This underscores the fact that observation of practical environmental problems can help us understand theoretical ecological principles. On the flip side, biological indicator organisms can provide a good record of the strength of pollution.

2. Natural selection ultimately results in the formation of races, subspecies, biotypes, and species. An understanding of the seemingly simple idea of a species is in fact quite complex. For example, there is the biological-species concept and the phylogenetic-species concept. The definition of *species* is very important because federal conservation legislation often emphasizes species over other categories.

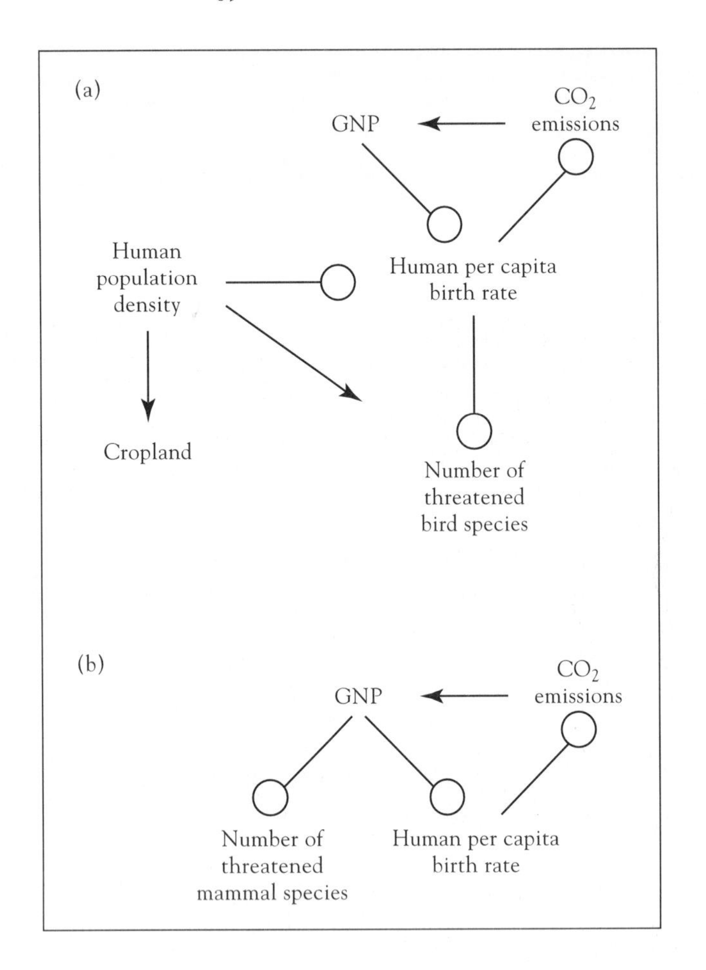

FIGURE 3.17 (**a**) Path analysis diagram of the factors influencing the numbers of threatened bird species per country (square-root transformed). An arrow from one factor to another indicates that the correlations are consistent with the hypothesis that the first variable directly influences the second (although other unobserved variables may intervene). A solid triangular head indicates a positive influence; an open circular head indicates a negative influence. (**b**) Path analysis diagram for threatened mammals. (*After Kerr and Currie 1995.*)

3. What causes the extinction of species? Such knowledge can be used to decrease current rates of extinction. At least seven life history traits may affect a species' sensitivity to extinction: rarity, dispersal ability, degree of specialization, population variability, trophic status, longevity, and intrinsic rate of population increase.

4. In order of importance, the three most important causes of extinction have been introduced species, habitat destruction, and direct exploitation. Introduced predators such as rats and cats have accounted for much extinction of ground-nesting birds and lizards on islands. In more modern times, habitat destruction, such as deforestation, is the prime cause of extinction. Indeed, of the factors that currently threaten living species with extinction, habitat loss threatens more than threequarters, overexploitation affects half, and introduced predators and competitors affect only 18 percent of threatened species.

DISCUSSION QUESTION

Which types of organisms deserve priority in conservation efforts and why?

SECTION

3
Behavioral Ecology

Do populations commonly self-regulate themselves or is control by biotic phenomenon more important? A study of behavior can help reveal whether biotic phenomenon like competition occur frequently. Here herring gulls are engaged in a territorial dispute on Long Island, New York. (*A. Morris, Visuals Unlimited ATR-282.*)

If ecology strives to explain the distribution patterns of plants and animals on earth, then an understanding of behavior is valuable (Berger 1996). Two schools of thought exist concerning the abundance of populations. One, viewed often as the "climate" school or "density independence," was championed by the Australian biologists Andrewartha and Birch in the 1940s and 1950s. Their view was that most population mortality was ultimately caused by the weather and was unpredictable in its effects. Populations of small insects, called thrips, which Andrewartha and Birch studied on plants, suffered violent changes from the effects of rainfall and other features of the weather. The alternative, often termed the "equilibrium" school or "density dependence," was championed by mathematicians such as Pearl and Lotka in the 1920s and 1930s and by the Australian entomologist Nicholson in the 1950s. The density dependence school of thought held that populations varied around some mean or equilibrium level. Any variations from that mean were "corrected" by biological phenomena. For example, a bonanza of food, which might promote high insect populations, would, in turn, be followed by an abundance of natural enemies that would reduce populations to their mean or equilibrium value. For many years, neither school of thought enjoyed the overwhelming support of evidence, and so it is even today—opinions sway back and forth about which theory is generally held to be more true. We shall return to this theme in the next section—Population Ecology.

Within the density dependence camp there have been discussions about which factors commonly cause populations to oscillate around the equilibrium value. These factors can be either intrinsic or extrinsic. Extrinsic factors include predators, parasites, and competition with other species—interspecific competition. Intrinsic factors center on control of population size at the level that is imposed by crowding or by competition with conspecifics for resources—intraspecific competition.

Some ecologists have argued that populations self-regulate themselves at a level that can be sustained without the ravages of starvation and disease—that is, at levels below which competition becomes important. In this way, wasteful squabbling and fighting for food is avoided. Nature is neat, tidy, and harmonious. Several pieces of evidence led early ecologists to believe that many populations did self-regulate their numbers. For example, many birds are territorial, and a population in a given area was often below that which could be supported by a given amount of prey. Second, any increase in organisms within a given area resulted in emigration, which again operated to maintain populations about a mean level. Experimental manipulations showed that some of these mechanisms operated at low densities in the absence of food or other resource limitations. Third, many species, in particular birds, showed high variation in reproductive rates. Some birds fledged more young than others. This was regarded as evidence of the adjustment of fecundity to balance mortality in the population. In the late 1940s, two ornithologists, David Lack and Alexander Skutch, brought the argument of self-regulation versus extrinsic regulation to a head. The argument was precipitated by data that showed that songbirds typically lay a clutch of four to six eggs in temperate regions of North America and Europe and only two or three in the tropics. Lack argued that birds in the tropics couldn't gather enough resources to fledge more than two to three young in the tropics. Skutch suggested that tropical populations were self-regulated in the tropics to ensure no resources were wasted. The trend is general, affecting virtually all groups of birds in all regions of the world. In the next twenty years the arguments were replayed many times with different data, but the theme was always the same: profusion of young or reproductive prudence?

In 1962 the self-regulation viewpoint was championed by British ecologist V. C. Wynne-Edwards who articulated the full concept of self-regulation in a book called *Animal Dispersion in Relation to Social Behaviour*. The premise was that most groups of individuals purposely controlled their rate of consumption of resources and their rate of breeding to ensure that the group would not go extinct—this was known as **group selection**. Individuals in successful groups would not tend to act selfishly. Groups that consisted of selfish individuals would over exploit their resources and die out. In concept the idea of self-regulation or group selection is straightforward, intellectually satisfying to many, and would seem to represent what nature ought to do—avoid the grisly clashes and potentially damaging confrontations of competition. Why fight to the death if a contest can be settled amicably or indeed if species never overstepped their limits so that confrontations never occurred? This is perhaps how humans would like to be. In the late 1960s and 1970s, the idea of self-regulation came under severe attack, and some of the data that supported the group selection camp was equally well explained by **individual selection**.

C H A P T E R

Group Selection and Individual Selection

Attractive though the idea of **group selection** is it has several flaws:

1. *Mutation:* Imagine a species of bird in which a pair lays only two eggs, and there is no overexploitation of resources. Suppose the tendency to lay two eggs is inherited as a group-selection trait. Now consider a **mutant** that lays six eggs, the equivalent of a "cheater" in this scenario. If the population is not overexploiting its resource there will be sufficient food for all the young to survive, and the six-egg genotype will become more common very rapidly. Gene frequencies in the population will change. This process would work for even larger brood sizes, and brood sizes would tend to increase until they became so large that the parents could not look after all their young, causing an increase in infant mortality. Thus, the clutch size in nature evolves so as to maximize the number of surviving offspring. Field studies of great tits in Wytham Woods, England, for example, show a median clutch size of eight-to-nine eggs, above which adult birds cannot reliably supply sufficient food for all chicks to survive.
2. *Immigration*: Even in a population in which all pairs laid two eggs and no mutations occurred to increase clutch size, "selfish" individuals that laid more could still migrate in from other areas. In nature, populations are rarely sufficiently isolated to prevent **immigration**.
3. *Individual selection:* For group selection to work, some groups must die out faster than others. In practice, groups do not go extinct fast enough for group selection to be an important force. Individuals nearly always die more frequently than groups, so individual selection will be the more powerful evolutionary force.

4. *Resource prediction:* For group selection to work, individuals must be able to assess and predict future food availability and population density within their own habitat. There is little evidence that they can.

Individual selfishness seems a more plausible result of natural selection. Group selection is probably a weak force and is only rarely very important (Maynard Smith 1976a) even though it may explain some phenomena in nature (Morrell 1996). For example, David Sloan Wilson of State University of New York, Binghampton, thinks group selection may explain why parasites seem to strike a balance with their hosts rather than maximize their virulence. Indeed, Wilson believes group selection operates at many levels (Wilson 1997). Thomas Seeley of Cornell suggests group selection can explain the selfless behavior of honey bees. But as Jerry Coyne of the University of Chicago argues, and as we shall see later, this is actually kin selection, and helping your offspring isn't really the same as group selection. Any reduction in population sizes from **self-regulation** is likely to come from intraspecific competition, of a selfish nature, in which individuals are still striving to command as much of a resource as they can (see Chap. 9). Indeed, we often see animals in nature acting in their own selfish interest. Male lions kill existing cubs when they take over a pride. The proximate cause may be the unfamiliar smell of the cubs. A similar Effect, known as the Bruce Effect, occurs in rodents, where the presence of a strange male prevents implantation of a fertilized egg or induces abortion in females (Bruce 1966). In the case of lions, the advantage of infanticide for the male lion is that, without their cubs, females come into the reproductive condition much faster, in nine months as opposed to twenty-five if the cubs are spared, hastening the day when males can father their own offspring. A male's reproductive life in the pride is only two-to-three years before he in turn is supplanted by a younger, stronger male. Infanticide ensures the male will father more offspring, and the tendency spreads by natural selection (Bertram 1975). There is no advantage to the mothers of infanticide of the killed cubs, but, being smaller, they are powerless to stop the males.

In human society, individuals also rarely act for the good of the group and instead tend to act selfishly. This causes many environmental problems such as overgrazing and overfishing, a phenomenon known as the Tragedy of the Commons (Applied Ecology: The Tragedy of the Commons.)

If individual selfishness is more common than group selection, we are obligated to provide explanations for phenomena like **altruism**, the existence of castes, and even the existence of sex itself, all of which smack of group selection.

4.1 Altruisic behavior does occur in nature and is often associated with kin selection.

Although natural selection favors individual rather than group selection, it is still common to see apparent cooperation. Animals of the same species groom one another, hunt communally, and give warning signals to each other in the presence of danger. How can this altruistic behavior be explained by natural selection?

We are not surprised to see a parent working hard in the caring for its young. All offspring have copies of their parents' genes, so parental care is genotypically selfish. Genes for altruism toward one's young will therefore become more numerous because

TABLE 4.1 **Brightly colored species of caterpillars of British butterflies are more likely to be aggregated in family groups than cryptic species. (From Harvey, Bull, and Paxton 1983)**

	NO. SPECIES OF CATERPILLARS	
DISPERSION	**APOSEMATIC**	**CRYPTIC**
Large family groups	9	0
Solitary	11	44

aggregate in kin groups (Table 4.1), so the death of one individual is most likely to benefit its relatives, such as siblings, and its genes will be preserved (Costa 1997). Even some solitary species have warning colors, suggesting a direct benefit as well—if the unlucky larva isn't killed by an attack, it will probably not be chosen again.

Another example of altruism again refers to lions. Lionesses tend to remain within the pride, whereas the males leave. As a result, lionesses within a pride are related, on average, by $r = 0.15$. Females all come into heat at the same time; one individual probably influencing the others' estrous cycles by means of pheromones. A similar phenomenon occurs in humans, where young women living in the same school dormitories may have synchronized menstrual cycles. In lionesses, the result is the simultaneous birth of cubs, and females exhibit the apparently altruistic behavior of suckling other females' cubs. Because the females are related, the selfish-gene hypothesis accounts for this behavior. Similarly, when male lions depart, they may act in concert to take over a pride; each one's genes will be perpetuated both through his own offspring and through those sired by his brothers (Bertram 1976; Bygott, Bertram, and Hanby 1979, Packer et al. 1991).

Although an altruistic act toward two sisters or brothers may seem to be genetically equivalent to a similar act toward an offspring, there are sometimes other ecological or proximate factors that tip the scales in favor of the offspring. Young may be more valuable in terms of expected future reproduction, having a higher potential reproductive output than older siblings or parents. Progeny may also benefit more from a given amount of aid. One insect fed to a nestling contributes more to its survival than the same insect would to the survival of a healthy adult sibling. Young may thus be thought of as super beneficiaries (West Eberhard 1975).

Not all acts of altruism result from behavioral extremes. A common example of altruism is the raising of an alarm call by "sentries" in the presence of a predator. The alarm maker is drawing attention to itself and risking increased danger by its behavior. For some groups, like ground squirrels, *Spermophilus beldingi*, individuals near an alarm maker bolt down their burrows, and those close neighbors are most likely to be sisters or sisters' offspring (Sherman 1977); thus, the altruistic act of alarm calling could be reasoned to be favored by kin selection. On the other hand, sentries can be subordinate individuals who are driven to the edge of the group where the risk is higher and they are forced to be alert for their own safety.

Working out the exact costs and benefits in a system of kin selection can be a nightmare and has often been a stumbling block in behavioral ecology. To get a precise measure of natural selection, one must be able to calculate an individual's contribution to the gene pool. This might be the contribution of an individual plus 0.5 times its number of brothers and sisters, 0.125 times its number of cousins, and so on. This genetical octopus has been termed **inclusive fitness** by Hamilton (1964). More commonly, behavior or **adaptation** is recognized as beneficial if it shows a closely designed fit to some problem presented by the animal's environment (Williams 1966). At least for the alarm makers, some selfish motive is involved because if a predator fails to catch prey in a particular area it is less likely to return, and the number of potential attacks over the alarm maker's lifetime is likely to be reduced (Trivers 1971).

Unrelated individuals may engage in altriuistic acts if there is a high probability that the altruism will be reciprocated.

Not all acts of altruism are directed to close relatives (Photo 4.1). In more than 200 species of birds and 120 species of mammals, individuals have been recorded helping others to reproduce. When a female olive baboon, *Papio anubis*, comes into estrus, a male forms a consort relationship with her, following her around and waiting for the opportunity to mate. Sometimes an unattached male enlists the help of another to engage the consort male in battle while the solicitor attempts to mate with the female. On a later occasion, the roles are reversed (Packer 1977). This is an example of reciprocal altruism (Trivers 1971). (Reciprocal altruism is common in human society where money is used to mediate its use.) Sometimes unrelated individuals will occupy the same territories as breeding individuals and help the parents raise offspring by foraging for additional food for the young. Although the helpers may be related to the parents, for example, in the well-studied Florida scrub jay (Woolfenden 1975; Woolfenden and Fitzpatrick 1984), sometimes they are not, for example, in mongooses (Rood 1978, 1990; Creel and Waser 1991) and dunnocks, *Prunella modularis* (Houston and Davies 1985; Davies et al. 1992). By comparing the nest or brood success of breeding pairs with helpers with those whose helpers were

■ **PHOTO 4.1** Grooming behavior in Japanese macaques. You scratch my back, I'll scratch yours, is an easy to understand maxim. Other types of altruism are much more difficult to explain. (*Steve Kaufman, Peter Arnold, Inc. AN-BE-25A.*)

removed, Brown et al. (1978, 1982) and Mumme (1992) were able to show that helpers do significantly increase parents' fitness. Again, the motive in most situations seems to be a sort of reciprocal altruism. In these situations, the habitat is usually saturated with breeders, and helpers could probably not obtain a territory for themselves. What they do is help increase territory size of a breeding pair; they are later able to carve off a fragment of this territory themselves, or they can take over from the breeding pair after one member dies.

For many predators, reciprocal altruism in the form of social hunting allows bigger game to be caught. The benefits of a large kill outweigh the cost of having to share the meat (Caraco and Wolf 1975). Many acts of altruism between unrelated individuals, whether in defense, attack, or mating, are exhibited by individuals living in social groups.

Of course, not all observed behavior can be explained by altruism. Many adaptive explanations are in fact more the result of an author's ingenuity than natural consequences of the facts. Great care must be taken to avoid such errors; otherwise, the resultant theories will have no more substance than Kipling's *Just So Stories* (Gould and Lewontin 1979). Often some individuals are simply manipulated by others. Cuckoos, for example, have used the devoted parental care of adult birds to rear their parasitic cuckoo young instead. Some authors have also suggested that in social insects some females could actually do better, in terms of **inclusive fitness**, by reproducing themselves rather than being sterile, but they have been manipulated by their parents. Often more than one factor has to be invoked to explain any apparent pattern in animal behavior. As an example, it would be easy to say that the long neck of the giraffe is an adaptation to feeding on high foliage, but it could equally well aid in predator detection. There are often such confounding variables in behavioral ecology, which make it tricky to decide which selective pressures cause a particular trait. Often it may be more than one.

Altruism in social insects may arise from the unique genetics of social insect reproduction.

Perhaps the most extreme altruism is the evolution of sterile castes in the social insects, in which some females, known as workers, rarely reproduce themselves but instead help others to raise offspring (Photo 4.2). The differentiation of one species into many

■ **PHOTO 4.2** In bees, wasps, ants, and termites, sterile castes or workers, like these paper wasps, exhibit the extreme altruistic act of forgoing reproduction. *(Michael L. Smith, Photo Researchers, Inc. OU 5589.)*

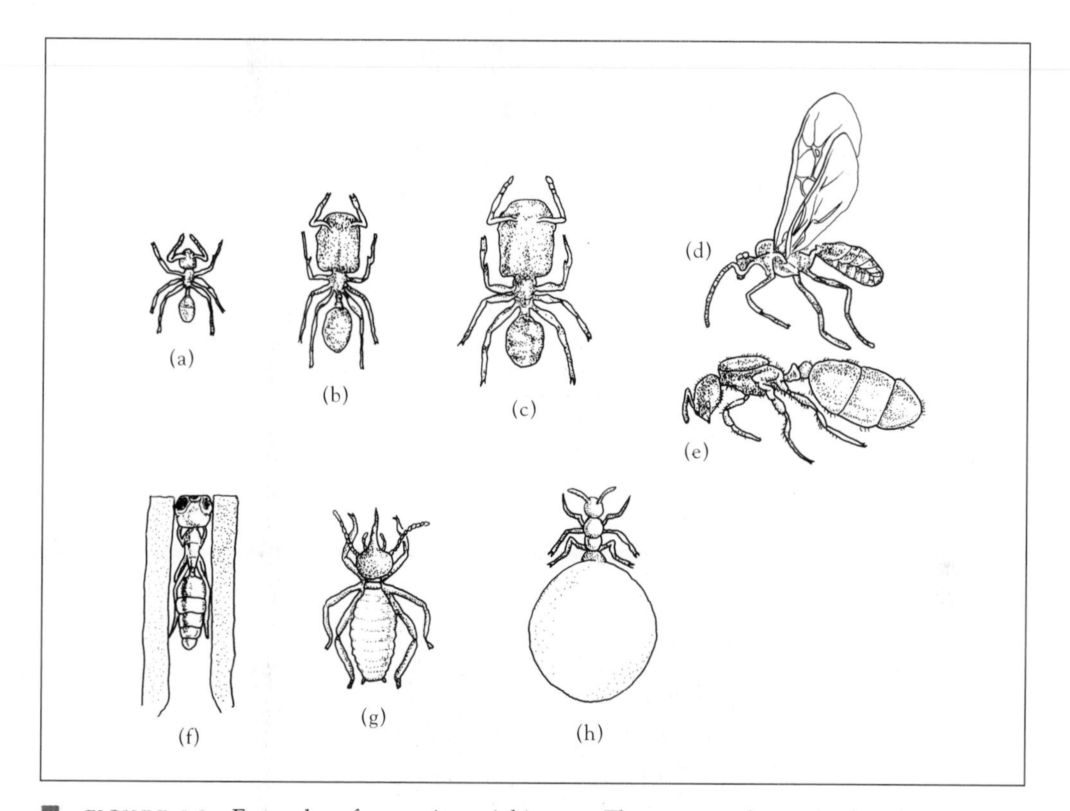

■ **FIGURE 4.2** Examples of castes in social insects. The top row shows the female castes and male of the myrmicine ant, *Pheidole kingi instabilis.* (**a**) Minor worker. (**b**) Media worker. (**c**) Major worker. (**d**) Male. (**e**) Queen. The bottom row shows various specialized castes in other species. (**f**) Soldier of the ant, *Camponotus truncatus*, blocking a nest entrance with its pluglike head which serves as a "living entrance" to the nest. (**g**) A sterile caste of the nasute termite, *Nasutitermes exitiosus*, which has a head shaped like a water pistol to spray noxious substances at an approaching enemy. (**h**) Replete worker of a *Myrmecocystus* ant, which lives permanently in the nest as a "living storage cask." *(From Wilson 1971.)*

different-sized castes in ants can be quite staggering (Fig. 4.2). Of the 263 living genera of ants known worldwide, 44 possess species with caste systems. The explanation of the peculiar system of castes was thought to lie primarily in the particular genetics of most (at least hymenopteran) social insect reproduction (Hamilton 1967) (Fig. 4.3). Males develop from unfertilized eggs and are **haploid**. Male gametes are formed without meiosis, so every sperm is identical. Thus, each daughter receives an identical set of genes from her father. Half of a female's genes come from her **diploid** mother, so the total relatedness of sisters is 0.75. Such a genetic system is called **haplodiploidy**. Thus, females are more related to their sisters than they would be to normal offspring. It is therefore advantageous to stay in the nest or hive and to try to produce new reproductive sisters.

The story is complicated a little bit when the interests of the queen are incorporated into the propensity for sterile castes. Queens are equally related to their sons and their daughters; $r = 0.5$ in each case. To maximize her reproductive potential, a queen should therefore produce as many sons as daughters, a 50:50 sex ratio. If she did, then sterile

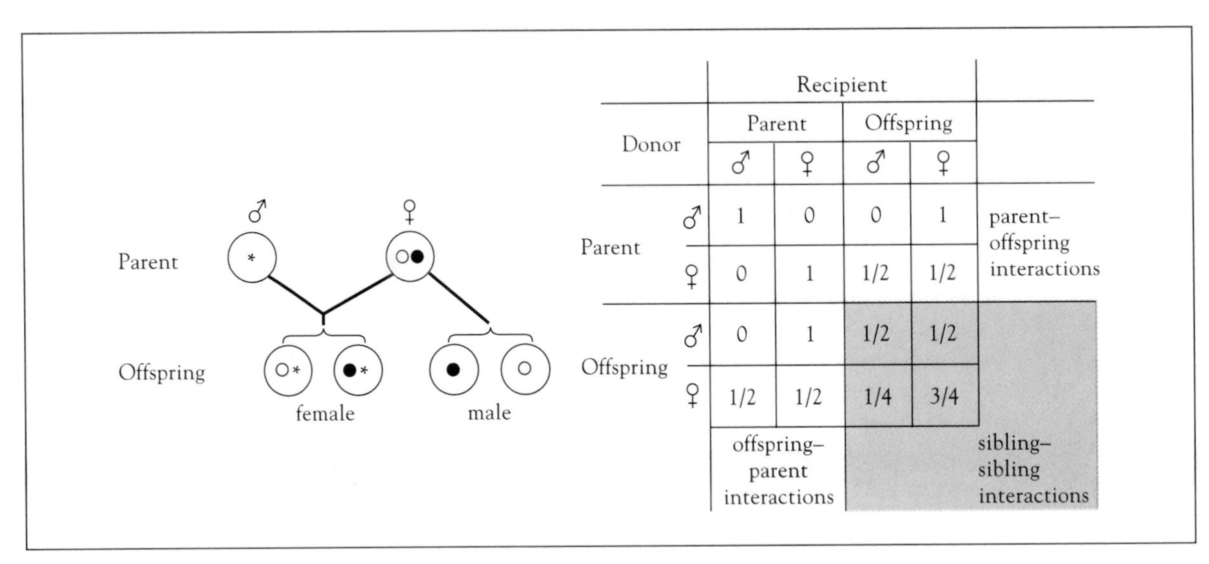

Donor		Recipient: Parent ♂	Recipient: Parent ♀	Recipient: Offspring ♂	Recipient: Offspring ♀
Parent	♂	1	0	0	1
Parent	♀	0	1	1/2	1/2
Offspring	♂	0	1	1/2	1/2
Offspring	♀	1/2	1/2	1/4	3/4

■ **FIGURE 4.3** Coefficients of genetic relationship in a haplodiploid mating system. (*After Ricklefs 1990.*)

worker females would spend as much time rearing brothers (to which they are related only by 0.25) as sisters. The average relatedness of a female to her siblings would then be 0.5, and she would do equally well to breed on her own. From the workers' viewpoint, it is far better to have more sisters, and in this conflict with the queen they appear to have won because in any colony there are more females than males, by a ratio of about 3:1 (Trivers and Hare 1976; but see also Alexander and Sherman 1977).

Elegant though these types of explanations are, they do not provide the whole picture: Large social colonies exist in termites too, but theirs is not a haplodiploid system, and colony males have only half their genes in common, on average. Furthermore, mole rats, mammals living underground in South Africa, have a division of labor based on castes and cooperative broods, but the animals are, of course, **diploid** (Jarvis 1981; Jarvis et al. 1994). There is only one breeding female, the queen; she apparently suppresses reproduction in other females by producing a chemical in her urine that is passed around the colony by grooming after visits to a communal toilet. The other castes perform different types of work in burrowing. One rat chisels away at the face of the burrow, and others shuffle the earth backward toward the burrow entrance. There, another individual kicks it out onto the surface. The earth carriers then return to the face of the burrow by leapfrogging over the crouched earth removers (Fig. 4.4). One caste, the "frequent workers," appears to do most of this work; another, the "infrequent workers," consists of heavier individuals that do some of the work; and even larger individuals, the nonworkers, rarely work at all. These are often male and may be a reproductive caste. There is even a special dispersing morph, large fat males, disinclined to mate with their own queen but with a strong tendency to leave home and mate with members of other colonies (O'Riain, Jarvis, and Faulkes 1996). These are equivalent to the dispersing morphs of ant colonies. Even before the case of the mole rats was known, Richard Alexander, curator of the Museum of Zoology at the University of Michigan, had sug-

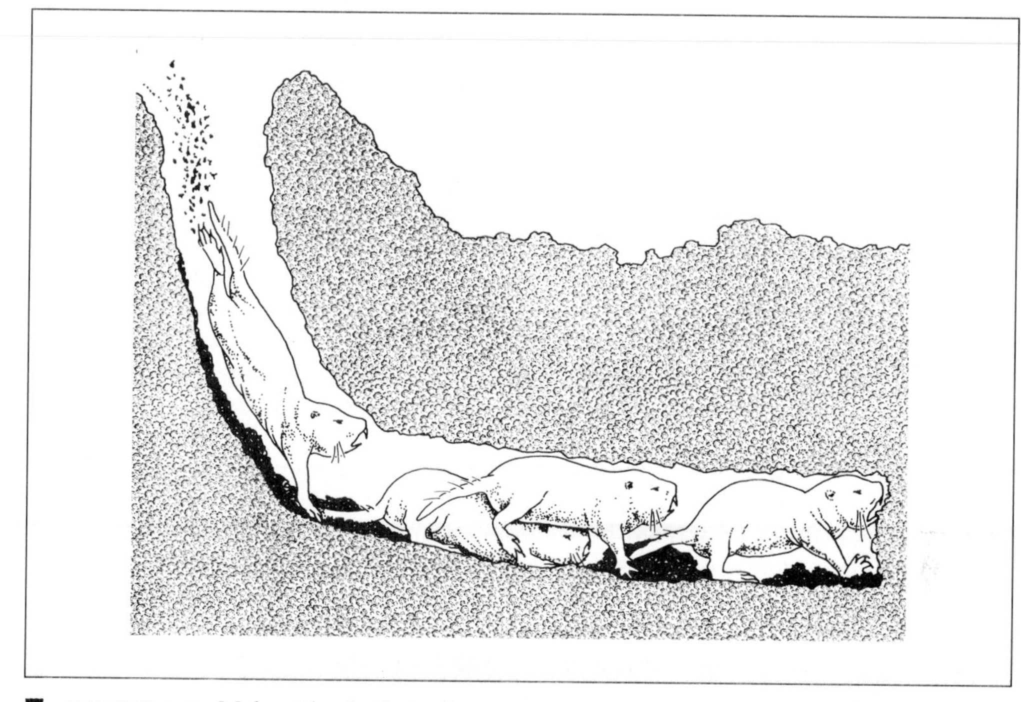

■ **FIGURE 4.4** Mole rat bucket brigade. (*Based on Jarvis and Sale 1971.*)

gested that it is the particular lifestyle of animals, not genetics, that promotes eusociality (Alexander 1974). He argued that in a normal diploid organism, females are related to their daughters by 0.5 and to their sisters by 0.5, so it matters little to them whether they rear sibs or daughters of their own. He predicted that mammals could exhibit a castelike society under certain conditions:

1. Where the individuals of the species are confined in nests or burrows.
2. Where food is abundant enough to support a high concentration of individuals in one place.
3. Where adults exhibit parental care.
4. Where there are mechanisms by which mothers can manipulate other individuals.
5. Where "heroism" is possible, whereby individuals give up their lives and by so doing can save the queen.

These factors can immediately account for eusociality in the termites as well. In mole rat colonies, where the burrows become as hard as cement, a heroic effort by a mole rat effectively stops a predator (commonly a snake) because predators cannot rip open the surrounding substrate. Self-sacrifice by a "worker" does translate into a genetic gain. In mole rats, the queen reigns supreme; all workers and nonworkers, whether male or female, develop teats during their pregnancies—testament to the power of their pheromonal cues. The superabundant food comes in the form of tubers of the plant

Pyrenacantha kaurabassana, which weigh up to 50 kg and can provide food for a whole colony. Often the tubers remain only half eaten and can regenerate in time. Why eusociality has developed in this but not in other systems of burrowing rodents such as prairie dogs is open to speculation.

Hamilton's genetic-relatedness theory seems even less appealing when one considers that the relatedness of workers in a hymenopteran colony is extremely close only if the colony is formed by a single queen who has mated once. When a queen mates twice and sperm mixes at random, the average relatedness between sisters is only 0.5. Evidence has come to light that, in a variety of hymenopterans, unrelated queens often initiate colonies together, dropping from mating swarms at the same time and engaging in pleiometrosis (cooperative nest digging and egg laying) (see references given by Strassman 1989). Some queens thus give up the opportunity to lay their own eggs, tending instead to the young of a co-foundress. In fire ants the number of co-foundresses commonly varies from two to five. How is this behavior explained? Females must not be able to predict which one will become the eventual egg layer when they begin a nest together. The benefits of nesting in a group must also be great. Part of the reason for cooperation apparently lies in the fact that many hymenopteran nests or colonies are clumped. Brood raiding by neighboring colonies is common (especially in ants), so attaining a large worker force quickly is critical to colony survival. In fire ants, especially, brood raiding is widespread, and in a study of newly initiated nests, Tschinkel (1990) documented the eventual merger of eighty nests into only two over the course of a month. Although this case was exceptional, brood raiding from four or five colonies over the space of a few hours is common, and the raids go back and forth from nest to nest until one colony eventually ends up with all the brood. Other advantages to large colony size may involve increased defense against predators and parasites and general lower adult mortality.

The evolution of sex has been difficult to explain in terms of individual selection.

One of the biggest stumbling blocks to the complete acceptance to individual selection over group selection is the existence of sex itself. The traditional explanation to "why have sex" was based on group selection because it demanded that an individual share its genes with those of another individual when making young. If this did not happen the species would not evolve and could, a few thousand generations later, be replaced by species that did. For this reason, sexual species were thought to be better off than asexual species. A sexual gene could spread only if it doubles the number of offspring an individual could have, which seems absurd. Imagine that an individual decided to forgo sex and instead pass on all its genes to its own offspring, taking none from its mate. It would then have passed twice as many genes on to the next generation as its rivals had. This would probably put it at a huge advantage since it would contribute twice as much to the next generation and would soon be represented most commonly in the species. Individuals that abandon sex could theoretically outcompete their sexual rivals in passing on genes. But in nature they do not, so sex must be beneficial in some way.

At first, a "lottery" model was popular. The logic here was that an asexual species could produce hundreds of equal quality offspring, but what counts in the evolutionary race is a handful of exceptional offspring. Breeding asexually is like having lots of lottery tickets with the same number. To stand a chance of winning the lottery you need lots of different tickets. But, to produce a future president of the United States, the lottery model suggested, one might need a few exceptional offspring not a lot of "average" offspring. Unfortunately, when scientists looked, there was no correlation between sexuality and ecological uncertainty. Sex should be commonest where in fact it is rarest—among highly fecund, small creatures in changeable environments. There sex is the exception. It is among big, long-lived, slow-breeding creatures in stable environments that sex is the rule.

In the 1980s sex was linked to a phenomenon known as "Muller's rachet." This theory proposed that sex exists to cleanse a species from accumulated damaging genetic mutations. When meiosis occurs and DNA is copied, some defects tend to accumulate as mutations occur. As successive mutations occur, it is likely that they will add more defects rather than repair previous ones. An analogy is that if you use a photocopier to make a copy of a copy of a copy of a document, the quality deteriorates with each successive copy. Only by using the unblemished original can you regenerate a clean copy. In the 1980s it was suggested that sex throws Muller's rachet into reverse. With sexual reproduction, new "perfect" individuals can arise from recombination where the damaging mutations are now absent. Again the field data do not conform to the theory. When two species, say an aphid that breeds asexually and one that breeds sexually, are competing for the same resource, the sexual population would probably be driven extinct by the asexual population's greater productivity, unless the asexual species' genetic drawbacks appear quickly—and this seems unlikely.

The most recent, and perhaps therefore the most accepted theory, is that sex is needed to fight disease (Ebert and Hamilton 1996). Diseases specialize in breaking into cells, either to eat them, as fungi and bacteria, or, like viruses, to subvert their genetic machinery for the purpose of making new viruses. To do that, the diseases use protein molecules that bind to other molecules on cell surfaces. There is necessarily a struggle between parasites and their hosts for the use of these binding proteins. Parasites invent new "keys" to use them, and hosts are constantly changing the locks, so that parasites are not able to bind with proteins. The logic is that if one so-called lock is common in one generation, the key that fits it will spread like wild fire. Thus, the individuals that have this lock will be quickly killed, and this type of lock will not be common a few generations later. Sexual species have a "library" of locks unavailable to a sexual species. Most notoriously variable, or polymorphic, genes are the very ones that affect resistance to disease—the genes for locks. Examples are genes that encode histocompatibility for antigens, proteins found on the surfaces of cells that help the immune system to distinguish "self" from foreign tissue or parasites. There is good evidence that asexuality is more common in species that are little troubled by disease, such as microscopic creatures and arctic or high-altitude plants and insects.

4.2 Organisms that live in dense groups may incur a cost in competition for resources, but group living can also provide a number of benefits.

Individual selection can explain the occurrence of altruism and the existence of sex within populations. But another common occurrence in nature is the observation that many species occur in dense shoals, flocks, or herds, which would surely only promote intense competition between individuals. If the central concern of ecology is to explain the distribution patterns of plants and animals, we must be able to provide a framework from which to understand the social behavior of animals. If dense congregations promote intense competition, there must be some high selective advantages of group living to compensate. In today's world, group living may be disastrous in that animals that organize into dense aggregations can easily be harvested by people and reduced to very low numbers. For example, the control of vampire bats in Latin America was greatly facilitated by the bats' habit of communal roosting and grooming (Mitchell 1986). Bats groom for two hours a day, and pesticides applied topically to only a few bats were soon transferred to the others. A ratio of fifteen to sixteen dead vampires to each one treated was obtained. The annual benefit to farmers in Nicaragua, resulting from increased milk production from host cattle, was U.S.$2,414,158 after vampire control. As annual costs were only U.S.$129,750, a favorable benefit-cost ratio of 18.6:1 was obtained.

Guppies, *Poecilia reticulata*, were first discovered in Trinidad in the nineteenth century by the Reverend P. L. Guppy. In their native habitat, guppies live in tighter groups when they are in streams in which predators are more common (Fig. 4.5), suggesting that being in a group helps an individual to avoid becoming a meal. When removed from a predator-rich part of the river to a predator-poor part, the guppies adjusted by growing bigger, living longer, and having fewer and bigger offspring (Resnick et al. 1977). Group living could reduce predator success in several different ways.

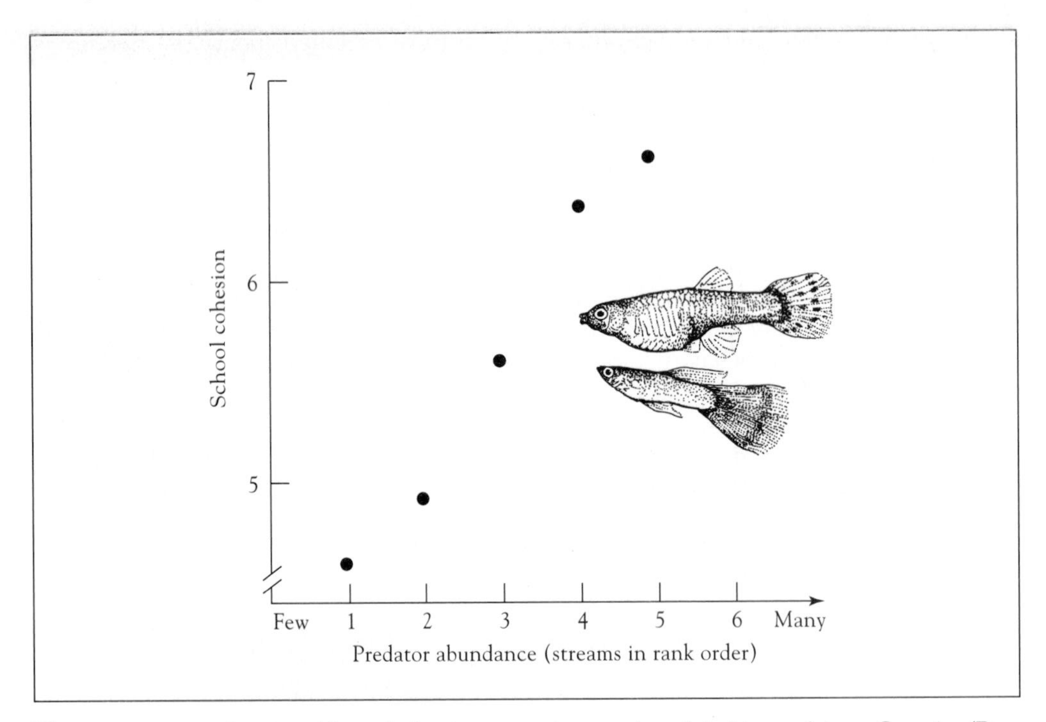

■ **FIGURE 4.5** Intraspecific variation in group size may be related to predators. Guppies (Poecilia reticulata) from streams with many predators live in tighter schools than those from streams with few predators. Each dot is a different stream, and "cohesion" was measured by a count of the number of fish in grid squares on the bottom of the rank. *(Modified from Seghers 1974.)*

■ **PHOTO 4.3** Why do animals live in groups, when groups may promote competition between individuals? For these snow geese, large flocks may be better able to detect predators, such as the bald eagle shown here just skyward of the flock. *(Johann Schumacher Design, Peter Arnold, Inc. AN-BE-50C.)*

Large groups may be more successful at detecting predators.

For many predators, success depends on surprise; if a victim is alerted too soon during an attack, the predator's chance of success is low. Goshawks (*Accipiter gentilis*) are less successful in attacks on large flocks of pigeons (*Columba palumbus*) mainly because the birds in a large flock take to the air when the hawk is still some distance away (Photo 4.3). If each pigeon occasionally looks up to scan for a hawk, the bigger the flock the more likely that one bird will spot the hawk early (known as the "many-eyes theory"). Once one pigeon takes off, the rest follow (Fig. 4.6) (Kenward 1978). Of course, cheating is a possibility because some birds might never look up, relying on others to keep watch while they keep feeding. However, at least in groups of Thompson's gazelle, the individual that happens to be scanning when a predator approaches is the one most likely to escape (Fitzgibbon 1989). This tends to discourage cheating.

An individual in a large group has a lower probability of being eaten if a predator attacks than does a solitary individual or one in a small group.

Normally, predators take only one prey item per attack. An individual antelope in a herd of a hundred has only a one in a hundred chance of being attacked, whereas a single individual has a one in one chance. Large herds may well be attacked more, but a herd is hardly likely to attract one hundred times more attacks than an individual (the dilution effect). In the Camargue, a marshy delta in the south of France, horses in large groups suffer less attacks per horse from biting tabanid flies than do solitary horses (Duncan and Vigne 1979).

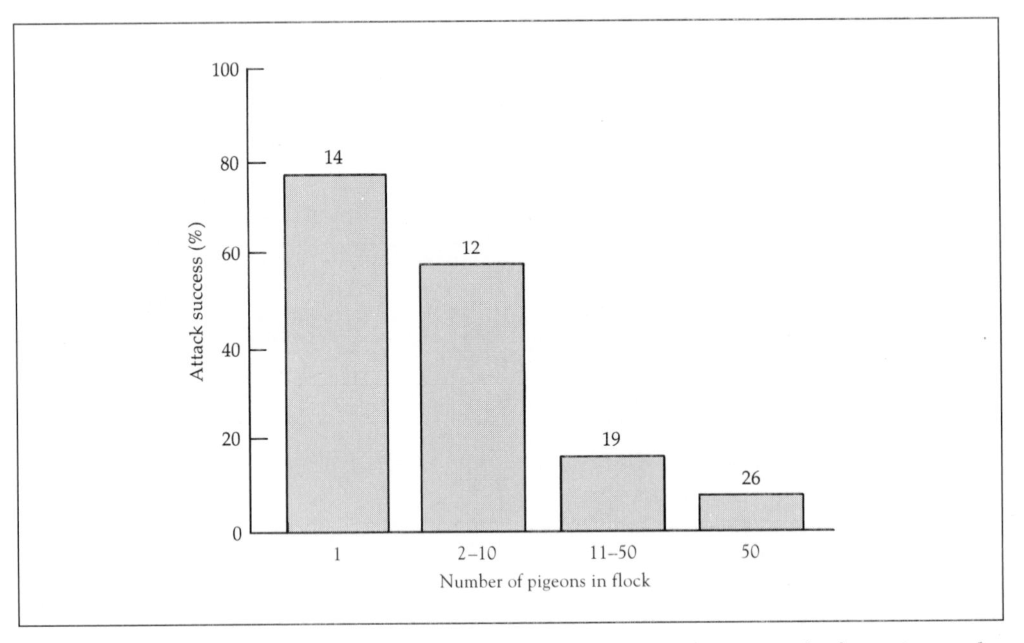

■ **FIGURE 4.6** The value of flocks of wood pigeons in deteting the approach of an avian predator, a goshawk. The graph shows the dependence of the goshawk's attack success on the size of the flock. *(Redrawn from Kenward 1978.)*

Associated with herds is a tendency to prefer the middle because predators are likely to attack prey on the edge of the group (this is often called the geometry of the selfish herd, Hamilton 1971). Part of the reason for this may be the difficulty of visually tracking large numbers of prey. Throw two tennis balls to a friend, and the chances are that he or she will drop them both, whereas one ball can be tracked much more easily. A similar phenomenon may operate in predator-prey interactions, explaining why predators take peripheral individuals (Neill and Cullen 1974). It may also be physically difficult to get to the center of a group, with many herds tending to bunch close together when they are under attack (Hamilton 1971). Furthermore, large numbers of prey are able to defend themselves better than single individuals, which usually flee. Nesting black-headed gulls mob crows remorselessly and reduce the success of the crows at stealing gulls' eggs (Kruuk 1964). Natural predators are rare in the Hawaiian Islands, and birds there seldom flock (Willis 1972). Who gets the best spots within the group? Perhaps the older, more experienced, or bigger individuals that commandeer the best positions. Perhaps the fastest individuals commandeer the positions at the edge of the group because they can flee more rapidly. It is not difficult to come up with more than one reason.

The spacing of individuals within a group may indicate what proceses are shaping group structure.

Within a group, individuals can space themselves different distances away from their neighbors. There are three basic kinds of spacing: random, clumped, and even (Fig. 4.7). We can visualize this pattern by imagining people in a meeting room. Many people would get together in social groups, creating a clumped pattern. If nobody knew each

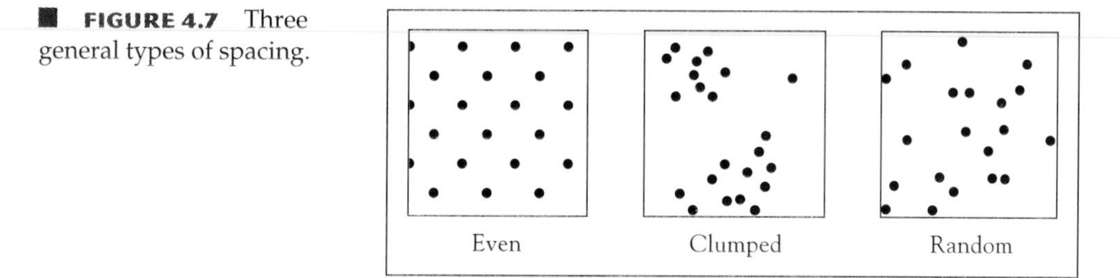

FIGURE 4.7 Three general types of spacing.

other they might maintain a certain minimum personal distance between themselves to produce an even distribution, or if nobody minded or cared about their position relative to anyone else we would get a random distribution.

The type of distribution observed in nature can tell us a lot about what processes shape group structure in nature. The most common spatial pattern is clumped, and this is probably because resources are clumped in nature. For example, certain plants may do better in moist conditions, and moisture is greater in low-lying areas. Social behavior between animals may also promote a clumped pattern. Competition may cause an even population pattern between trees in a forest. At first, the pattern of trees and seedlings may appear random as seedlings develop from seeds dropped at random, but competition between roots may cause some trees to be outcompeted by others, causing a thinning out or self-thinning and resulting in an even distribution. Thus, the population pattern starts random but ends up even.

Large groups may be more successful at foraging for food than solitary individuals or small groups.

Living in groups may confer advantages for predators as well as prey. Predatory groups often capture prey that are difficult for a single individual to overcome, either because the prey is too large (lions hunting adult buffalo) or because it is too elusive (killer whales hunting dolphins). For species that feed on large ephemeral food clumps such as seeds or fruits, the limiting factor is often the location of a good site or tree. Once it has been found, there is usually plenty of food. It has been proposed, with good evidence from *Quelea* birds in Africa, that in large roosts successful foragers may be followed by birds that had been previously unsuccessful. This "mutual parasitism" supposes that poor foragers are in some way able to distinguish successful birds (Ward and Zahavi 1973) and that groups are centers of information transfer (the information center hypothesis). Brown (1988) has shown that cliff swallows (*Hirundo pyrrhonota*) in southwestern Nebraska nest in colonies that serve as information centers in which unsuccessful individuals locate and follow successful individuals to aerial insect food resources. Brown was able to factor out the confounding effects of increased ectoparasitism (by nest fumigation) on larger colonies and colony location (by reducing certain colony sizes) to show that increased nestling weight at larger colonies was due to more successful foraging, which was attributable to more efficient transfer of information among colony residents.

There are additional benefits and further drawbacks to group living (Table 4.2). For example, large groups may attract more parasites, but small groups' individuals may not be able to find suitable mates. It is clear that conflicting selective pressures operate on group size to determine the eventual number of individuals in a herd or flock. The operation of just two conflicting variables, competition for food and the presence of

TABLE 4.2 Examples of studies in which possible costs and benefits of group living other than those mentioned in the text have been measured. (After Krebs and Davies 1993)

HYPOTHESIS	EXAMPLE
1. Saving of energy by warm-blooded animals as a result of thermal advantage of being close together.	Pallid bats (*Antrozous pallidus*) roosting in groups use less energy than solitary neighbors.
2. Chance for small to overcome competitive superiority of a large species by being in a group.	Groups of striped parrot fish (*Scarus croicus*) can feed successfully inside the territories of the competitively superior damselfish (*Eupomacentrus flavifrons*).
3. Hydrodynamic advantage for fish swimming in a school. They save energy by positioning themselves to take advantage of vortices created by others in the group.	Measurements of distances and angles between individuals show that they are not correctly positioned to benefit according to the predictions of the theory.
4. Increased incidence of disease as a result of close proximity to others.	Number of ectoparasites in burrows of prairie dogs (*Cynomys spp.*) increase in larger colonies. Number of ectoparasitic bugs and fleas increases with size of cliff swallow (*Hirundo pyrhonota*) colonies.
5. Risk of cuckoldry by neighbors.	In colonially nesting red-winged blackbirds (*Agelaius phoeniceus*), the mates of vasectomized males laid fertile eggs. They must have been fertilized by males other than their mates.
6. Risk of predation on young by cannibalistic neighbors.	In colonies of Belding's ground squirrel (*Spermophilus beldingi*), females with small territories are more likely to lose their young to cannibalistic neighbors than are females with large territories around their burrows.

predators, on the size of bird flocks is illustrated in Fig. 4.8. Increased predation rates increase flock size while increased food levels decrease flock size. Many other variables are likely to influence flock size at the same time.

To explain why some animals live in groups of fifty and others in groups of ten is a little harder. Even if there is a best or "optimal" group size, say seven, neighboring individuals would probably join the group because they could do better than they could alone. Fig. 4.9 illustrates that, for this example, even where the group size becomes twelve, individuals would still do better than they would alone (Sibly 1983). Thus, although the optimal group size might be seven, a range of two to fourteen would be expected in nature.

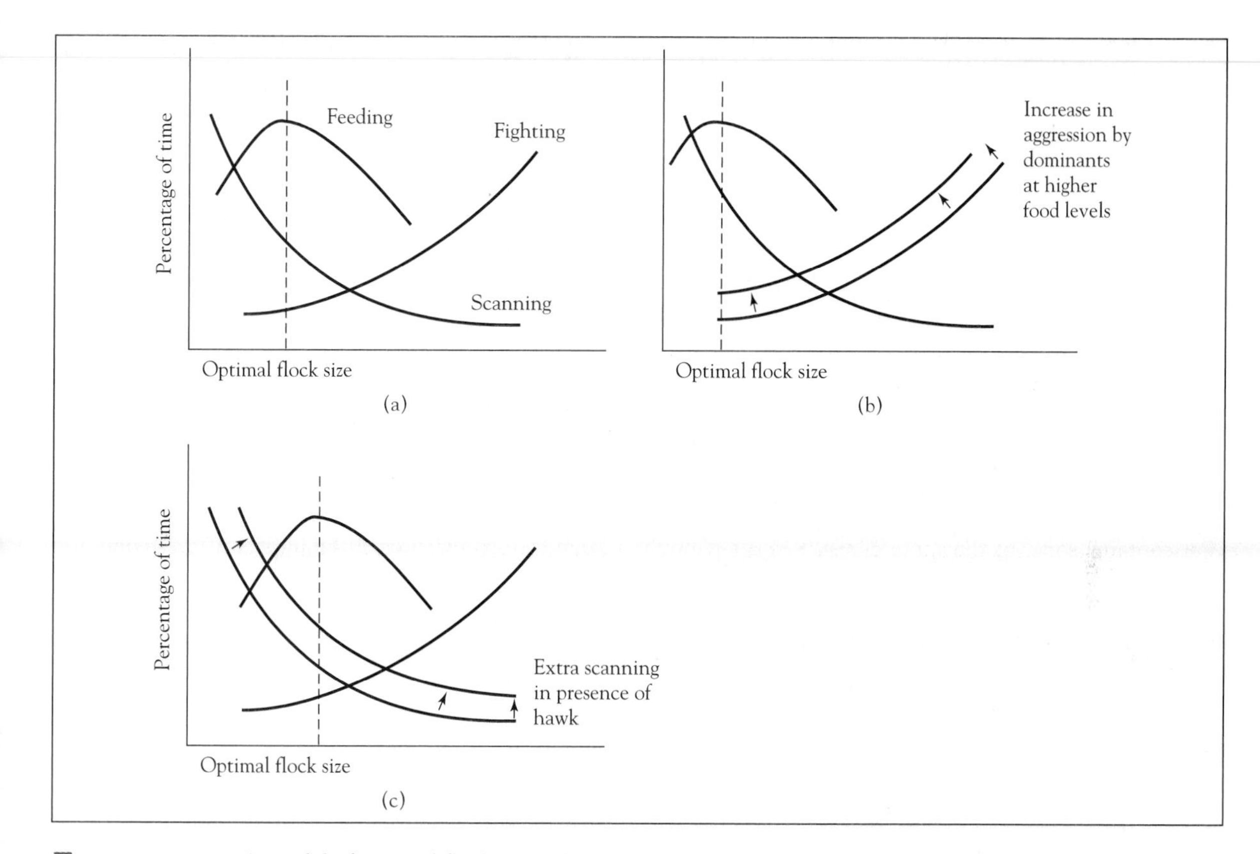

FIGURE 4.8 A model of optimal flock size. This is a time budget model that assumes birds do only one of three things: feed, fight, or scan. (**a**) The trade-off between squabbling or fighting and scanning for predators. As flock size increases, birds spend more time fighting and less time scanning. An intermediate flock size gives the maximum proportion of time feeding. (**b**) The effect of an increase in resources on flock size. When food is more plentiful, dominant birds can afford to spend more time attacking subordinates. The optimal flock size for the average bird therefore decreases. (**c**) The effect of an increase in predation on flock size. When predation risk is increased by the flight of a hawk over the flock, the scanning level should go up, and the optimal flock size is increased. (*Modified from Krebs and Davies 1981.*)

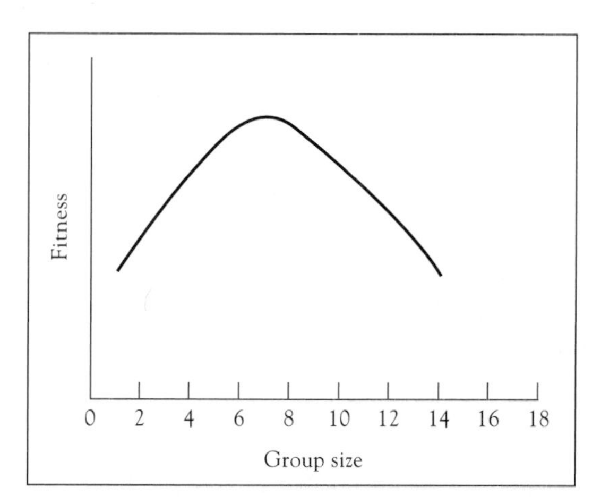

FIGURE 4.9 Sibly's model of optimal and stable group size. Each individual joins the group that maximizes its fitness so that the optimal size of seven is not necessarily stable—it will be joined by solitary individuals for example. (*After Sibly 1983.*)

SUMMARY

1. Some ecologists believed that populations of species in nature are maintained at constant or equilibrium levels by group selection—self-regulation by individuals in a population to prevent overexploitation of resources. Attractive though the idea of group selection is, it has several flaws, such as mutation, immigration, and resource prediction. Individual selfishness seems a more plausible result of natural selection.

2. If individual selection is more likely than group selection, how can we explain acts of altruism in nature between individuals? One popular explanation that was invoked was kin selection—selection for behavior that lowers an individual's own chances of survival and reproduction but raises that of a relative that contains some of the same genes. Kin selection was invoked to explain the unusual caste systems of social insects, especially those with a haplodiploid mating system.

3. Eusociality and cooperation are not only found in haplodiploid organisms. Termites and naked mole rats exhibit a caste system too. In 1974, Richard Alexander predicted that certain ecological reasons could be sufficient to cause a castelike society. These reasons were confinement to burrows, high food concentrations, parental care of offspring, mechanisms for mothers to manipulate other individuals, and the opportunity for heroism.

4. While ecological or evolutionary reasons can explain the existence of altruism and social behavior, group size can often be seen to be a compromise between large groups that lessen the impact of predation and small groups that lessen the impact of competition.

DISCUSSION QUESTION

One often reads that a change in the attitudes of humans is necessary to reduce environmental problems such as pollution or the overuse of resources. Such ideas are often based on altruism. This chapter has argued that individual selfishness is much more likely than altruism. Can humans act selfishly and still reduce environmental problems?

CHAPTER

5

Life History Variation

An organism's life history largely concerns its lifetime pattern of growth and reproduction. For example, why is it that some plants, especially those that we regard as weeds, produce vast numbers of tiny seeds while some trees produce fewer, bigger ones? Why do many species have a 50:50 sex ratio? Why is it that the ratio of age-at-maturity:average-life span is often roughly constant within a group of organisms but different between groups (e.g., mammals 1.3, fish 0.45)? Many ecologists believe that an understanding of such differences in life history strategies is fundamental (Lessels 1991; Stearns 1992). A knowledge of an organism's age structure, sex ratios and reproductive strategies, and dispersal capabilities is certainly important in understanding how populations grow.

Individual size is the most apparent aspect of an organism's life history. Large size may increase an organism's competitive ability or its success in deterring predators. Larger organisms are less vulnerable to environmental variation because of their small surface:volume ratio. In addition, within a species, large individuals usually produce more offspring (Fig. 5.1). On the other hand, large individuals require more resources to maintain themselves, and they often attract more attention from predators. As a result, intermediate size may be optimal in terms of survivorship and/or reproductive output.

5.1 Reproductive strategies differ among organisms, with some breeding continuously, others in discrete intervals, and some just once in their lifetime.

A particular size may achieved by starting large, growing fast, growing for a long time, or any combination of these. In addition, development of reproductive status may be quick or long. Organisms can produce all of their offspring in a single reproductive event. This is called *semelparity* and occurs in organisms like salmon, bamboo, and

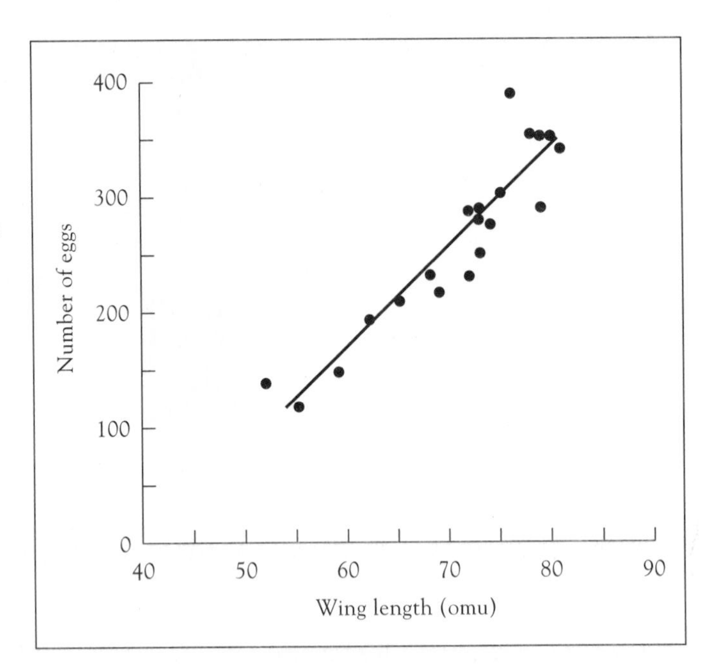

FIGURE 5.1 Relationship between fecundity (number of eggs) and wing length (ocular micrometer units) for *Asphondylia*, a phytophagous gall-making fly reared from a seaside plant *Borrichia frutescens*. (From Stiling and Rossi 1997.)

yucca plants where individuals reproduce once only and die. This is in contrast to repeated reproduction or *iteroparity*, where organisms, like tropical trees, reproduce continuously. Semelparous organisms may live for many years before reproducing, like the yuccas, or they may be annual plants that develop and drop seed within a year. Many annuals actually have seeds that can lay dormant in a seed bank in the soil for over a year, so they are not really annuals at all.

Among iteroparous organisms there is much variation in the number of separate clutches and in the number of offspring in a clutch. Many species have distinct breeding seasons that lead to overlapping but distinct generations, such as in temperate birds or temperate forest trees. For a few species, individuals reproduce repeatedly and at any time of the year. This is continuous iteroparity and is exhibited by some tropical species, many parasites, and, of course, humans.

Why do species have either semelparous or iteroparous reproductive strategies? Cole (1954) showed mathematically that for a hypothetical organism the two strategies produced very similar numbers of offspring over the same number of years. Why then do species bother to have repeated reproduction? The answer may lie in environmental uncertainty. If survival of juveniles is very poor and unpredictable selection favors repeated reproduction and long reproductive life, repeated reproduction spreads the risk of reproducing over a longer time period and acts as an adaptation to thwart environmental fluctuations. This is often referred to as "bet-hedging" (Stearns 1976).

5.2 All continuously breeding populations tend toward a stable age structure.

The reproductive strategy employed by an organism has a strong effect on the subsequent age structure of a population. The age structure of populations can be characterized by specific age categories, such as years or months, or other categories like eggs, larvae, or pupae in insects or size classes in plants.

All continuously breeding populations tend toward a stable age distribution. The ratio of each age group in a growing population remains the same if the birth rate and the death rate at any age do not change. If the stable age distribution is disrupted by any cause, such as natural catastrophe, disease, starvation, or emigration, the age composition will tend to restore itself upon return to the previous conditions, provided of course the rates of birth and death are still the same (see Chap. 6). When the population is not growing and its age structure remains the same it is said to have a stationary age distribution. The ratio of young to adults in a relatively stable population of most mammals and birds is approximately 2:1 (Alexander 1958) (Fig. 5.2). A normally increasing population should have an increasing number of young, whereas a decreasing population should

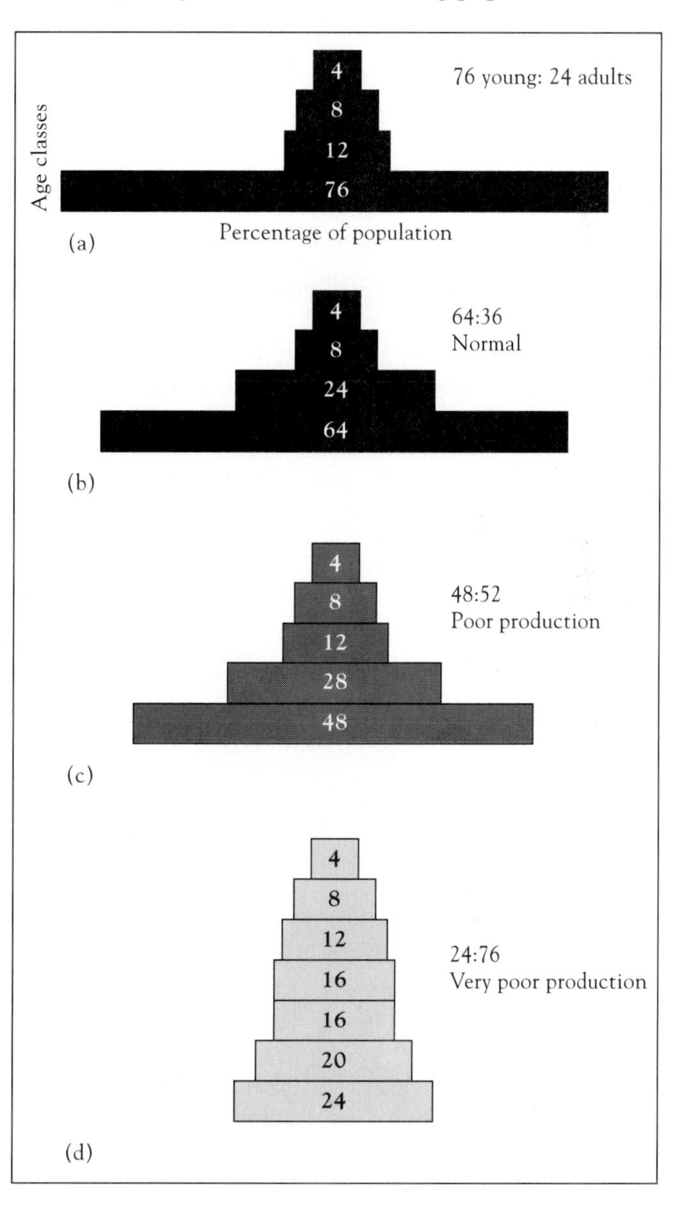

FIGURE 5.2 Theoretical age pyramids, with younger age classes on the bottom and older ones at the top. The length of the bar represents the percentage of the population in the class. (**a** ,**b**) Growing populations in general are characterized by a large number of young, giving the pyramid a broad base. (**c** ,**d**) Populations with a high proportion of individuals in the older age classes; these populations are dominated by older individuals and the population is in decline.

have a decreasing number of young; however, many variations on this theme are possible. Whatever the relationships, the number in one age class that enters the next age class influences the age structure of a population from year to year. The loss of age classes can have a profound influence on a population's future. In an overexploited fish population, the older reproductive age classes are often removed. If the population experiences reproductive failure for one or two years, there will be no young fish to move into the reproductive age class to replace the fish removed, and the population can collapse.

For humans, such data can be combined with historical and sociological information to provide an intelligible picture of a country's population (Fig. 5.3). We can also contrast the age distribution of a "developing" (i.e., poor) country with a developed country (Fig. 5.4). The age distribution for the developing nations tapers sharply from its base to its apex, while that for the developed nations has near vertical or even overhanging sides until the older age classes. This contrast is partly due to the higher birth rates and lower survivorship rates in the poorer countries.

5.3 Mating systems may be monogamous or polygamous.

Most sex ratios are 50 percent males to 50 percent females.

Sex ratio refers to the proportions of the two sexes, male to female, in a population. Sex ratios are of interest to pure and applied ecologists alike. For example, hunters prefer to take male deer—bucks—and would be pleased to have a deer population skewed toward males. However, too many males would severely limit herd growth, making the job of game managers very difficult.

In nature, most males seem superfluous because one male could easily fertilize all the females in a local area. For example, five milliliters (ml) of human semen contain

■ **FIGURE 5.3** A population pyramid for France on January 1, 1992. (1) Shortfalls in births due to the 1914-1918 war. (2) Passage of empty classes to reproductive age. (3) Shortfalls in births due to the 1939-1945 war. (4) "Baby boom." (5) Population in decline? *(Based on French data from the Institut National d'Ètudes Démographiques.)*

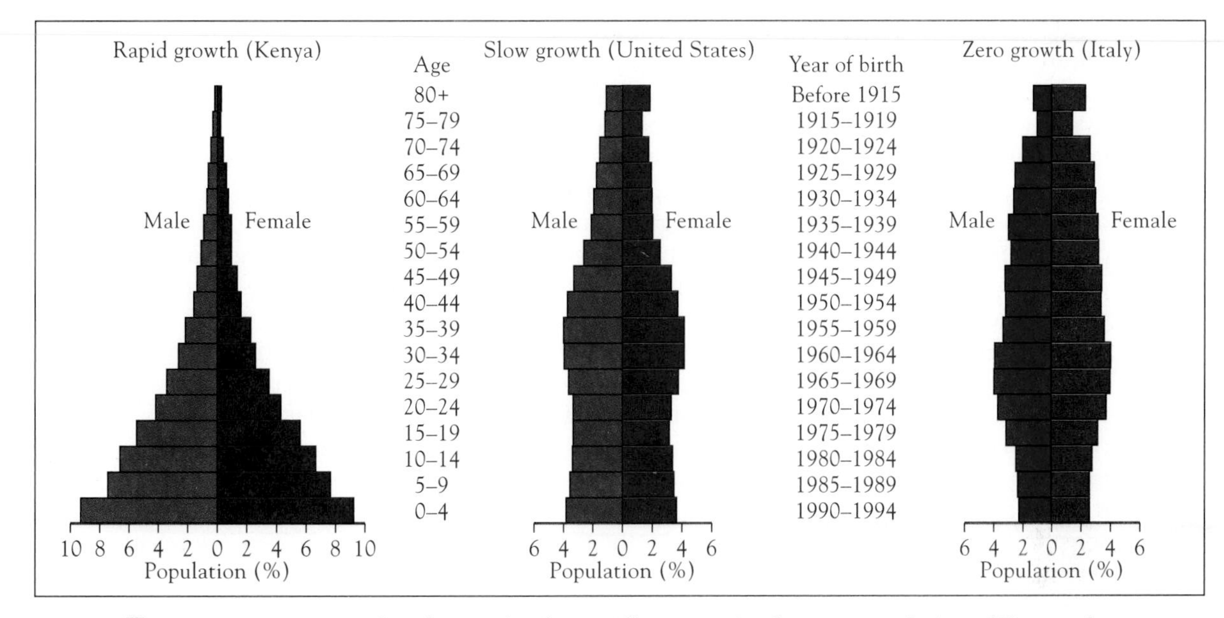

■ **FIGURE 5.4** Age distribution for the rapidly increasing human population of Kenya, the slowly increasing population of the United States, and the stationary population of Denmark in 1990. *(From McFalls, 1995.)*

enough sperm to fertilize twice the female population of the United States. If one male can fertilize dozens of females, why are there few species with a sex ratio of, say, one male to twenty females? The sex ratio is more often about 1:1. The answer again lies with the selfish genes (Fisher 1930). If a population contained twenty females for every male, then a parent whose children were exclusively sons could expect to have twenty times the number of grandchildren produced by a parent with mostly daughters. Such constraints operate on the numbers of both male and female offspring, keeping the sex ratio at about 1:1. In humans in the United States, the sex ratio is about 1:1.05 (males to females) at birth, dropping to 1:1.03 at age 15, 1:1.00 between ages 20 and 45, and 1:0.95 thereafter.

The most common exception to the 1:1 sex ratio is the situation in which it costs more to produce sons or daughters. Suppose sons are twice as costly to produce as daughters because they are bigger and eat more. Sons are a bad investment because each grandchild is twice as costly to produce via a son as via a daughter. The sex ratio would then swing toward a female bias until a ratio of 1:2 (males:females) was reached. Another exception to the 1:1 sex ratio is due to local mate competition, in which one male dominates in breeding, and other males therefore become superfluous because they are never likely to leave any offspring. This phenomenon is most likely to occur in species with low powers of dispersal in which brothers stay in the same place and inbreeding is frequent. In such situations, for example, in parasitic Hymenoptera, whose broods develop and mate in or around patches of host insect larvae, biased sex ratios are common (Werren 1980). Similar biased sex ratios occur in Old World monkeys. Groups consisting of more than ten females contain many breeding males, whereas those consisting of five or fewer females contain one adult male. In the viviparous mite *Acarophenox*, brood size is about twenty, and each brood contains only one son. The male mates with his sisters inside the mother and dies before he is born (Hamilton 1967). Where local mate competition is not so important or dispersal is more prevalent, more males may be produced.

The effects of local mate competition on sex ratios can also be seen where the effective patch size varies. That is, if the number of hosts, say, eggs in a clutch, were small, local mate competition would be increased. A similar effect would result if the number of foundresses, female parasites attacking the patch, were large. In accord with these theories, Waage (1982) reviewed the biologies of different species of Scelionidae, parasites of insect eggs. In small patches, female-biased sex ratios were produced. In large patches, presumably with little local mate competition, normal (1:1) sex ratios were produced, as they were for species attacking isolated eggs, for in these cases dispersal was very common. Sex ratios in mammals may also change due to local mate competition though the alteration of sex ratio necessarily follows fertilization, and the mechanism is yet unknown. Among some species of primates, high-ranking females have access to more resources than low-ranking females, and males disperse. Because daughters inherit their mother's rank and its associated access to resources, high-ranking females produce more daughters, whereas low-ranking females, with access to fewer resources, produce more sons. A similar phenomenon is found in the Cape Mountain Zebra *(Equis zebra zebra)* (Lloyd and Rasa 1984).

In polygynous mating systems, one male mates with many females.

Physiological constraints often dictate that female organisms must care for the young because they are the ones most often left "holding the baby." Because of these constraints, at least in mammals, males are able to desert, and most mammals have polygynous mating systems in which each male mates with several females, but each female mates with only one male. In cases in which females are able to shed their young at an early stage and in which there is no parental care—for example, in fish, amphibians, and, to some extent, birds—different strategies are adopted. As an example, it is the male sea horse, *Hippocampus*, who carries the fertilized eggs around in his brood patch; females desert to form reserves for more eggs (Photo 5.1). In most fish, the female deposits her eggs first, and then the male fertilizes them. Females are thus able to desert, leaving the male "holding the babies," sometimes literally in mouth breeders. In those fish with internal fertilization, however, it usually falls to the female to exhibit parental care. Lack (1968) concluded that over 90 percent of bird species are monogamous, a trait that may well engender empathy in many bird-watchers. However, recent DNA evidence suggests many monogamous species may not be a faithful as was once thought.

Polygyny can be influenced by the spatial or temporal distribution of breeding females. In cases where all females are sexually receptive at the same time, there is little opportunity for a male to garner all the females for himself. Monogamous relationships are more common in these situations, and Knowlton (1979) has suggested that female synchrony has evolved specifically to enforce monogamy on males. Where female reproductive receptivity is spread out over weeks or months, there is much more opportunity for males to mate with more than one female. For example, females of the common British toad, *Bufo bufo*, all lay their eggs within a week, and males generally have time only to mate with one female. In contrast, bullfrog females, *Rana catesbeiana*, have a breeding season of several weeks, and males may mate with as many as six

PHOTO 5.1 Fish are one of the few vertebrates where females deposit their eggs first and males fertilize them. Females are thus able to desert and leave the male to rear the babies. In male sea horses, *Hippocampus* sp., the male carries the fertilized eggs around in his brood patch. (*Jeff Foott, DRK Photo, 136026.*)

females in a season. Also, where some critical resource, say, available breeding or nesting sites, is in short supply and is patchily distributed, not uniformly spread out over the habitat, there is great opportunity for certain males to dominate it and to breed with more than one visiting female. Male orange-rumped honeyguides, *Indicator xanthonotus*, defend bees' nests, and when a female comes to feed the male mates with her, exchanging food for sex. The more nests he can defend, the more females he will attract (Cronin and Sherman 1977).

In the lark bunting (*Calamospiza melanocorys*), which mates in North American grasslands, males arrive first, compete for territories, and then display with song flights to attract females. The major source of nestling death in this species is overheating from overexposure to the sun. Prime territories are those with abundant shade, and some males with shaded territories attract two females, even though the second female can expect no help from the male in the process of rearing young. Males in some exposed territories remain bachelors for the season. Pleszczynska (1978) was able to predict with good success the status of males (bigamous, monogamous, or bachelor) on the basis of territory quality before the arrival of females. Furthermore, supplementing open areas with plastic strips to provide shade turned them from bad to good territories. Predation is a strong selective pressure that acts in a similar manner to force polygynous relationships on birds where females choose males with safe territories (Rubenstein and Wrangham 1986).

From the male's point of view, territory-based polygyny is advantageous; from the female's point of view, there are drawbacks. Although by choosing dominant males a

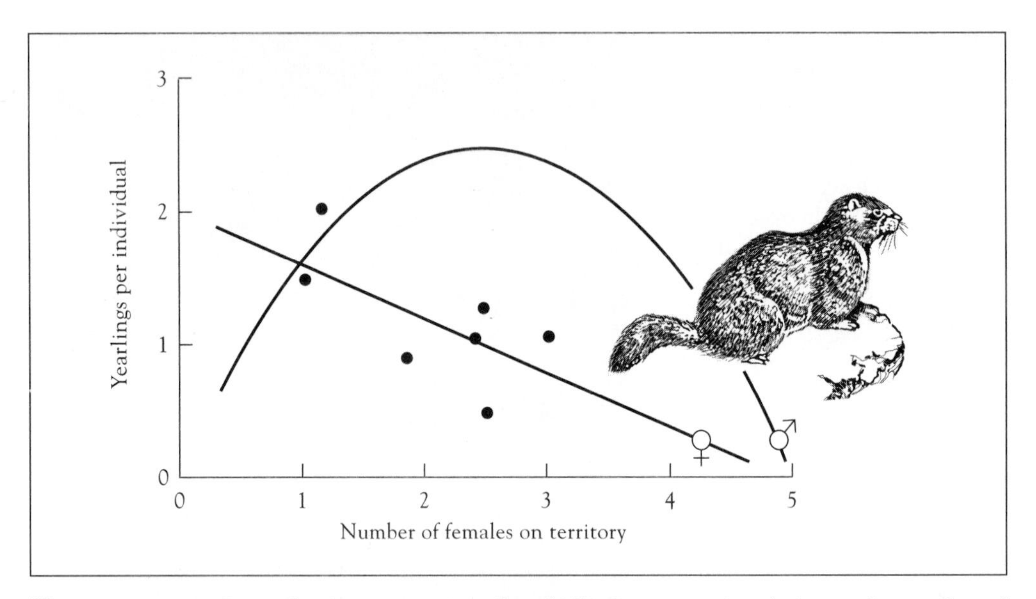

FIGURE 5.5 Reproductive success in yellow-bellied marmots in relation to the number of females on a male's territory. The success per female declines with number of females (solid dots). The success of the male, which is simply the success per female multiplied by the number of females, peaks at 2-3 females. *(Modified from Downhower and Armitage 1971.)*

female may be gaining access to good resources, she may also have to share these resources with other females. In the yellow-bellied marmot (*Marmota flaviventris*), males attract more females if they defend the best burrow sites. From the female's point of view, more females in a burrow means lower success per female (Fig. 5.5) Although it is best for a female to be with a monogamous male, it is best for the male to mate with two-to-three females. A compromise is often evident in which about two females are usually observed per territory (Downhower and Armitage 1971). Orians (1969b) formally modeled this type of optimization of number of females per male (Fig. 5.6).

Sometimes males simply defend females as a harem without bothering to command a conventional resource-based territory. This pattern is more common when females naturally occur in groups or herds, perhaps to avoid predation. Usually the largest and strongest males command all the matings, but being a harem master is usually so exhausting that males may only manage to remain at the top for a year or two. In the elephant seal, *Mirounga* sp., named for the enlarged proboscis of the male, males constantly lumber across the beach, squashing pups in the process. Because the offspring are likely to have been fathered by the previous year's dominant bulls, the havoc matters little to the present males (Cox and Le Boeuf 1977; McCann 1981).

Sometimes, polygynous mating occurs where neither resources nor harems are defended. In some instances, particularly in birds and mammals, males display in specific communal courting areas called **leks**. The females choose their prospective mates after the males have performed often elaborate displays. A few males may perform the vast majority of the matings (Wiley 1973). Perhaps the largest lek in the world is in Lake Malawi, Africa, where as many as fifty thousand male cichlid fish may display on a sand

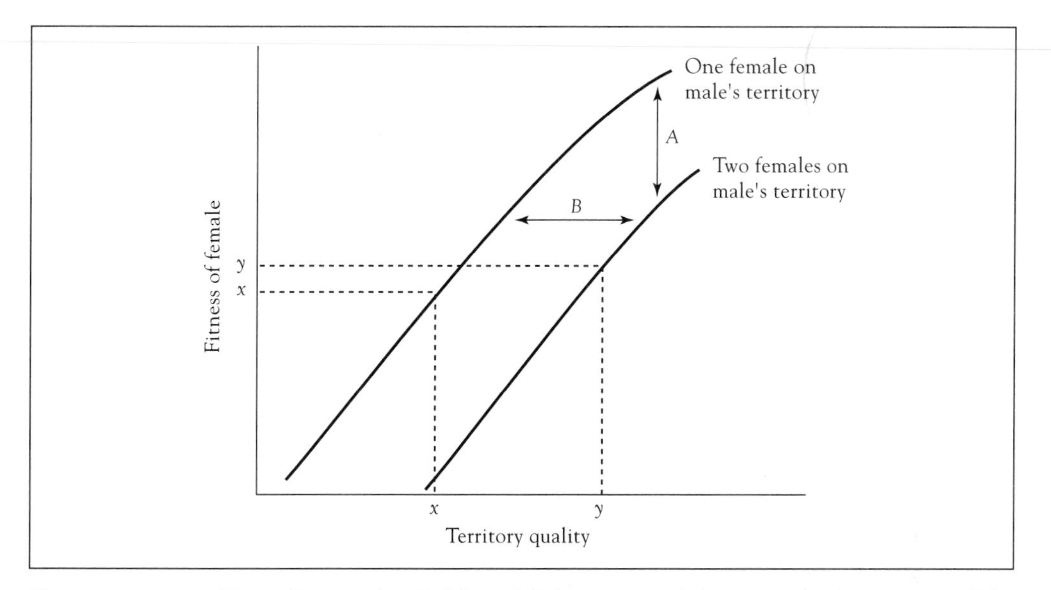

FIGURE 5.6 The polygyny threshold model. It is assumed that reproductive success of the female is correlated with environmental factors, such as the quality of the territory in which she breeds, and that females choose mates from the available males. In the model, a female suffers a decrease in fitness, A, by going to an already mated male from the fitness she could expect if she had a male all to herself. Despite this drop in fitness, provided the difference in quality between the territories is sufficient (B = the polygyny threshold), a female may expect greater reproductive success if she breeds with an already mated male. For example, a female who shared a male on territory y would do better than a female who had a male all to herself on territory x. *(Redrawn from Orians 1969b.)*

bar 4 km long (McKaye 1983). Five main hypotheses have been proposed to explain leks (Bradbury and Gibson 1983; Bradbury et al. 1989). Males are thought to aggregate

- To decrease predation
- To increase the efficiency of attracting females
- To be close to "hot spots" through which the largest females pass
- To make use of limited display sites
- To take advantage of female's preference for clumped males

As yet, insufficient evidence is available to distinguish between these theories.

On the grounds that competition for mates is more intense between males of polygynous species than in monogamous ones and that large size increases an individual's fighting ability, it is not unrealistic to expect a relationship between sexual **dimorphism** and degree of polygyny, and such relationships have been observed in reptiles, amphibians, ungulates, carnivores, and primates (Clutton-Brock and Harvey 1984). Males are larger where they have the opportunity to monopolize more females. If this line of argument is taken one step further, it might be expected that the weaponry of males (horns, antlers, or teeth) might increase relative to body size in strongly polygynous species relative to monogamous ones. Such has been found to be the case in primates (Fig. 5.7). Further, competition occurs not only as fighting between males prior to copulation but also, in systems where females mate more than once, between the sperm of males inside the females. Vol-

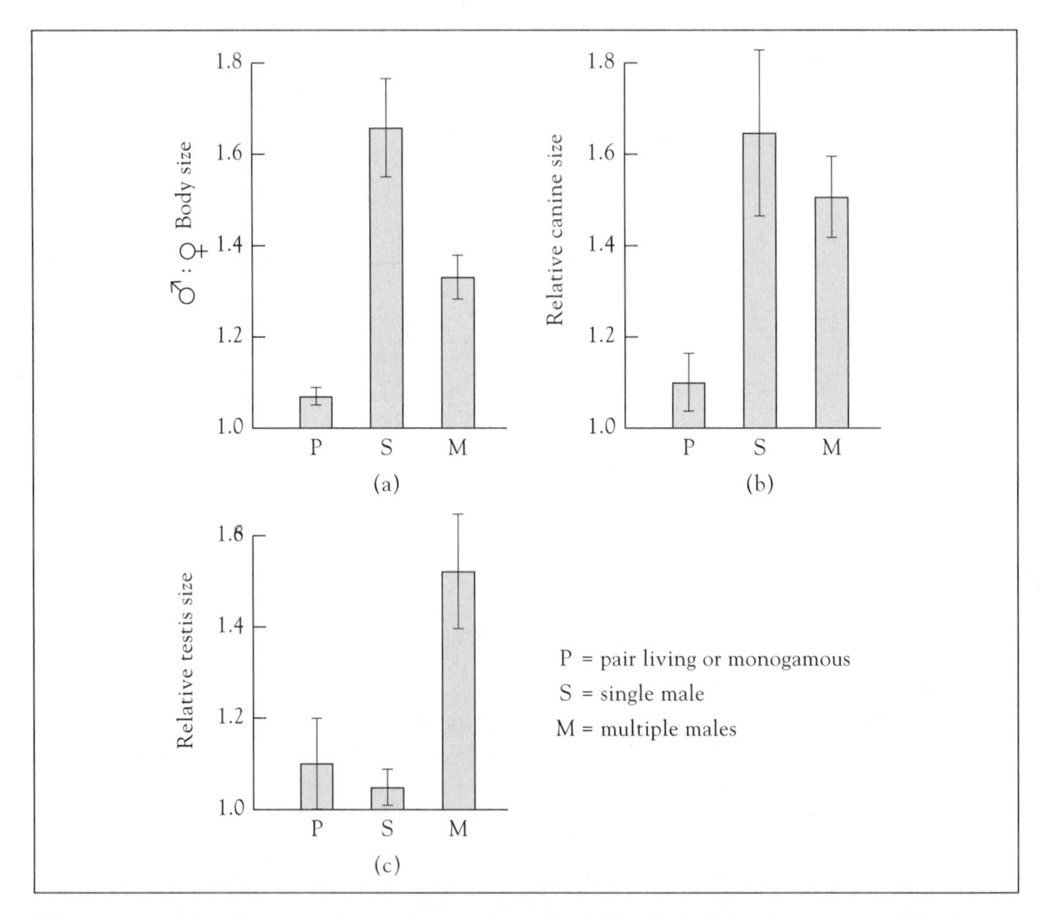

■ **FIGURE 5.7** Features of primate genera with different breeding systems. (**a**) Body size dimorphism (adult male divided by adult female weight). (**b**) Relative canine size (a measure of canine size dimorphism). (**c**) Relative testis size (a measure of testis size after body-size effects have been removed). Bars indicate one standard error in each direction from the mean. *(Based on data of Harvey, Kavanagh, and Clutton-Brock 1978 and Harcourt et al. 1981.)*

ume of ejaculate is related to testis size, and it is also true in primates that males from species in which females copulate with more than one partner have larger testes relative to body size (Clutton-Brock, Guinness, and Albon 1982; Harvey and Harcourt 1982).

It is only in birds and mammals, however, that such large males are commonly found. In the great majority of other vertebrates (frogs and snakes) and in invertebrates, it is the female that is the larger sex (Greenwood and Wheeler 1985). Male spiders are often much smaller than the female and have to use extreme caution when approaching a potential mate. In some deep-sea anglerfish, the mate is merely a tiny fused appendage on the body of the female. In most cases, this situation is not surprising, given the number and volume of eggs females produce together with the attendant food reserves. There are, of course, exceptions, with large males, horned beetles, and some crabs being prominent among them. The reasons for these differences in sexual dimorphism across the animal kingdom are not yet clear.

In Polyandrous mating systems, one female mates with many males.

In most **polygamous** systems, those in which one individual mates with more than one individual of the opposite sex, the polygamous sex is the male, and such systems are termed **polygynous**. The opposite of polygyny, **polyandry**, in which the female is polygamous, is much rarer. Nevertheless, it is practiced by a few species of birds (Oring 1981). In Arctic waders such as sanderling, *Calidris alba*, and Temminck's stint, *Calidris temminckii*, the males defend territories. The female lays one clutch of eggs that the male incubates and another that she herself incubates. In the Arctic tundra, the season is short but productive, providing a sudden but short-lived wealth of food. In the spotted sandpiper, *Actitis macularia*, the productivity of breeding grounds is so high that the female becomes rather like an egg factory, laying five clutches of twenty eggs in forty days. Her reproductive success is limited by the number of males she can find to incubate the eggs, and females compete for males, defending territories where the males sit (Lank, Oring, and Maxson 1985). Many more examples of monogamous and polygynous patterns are discussed by Rubenstein and Wrangham (1986).

5.4 When individuals select their habitats, they may or may not divide it up into territories.

Mating systems and male and female reproductive success are strongly influenced by the quality of the habitat that they inhabit. Habitat selection is of critical importance to many species. Organisms do not always occupy all their potential range even though they could live there. Such individuals choose to live in certain habitats. The selection may be based to a large extent on food availability and the presence of competitors (Orians and Whittenberger 1991; Wiens 1985; Bazzaz 1991).

Imagine two habitats, one rich and one poor in resources. All individuals are free to go where they like, a situation referred to as "ideal free conditions" (Fretwell and Lucas 1970). As individuals arrive at the rich habitat, resources will gradually be depleted. Eventually a point will be reached where the next arrivals will do better by occupying

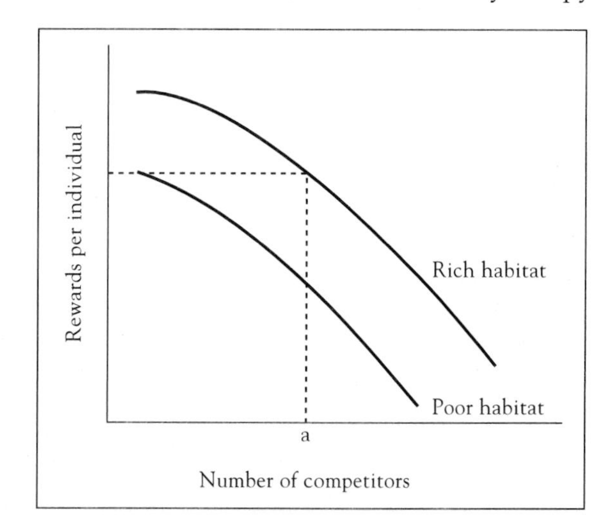

FIGURE 5.8 The ideal free distribution. Every individual is free to choose where to go. The first arrivals will go to the rich habitat. Because of resource depletion, the more competitors, the lower the rewards per individual, so at point *a* the poor habitat will be equally attractive. Thereafter, the habitats should be filled so that the rewards per individual are the same in both.

the poor habitat, where, although resources are scarcer, competition is less (Fig. 5.8). A human social equivalent occurs at grocery stores, where customers distribute themselves among lines according to an ideal free distribution so that everyone has about the same waiting time. The ideal free distribution was demonstrated in an experiment using six sticklebacks by Milinksi (1979) who added *Daphnia* prey to one end of a tank at twice the rate of the other end. The stickleback distributed themselves with two at the "slow feed" end and four at the "fast" end, so that each animal was feeding at the highest rate. When the feeding regimes were reversed, the fish redistributed themselves accordingly.

For many animals, peaceful resource allocation does not occur. Individuals compete and fight over resources (Photo 5.2) so that richer habitats always get filled up first before animals exploit the poorer resource (Fig. 5.9). Many birds such as great tits (*Parus major*) and red grouse (*Lagopus lagopus*) are territorial and defend the richest territories (Krebs 1971; Watson 1967). After these are completely filled up, excluded birds occupy poor habitats, and when these too are filled birds form floating groups where their chances of survival are low. Thus, the strongest individuals are despots, who force others into low-quality areas.

■ **PHOTO 5.2** Fighting strategy, as between these mule deer bucks, may depend on features such as relative size and whether or not an animal is a territory owner. (*Irene Vandermolen, Stock Boston, LOR 0066B.*)

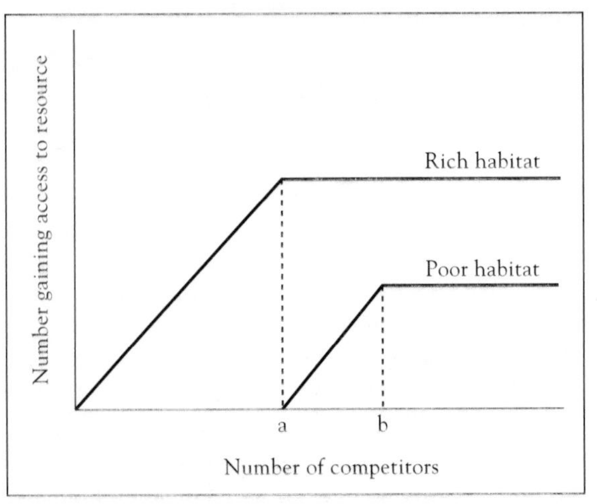

■ **FIGURE 5.9** Despotic behavior. Competitors occupy the rich habitat first of all. At point *a* this becomes full, and newcomers are forced to occupy the poor habitat. When this is also full (point *b*), further competitors are excluded from the resource altogether and become "floaters." (*Redrawn from Brown 1969.*)

In nature, despotism and resource sharing are not always mutually exclusive but can be combined in an overall behavioral strategy. In cottonwood aphids (*Pemphigus betae*), adult females settle on cottonwood (*Populus angustifolia*) leaf veins in the spring, sucking the sap and reproducing **parthenogenetically**. Eventually, the expanding tissue forms a gall around the aphid. Large leaves provide the richest supplies of sap and allow formation of up to seven times the number of progeny produced by mothers on small leaves. As a result, large leaves are quickly colonized, often by more than one aphid per leaf. Other colonizing females reduce the availability of sap, and less progeny are produced by each. In fact, the system fulfills the ideal free conditions such that, on large leafs with three aphids, the average number of progeny per female is equal to that on a medium leaf with two mothers or on a small leaf with a single mother. However, although the *average* success on different leaves is in accordance with an ideal free situation, not all competitors get equal rewards. The best place to be on a leaf is nearest the leaf petiole, in a position to tap the incoming sap first. Basally positioned mothers produce more progeny than more apical ones, so females compete for this optimal position in a despotic fashion, jostling for position; the bigger individuals usually win (Whitham 1978, 1979, and 1980).

In disputes over resources some individuals stay and fight, others turn and flee.

In the aphid example, the females face the problem of whether to fight for a good leaf position or to retreat and look for a new leaf. How does an individual solve this dilemma? Why do some animals, like the *Pemphigus* aphids, fight and risk serious injury, whereas other animals settle disputes by display patterns?

John Maynard Smith and his associates (see Maynard Smith 1976b, 1979, and 1982) tried to elucidate fighting in animals by considering contests in which there are different sorts of strategies. Imagine a game in which "Hawks" always fight to injure and kill their opponents (risking injury themselves) and "Doves" simply display and never engage in serious fights. These two strategies are chosen to represent the two possible extremes that can be seen in nature.

In this evolutionary game, we give the contestants rewards—the winner of a contest scores +50 and the loser 0. The cost of serious injury is −100 because the injured player may not be able to compete again for a long time; the cost of wasting time in a display is −10. These scores are arbitrary measures of fitness, and it will be assumed that Hawks and Doves reproduce their own kind faithfully in proportion to their scores. The next step is to draw up a two-by-two matrix with the average scores for the four possible types of encounter (Table 5.1). Consider what happens if all individuals in the population are Doves. Every contest is between a Dove and another Dove and the score, on average, is +15. In this population, any mutant Hawk could do very well, and the Hawk strategy would soon spread because when a Hawk meets a Dove it gets +50. It is clear that Dove is not an evolutionarily stable strategy (ESS), that is, a strategy that, if adopted by most of a population, cannot be bettered by any other strategy and will therefore become established by natural selection.

However, the Hawk strategy would not spread to take over the entire population. In a population of all Hawks, the average score is −25, and any mutant Dove would do better because when a Dove meets a Hawk it gets 0 (which is not very good but still better than −25!). The Dove strategy would spread if the population consisted mainly of Hawks. Therefore, Hawk is not an ESS either.

TABLE 5.4 Some of the correlates of r and k selection.

	R SELECTION	K SELECTION
Favorable climate	Variable and/or unpredictable; uncertain.	Fairly constant and/or predictable; more certain
Mortality	Often catastrophic, nondirected, density independent	More directed, density dependent
Survivorship	Often type III	Usually types I or II
Population size	Variable in time, nonequilibrium; usually well below carrying capacity; some recolonization each year; communities not saturated with species; colonization each year.	Fairly constant in time, equilibrium; at or near carrying capacity of the environment; no recolonization necessary; communities saturated with species
Intra– and interspecific competition	Variable, often lax	Usually keen
Selection favors	1. Rapid development 2. High reproductive rate 3. Early reproduction 4. Small body size 5. Single reproduction	1. Slower development 2. Great competitive ability 3. Delayed reproduction 4. Larger body size 5. Repeated reproduction
Length of life	Short, usually less than one year.	Longer, usually more than one year

to cope with habitat disturbance (especially human made but also by herbivores); *C* strategists (competitors) are adapted to live in supposed highly competitive but benign environments like the tropics; and *S* strategists (tolerators) are adapted to cope with severe environmental conditions (Fig. 5.12).

More recently, Silvertown (Silvertown, Franco, McConway 1992; Silvertown et al. 1993) provided a demographic interpretation of Grime's ideas, known to some as Grime's triangle (Fig. 5.13). The contribution of (i) fecundity, *F*, (ii) growth, *G*, and (iii) survival, *L*, to the overall reproductive rate, *R*, was determined. Individuals could then be placed on triangular diagrams with *F*, *G*, and *L* as axes, from which it became apparent that species of particular life-forms (e.g., woody plants) and from particular environments (e.g., forests) tend to be found together on the diagrams (Fig. 5.13).

Useful though each of these schemes is, the *r* and K concept remains the rock on which each is based. Sometimes, however, the MacArthur-Wilson foundation seems a little shaky. Despite the apparently broad array of support for the *r* and K concept from a wide variety of taxa, on closer examination the actual empirical evidence for this idea is equivocal. Stearns (1977) in an extensive review of the data available at the time, found that of thirty-five thorough studies on life history strategies, eighteen conformed to the

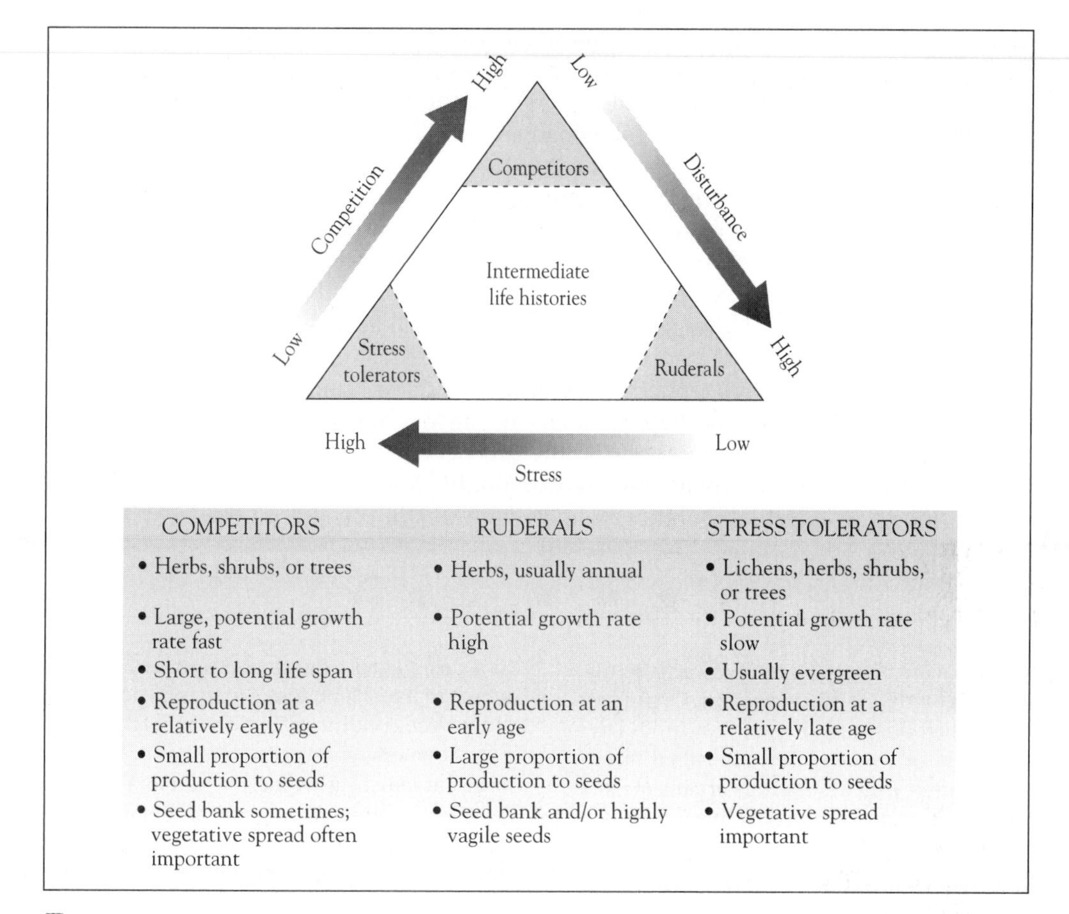

FIGURE 5.12 Expected plant life histories based on a model in which stress, disturbance, and competition are the important selective factors. (*Based on data in Grime 1979*).

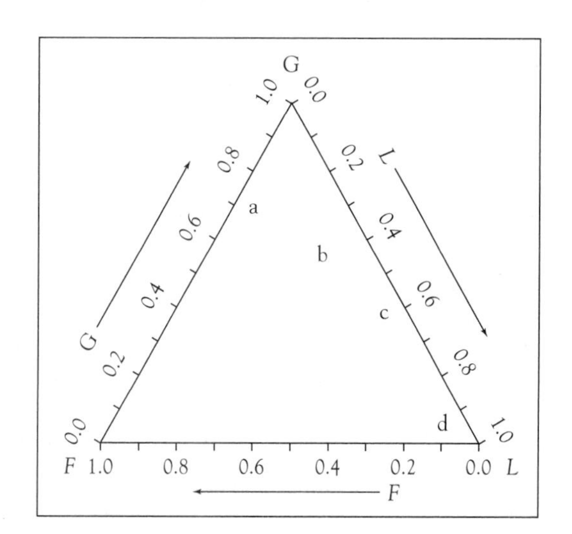

FIGURE 5.13 The distribution of species of perennial plants in the *G-L-F* (growth, survival, fecundity) triangle. Note that *G*, *L*, and *F* values are defined such that they always sum to 1. (**a**) *Semelparous* herbs. (**b**) *Iteroparous* herbs of open habitats. (**c**) *Iteroparous* forest herbs. (**d**) Woody plants.

scheme while seventeen did not. In addition, different authors have used the terms *r* and K selection too loosely and in different senses (Parry 1981, but see Boyce 1984). It has thus become an "omnibus" term (Milne 1961), a term with so many different intuitive definitions as to be ambiguous. For many people, *r* and K selection is what Hardin (1957) termed a **panchreston**—something that can explain almost anything. Furthermore, the more modern triangular schemes of Grime and Silvertown imply a trade-off between strategies. That is, a thickly trunked tree that is a good competitor and resistant to herbivores cannot also be a fast grower and have early reproduction. Although Loehle (1987) found trade-offs in North American hardwood trees, they were scarce among conifers. For example, in slow-growing hardwoods such as hickory and oaks, median tree age was just under two hundred years, whereas for fast-growing ones, such as box elder and cottonwood, trees tended to die at well under a hundred years. For conifers such trade-offs were absent perhaps because one big loss may result in myriads of small undetectable gains, not one major one. Trade-offs need not always be one to one. However, while the *r*/K dichotomy may be an oversimplification of what goes on in nature it has provided a wonderful theoretical framework and stimulus for much research.

SUMMARY

1. An organism's life history largely concerns its lifetime pattern of growth and reproduction. Organisms that produce all their offspring in a single reproductive event, like salmon, or yucca plants, are called semelparous. This contrasts with iteroparous organisms, like trees and mammals, where organisms live a relatively longer time and reproduce continuously.

2. Reproductive strategy has a strong effect on the age structure of a population. All continuously breeding populations tend toward a stable age distribution with the ratio of each age group in a growing population remaining the same. The ratio of young to adults in a relatively stable population of most mammals and birds is approximately 2:1. Lower ratios indicate populations in decline.

3. A 1:1 ratio of males to females is the expected sex ratio in most populations except where the cost to produce sons is different from that of daughters or there is local male competition. In polygynous mating systems, males mate with more than one female; in polyandrous systems females mate with more than one male.

4. The reproductive success of an organism depends to a large extent on its choice of the right habitat in which to live. Habitat selection may occur according to "ideal free conditions," or it may be influenced by territory owners or despots. Whether or not to fight over a habitat may be governed by certain rules. These rules have been modeled as game strategies by John Maynard Smith using the simple strategies of hawks and doves. What strategy to play depends very much on what strategy other organisms are playing.

5. Habitat selection may be affected by the ability to disperse over different spatial scales. Some suitable habitats may be underutilized because organisms cannot disperse into them.

6. The effects of mating systems, habitat selection, and dispersal can be synthesized under the banner of the *r-K* continuum. Organisms that are r-selected are poor competitors but have high per-capita population growth rate (*r*). Such species are often highly dispersive species that colonize new habitats before being outcompeted by *K*-selected species. *K*-selected organisms are good competitors and often exist in mature habitats, close to the carry capacity (*K*), where they can outcompete most species.

DISCUSSION QUESTION

What happens to the proportion of hawks and doves in a population as the payoffs vary?

S E C T I O N

4 Population Ecology

Wild lupines bloom in Acadia National Park, Maine. What is the most important source of mortality for them—competition with other plants, herbivory by insects or vertebrates, or a lack of soil nutrients or moisture? Such questions abound in nature and understanding what limits populations is the province of population ecology. *(Tom Till, DRK Photo 211315.)*

What determines the abundance of species in nature? This question has continued to intrigue ecologists ever since 1957, the date of the famous Cold Spring Harbor Symposium in North America at which the proponents of various views vigorously aired their opinions. The theme of the symposium was the debate over density dependence versus density independence (see Chap. 13). On the one hand, there were the correlations of abundance and population changes of organisms with aspects of weather (Davidson and Andrewartha 1948b). There was also a good demonstration that the intrinsic rate of natural increase of a species could be influenced by laboratory-controlled temperature and moisture (Birch 1953). On the other hand, there were many laboratory populations that showed a reasonably good fit to a model known as the logistic (which suggests competition acts so as to slow the growth rates of populations at high densities). There were also some long-term field studies showing population fluctuations but a fairly constant mean level of abundance. Finally, there existed the apparently logical and mathematical argument that without density dependence, populations would fluctuate with increasing amplitude, either going extinct or reaching completely unrealistic numbers.

Since that time, the same types of questions have been posed many times over but in a more sophisticated manner. Most ecologists accept that there are upper limits within which population densities are constrained, but that still doesn't necessarily mean that density dependence occurs (see Chap. 13). In order to address fully the density-dependent versus density-independent debate, we must examine how populations grow, whether they are affected much by the weather or by other organisms such as competitors, parasites, predators, and, in the case of plants, herbivores. Only after an examination of these factors can we begin to construct a synthesis about which type of population is controlled by which type of factors.

CHAPTER

6

Population Growth

Within their areas of distribution, plants and animals occur in varying densities. We recognize this pattern by saying an animal is "rare" in one place and "common" in another. There are, however, at least two ways to think about rarity: a species we know as rare can be common in at least one part of its range and rare everywhere else (globally restricted) or rare throughout its range (of limited abundance). Rarity can be strongly influenced by human activities, but it can also be produced naturally; some species are rare because they are endemic to limited areas, such as islands (globally restricted). For example, more than 50 percent of the Canary Islands' plant species, which are about 95 percent endemic, and about 66 percent of the 155 endemic plants on Crete are considered endangered (Lucas and Synge 1978). Habitat destruction in these cases adds heavily to the perils of an already restricted range. Rarity can also be the result of the dynamics of food chains (Chap. 20); top predators are normally less abundant than their prey, and only about fourteen hundred panthers are thought to have occupied the whole state of Florida even before the arrival of Europeans (of limited abundance) (Cristoffer and Eisenberg 1985).

For more precision than a statement of rare versus common, and especially for management purposes, it is desirable to quantify population density and more precisely to determine what fractions of a population consist of juveniles and adults. Normally, density is calculated for a small area, and total abundance over a bigger area is estimated from these figures. Apart from pure visual counts of organisms, sampling methods include the use of the following:

Traps: Live traps, snap traps, light traps (for night-flying insects), pitfall traps (for crepuscular species), suction traps, pheromone traps. The mark and recapture of animals is a particularly good way to get information on population sizes.

Fecal pellets: For hare, mice, rabbits, and so on.

Vocalization frequencies: For birds or frogs.

Pelt records: Taken at trading stations for large mammals.

Catch per unit effort: Especially useful in fisheries, where catch is often given per one hundred trawling hours.
Percentage ground cover: For plants.
Frequency of abundance along transects or in quadrants of known area: For plants and sessile animals, which remain in place to be counted.
Feeding damage: Useful for estimating the relative numbers of herbivorous insects.
Roadside spottings in a standard distance: Often used in bird counts.

Southwood (1977) provides an exhaustive review of these techniques and many more. From such data one can estimate not only the density of a population but also the relative frequency of juveniles and adults, larvae and nymphs, or other types of immatures. Great care must be taken in data collection; in particular, one must always consider whether the sampling regime is likely to bias results. For example, Mallet et al. (1987) showed how mark-recapture techniques strongly influenced butterfly behavior and hence population-size estimates. *Heliconius* butterflies avoid specific sites where they have been handled. In a fascinating article, Spear (1988) showed that nesting gulls can recognize and distinguish different individual human investigators, mostly on the basis of facial features! Some familiar faces do not elicit the disruption of nesting areas as much as unfamiliar faces do.

With data, it is possible to construct life tables that show precisely how a population is age structured. Life table construction is termed *demography*. An accurate measure of age is essential here; the use of size as an indicator of age is tenuous at best and at worst leads to underestimates of juveniles and overharvesting. A knowledge of age structure and of age-specific fertilization can then lead to accurate predictions about population growth. Basically, there are two different types of life table: age-specific (cohort analysis) and time-specific (static).

6.1 Life tables provide a tool for characterizing a species' life history strategies by estimating how the likelihood of mortality and reproduction change with age.

Time-specific life tables provide a snapshot of a population's age structure from a sample at a given time.

The basic data used in a life table are the number of individuals in each age class. These data can either be obtained by following a group of individuals (a cohort) from birth until all members of the cohort have died (a dynamic or cohort life table), or from a snapshot of the age structure of the population at a single point in time (a time-specific or static life table).The latter situation is more common with long-lived organisms where it is impractical to follow a single cohort from birth to death.

Time-specific life tables are useful in examining populations of long-lived animals, say, herds of elephants, where following a cohort of individuals from birth to death would be impractical. An example of such a time-specific life table, prepared from a collection of skulls of known ages, for Dall mountain sheep (shown in Photo 6.1) living in Mount McKinley (now Denali) National Park, Alaska, is shown in Table 6.1. The values given in the columns of life tables are symbolized by letters:

▪ **PHOTO 6.1** Dall Mountain sheep, *Ovis dalli*, in McKinley Park, Alaska. Collections of skulls, together with accurate age estimates of the skulls, permitted construction of an accurate life table for this species. *(Bean, DRK Photo, 116027.)*

x = age class or interval (years)

n_x = the number of survivors at beginning of age interval x

d_x = number of organisms dying between the beginning of age interval x and the beginning of age interval $x + 1$

l_x = proportion of organisms surviving to beginning of age interval x

q_x = rate of mortality between the beginning of age interval x and the beginning of age interval $x + 1$

e_x = mean expectation of life for organisms alive at beginning of age x (the goal of actuaries)

Beginning with columns x *(age class)* and n_x (number of individuals at the beginning of that age class), one can begin to calculate the other variables in the life table. From n_x, calculate the number of deaths occurring during age class $x(d_x)$, as follows:

$$d_x = n_x - n_{x+l}$$

For example, in the first age class, n_{0-1}, = 1000, indicating that there were one thousand individuals alive at the beginning of the age class. Only 801 individuals were alive at the beginning of the next age class $(n_{l-2} = 801)$, so we therefore know that 199 individuals died during the first age class.

TABLE 6.1 **Time-specific life table for the Dall Mountain Sheep (*Ovis dalli*) based on the known age of death of 608 sheep dying before 1937 (both sexes combined). Data are expressed per 1,000 individuals.**

x	n_x	d_x	l_x	q_x	L_x	T_x	l_x
0–1	1000	199	1.000	.199	900.5	7053	7.0
1–2	801	12	0.801	.015	795	6152.5	7.7
2–3	789	13	0.789	.016	776.5	5357.5	6.8
3–4	776	12	0.776	.015	770	4581	5.9
4–5	764	30	0.764	.039	749	3811	5.0
5–6	734	46	0.734	.063	711	3062	4.2
6–7	688	48	0.688	.070	664	2351	3.4
7–8	640	69	0.640	.108	605.5	1687	2.6
8–9	571	132	0.571	.231	505	1081.5	1.9
9–10	349	187	0.439	.426	345.5	576.5	1.3
10–11	252	136	0.252	.619	174	231	0.9
11–12	96	90	0.096	.937	51	57	0.6
12–13	6	3	0.006	.500	4.5	6	1.0
13–14	3	3	0.003	1.00	1.5	1.5	0.5

Note: A small number of skulls without horns, but judged by their osteology to belong to sheep nine years old or older, were apportioned pro rata among the older age classes. Mean length of life 7.09 years.

Next, values of n_x are used to calculate survivorship, which is the proportion of individuals in a cohort that survive to each age class. Survivorship, or l_x, is calculated by dividing the number of individuals in an age class by the number in the original cohort:

$$l_x = \frac{n_x}{n_0}$$

Survivorship to the fourth age class (3 to 4 years) is therefore 776/1000, or 0.776. This value tells us that over three-quarters of those individuals born will survive to be at least three years old.

The mortality rate within each age class is estimated with the variable q_x, which is calculated by

$$q_x = \frac{d_x}{n_x}$$

This simply expresses the number of deaths during an age class (d_x) as a proportion of the number of individuals that were alive at the beginning of that age class (n_x) and so indicates the fraction of individuals that died during that age class. For example, in the sixth age class (5 to 6 years), there were 46 deaths out of 734 individuals, so $q_x = 46/734$ or 0.063. This mortality rate allows us to determine whether mortality increases with age, decreases with age, or is independent of age.

The next parameter that is commonly estimated is life expectancy, which is the average number of additional age classes an individual can expect to live, on average, at each age. This calculation requires two intermediate steps. First, the average number of individuals alive during each age class must be calculated. Since n_x is the number of individuals alive at the beginning of each age class, the number alive during an age class can be estimated by:

$$L_x = \frac{n_x + n_{x+1}}{2}$$

A numerical example, from the fourth and fifth rows of Table 6.1, is

$$L_3 = \frac{n_3 + n_4}{2} = \frac{776 + 764}{2} = 770$$

The second step in determining life expectancy is to calculate a quantity called T_x. This is a purely intermediate step and unlike the other columns in a life table is without real biological meaning. T_x is calculated by summing the values of L_x in age class x and all subsequent (older) age classes. Thus,

$$T_x = \sum_{i=1}^{x} L_x$$

The value of T_x for the eleventh age class (11 to 12 years) is therefore calculated as follows:

$$T_{11} = L_{11} + L_{12} + L_{13} = 57$$

$$= 51 + 4.5 + 1.5 = 57$$

Life expectancy, or e_x, can now be calculated as follows:

$$e_x = \frac{T_x}{n_x}$$

The value of e_x indicates the average number of additional age classes an individual can expect to live at each age, so

$$e_{10} = \frac{174 + 51 + 4.5 + 1.5}{252} = 0.92$$

Therefore, ten-year-old sheep can be expected to live, on average, a further 0.92 years.

Taken as a whole, these calculations allow important features of a population to be quantified: What is the maximum age of individuals in this population? How does survivorship change with age? When in an individual's life is mortality rate highest?

The n_x data are commonly plotted against age to give a survivorship curve, as shown in Fig. 6.1 (Pearl 1928). For the Dall mountain sheep there is an initial decline in survivorship as young lambs are lost, then the curve flattens out, indicating sheep survive well through about age 7 or 8. Then old age takes its toll, and sheep numbers decline rapidly.

A second type of life table examines reproductive output of individuals as a function of age. This table is typically called a fecundity schedule. Fecundity is defined as the number of offspring produced by each breeding individual. Fecundity schedules describe reproductive output and survivorship of breeding individuals only: in a species with separate sexes, they include only females. Table 6.2 shows a fecundity schedule for a pride of female lions. Survivorship, l_x, is calculated as before from the number of individuals surviving to each age class (n_x, data not shown). Fecundity, m_x, is the average number of offspring produced per capita at each age class. Fecundity is less than one in some age classes because not all females reproduced.

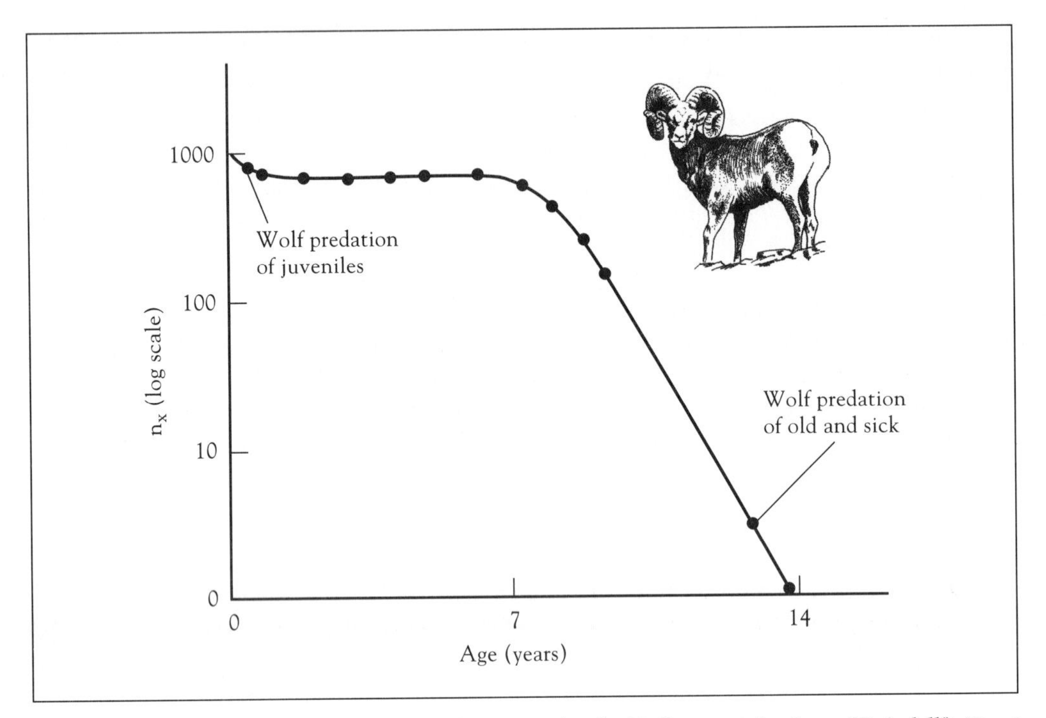

■ **FIGURE 6.1** Time-specific survivorship curve for the Dall mountain sheep (*Ovis dalli*). (*Based on data from Murie 1944.*)

Fecundity schedules not only allow examination of age-specific trends in reproduction but also allow calculation of the population's net reproductive rate, R_0. R_0 is calculated by multiplying l_x by m_x and summing over all age classes. Thus, the net reproductive rate, R_0, is given by the formula

TABLE 6.2 Fecundity schedule for female lions.

x (YEARS)	l_x	m_x	$l_x m_x$
0–1	1.00	0.00	0.00
1–2	0.75	0.00	0.00
2–3	0.58	0.14	0.08
3–4	0.58	0.57	0.33
4–5	0.50	0.50	0.25
5–6	0.42	0.00	0.00
6–7	0.33	1.25	0.42
7–8	0.25	0.67	0.17
8–9	0.25	0.33	0.08
9–10	0.17	0.00	0.00
10–11	0.08	0.00	0.00
			$R_0 = 1.33$

$$R_0 = \sum_{i=1}^{x} l_x m_x$$

The net reproductive rate is defined as the number of breeding individuals that will be produced by each breeding individual in a population. As its calculation indicates, it takes into account not only the number of offspring that a female will produce at each age but also the probability that each female will survive to that age. If each individual exactly replaces herself, then $R_0 = 1$ and the population is stationary. If each individual more than replaces herself, $R_0 > 1$ and the population is increasing. If each individual does not replace herself, then $R_0 < 1$ and the population is declining. In the example illustrated in Table 6.2, R_0 is 1.33, indicating that on average each female will produce 1.33 female offspring and that the population is therefore increasing.

Usually the calculation of m_x is facilitated by a knowledge of F_x, the total number of eggs, seeds, or young deposited divided by the total number of reproducing individuals such that

$$m_x = \frac{F_x}{n_x}.$$

A worked example is shown in Table 6.3 for the plant *Phlox drummondi*. In the case of the *Phlox* population R_0 was 2.4, meaning there was a 2.4-fold increase in the size of the population over one generation. If this were true every year then we would be knee-deep in *Phlox*. This was evidently a good year. In bad years R_0 would be much less than

TABLE 6.3 Fertility schedule for *Phlox drummondi*. (From data in Leverich and Levin 1979)

AGE INTERVAL (DAYS) x	NUMBER SURVIVING TO DAY x n_x	PROPORTION OF ORIGINAL COHORT DYING DURING INTERVAL l_x	PROPORTION OF ORIGINAL COHORT SURVIVING TO DAY x d_x	F_x	m_x	$l_x m_x$
0–63	996	0.329	1.000	—	—	—
63–124	668	0.375	0.671	—	—	—
124–184	295	0.105	0.296	—	—	—
184–215	190	0.014	0.191	—	—	—
215–264	176	0.004	0.177	—	—	—
264–278	172	0.005	0.173	—	—	—
278–292	167	0.008	0.168	—	—	—
292–306	159	0.005	0.160	53.0	0.33	0.05
306–320	154	0.007	0.155	485.0	3.13	0.49
320–334	147	0.043	0.148	802.7	5.42	0.80
334–348	105	0.083	0.105	972.7	9.26	0.97
348–362	22	0.022	0.022	94.8	4.31	0.10
362–	0	—	0.000	—	—	—
				2408.2		2.41

Note: The columns are explained in the text.

1, but many year's data reveal a more balanced picture. The reasons for good and bad years may include bad weather, herbivory, or disease—these will be addressed later.

Despite the value of time-specific life tables, there are some assumptions that limit their accuracy. Paramount among these is that equal numbers of offspring are born each year. For example, if the rate of mortality of two-year-old Dall mountain sheep were identical to the rate of four year-old sheep but there were more two-year-old sheep born because of favorable climate in that particular year, then more skulls of two-year-old sheep would be found later on and a higher rate of mortality of two-year-olds would be implied. There is often no independent method for estimating the birth rates of each age class.

Age-specific life tables follow an entire cohort of individuals from birth to death.

For organisms with short life spans, usually completed within a year or those with distinct breeding cycles, a snapshot, time-specific life table may not give the correct picture. It will be severely biased toward the juvenile stage common at that moment. In these cases age-specific tables are used, which follow one cohort or generation. Population censuses must be conducted frequently but only for a limited time (usually less than a year). An age-specific life table for the spruce budworm (Photo 6.2) is shown in Table 6.4 and represented graphically in Fig. 6.2. (See also Ecology in Pratice: Tomo Rayama.) Spruce budworms are larvae of tortricid moths.

They excavate the terminal and lateral buds of conifers and are an economically

■ **PHOTO 6.2** Spruce budworms feeding on Jack Pine in Ontario can cause substantial economic loss to planted forests. For this reason there is much impetus to understand what influences their numbers. Because the life cycle is passed so quickly, it is possible to follow an entire cohort of budworms, from egg to moth, and construct an age-specific life table. (*Lynch, DRK Photo 147379.*)

TABLE 6.4 Age-specific table for the 1952–1953 generation of the spruce budworm on the plot G4 in the Green River watershed of New Brunswick. (After Morris and Miller 1954)

AGE INTERVAL x	NO. ALIVE AT START OF AGE INTERVAL n_x	FACTOR RESPONSIBLE FOR MORTALITY	NO. DYING DURING AGE INTERVAL d_x	MORTALITY RATE q_x
Eggs	174	Parasites	3	0.017
		Predators	15	0.086
		Other	1	0.006
		Total	19	0.109
Instar I	155	Dispersal	74.4	0.480
Hibernacula	80.6	Winter	13.7	0.170
Instar II	66.9	Dispersal	42.2	0.631
Instar III-VI	24.7	Parasites	8.92	0.360
		Disease	0.54	0.040
		Birds	3.39	0.137
		Other	10.57	0.428
		Total	23.42	0.948
Pupae	1.28	Parasites	0.10	0.078
		Predators	0.13	0.101
		Other	0.23	0.180
		Total	0.46	0.360
Moths	0.82		0.82	1.000
Total for generation (egg to adult)			173.18	0.995

Note: All numbers expressed per 10 sq. ft. of branch surface.

important pest, especially in eastern Canada. Eggs are laid in August, and young caterpillars feed until the fall, when they pupate in hibernacula. In the spring the caterpillars resume feeding and pupate, and adult moths emerge in the early summer to lay new eggs, starting the cycle over again. Various sources of mortality operate at different stages in the life cycle, so, for example, eggs or larvae may be parasitized by small wasps, caterpillars may be eaten by birds or infected by fungi, and pupae may be eaten or parasitized. In fact, one of the biggest decreases in density is caused by caterpillar dispersal—they balloon away on silken threads to try their luck on new trees whose foliage may be at a more appropriate stage for consumption. The fact that most caterpillars do not land on a suitable tree is a big source of mortality.

As a result of both time-specific and age-specific demographic techniques, three general types of survivorship curves can be recognized (Fig. 6.3): Type III, in which a large fraction of the population is lost in the juvenile stages; Type II, in which there is an almost linear rate of loss; and Type I, in which most individuals are lost when they are older. Type I curves are often observed in vertebrates or organisms that exhibit parental care and protect their young. Type III curves are often exhibited by invertebrates such as insects, many plants, especially weeds, and marine invertebrates that do not exhibit parental care. For

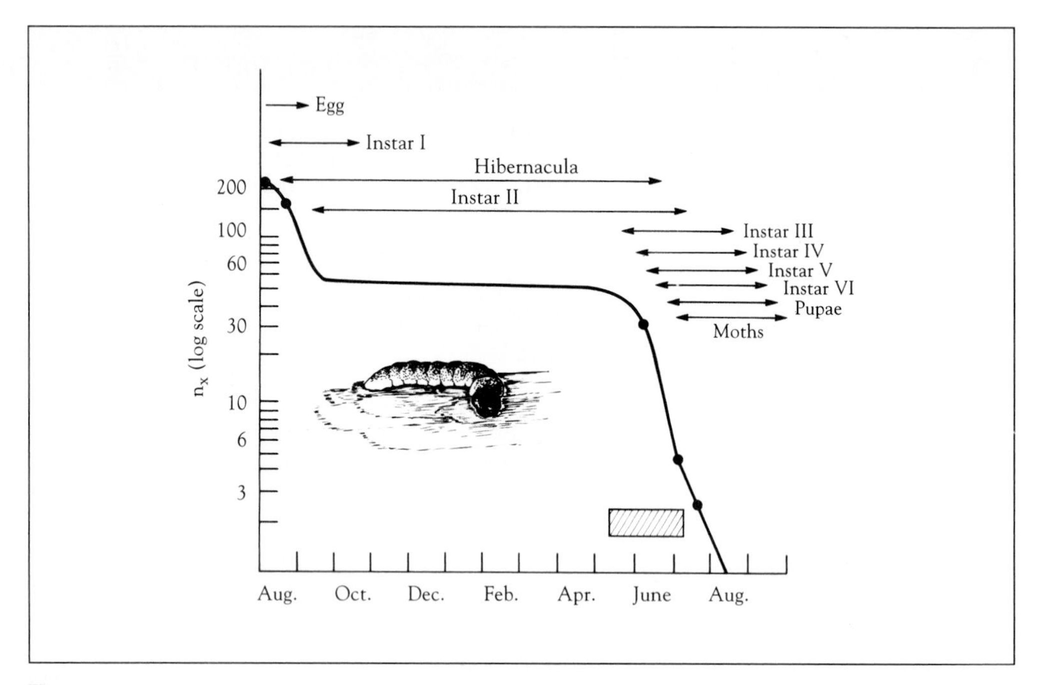

■ **FIGURE 6.2** Survivorship curve for the spruce budworm during the 1945–1960 outbreak in New Brunswick. This insect has one generation per year, and females can lay 200 eggs. Regular census points on the curve are indicated by dots, and the critical large larval stage in early summer is indicated by cross-hatching. (*Drawn after Morris 1963.*)

example, barnacles release millions of young into the sea, but most drift off and are eaten by predators. Only a few survive and settle in the rocky intertidal (although, once there, they show excellent survivorship). Some examples are shown in Table 6.5.

What's the difference between a time-specific and a cohort life table? Not much, but there is some difference. It is very difficult to get data for both types of life table for most populations, but for humans it is possible because we can use birth and death records for people born in say 1880 and follow them for their entire lives. We can also take a snapshot or cross-section of the population at 1880 to get a time-specific survivorship curve. Comparisons of the two reveal what the population would have looked like had it continued surviving at the rates observed in 1880 (Fig. 6.4). But static curves ignore improvements in medicine and nutrition that increased survival rates. In the natural world static curves ignore environmental variation, that is, good years and bad years.

6.2 Deterministic models of population growth predict changes in population size based on intrinsic properties of the population such as growth rate, current size, and carrying capacity.

It might seem that before we can determine how a population grows we must know the age structure of the population and age-specific survival rates and birth rates. Surprisingly, this is not the case. Lotka's theorem (1922) shows that if a population is subject to

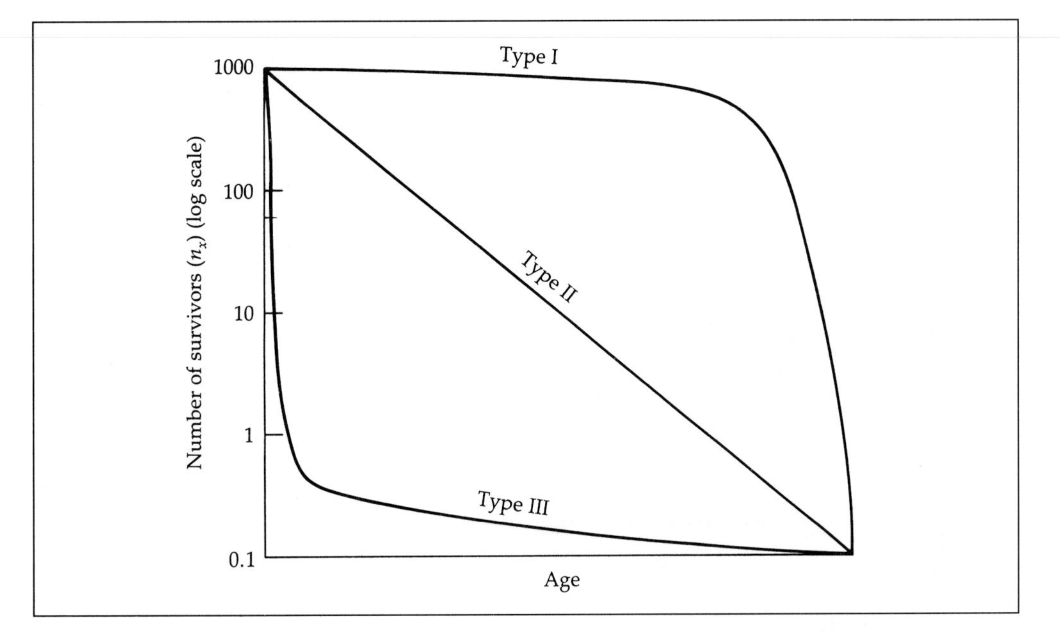

■ **FIGURE 6.3** Hypothetical survivorship curves. Type I includes man, mammals, and higher animals, often with parental care of young. Type II includes some birds and some invertebrates such as hydra. Type III includes some insects and many lower organisms with pelagic juvenile stages such as benthic invertebrates, mollusks, oysters, and fishes.

TABLE 6.5 Species survivorship curves. (Data from many sources)

SPECIES	TYPE OF ORGANISM	SURVIVORSHIP CURVE
Phlox drummondi	Phlox plant	I
Cervus elophus	red deer	I
Cervus canadensis	elk	I
Ranuculus acris	buttercup	II
Geospiza scandens	ground finch	II
G. fortis	medium ground finch	II
Ambostyma texanum	salamander	II
Lepidosaphes ulma	scale insect	III
Melilotus alba	sweet clover	III
Balanus glandula	barnacle	III
Chorthippus brunneus	grasshopper	III
Lymantria dispar	gypsy moth	III

ECOLOGY IN PRACTICE

Tomo Royama, Canadian Forest Service

Years ago, as a graduate student in Tokyo, I was browsing through *The Canadian Entomologist* and Canadian *Journal of Zoology* and came across a series of papers on spruce budwom in eastern Canada, called the Green River Project. It was interesting because the field study of an insect population based on life tables was a novelty then. So I wrote a review article about it just before I was leaving for Oxford for further study, not at all realizing, of course, that I was going to take over the budworm project in Fredericton later in my life.

I left Japan because my professor, a forest entomologist, was reluctant to read my doctoral dissertation: "The Breeding Biology of the Great Tit." So I sent it to Dr. David Lack of Oxford for his evaluation. He was interested in it and suggested that I continue my study under him. In these studies, I designed and built an automatic 16mm camera to record food items that adult great tits brought to nests to feed their young. In all, I took fourty one thousand pictures at ten nesting boxes over five breeding seasons. The list of the food items published in the 1970 *Journal of Animal Ecology* paper, I believe, is the most complete record, thus far, of a wild bird's menu.

Four years later, I was finishing my work at Oxford and gave a seminar on predation. In the audience, there happened to be a visitor from the Canadian Forestry Service, Chuck Buckner. After the seminar, he asked if I was interested in joining the Canadian Forest Service (CFS). And that is where I ended up.

I doubted the credibility of the conventional view of budworm outbreak processes because I noticed that the major assumptions were not deduced from a thorough analysis of data. It was more likely that the data were used to justify the hypotheses that had already been intuitively formulated. I decided to reanalyze the Green River data.

After the reanalysis, I was convinced that the rise and fall of a budworm population were dictated by changes in the survival of feeding larvae and pupae. Thus, to understand the budworm outbreak processes properly, it is crucial to know precisely what causes their mortality. Toward the end of the 1970s, I proposed that we resurrect the budwom survivorship study. In particular, I proposed the daily observations of budworm mortality throughout the summer season. In the Green River study, so many plots were sampled but only a few times in each plot. I proposed, instead, to sample only a few plots but as frequently as we could.

I suppose a valid hypothesis could be conceived through a logical consideration alone. The idea thus conceived could be published as a viable hypothesis that could be left to someone else for testing. This seems to be a major pattern in ecological research in the past few decades. Nothing is wrong with this approach, of course; if the basic assumption is appropriate and the conclusion is properly deduced it might be worth testing against observations and experiments designed for that specific purpose. Nonetheless, there is an inherent danger in this approach. The natural system might be viewed through a narrow window, and the gathering of information tends to be fragmented. But it might be tempting to follow this pattern in the era of publish or perish with the meager R&D money granted. The idea producer can live on a comparatively small grant and the tester can publish the result, whether supportive or negative, so that a publication is almost guaranteed.

I personally have resisted following this pattem; I tried to deduce a theory from data that contained comprehensive information. This is not a lucrative way of doing science nowadays because I could publish, if at all, only when I succeeded in arriving at a positive, coherent result. I could afford to do this because I worked in the govemment. A university professor once enviously remarked that I was sitting in the ivory tower. Money used to be so plentiful that while staying at the Green River summer camp, with free accommodation and free meals, I remember one student once ate six T-bone steaks.

So I took time to do my research—spending a lot of time fishing too. But I always had data with me, even if they were not mine; I collected as much information on the same system from as many different sources as I could find. Many of them at first sight contradicted each other. However, the real advantage of this approach is that once a coherent explanation was found its credibility would be high. Furthemore, once conceived, the idea need not be tested because it has already come out of a large collection of data; the idea only needs to be applied to a new system or situation for further generalization.

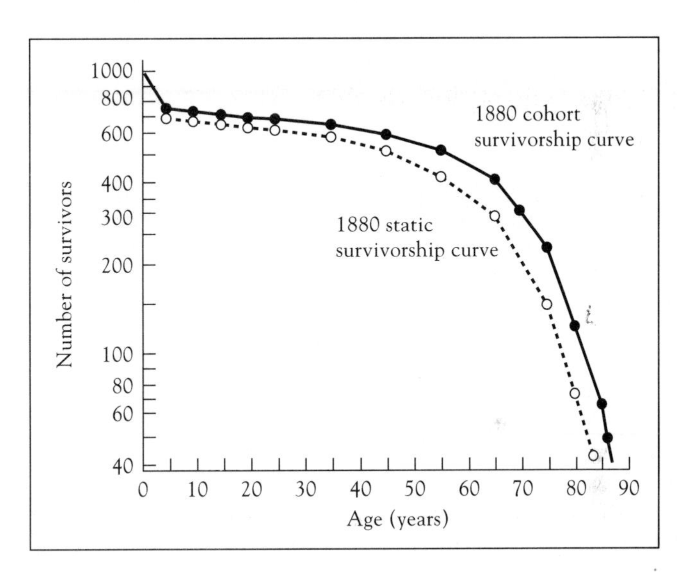

FIGURE 6.4 Comparison of cohort or generation survivorship of males born in 1880 in England and Wales with static or time-specific survivorship of males for 1880.

a constant schedule of birth and death rates, it will quickly approach a fixed or stable age distribution, whatever the initial age distribution may have been to start with. Krebs (1994) has provided a simple example to illustrate this point.

Suppose that we have a parthenogenetic organism that lives three years and then dies. It produces two young at exactly one year of age, one young at exactly two years of age, and no young at year three. The life table for this hypothetical organism is thus simple:

x	l_x	m_x	$l_x m_x$
0	1	0	0
1	1	2	2
2	1	1	1
3	1	0	0
4	0	—	—

TABLE 6.6 A population growing geometrically develops a stable age distribution.

YEAR	NUMBER AT AGES 0	1	2	3	TOTAL POPULATION SIZE	% AGE 0 IN TOTAL POPULATION
0	1	0	0	0	1	100.0
1	2	1	0	0	3	66.67
2	5	2	1	0	8	62.50
3	12	5	2	1	20	60.00
4	29	12	5	2	48	60.42
5	70	29	12	5	116	60.34
6	169	70	29	12	280	60.36
7	408	169	70	29	676	60.36
8	985	408	169	70	1632	60.36

$$R_0 = \sum_{0}^{4} l_x m_x = 3$$

If a population of this organism starts with one individual aged 0, the population growth will be as shown in Table 6.6.

Note that the age distribution quickly becomes fixed or stable, with about 60 percent (985/1632) at age 0, 25 percent (408/1632) at age 1, 10 percent (169/1632) at age 2, and 4 percent (70/1632) at age 3. This demonstrates Lotka's (1922) conclusion that a population growing geometrically develops a stable age distribution. In considering population growth we don't have to worry much about population age structure. We can rely on R_0, which in this case is three.

Geometric or exponential growth occurs in populations unlimited by competition for resources.

A population released into a favorable environment will begin to increase in numbers. Consider a **univoltine** insect with one breeding season and a life span of one year. The population size at time $t + 1$ is given by

$$N_{t+1} = R_0 N_t$$

where N_t is the population size of females at generation t, N_{t+1} is the population size of females at generation $t + 1$, and R_0 is the net reproductive rate or number of females produced per female per population. Clearly, much depends on the value of R_0; when $R_0 < 1$, the population goes extinct, when $R_0 = 1$ the population remains constant, and when $R_0 > 1$ the population increases. When $R_0 = 1$, the population is often referred to as being at **equilibrium**, where no changes in population density will occur. Even if R_0 is only fractionally above 1, population increase is rapid (Fig. 6.5).

Northern elephant seals were nearly hunted to extinction in the late nineteenth century because of demand for their blubber. About twenty surviving animals were found off Mexico on Isla Guadalupe in 1890, and the population was protected. The actual growth of the elephant-seal population and recolonization of old habitats matched well

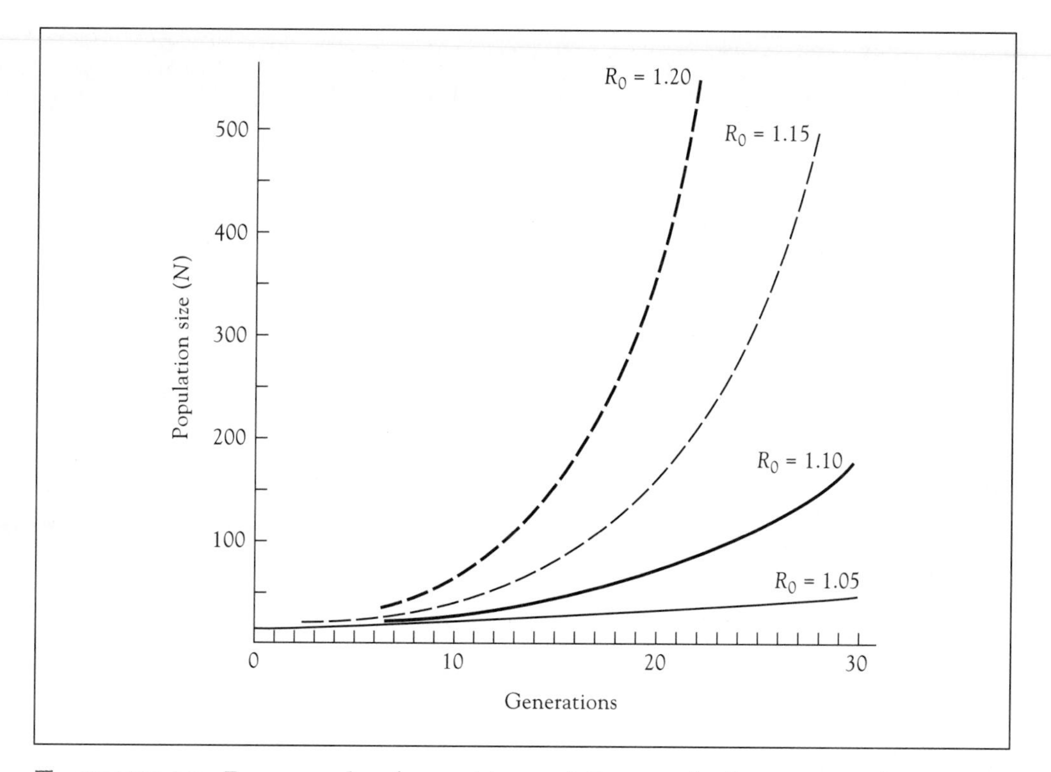

■ **FIGURE 6.5** Four examples of geometric population growth, discrete generation, constant reproductive rate. $N_0 = 10$.

the growth predicted when $R_0 = 2$ and generation time was eight years (Le Boeuf and Kaza 1981). Predicted numbers were about 80 in 1906 and 40,960 in 1978. Censuses showed that actual population numbers were 125 in 1911 and 60,000 in 1977. The growth of protected annually breeding populations recovering from overexploitation in the past often provides some of the best fits of actual population growth curves to this kind of model. The growth of some exotic species introduced into novel habitats also seems to fit the geometric pattern. The rapid expansion of rabbits after their introduction into South Australia in 1859 is a case in point. Sixteen years later, rabbits were reported on the west coast, having crossed an entire continent, over eleven hundred miles, despite the efforts of Australians to stop them by means of huge, thousand-mile-long fences (Fenner and Ratcliffe 1965).

The global human population also seems to fit a geometric pattern and is a source of concern to many (see Applied Ecology: Human Population Growth). In general, however, in over two hundred cases of mammalian introductions, most have failed to grow in an explosive or geometric pattern of increase (De Vos, Manville, and Van Gelder 1956). Of course, the Earth is not overrun with animals or plants, usually because R_0 decreases at high densities because of an increase in death rates and decrease in birth rates brought about by food shortage or epidemic disease. The value of R_0 usually changes as resources run short. (For humans this doesn't seem to have happened yet.) However, the effect of changing reproductive rates is usually discussed with reference to a different set of equations (later in this section) in logistic growth.

APPLIED ECOLOGY

Human Population Growth

Up until the beginning of agriculture and the domestication of animals, about 10,000 B.C., the average rate of population growth has been very roughly estimated at about 0.0001 percent per annum. After the establishment of agriculture, the world's population grew to about 300 million by A.D. 1, and 800 million by the year 1750, but still the average growth rate was well below 0.1 percent per annum. The modern period of rapid population growth may be regarded as having started at about 1750. Average annual growth rates climbed to about 0.5 percent between 1750 and 1900, 0.8 percent in the first half of this century, and 1.7 percent in the second half. Advances in medicine and nutrition are certainly responsible for a large part of the growth in reproductive rates and the longevity of humans. In this relatively tiny period of human history (equivalent to the last one minute on a twenty-four-hour clock), the world's human population increased from 0.8 billion in 1750 to 5.6 billion in 1995. Thus, in less than 0.1 percent of human history more than 80 percent of the increase in human numbers has occurred. This rapid increase in human growth rates in recent times is illustrated in Figure 1.

It was estimated that in 1995 the world's population was increasing at the rate of three people every second. In 1984, United Nations' projections estimated that the most likely course of world population was one of growth toward an eventual total stabilizing at around 10 billion people by the year 2100. However, human growth rates have not been reduced as quickly as was anticipated. The United Nations' 1995 projections point to a world population stabilizing at around 11.6 billion toward the year 2150, though nobody knows for sure (Pulliam and Haddad 1994; Cohen 1995; Lutz, Sanderson, and Scherbov 1997).

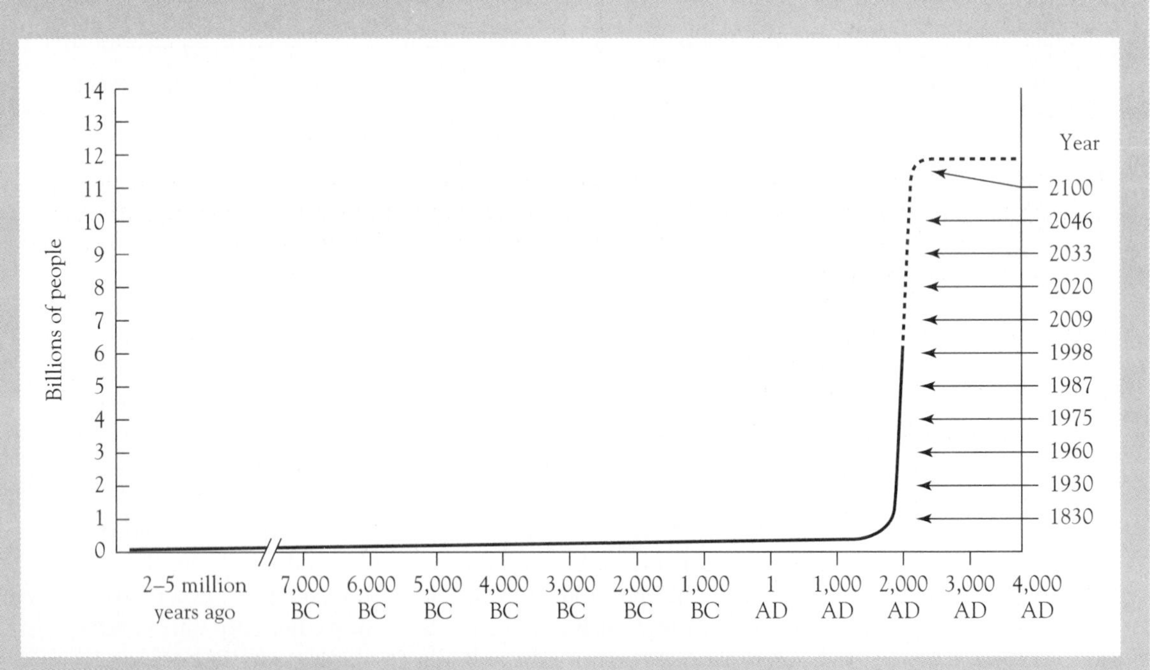

■ **FIGURE 1** The world population explosion. For most of human history, the population grew very slowly, but in modern times it has suddenly "exploded." Where and when it will level off is the subject of much debate.

TABLE 1 Trends in world population size, rate of increase, and density of people in 1990.

REGION	POPULATION (10^6) 1960	1970	1980	1990	RATE OF INCREASE (%) 1960–1965	1970–1975	1980–1985	1990–1995
WORLD	3,019	3,698	4,450	5,292	1.99	1.96	1.74	1.71
Developed regions	945	1,049	1,136	1,205	1.19	0.86	0.65	0.48
Developing regions	2,075	2,649	3,314	4,087	2.35	2.38	2.10	2.06
Africa	281	363	481	648	2.48	2.69	2.95	3.01
Latin America	218	285	362	448	2.80	2.48	2.19	1.94
Northern America	199	226	252	276	1.49	1.06	1.00	0.71
Asia	1,667	2,101	2,583	3,108	2.19	2.27	1.86	1.82
Europe	425	460	484	498	0.91	0.58	0.32	0.22
USSR	214	243	266	288	1.49	0.96	0.84	0.68
Ocenia	16	19	23	26	2.09	1.78	1.55	1.34

In the developed countries, the average annual rate of population increase has slowed considerably, from 1.19 percent in 1960–1965 to 0.48 percent in 1990–1995. In the developing world, however, average annual growth rates rose from 2.35 percent in 1960–1965 to 2.38 percent in 1970–1975 before dropping to 2.1 percent in 1980–1985. The results of these growth rates are that in developed nations the population has seemed to stabilize at around 1 billion people, but the number of people in developing countries is still increasing (Table 1). The result is that the world population growth is still rising and is expected to rise to over 6 billion people by the year 1998. If human population growth fuels pollution and resource use then developing countries, particularly those in Africa, are clearly most at risk. However, even though African countries have a high rate of population increase, Africa's total population, at about 650 million people, is far below the 3 billion that inhabit Asia.

Thus, in terms of the global population level, even a small decrease in the rate of population increase in Asia would reduce the absolute numbers of humans more than would a substantial decrease in the rate of population increase in Africa.

Global population growth can also be examined by looking at fertility rates, the average number of live births that would typically be borne by a woman of childbearing age (Table 2).

Again, the fertility rate differs considerably between geographic areas. In Africa, the total fertility rate of 6.0 in 1990 has scarcely declined since the 1950s when it was around 6.6 children per woman. In Latin America and Southeast Asia, fertility rates have declined considerably from the 1970s and are now at around 3.3 and 3.2, respectively. In 1960–1965, the global average number of children born per woman was almost five. In 1990, the average was 3.3, but this is still greater than the 2.1 needed for zero population growth. In more developed regions, the fertility rate

■ **PHOTO 1** Chandini Chowk, India, once a grand avenue for imperial procession in the 1600s, now a crowded bazaar. Curbing human population growth is the key to minimizing environmental degradation. *(Porterfield/Chickering, Photo Researchers Inc. 6W 9756.)*

has declined steadily in recent times and is now expected to remain fairly stable, at around 1.9, for the next few decades. Part of the reason for the differences among developed and developing countries has been linked to contraceptive use. For example, a 1992 Johns Hopkins University study indicated that in developed countries over 70 percent of couples, where the woman was of reproductive age, were using some form of contraceptive. In developing countries, this level was estimated to be around 45 percent. In Africa, only an estimated 14 percent of couples used contraceptives, whereas the level of use was around 50 percent in Asia and about 57 percent in Latin America. Some of the most startling reductions in fertility have been experienced in China. In the 1950s and 1960s, the total fertility rate in China was six children per woman. During the 1970s, family planning services were offered by the government, and incentives were also offered to couples who agreed to limit themselves to one child. In 1990

TABLE 2 The net reproductive rate, R_0, (sometimes called the total fertility rate), the average number of children per woman in her reproductive lifetime, 1990 data.

REGION	FERTILITY RATE	REGION	FERTILITY RATE
Africa	6.0	Former USSR	2.3
South America	4.0	Europe	2.0
Asia	4.0	United States	1.9
Oceania	3.0	World	3.0
North America	3.0		

an estimated 75 percent of the population used birth control, and the fertility rate had dropped to 2.2.

Government-supported family planning programs have been implemented in an increasing number of countries during the past two decades. In 1976 only 97 governments provided direct support for family planning, compared with 125 in 1988. In contrast, the number of governments limiting access to family planning fell from 15 in 1976 to 7 in 1988. However, as of 1989, there were still 31 countries in the developing world where couples have virtually no access to modern family planning methods. One could also argue that reductions in birth rates have not occurred in some developing countries because women wish to have large families, but there are many surveys that show that women in developing countries would like to have fewer children than they already have. For example, for young women (15 to 19 years) in virtually every country outside sub-Saharan Africa, the desired number of children is below three: Thailand 1.8, Colombia 2.1, and Brazil 2.2. However, in sub-Saharan Africa the desired number in most countries surveyed is five or more. Surveys of older women (45 to 49 years) often reveal an even higher number of desired children. However, it has been shown that these higher values are biased by rationalization, a post-facto revision of desired family size to make it closer to the actual number.

It is interesting that despite the concerns in many parts of the world about the increase in human population growth rates, in some countries the concern is that the growth rates are not enough. This attitude has been prevalent in some Western European countries. Since about 1965 the "baby boom" in these countries, together with other more developed countries like Australia, Canada, New Zealand, and the United States, has turned into a "baby bust," and total fertility rates have fallen below the national replacement level of 2.1 children per woman. This rate of replacement, 2.1 children per woman, allows for natural mortality and, in the end, produces a natural increase of about 2.0, ensuring that the population neither decreases nor increases. A negative growth rate could eventually severely affect the political and economic structure of some nations, and that alarms some politicians.

For many species, including bacteria, fungi, insects, invertebrates, and many plants and animals in warm climates, reproduction occurs not seasonally but year round and generations overlap. For such species the rate of increase is essentially the same as in Figure 6.5, i.e., geometric increase that, however, because of its constant nature has to be described by a differential equation. We can best see why this is with an analogy to interest rates. If we have a population of one thousand organisms increasing at a finite rate of 10 percent per year, then the population size at the end of year 1 will be 1000+ (10% of 100) = 1100. The following tables give the corresponding values for subsequent years.

YR	N
0	1000
1	1110
2	1210
3	1331
4	1464
5	1610

If we repeated these calculations with a finite rate of increase of 5 percent every six months, our numbers would be a little different. After six months, the population size would be 1,050, after one year 1,102, two years 1,215, three years 1,340, and so on. If we repeated the calculations every day or every hour our values would be slightly different again. Because many populations are continuous breeders this is quite appropriate. Dividing a year into one thousand short time periods, we have a population size of 1000.1 in the first thousandth of the year, 1000.2 in the second thousandth and so on. If we repeat this for one thousand time intervals we will end up with 1,105 organisms at the end of the first year.

The differential equation that describes this type of growth is

$$\text{rate of increase } \frac{dN}{dt} = rN = (b - d)N$$

where N = population size, t = time, r = **per-capita rate of population growth**, b = instantaneous birth rate, and d = instantaneous death rate.

In this equation r is analogous to R_0. In populations with a stable age distribution, that is where the proportions of different age classes remain the same from year to year,

$$r \simeq \frac{(\ln R_0)}{T_c}$$

where T_c is the generation time (Southwood 1976)

This essentially means that, for a continuously breeding population, the net reproductive rate divided by the generation time would give an approximation of the instantaneous rate of population growth. To express this rate per individual we use the natural log, ln, of R_0. To convert this instantaneous rate into a finite rate, we can use the formula

$$\text{finite rate of increase } \lambda = e^r$$

Thus for, $r = 0.881$, $\lambda = 2.413$ individuals per year.

We can also use the geometric growth model to estimate the doubling time for a population growing at a certain rate. Following the rules of calculus, the integral form of

$$\frac{dN}{dt} = rN$$

is

$$N_t = N_0 e^{rt}$$

Thus,

$$\frac{N_t}{N_0} = e^{rt}$$

If the population doubles, then $N_t/N_0 = 2$. Thus

$$2.0 = e^{rt}$$

or

$$\ln (2.0) = rt$$

and

$$t = \frac{0.69315}{r}$$

where

t = time for population to double its size

r = rate of population growth per capita

A few values for this relationship are given for illustration:

r	t
0.01	69.3
0.02	34.7
0.03	23.1
0.04	17.3
0.05	13.9
0.06	11.6

Thus, if a human population is increasing at an instantaneous rate of 0.03 per year (finite rate = 1.0305), its doubling time would be about 23 years, if geometric increase prevails. Even very low rates of population increase give very short doubling times. In other organisms doubling times can vary from as little as 17 minutes for an *E. coli* bacteria to as much as 25.3 years for a southern beech tree, *Nothofagus*.

Logistic growth occurs in populations in which resources are or can be limiting.

For many species resources become limiting as populations grow. The intrinsic rate of natural increase decreases as these resources are used up. Thus, a more appropriate equation to explain population growth under these conditions is:

$$\frac{dN}{dt} = rN\frac{(K - N)}{K} \quad \text{or} \quad \frac{dN}{dt} = rN\left(1 - \frac{N}{K}\right)$$

where K is the upper asymptote or maximal value of N, commonly referred to as the car-

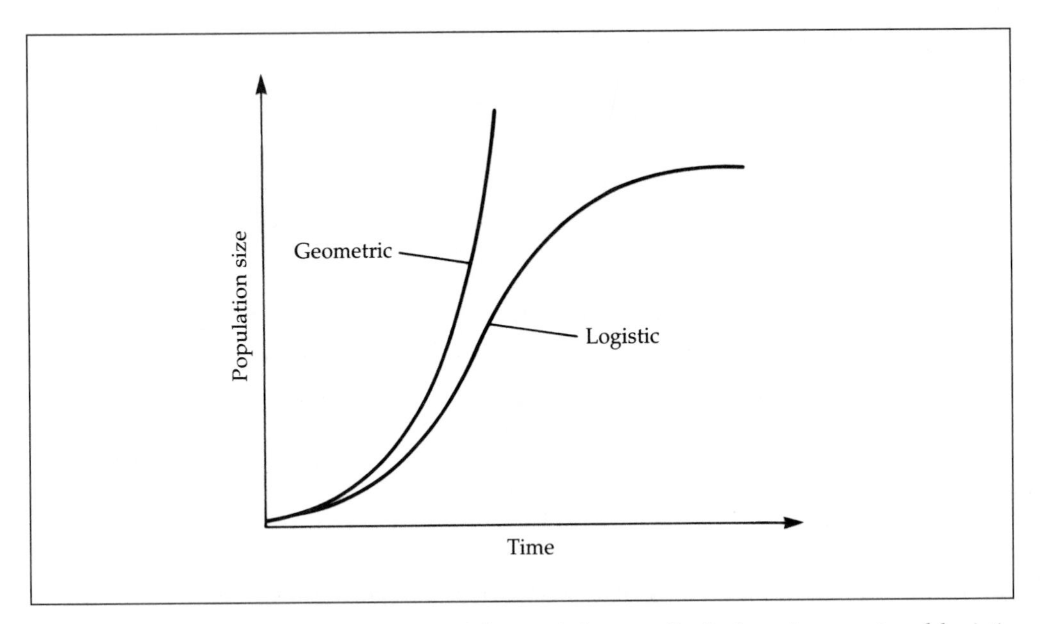

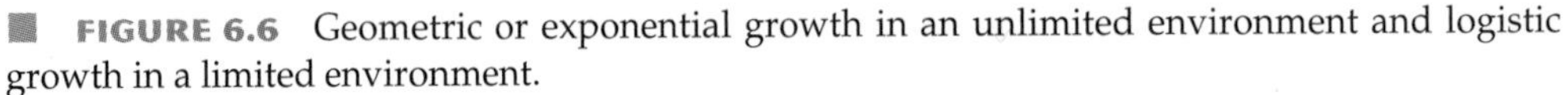

■ **FIGURE 6.6** Geometric or exponential growth in an unlimited environment and logistic growth in a limited environment.

rying capacity of the environment at the equilibrium level of the population. In essence this equation means

$$\begin{array}{lcccl} \text{rate of increase of population per unit time} & = & \text{rate of population increase} & \times & \text{population size times unused opportunity for population growth} \\ & = & r & \times & N\dfrac{(K-N)}{K} \end{array}$$

When this type of growth is represented graphically, a sigmoidal, S-shaped, or so-called *logistic curve* results (Fig. 6.6). This equation was first described by Verhulst (1838) and was derived independently by Pearl and Reed (1920) to describe human population growth in the United States.

Logistic growth entails many important assumptions. Five of the most important are listed below:

1. The relation between density and rate of increase is linear.
2. The effect of density on rate of increase is instantaneous.
3. The environment (and thus K) is constant.
4. All individuals reproduce equally.
5. There is no immigration or emigration.

For many laboratory cultures of small organisms, such assumptions are easily met. Thus, early tests of these models using laboratory cultures of yeast or bacteria suggested they were valid (Gause 1934). For field populations of larger animals, assumptions are not so easily met.

1. In nature *each* individual added to the population probably does not cause an incremental decrease to r .
2. In nature there are often time lags, especially in species with complex life cycles. For example, in mammals it may be months before pregnant females give birth even when resources have been favorable for months.
3. In nature K may vary seasonally or with climate.
4. In nature often a few individuals command many matings.
5. In nature there are few barriers preventing dispersal.

Thus, it is not surprising that there are few good examples of population data fitting the logistic. For example, in 1911 reindeer were introduced onto two islands, St. Paul and St. George, in the Bering Sea off Alaska. About twenty reindeer were introduced onto each island, both of which were completely undisturbed, having no predators and no hunting pressure (Fig. 6.7) The St. George population reached a low ceiling of 222 in 1922, then subsided to a herd of about forty. The St. Paul population grew enormously, to about two thousand in 1938 but then crashed to eight animals in 1950. There appeared

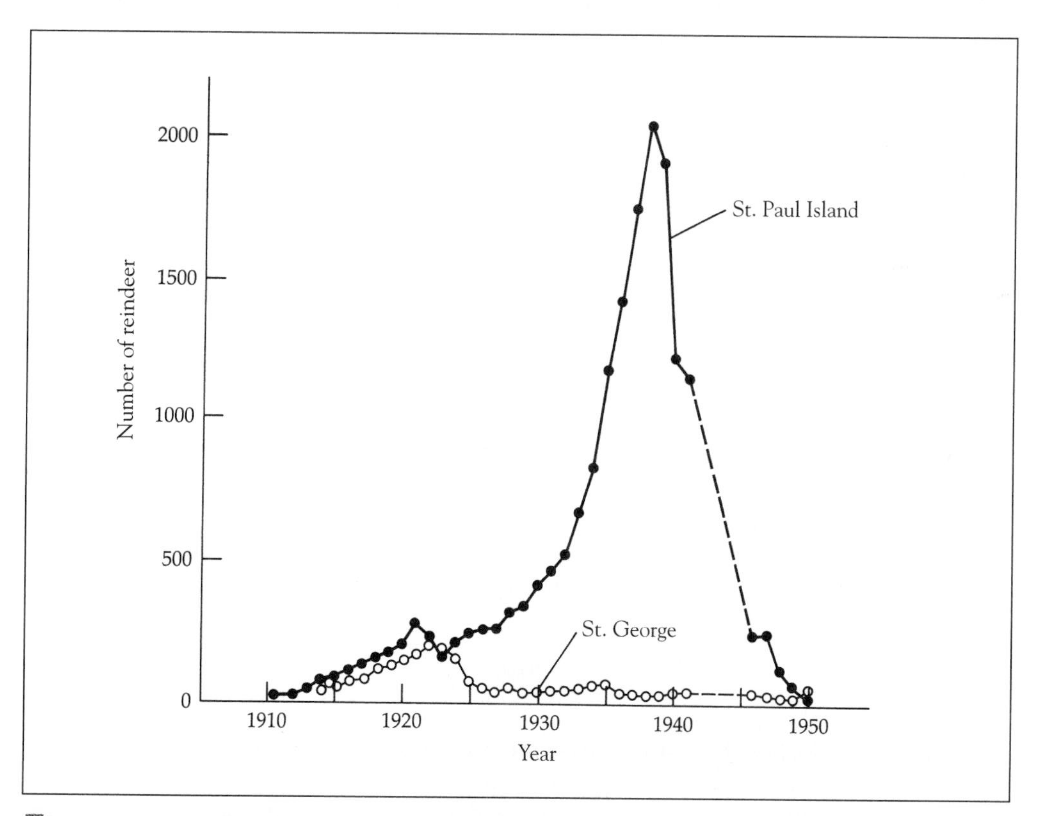

■ **FIGURE 6.7** Reindeer population growth on the Pribilof Islands, Bering Sea, from 1911, when they were introduced, until 1950. *(Redrawn from Scheffer 1951.)*

to be no ecological differences between the islands and no reason for the differences in population growth observed. Neither population had fit the pattern of logistic growth (Klein 1968). Winter food shortage, particularly of lichens, was seen as the major cause of the population crashes, but there seemed to be an overshoot of the population beyond the carrying capacity before this kicked in. This suggests a lag in response time between reproduction and food supply (see next section). Even Pearl and Reed's (1920) own prediction on human growth in the United States has not held true. On the basis of census data taken from 1790 to 1910, they projected that an asymptote of 197 million would be reached around the year 2060. The census data for 1920 to 1940 fit the curve well, but since then the population has increased geometrically rather than logistically, resulting in projections of 260 to 350 million in the year 2025. To be fair to them, immigration rates have greatly increased and now contribute the same number of new people to the U.S. population as births. At least five hundred thousand immigrants enter the country each year legally, and probably one million illegally.

Time lags in a population's response to changing resource availability can influence the rate at which a population reaches an equilibrium at carrying capacity.

If there is a time lag of length τ between the change in population size and its effect on population growth rate, then the population growth at time t is controlled by its size at some time in the past, $t - \tau$. If we incorporate the time lag into the logistic growth equation

$$\frac{dN}{dt} = rN\left(1 - \frac{N_{t-\tau}}{K}\right)$$

then population growth is affected by the length of the time lag τ.

Say, for example, in an unlagged population where

$$\frac{dN}{dt} = rN\left(1 - \frac{N_t}{K}\right)$$

then if $r = 1.1$, $k = 1000$ and $N = 900$

$$\frac{dN}{dt} = 1.1 \times 900\left(1 - \frac{900}{1000}\right) = 990 \times 0.1 = 99$$

so the new population size is $900 + 99 = 999$.

If there is a time lag such that at the time the population is nine hundred the effects of crowding are only being felt as thought the population were eight hundred, then

$$\frac{dN}{dt} = 1.1 \times 800\left(1 - \frac{800}{1000}\right) = 880 \times 0.2 = 176$$

so the new population size is 1,076.

The effect of the time lag is to increase the population growth rate over the ordinary logistic. It is also now possible for a population to overshoot its carrying capacity.

The effects of time lags on populations also depend on the response time of the population, which is inversely proportion to r. Populations with high growth rates have short response times ($1/r$). The ratio of the time lag (τ) to the response time ($1/r$), or $r\tau$,

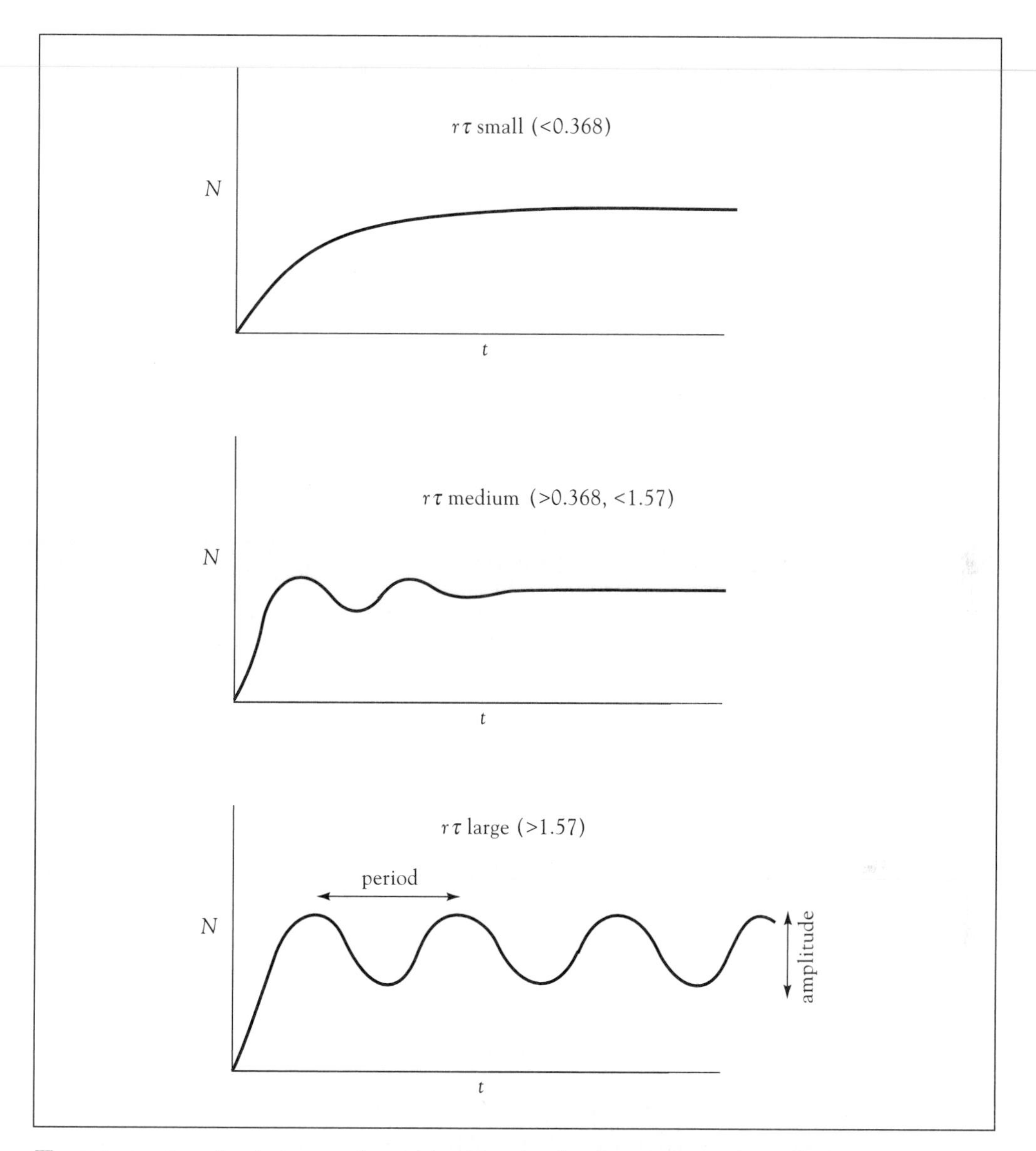

■ **FIGURE 6.8** Logistic growth curves with a time lag for species with overlapping generations. Growth depends on the value of $r\tau$, the product of the intrinsic rate of increase and the time lag. In this figure, $r\tau$ increases from top to bottom. At a certain value of $r\tau$, the population oscillations become stable (stable limit cycle) and no longer converge on the carrying capacity. *(After Gotelli 1995.)*

affects population growth. If $r\tau$ is small, < 0.368, the population increases smoothly to the carrying capacity (Fig. 6.8). If $r\tau$ is large, > 1.57, the population enters into a stable oscillation called a limit cycle, rising and falling around K but never settling on the equilibrium value. We can think of a population at just below the carrying capacity having enough food for every individual to reproduce. After reproduction, however, few of the resulting juveniles will command enough resources to breed, and the population will crash again. In this scenario, territorial species might be less likely to undergo limit cycles because territory holders will usually always command enough resources to breed. On the other hand, invertebrate species that often scramble for resources may be more likely to undergo a limit cycle. When values of $r\tau$ are intermediate, the population undergoes oscillations that dampen with time until K is reached.

In stable limit cycles the period of the cycle is always about 4τ, so a population with a time lag of 1 year (annual breeders) may expect to reach peak densities every four years. This may explain the observation of a four-year cycle in high latitude annually breeding mammals, like lemmings.

For species with discrete generations, the logistic equation becomes

$$N_{t+1} = N_t + rN_t\left(1 - \frac{N_t}{K}\right)$$

Although we know $r\tau$ controls population growth, with discrete generations the time lag is always 1.0, thus the value of r alone controls the dynamics. Again, if r is small, < 2.0, the population generally reaches r smoothly. At values of r between 2 and 2.449, the population enters a stable two-point limit cycle with sharp "peaks" and "valleys" rather than smooth ones (Fig. 6.9). We can imagine that with continuously breeding species there is more chance for these rough edges to be "smoothed" out, but with annual breeders there is less chance for this to happen. Between r values of 2.449 and 2.570, more complex limit cycles result. One illustrated in Fig. 6.9 has two distinct peaks and valleys before it starts to repeat. At values of r larger than 2.57 the limit cycles break down and the population grows in complex, nonrepeating patterns often known as "chaos." It is important to note that chaos doesn't mean random, although random population growth and chaos certainly look similar. The important point is that with chaos, the same chaotic population growth patterns would repeat every time the model is run, as long as the specified values of N_t, r, and K are the same. With a "messy" data set it is sometimes hard to know if a population is behaving in a chaotic manner or random or stochastic changes are driving the dynamics. The effects of these random or stochastic changes are explained a little more in the next section.

6.3 Stochastic models of population growth incorporate the effects of genetic variability and extrinsic factors like climate on population dynamics.

It may be too simplistic to expect organisms in real life to behave exactly like numbers in an equation. Part of the reason is the genetic variability between individuals that causes some females to produce more offspring, on average, than others, or some animals to be more resistant to climatic stress, predator pressure, and other factors. Part of the reason is environmental variation itself. Given this variability, termed *stochasticity*, can we ever hope to predict population processes accurately or to incorporate such variability into a model?

Stochastic models of population growth are based largely on probability theory. Rather than exactly two offspring, one might assume that each female has a 0.5 probability of giving birth to two offspring, a 0.25 chance of producing three progeny, and a

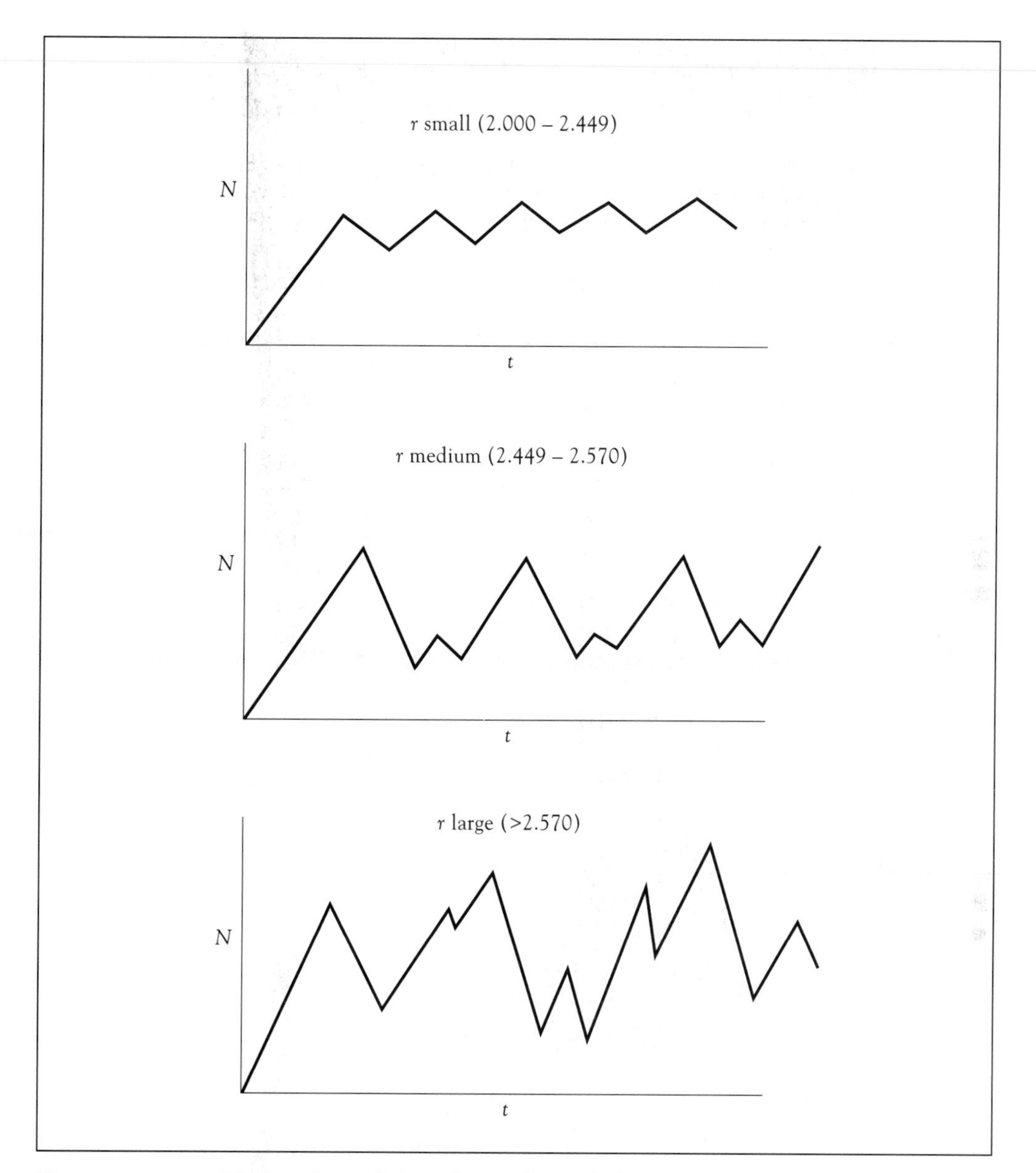

■ **FIGURE 6.9** Limit cycles and chaos for species with discrete generations. The discrete logistic growth curve is effected by the size of r. Values of r increase from top to bottom in this figure, and the number of points in the limit cycle also increases. In the end, a chaotic pattern results that is difficult to distinguish from randomness. (*After Gotelli 1995.*)

0.25 chance of producing one. For example, reproductive rates might vary according to environmental conditions or parental genotypes.

For the geometric **deterministic model** discussed previously, if $R_0 = 2$ and $n = 5$, then

$$N_{t+1} = R_0 N_t$$
$$= 2 \times 5 = 10$$

For a stochastic model, a coin can be flipped to mimic the probability of the outcome, where tails/heads or heads/tails = two offspring, two tails = one offspring, and two heads = three offspring. Here are the results I got one night (for the second edition of this book):

	OUTCOME OF TRIAL			
PARENT	**1**	**2**	**3**	**4**
1	2	3	3	2
2	3	1	1	1
3	3	1	2	2
4	1	1	3	3
5	3	1	1	1
Total population in next generation	12	7	10	9

Some of the outcomes are above the expected value of 10, and some are below. If this technique is continued, a frequency histogram can be constructed (Fig. 6.10).

Stochastic models can also be developed for geometric growth. Again, such a model is best explained by referring to the corresponding geometric equation, where

$$dN/dt = rN = (b - d)N$$

and if $b = 0.5$, $d = 0$ (often true in new populations), and $N_0 = 5$, then using calculus, the integral form of this equation is

$$N_t = N_0 e^{rt} = 8.244$$

so for one generation $N_t = N_0 e^r = 8.244$.

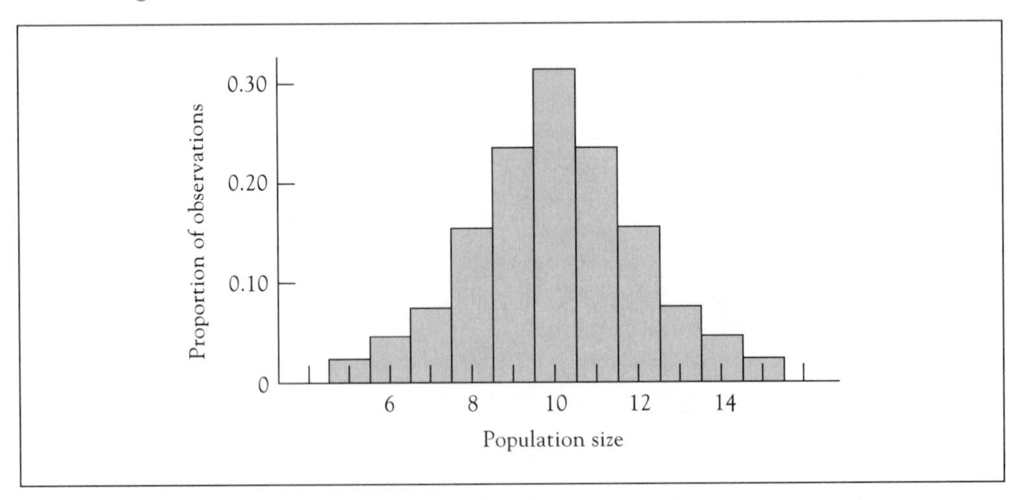

■ **FIGURE 6.10** Stochastic frequency distribution for size of a female population after one generation; beginning with five females. In this case, probability of having two female offspring = 0.5, probability of having three female offspring = 0.25, and probability of having one female offspring = 0.25.

With a stochastic model, the probability of one individual not reproducing in one time interval must be calculated as $e^{-b} = 0.6065$, and likewise the probability of one individual reproducing once in one time interval is

$$1 - e^{-b} = 0.3935$$

Then for $N = 5$, the chance that no individuals will reproduce is $0.6065^5 = 0.082$. Similar calculations for other combinations eventually produce a frequency histogram, from which a population growth curve can be constructed (Fig. 6.11). With stochastic model population growth could vary dramatically, although there is still a most likely trajectory for it to follow.

If death occurs in the population, there will be a chance that the population will become extinct. If birth rate is greater than death rate, then

$$\text{probability of extinction} = \left(\frac{d}{b}\right)^{N_0} \text{ as time} \to \infty$$

Thus, if $b = 0.75$, $d = 0.25$, and $N_0 = 5$, then

$$\left(\frac{d}{b}\right)^{N_0} = 0.0041$$

but if $b = 0.55$, $d = 0.45$, and $N_0 = 5$, then probability of extinction $= 0.367$.

The larger the initial population size and the greater the value of $b - d$, the more resistance to extinction population becomes. In reality $b - d$ is often zero, so $(d/b) = 1.0$ as time $\to \infty$. In other words, extinction is a certainty for a population given a long enough

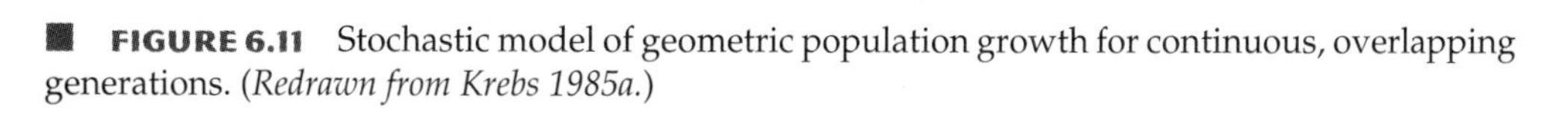

■ **FIGURE 6.11** Stochastic model of geometric population growth for continuous, overlapping generations. (*Redrawn from Krebs 1985a.*)

time span and is likely to occur more quickly for a small population. Fischer, Simon, and Vincent (1969) believed that probably 25 percent of the species of birds and mammals that have become extinct since 1600 may have died out naturally. Such stochastic effects are particularly important when the conservation of small populations of rare species is considered. For example, Schaffer and Samson (1985) have predicted that if N_e (effective population size) = 50 for grizzly bears, demographic stochasticity alone would cause extinction on average once every 114 years. A model of the spotted owl *(Strix occidentalis caurina)* suggests that demographic stochasticity is more likely than genetic factors to extinguish local subpopulations over the short term of decades (Simberloff 1986b).

Stochastic models introduce biological variation into population growth and are much more likely to represent what is happening in the field. The price paid is complicated mathematics as many new factors must be incorporated, such as the probability that a predator will kill a certain number of individuals or that there will be enough food available. Carrying capacity, K, may vary seasonally or randomly. For example, the number of insect prey available to birds may vary from spring to summer and may also be depressed by unusual cold snaps or wet weather. In general, the more variable the environment, the lower the population size. Stochastic models become more important as population sizes get smaller. If all populations were in the millions, one could throw away stochastic models—deterministic ones would do. These days, of course, populations of many mammals, except humans, tend toward the hundreds of thousands rather than the millions.

Although some readers may have by now despaired of ever producing a robust model of population growth that incorporates all the necessary factors, some closing generalizations can be made about models of population growth:

1. Models can provide simplified ideas about how nature operates. Often a simple model can accurately predict how a population will change in nature.
2. Models are developed from basic principles about what should be happening in nature.
3. Models can help us figure out what things we should be measuring in nature and perhaps where we need to go.
4. There are two inherent dangers in models. One is they lose their utility if they are made complex or specific. The second is that they are difficult to test if they are too complex.

Some generalities that have been made about population growth are as follows:

1. There is a strong correlation between size and generation time in organisms ranging from bacteria to whales and redwoods (Bonner 1965),
2. Organisms with long generation times have lower per-capita rates of population growth (Heron 1972). Therefore,
3. Larger animals have lower rates of increase, r (Fig. 6.12). For any given size, warm-blooded animals, endotherms, have a higher rate of increase than ectotherms, which in turn are more fecund than unicellulars.

Such generalizations are particularly important to conservation efforts because they underline how long it takes for populations of large animals to rebound after ecological disasters or for large trees to reappear after forest clear-cutting. They may also point to reasons why larger species are more prone to extinction than are smaller ones.

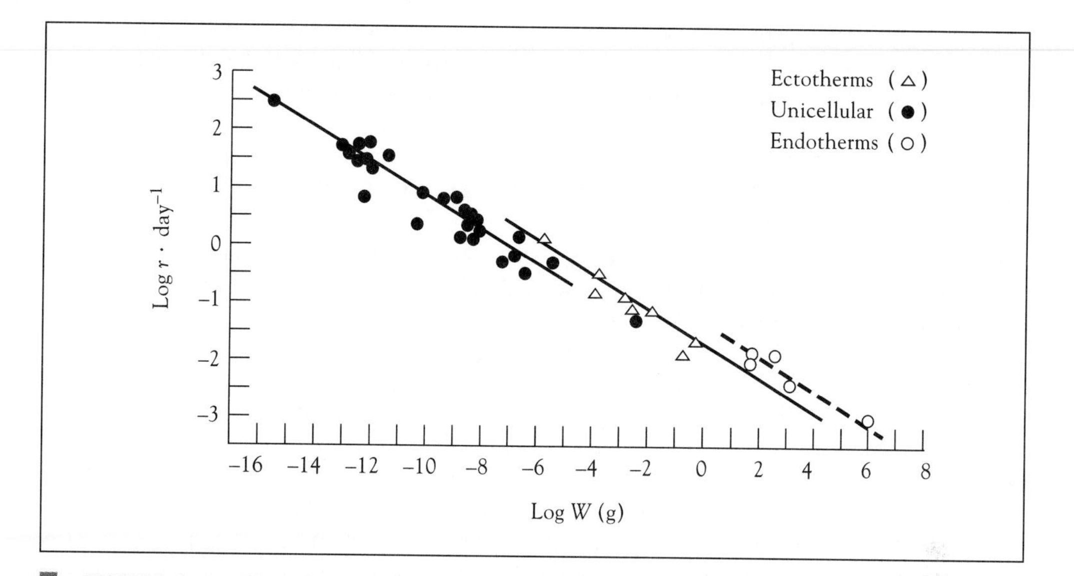

FIGURE 6.12 The relationships of the intrinsic rate of natural increase to weight for various animals. (*Redrawn from Fenchel 1974.*)

SUMMARY

1. For longer-lived organisms, time-specific life tables provide a snapshot of a population's age structure at a given time. Such tables often contain information on the number of organisms dying at any given age interval, the proportion surviving, and the mean expectation of life for organisms at any age interval. Data in most any column can be calculated from any other column.

2. For organisms with short life spans, usually completed within a year, age-specific life tables, which follow one cohort or one generation, are used.

3. Three types of survivorship curve can be recognized: Type III, where most individuals are lost as juveniles (invertebrates, fish, frogs, and plants); Type II, in which there is an almost linear rate of loss (some plants and birds); and Type I, where most individuals die after they have reached sexual maturity (mammals).

4. Population growth may be discussed under the banners of deterministic models or stochastic models. In deterministic models, rates of population increase are given precisely determined values. In stochastic models, variation is allowed around this value—as would be the case in nature. For small populations, stochastic events are important, and stochastic models are therefore often used in conservation theory. Stochastic models are more difficult to develop than deterministic ones. For large populations, deterministic models give good approximations of population growth.

5. In both deterministic and stochastic population growth, where environments are favorable populations may undergo geometric or exponential growth, which results in J-shaped population-growth curves.

6. In the real world, there are often limits set on population growth. Such limits might be the exhaustion of space or resources or limitation by waste production. As a result, most population-growth curves are logistic and result in an S-shaped growth curve, where the upper level or carrying capacity, K, represents the total number of individuals or the total biomass that can exist in an area.

7. Population growth curves are not always smooth in trajectory. They may oscillate about the carrying capacity or exhibit what appears to be random fluctuations. Such fluctuations may actually be the result of time lags and may result in chaotic growth curves—curves that look as if they are generated by random fluctuations but are actually the result of precise time lags and rates of reproduction.

DISCUSSION QUESTION

Human populations appear to be growing geometrically. Do you think there will be a limit to human population growth? If so, what will it be, and will it be set by space, food, water, or pollutants? What would happen to human population growth if the carrying capacity, K, and the net reproductive rate, r, are not constant but vary with time?

C H A P T E R

Abiotic Factors

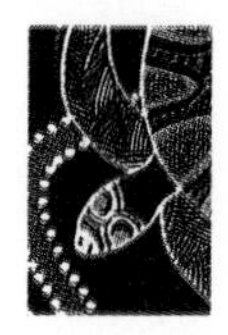

The local distribution patterns of many species are limited by certain physical or abiotic factors of the environment such as temperature, moisture, light, pH, soil quality, salinity, water current, and so on. Leibig's law of the minimum, coined by Justus Leibig in 1840, states that the distribution of species will be controlled by that environmental factor for which the organism has the narrowest range of adaptability or control. For example, forest trees differ in their tolerance to shade. Sun-loving species soon overtop the shade-loving species because of their faster rate of growth in full sun (see Fig. 7.1). However, they cast shade that is too dark for their own seedlings and that inhibits their growth. Light levels below their crowns are still sufficient for the growth of those species that do better in intermediate light levels and in turn for species that are the most tolerant of shade. Thus, a stratified canopy may develop

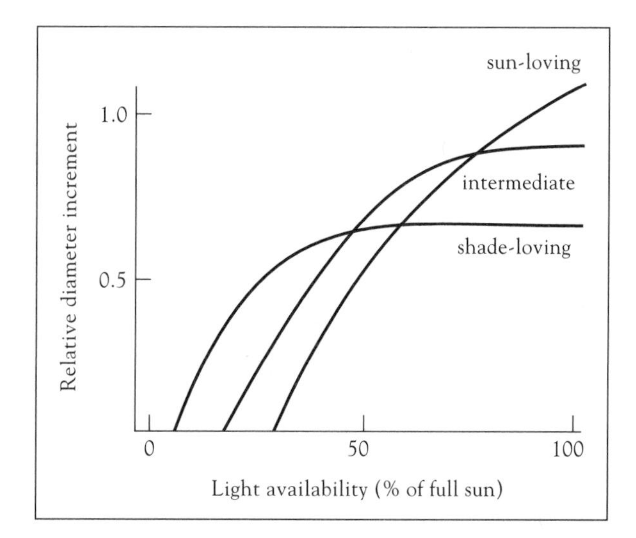

■ **FIGURE 7.1** Relative rates of tree growth as a function of light availability for sun-loving, intermediate, and shade-loving species.

TABLE 7.1 Separation of some deciduous tree species in the northeastern United States by tolerance classes for shade and low nitrogen availability.

		TOLERANCE CLASS FOR SHADE		
		TOLERANT	INTERMEDIATE	INTOLERANT
Tolerance class for low nitrogen availability	*Tolerant*	Closed canopies—poor sites Hickory	Gaps—poor sites White oak Chestnut oak	Open—poor sites Bigtooth aspen
	Intermediate	Closed canopies—moderate sites Beech Red maple	Gaps—moderate sites Basswood	Open—moderate sites Trembling aspen
	Intolerant	Closed canopies—rich sites Sugar maple Black gum	Gaps—rich sites White ash Northern red oak	Open—rich sites Tulip poplar

with sun-loving species at the highest level, intermediate species below them, and shade-loving species lower still. Often rather than just one limiting factor there are many factors, all interacting. Plant growth rates differ not only because of shade but also for other resources such as the availability of water and nutrients (nitrogen, phosphorus, potassium, calcium, magnesium, and sulfur). A lack of nitrogen is by far the most critical factor limiting the growth of certain species. Species' responses to nitrogen availability can be described in much the same way as their response to light. However, species that are tolerant of low light levels are not necessarily also tolerant of low nitrogen availability. Table 7.1 lists some species of eastern deciduous tree forests and their requirements with respect to tolerance to shade and nitrogen. The data in the table indicate where each species might be common within a large forested area containing young, disturbed and old, undisturbed stands. For example, white ash is intermediate for light levels and intolerant for lack of nitrogen, and it tends to occur in small disturbances on sites with rich soil. It may now be apparent why there are so many species of tree in a given forest region. Nine distinct types of environment can be delineated by their light and nitrogen availability alone.

Some species are tolerant of a wide range of environmental conditions, others of only a narrow range, but each functions best only over a limited part of the gradient, and this is termed a species' optimal range (Fig. 7.2). It must also be remembered that part of a preferred optimal range may already be occupied by a competitively superior species. In the field, species may not occupy their full ranges, as measured in the laboratory in terms of abiotic factors, because of competition with other organisms (Fig. 7.3) (see also Chap. 9).

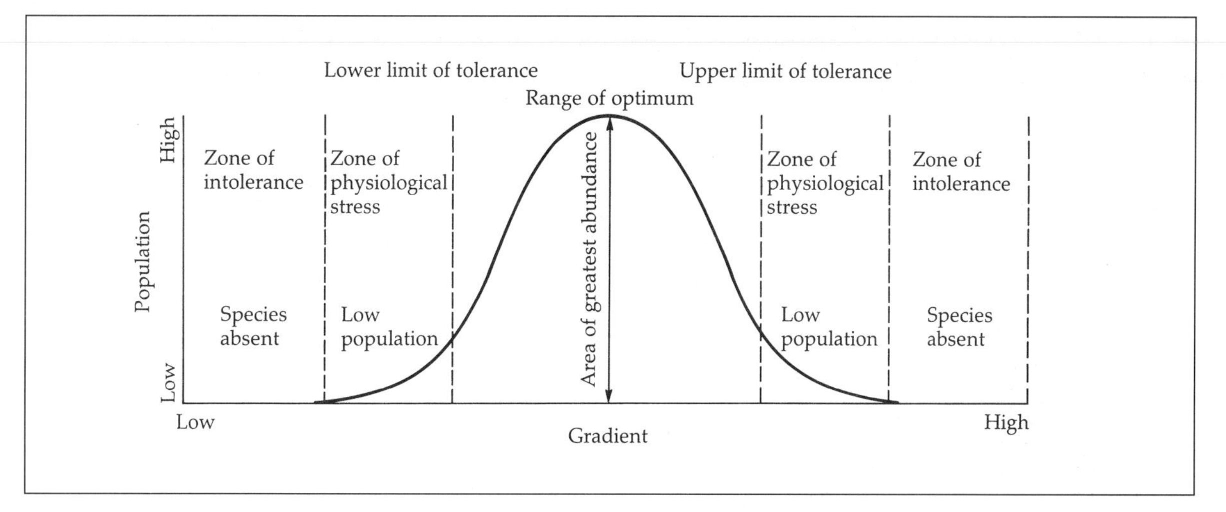

FIGURE 7.2 Organismal distribution along a physical gradient. *(Modified from Cox, Healey, and Moore 1976.)*

This whole line of thinking is important in evolutionary ecology for it suggests there are trade-offs in life history strategies. A species adapted to an arid environment could not also do well in flooded habitats. A species that does well in full sunlight would not do well in heavy shade. There are two broad classes of abiotic variables to consider, those that directly affect organisms, temperature, wind, salinity, and pH, and those that act as resources, nutrients, soils, light, and moisture.

7.1 The abiotic factors that are most commonly found to limit the distribution and productivity of terrestrial organisms are temperature, moisture, and nutrient availability. These are all affected by climate.

There are substantial temperature differentials over the Earth, a large proportion of which are due to variation in the incoming solar radiation. In higher latitudes, the sun's rays hit the Earth obliquely and are thus spread out over more of the Earth's surface than they are in the equatorial regions (Fig. 7.4). More radiation is also dispersed in the higher latitudes because the sun's rays travel a greater distance through the atmosphere. The result is a much smaller (40 percent) total annual insolation in polar latitudes than in equatorial areas (Fig. 7.5). In the summer, increased day length in high latitudes increases insolation, but shorter day length in winter decreases the daily total. The reason is that the Earth's axis of rotation is inclined at an angle of 23.5° (Fig. 7.6); the Northern Hemisphere is treated to long summer days while the Southern Hemisphere has winter and vice versa. At the summer solstice in the Northern Hemisphere (June 22), light falls perpendicularly on the Tropic of Cancer; on December 22 it shines perpendicularly on the Tropic of Capricorn. On March 22 and September 22 (the equinoxes), the sun's rays fall perpendicularly on the equator, and every place on Earth receives roughly

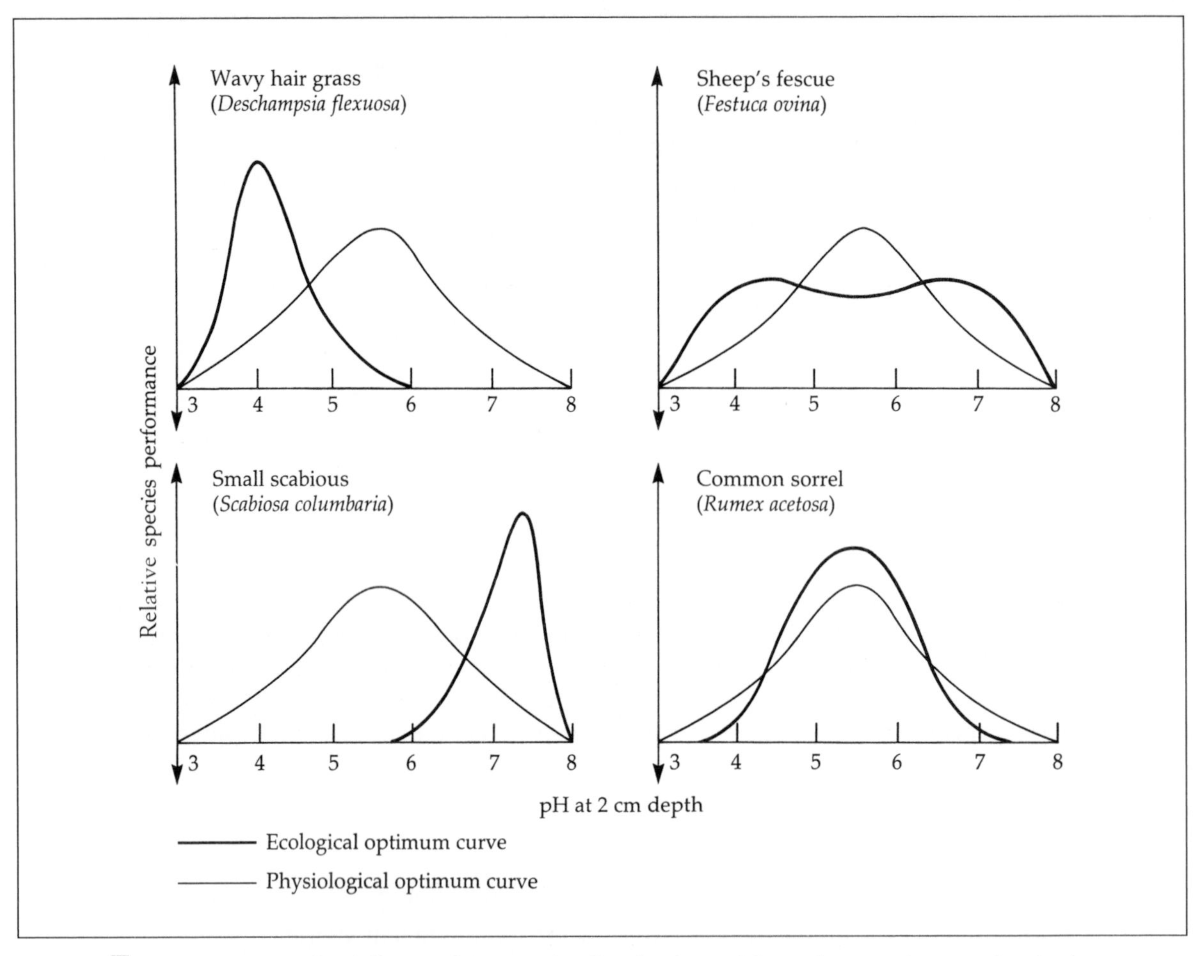

■ **FIGURE 7.3** The difference between the distributions of four plant species growing in the field (ecological optimum curve) and under noncompetitive conditions in controlled laboratory plots (physiological optimum curve). *(Redrawn from Collinson 1977.)*

the same day length. These effects do not translate into a linear relationship between temperature at the surface and latitude—at the tropics both cloudiness and rain reduce mean temperature, and relatively cloud-free areas beyond this zone increase mean temperature relative to isolation (Fig. 7.7).

Global temperature differentials create winds and drive atmospheric circulation. The first contribution to a classical model of general atmospheric circulation was made by George Hadley in 1735. Hadley proposed that solar energy drove winds that in turn influenced the circulation of the atmosphere. He proposed that the large temperature contrast between the very cold poles and the hot equator would create a thermal circulation. The warmth at the equator caused the surface equatorial air to become buoyant and rise vertically into the atmosphere. As it rose away from its source of heat, it cooled and became less buoyant but was unable to sink back to the surface because of the warm air behind

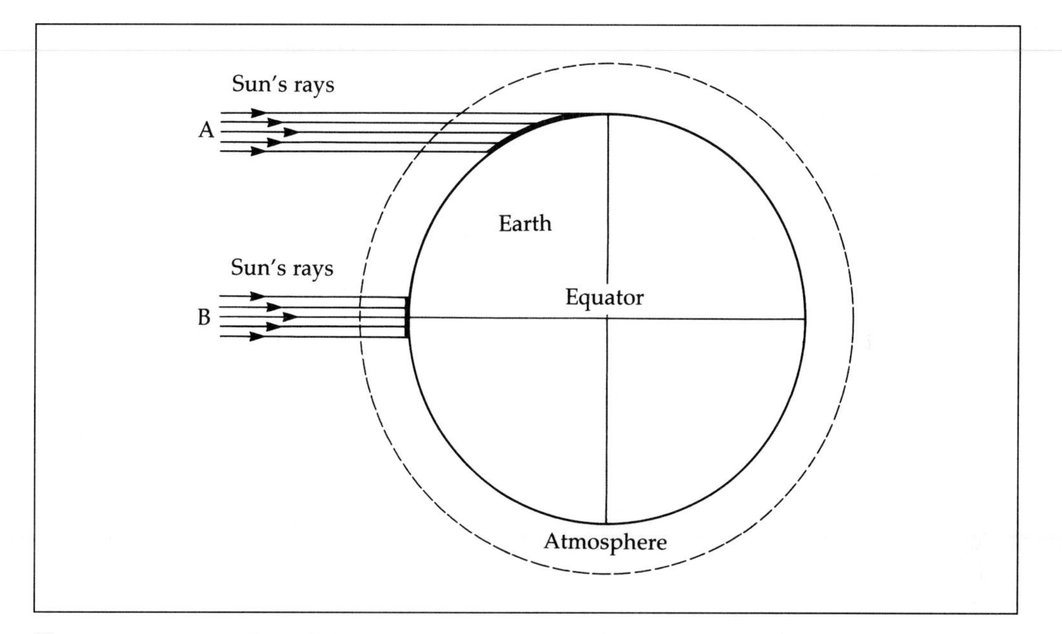

■ **FIGURE 7.4** Effect of the Earth's shape and atmosphere on incoming radiation. In polar areas the sun's rays strike the Earth in an oblique manner (A) and deliver less energy than at tropical locations (B) for two reasons: (1) because the energy is spread over a larger surface in A and (2) because it passes through a thicker layer of absorbing, scattering, and reflecting atmosphere.

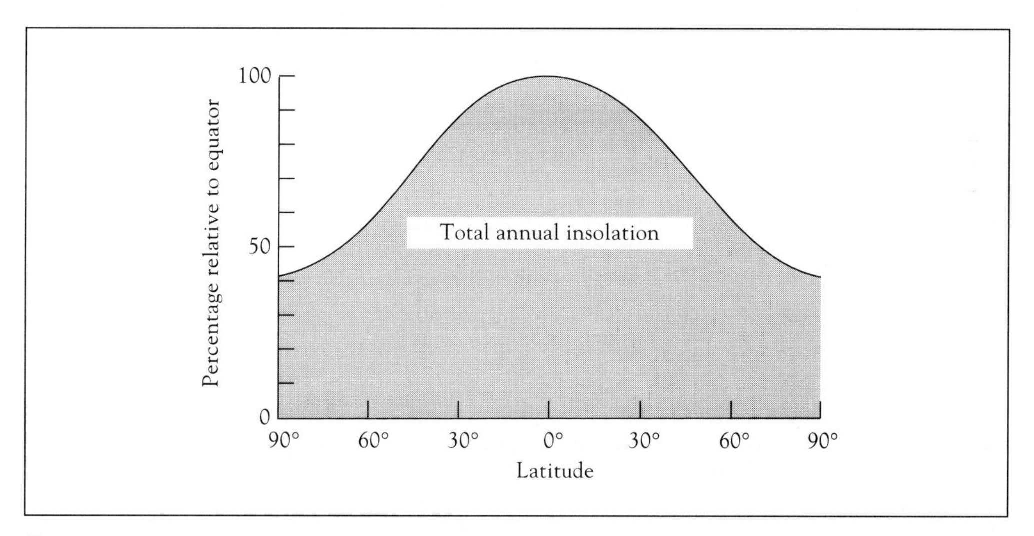

■ **FIGURE 7.5** Insolation at different latitudes during the year. The amount of solar energy is expressed as a percentage of the amount at the equator.

it. Instead it spread north and south away from the equator, eventually returning to the surface at the poles. From there it flowed back toward the equator to close the circulation loop. Hadley suggested that on a nonrotating earth this air movement would take the form of one large convection cell in each hemisphere, as shown in Fig. 7.8.

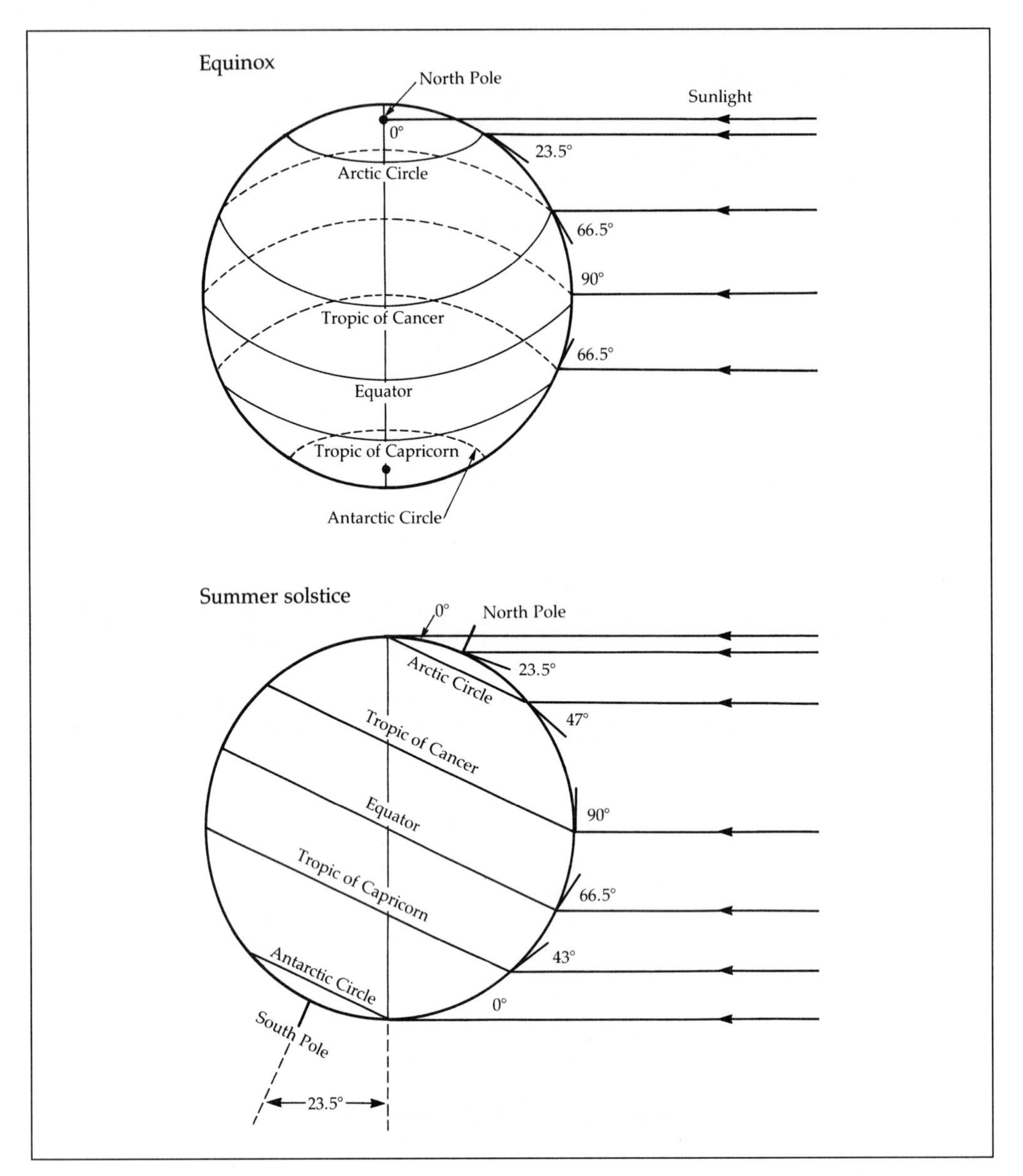

■ **FIGURE 7.6** Effects of the Earth's inclined axis of rotation on amount of insolation. The Earth's axis of rotation is inclined at an angle of 23.5 degrees, which causes increasing seasonal variation in temperature and day length with increasing latitude.

When the effect of the earth's rotation is added, the surface flow becomes somewhat easterly (toward the west). This is a consequence of the so-called Coriolis effect. In any one revolution of the Earth, any point on the equator circumscribes a greater circle than any point north or south. Thus, the equator of the Earth is moving fastest during rotation. Imagine a rocket fired from the north pole toward the equator. In the time it took the rocket to reach the equator, hours later, that point on the equator may have moved

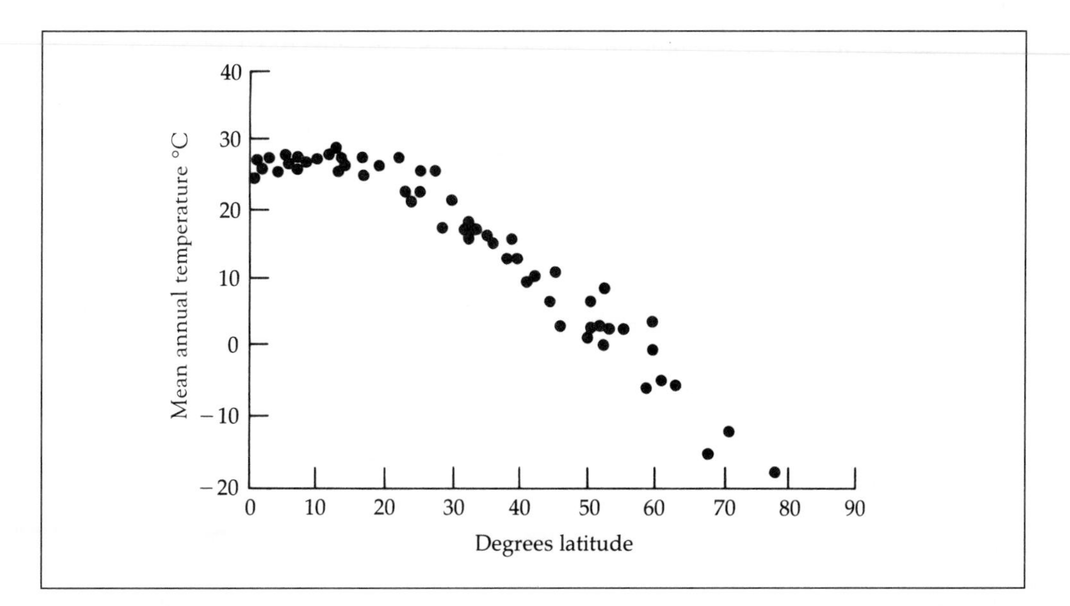

■ **FIGURE 7.7** Mean annual temperature (degrees C) of low-elevation, mesic, continental locations on latitudinal gradient. Note the wide band of similar temperatures at the tropics. *(Redrawn from Terborgh 1973.)*

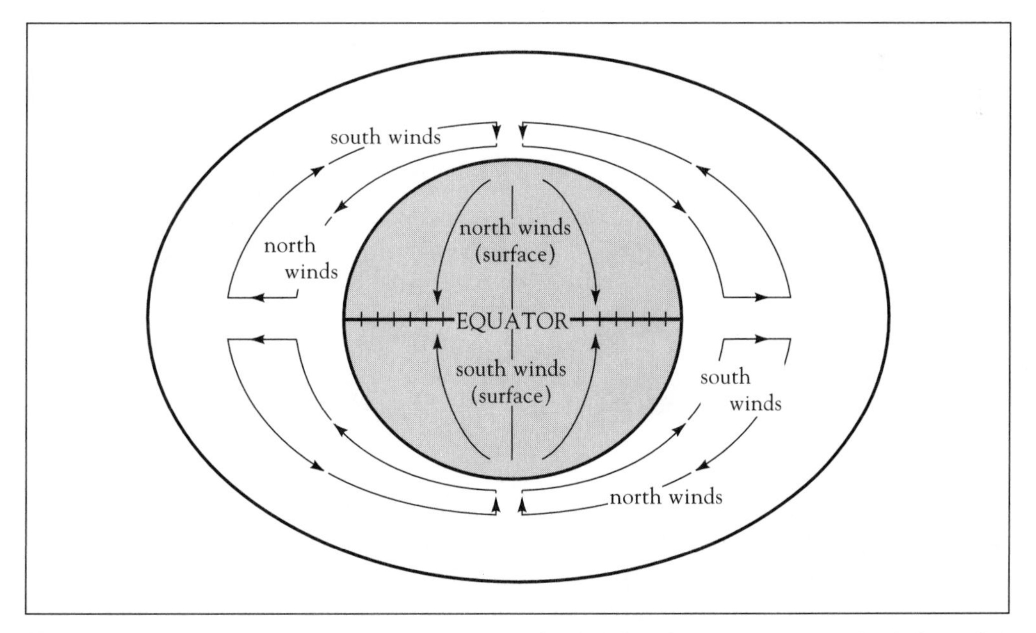

■ **FIGURE 7.8** Simple convective circulation of air on a uniform, nonrotating Earth, heated at the equator and cooled at the poles, according to the scientist George Hadley in 1735.

15° to the east. Thus, the straight line of the rocket's path effectively becomes curved toward the west. A similar phenomenon occurs with winds.

Attractive though this simple theory is, we have to modify it to fit the data. In the 1920s a three-cell circulation in each hemisphere was proposed to fit the earth's heat balance (Fig. 7.9).

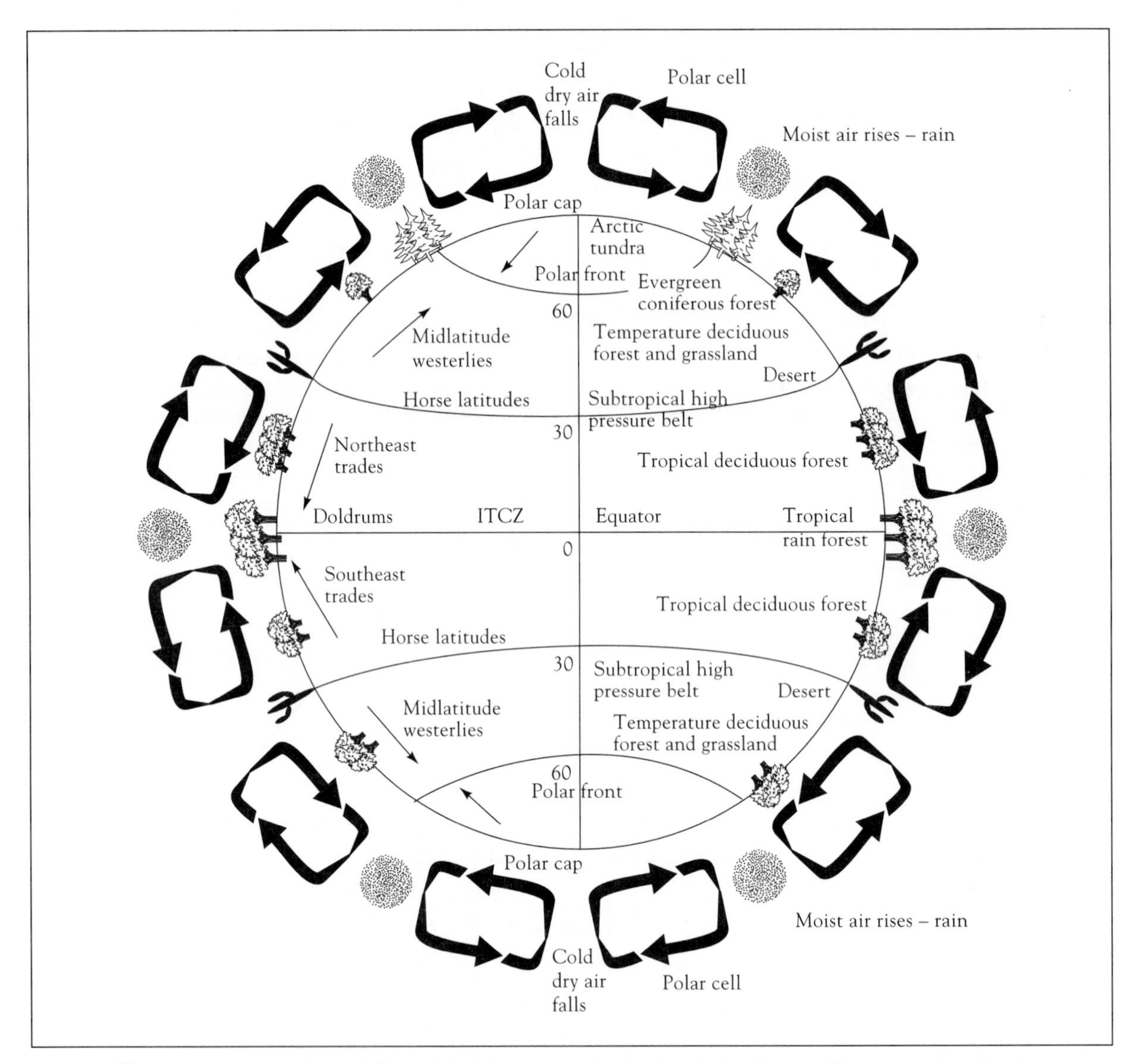

■ **FIGURE 7.9** Three-cell model of the atmospheric circulation on a uniform, rotating Earth heated at the equator and cooled at the poles. The direction of air flow and the ascent and descent of air masses in six giant convection cells determine Earth's general climatic zones. This uneven distribution of heat and moisture over different parts of the planet's surface leads to the forests, grasslands, and deserts that make up the planet's biomass.

The contribution of George Hadley is still recognized in that the most prominent of these three cells, the one nearest the equator, is called the Hadley cell. In the Hadley cell, the warm air rising near the equator forms towers of cumulus clouds that provide rainfall, which in turn maintains the lush vegetation of the equatorial rainforests (Fig. 7.10). As the upper flow in this cell moves poleward, it begins to subside in a zone between 20° and 35° latitude. Subsidence zones are areas of high pressure and are the sites of the world's tropical deserts. This is because the subsiding air is relatively dry, having released all its moisture over the equator. In the subsidence zone, winds are generally weak and variable near the center of this zone of descending air. The region has popularly been called the horse latitudes. The name is said to have been coined by Spanish sailors, who, crossing the Atlantic, were sometimes becalmed in these waters and reportedly were forced to throw horses overboard as they could no longer water or feed them.

From the center of the horse latitudes, the surface flow splits into a pole branch and an equatorial branch. The equatorial flow is deflected by the Coriolis force and forms the reliable trade winds. In the Northern Hemisphere the trades are from the northeast, where they provided the sail power to explore the New World; in the Southern Hemisphere the trades are from the southeast. The trade winds from both hemispheres meet near the equator in a region that has a weak pressure gradient, the intertropical convergence zone. This region is also called the Doldrums. Here the light winds and humid conditions provide the monotonous weather that may be the basis for the expression "down in the doldrums."

In the three-cell model the circulation between 30° and 60° latitude is just opposite that of the Hadley cell. Net surface flow is poleward and because of the Coriolis effect the winds have a strong westerly component. These prevailing westerlies were known

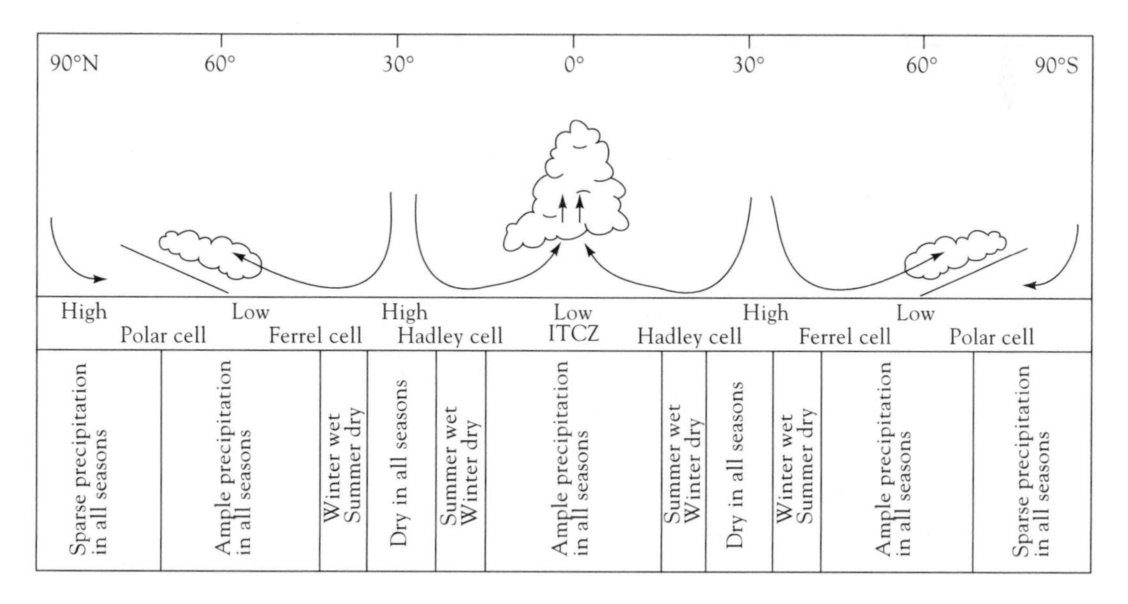

■ **FIGURE 7.10** Schematic illustration of zonal precipitation patterns. High and low refer to prevailing atmospheric pressures.

to Benjamin Franklin, perhaps the first American weather forecaster, who noted that storms migrated eastward across the colonies.

Winds, together with the rotation of the Earth, create currents. The major currents act as "pinwheels" between continents, running clockwise in the ocean basins of the Northern Hemisphere and counterclockwise in those of the Southern Hemisphere. Thus, the Gulf Stream, equivalent in flow to fifty times all the world's major rivers combined, brings warm water from the Caribbean and the U.S. coasts to Europe, the climate of which is correspondingly moderated. The Humboldt current brings cool conditions almost to the equator along the western coast of South America (Fig. 7.11). The climates of coastal regions may differ markedly from those of their climatic zones; many never experience frost, and fog is often evident.

Superimposed on the broad geographical temperature trends are the influences of altitude and land mass. There is a drop of 1°C for every 100 m increase in altitude in dry air as a result of the "adiabatic" expansion of air as atmospheric pressure falls with the increase in altitude. The sinking of dense cold air into the bottom of a valley at night can make it as much as 3°C colder than the side of the valley only 100 m higher. The effects of land mass are attributable to different rates of heating and cooling of the land and the sea. The land surface reflects less heat than the water, so the surface warms more quickly

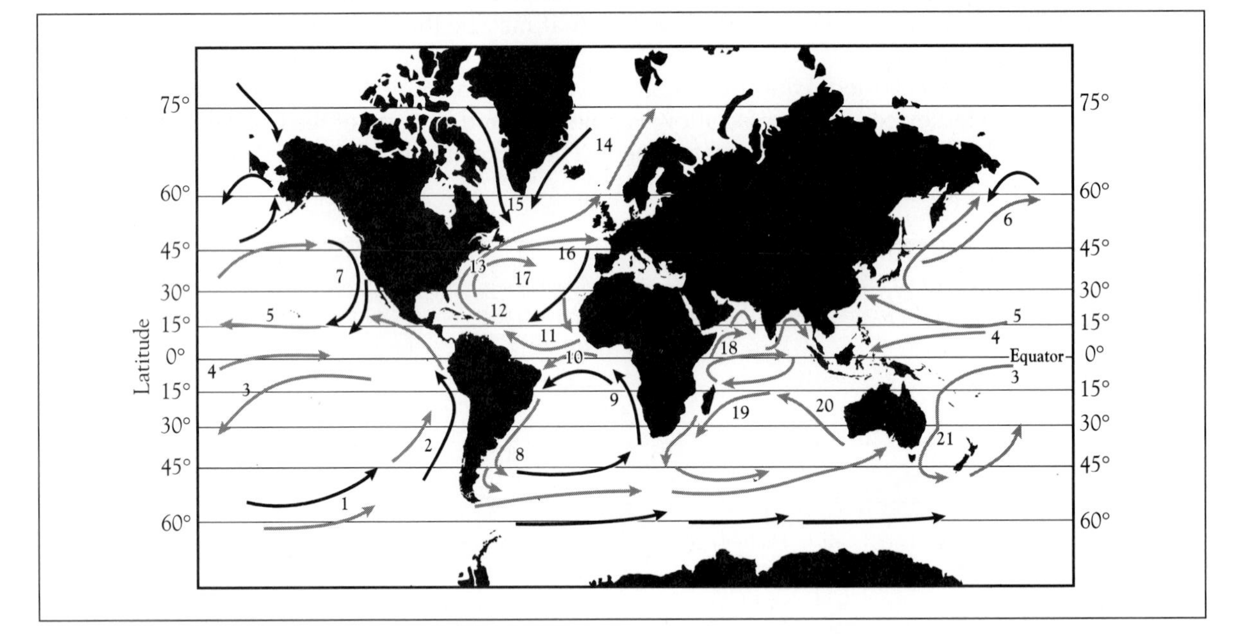

■ **FIGURE 7.11** Ocean currents of the world: (1) Antarctic West Wind Drift; (2) Peru Current (Humboldt); (3) South Equatorial Current; (4) Counter-Equatorial Current; (5) North Equatorial Current; (6) Kuroshio Current; (7) California Current; (8) Brazil Current; (9) Benguela Current; (10) South Equatorial Current; (11) Guinea Current; (12) North Equatorial Current; (13) Gulf Stream; (14) Norwegian Current; (15) North Atlantic Current; (16) Canaries Drift; (17) Sargasso Sea; (18) Monsoon Drift (summer east; winter west); (19) Mozambique Current; (20) West Australian Current; (21) East Australian Current. Dashed arrows represent cold water.

and loses heat more quickly. The sea therefore has a moderating "maritime" effect on the temperatures of coastal regions and especially islands.

7.2 Low temperatures impose significant ecological constraints by limiting the amount of time in which physiological activity is possible.

Environmental temperature is an important factor in the distribution of organisms because of its effect on biological processes and the inability of most organisms to regulate body temperature precisely. In plants, temperature is particularly important because cells may rupture if the water they contain freezes. High temperature is critical because the proteins of most organisms denature, that is, are destroyed at temperatures above 45°C. In addition, few organisms can maintain a sufficiently high metabolic rate at very low or very high temperatures. Some organisms, of course, have extraordinary adaptations to enable them to live outside this temperature range. Tardigrades can freeze in ice yet miraculously return to life at the next spring thaw. Penguins can survive the coldest regions. As so-called warm-blooded organisms or endotherms that generate heat within their own bodies, mammals and birds can cope with temperature extremes better than many organisms because they can regulate their own temperature. But even mammals and birds function best within certain environmental temperature ranges. Ecotherms, by contrast, rely on external sources of heat and include plants and other animals.

Temperature resistance in plants, though poorly understood, is often critical to their distribution patterns. Many tropical plants are sensitive to "chilling" and are damaged by cooler temperatures, even above freezing. The exact mechanism is unknown but may be related to the breakdown of membrane permeability and the leakage of ions such as calcium (Minorsky 1985). Under colder conditions, water must be bound up in such a chemical form that it cannot change to damaging ice. Usually the water within the cells does not freeze, but the extracellular water does. This draws water out from the cell, damaging its osmoregulatory machinery, in much the same way as drought or salinity. Injury by frost is probably the single most important factor limiting plant distribution. As an example, the saguaro cactus can easily withstand frost for one night as long as it thaws in the day, but it will be killed when temperatures remain below freezing for thirty-six hours. In Arizona the limit of the cactus's distribution corresponds to a line joining places where, on occasional days, it fails to thaw (Fig. 7.12). For some plants, general coldness, not freezing, limits distribution. The northern boundary of the wild madder, *Rubia peregrina*, in Europe coincides with the January 4.5°C isotherm (Fig. 7.13), and it has been suggested that this temperature is critical for the early growth phases of new shoots.

Frost injury has caused losses to agriculture of over $1 billion annually in the United States and has been considered an unavoidable result of subfreezing temperatures, but genetic engineering could slightly modify the trend. Frost injury is precipitated by the ice-nucleation activity of just five species of bacteria that live on plant surfaces. Recently, the DNA sequences conferring ice nucleation have been identified, isolated, and prevented from working in an engineered strain of one of them, *Pseudomonas syringae* (Lin-

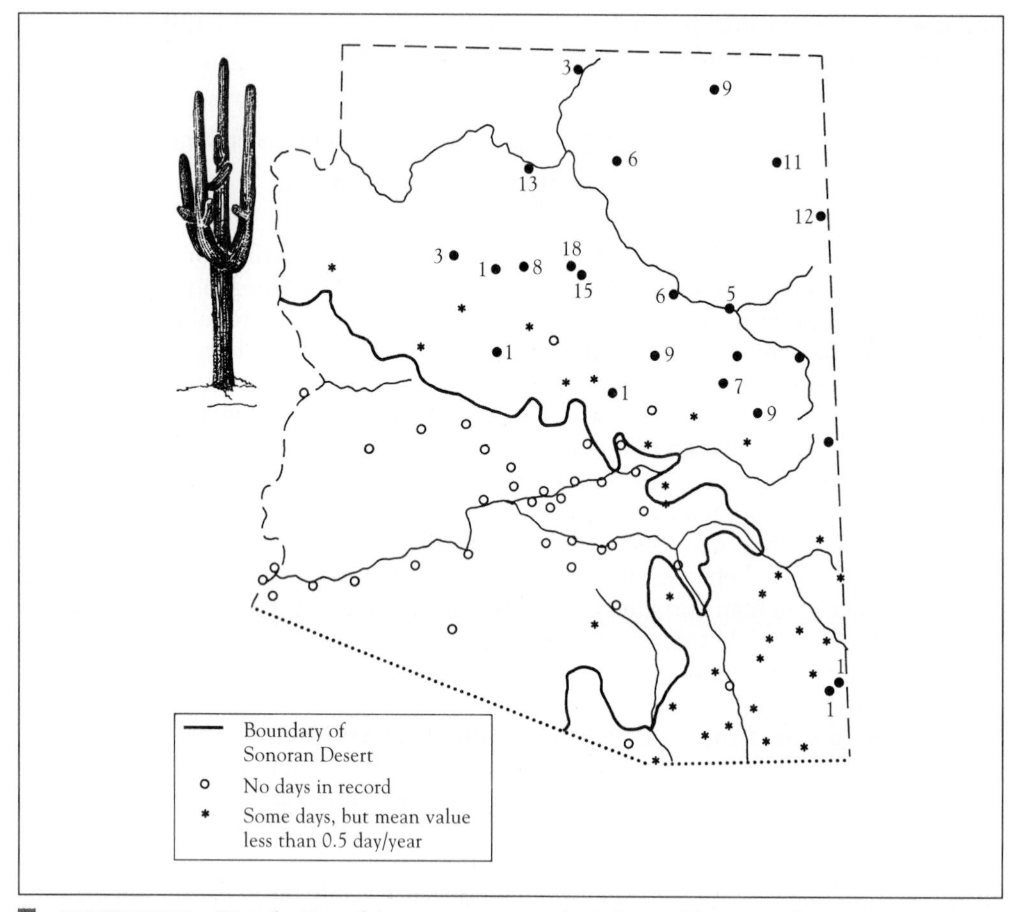

■ **FIGURE 7.12** Distribution of the saguaro cactus in Arizona. There is a close correspondence between the northern and eastern edges of the cactus's range in the Sonoran Desert and the line beyond which it occasionally fails to thaw during the day. The numbers are mean numbers of days per year with no rise above freezing. *(Modified from Hastings and Turner 1965.)*

dow 1985). When such a strain is allowed to colonize plants, frost damage is greatly reduced, and plants can withstand approximately 5°C cooler temperatures before frost forms. The promise of this technique for the increase of agricultural yields and the alteration of normal plant distribution patterns is staggering.

The range limits of even warm-blooded animals may correlate with temperature too. For example, the Eastern phoebe has a northern winter range limit that coincides well with the −4°C isotherm of average minimum January temperatures (Fig. 7.14, Root 1988). Such limits are probably associated with the energetic demands associated with cold temperatures. Colder temperatures mean higher metabolic costs that are in turn dependent on high feeding rates. The distribution of the vampire bat (*Desmodus rotundus*) (Photo 7.1) is from central Mexico to northern Argentina (Fig. 7.15). Both the northern and southern limit of its range parallels the 10°C minimal isotherm of January because of the bat's poor capacity for thermal regulation below 10°C (McNab 1973). There is a tendency for endotherms to be bigger in colder areas (Bergmann's Rule) and

■ **FIGURE 7.13** The distribution of wild madder (*Rubia peregrina*) in Europe (shaded) and the location of the January isotherm for 4.5 degrees C. (*Modified from Cox, Healey, and Moore 1976.*)

have shorter limb extremities (Allen's Rule). In both cases, the result is that organisms have a relatively smaller surface area across which to lose heat.

Small-scale temperature changes can be important too. The winter sun, shining on a cold day, can heat the south-facing side of a tree (and the habitable cracks and crevices within it) to as high as 30°C; and the air temperature in a patch of vegetation can vary by 10°C over a vertical distance of 2.6 m from the soil surface to the top of the canopy. These local variations in temperature, create small-scale micro habitats that may be very important to an organism's distribution. For example, the rufous grasshopper (*Gomphocerripus rufus*) is distributed widely in Europe but in Great Britain reaches its northern limit only 150 km from the south coast where it is restricted to steep south-facing and therefore relatively sun-drenched and warm, grassy slopes. For some organisms, like insects, there is

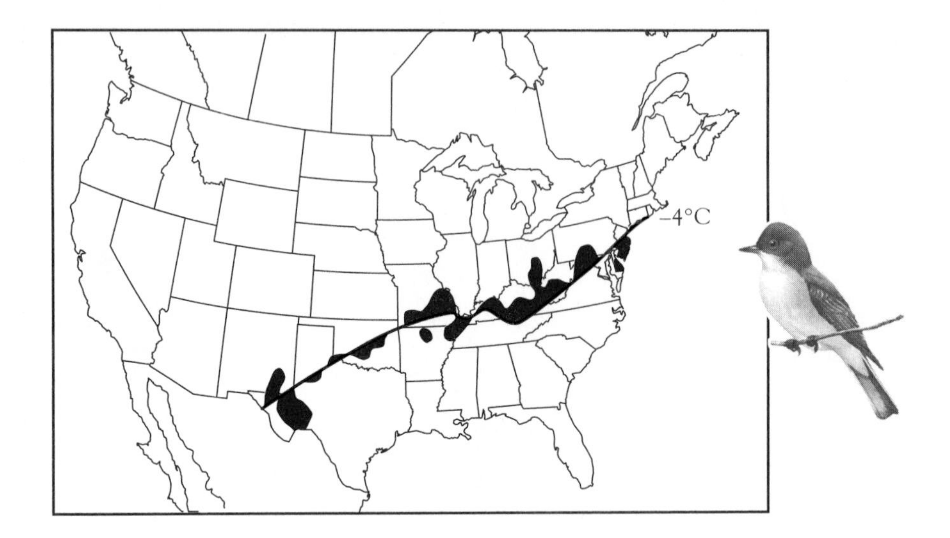

■ **FIGURE 7.14** Contour map of the winter distribution of the Eastern phoebe (*Sayornis phoebe*). The bold line marks the edge of the winter range, which is associated with the −4°C isotherm of average minimum January temperatures. The area of deviation between the range boundary and the isotherm is shaded. *(From Root 1988.)*

■ **PHOTO 7.1** The distribution of the vampire bat (*Desmodus rotundus*) is limited by the 10° C minimal isotherm of January because of the bat's poor capacity to keep itself warm below that temperature. In theory the bat should be able to exist in extreme southern Texas, or in South Florida if introduced there. *(Zig Leszczynski, Animals Animals M-33231.)*

a linear relationship between temperature and development. If the temperature threshold for the development of a grasshopper is 16°C, it might develop in say 17.5 days at 20°C or 5 days at 30°C. The absolute length of time is not important but rather a combination of both—called degree days. Thus, 17.5 days × 4 degrees above the threshold = 70 day degrees, and 5 days × 14 degrees = 70 also (Davidson 1944).

High temperatures impose equally severe ecological constraints on organisms.

High temperatures also can limit species abundance and distribution. Relatively few species can survive temperatures more than a few degrees above their metabolic optimum. Organisms effectively cool themselves by water loss by evaporation, but they soon dry out if an abundant supply of water is not available. Some of the most resistant life

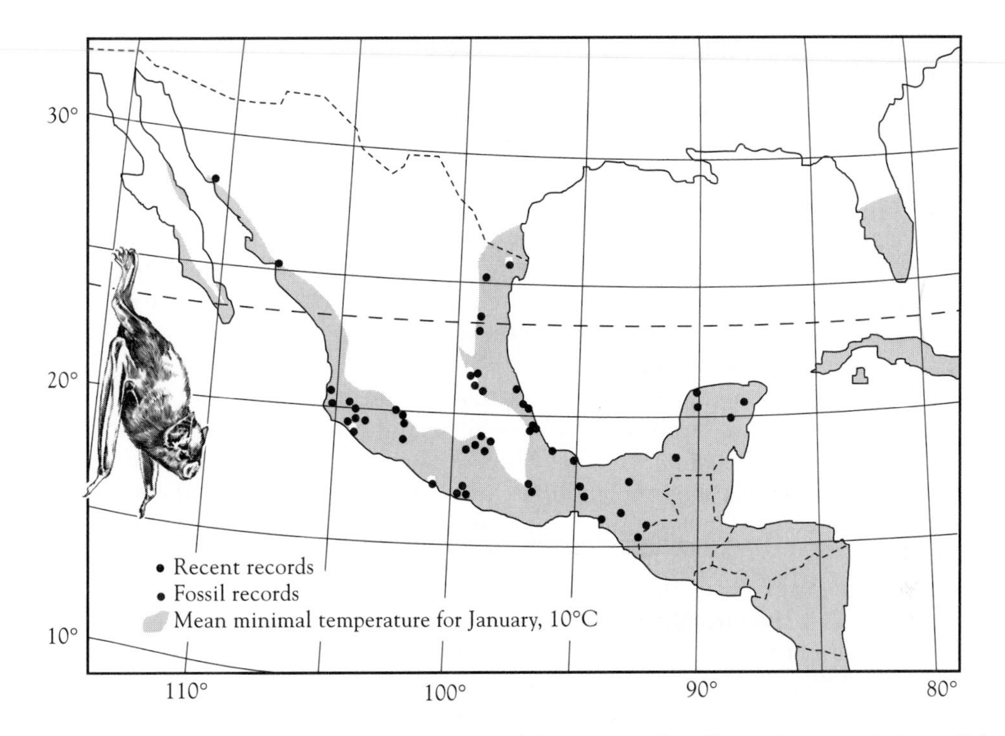

■ **FIGURE 7.15** The northward distribution of the vampire bat (*Desmodus rotundus*) parallels closely the 10°C minimal isotherm for January. Its southward distribution in Argentina and Chile is also limited by the same isotherm. (*After McNab 1973.*)

history stages are the resting spores of fungi, cysts of nematodes, and seeds of plants. Dry wheat grains can withstand temperatures of 90°C for short periods of up to ten minutes. Decomposing organic matter may reach temperatures of 75°C driven by the metabolic heat of fungi, like species of *Mucor asperigillus* and *Hamicola*. Natural hot springs are home to *Thermus aquaticus*, which grows at temperatures of 67°C, and some thermophilic bacteria collected from deep-sea vents have been cultured at temperatures of 100°C, much higher than those originally thought to place limits on life.

Many organisms are even adapted to withstand the high temperatures of fires.

The ultimate high temperatures that organisms face are brought about by fire. Before the arrival of Europeans in North America, fires started by lightning were a frequent and regular occurrence in some areas (Beaufait 1960), for example, in the pine forests of what is now the southeastern United States. These fires, because they were so frequent, consumed leaf litter, dead twigs and branches, and undergrowth before they accumulated in great quantities. As a result, no single fire burned hot enough or long enough in one place to damage large trees—each one swept by quickly and at a relatively low temperature. The dominant plant species of these areas came to depend both directly and indirectly on frequent, low-intensity fires for their existence. The jack pine, *Pinus banksiana*, and the long-leaf pine, *P. palustris*, have serotinous cones, which remain sealed by resin until the heat of a fire melts them open and releases the seeds, and therefore they depend directly on fire for their reproductive success. Much of the rest of the fire-adapted vegetation would be supplanted by other species if fires did not suppress those species periodically (Wade, Ewel, and Hotstetler 1980; Christensen 1981). For example, in the midwestern United States, eastern prairies are maintained by fire because in its absence, deciduous forest would take over (Collins and Wallace 1990).

Management practice that attempts to maintain forests in their natural state by preventing forest fires completely often has exactly the opposite result. First, trees like the jack pine simply stop reproducing in the absence of fire. Second, species like the longleaf pine and wiregrass that depend on fire to suppress their competitors are soon replaced by species characteristic of other communities. Finally, when a fire does occur, fuel has had a much longer period in which to accumulate on the forest floor and the result is an inferno—a fire that is so large and burns so hot that it consumes seeds, seedlings, and adult trees, native and competitor alike. Photo 7.2a shows a longleaf pine seedling; part b shows a small, rapidly moving "natural" fire of the sort that prevailed before current management practices were instituted; and part c shows a destructive fire of the sort that results when litter is allowed to accumulate for long periods.

Management practices that prevent fire arise from the mistaken assumption that forests evolved in the virtual absence of fire. Lightning-caused fires are particularly frequent in the southeastern United States, but even in other areas many more fires are naturally caused than most people believe. For example, in the western half of the United States, nearly half of the yearly average of over ten thousand fires are thought to be started by lightning (Brown and Davis 1973).

■ **PHOTO 7.2A** Adaptation to fire. Longleaf pine seedling, showing the dense cluster of long green needles that protects its growing tip from the low temperature, fast-moving fires that suppress in competitors. *(Photo by Dana C. Bryan, Florida Park Service.)*

■ **PHOTO 7.2B** A "natural" fire in a stand of longleaf pine. (*Photo by Florida Park Service.*)

■ **PHOTO 7.2C** The results of unsound management practices—a destructive fire. Natural fires do not burn very hot because they are frequent and not much litter accumulates before each burn. When natural burning is stopped, litter accumulates rapidly; subsequent fires, of whatever origin, quickly get out of control, not only killing the seedlings but leaping into the forest canopy and devastating the mature trees. (*Photo by U.S. Department of Agriculture.*)

Despite the correlation between the distribution of some species and temperature we need to be cautious about reading too much into species distributions and temperature maps. This is because the temperatures measured for constructing maps are only rarely those that the organism's experience. In nature, an organism may choose to lie in the sun or hide in the shade and, even in a single day, may experience a baking midday sun and a freezing night. Moreover, temperature varies from place to place on a far finer scale than will usually concern a geographer, but it is these local conditions that will be important for a particular species.

Most often it is not the mean average temperature that limits species but the frequency of occasional extremes—like freezes for the saguaro cacti. Agriculturalists know this only too well. It is the frequency and strength of the periodic freezes that limits the northern distribution of oranges in Florida and the southern distribution of coffee (*Coffea arabica* and *C. robusta*) in Brazil.

In these cases, agriculturalists have provided three important take-home messages for ecologists, messages that should be heeded in the study of biotic factors as well:

1. Moving organisms outside their normal range (experimentation) is a useful way to establish what controls the natural limits.
2. Biotic (or abiotic) extremes may be more useful than averages in determining distribution patterns.
3. Abiotic extremes may only be apparent in a limited number of years, e.g., severe freezes or floods, so we may have to wait several years to get useful information.

It is also worth noting that even if adverse temperatures, or other factors, do not kill directly they may interact with other variables such that attack by parasites or predators may be increased.

Because so many species are limited in their distribution patterns by global temperatures, there is concern that if global temperatures change then some species will be driven to extinction or that their geographic range will change and that the geographic location of many centers of agriculture and forestry will be altered. This is the fear of global warming (Applied Ecology: Global Warming).

7.3 Wind can amplify temperature gradients and be important an mortality factor in its own right.

As mentioned earlier, temperature gradients can create winds. Wind can be important because it amplifies the effects of environmental temperature on organisms by increasing heat loss by evaporation and convection (the wind-chill factor). Wind also contributes to water loss in organisms by increasing the rate of evaporation in animals and transpiration in plants. In addition, wind aids in the pollination of many plants by blowing pollen from flower to flower, and it disperses plant seeds, so the strength of the wind is important.

King (1986) has analyzed the growth of tree form and height (using *Acer saccharum*, sugar maple) and its relation to susceptibility to wind damage. Trees should grow as rapidly as possible to escape shade conditions. There is therefore a trade-off between trunk height and diameter—that needed to keep the tree upright. Support efficiency was analyzed in terms of the ratio of actual trunk diameter to the minimum required to keep the tree erect (the stability safety factor). A trade-off was expected between height-growth efficiency (maximized by a minimally designed trunk) and ability to resist storms. The lowest stability safety factors (1.8) were observed in saplings. The maxima were observed in mature canopy trees, which had trunks two to six times the minimum needed to keep trees erect in the absence of winds. However, large trees snap more frequently in high winds than do the more supple saplings. Wind speed is higher in the canopy than closer to the ground, and large trunks are less flexible.

Wind can be an important mortality factor (Photo 7.3). Fifteen million trees in the south of England perished on October 16, 1987, in the wake of a mighty storm (Kerr 1988). Records suggest that such winds had not hit the region for at least three hundred years, so this may have been a rare event. Others suggest that such severe weather will become more common (Thompson 1988) because the world's climate is changing. Five of England's biggest freezes have come since 1978; the frequency of disastrous hurricanes in the South Pacific, especially over Fiji, has increased from one every twelve years to one every seven years, and six storms were recorded between 1981 and 1985 alone. Woodley et al. (1981) present data to show that hurricanes can devastate coral reefs. Wind extremes can undoubtedly play a big part in the distributions of plants and animals.

■ **PHOTO 7.3** This huge live oak tree, *Quercus virginiana*, was felled by strong winds in North Florida. Abiotic factors can clearly have a big influence on the population densities of many forms of life. (*Photo by Peter Stiling.*)

APPLIED ECOLOGY

Global Warming

The scientific debate on global warming centers on two issues: how fast is global warming occurring and how much do human-made "greenhouse" gases contribute to such warming? Political and economic debate centers upon what economic tools are available to control the anthropogenic emissions of greenhouse gases and how vigorously should they be applied.

Increased global warming is caused by something known as the **greenhouse effect**. The greenhouse effect is brought about by the ability of the atmosphere to be selective in its response to different types of radiation. The atmosphere readily transmits solar radiation, allowing about 50 percent of it to pass through unaltered to heat the Earth's surface. The energy absorbed by the Earth is radiated back into the atmosphere, but this terrestrial radiation is long-wave infrared and instead of being transmitted through the atmosphere much of it is absorbed by clouds, causing the temperature of the atmosphere to rise. A lot of the energy absorbed in the atmosphere is then returned to the Earth's surface, causing its temperature to rise also. This is considered similar to the way in which a greenhouse works—allowing sunlight in but trapping the resulting heat inside—hence, the use of the word *greenhouse effect* (Fig. 1.).

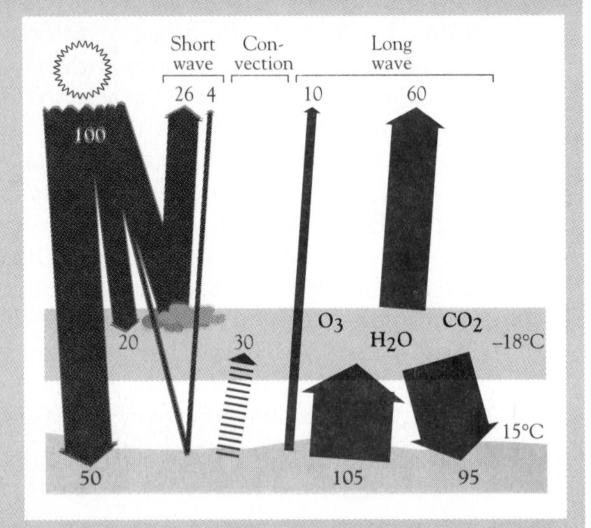

■ **FIGURE 1** There are two controls on the temperature of the Earth, the incoming solar radiation and the insulating effect of the atmosphere. The energy from the sun is almost constant, and the main climatic changes are the result of changes in the composition of the atmosphere. The diagram shows the incoming short-wave solar radiation set at 100 units, and 30 percent of this is reflected by clouds or the Earth's surface. Twenty percent is absorbed by water vapor, ozone, and dust. Half the incoming radiation warms the Earth's surface, and this then radiates long-wave radiation (infrared), which passes less readily through the atmosphere. If the absorption by the greenhouse gases increases, a warmer surface temperature (including the oceans) may result.

Without some type of greenhouse effect global temperatures would be much lower than they are, perhaps averaging only −17°C compared to the existing average of +15°C. The greenhouse effect helps explain the hot temperatures on Venus, which is blanketed in CO_2, and the frigid conditions of Mars, which has very little atmosphere.

The greenhouse effect is an important characteristic of the atmosphere, yet it is made possible by a group of gases, including water vapor, which together make up less than 1 percent of the total volume of the atmosphere. There are about twenty of these greenhouse gases. After water vapor, the five most important are carbon dioxide, methane, chlorofluorocarbons, nitrogen oxide, and ozone (Table 1).

All of these gases have increased in atmospheric concentration since the industrial times (Fig. 2), and the rate of increase is accelerating. The most important of these is carbon dioxide. It has a lower global warming potential per unit of gas than any of the others, but it is far more common in the atmosphere. There is no doubt that the amount of greenhouse gases in the atmosphere is hugely influenced by natural causes. For instance, nearly two-thirds of the nitrous oxide in the atmosphere comes from natural soils and the oceans, and one-third of the methane comes from bogs, swamps, and the action of termites. Volcanoes can emit huge quantities of dusts, carbon, and also sulfur into the atmosphere. However, the steady increase in greenhouse gases over the past two

TABLE 1 Atmospheric concentrations, increase, residence time, sources, and sinks for the major greenhouse gases and their contribution to global warming.

	CO_2	CH_4	N_2O	O_3	CFCS
Radiative absorption per ppm of increase	1	32	150	2000	>10000
1990/1991 concentration	335 ppmv	1.72 ppmv	310 ppbv	na	0.28-0.48 ppbv
Contribution (%) to global warming	50	19	4	8	15
Annual increase (%)	0.5	1	0.2-0.3	2	3
Residence time (year)	100	8.0-12.0	100-200	0.1-0.3	65-110
Total source	6.5-7.5 bt C	400-640 mt CH_4	11-17 mt N	-	-
% biotic	20-30	70-90	90-100	-	0
Major anthropogenic sources	Fossil fuel use (75%), deforestation, and shifting cultivation (25%)	Rice paddies 17% ruminants 22% landfill sites 8% biomass burning 11% coal mining + gas exploitation 28% animal wastes 7% sewage treatment 7%	Cultivated soils 52% fossil fuel 25% mobile sources 11% industrial 12%	Atmospheric	Industrially manufactured (as aerosol propellants, foam insulators)

Notes:
ppmv = parts per million by volume,
ppbv = parts per billion by volume
bt = billion tons
mt = million tons

hundred years cannot easily be explained by natural causes. There is little doubt that greenhouse gases are influenced by human activities. Paramount among these is the mining and burning of fossil fuels, which accounts for 75 percent of the increase in CO_2 emissions, about 39 percent of methane output, and 36 percent of nitrous oxide emissions—close to half of all the greenhouse gas emissions (Photo 1). Land alteration patterns via deforestation, rice paddies, domestic animals, and agricultural soils account for almost 25 percent. Thus, anthropogenic effects cause nearly 75 percent of the increase in greenhouse gases and natural effects only 25 percent. It is very difficult to get accurate estimates of the direct output of greenhouse gases from human causes—after all, methane and nitrous oxide release and deforestation occur globally over a large scale. However, much of the CO_2 output is caused from the industrial burning of fossil fuels and from the manufacture of cement, and these can be more easily documented (Fig. 3).

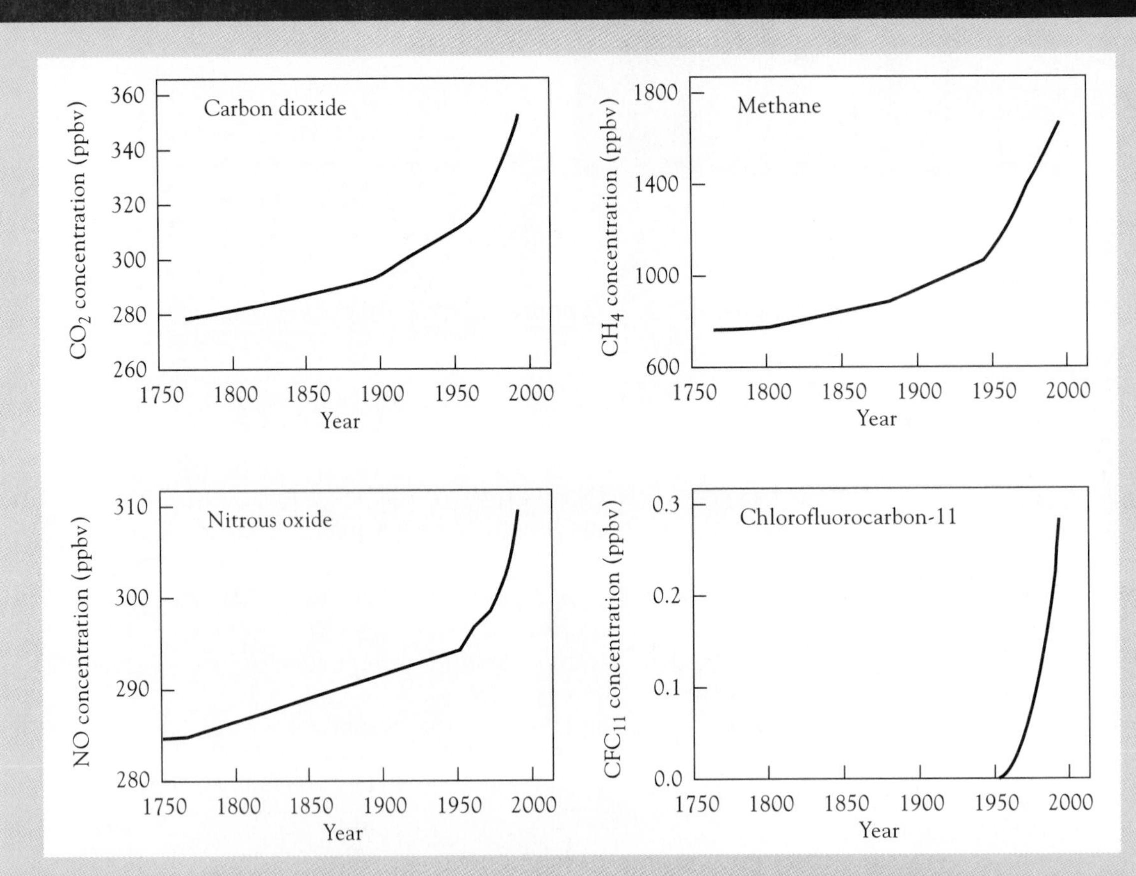

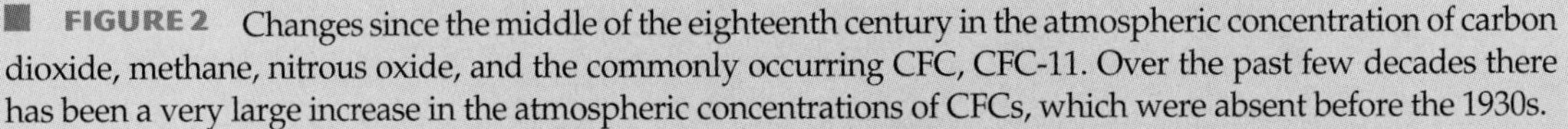

FIGURE 2 Changes since the middle of the eighteenth century in the atmospheric concentration of carbon dioxide, methane, nitrous oxide, and the commonly occurring CFC, CFC-11. Over the past few decades there has been a very large increase in the atmospheric concentrations of CFCs, which were absent before the 1930s.

Because of the huge social upheaval involved in lessening the use of fossil fuels and cement production it has not yet been possible to reduce CO_2 output. CFC (chlorofluorocarbon) production, however, has been cut dramatically, but this is largely because of another problem—CFCs reduce atmospheric ozone. Also, it has not caused undue economic hardship to replace CFCs with other chemicals. Replacing fossil fuels is a very sticky issue. Politicians are loathe to take this step, unless the evidence is strong and the economic damage (to their state) can be minimized.

What is the evidence for global warming? Instrumental temperature readings have been available for a large number of locations since the middle of the nineteenth century. From these it is possible to construct records of mean global surface temperature. Typically, the records show an irregular global warming amounting to about 0.5°C since 1880 (Fig. 4). But there are several problems with the records. For example, although the number of recording stations may be large, their geographic distribution is not truly global. The majority of stations lie in populated

PHOTO 1 The Baikalski Paper Mill on the south side of Lake Baikal in Russia. Cutting down forests to make paper may reduce the absorption of CO_2 by plants from the atmophere while burning wood or fossil fuels can increase the already existing atmospheric CO_2. Both processes exacerbate global warming. *(D. Allan, Animals Animals/Earth Scenes, ECO 3100AD00101.)*

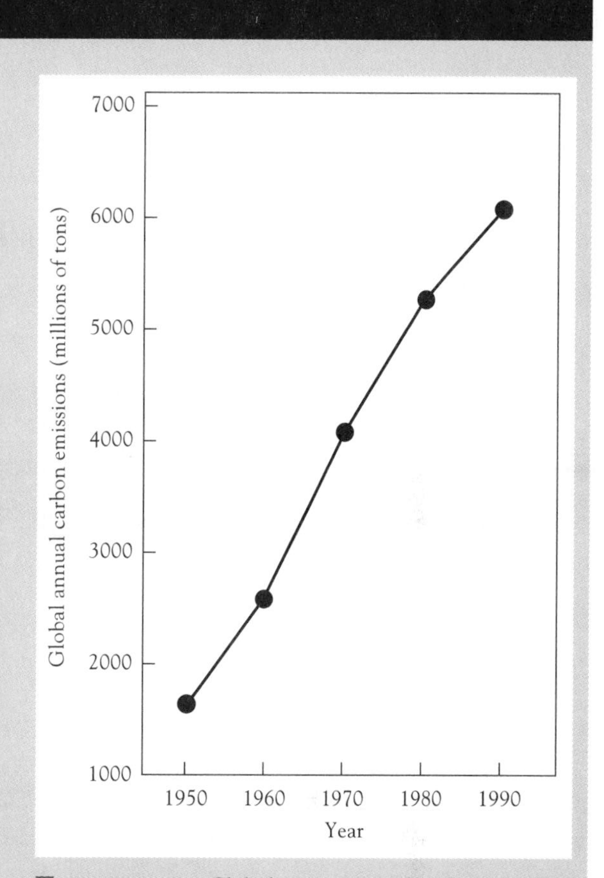

FIGURE 3 Global emissions of carbon dioxide from fossil fuel combustion and cement manufacture, 1950-1990.

areas of the Northern Hemisphere, particularly in Europe and North America. Some stations may have experienced substantial warming due to changes in land use and population density—the urban heat island effect. This term refers to the fact that the temperatures within cities are generally higher than in rural areas. The average urban temperature can be raised by 1 to 2°C and summer rainfall also increases. Despite these problems, it is not denied that some global warming has definitely occurred. We need to know how much warming will occur in the future.

Computer Models and Predictions

It is difficult to design accurate computer models of global warming so that they include all the necessary mechanisms that may influence the degree of warming. There are simply too many variables to incorporate into models (Kerr 1997; Trenberth 1997) (Fig. 5). For instance, higher temperatures associated with an intensified greenhouse effect would bring about more evaporation from the Earth's surface. This could lead to increased cloudiness as the rising water vapor condensed. The clouds in turn would reduce the amount of solar radiation reaching the surface and therefore cause a temperature reduction, which

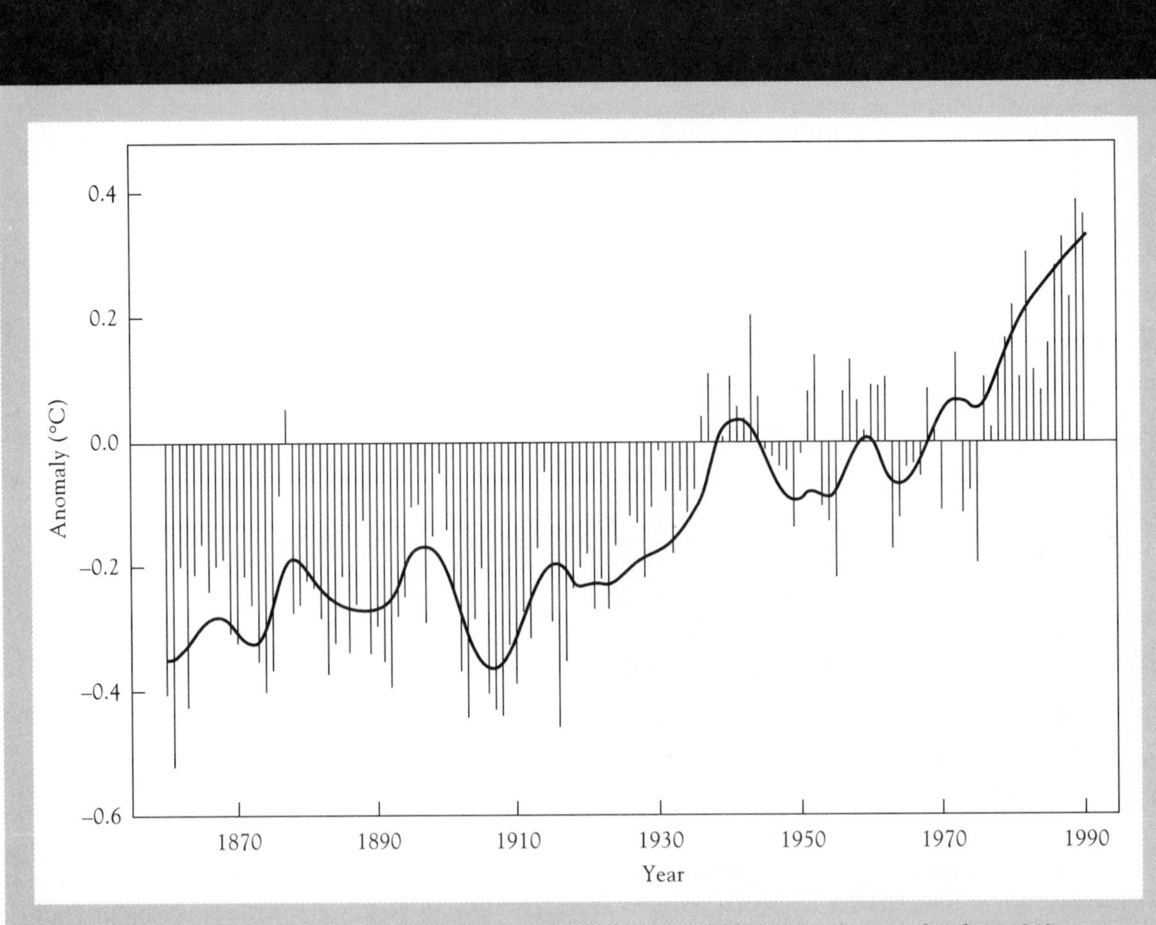

■ **FIGURE 4** Estimated mean global surface temperature, 1860-1990, relative to levels in 1940.

might moderate the increase caused by the greenhouse effect. This is a negative feedback mechanism that could prevent escalation in global warming. There are also positive feedback mechanisms that augment the change. The colder northern waters of the world's oceans, for example, act as an important sink for CO_2, but their ability to absorb gas decreases as temperatures rise. Global warming would reduce the ability of the oceans to act as a sink. Instead of being absorbed by the oceans, CO_2 would remain in the atmosphere, thereby adding to the greenhouse effect. Another positive feedback mechanism is permafrost melting resulting in increased global warming and the production of more methane from bogs and swamps. Neither positive nor negative feedback mechanisms have adequately been dealt with in most models of global warming. As a result, models showed more warming than had actually occurred.

The April 1996 report from the UN's Intergovernmental Panel on Climate Change (IPCC), in its first full report for five years, was able to finally explain the perplexing lack of fit of the models to the data. Although the chief influence on climate was the buildup of CO_2 and other greenhouse gases in the atmosphere, in many parts of the world, including much of Europe and North America, warming has been masked by another form of pollution from burning fossil fuel—aerosols of sulfates and soot that form a thin haze that reduces solar heating. The IPCC concluded that, globally,

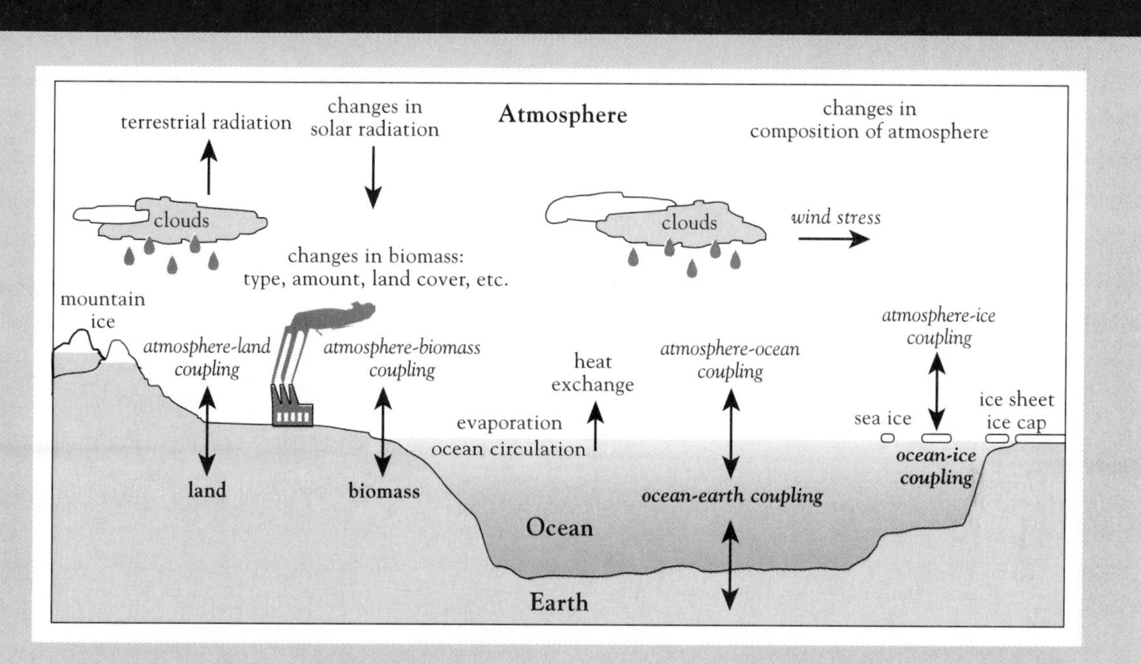

■ **FIGURE 5** Schematic illustration of some of the processes simulated in global climate models. The figure is incomplete as the effects such as added methane from thawing permafrost are not included.

aerosols have a cooling effect of around 0.5 watts per m^2, compared with the global warming from greenhouse gases of 2 to 5 watts. By combining the impacts of greenhouse gases and sulfate aerosols the IPCC produced a model of the world climate that matched well the observed surface air temperature data. In a key section of the report, Ben Santer of the Lawrence Livermore National Laboratory in California and Tom Wigley of the U.S. National Center for Atmospheric Research in Colorado argued "that the pattern was distinguished from natural variability with a high level of statistical confidence." In the strongest wording ever from the group they ended by stating that "the balance of evidence suggested a discernible human influence" on climate. By the IPCC's latest predictions, a doubling of atmospheric CO_2 levels, predicted for the end of the next century, would push temperatures up worldwide by between 1 and 3.5°C. Naturally, there were some dissenters. Loudest among these were representatives of Saudi Arabia, Kuwait, and Dow Chemicals.

The Environmental Impact of Global Warming

It is clear that a knowledge of the speed and extent of global warming is critical. To try and predict the effect of global warming most scientists focus in on the future point when the effects of CO_2 have doubled—that is, about 560 parts per million (ppm) compared to the preindustrial level of 280 ppm. This is usually taken to be about 2,100. Although CO_2 levels will be only about 500 ppm at that time, the additional effects of CFCs, methane, and the like will bring the "CO_2 equivalent" concentration to 560 ppm. At that time, it is argued that global temperatures will be about 1 to 3.5°C (about 6°F) warmer than present and will increase a further 0.3°C each decade. This increase in heat is comparable to the warming that ended the last Ice Age! Much depends on the rate of greenhouse gas emissions. If they accelerate the warming could be more severe; if they slow down the global warming will not be so catastrophic.

If we accept the typical scenario of gradual global warming we can ask, "What are the environmental consequences on natural and human-made ecosystems"?

Natural Ecosystems

The main drawback associated with global warming will be the severe changes in the communities of organisms other than humans. Some authors point out that many species can adapt to slight changes in their environment. However, evolution is usually assumed to be irrelevant in a global warming scenario because it is thought that most species cannot evolve significantly or rapidly enough to counter climate changes. The anticipated changes in global climate are expected to occur at a rate that will probably be too rapid to be tracked by evolutionary processes such as natural selection. It is also not likely that all species can simply disperse and move north or south into the newly created climatic regions that will be suitable for them.

Many tree species take hundreds, even thousands, of years for substantial dispersal via progeny. The tree with one of the fastest known dispersal rates is spruce, which expanded its range an average of only 20 km every ten years, beginning about nine thousand years ago. Because plants can "move" only slowly, the resultant changes in the floras will have dramatic effects on animals, both on the large biogeographic scale and on the local community scale. Herbivores cannot disperse without the foliage that they feed on first being present. As an example, it has been shown by paleobotanist Margaret Davis from the University of Minnesota that in the event of a CO_2 doubling, beech trees, presently distributed throughout all of the eastern United States and southeastern Canada would die back in all areas except in northern Maine, northern New Brunswick, and southern Quebec (Fig. 6). Of course, this could possibly be offset by the creation of new favorable habitats in central Quebec. However, most scientists believe that the rates of the poleward shifts of climatic zones would outstrip migration rates of trees, hence extinctions would be increased.

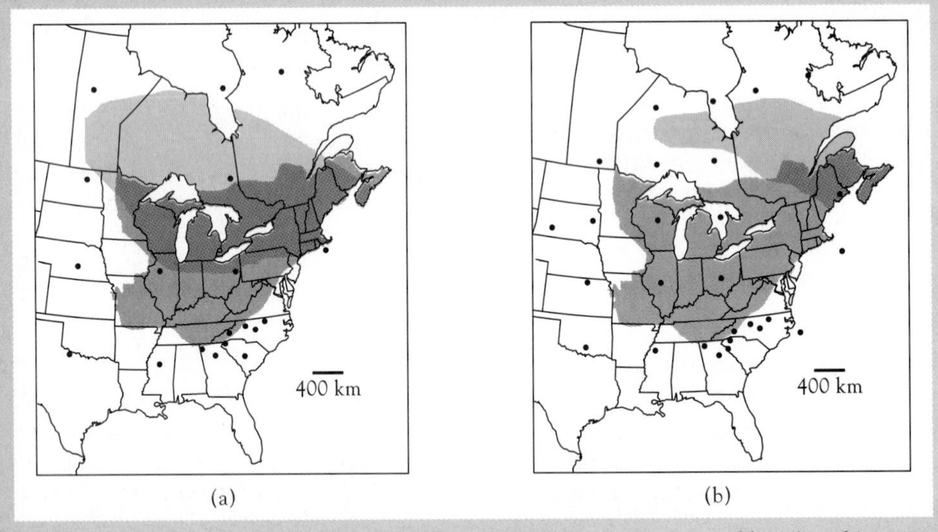

■ **FIGURE 6** The present geographic range of sugar maple (horizontal lines) and its potentially suitable range under doubled CO_2 (vertical lines). Cross-hatching indicates the region of overlap. Note that different models give different predictions: **(a)** uses a scenario from the Goddard Institute for Space Studies; **(b)** uses a scenario derived from the Goddard Fluid Dynamics Laboratory. Grid points are sites of climatic data output for each model. (*From Davis and Zabinski 1992*).

Even more damaging changes could occur on some isolated islands where northerly or southerly range migration is not possible, thus dooming some species to certain extinction. This has led some people into a debate over whether humans need to remove some species from the field and hold them in captive breeding programs for reintroduction later when the climate reverts back to "normal." Alternatively, translocating wild individuals into potentially more hospitable environments is an option.

One of the main concerns voiced is that the delicate balance between diseases, their vectors, and humans might be upset as tropical climates that are so hospitable to spawning and spreading diseases move poleward. The spread of infectious diseases is controlled by the range of their vectors—mosquitoes and other insects. Because these are all cold-blooded organisms, they are susceptible to subtle changes in temperatures. Increases in temperatures mean increases in the activity and ranges of these vectors. Data on recent trends support this observation. A 1°C increase in the average temperature in Rwanda in 1987 was accompanied by a 337 percent rise in the incidence of malaria that year as mosquitoes moved into mountainous areas where they had previously been absent. Also, *Aedes aegypti*, a mosquito that carries dengue and yellow fever, has extended its range high into the mountain areas of such diverse areas as Colombia, India, and Kenya. Although global warming is expected to deliver its most deadly punch in the tropical areas of the world (see Table 2) where over six hundred million people are affected (and two million die), the United States is not immune. A 1993 outbreak of hantavirus respiratory illness in the southwest killed twenty-seven people and was linked to an El Niño event. An explosion in the deer mouse population (the viral host) followed from an El Niño that precipitated heavy rains and increased the animals' food supply. A computer model by a Dutch public health team proposed that an average global

TABLE 2 Major tropical diseases likely to spread with global warming.

DISEASE	VECTOR	POPULATION AT RISK (MILLIONS)	PREVALENCE OF INFECTION	PRESENT DISTRIBUTION	LIKELIHOOD OF ALTERED DISTRIBUTION WITH WARMING#
Malaria	mosquito	2100	250 million	(sub)tropics	+++
Schistosomiasis	water snail	600	200 million	(sub)tropics	++
Filariasis	mosquito	900	90 million	(sub)tropics	+
Onchocerciasis *(river blindness)*	black fly	90	18 million	Africa/Latin America	+
African trypanosomiasis *(sleeping sickness)*	tsetse fly	50	25,000 new cases/year	tropical Africa	+
Dengue	mosquito	*estimates unavailable*		tropics	++
Yellow fever	mosquito	*estimates unavailable*		tropical South America & Africa	+

Notes: # As assessed by WHO: + = likely; ++ = very likely, +++ = highly likely

temperature increase of 3° in the next century could result in 50 to 80 million new malaria cases per year.

Another drawback associated with global warming is often seen to be an increase in dryness likely to accompany the rising temperatures. Modern droughts are frequently blamed on global warming. However, average global precipitation should increase (Fig. 7). Droughts are more likely a consequence of the normal variability in patterns of rainfall. Departures from normal are more common in arid areas, thus compounding problems of already inherent aridity. Nevertheless, despite the overall increase in rainfall, a decrease would be apparent in some areas, and these areas would likely be over already dry areas. Turning up the temperature there would not increase evaporation—the area is like a dry sponge already. The areas with reduced rainfall would be the midcontinental areas of America and Asia. Increased droughts here would probably cause the widespread extinction of plant and animal life in existing ecosystems. On the other hand, rainfall would increase over moist areas and coastlines where increased evaporation would lead to increased rainfall. Thus, the world's current grain-producing areas would become drier than they are now following the global warming. Corn yields would be

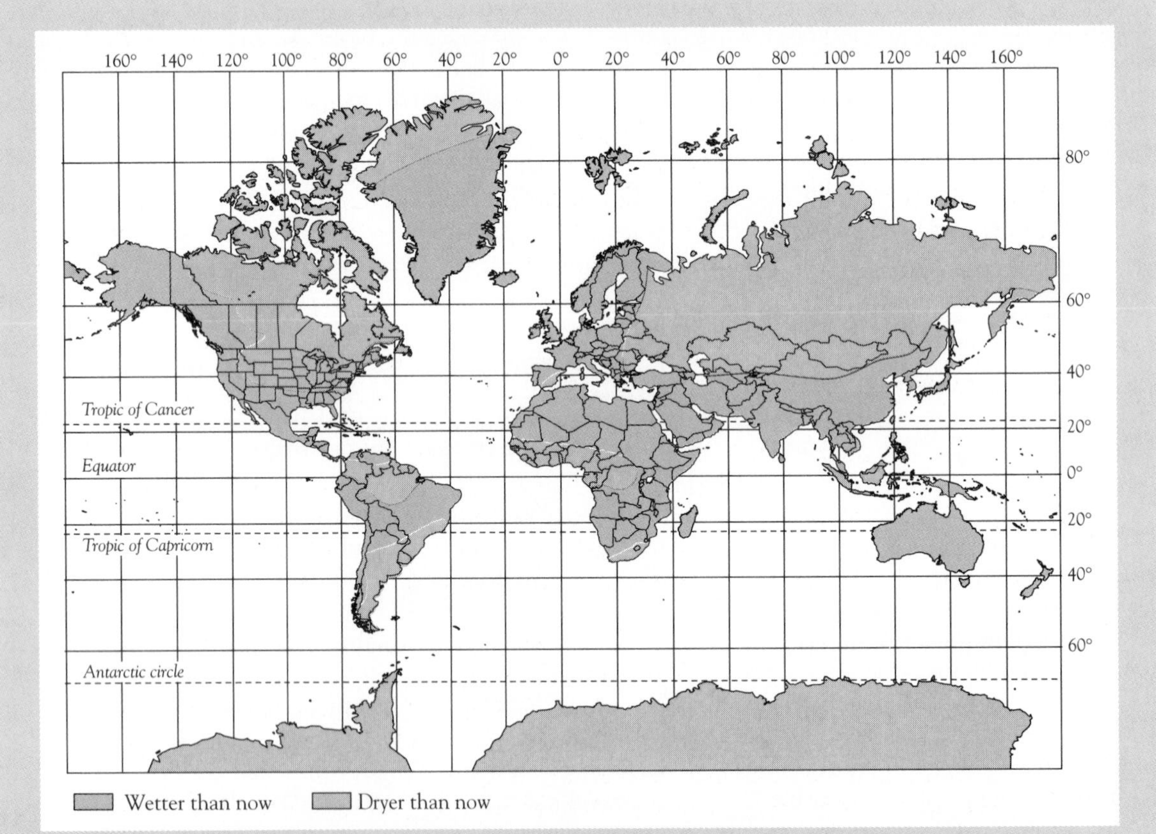

■ **FIGURE 7** Predicted changes in precipitation patterns caused by global warming. Mid-continent regions will probably be drier than now, but coastal areas and much of the tropics could be wetter.

reduced in the midwestern plains of the United States, and a major increase in the frequency and severity of drought would lead to more frequent crop failure in the wheat-growing areas of Canada and Russia. This would be partly offset by increased rainfall in some tropical countries. This high rainfall in the tropics could lead to increased rice yields. Some of the grain-growing areas in Australia would also experience increased precipitation and higher temperatures. Thus, the pattern of the world's grain trade, which depends heavily on the annual North American surplus, would probably be disrupted and take a while to become adjusted. Furthermore, although more rainfall could fall in many areas, it is not at all certain that this would result in increased soil moisture. Higher temperatures could increase evaporation, and increased plant growth could increase transpiration.

As well as the direct consequences of global warming on the Earth's biota, there are potentially a myriad of other indirect effects such as sea-level rises and consequences for water supplies, leisure activities, and electricity requirements.

Winds can also modify wave action in the sea. On the rocky intertidal, algae such as fucoids and kelps survive repeated pounding by a combination of holdfasts and flexible structures. The animals of the intertidal have powerful organic glues and muscular feet to hold them in place. Current can also be critical in streams where in turbulent upstream regions, plants and animals are in danger of being washed away. Here rock-encrusting forms like algae and mosses grow, and animals have flattened bodies to hide under crevices.

7.4 Salt concentrations can affect water uptake.

For organisms in both terrestrial and aquatic habitats, the concentration of salts in the soil or in water is important because higher concentrations increase osmotic resistance to water uptake. On land, this occurs in arid conditions where water evaporates and crystalline salt accumulates. It can be of critical importance to agriculture where continued watering in arid environments greatly increases salt concentration and reduces crop yields.

Along the salt marshes of sea coasts, high salt concentrations occur, and the vegetation consists largely of halophytes, species that can tolerate higher salt concentrations in their cell sap than nonhalophytes. Many plants also have ways of overcoming the excess salt (Adam 1990). Plants like mangrove and *Spartina* grasses have salt glands that excrete salt to the surface of the leaves where it forms tiny white salt crystals.

Windborne salt can affect the distribution of plants on sand dunes too. On Atlantic coasts of the United States, sea oats, *Uniola paniculata*, can withstand high atmospheric salt and occur near the shoreline while another dune grass, *Andropogon littoralis*, is less tolerant and occurs in protected areas back from the shoreline (Oosting and Billings 1942).

7.5 Acidity or alkalinity of the environment affects plant and animal distributions.

The pH of soil or water can exert a powerful influence on the distributions of species. For plants, most roots are damaged below pH 3 or above pH 9. Only a minority of organisms can grow below pH 4.5. Even between extremes, soil or water pH can have indirect effects limiting the availability of nutrients to organisms. Generally, the number of species decreases in acid waters. Edmondson (1944) found that the distribution of eight species of rotifer was limited by the water pH so that they did not occur in all the possible lakes available. Lake trout disappear from lakes in Ontario and the eastern United States when the pH falls below about 5.2 (Mills et al. 1987). Although this pH does not affect survival of the adult fish, it affects the survival of juveniles. The deposition of acid rain can greatly exacerbate this problem (see Applied Ecology: Acid Rain in chapter 22). Chalk and limestone grasslands (so called high lime soils) carry a much richer flora (and associated fauna) than do acid grasslands.

Plants have been broadly classified as calcicole (lime-loving), calcifuge (lime-hating), and neutrophiles, tolerant of either condition (Larcher 1980). Acidity, however, usually depends on more than just calcium availability, and involves the presence of iron and aluminum. True calcifuges, such as rhododendrons and azaleas, have a low lime requirement and can live in soils with a pH of 4.0 and less. They are also tolerant of aluminum and have a high demand for iron. True calcicolous plants, such as alfalfa (*Medicago sativa*), blazing star (*Chamaelirium luteum*), and southern redcedar (*Juniperus silicicola*) are restricted to soils of high pH, not because they have any particular demand for calcium, but because they are susceptible to aluminum toxicity, acidity, and other factors influenced by calcium.

7.6 The availability of mineral nutrients like nitrogen and phosphorus is a significant limit on primary productivity in many areas.

Many abiotic factors such as nutrients, light, water, and carbon dioxide are important for the distribution of organisms because they are used by them. Organisms require a wide variety of chemical elements, in addition to the basic building blocks of hydrogen, carbon, and oxygen (Table 7.2). The elements required in the greatest amounts, nitrogen, phosphorus, sulfur, potassium, calcium, magnesium, and iron are known as macronutrients. In addition, there are a whole host of nutrients needed in so-called trace amounts. Animals get most of their nutrients from their prey-plants or other animals. Plants get theirs from water, often in the soil. The availability of each of these elements varies with the temperature and acidity of the soil. A knowledge of soils is therefore critical for determining nutrient supply. Generally, a widely spaced, extensive root system tends to maximize access to nitrate, while a narrowly spaced, intensively branched system maximizes access to phosphates (Nye and Tinker 1977). A comparison of elements in the soil and those accumulated by the vegetation (Table 7.3) suggests that certain elements like nitrogen are likely to be limiting. On the other hand, some nutrients in aquatic systems far exceed levels necessary for organisms (Table 7.4).

Animals also face problems acquiring nutrients. In plants, the carbon-to-nitrogen ratio is about 40:1. In mammals, it is about 14:1 (Mattson 1980) so herbivores are under selective pressure to find nutrient-rich plants (see also Chap. 11, "Herbivory"). One

TABLE 7.2 Macronutrients and trace elements required by organisms and a partial list of their functions.

NUTRIENT		USED IN
Macronutrients		
Nitrogen	(N)	Proteins and nucleic acids
Phosphorus	(P)	Nucleic acids, phospholipids, and bone
Sulfur	(S)	Proteins
Potassium	(K)	Solute in animal cells, sugar formation in plants, protein synthesis in animals
Calcium	(Ca)	Bone, muscle contraction, blood clotting, plant cell walls, regulation of cell permeability
Magnesium	(Mg)	Chlorophyll and many enzyme activation systems
Iron	(Fe)	Hemoglobin and enzymes
Sodium	(Na)	Extracellular fluids of animals, osmotic balance, nerve transmission
Chlorine	(Cl)	Chlorophyll, osmotic balance
Trace elements		
Manganese	(Mn)	Chlorophyll, fatty acid synthesis
Zinc	(Zn)	Auxin production in plants, enzyme systems
Copper	(Cu)	Chloroplasts, enzyme activation
Boron	(B)	Vascular plants and algae
Molybdenum	(Mo)	Enzyme activation systems
Aluminum	(Al)	Nutrient for ferns
Silicon	(Si)	Nutrient for diatoms
Selenium	(Se)	Nutrient for planktonic algae
Cobalt	(Co)	Nutrient for mutualistic association of legumes and nitrogen-fixing bacteria in root nodules and in ruminants
Vanadium	(V)	Tunicates, echinoderms, and some algae
Iodine	(I)	Higher animals, thyroid metabolism
Fluorine	(F)	Bone and teeth formation

nutrient in particularly short supply for herbivores may be sodium, and many species, like moose, may deliberately chosen to feed on sodium-rich vegetation (Belovsky 1981). Such large herbivores may even eat mineral-rich soil to satisfy their needs or seek out mineral licks. Deer on diets low in calcium and phosphorous show stunted growth, small antlers, and low reproductive success.

Low nutrient availability is often associated with low soil temperatures and anaerobic soil conditions.

Soil is the complex loose terrestrial surface material in which plants grow. It is composed of three things, (1) weathered fragments of parent material in organic matter in various stages of breakdown, (2) soil water, and (3) the minerals and organic compounds dissolved in it. Soil is also teeming with life, much of it microscopic. When a plant or any organism

TABLE 7.3 Typical concentrations of elements in soils and annual uptake by plants. (After Bohn, McNeal, and O'Connor 1979)

ELEMENT	SOIL CONTENT (WEIGHT %)	ANNUAL PLANT UPTAKE ($KG\ HA^{-1}\ YR^{-1}$)	SOIL CONTENT ANNUAL PLANT UPTAKE (YEARS)
Silicon (Si)	33	20	21,000
Aluminum (Al)	7	0.5	180,000
Iron (Fe)	4	1	52,000
Calcium (Ca)	1	50	260
Potassium (K)	1	30	430
Sodium (Na)	0.7	2	4600
Magnesium (Mg)	0.6	4	2000
Titanium (Ti)	0.5	0.08	62,000
Nitrogen (N)	0.1	30	40
Phosphorus (P)	0.08	7	150
Manganese (Mn)	0.08	1	1000
Sulfur (S)	0.05	2	320
Fluorine (F)	0.02	0.01	26,000
Chlorine (Cl)	0.01	0.06	220
Zinc (Zn)	0.005	0.01	6500
Copper (Cu)	0.002	0.006	4200
Boron (B)	0.001	0.03	400
Molybdenum (Mo)	0.0003	0.0003	13,000
Selenium (Se)	0.0000001	0.0003	40

Note: Carbon, oxygen, and hydrogen make up most of the remaining nutrients in the soil.

TABLE 7.4 Percentage composition of dissolved minerals in rivers (freshwater) in seawater and in the blood plasma and cells of frogs. (From material in Reid 1961 and Gordon 1968)

MINERAL ION	DELAWARE RIVER	RIO GRANDE RIVER	SEAWATER	FROG PLASMA	FROG CELLS
Sodium	6.7	14.8	30.4	35.4	1.3
Potassium	1.5	0.9	1.1	1.3	77.7
Calcium	17.5	13.7	1.2	1.2	3.1
Magnesium	4.8	3.0	3.7	0.4	5.3
Chlorine	4.2	21.7	55.2	39.0	0.8
Sulfate	17.5	30.1	7.7	—	—
Carbonate	33.0	11.6	0.4	22.7	11.7

Note: Sulfate (SO_4^{-2}) contains sulfur and carbonate (CO_3^{2-}), carbon. The sums of the columns do not equal 100 because all dissolved substances are not included.

dies, the process of decay occurs in the soil through the activities of soil organisms, especially bacteria and fungi. Humus, finely ground organic matter mixed with the mineral part of the soil, is produced, and eventually the minerals are taken in by the plant roots.

One of the best examples of soil effects on the distribution of plants is the serpentine soils that occur in scattered areas all over the world, including California. Serpentine rock is basically a magnesium iron silicate. Plants that grow there must be tolerant of low calcium; nitrogen and phosphorus; and high nickel, chromium, and magnesium. Such conditions are lethal to many plants, hence, serpentine soils have little value for agriculture or forestry. However, many species have adapted to these conditions, forming stunted, endemic communities not found on normal soils (Whittaker 1954).

Because most soil-forming processes, like litter fall and weathering of rocks, tend to act from the top down, soil develops a vertical structure referred to as the soil profile (Table 7.5; Fig. 7.16). Soil scientists recognize five main layers, or horizons. The O horizon is the humus layer. It may be subdivided into Oi, the litter layer, and Oa, the humus layer proper. The leaf litter layer may fluctuate seasonally, whereas the zone of actual decomposition does not. The next layer is the topsoil, or the A horizon, where most of the plant roots are located. Dead organic matter is added to this layer as it mixes with the mineral soil below. The E layer is the zone of maximum leaching or eluviation, hence the label. Soils have either an A or E layer or both. Organic material is largely absent from the B horizon, and the mineral parent material is less thoroughly weathered. Materials leached from the A horizon may be deposited here (illuviation). Below the B horizon in some soils may be compact, slowly permeable layers like claypans. Below this is the C horizon, generally consisting of weathered parent material such as glacial drift or bedrock.

Soil texture is based on the sizes of mineral particles making up the soil, classified as gravel, >2 mm sand <2 mm, silt <0.05 mm, and clay <0.002 mm (Fig. 7.17). Soils made up mainly of small particles like clay are called heavy soils. Water soaks into heavy soils slowly but is retained well; such soils have the potential for being very fertile. Light soils, such as sandy soils, are well aerated and allow free movement of roots and water but are relatively infertile because water and nutrients leach out of them.

The difference in potential fertility between light and heavy soils results from the way minerals are retained. Such minerals as calcium and magnesium (specifically, most elements that form positive ions, or cations, when dissolved) are stored on the surface of par-

TABLE 7.5 Major soil horizons and their divisions.

O	Surface layer—the organic layer on top of the mineral soil—litter and humus.
(A)	Mineral soil in which organic matter is accumulating from above and from which clay, iron, or aluminum may be leached. Sometimes subdivided into an E layer.
(B)	Zone of deposition of clay, iron, or aluminum. Often blocky or columnar structure. In many soils, reddish from iron oxides.
(C)	Little weathered. In areas where precipitation/evapotranspiration ratio <1, calcium or magnesium carbonate layer (ca) usually present.

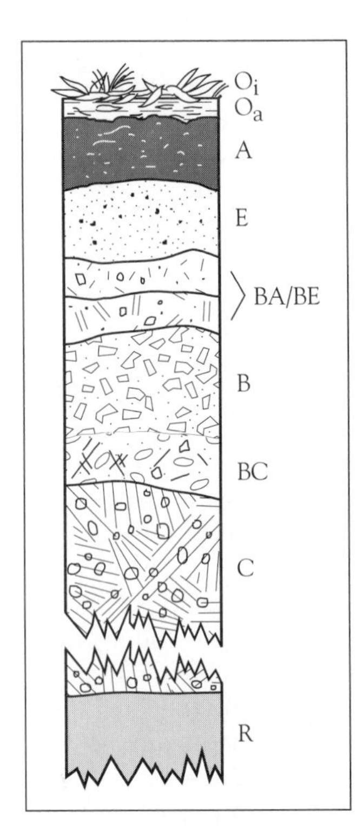

■ **FIGURE 7.16** A generalized profile of the soil. Rarely does any one soil possess all of the horizons shown. *Oi*, loose leaves and organic debris. *Oa*, organic debris partially decomposed (fermented) or matted. *A*, dark-colored horizon with high content of organic matter mixed with mineral matter. *E* lighter-colored horizon of maximum leaching. Prominent in spodosols, it may be faintly developed in other soils. *B*, zone of maximum accumulation of clay minerals or of iron and organic matter. Subhorizons include *BA* or *BE*, transitional to *B*, but more like *B* than *A*, *BC* or *CB*, transitional to *C*. *C*, the weathered material, either like or unlike the material from which the soil presumably formed. A gley layer may occur, as well as layers of calcium carbonate, especially in grassland. *R*, consolidated bedrock.

ticles. Anions, by contrast, are dissolved in soil water. Both are more prevalent in heavy soils. Plants' roots remove cations and replace them with hydrogen ions. The potential fertility of a soil depends primarily on cation exchange capacity, a measure of the number of sites per unit of soil on which hydrogen can be exchanged for mineral cation. Clay soil may have a cation exchange capacity from two to twenty or more times that of sand. A soil with high cation exchange capacity and consequently high potential fertility may be actually infertile if most of the sites on the particles are already filled with hydrogen ions. Such a soil will be acid and will have little calcium to supply to the growing plant. Percentage base saturation, the percentage of the exchange capacity satisfied by calcium, magnesium, and similar elements, measures this aspect of fertility. If percentage base saturation is 60 percent, then 60 percent of the sites are filled by basic ions and 40 percent by hydrogen ions.

Soil development is strongly influenced by rainfall, evaporation, temperature, and vegetation. In the arctic, tropic, temperate, and arid regions different soils are largely a result of different climates (Table 7.6; Fig. 7.18). However, in the northern Midwest, coniferous forest, deciduous forest, and grassland grow within a few miles of one another, and the soil differences are entirely due to the differing effects of the vegetation. Coniferous forest produces strongly acid soils with a heavy layer of undercomposed litter and hardpan of leached clay in the B horizon. Deciduous forest produces a thinner, less acid litter, and this grades into the mineral soil below. In grassland the soil is less acid still, and each year the entire aboveground part of the plant dies back, allowing decomposing organisms access to the nutrients at or near the soil surface. The grassland soil is dark because of the addition of this organic material in addition to that added to the soil at various depths by the death each year of many fine roots. The general range of litter fall in temperate ecosystems is 250 to 450 g/m^2 per year or one to two tons per acre, although

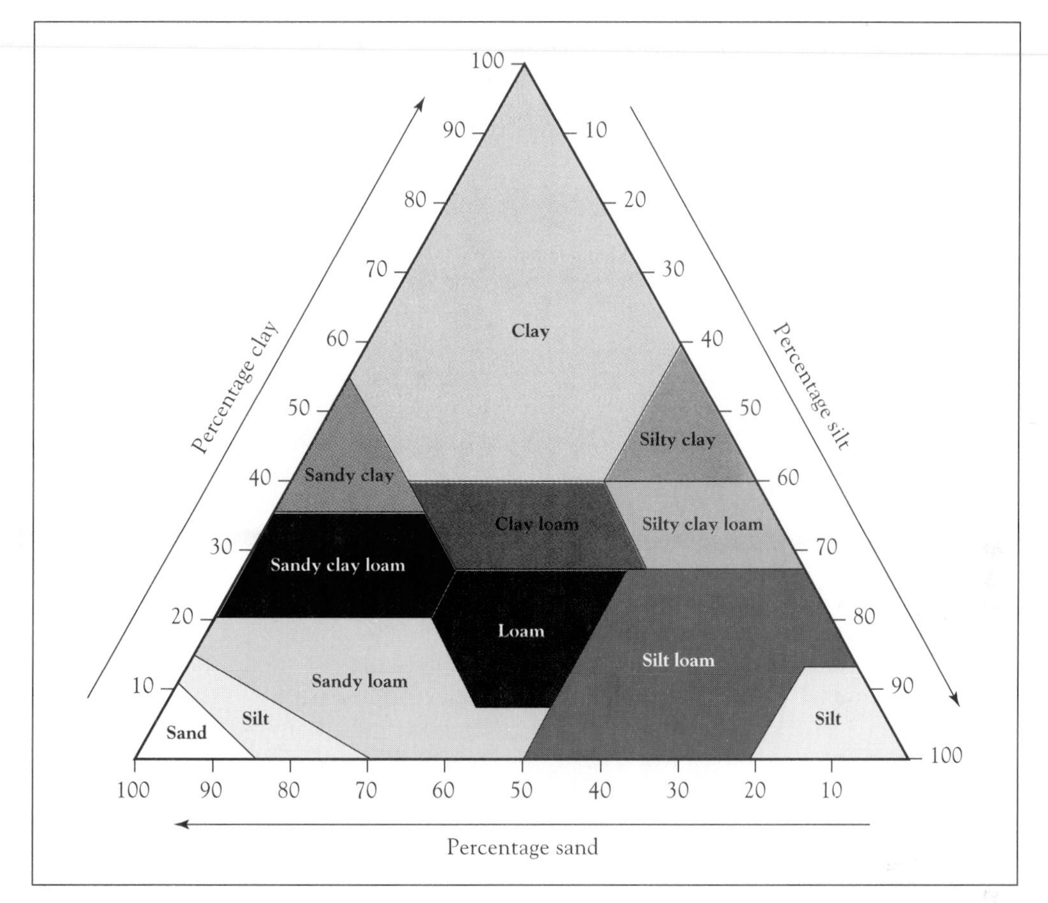

■ **FIGURE 7.17** A soil texture chart showing the percentages of clay (below 0.002 mm), silt (0.002 to 0.05 mm), and sand (0.05 to 2.0 mm) in the basic soil textural classes. For example, a soil with 60 percent sand, 30 percent silt, and 10 percent clay would be considered as sandy loam.

ranges from almost 0 (deserts) to 72,000 g/m^2 (tropical forests) have been reported. The rate at which the litter decays varies from a half-life (i.e., 50 percent decay) of 0.36 years in tropical forest to 1.01 years in temperate coniferous forest to 11.23 years in boreal coniferous forest. The deep acid humus of coniferous forests is sometimes known as mor humus and the thin, less acid humus of temperate deciduous forest as mull humus after the Danish forester P. E. Muller who described the phenomenon in 1879.

7.7 Light is an important limiting resource for plants.

Light is important to organisms for two different reasons. First, it affects the timing of daily and seasonal rhythms in plants and animals. The activity patterns of nocturnal organisms use light as a cue for activity cycles, and the breeding season of animals and plants are set by the organism's response to day length changes. In these cases, however, light is not a limiting resource or a limiting physical variable; it is merely acting as a cue.

TABLE 7.6 Main types of climatic-vegetational soil development.

TYPE OF SOIL DEVELOPMENT	CLIMATE	VEGETATION	IMPORTANT PROCESSES	KINDS OF SOILS
Gleization	Cold, little rainfall	Tundra	In summer, soils remain wet due to poor drainage caused by permanently frozen lower layers (permafrost) or clay pans. In winter, freezing at top compresses middle layers (glei) which are sticky and blue-gray because of iron in the reduced state.	Tundra or inceptisols
Podzolization	Cool, fairly moist	Coniferous or deciduous forest	Percolation of acid water downward leaches out carbonates. A horizon becoming acid and ash-colored (podzol). If extremely acid, clays may be leached and deposited in the B horizon and form a hard pan. Decomposition is slow under acid conditions, and there is often a sharp break between the humus layer and the upper mineral layer.	Podzols or spodisols (coniferious forests). Gray-brown podzols or Alfisols (deciduous forests; podzolization less extreme so that soils are less acid and more fertile). Red-yellow podzols or ultisols (southern forests; transitional to lateritic soils) Lateritic soils or oxisols

(continued)

The second reason that light is important is that it can be a limiting resource for photosynthesis. Some plants, like oak and maple trees, reach maximal photosynthesis at one quarter of full sunlight. Others, like sugarcane never reach a maximum but continue to increase their photosynthetic rate as light intensity increases (Fig. 7.19). As we mentioned at the beginning of the chapter, there are thus shade-tolerant and shade-intolerant species.

One reason photosynthetic rate varies among plants is that there are three different

TABLE 7.6 (continued) Main types of climatic-vegetational soil development.

TYPE OF SOIL DEVELOPMENT	CLIMATE	VEGETATION	IMPORTANT PROCESSES	KINDS OF SOILS
Laterization	Warm, moist	Tropical forest	Clay minerals decompose rapidly and release bases, keeping soil from becoming highly acidic. Humus decomposes rapidly and does not accumulate. High rainfall ensures heavy leaching. Silica fraction is leached, but oxides of iron, aluminum, and manganese remain, giving solid reddish or yellowish color.	Lateritic soils or oxisols.
Calcification	Warm to cool, not moist	Grassland, savanna desert	Leaching not sufficient to carry away calcium carbonates that may accumulate as hardpan; soil remains neutral to alkaline, and clay is not leached. Dense roots and tops add organic matter and nutrients. Darkness is related to amount of organic matter and thus to rainfall. In desert, wind erosion may remove finer particles.	Prairie soils or mollisols.

Note: In addition some soils develop in other ways. Aridosols in deserts are light in color and low in organic matter; vertisols are heavy in clay and swell and crack in wet and dry periods, respectively; histosols have high organic context and are very wet, like peat and bogs; andisols develop from volcanic ash; and entisols are very young with weakly developed horizons. In addition to these 11 orders are soil series. In the United States there are about 7,000 of these, with names like Hagerstown series. These may be attached to structural classes to produce names like Hagerstown sandy-loam.

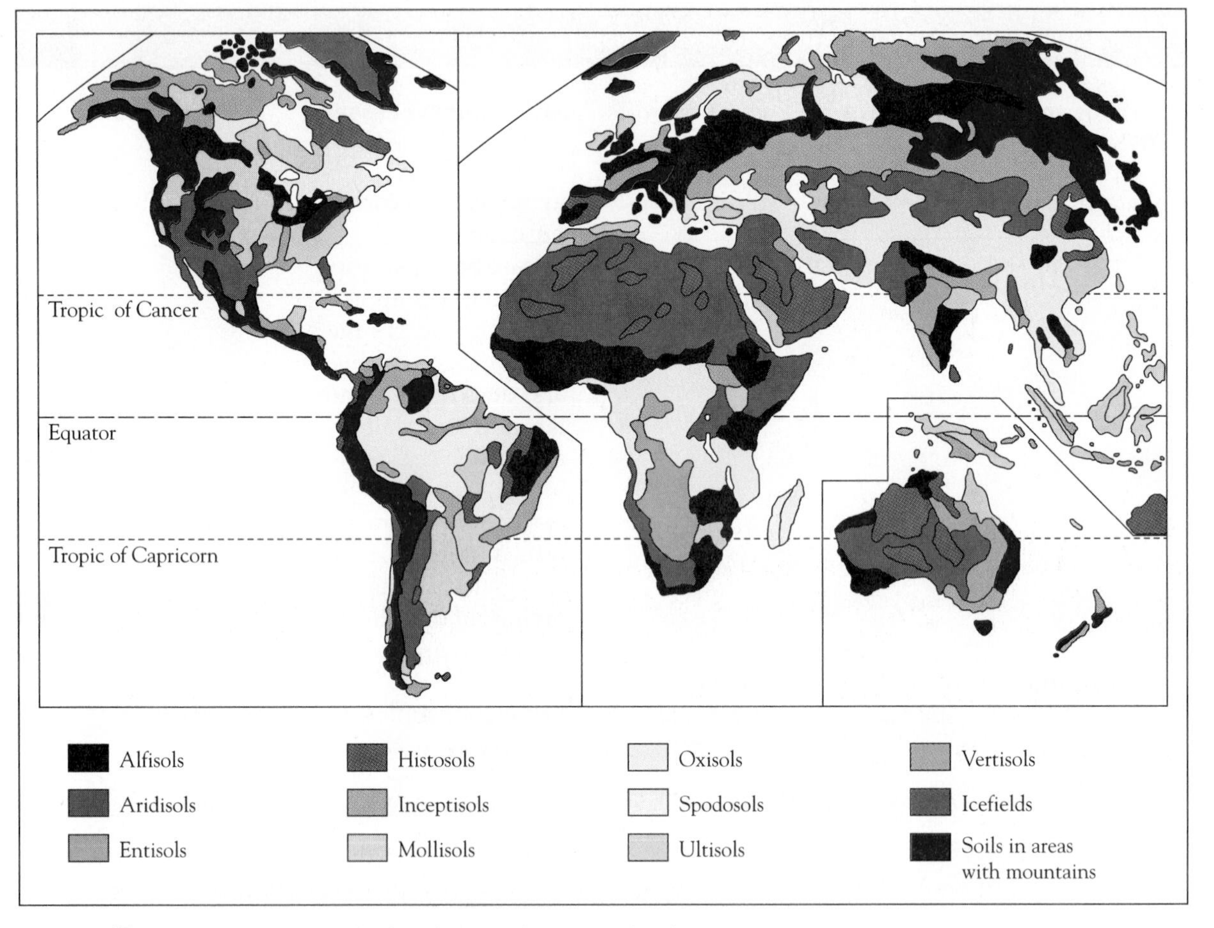

▪ **FIGURE 7.18** World distribution of major soil orders.

biochemical pathways by which the photosynthetic reaction can occur. Most plants use the C_3 pathway (or Calvin-Benson cycle); CO_2 from the air is first converted to 3-phosphoglyceric acid, which is a three-carbon molecule (hence, the name C_3) by the enzyme Rubisco, which is present in massive amounts in leaves (25.3% of total leaf nitrogen). Until the mid-1960s, this pathway was believed to be the only important means of fixing carbon in the initial steps of photosynthesis. In 1965, sugarcane was discovered to fix CO_2 by first producing malic and aspartic acids (four-carbon acids), and a new C_4 pathway of photosynthesis (Hatch-Slack) was uncovered (Ehleringer and Manson 1993). Such C_4 plants have all the biochemical elements of the C_3 pathway as well, so they can use either method to fix CO_2. Typical C_4 plants do not reach saturation light levels even under the brightest sunlight, and they always produce more photosynthate per unit area of leaf than do C_3 plants. Therefore, C_4 plants are more efficient than C_3 plants (Fig. 7.20).

C_4 species are more common in tropical areas than in temperate or polar areas (Fig. 7.21). Most C_4 species are grasses and sedges. Some desert succulents, such as cacti of the genus *Opuntia*, have evolved a third modification of photosynthesis, crassulacean

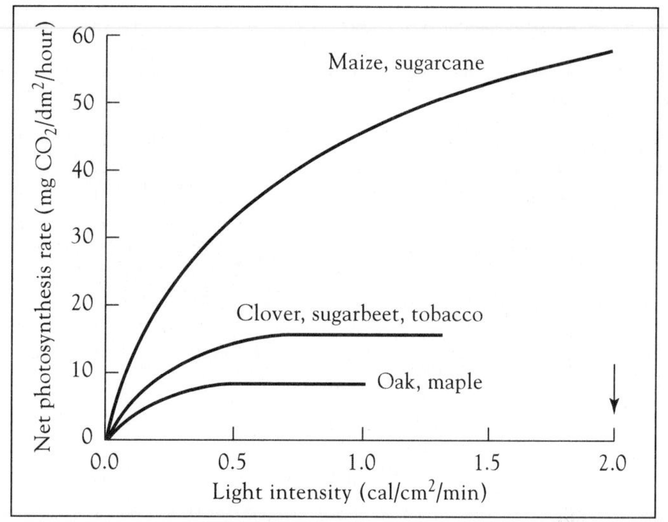

■ **FIGURE 7.19** The effect of light intensity on the rate of photosynthesis in several species of plants. Photosynthesis was measured by CO_2 uptake at 30°C and 300 ppm CO_2 in air. The arrow on the light axis marks the approximate equivalent of full summer sunlight. *(Redrawn from Zelitch 1971.)*

acid metabolism (CAM) These plants are the opposite of typical plants because they open their stomata to take up CO_2 at night, presumably as an adaptation to minimize water loss in the day. This CO_2 is stored as malic acid, which is then used to complete photosynthesis during the day. The CAM plants have a very low rate of photosynthesis and can switch to the C_3 mode during daytime. They are adapted to live in very dry desert areas where little else can grow.

The C_3-C_4 dichotomy does not just concern light. C_4 plants can absorb CO_2 much more effectively than C_3 plants. Consequently, they do not have to open their leaf stomata as much, and so they lose less water than C_3 plants. C_4 plants are therefore more frequent in arid conditions. Also, Rubisco concentration, and leaf nitrogen content, is generally lower in C_4 plants. As a consequence, C_4 plants are nutritionally less attractive

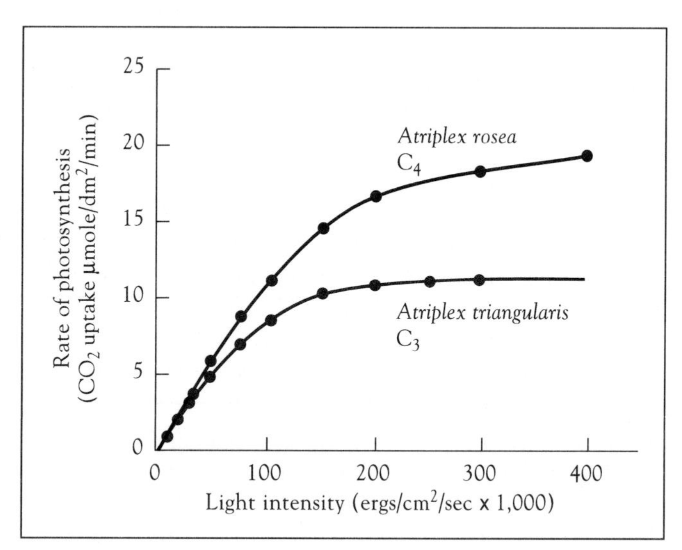

■ **FIGURE 7.20** Comparative photosynthetic production of the C_3 species *Atriplex triangularis* and the related C_4 species *Atriplex rosea*. The plants were grown under identical conditions of 25°C day/20°C night, 16-hour days, and ample water and nutrients. *(Redrawn from Bjorkman 1973.)*

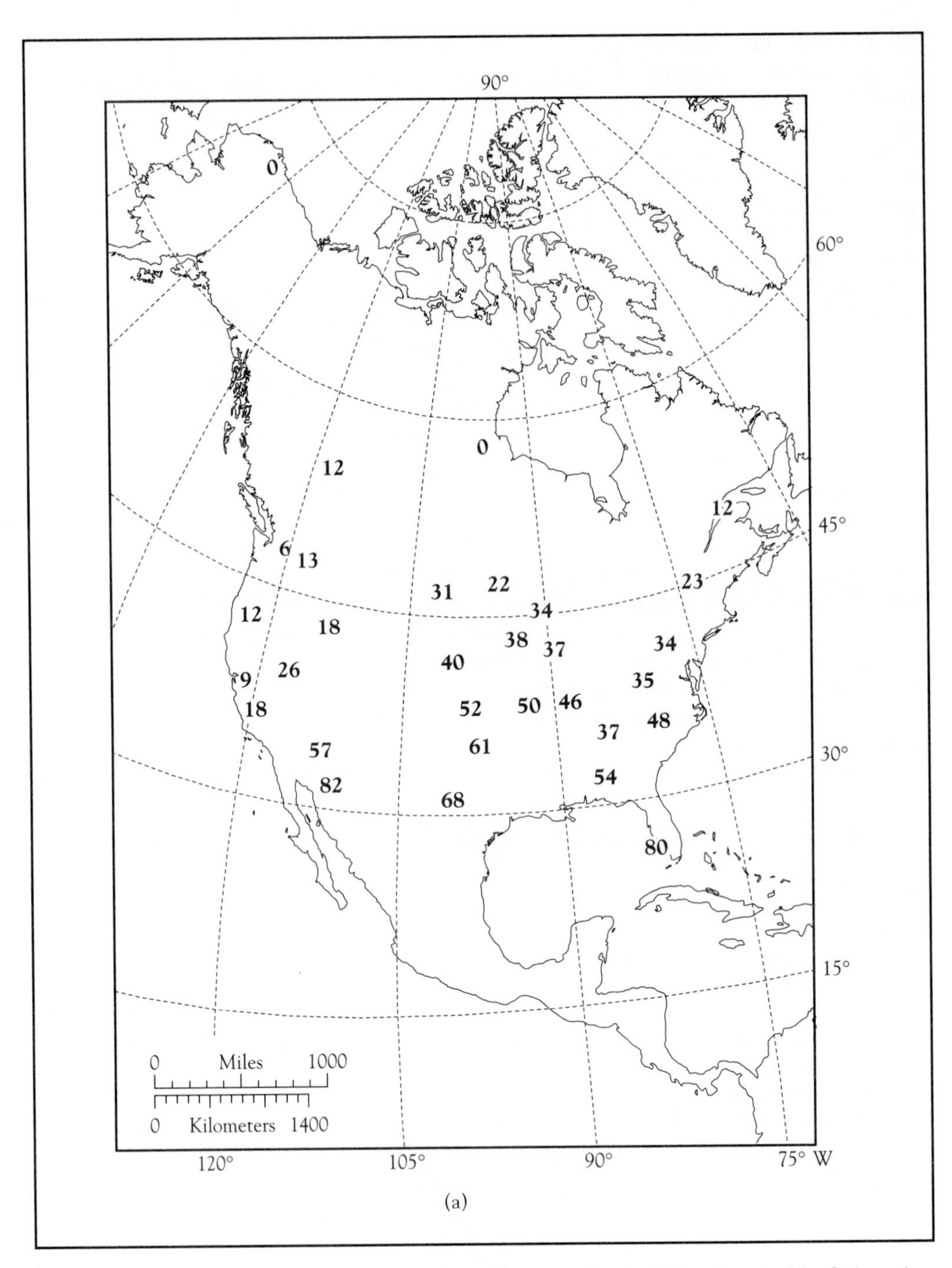

■ **FIGURE 7.21** (**a**) Percentage of C_4 species in the grass floras of 32 regions in North America. *(From Teeri and Stowe 1976.)* (**b**) Approximate contour map of C_4 native grasses in Australia. Lines give percentages of C_4 species in total grass flora for 75 geographic regions. *(From Hattersley 1983.)*

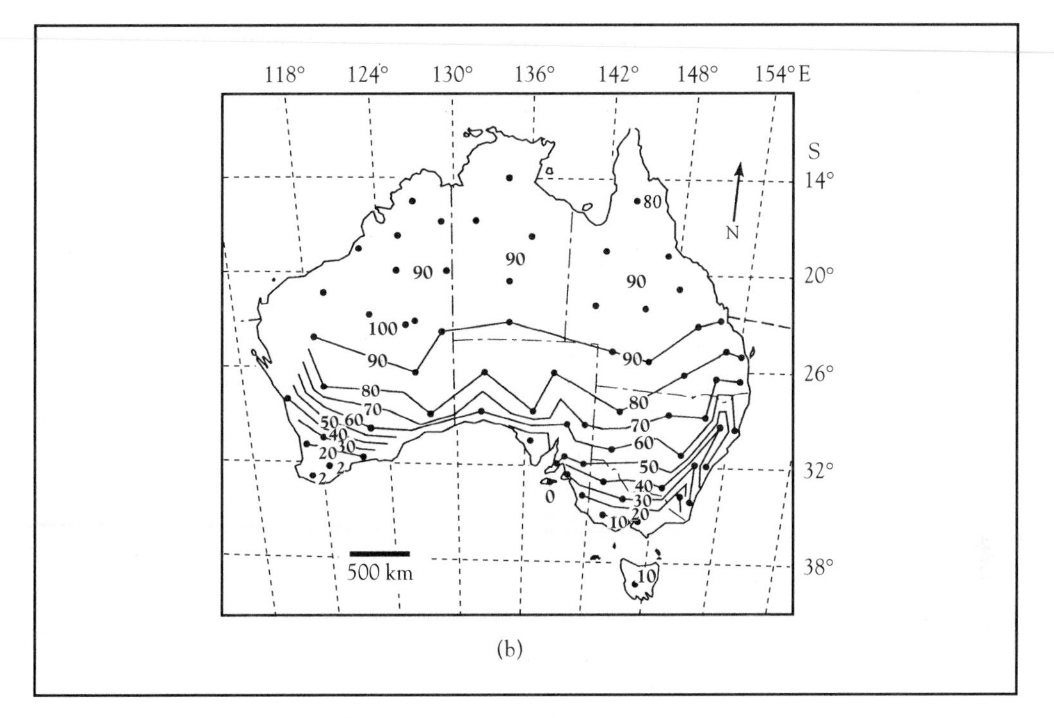

■ **FIGURE 7.21** (continued)

to some herbivores. Why haven't C_4 plants come to dominate world vegetation? Probably because they do not perform so well in the shady wet environments of northern latitudes or tropical forests. The few C_4 plants that have penetrated into temperate regions are often found in stressful environments where osmotic conditions limit water availability, e.g., *Spartina* grasses in salt marshes.

7.8 Water availability limits the distribution and abundance of many species.

Soils are critical not only because of the nutrients they provide but also because of the water they hold. Although temperature alone is an important limiting factor, water has an equally important effect on the ecology of terrestrial organisms. Protoplasm is 85 to 90 percent water, and without moisture there can be no life. Globally, there is a belt of high precipitation around the equator, broadly corresponding to the tropics, and a secondary peak between latitudes 45° and 55°. Rates of evaporation and transpiration are primarily dependent on temperature, hence, the importance of water and temperature combined. Over about one-third of the Earth, evaporation exceeds precipitation—these areas are the deserts.

The distribution patterns of most plants are limited by available water. Some, for example, the water tupelo tree in the United States, do best when completely flooded and are thus predominant only in swamps. For many plants, the limiting amount of moisture is much lower. In cold climates, water can be present but locked up as permafrost and, therefore, unavailable—a frost-drought situation. Alpine timberlines may be affected by winter desiccation and frost drought.

Animals face problems of water balance, too, and the distribution of organisms on the intertidal rock face can be strongly affected by desiccation (Connell 1961 and Fig. 9.7). However, many animals can move away from dry or intolerable environments. Many desert animals are small and can hide underground in the heat of the day. This brings home the point that some organisms have two strategies for coping with extremes—tolerance or avoidance. For plants, of course, avoidance is difficult. Larger animals too cannot avoid environmental extremes so easily, and because most depend ultimately on plants as food their distributions are intrinsically linked to those of their food sources. The distributional boundary of the red kangaroo, *Macropus rufus*, in Australia coincides with the 400 mm rainfall contour because the kangaroos are dependent on arid-zone grasses that grow there (Fig. 7.22). In the wake of an extraordinary El Niño event (an irregular increase in water temperature in the eastern Pacific Ocean) in 1982–1983, the rainfall on Isla Genovesa, Galapagos, increased from its normal 100 to 150 mm during the rainy season to 2,400 mm from November 1982 through July 1983. Plants responded

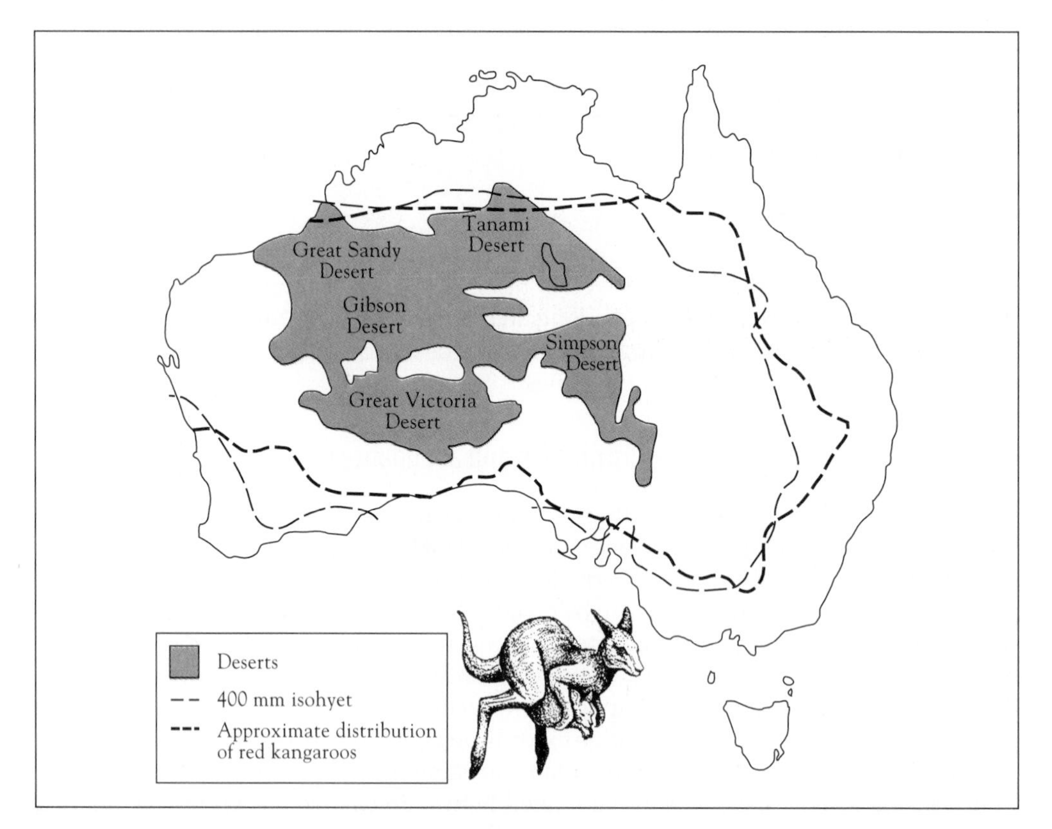

■ **FIGURE 7.22** Distribution of the red kangaroo in the arid regions of Australia and the 400 mm (15 inch) rainfall line. Red kangaroos are relatively rare in the large desert areas shown. (*Modified from Krebs 1985a.*)

with prodigious growth, and Darwin's finches bred up to eight times rather than their normal maximum of three (Grant and Grant 1987).

7.9 Abiotic factors can affect species abundance as well as distributions.

Not only does climate play an important role in the global distribution of species, it can affect their densities within these ranges too. The most famous proponents of this idea were the Australian entomologists Davidson, Andrewartha, and Birch who studied densities of small insects, called thrips, in rose bushes. Like many insects, the thrips underwent large fluctuations in density. These scientists found that 78 percent of the variation in population maxima was accounted for by variations in weather (Andrewartha and Birch 1954).

Enright (1976) suggested that density limitation by climate may be more common than we think. He argued that many ecologists are preoccupied with the effects of competitors, predators, and parasites on populations and that the effect of climate is often overlooked. Biogeographers have described whole arrays of species, or communities, that are restricted to certain climatic zones. Why should we deny that weather and climate can profoundly influence densities of species within their range? Often abundance declines toward the boundaries of a species range and is correlated with environmental variables. Of course, in the center of a range climatic variables may be less likely to cause observed density differences, if they occur. Again, the issue of scale arises. Over small scales biotic variables may be important; over larger scales climate may be more important. Even over small scales the abundance of refuges from climate may be important, and the indirect effects of climate, by reducing foliage quality, may affect the density of herbivores.

7.10 Abiotic factors may determine how many species occur in a community.

Abiotic factors not only affect the distribution patterns and abundances of individual species, they also play an important part in community richness—the number of species in any given locale can be affected by climate. Currie and Paquin (1987) showed that, of all the environmental data available, evapotranspiration rates (strongly correlated with primary production and hence available energy) are the best predictors of tree species richness in North America (Fig. 7.23).

If abiotic factors can affect species distributions, they also could potentially affect the composition of species that form a community (see also Chap. 15). The fact that tree species composition at high elevations in Georgia matches species composition at low elevations in Quebec province is suggestive of this.

It was Robert Whittaker who first formalized the concept of community structure as being governed by physical variables. Whittaker (1967) invited us to consider an environmental gradient, which could be a long, even, uninterrupted slope of a mountain. He considered four hypotheses to explain the distribution patterns of plants and animals on the gradient (Fig. 7.24).

1. Competing species, including dominant plants, exclude one another along sharp boundaries. Other species evolve toward close association (perhaps mutualisms) with the dominants. There thus develop distinct zones along the gradient, each zone having its own assemblage of species, giving way at a sharp boundary to another assemblage of species (community).

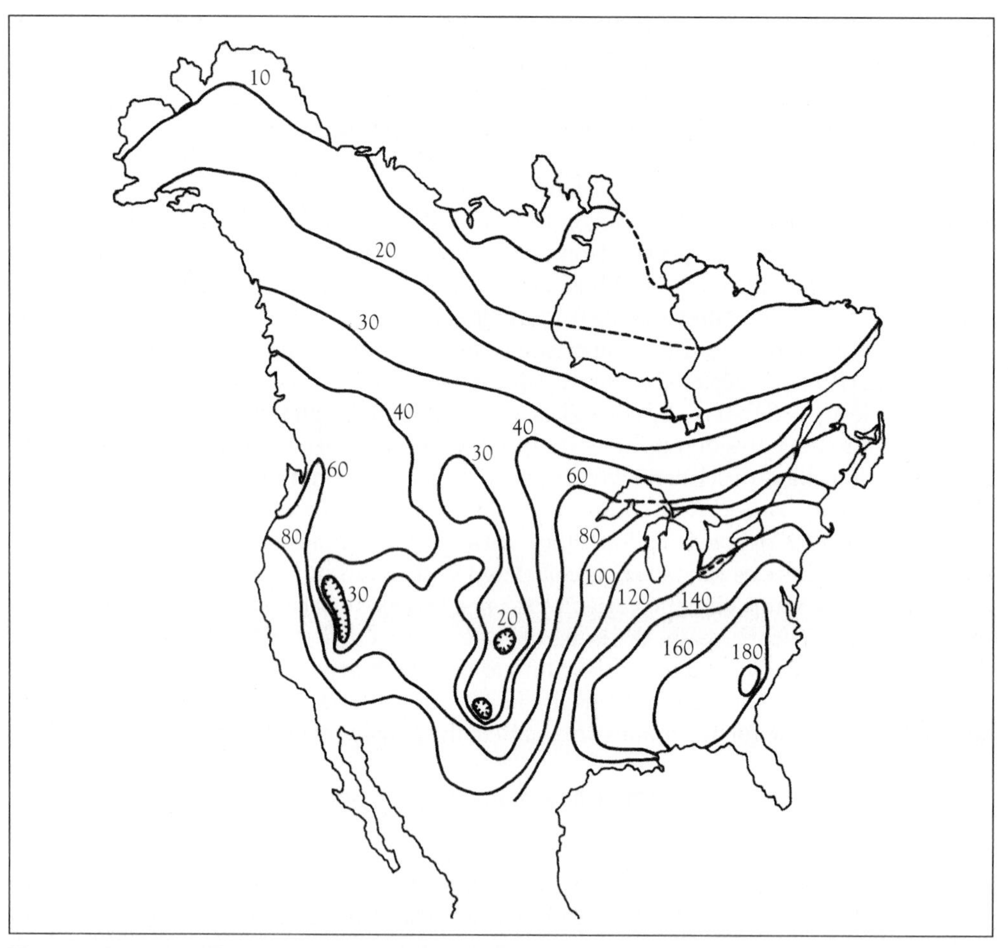

■ FIGURE 7.23 Tree species richness in Canada and the United States. Contours connect points with the same approximate number of species per quadrat. *(Redrawn from Currie and Paguin 1987.)*

2. Competing species exclude one another along sharp boundaries but do not become organized into groups with parallel distributions.
3. Competition does not, for the most part, result in sharp boundaries between species populations. Evolution of species toward adaptation to similar physical variables will, however, result in the appearance of groups of species with similar distributions.
4. Competition does not usually produce sharp boundaries between species populations, and evolution of species to similar physical variables does not produce well-defined groups of species with similar distributions. Centers and boundaries of species populations are scattered along the environmental gradient, and communities are not easily recognized.

To test these possibilities Whittaker (1967) examined the mountain vegetation in various mountain ranges of the western United States. Vegetation samples of plant popu-

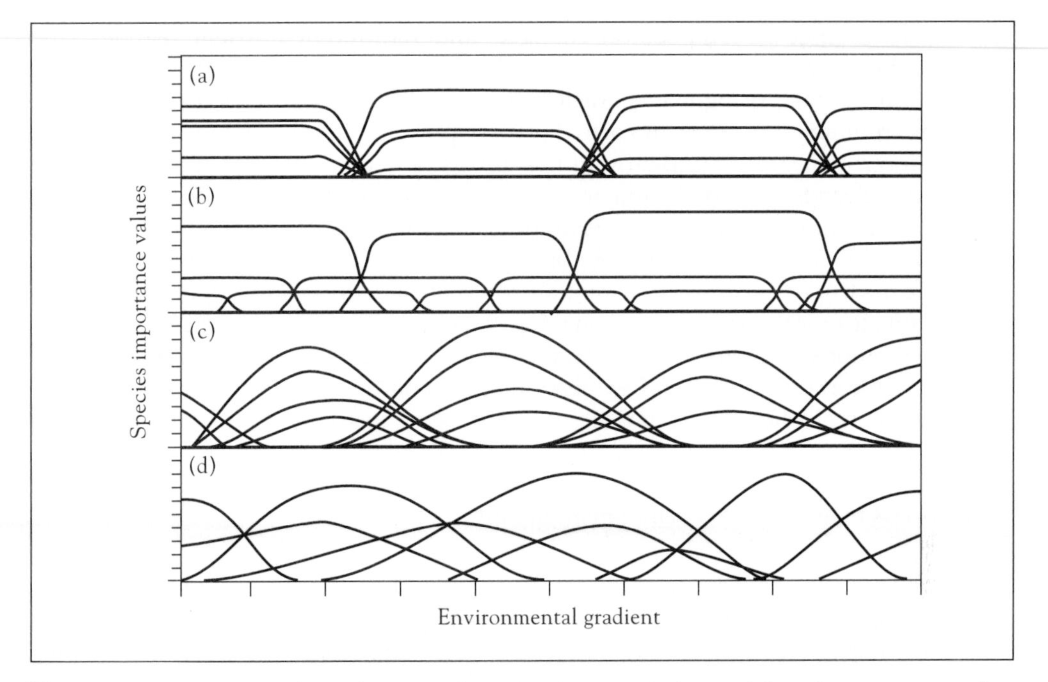

■ **FIGURE 7.24** Four hypotheses on how species populations might relate to one another along an environmental gradient. Each curve in each part of the figure represents one species population and the way it might be distributed along the environmental gradient. (*After Whittaker 1975.*)

lations were taken along an elevation gradient from the tops of the mountains down to the bases along with data on various abiotic variables like soil moisture. The results supported hypothesis D. Whittaker concluded that his observations agreed with the "principle of species individuality" (asserted by Gleason 1926, see Section 5) and that each species was distributed in its own way, according to its own genetic, physiological, and life-cycle characteristics. The broad overlap and scattered centers of species populations along a gradient implied that most communities intergrade continuously along environmental gradients, rather than forming distinct, clearly separated zones, and that species composition at any one point was largely determined by abiotic factors.

Nowhere has the link between abiotic factors and species richness been better demonstrated than on long-term grassland fertilizer trial plots at Rothamstead Experimental Farm in England, established in 1856 and continually monitored to the present day. The unfertilized plots supported some sixty species of higher plants, including all of the species of plants found on the fertilized plots (Thurston 1969). Species diversity on the unfertilized plots was high, and no species was clearly dominant. The vegetation was short and primary productivity low. On plots that received applications of phosphorus, potassium, sodium, and magnesium, but no nitrogen, legumes such as clover became dominant at the expense of other species. This is because legumes can fix their own nitrogen (see Chap. 22). The addition of nitrogen discouraged legumes, reduced their growth, and encouraged grasses. Nutrient addition to grassland enhanced the growth of a few highly competitive and lush-growing grasses at the expense of creeping and rosette-type species. Thus, species composition changed under different nutrient regimes, and species richness was much reduced (see also McGraw and Chapin 1989).

One can, in fact, recognize characteristic assemblages of species for many of the major environmental settings on Earth. For example, tall broadleaf evergreen trees and lush undergrowth grow rampantly in the tropics, whereas cacti are sparse but most prevalent in deserts. Each of the major types of floral and faunal assemblages is referred to as a **biome**. Thus, cooler, drier areas dominated by tall, deciduous trees form the temperate forest biome. A number of major biome types are recognized by ecologists, among which are the tropical rain forest, temperate forest, desert, grassland, taiga, tundra, tropical savanna, tropical scrub and seasonal forest, temperate rain forest, and chaparral. In aquatic situations there are coral reefs, the open ocean, estuaries, freshwater environments, and the intertidal biome. The meteorological conditions necessary for certain biomes and their physical characteristics are discussed in the section on communities (Section 5).

7.11 Species may adapt to specific environments.

Although many species are limited by physical or abiotic variables, even Darwin recognized that species could extend their distribution by local adaptation to limiting environmental factors. It was a Swedish botanist, Göte Turesson, who first began looking seriously at adaptations to local environmental conditions in plants. Turesson (1922) coined the word *ecotype* to describe genetic varieties within a single species. He experimented with varieties by collecting plants from a variety of areas and growing them together in field or laboratory plots at one site. The type of result he obtained in the early work can be illustrated with one example. *Plantago maritima* grows as a tall robust plant in marshes along the coast of Sweden and also as a dwarf plant on exposed sea cliffs in the Faeroe Islands. When plants from marshes and from sea cliffs are grown side by side in a an experimental garden, the height difference is not as extreme but remains significant, with over 50 percent difference between types (Turesson 1930). This method of experimentation is known as a **common garden** and is an attempt to separate the environmental and genetic components of variation. Plants of the same species growing in such diverse environments as sea cliffs and marshes can differ in morphology and physiology in three ways: (1) All differences are phenotypic, and if seeds are transplanted from one environment to the other they will respond in exactly the same way and the same as the resident individuals; (2) all differences are genotypic, and if seeds are transplanted between areas the mature plants will retain the form typical of their original habitat; or (3) some combination of phenotypic and genotypic determination produces an intermediate result. In natural situations, the third case is most usual (see review in Heslop-Harrison 1964).

Good examples of adaptation to local environments are provided by some crops. Six hundred years ago, all cotton plants were perennial shrubs confined to the frost-free tropics. Gradually, forms were selected to fruit early and produce a sizable crop in the first growing season. Early-fruiting varieties were suitable for cultivation in temperate regions where they could produce cotton before winter. Cold winters and hot summers imposed an annual growth habit, and now all commercial cottons are obligate annuals that can be grown in cold-winter areas and semiarid climates. All this—adaptation to previously lethal environments (Hutchinson 1965)—has been achieved in a maximum of six hundred generations. In less than fifty years, the grass *Agrostis tenuis* has evolved populations that live on spoil tips in Great Britain, the areas of mine wastes often rich in noxious elements like lead, copper, or zinc (Antonovics, Bradshaw, and Turner 1971). In this case, natural selection has favored the very few individuals tolerant to such areas, and these have prospered. Races of tolerant grasses are now being used to revegetate toxic soils (Smith and Bradshaw 1979).

SUMMARY

1. The local distribution patterns of most species are limited by certain physical or abiotic factors of the environment such as temperature, moisture, light, pH, soil, salinity, and water current. There are two broad classes of abiotic variables—those that directly affect organisms, such as temperature, wind, pH, and salinity, and those that act as resources, such as nutrients, light, and water.

2. One of the most important environmental variables in terms of its influence on plant and animal distribution patterns is temperature. Temperature resistance in plants, especially frost resistance, is critical to their distributions. High temperature resistance, especially to fire, is also important.

3. Wind can amplify temperature gradients and affect species distributions directly.

4. Soil salt concentrations and pH both can affect water uptake in plants and be important for aquatic organisms.

5. Nitrogen and phosphorus are two of the most limiting nutrients to plants and animals.

6. The amount of available light strongly affects plants' photosynthetic rates. Different metabolic pathways, C_3, C_4, or CAM, exist in plants in different light regimes.

7. Plants and animals face problems of water balance, and water limits the distribution of many species.

8. If abiotic factors can limit where species occur, they probably have a strong influence on population densities within these areas of distributions too.

9. The influence of humans on abiotic variables may be substantial. Increasing CO_2 output from the burning of fossil fuels and increasing deforestation may cause global CO_2 levels to increase so much that the world's climate might change slightly (so-called global warming). Other anthropogenic changes include increased SO_2 output from fossil-fuel burning, leading to acid rain, acid lakes, and acid streams.

DISCUSSION QUESTION

Do you think abiotic factors, such as soil nitrogen or available water, influence plant growth more than biotic variables, such as the presence of herbivores or competition from other plants that compete for the same resources?

CHAPTER

Mutualism

In addition to their own internal dynamics, populations are affected by the other species with which they interact. Species interactions can take a variety of forms, summarized in the following table.

NATURE OF INTERACTION		SPECIES 1	SPECIES 2
Mutualism*	(Chapter 8)	+	+
Commensalism*	(Chapter 8)	+	0
Predation	(Chapter 10)	+	−
Herbivory	(Chapter 11)	+	−
Parasitism*	(Chapter 12)	+	−
Allelopathy	(Chapter 9)	−	0
Competition	(Chapter 9)	−	−

+ = positive effect; 0 = non effect; − = deleterious effect.
*Examples of a symbiotic relationship, in which the participants live in intimate association with one another.

Herbivory, predation, and parasitism all have the same general effects, a positive effect on one population and a negative effect on the other. Competition affects both species negatively. **Mutualism** and **commensalism** are less commonly discussed in ecology but are tied together, with **parasitism**, under the banner of **symbiotic** relationships. In symbiotic relationships, the partners in the association live in intimate association with one another; they are always found in close proximity (Saffo 1992; Bronstein 1994). The effects of mutualism and commensalism are different from those of parasitism, however, and are discussed separately here.

In mutualistic arrangements (reviewed by Bronstein 1994a), both species benefit. Photos in the color inserts show members of some mutualistic relationships. In mutualistic pollination systems both plant and pollinator (insect, bird, or bat) benefit, one usually by a nectar meal and the other by the transfer of pollen. In one extraordinary case, male euglossine bees visiting orchids in the tropics do not collect nectar or pollen but are rewarded instead with a variety of floral fragrances, which they modify to attract females (Williams 1983). The tightness of the relationship in some pollination systems is underlined by the phenomenon of *buzz pollination* (Erickson and Buchmann 1983). Certain flowers with anthers opening through pores at the top shed their pollen when subjected to vibrations emanating from the buzzing of bee wings. To ensure that some pollen falls on its target bee below, the pollen is negatively charged. As the pollen rains down, it is attracted electrostatically to the bees, which tend to have positive charges.

Both parties also benefit in mutualistic fashion from seed dispersal when fruits are eaten by frugivorous birds, bats, or mammals; the consumer receives a meal and the plant receives an effective means of progeny **dispersal**. The lack of plant mobility has made many of them dependent on animals for pollination and seed dispersal. Temple (1977) argued that the tree *Calvaria major* on the island of Mauritius has produced no seedlings for the past three hundred years because its seeds do not germinate unless they have first passed through the digestive system of the now-extinct dodo. However, Witmer (1991) exploded this myth by noting that, (a) seeds can geminate without abrasion in bird guts and (b) some living trees less than three three hundred years old exist, whereas the dodo went extinct in the 1660s, more than three hundred years ago. Corals are mutualists too. The animal polyps contain unicellular algae. Different species of coral reef sponges themselves live in a mutualistic arrangement protecting one another from storm breakages and predation (Wulff 1997). The mutualistic association of humans with domestic animals or crops has permitted the most far-reaching ecological changes on each (Applied Ecology: Humans in Mutualistic Relationships).

Several recent articles (Cherif 1990; Keddy 1990) have commented that mutalistic interactions are not covered in sufficient detail in modern ecology texts. This is in contrast to ecological textbooks of the 1920s to the 1940s where positive interactions were hypothesized to be important driving forces in communities (Clements, Weaver, and Hansson 1926; Allee et al. 1949). However, even in the ecological literature the frequency of research articles on mutualism (14%) is much less than that for other interaction types such as competition (31%), predation (24%), or herbivory (30%) (Bronstein 1991). So textbooks may merely reflect the state of the ecological literature. This low level of representation could be because many mutualism studies have been descriptive and have focused on particular adaptations or life history characteristics of organisms and not on theory, whereas studies on other interactions focus on the interaction itself and the mechanisms involved. Bertness and Hacker (1994) also suggest that studies of mutualism are not often experimental and are paid little attention to by theorists.

There are few appropriate models for mutualistic interactions. We could incorporate the positive effect of one species on the other by modifying the Lotka-Volterra equations such that

$$\frac{dN_1}{dt} = r_1 N_1 \frac{(K_1 - N_1 + \partial_{12} N_2)}{(K_1)}$$

ECOLOGY IN PRACTICE

Judith Bronstein, University of Arizona

A glance at this textbook should be enough to convince you that much of ecology deals with antagonistic interactions between organisms, such as competition, predation, parasitism, and herbivory. There is no question that in nature, mutually beneficial interactions within and between species are also very common. Ecologists only rarely study them, though. At the moment, our understanding of cooperation mostly exists through a set of well-known case studies. The goal of my research is to help move the study of cooperation away from the usual storytelling approach and toward the kind of rigorous hypothesis testing that ecologists have long used to study antagonisms.

I first became interested in community ecology and species interactions through my undergraduate courses at Brown University. As a beginning graduate student at the University of Michigan, a summer spent assisting my professor Beverly Rathcke in her field studies of plant/pollinator communities helped to focus my interests on mutualisms (cooperative interactions between species). The pivotal experience that shaped my future research came when I had the opportunity to take a semester-long course in tropical ecology in Costa Rica. These wonderful courses are offered to graduate students several times each year through the Organization for Tropical Studies, a group of universities that cooperate to provide tropical research and educational opportunities.

The instructors of our course introduced us to some of the most spectacular of the stories that have shaped our limited understanding of mutualism. The one that really piqued my interest was the fig pollination mutualism. There are about eight hundred species of figs in the tropics, each of which is pollinated by a different species of fig wasp. These tiny wasps (one of my mentors jokes that they are "about the size of cosmic dust") deliberately pollinate figs, then lay eggs in some of the fig's flowers; their offspring feed on many of the fig's seeds. What intrigued me was that for the fig trees, fig wasps are simultaneously beneficial (since adults are pollinators) and harmful (since the pollinators' offspring are seed predators). I wondered, how did these trees end up with such destructive insects as their only possible pollinators? Is there anything the trees could do to lower the damage that the wasps inflict, without killing them off entirely (since if they did, the trees would never reproduce again)? More generally, how can this mutualism persist over the long term, in light of the apparent conflict between the partners? None of the answers to these questions were known. My graduate research eventually focused on measuring those conflicts and some of their ecological consequences in one Costa Rican fig/fig wasp mutualism.

It would have been easy to have used my findings simply to add more facts to the large database on these odd interactions. Instead, my professors pushed me, and I pushed myself, to take the tougher but much more scientifically significant route of looking beyond the specifics of my study system to address big, general questions about mutualism. But what were those big questions? To my surprise, I discovered that the study of mutualism was so new that researchers were barely beginning to suggest what they were. I chose to focus on the hypothesis that in all mutualisms, not just the fig/fig wasp interaction, there are serious conflicts between the partners. (The basic idea is that mutualism is a kind of "reciprocal parasitism": each partner is out to do the best it can, by obtaining what it needs from its mutualist at the lowest possible cost to itself.) Ultimately, by taking the risk of "thinking big" in my graduate research, I found myself playing a major role in developing the very young field of mutualism studies. If instead I had been content to see my goal as simply to discover more facts about figs, I'm sure that today I would be an unemployed fig specialist rather than a professor.

More recently, the challenge has become to convince other ecologists that my chosen specialty is an important and interesting one. Many suggestions have been offered over the years for why ecologists have studied mutualism so little, especially compared to competition and predation. Some have attributed it to Western male scientists' supposed biases toward seeing aggression in nature. I tend to disbelieve this (though there does seem to be a trend for mutualism to be studied by women). Others argue that cooperation is simply not a biological phenomenon important enough to deserve much attention. I strongly disagree. But, as a scientist, it is not sufficient for me to simply believe that this view is wrong. My responsibility is to convince others that I am right by conducting good ecological studies that critically test my ideas.

I am currently associate professor of ecology and evolutionary biology at the University of Arizona. Like all university professors, I must find a way to balance research with my equally important (and enjoyable!) role as a teacher and mentor of graduate and undergraduate students. In my case, I must also set aside large blocks of time to spend with my husband and new son. Trying to balance these conflicting demands on my time has certainly proved to be the biggest challenge of my career. My primary role models these days are the increasing numbers of professionals, especially in university positions, who are successfully managing the delicate balancing act between family and career. I would like to think that the days are over when having both a thriving career and a happy home life are thought to be mutually exclusive for women.

and

$$\frac{dN_2}{dt} = r_2N_2\frac{(K_2 - N_2 + \partial_{21}N_1)}{(K_2)}$$

where ∂_{12} and ∂_{21} are the positive effects of species 2 on species 1 and species 1 on species 2, respectively. (See Chap. 9 for a fuller discussion of similar modifications to the Lotka-Volterra equations for competition theory.) Such modifications often lead to unrealistic solutions in which both populations increase to unlimited size. We could allow each species to increase the carrying capacity of the other but place a limit on the interaction such that $\partial_{12}\partial_{21} < 1$. Again, the models are unstable and often lead to the extinction of one species (May 1976). However, some authors have suggested that is an accurate finding: few obligate mutualisms in nature are very stable in the face of environmental change.

Some generalizations of mutualisms have been advocated:

1. The need for mutualism decreases with increased resource availability. For example, mycorrhizal fungi augment certain plant's limited supply of nutri-

APPLIED ECOLOGY

Humans in Mutualistic Relationships

The mutualisms with perhaps the most important implications involve those associated with human agriculture (Photo 1). Most introduced crops will always remain dependent on people for their survival—requiring that water or competing weeds be removed. Most of the world's crops (whose origins are shown in Table 1) have been moved and only recently. Introduced animals such as sheep, probably derived from Asiatic mouflon (*Ovis orientalis*), and cows, from the European auroch, are also recent animals. The population size of humans is increased in the presence of these crops and domesticated animals and, likewise then, populations of the crops and domesticated animals would be reduced in the absence of humans. Fully 35 percent of the world's land area is given over to crops and grazing pastures (Table 2), and the head of livestock in the world runs into the billions (Table 3).

Unfortunately, the side effects of this mutualism on natural systems are immense:

■ **PHOTO 1** Modern agriculture illustrates one of the most far-reaching kinds of mutualism. These radishes, near Perrydale, Oregon, could not exist without human help and, at the same time, crops sustain human populations. Such mutualisms between humans and crops have radically altered the ecological landscape. (*Bob Pool, Tom Stack & Associates 3180-231.05.*)

1. Pollution of water bodies. Runoff water often contains high levels of nitrogen and phosphorus from fertilizers as well as residual pesticides and manure, all of which contaminate streams. These can all affect communities of aquatic plants and animals.
2. Loss of topsoil. Some four billion tons of topsoil are washed into U.S. waterways each year. Topsoil can also be whipped away by wind.
3. Continued irrigation for agriculture can cause salting of the land, as was being discovered in California in the 1970s. Even "fresh" water contains minute quantities of dissolved salts, and as water evaporates or is used by plants, the salt is left behind and accumulates in the soil. Mesopotamia, probably the world's oldest irrigated area, on the plains of the Tigris and Euphrates, had a fertility legendary throughout

TABLE 1 Geographic origin of some important crops.

CROP	AREA OF ORIGIN
Apple	Western Asia
Banana	Malaysia
Carrot	Eastern Mediterranean
Coffee	Ethiopia
Cucumber	Egypt, India
Eggplant	India
Lettuce	Egypt
Orange	Indochina
Peach	China
Peanut	Central and South America
Pineapple	Amazonia
Potato	Andes
Rhubarb	Russia
Rice	Asia
Sugarcane	India
Tea	India
Tomato	Mexico
Wheat	Nile Valley

TABLE 2 Agricultural and grazing lands, in thousands of hectares.

	CONTINENTAL AREA EXCLUDING INLAND WATER	ARABLE LAND AND PERMANENT CROPS	%	PERMANENT PASTURES AND MEADOWS	%
World	1,3078,873	1,476,483	11.3	3,170,822	24.2
Africa	2,964,595	184,869	6.2	788,841	26.6
North and Central America	2,242,075	274,626	12.2	367,062	16.4
South America	1,781,851	140,638	7.9	458,364	25.7
Asia	2,757,252	454,253	16.5	644,669	23.4
Europe	487,067	139,625	28.7	84,260	17.3
Oceania	850,967	50,285	5.9	453,026	53.2
Former U.S.S.R.	2,240,220	232,187	10.4	374,600	16.7

Note: Permanent crops include tree crops such as rubber, citrus, and cocoa.

TABLE 3 Number of the seven species of food livestock in the world in 1992.

LIVESTOCK	NO. (MILLIONS)
Turkeys	259
Goats	574
Ducks	580
Pigs	864
Sheep	1,138
Cattle	1,284
Chickens	11,279

the Old World. The area now evidences some of the world's lowest crop yields (Eckholm 1976).

4. Severe grazing by livestock can induce desertification. The cattle population in the Sahel, the region immediately south of the Sahara desert in Africa, increased fivefold in the period 1940–1968.
5. Finally, agriculture can result in a severe loss of wildlife through habitat removal due to deforestation, drainage of swamps, and hedgerow removal.

ents, when phosphorus is added to the soil plants reduce their mycorrhizal infections (Bowen 1980).

2. Mutualisms are more common in stressful habitats (see Section 8.3).
3. Per-capita benefits of mutualism in large populations may be reduced. For example, species inhabited by nutrient-providing cellular organisms often have mechanisms that limit population growth of these organisms (Falkowski et al. 1993).

8.1 Interactions between plants and their pollinators are commonly, although not always, mutualistic.

Nearly 45 percent of all studies of mutualism involve pollination systems (Bronstein 1991). This may be partly because of the great diversity of apparently tight, **coevolved**, and interesting systems to study and partly because money is available to study them. Over ninety crops in the United States alone are pollinated by insects.

There is strong selective pressure for plants to develop intimate relationships with their pollinators. Up to 37 percent of the photosynthate that the milkweed, *Asclepias*, produces is used to produce nectar (Southwick 1984). One of the finest studies in obligate mutualism involves figs and fig wasps as studied by Janzen (1979a). Over nine hundred species of *Ficus* exist, and virtually every one must be pollinated by its own species of agaonid wasp. The fig that we actually eat has an enclosed inflorescence containing many flowers. A female wasp enters through a small opening, pollinates the flowers, lays eggs in their ovaries, and then dies. The progeny develop in tiny galls and hatch inside the fig. Males hatch first, locate female wasps within their galls, and thrust their abdomens inside the galls to mate with them. The males then die, without ever having left the fig. Females collect pollen from the fig and then leave to search out new figs in which to lay their progeny. Both fig and wasp benefit—the flower is pollinated and fruits form and so do seeds, although some are eaten by the developing wasp. There is clearly a necessary balance between mutualism and over-exploitation. A similar mutualistic relationship occurs between yucca plants and yucca moths (Addicott 1986). Both mutualisms are highly coevolved. The distribution of each species of yucca or fig is controlled by the availability of its pollinator and vice versa. In the late nineteenth century, Smyrna figs were introduced into California, but they failed to produce fruit until the proper wasps were introduced to pollinate them.

It is interesting that such highly coevolved systems arose because on superficial examination the needs of the plants and those of the pollinators seem to conflict sharply. From the plant's perspective, an ideal pollinator would move quickly among individuals but retain a high fidelity to a plant species, thus ensuring that little pollen is wasted as it is inadvertently brushed onto the pistils of other plants. The plant should provide just enough nectar to attract a pollinator's visit. From the pollinator's perspective, it would probably be best to be a generalist and to obtain nectar and pollen from flowers in a small area, thus minimizing energy spent on flight between patches. This casts doubt on whether such relationships are truly mutualistic in nature or whether both species are actually "trying to win" in an evolutionary arms race. One way in which the plants encourage the pollinator's species fidelity is by sequential flowering through the year of different plant species and by synchronous flowering within a species (Heinrich 1979). Mutualistic interactions are evolutionarily stable only when both interacting species possess mechanisms to prevent excessive exploitation. For example, in the yucca moths, flower abortion results if too many moth eggs are laid (Pellmyr and Huth 1994).

There are cases where both flower and pollinator try to cheat. In the bogs of Maine, the grass pink orchid *(Calopogon pulchellus)* produces no nectar, but it mimics the nectar-producing rose pogonia *(Pogonia ophioglossoides)* and is therefore still visited by bees. Bee orchids have even gone so far as to mimic female bees; males pick up and transfer pollen while trying to copulate with the flowers. So effective are the stimuli of flowers of the orchid genus *Ophrys* that male bees prefer to mate with them even in the presence of real female bees! Conversely, some *Bombus* species cheat by biting through the petals at the base of flowers and robbing the plants of their nectar without entering through the tunnel of the corolla and picking up pollen. Recent studies of mutualism in yuccas has revealed a slew of bogus yucca moths that oviposit into fertilized flower and fruit, so depending on the real pollinating moths for the subsistence of their larvae (Pellmyr et al. 1996). However, in a clever study in which he put plastic "anti-theft" collars on bluebell flowers in Alaska, Morris (1996) showed that the "cheaters" actually often switched between nectar-robbing and pollen collection, depending on flora development stage. Nectar reward in mature flowers acts as an enticement to pollinators, which then enhance plant reproduction by legitimately visiting early-stage flowers.

Finally, it is interesting to speculate about why ants, usually the most abundant insects in a given area, are so rarely involved in pollination. One reason might be that the subterranean nesting behavior of many ants exposes them to a wide range of pathogenic fungi and other dangerous microorganisms, to which they respond by producing large amounts of antibiotics. These antibiotics inhibit pollen function (Beattie et al. 1984; Beattie 1985). Peakall, Beattie, and James (1987) demonstrated that ants without metapleural glands, and therefore without these secretions, do successfully pollinate orchids in Australia.

8.2 Plant seeds and fruits provide highly nutritious rewards to animals that disperse them.

Mutualistic relations are highly prevalent in the seed-dispersal systems of plants. Studies on these dispersal systems are also highly prevalent in the mutualism literature, with 38 percent of studies focusing on dispersal. In the tropics, some fruits are dispersed by birds that are strictly frugivorous. These fruits provide a balanced diet of proteins, fats, and vitamins (Proctor and Proctor 1978). In return for this juicy meal, birds unwittingly disperse the enclosed seeds, which pass unharmed through the digestive tract. Some plants, instead of producing highly nutritious fruits to attract an efficient disperser, simply produce abundant mediocre fruit in the hope that some of it will be eaten by generalists. Fruits taken by birds and mammals often have attractive colors—red, yellow, black, or blue; those dispersed by nocturnal bats are not brightly colored but instead give off a pungent odor to attract the bats. (In contrast, because birds do not have a keen sense of smell, fruits eaten by them are generally odorless.)

In general, the relationships are not as obligately mutual as are plant-pollinator systems, because seed dispersal is performed by more generalist agents. Nevertheless, a wide array of adaptations exists; one has only to look at the impressive specialization of parrot beaks, strong and sharp to crack and peel fruits, to see that the mutualistic relationship between plant and seed disperser is strong in this case. Some bizarre strategies also exist; for example, in the floodplains of the Amazon, fruit-and seed-eating fish have evolved that disperse seeds (Goulding 1980). Microbes appear to be good at cheating the plant in this system, for they will readily attack the fruit without dispersing it. Janzen (1979c) has suggested that microbes deliberately cause fruit, and other resources like carcasses, to "rot," reserving it for themselves, by manufacturing ethanol and rendering the medium distasteful to vertebrate consumers. Animals eating such food are selected against because they become drunk and are easy victims for predators. As a countermeasure, some vertebrates have "learned" evolutionarily to tolerate these microorganisms and even to use them in their own guts to digest food. Alternatively, they often possess the enzyme alcohol dehydrogenase to break down alcohol.

8.3 In addition to the well-studied pollination and dispersal mutualisms, mutualisms can take a variety of other forms.

Some mutualisms decrease the susceptibility of one partner to parasites.

On coral reefs "cleaner" fish nibble parasites and dead skin (which might otherwise cause disease) from their customer fish at specific cleaning stations (Ehrlich 1975). Such systems often leave their participants open to cheaters. Saber-toothed blennies of the genus *Aspidontus* bear a striking resemblance to the common cleaner wrasse, *Labroides dimidiatus*. Instead of performing a cleaning function, however, the blenny bites chunks out of the customers. Saber-tooths are protected from attack by their resemblance to the cleaners, though customers, in time, learn to avoid the cleaning stations that blennies frequent. Even the cleaner fish themselves can damage the scales of their hosts, and some authors question whether this interaction is ever mutualistic (Gorlick, Atkins, and Losey 1978).

Some mutualisms increase the availability of resources to both partners.

One of the oldest ideas about mutualisms is that they are more common under harsh physical conditions when neighbors buffer one another from limiting physical stresses (Hay 1981; Wood and Del Moral 1987; Bertness 1987; Callaway 1997; Hacker and Gaines 1997).

Pugnaire, Haase, and Puigolefabregas (1996) showed how a strong facultative mutualism exists between two plants, the leguminous shrub *Retama sphaerocarpa* and understory plant *Marrabium vulgare* in a semiarid region in southeastern Spain. In these almost desertlike conditions, the interaction between these two species was indirect: the soil nitrogen conditions and moisture were much improved in the "islands of fertility" that resulted from their association. *Retama* shades *Marrubium* providing a favorable microclimate while *Marrabium* somehow enhances the availability of water to *Retama*. Nutrient cycling is probably increased due to litter accumulation. As a result, both species had greater specific leaf area, leaf mass, shoot mass, more flowers, and higher leaf nitrogen content when growing together than when growing alone (Fig. 8.1).

Bertness and his colleagues (Bertness and Leonard 1997) have detected positive interactions in intertidal habitats involving organisms on the intertidal rock face and between plants in intertidal salt marshes in New England (see page 238).

There are a number of well-studied cases of insects providing protection against herbivory to their plant hosts.

In terrestrial systems one of the most commonly observed mutualisms is between ants and aphids. Aphids are fairly helpless creatures, easy prey to marauding ants. Yet in general ants tend to farm aphids like so many cattle. The aphids secrete honeydew, a sticky exudate rich in sugars, which the ants enjoy. In return, ants protect aphids from

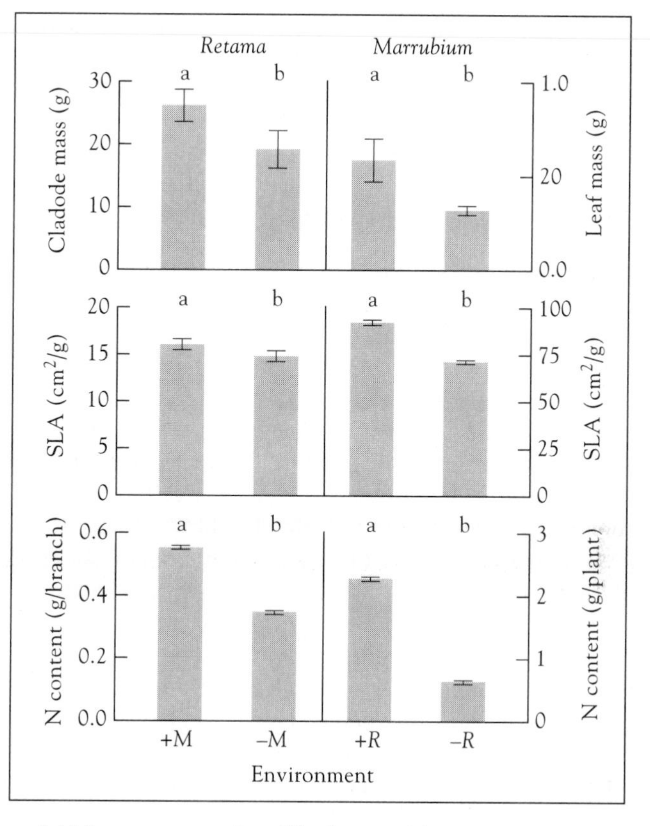

■ **FIGURE 8.1** Dry mass of cladodes or leaves per branch, specific leaf area (SLA), and nitrogen content in *Retama sphaerocarpa* (R) and *Marrubium vulgare* (M) growing in associaton (+R, +M) or alone (−R, −M) at Rambla Honda, Almerìa, Spain. Data represent means; error bars represent 1 SE; those with different letters are signficantly different at $P < 0.05$ (t test). (*After Pugnaire et al. 1996*).

an array of predators, such as syrphid larvae, parasites like braconid wasps, and other competing insects, by vigorously attacking them (Fowler and MacGarvin 1985). These three examples illustrate another common feature of mutualisms—often a third (or a fourth) species is involved in the relationship, like a competitor or predator of one of the mutualists. This is another reason why modeling mutualisms is so hard.

Obligatory mutualisms are those in which the interaction is necessary to the existence of both partners.

In some cases, the mutualistic relationship is so tight that neither participant could exist without the other, and this is called *obligatory mutualism*. Such is the case for many lichens, which are combinations of algae (which provide the photosynthate) and fungi. The "lichenized" fungi include within their bodies and near the surface a thin layer of algal cells, forming only 3 to 10 percent of the weight of the thallus body. Of the fifty thousand or so species of fungi, 25 percent are lichenized. Lichenized forms occur in deserts, in alpine regions, and across a wide range of habitats. Nonlichenized fungi are usually restricted to being parasites of plants or animals or to being involved in decomposition.

Many ruminants shelter symbiotic bacteria in their guts, which break down plant tissue to provide energy for their hosts; cellulose is otherwise indigestible for mammals (Hungate 1975). Likewise, the roots of most higher plants (except the Brassicaeceae) are actually a mutualistic association of fungus and root tissue—the mycorrhizae. The fungi require soluble carbohydrates from their host as a carbon source (up to 40 percent of the photosynthate produced), and they supply mineral resources, which they are able to extract efficiently from the soil, to the host (Allen 1992; Harley and Smith 1983; Lynch 1990).

Mutualisms can have communitywide effects.

Although most studies of mutualism focus primarily on the two species directly involved, mutualisms can have strong indirect effects. Bertness and Hacker (1994) have described an interesting mutualism in the salt marshes of New England between the perennial shrub, Marsh elder, *Iva frutescens*, and black-grass, *Juncus gerardi*. Bertness and Hacker experimentally removed *Juncus* from around some *Iva* plants and found soil salinities doubled and soil oxygen levels deceased. The photosynthetic rate of *Iva* went down and, fourteen months later, the *Iva* were dead. *Juncus* neighbors clearly have strong positive effects on adult marsh elders by reducing the soil salinity and enabling *Iva* to survive at places in the salt marsh where, alone, they would die. Hacker and Bertness (1996) went on to show how plant mutualisms had communitywide ramifications. Population growth rates of aphids on *Iva* in the absence of *Juncus* were so low that the aphids were unable to produce enough offspring to replace themselves. In addition, the number of ladybird beetle predators of the aphids changed on *Iva* and *Juncus* so the mutualism had an effect one and even two trophic levels removed from the plants themselves.

Mycorrhizal fungi can also affect herbivore loads on host plants. Ectomycorrhizal mutualists found on the roots of pinyon pines (*Pinus edulis*) can positively or negatively affect densities of the needle scale insect *Matsucoccus acalyptus* by either improving plant vigor or increasing plant investment in antiherbivore defenses. The relationship between systemic fungal endophytes and vascular plants, usually thought to be a parasitic infection, has also been viewed as mutualism (Clay 1990; Wilson 1995). The fungi are thought to aid their hosts in defending against herbivory. For example, fungal endophytes of fescue grasses, *Festuca* sp, reduce grazing by herbivores and cause widespread losses to the U.S. livestock industry (Bazely et al. 1997).

However, the few studies of interactions of endophytes with insect herbivores of woody plants show a wide range of effects from negative to positive (Gange 1995). For example, Faeth and Hammon (1997a, b) did not find strong correlative evidence that endophytes interact mutualistically with Emory oak in Arizona by increasing resistance to a dominant herbivore, the leafminer *Cameraria*. Higher infections on mined leaves were likely caused by leafmining damage that enhanced fungal colonization and penetration into oak leaves. Experimental infestation of fungal endophytes using spore suspensions did not affect leafminer survival or size at pupation.

8.4 Commensalistic relationships are those in which one partner receives a benefit while the other is unaffected.

In commensal relationships one member derives benefit while the other is unaffected. Such is the case when sea anemones grow on hermit-crab shells. The crab is already well protected in its shell and gains nothing from the relationship, but the anemone gains continued access to new food sources. The same benefits accrue to members of a phoretic relationship (**phoresy**), in which the association involves the passive and more temporary transport of one organism by another, as in the transfer of flower-inhabiting mites from bloom to bloom in the nares of hummingbirds (Colwell 1973). Some of the most numerous examples of commensalism are provided by plant mechanisms of seed dispersal. Many plants have essentially cheated their potential mutualistic seed-dispersal agents out of a meal by developing seeds with barbs or hooks to lodge in the animals' fur rather than their stomachs. In these cases, the plants receive free seed dispersal, and the animals receive nothing. This type of relationship is fairly common; most hikers have been plagued by "burrs" and "sticktights." However, sometimes these barbed seeds can cause great discomfort to the animal in whose fur they become entangled. Fruits of the genus *Pisonia* (cabbage tree), which grows in the Pacific region, are so sticky that they cling to bird feathers. On some islands, birds and reptiles can become so entangled with *Pisonia* fruits that they die. This relationship may have crossed the line from commensalism to an antogonistic relationship such as parasitism.

SUMMARY

1. Organisms do not exist alone but instead co-occur with many species where many different interactions such as mutualism, commensalism, herbivory, predation, parasitism, and competition are possible. Mutualism, commensalism, and parasitism are all symbiotic relationships where the partners live in close association with one another. However, in mutualism and commensalism, there are no negative effects of one species on the other, so discussions of these phenomena are usually treated together.

2. In mutualisms, both species benefit. A common example is pollination, where plants benefit from the transfer of pollen and where the pollinator, often an insect, gains a nectar meal. In seed-dispersal systems, the plant provides a fruit meal for birds and mammals and, in turn, benefits from dispersal of seeds into new areas.

3. Obligatory mutualisms are mutualisms in which the species cannot live apart. Examples include lichens, a mutualism between fungi and algae; mycorrhizae, an association between fungae and plant roots; and the symbiotic bacteria in ruminant guts.

4. In commensalisms, one member derives benefit while the other is unaffected. One of the most common examples of such phenomena is phoresy, the passive transport of one organism by another.

DISCUSSION QUESTION

At least six types of mutualism have been recognized in nature: those that deter predation, increase the availability of prey or resources, feed on (or compete with) a predator, increase the competitiveness of one partner in the mutualism with other species outside the mutualism, decrease the vigor of a competitor, or feed on (or compete with) a competitor (Addicott and Freedman 1984). Discuss examples of these. Where might mutualisms be more common: in more stressful temperate areas or in the more stable tropics?

C H A P T E R

Competition and Coexistence

For many biologists, implicit in Darwin's theory of natural selection was a view of nature as "red in tooth and claw" in which species scrambled to outcompete each other and leave the most offspring. How true a picture is this?

It is first worthwhile to be specific about what types of competitive event may occur in nature (see photos in color insert). Competition may be intraspecific, between individuals of the same species, or interspecific, between individuals of different species. Competition can also be characterized as resource competition or contest competition. In resource (scramble) competition, organisms compete directly for the limiting resource, each obtaining as much as it can. Under severe stress, for example, when fly maggots compete in a bottle of medium, few individuals can command enough of the resource to survive or reproduce. Such competition is most evident between invertebrates. In contest (interference) competition, individuals harm one another directly by physical force. Often this force is ritualized into threatening behavior associated with territories. In these cases strong individuals survive and take the best territories, and weaker ones perish or at best survive under suboptimal conditions. Such behavior is most common in vertebrates.

9.1 The effect of competition on the growth of a population can be predicted using mathematical models.

The Lotka-Volterra competition models are based on the logistic equation of population growth.

The growth rate of populations of two species growing independently (i.e., not interacting with each other) can be described using the logistic growth equations (Chap. 6). As before, r is the per-capita rate of population growth, N is the population size, and K is the carrying capacity. The subscripts in these equations refer to the species, so r_1 is the per capita population growth rate of species 1 and r_2 is the per capita population growth rate of species 2.

$$\frac{dN_1}{dt} = r_1 N_1 \frac{(K_1 - N_1)}{K_1}$$

and

$$\frac{dN_2}{dt} = r_2 N_2 \frac{(K_2 - N_2)}{K_2}$$

In the single-species logistic equation, the growth of a population is affected by the population size relative to carrying capacity. The quantity $K - N$ defines how far below (or above) carrying capacity a population is, and it can easily be modified to account for the presence of a second competing species. To do this, it is necessary to define a "conversion factor" that quantifies the per capita competitive effect on species 1 on species 2 and vice versa. For example, if two individuals of species 2 take up the same amount of resources as one individual of species 1, then the per capita competitive effect on species 1 of species 2 is 0.5. This conversion factor is defined by the term ∂, in which

∂_{12} = per capita competitive effect *on* species 1 *of* species 2
∂_{21} = per capita competitive effect *on* species 2 *of* species 1

The conversion factor allows the logistic equation to be modified to take into account the effects of both species on the growth rate of each species:

$$\frac{dN_1}{dt} = r_1 N_1 \frac{(K_1 - N_1 - \partial_{12} N_2)}{K_1}$$

$$\frac{dN_2}{dt} = r_2 N_2 \frac{(K_2 - N_2 - \partial_{21} N_1)}{K_2}$$

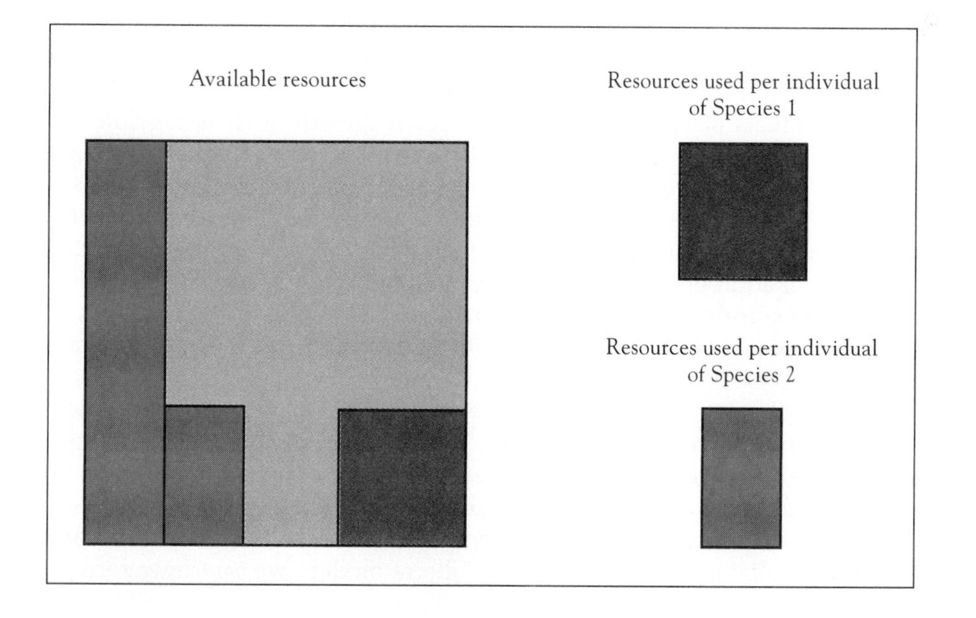

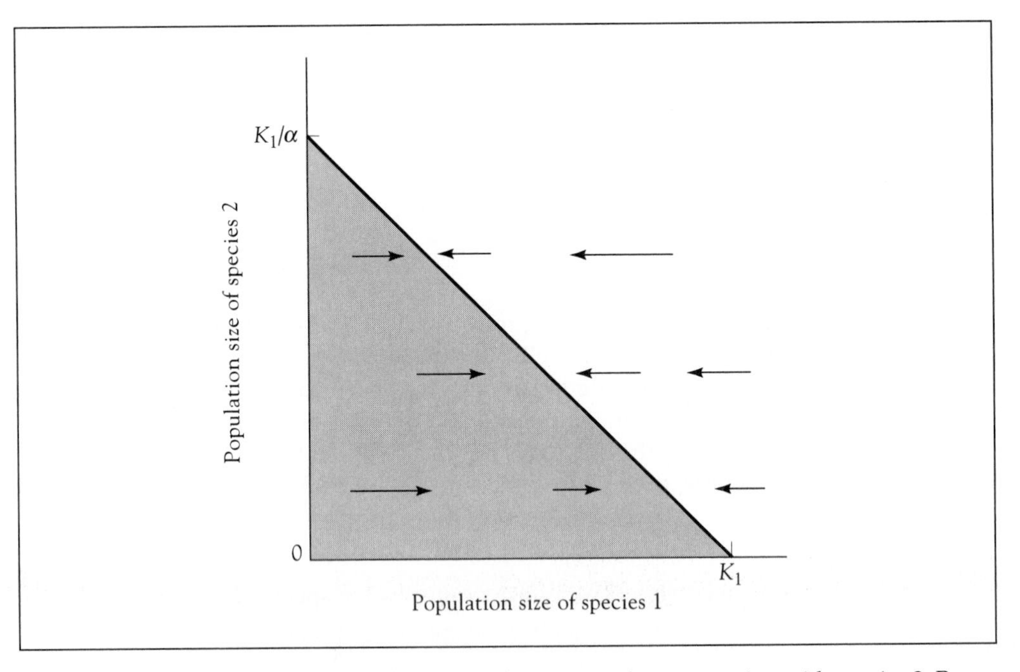

■ **FIGURE 9.1** Changes in population size of species 1 when competing with species 2. Populations in the shaded area will increase in size and will come to equilibrium at some point on the diagonal line. Along the diagonal, $dN/dt = 0$.

The population size relative to carrying capacity is now defined by the abundance of both species, and the growth rate of each population is determined by per-capita population growth rate (r) carrying capacity for that species (K) and the abundance of species 1 (N_1) and species 2 (N_2). These equations can be used to describe the joint dynamics of the two populations (e.g., how the abundance of both species changes together).

Such relationships can be expressed graphically (Fig. 9.1). Population growth of N_1 continues to the carrying capacity of the environment K_1 in the absence of N_2. If there are K_1/∂ individuals of N_2 present, no population growth of N_1 is possible because species 2 has "filled up" the available space. Between these two extremes are many combinations of N_2 and N_1 at which no further growth of N_1 is possible. These points fall on the diagonal $dN_1/dt = 0$, which is often called the *zero isocline*. Population growth of N_2 can be represented by a similar diagram. Combining the two figures and adding the arrows by vector addition illustrates what happens when the species co-occur (Fig. 9.2). Essentially, there are four possible outcomes: species 1 goes extinct, species 2 goes extinct, either species 1 or species 2 goes extinct, depending on the initial densities, or the two species coexist.

A drawback to the Lotka-Volterra model of competition is that no mechanisms are specified that drive the competitive process. Tilman (1982, 1987) criticized this approach and emphasized that we need to know the mechanism by which competition occurs. Knowing the mechanism will enable better predictions of the outcome. Tilman began by considering the responses of two competing plant species to environmental variables, say nitrogen and light. As with the Lotka-Volterra models, we can draw zero-growth

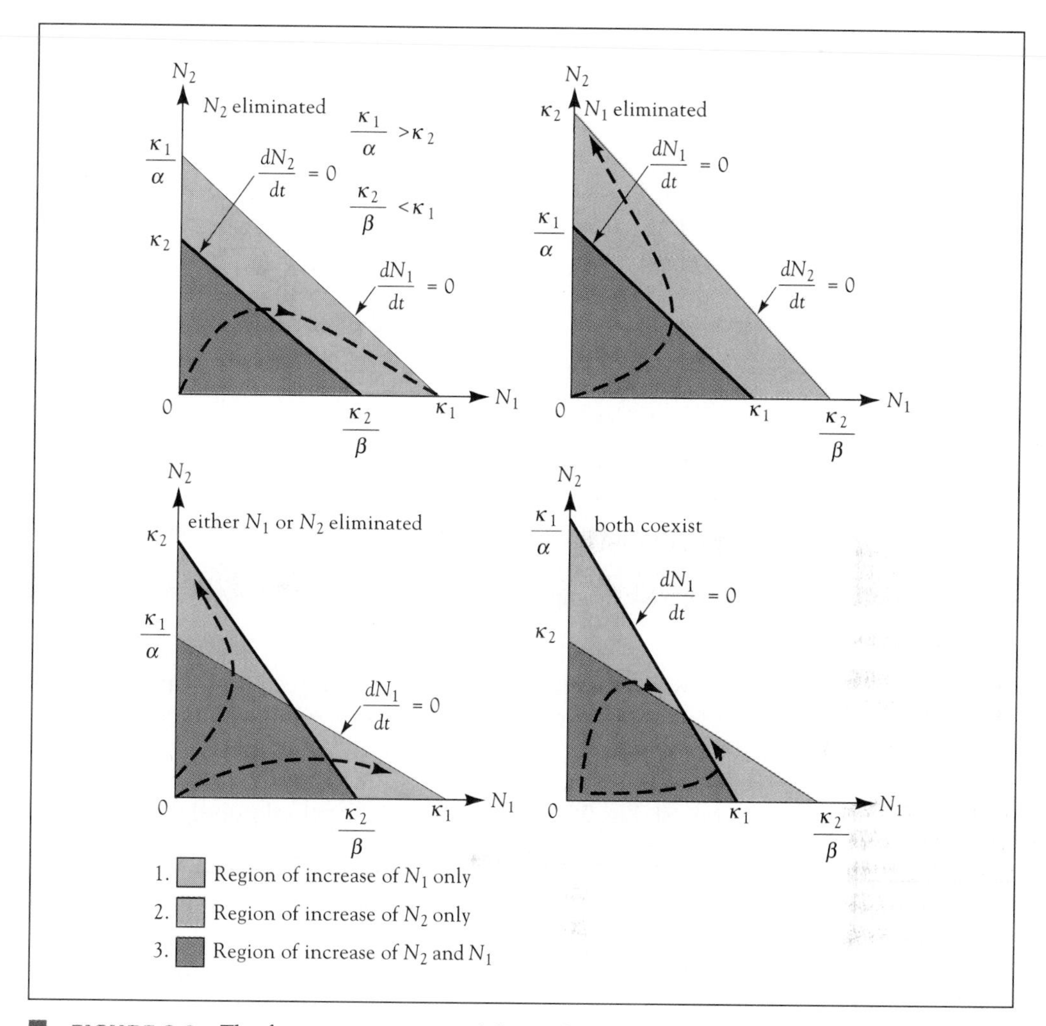

■ **FIGURE 9.2** The four consequences of the Lotka-Volterra competition equations. Axes are population sizes of N_1 and N_2.

isoclines for both species, this time based on their responses to light and nitrogen (the zero isocline for one species is shown in Fig. 9.3). If light levels are too low, a species will not grow; above a certain light level, growth proceeds. The same happens for nitrogen levels. A sort of all-or-nothing response is envisioned. We can do the same for the second species. By superimposing the two zero-growth isoclines we can determine the outcome if the two species coexist in the same habitat (Fig. 9.4). Once again, four different outcomes are possible. In the first instance, species B will always need more of both resources than species A, and species A wins out while species B goes extinct. In the second instance, the roles are reversed. In the two remaining cases, the zero-growth isoclines cross and there is an equilibrium point. To determine what happens in these other two scenarios an extra piece of information is needed: the consumption curves of each species. In Fig. 9.4 (c) and (d) species A is limited by resource 2 and species B is limited

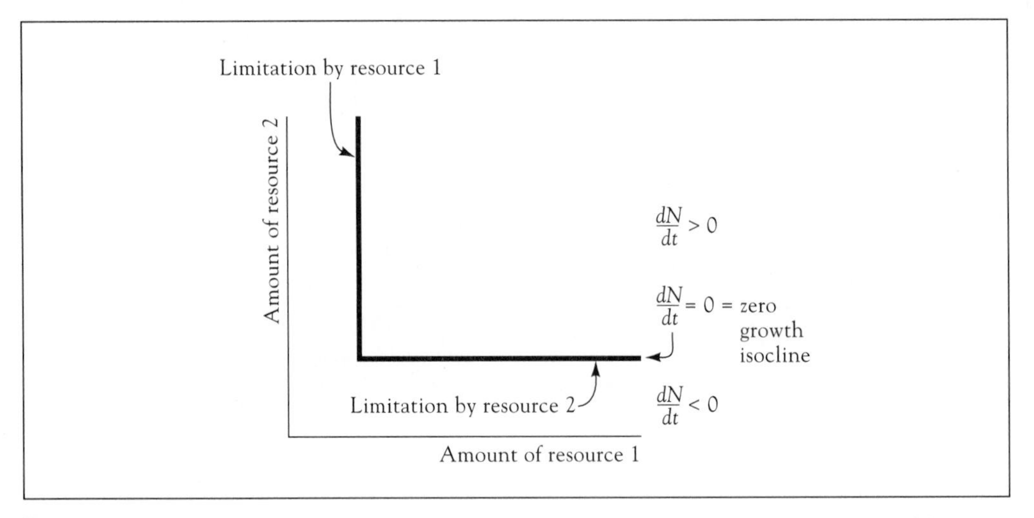

FIGURE 9.3 The response of an organism to variations in two essential resources (like nitrogen and light for plants). The thick line represents the zero-growth isocline. Above this line the population will increase in size. Below this line, the population will decline. (*After Tilman 1982.*)

by resource 1. If species A consumes relatively more of resource 1 than species B, the equilibrium point is unstable and one or other of the species will go extinct depending on which is more limited by resource 1, in this case species B. If, however, each species consumes more of the resource that limits its own growth, then the species reach a stable equilibrium and both coexist. Fig. 9.5 illustrates what the actual population densities of species A and B look like given scenario (d) in Fig. 9.4.

Tilman tested this model by growing two species of diatoms, *Asterionella formosa* and *Cyclotella meneghiniana*, in chemostats under controlled rates of nutrient supply. *Asterionella* required higher levels of silicon and *Cycloteda* higher levels of phosphorus. In Tilman's experiments there were four levels (two levels of two factors) of supply: low phosphate, high silicate to high phosphate, low silicate. As predicted by his theory, Tilman could get coexistence or get one or the other species to go extinct by varying the nutrient supply. Although this approach is encouraging, it is difficult to transfer it from the laboratory to the field, where estimation of supply points is extremely difficult (Sommer 1990).

9.2 Laboratory studies of competition have both supported and contradicted the predictions of the Lotka-Volterra competition models.

One must ask whether mathematical formulations represent real biological systems. One of the first and most important tests of these equations was performed in 1932 by a Russian microbiologist, who studied competition between two species of yeast, *Saccharomyces cervisiae* and *Schizosaccharomyces kephir* (species renamed since 1932) (Gause 1932). Alone, each species grew according to the logistic curve; the asymptote reached was a function of ethyl alcohol concentration. Ethyl alcohol is a by-product of sugar breakdown under anaerobic conditions and can kill new yeast buds just after they separate from the mother cell. In

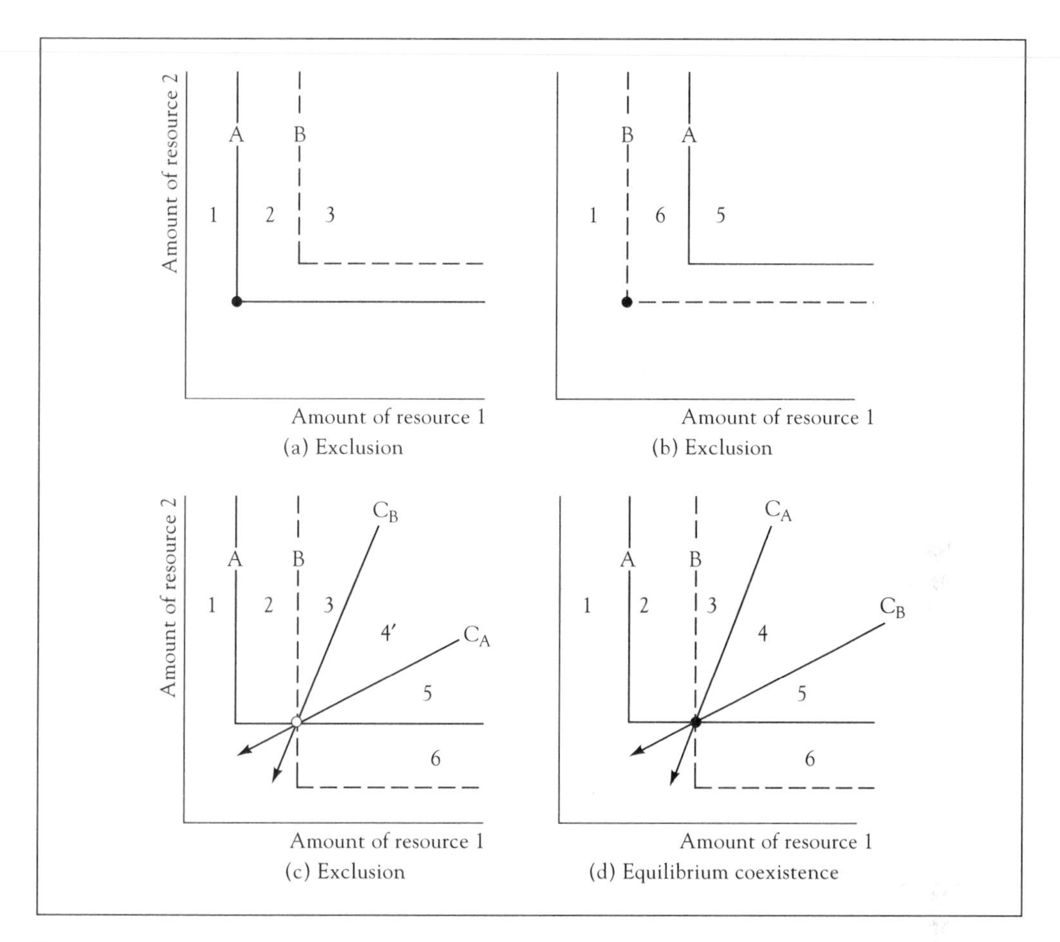

FIGURE 9.4 Tilman's model of competition for two essential resources. The zero-growth isoclines for species A and B are shown, along with the consumption rates for each species (C_A and C_B). For all four cases, the regions are labeled as follows: 1 = neither species can live; 2 = only species A can live; 3 = species A wins out; 4 = stable coexistence; 4^1 = unstable coexistence, one or the other species wins; 5 = species B wins; 6 = only species B can live. • = stable equilibrium, o = unstable equilibrium. In (c) and (d), species B is limited more by resource 1 and species A by resource 2. In (c), species A consumes relatively more of resource 1 than species B and wins out in competition. In (d), each species consumes more of the resource that more limits its own growth, and the species come to a stable equilibrium with both species coexisting. *(From Tilman 1982.)*

cultures where the two yeasts grew together, population densities were lower than they were under single-species conditions (Fig. 9.6). From these data, Gause was able to calculate that $\partial_{12} = 3.15$ and $\partial_{21} = 0.44$; that is, 1 volume of *Saccharomyces* = 3.15 volumes of *Schizosaccharomyces*. Because alcohol is the limiting factor, Gause argued that he could determine ∂_{12} and ∂_{21} by measuring alcohol production of the two yeasts, which turned out to be 0.113 percent EtOH/cc yeast for *Saccharomyces* and 0.247 percent for *Schizosaccharomyces*. Thus, $\partial_{12} = 0.247/0.113 = 2.18$, and $\partial_{21} = 0.113/0.247 = 0.46$. The values of ∂_{12} and ∂_{21} obtained from the Lotka-Volterra equations were indeed in general agreement with those

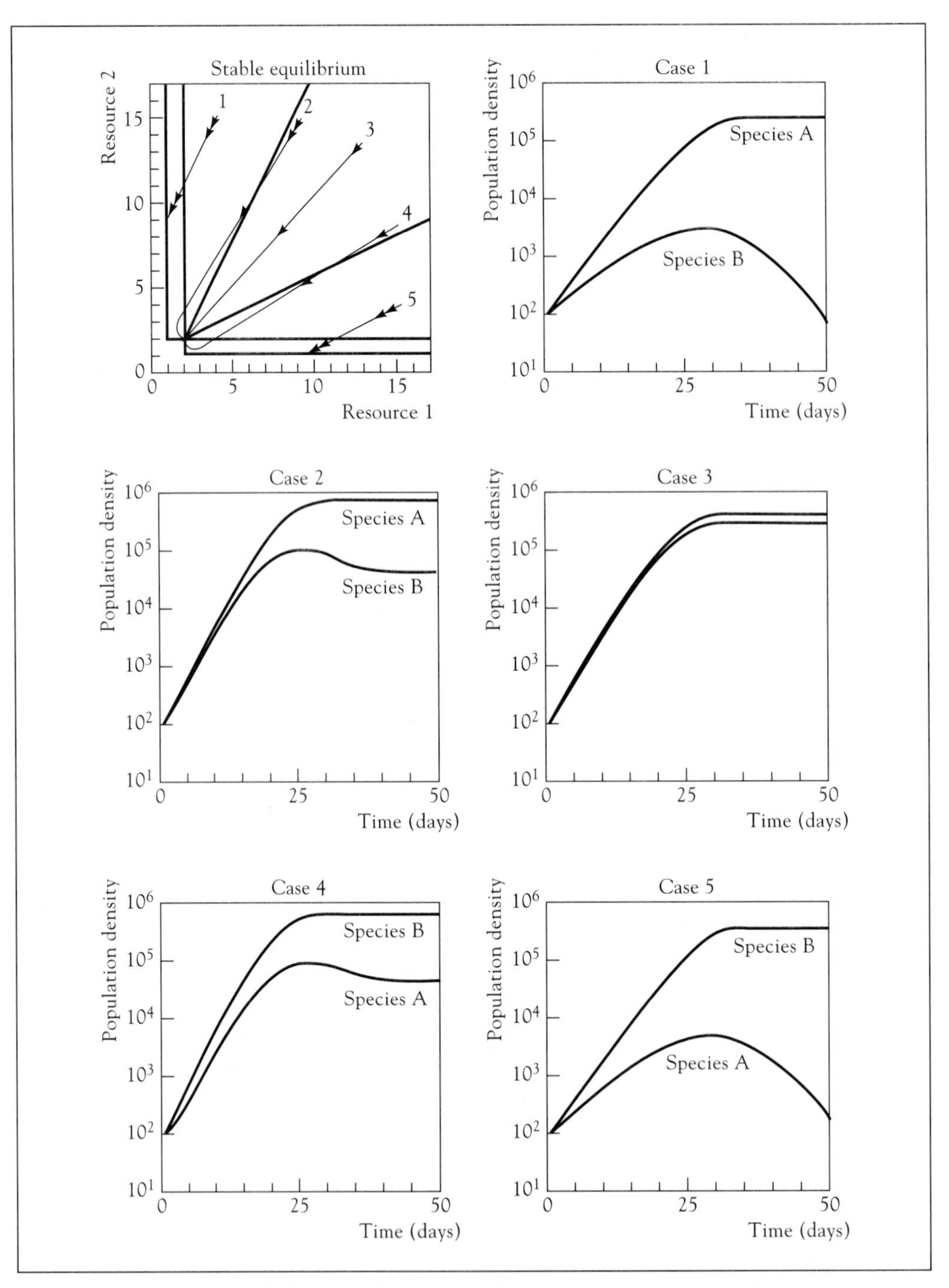

FIGURE 9.5 The outcome of competition for Tilman's model for five different resource supply points for Figure 9.4, equilibrium coexistence. The resulting five population curves for the two species show the time course of competition. (*From Tilman 1982.*)

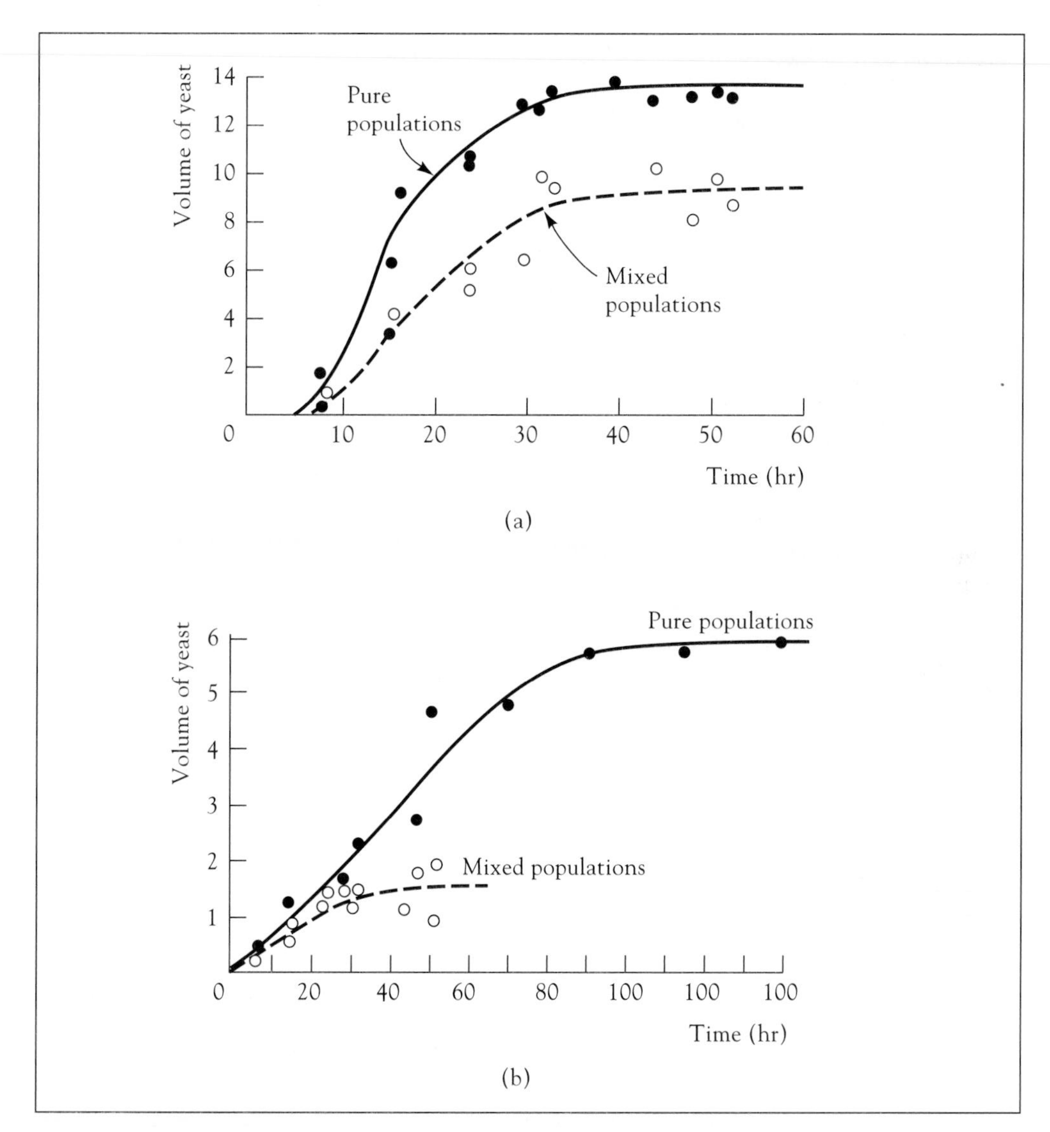

FIGURE 9.6 (**a**) Population growth of the yeast *Saccharomyces* in pure cultures and in mixed cultures with *Schizosacchoromyces*. (*After Gause 1932.*) (**b**) Population growth of the yeast *Schizosacchoromyces* in pure cultures and in mixed cultures with *Sacchoromyces*. (*After Gause 1932.*)

obtained independently by a physiological method. It is interesting that in this situation, population growth was limited by pollution of the environment by alcohol, not limiting responses. Many people think the same will be true for human populations.

In the late 1940s, Thomas Park and his students at the University of Chicago began a series of experiments examining competition between two flour beetles, *Tribolium confusum* and *Tribolium castaneum* (Photo 9.1). *Tribolium confusum* usually won, but in initial experiments the beetle cultures were infested with a sporozoan parasite, *Adelina*, that killed some beetles, particularly individuals of *T. castaneum*. In these early experiments

■ **PHOTO 9.1** In the 1940s, confused flour beetles, *Tribolium confusum*, were used as a model system by University of Chicago researcher Thomas Park to study competition. (*Degginger, Photo Researchers, Inc. 7W1467.*)

(Park 1948), *T. confusum* won in sixty-six out of seventy-four trials because it was more resistant to the parasite. Later, *Adelina* was removed, and *T. castaneum* won in twelve out of eighteen trials. Most importantly, with or without the parasite there was no absolute victor; some stochasticity was evident. In general, the species were mutually antagonistic, that is, they ate more eggs and pupae of the other species than they did of their own. Park then began to vary the abiotic environment and obtained the results shown in Table 9.1. It was evident that competitive ability was greatly influenced by climate; each species was

TABLE 9.1 Results of competition between the flour beetles *Tribolium castaneum* and *T. confusum*. (*After Park 1954*)

TEMP. °C	RELATIVE HUMIDITY %	CLIMATE	SINGLE SPECIES NUMBERS	MIXED SPECIES (% WINS)	
				T. confusum	*T. castaneum*
34	70	Hot-moist	conf = cast	0	100
34	30	Hot-dry	conf > cast	90	10
29	70	Temperate-moist	conf < cast	14	86
29	30	Temperate-dry	conf > cast	87	13
24	70	Cold-moist	conf < cast	71	29
24	30	Cold-dry	conf > cast	100	0

a better competitor in a different microclimate. However, single-species rearings in a given climate could not always be relied on to predict the outcome of mixed-species rearings (examine the entry for cold-moist climate). Later, it was found that the mechanism of competition was largely predation on eggs and pupae by larvae and adults. Park then varied the predatory tendencies of the beetles by selecting different strains; he obtained different results according to the strain of each beetle used (Table 9.2) although again the results could not always be predicted from the particular strains used (examine results for *T. Castaneum* CI). However, Park had demonstrated a complete reversal of competitive outcome as a function of temperature, moisture, parasites, and genetic strains.

9.3 Studies of competition in natural settings reveal that the prevalence and strength of competition varies among habitats and among years.

What of systems in nature where far more variability exists? One view holds that competition in nature is rare because by now, of all potential competitors, one has displaced the other. An alternative view holds that competition is a common enough force in nature to be a major factor influencing evolution. A third alternative is that predation and other factors hold populations below competitive levels.

The question is important in applied situations, for example, in biological-control campaigns, because it is vital to know whether releasing one natural enemy against a pest is likely to be more effective than the release of many, where competition between enemies might reduce their overall effectiveness (Rosenheim et al. 1995; Ferguson and Stiling 1996).

TABLE 9.2 Results of competition experiments between the flour beetles *Tribolium castaneum* and *T. confusum*. (*After Park, Leslie, and Metz 1964*)

T. castaneum STRAIN	*T. confusum* STRAIN	NO *castaneum* WINS	NO *confusum* WINS
CI	bI	10	0
	bII	10	0
	bIII	10	0
	bIV	10	0
CII	bI	1	8
	bII	0	10
	bIII	0	10
	bIV	4	6
CIII	bI	0	10
	bII	0	9
	bIII	0	10
	bIV	0	10
CIV	bI	9	1
	bII	9	0
	bIII	9	1
	bIV	8	2

Note: Predatory tendency of strain does low → high as CI → CIV and bI → bIV.

Circumstantial evidence suggests fewer natural enemies become established where many are released (Ehler and Hall 1982), yet sometimes this phenomenon does little to reduce overall effectiveness of control (Ehler 1979). Competitive effects between plants are also often thought to be of paramount importance in influencing crop yields, and many applied ecologists immediately assume all plants compete if resources are limiting (for example, Reynolds 1988). Again, this is important in applied ecology because while agronomists may strive to reduce competition, entomologists argue that more than one crop is valuable in order to encourage a wide variety of natural enemies and thus so reduce insect pest densities.

The most direct method of assessing the importance of competition is to remove individuals of one species A and to measure the responses on species B (Wise 1981). Often, however, such manipulations are difficult to make outside the laboratory. If individuals of species A are removed, what's to stop them from migrating back into the area of removal? If cages are used to stop immigration of species A, emigration of species B is prevented and the numbers of species B may go unnaturally high (the so-called cage effect or "Krebs effect" [Strong and Stiling 1983]). Three of the most cited examples of competition in nature involve barnacles, parasitic wasps, and chaparral shrubs.

Two barnacles, *Chthamalus stellatus* and *Balanus balanoides*, dominate the British coasts. Their distribution on the intertidal rock faces is often well defined (Fig. 9.7). Joe Connell (1961) showed that *Chthamalus* could survive in the *Balanus* zone when *Balanus* was removed. In nature, *Balanus* grows faster on rocks of the middle intertidal zone, squeezing *Chthamalus* out by undercutting *Chthamalus* and eventually levering them off. The limits of *Balanus* distribution are determined in the upper zone by desiccation and

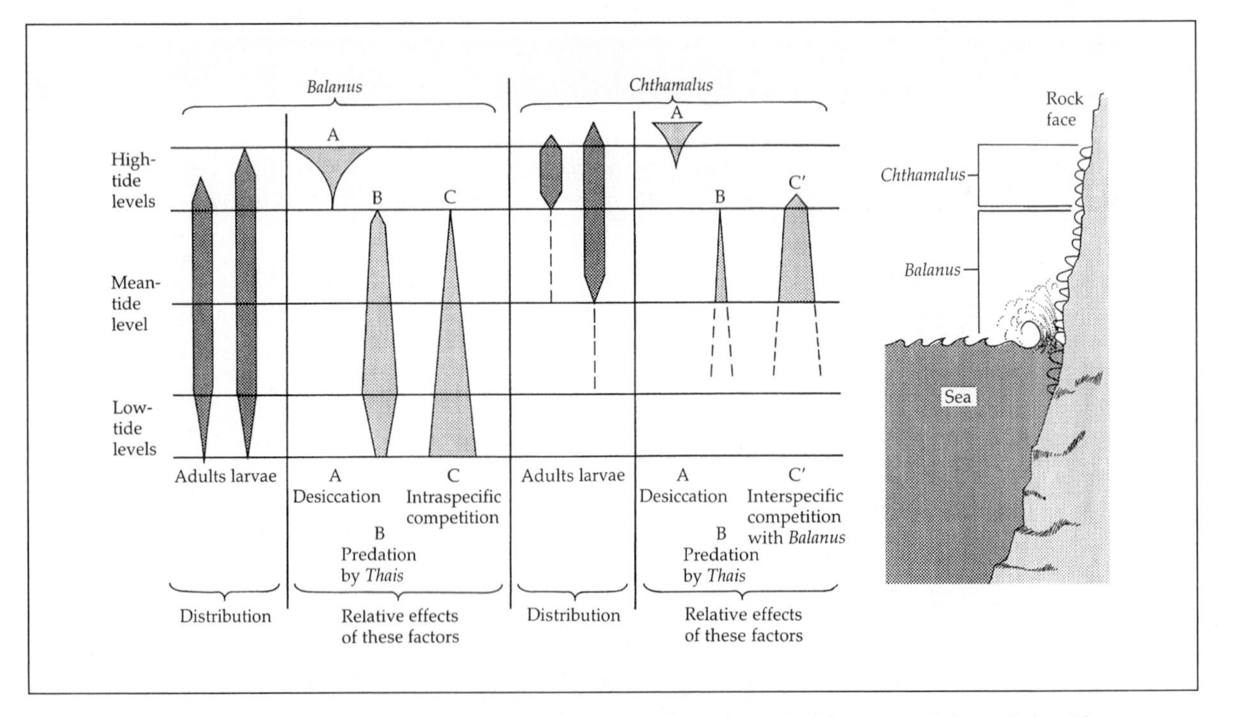

■ **FIGURE 9.7** Intertidal distribution of adults and newly settled larvae of *Balanus balanoides* and *Chthamalus stellatus* at Millport, Scotland, with a diagrammatic representation of the relative effects of the principal limiting factors. (*Modified from Connell 1961.*)

in the lower zone by predation and competition for space with algae. *Chthamalus* is more resistant to desiccation than *Balanus* and is normally found only high on the rock face. Thirty-four years later, in 1988, Connell repeated his competition experiments in the same area of the Scottish coast and, once more, observed strong evidence for competition (Connell 1990). Overgrowth of one organism by another is quite common in marine encrusting organisms and has also been shown to occur between competing lichens on rock faces in the Appalachians (Harris 1996).

Pest control is big business in the United States, and for some biological control projects there is much labor available, in the form of trained "scouts," to survey an area for evidence of successfully reproducing released enemies. Thus, when three species of parasitic wasps of the genus *Aphytis* (Photo 9.2) were introduced into southern California to help control the red scale *(Aonidiella aurantii)*, an insect pest of orange trees, there were unprecedented amounts of data on the results. *Aphytis chrysomphali* was introduced accidentally from the Mediterranean in 1900 and became widely distributed. In 1948 *A. lignanensis* was introduced from south China and began to replace *A. chrysomphali* in many areas (Debach and Sundby 1963), such as Santa Barbara County and Orange County:

	PERCENTAGE OF INDIVIDUALS	
	A. CHRYSOMPHALI	**A. LIGNANENSIS**
Santa Barbara County		
1958	85	15
1959	0	100
Orange County		
1958	96	4
1959	7	93

In 1956-1957 another species, *A. melinus*, was imported from India and immediately displaced *A. lignanensis* from the hotter interior areas:

	PERCENTAGE OF INDIVIDUALS	
	A. IGNANENSIS	**A. MELINUS**
Coastal, Santa Barbara County		
1959	100	0
1960	95	5
1961	100	0
Interior, Santa Barbara County		
1959	50	50
1960	6	94
1961	4	96

ECOLOGY IN PRACTICE

Joe Connell, University of California, Santa Barbara

It's hard to know whom one should credit with helping in any of the successes that occur in one's career, but I'm sure that the nuns in my first eight elementary school years had a lot to do with teaching me the discipline to get a job done. I certainly learned that I'd better always have my homework done (and done correctly) when I got to school in the morning!

I had a hard time deciding what to do with my life. I lived in a small town where the only professions I saw practiced were doctor, lawyer, or engineer. So I began college in an engineering school. Only after I got in the army (World War II), did I meet people who had studied to be biologists, usually wildlife managers. Since I had always loved watching birds and identifying trees, it was a revelation to discover that somebody would actually pay you to do it! I've never worried about the size of my paycheck because I always reckoned that since doing ecology was such fun, one shouldn't really be getting paid for it.

In the war I was trained in meteorology, and as soon as I got out of the army I finished a B.S. degree in meteorology at the University of Chicago. Then I finally decided what to do with my life: to become a biologist. I went to U. C. Berkeley, where my advisor was A. C. Leopold. He advised me to go for an M. A. degree and to do my research on a potential game animal that was very common but completely unstudied, the brush rabbit. This choice had a great influence on my future research career; I learned that one shouldn't choose a research species solely because it had never been studied or a topic solely because it might some day be a benefit to mankind. The reason why this animal had never been studied soon became clear; it was very difficult to capture, and even when I learned how (I caught one nineteen times!), my results were pretty dull: I knew where I had caught them but not much more.

It was discouraging, dull research, and at the finish I decided that this wasn't for me and became a high school teacher. This was rewarding, hard work, and I would probably have stayed with it but was stimulated to go back to college when the government announced that any veteran who was on a G.I. Bill scholarship must use it up immediately or lose it. So I decided to go back to university but to do so in a foreign country. I went to the University of Glasgow, in Scotland, because I had met the professor of zoology, C. M. Yonge, when he visited Berkeley. (Here I'd like to acknowledge the G.I. Bill, which gave me four years of college and the opportunity to become an ecologist; it may be the best "entitlement" our government ever provided its citizens).

Having learned the hard way what *not* to choose as a study animal, I remembered a wonderful paper, by E. S. Deevey in the *Quarterly Review of Biology* in 1947, that I had had to review in a graduate seminar. In it he described a 1938 paper by H. Hatton in an obscure (to Americans) French journal. It was in French, but I slogged through it and discovered the idea of controlled field experiments, which this fellow had conceived and carried out in 1929–1932, on three marine intertidal animals and three algae. I decided that I would study one of the same species he did, a barnacle, and since Hatton had concentrated on physical factors I would concentrate on biological ones, namely, the effects of population density and predators, using field experiments as Hatton's paper had taught me. (Research is usually regarded as the alternative to teaching by university faculty, but some of the most important teaching is done through one's papers.) So Deevey and Hatton were my early role models both through their publications; I met Deevey later, and I went to France to try to find and thank Hatton, but couldn't; his Professor thought he might have died in the war.

One aside about doing a Ph.D.: my professor wisely advised me not to try to spread my Ph.D. research too widely but to confine it to the effects of density and predators. However, I saw what looked like a neat interaction between two competing species of barnacles, so, without telling him, I went ahead and did a field experiment on competition. I didn't include it in my final Ph.D. thesis, so he never found out about my disobedience! It turned out to be one of my most-quoted papers, so my advice is: be kind to your professor, but don't always take his advice!

I find dreaming up new ideas and getting and analyzing data, easy and fun, but writing is very difficult for me. I have to force myself to stop reading a mystery novel and get down to writing; here is where the training in discipline by my elementary school nuns has its effect! (After what I put them through, they just have to be in heaven!) I usually write five to ten (sometimes more) drafts of a paper before I'm satisfied and inflict many of these drafts on my long-suffering friends to criticize.

However, sometimes data analyses were the hardest part, particularly in two papers in which I reviewed and analyzed the data from papers written by others: one in 1983 for *American Naturalist* and one in 1997 in *Coral Reefs* . Here the problem was to understand the data from the author's published tables or figures. In a few cases, I had to write the author and ask her or him to clear up my difficulty. In two cases this resulted in the author discovering errors in the published results, so the only accurate analyses are in my papers! In one paper I simply couldn't figure out the results, so I had to use only a part of the data. I also discovered (in the 1983 paper) that papers containing only negative results apparently seldom get published.

I've not been embroiled in much controversy, though my 1983 review paper was stimulated by a particularly acrimonious criticism that I felt I had to answer. Critical comments are usually very helpful, provided they are meant to be constructive; they are routinely given by reviewers of grant proposals or papers one has submitted for publication. I certainly need them since I seem to be unable to recognize my own shortcomings. Having colleagues who don't hesitate to point them out is essential, and I've been very lucky in this; in our group, we all realize that it's better to have one's friends, rather than one's rivals, correct your errors!

I was very lucky that my career coincided with the golden age of funding support for science, 1950 to 1980. When competition for grants got really heavy in the 1980s, I had already established a track record, so I continued to be able to get National Science Foundation funding. I got turned down several times on first drafts of a proposal but usually managed to rewrite and get the funding the second time. Young scientists just starting now have a much tougher time than I did when I got my first grant in 1959. Luck and chance play a part in ecology and in one's life, and I've been very lucky.

The mechanism by which competitive displacement occurred was that female *A. melinus* could use smaller- scale insects as hosts and could lay a higher proportion of "female" eggs in them than the other two species (Luck and Podoler 1985; Murdoch, Briggs, and Nisbet 1996). Thus, *A. melinus* preempted most scales as hosts before the other parasites could use them. There are probably very few other examples of such well-documented competitive displacement in nature. However, the caveat here is that all species were exotic and may have been expected to compete more than native species

■ **PHOTO 9.2** What is a parasitoid?—a parasite that lays its egg in another organism and whose resultant larva kills its host. Here a parasitic *Aphytis* wasp lays its egg in a *Dactynotus* aphid. Parasitoids can be very effective agents in biological control; the interaction of three species of *Aphytis* wasp parasitoids in California provided a good example of competition in the field. (*Kuhn, DRK Photo 140701.*)

that have evolved together over millions of years. Nevertheless, this underscores the point that the study of competition is important in real-world situations. An understanding of competition between native and exotic species, and what the likely outcomes will be is also important (Applied Ecology: The Effects of Exotic Competitors on Native Fauna).

Plants are often thought to suffer more from competition than animal populations because plants are rooted in the ground and cannot move to escape competitive effects. In Southern California chaparral, grassland shrubs such as the aromatic *Salvia leucophylla* and *Artemisia california* are often separated from adjacent grassland by bare sand one to two meters wide. Volatile terpenes are released from the leaves of the aromatic shrub; these inhibit the growth of nearby grasses (Muller 1966). Some plants, such as black walnut, *Juglans nigra*, produce similar chemicals (here, juglone) from their roots, which leach into the soil, killing neighboring roots (Massey 1925). This phenomenon is termed **allelopathy**; the action of penicillin among microorganisms is a classic case. In many cases, such allelopathic chemicals are toxic to some competitors but not to others. Competitive interactions between plants are of paramount importance in **agroforestry**, a relatively new concept in which crops are grown under forest cover so that the land will yield both food and timber. Many species of tree, especially *Eucalyptus* in tropical regimes, are not suited to the practice of agroforestry because of their adverse effects on plants growing beneath them (Young 1988).

Despite the difficulties of demonstrating competition, it is valuable to know how frequent it might be in nature. Two reviews of the literature attempted to answer this question. Connell (1983) reviewed 72 studies on active competition as reported in the literature. Competition was found in 55 percent of the 215 species and in 40 percent of 527 experiments involved. Connell suggested that his result appears logical if one takes the following view. Imagine a resource set with, say, four species distributed along it. Then if only adjacent species competed, competitive effects would be expected in only three out of the six species pairs (50 percent) (Fig. 9.8). Of course, the mathematics would be drastically different according to the number of species on the axis. For any given pair of adjacent species, however, competition would be expected, and indeed Connell found that, in studies of single pairs of species, competition was almost always reported (90 percent), whereas in studies involving more species the frequency was only 50 percent.

In a parallel but independent review of 150 field experiments, Schoener (1983) reported competition in more than 90 percent of 164 studies and in 75 percent of the species studied. Why the difference in Connell's and Schoener's studies? Because of slightly different samples of studies, methods of analysis, and, perhaps, predispositions of the authors. Connell, perhaps believing predation to be a more important force in nature, was more rigorous in what he accepted as a satisfactory experiment. For example,

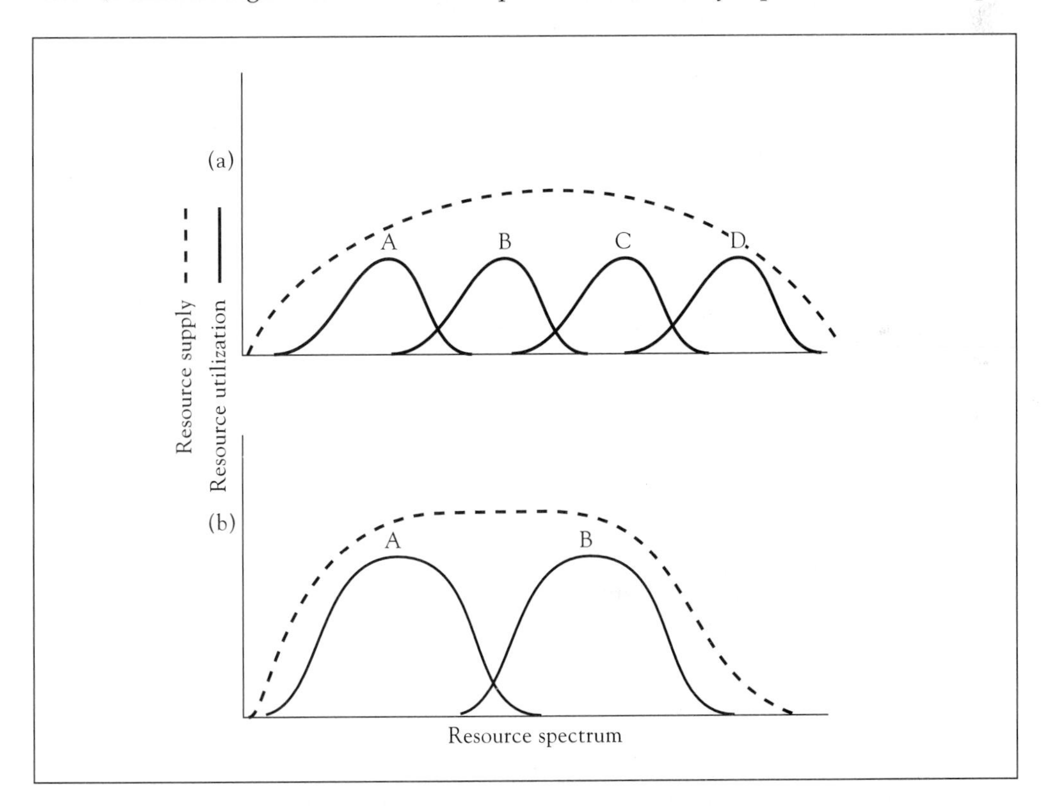

FIGURE 9.8 (**a**) Resource supply and utilization curves of four species, A, B, C, and D along a resource gradient. If competition occurs only between species with adjacent resource utilization curves, the competition will be expected between A and B, B and C, and C and D—three of the six possible pairings. (**b**) When only two species utilize a resource set, competition would nearly always be expected between them.

The Effects of Exotic Competitors On Native Fauna

Worldwide, humans have succeeded in establishing at least 330 species of nonnative birds and mammals in more than fifteen hundred separate cases. In many cases, exotics compete with natives for resources, often with serious consequences. In New Zealand there is a large overlap in food choice between introduced red deer, feral goats, bushy-tailed possums, and the native kokako bird. Species share resources, but this does not prove competition because it is not known if any of the populations are affected. Similarly, in England the introduced grey squirrel and the native red squirrel show more-or-less exclusive distributions that may be a result of competition, but it has recently been argued that the red squirrel was declining in certain areas, and the grey squirrel simply took over these vacant habitats.

The introduction to Gatun Lake, Panama, of the cichlid fish *Cichla ocellaris* (a native of the Amazon) is thought to have led to the elimination of six of the eight previously common fish species within five years (Zaret and Paine 1973; see also Payne 1987). Also, the introduction of the game fish *Micropterus salmoides* (largemouth bass) and *Pomoxis nigromaculatus* (black crappie) led to the diminution of local fish and crab populations. Similarly, after construction of the Welland Canal linking the Atlantic Ocean with the Great Lakes, much of the native fish fauna was displaced by the alewife (*Alosa pseudoharengus*) through competition for food (Aron and Smith 1971). Introduced African dung beetles were successful in reducing the numbers of pest flies that competed for dung in Australia (Moon 1980).

Competition can be of two types, resource or contest. Contest competition involves aggressive behavior between two species. For example, on the Japanese Island of Oshima, contest competition has been observed between the introduced gray-bellied squirrel and the oriental white-eye bird. The squirrel chases the bird away from the flowers of camellia, which not only affects the bird but also the plant since it is pollinated by white-eyes. In exploitative, or indirect competition, species compete indirectly for a common resource. For example, nesting boxes made for eastern bluebirds on Bermuda are used by introduced great kiskadees as perches, making it impossible for the bluebirds to nest in them. Table 1 summarizes some data regarding the extent of competition from introduced species. Competitive interactions from introduced birds seem more common (18%) than those due to introduced mammals (9%). Nevertheless, mammals have been found to be serious competitors in some instances. This is especially true in Australia, where one single order, the Marsupialia, has become highly diversified to fill the roles of insectivores, predators, grazers, fruit eaters, and so on. The rabbit is a serious competitor for burrows with an

TABLE 1 Number of introductions of mammals and birds resulting in potential competition. (*After Ebenhard 1988*)

CATEGORY	CONTINENTS	CONTINENTAL SHELF ISLANDS	OCEANIC ISLANDS	TOTAL
Mammals	24/60 (29%)	6/108 (5%)	15/291 (5%)	45/109 (9%)
Herbivores	15/34 (31%)	3/54 (5%)	13/119 (10%)	31/207 (13%)
Birds	11/29 (28%)	1/40 (2%)	22/88 (20%)	34/157 (18%)

Note: Data indicate number of introductions with/without potential competition with percentage introductions with potential competition in parentheses.

■ **PHOTO 1** Introduced animals can sometimes out-compete native ones. Sheep in Australia compete with Kangaroos, such as these western Grey Kangaroos in Victoria, Australia, for forage, often reducing the kangaroo population. *(John Cancalosi, Peter Arnold 39554.)*

animal called the boodie, the only burrowing kangaroo, and with the common rabbit bandicoot. Among the predators, the cat is probably competing with dunnarts, and the dingo (feral dog) has been thought to have excluded the thylacine (native marsupial wolf like animal) from mainland Australia. Goats and sheep probably compete with a range of kangaroo species, especially the brush-tailed rock wallaby and larger species, such as the red kangaroo and the western gray kangaroo (Photo 1). Sometimes competition has been suggested to be more intense between introductions of species closely related to native species than between distantly related species. For example, the American beaver has been introduced into Finland, where it excludes the European beaver, with the result that the two species now have separate geographic distributions. What generally is being competed for? When mammals are introduced, competition is strongest over food, while for introduced birds, competition is strongest for nest sites (Table 2).

TABLE 2 Competition objects (only cases involving an introduced bird or mammal species). *(Data from Ebenhard 1988)*

	INTRODUCED MAMMALS		INTRODUCED BIRDS	
OBJECTS	NO. OF CASES	%	NO. OF CASES	%
Food	31	63	8	18
Nest-site	3	6	29	64
Space	7	14	1	2
Not known	8	16	7	16
Total	49	100	45	100

Hairston (1989) points out that at least one of the experiments accepted as evidence of competition by Schoener did not meet the necessary requirements of experimental design.

Both reviews may well overrepresent the actual frequency of competition in nature because of some common flaws:

1. "Positive" results demonstrating a phenomenon (here competition) may tend to be more readily accepted into the literature than "negative" results

demonstrating patterns indistinguishable from randomness (Connell 1983; see also Csada, James, and Espie 1996, Chap. 1).
2. Scientists do not study systems at random; those interested in competition may well choose to work in a system where competition may be more likely to occur.

On the other hand, the reviews may fail to reveal the true importance of competition because

1. By now most organisms have evolved to escape competition and the lack of fitness it may confer.
2. Competition may only occur in certain "crunch" years where resources are scarce (Weins 1977). Nevertheless, this competition is severe enough to structure the community. If a crunch year occurs only one in five years and a researcher does experiments in any of the other four, competition may go undetected.

Despite these drawbacks some general patterns were evident from Connell's and Schoener's work if one assumed there are no taxonomic biases in reporting the frequency of competition (Table 9.3). Herbivorous insects (especially leaf feeders) and filter feeders (such as clams) showed less competition than plants, predators, scavengers, or grain feeders. Marine intertidal organisms tended to compete more than terrestrial ones and large organisms more than small ones. Some patterns like this seemed logical; for example, given limited intertidal space, it would not seem odd to detect competition for space between sessile organisms. Seeds and grains also provide a limiting but very important nutrient-rich resource for desert granivores. Brown and co-workers (Brown et al. 1986; Heske, Brown, and Mistry 1994) provide good evidence that all types of grain feeders, from rodents and birds to ants, compete for this resource.

In addition, botanists (e.g., Grime 1979) have argued that competition is relatively unimportant for plants in unproductive environments because plant biomass and therefore resource depletion is low. Instead, such environments are dominated by stress-tolerant species. However, others (e.g., Tilman 1988) have argued that competition occurs across all productivity gradients but that the resources concerned may differ. Thus, in unproductive environments competition is primarily for below-ground resource (i.e., nutrients

TABLE 9.3 Percentage of experimental studies showing interspecific competition. (*After Connell 1983*)

	TERRESTRIAL		MARINE		FRESHWATER		TOTAL	
	NO. EXP.	%	NO. EXP.	%	NO. EXP.	%	NO. EXP.	%
Plants	205	30	31	68	2	50	238	35
Herbivores	45	20	13	69	0	—	58	31
Carnivores	36	11	5	60	3	67	44	20
Total	286	26	49	67	5	60	340	32
Invertebrates	57	16	37	32	0	—	94	22
Vertebrates	47	23	10	90	3	67	60	37

and water) while in productive environments competition is primarily for light. No census on the "Grime-Tilman" debate has yet emerged (Goldberg and Novoplansky 1997).

In support of the "competition-between-herbivores-is-feeble" theory Lawton (1984) has argued that there is much evidence of vacant space on the plants of the world. As evidence, he showed that bracken fern (*Pteridium aquilinum*) in Europe has a large array of chewing, sucking, mining, and galling insects in a wide range of habitats, but in the United States and especially in Papua, New Guinea, whole types of feeders, for example, gall formers, are missing (Fig. 15.9). With such vacant space available, insects cannot be expected to compete so fiercely. The fact that so many introduced insects, other animals, and plants have become established and thrive when introduced into new countries and novel habitats suggests that few niches in natural ecosystems are filled.

Lawton and McNeill (1979) also suggested that insect herbivores often "lie between the devil and the deep blue sea," that is, between a huge array of predators and parasites on the one hand and a deep blue sea of abundant but low-quality food on the other. As a result, they could scarcely become abundant enough to compete. The foundations of this argument had been laid years earlier by Hairston, Smith, and Slobodkin (1960), who had essentially argued that the "Earth is green" and that the phytophagous insects that could eat this greenery must therefore be held in check by animals from the next trophic level up.

However, the latest review (Denno, McCLure, and Ott 1995) suggests that, in the mid-1980s to early 1990s, competition in insects was actually found more frequently than in previous decades, so now the pendulum has swung back the other way. Their examination of 193 pair-wise species interactions showed interspecific competition in 76 percent of the cases. Why the difference? Insect ecologists had perhaps underestimated the ability of plants to respond chemically and physiologically to herbivore attack, including how far such effects could spread through the plant's transport system and how long they would last, discouraging feeding by other species. That is, herbivores occupying different parts of the plant may still compete by exploiting a common resource such as phloem sap. Aphids feeding on roots compete with other aphid species galling the leaves without ever coming directly into contact with them (Moran and Whitham 1990). Likewise, even herbivores that are separated in time are potential competitors too, if they induce a long-lasting response in the plant. Larvae of the winter moth (*Operophtera brumata*) feeding on young oak leaves have severe adverse effects on leaf miners that appear later in the season (West 1985). If competition is frequent even between insects, it is likely a pervasive force in nature.

By what mechanism does competition most commonly operate? Schoener (1985) divided the mechanisms into six categories.

1. *Consumptive or exploitative:* using resources.
2. *Preemptive:* using space.
3. *Overgrowth:* one species growing over another and blocking light or depriving the other of a resource.
4. *Chemical:* by-production of toxins.
5. *Territorial:* behavior or fighting in defense of space.
6. *Encounter:* transient interactions directly over specific resources.

Exploitative competition is by far the most common, occurring in $71/188 = 37.8$ percent of cases (Table 9.4). Some of Schoener's findings about mechanisms are easy to inter-

TABLE 9.4 Mechanisms of interspecific competition in experimental field studies. (*After Schoener 1985*)

	MECHANISM						
GROUP	CONSUMPTIVE	PREEMPTIVE	OVERGROWTH	CHEMICAL	TERRITORIAL	ENCOUNTER	UNKNOWN
Freshwater							
Plants	0	0	1	1	0	0	0
Animals	13	1	0	1	1	5	2
Marine							
Plants	0	6	4	1	0	0	0
Animals	9	10	6	0	7	6	0
Terrestrial							
Plants	28	3	11	7	0	1	9
Animals	21	1	0	1	11	15	6
Total	71	21	22	11	19	27	17

pret—preemptive and overgrowth competition appear among sessile space users, primarily terrestrial plants and marine macrophytes and animals living on hard substrates. Territorial and encounter competition occur among actively moving animals, especially birds and mammals. Chemical competition occurs among terrestrial plants; toxins become too diluted in aquatic systems.

A final tidbit from both Connell's and Schoener's studies is that most often only one member of a species pair responded to the addition or removal of individuals of the other. The logic here is that such asymmetric competition should be expected—the superior competitor will probably be more strongly limited by some other factor—environmental tolerances or predators. Usually, the larger organism has the competitive advantage (Persson 1985).

More recently, Gurevitch et al. (1992) conducted a meta-analysis of competition in field experiments. This type of analysis uses statistical techniques to the compare the strength of effects of competition in various studies, in contrast to the perceived "vote-counting" procedures of both Connell and Schoener where studies are simply counted yes or no as to whether they show competition or not. The meta-analysis examined field competition experiments on ninety-three species in a wide variety of habitats. There were very strong effects of competition on body size, that is, biomass, but effects on population density were not investigated. Surprisingly, primary producers, plants, and carnivores did not show such strong effects of competition as filter feeders or herbivores (Fig. 9.9). Competition was strong for all herbivore groups except for terrestrial arthropods. There were no differences in the effects of competition in terrestrial, freshwater, or marine systems, for plants or carnivores, nor in high-productivity (fields, prairies, meadows) versus low-productivity systems (deserts, arctic tundra).

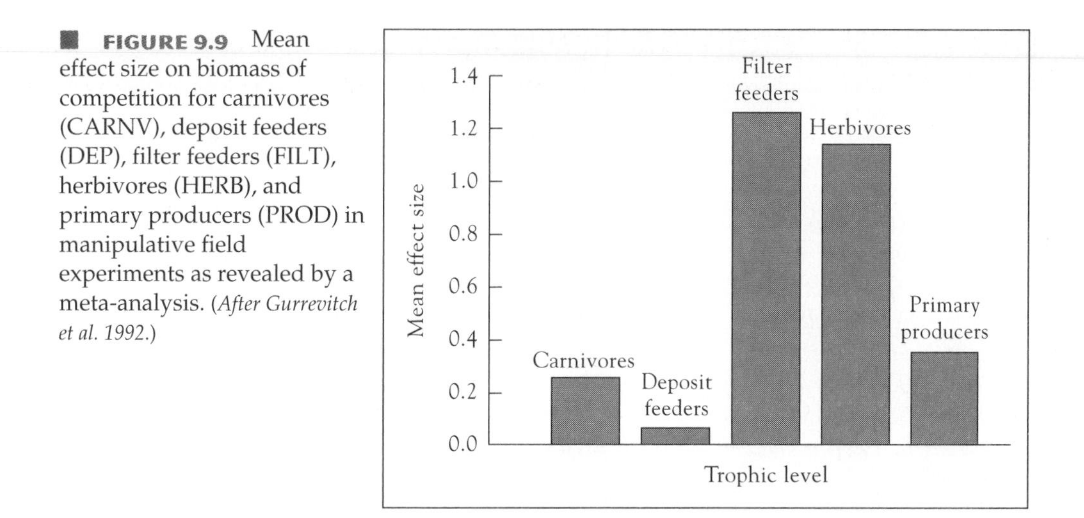

■ **FIGURE 9.9** Mean effect size on biomass of competition for carnivores (CARNV), deposit feeders (DEP), filter feeders (FILT), herbivores (HERB), and primary producers (PROD) in manipulative field experiments as revealed by a meta-analysis. (*After Gurrevitch et al. 1992.*)

9.4 Competition can either lead to the displacement of the competitively inferior species or to coexistence.

In 1934 Gause wrote, "As a result of competition two similar species scarcely ever occupy similar niches, but displace each other in such a manner that each takes possession of certain peculiar kinds of food and modes of life in which it has an advantage over its competitor." What exactly is a niche? Perhaps the most useful way to think of a niche is the way Joseph Grinnell described it in 1917, as a subdivision of the habitat. Hutchinson (1958) suggested one could more precisely define a niche in terms of places that contained an organisms' diet needs, its temperature and moisture requirements, pH and so on, and this would be the *fundamental niche* of a species. So Gause was saying that two species with similar requirements cannot live together in the same place. Hardin (1960) stated that this be known as the *competitive exclusion principle*, i.e., complete competitors cannot coexist, just as the Lotka-Volterra equations suggested. But, as anyone who has ever spent time in the field knows, we see myriads of different species apparently coexisting side by side all the time. How can this be?

Active competition may not always lead to competitive displacement. It is conceivable, for example, that, in areas of overlap, species change their lifestyles or feeding habits so that competition is minimized. If species do compete in nature, the important question is perhaps not how much competition goes on but how similar competing species can be and still live together. This question has received more attention in ecology than any other single topic (Schoener 1974), but Lewin (1983) suggests that it has led ecologists into futile works and blind alleys.

One hypothesis explaining the coexistence of competing species states that morphological differences may allow coexistence.

In a seminal paper entitled "Homage to Santa Rosalia, or why are there so many kinds of animals?" G. Evelyn Hutchinson (1959) looked at size differences, particularly in feeding apparatus, between congeneric species when they were **sympatric** (occurring

TABLE 9.5 **Size relationships, the ratio of the larger to the smaller dimension, between congeneric species when they are sympatric and allopatric. Culmen is beak length.** (*From Hutchinson 1959*)

ANIMALS AND CHARACTER	SPECIES	MEASUREMENT (MM) WHEN "SYMPATRIC"	"ALLOPATRIC"	RATIO WHEN "SYMPATRIC"	"ALLOPATRIC"
Weasels	*Mustela nivalis*	39.3	42.9	1.28	1.07
(skull)	*M. erminea*	50.4	46.0		
Mice	*Apodemus sylvaticus*	24.8		1.09	
(skull)	*A. flavicollis*	27.0			
Nuthatches	*Sitta tephronota*	29.0	25.5	1.24	1.02
(culmen)	*S. neumayer*	23.5	26.0		
Darwin's	*Geospiza fortis*	12.0	10.5	1.43	1.13
Finches	*G. fuliginosa*	8.4	9.3		
(culmen)					

together) and **allopatric** (occurring alone) (Table 9.5). (The conceptual basis and explicit discussion of this approach had actually been laid out by Julian Huxley seventeen years earlier in 1942 [Carothers 1986]).

Ratios between characters studied when species were sympatric ranged between 1.1 and 1.43, and Hutchinson tentatively argued that the mean value of 1.28 could be used as an indication of the amount of difference necessary to permit coexistence at the same trophic level but in different niches. Some authors extrapolated that ratios of between 1.3 and 2.0 indicated sufficient differences to permit coexistence since weight varies to the third power of length and $1.3^3 = 2.2$. Despite its appeal, Hutchinson's idea came under heavy fire for the following reasons:

1. In a large series of examples purporting to support this hypothesis, statistical analysis was not appropriate. Further tests showed no more differences between species than would occur by chance alone (Simberloff and Boecklen 1981).
2. Size-ratio differences have too loosely been asserted to represent the ghost of competition past (Connell 1980) when in fact they could have evolved for other reasons.
3. Biological significance cannot always be attached to ratios, particularly those of structures not used to gather food: ratios of 1:1.3 have been found to occur between members of sets of kitchen skillets, musical recorders, and children's bicycles (Horn and May 1977) (Photo 9.3). Maiorana (1978) argued that 1:1.3 ratios may simply reflect something about our perceptual abilities.

Needless to say, followers of Hutchinson's ideas have been swift to rebutt some of these ideas. Losos, Naeem, and Colwell (1989) argued that the Simberloff-Boecklen statistics were deficienct in statistical power. In other words, they erred too strongly on the side of accepting a null hypothesis of no effect of competition when such an hypothesis was false. Doubtless, variations of these arguments will continue to be tossed back and forth

PHOTO 9.3 Should biological significance be attached to particular size ratios between organisms? Ratios of 1 : 1.3 have been found to occur between members of sets of kitchen knives, skillets, musical recorders and children's bicycles. (*Leonard Lessin, Peter Arnold Inc. 155264.*)

in the future. It is noteworthy, however, that even one of the chief protagonists in the debate, Daniel Simberloff, has recently published articles that support the idea of character displacement in mammals (see, for example, Dyan and Simberloff 1994)

There are some authors who have tried to use natural experiments to support the validity of Hutchinson's ratios. In the primeval forests of Canada, four indigenous parasitoids attacked the wood-boring siricid larva *Tremex columba*. Each had a different ovipositor length and laid an egg only when the ovipositor was fully extended. The ovipositor can be regarded as a food-provisioning apparatus for the larva; each species laid eggs in *Tremex* cocoons at a different depth in logs. In the 1950s a fifth species, *Pleolophus basizonus*, was introduced into the area in an attempt to control another pest, the European sawfly. Its ovipositor length was intermediate between those of two of the existing species. Even before the introduction of *P. basizonus*, the first three species were tightly packed but still maintained the minimum separation of about 1:1.1 noted by Hutchinson. However, when *P. basizonus* was introduced, strong competition ensued. *Pleolophus basizonus* or either of the two other species could have been displaced. In fact, *M. aciculatus* and *M. indistinctus* were forced out of the more favorable high-host-density sites (Price 1970) (Fig. 9.10).

Species may also coexist by partitioning the resources to avoid competition.

Besides separation in size, species may also differ in their use of a particular set of resources, such as food or space. Consider three species normally distributed on a resource set (Fig. 9.11), where $K(x)$ is the resource availability of x or carrying capacity, d is the distance between abundance maxima, and w represents one standard deviation, approximately 68 percent of the area on one side of the curve. It has been argued that if $d/w < 1$, species cannot coexist; if d/W is < 3, there will be some interaction between species, and if $d/w > 3$, species coexist harmoniously (see Southwood 1978). One problem is that species abundances are often not normally distributed or have a discontinuous distribution.

Where the resource has a discontinuous distribution or occurs in distinct units, like leaves on a shrub, resource utilization can be illustrated graphically as in Fig. 9.12. The niche breadth of a species can then be quantified by Levins's (1968) formula

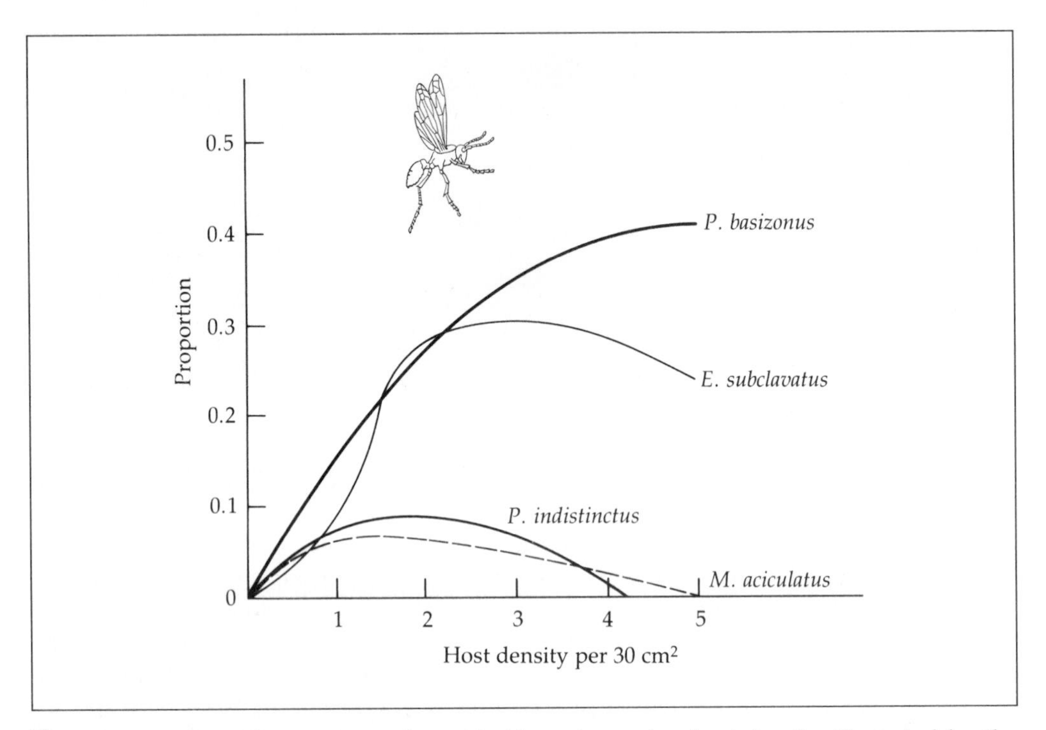

■ **FIGURE 9.10** The response of parasitoids to increasing host density illustrated by the change in the proportion of each species in the total parasitoid complex. As the host density increases, competition between parasitoids becomes more severe. Note that as the introduced *Pleolophus basizonus* increases, the first to be suppressed are those closest in ovipositor lengths. *(Redrawn from Price 1970.)*

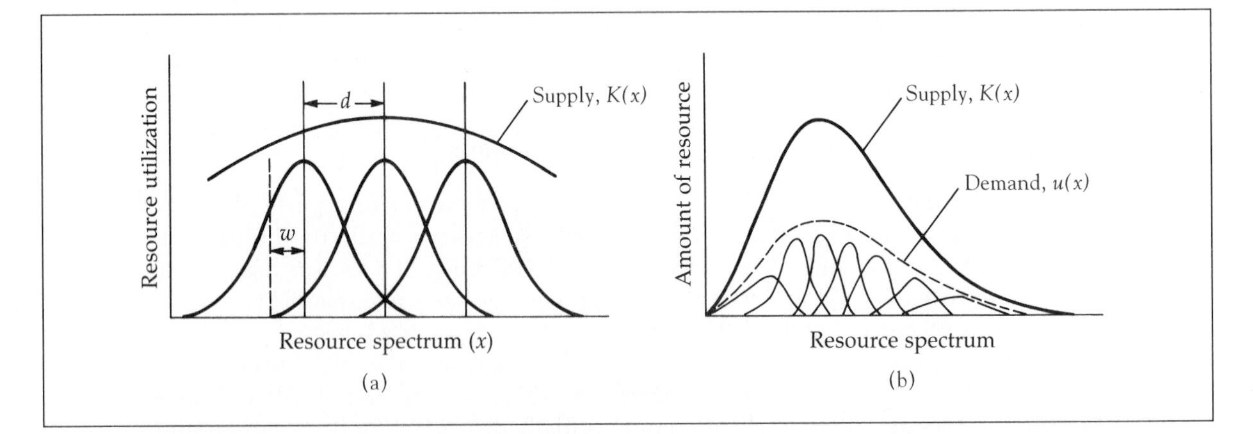

■ **FIGURE 9.11** Theorethical resource-utilization relationships. (**a**) The "simplest case" of three species with similar (and normal) resource-utilization curves. d = distance apart of means, w = standard deviation of utilization, and d/w = resource separation ratio. (**b**) The more typical case with varying resource-utilization curves broadest in the region of fewer resources and less interspecific competition. ([a] *Modified from May and MacArthur 1972,* [b] *redrawm from Pianka* 1976.)

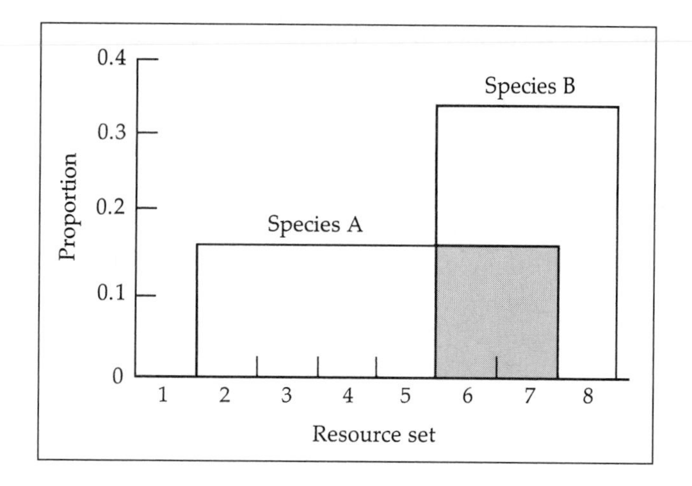

FIGURE 9.12 Hypothetical distributions of a species, A, with a broad niche and a species, B, with a narrower niche, on a resource set subdivided into eight resource units. The species have the same proportional similarity (shaded zone), but species A overlaps B more than B overlaps A.

$$\text{niche breadth} = \frac{1}{\sum_{i=1}^{s} p_i^2 \ (S)}$$

where p_i = proportion of species found in the ith unit of a resource set of S units, such that B_{max} (the maximum width breadth) = 1.0 and B_{min} (the minimum width breadth) = $1/S$. Proportional similarity between species is then given by

$$PS = \sum_{i=1}^{n} p_{ni}$$

where p_{ni} is the proportion of the less abundant species of the pair in the ith unit of a resource set with n units.

Thus, in Fig. 9.12 the P_{ni} for each of the seven units of resources shown where at least one species occurs is 0, 0, 0, 0, 0.166, 0.166, and 0. The proportional similarity is then 0.333.

In accordance with Hutchinson's ideas, *PS* values of less than 0.70 have been taken to indicate possible coexistence, those greater than 0.70 competitive exclusion. However, species may differ not only along one resource axis but along many, such as food, temperature, and moisture. For two resource axes, proportional similarity indices can be combined—proportional similarity values of 0.8 and 0.6 on two axes combine to give an overall *PS* of 0.48. Theoretically, coexistence would be permitted in cases where combined *PS* values $0.7 \times 0.7 = 0.49$ or less. Such analyses become more complex and more subject to error as new axes are included.

Perhaps the most serious criticism of both d/w and *PS* treatment is that resource axes identified by the researcher as important may not accurately reflect limiting resources for organisms. Can sweep-net samples of insects on foliage reliably indicate food availability for birds? Furthermore, faced with the apparent contradiction that many ecologically similar species coexist with no apparent differences in biology, many researchers would argue that the correct niche dimensions had not yet been examined.

It is worth noting that, in some situations, competing species have been found to differ hardly at all in morphology and yet still are found together (Laurie, Mustart, and Cowling 1997). Two reasons have been proposed. First, in the presence of high levels of predation, competitively dominant species may be selected by predators over less abundant prey. In this situation, good competitors will probably never be able to eliminate poor competitors totally if predation occurs; this is the idea of predator-mediated coexistence, and it will be addressed again later (as a method to explain the high diversity of some communities; see Section 5). Second, it is important to realize that many real populations in nature exist not in closed systems but in open areas where migration is possible and there is good connectedness between populations. Caswell (1978) has theorized that, given good connectedness between areas, immigration into an area of competitively inferior species from areas where they do well will be sufficient to maintain reasonable population sizes for both competitors for an indefinite time. Thus, if predators open up resources in an environment by killing members of the competitively dominant species, high connectedness between populations means that competitively inferior species may repeatedly first appear there by immigrating from other areas.

SUMMARY

1. Competition may be intraspecific (between individuals of the same species) or interspecific (between individuals of different species). It may also be viewed as scramble competition for a limiting resource, or contest competition where individuals compete directly with one another.

2. The first attempts to model how competing species interacted were originated by Lotka (1925) in the United States and by Volterra (1926) in Italy. In a two-species interaction, four outcomes are possible: species 1 goes extinct; species 2 goes extinct; either species 1 or 2 goes extinct, depending on starting conditions; or both species coexist.

3. Thomas Park's laboratory studies of competition between four beetles showed that the outcome of competition could be changed by environmental conditions, by the presence or absence of natural enemies, and by the genetic strain of the competitors involved.

4. In nature, there are many experimental studies that show competition between different type of organisms, such as between barnacles, between parasitic wasps, and between chaparral plants.

5. Reviews on the frequency of competition in nature demonstrate that competition between species has been found in 55 percent to 75 percent of the species involved.

6. Schoener (1985) recognized at least six mechanisms of competition: consumptive (exploitative), preemptive, overgrowth, chemical, territorial, and encounter. Exploitative competition is by far the most common, occurring in at least 37.8 percent of cases.

7. It has frequently been argued that competition is minimized and that species can coexist if they utilize different resources. What is the amount by which competitors can overlap in resource utilization but still coexist? Hutchinson (1959) suggested that a ratio of 1:1.3 in the morphological sizes of feeding apparatus was necessary. Other authors have studied resource use more directly and have suggested that proportional similarity values of no more than 70 percent are necessary.

DISCUSSION QUESTION

Which type of competition would you expect to be the more important in nature—intraspecific or interspecific? Does intraspecific competition have to be weak in gregarious species? If interspecific competition between species can cause species to partition resources or to have different morphological characteristics, why wouldn't intraspecific competition cause even greater character displacement or resource partitioning between individuals of the same species?

CHAPTER

10

Predation

Several types of predation can be recognized:

Herbivory:	Animals feeding on green plants.
Carnivory:	Animals feeding on herbivores or other carnivores.
Parasitism:	(a) Animals or plants feeding on other organisms without killing them and (b) parasitoids, usually insects, laying eggs on other insects, which are subsequently totally devoured by the developing parasitoid larvae. Usually, one or more parasitoids can complete development on a single host.
Cannibalism:	A special form of predation, predator and prey being of the same species.

Although herbivory and parasitism could be viewed widely in the context of predation, each has special characteristics that tend to separate it out into a special subdivision of predation, and each subject will be discussed separately. Herbivory, for example, is often nonlethal predation on plants. Many leaves or parts of leaves of a plant can be eaten without serious damage to the "prey." In the animal world, on the other hand, predation generally means death for the prey. Parasitism, too, is a unique type of event, in which one individual prey is commonly utilized for the development of one or more predators. In the special case of insect herbivores on plants, there is great debate about whether such insects can be called plant *parasites* because they seem to fit such a definition fairly well (Price 1980, see Chap. 12). Carnivory embodies the traditionally held view of predation, and cannibalism can be seen as a special case of it. The degree of lethality of these interactions on their hosts/prey and the intimacy with them are summarized in Fig. 10.1.

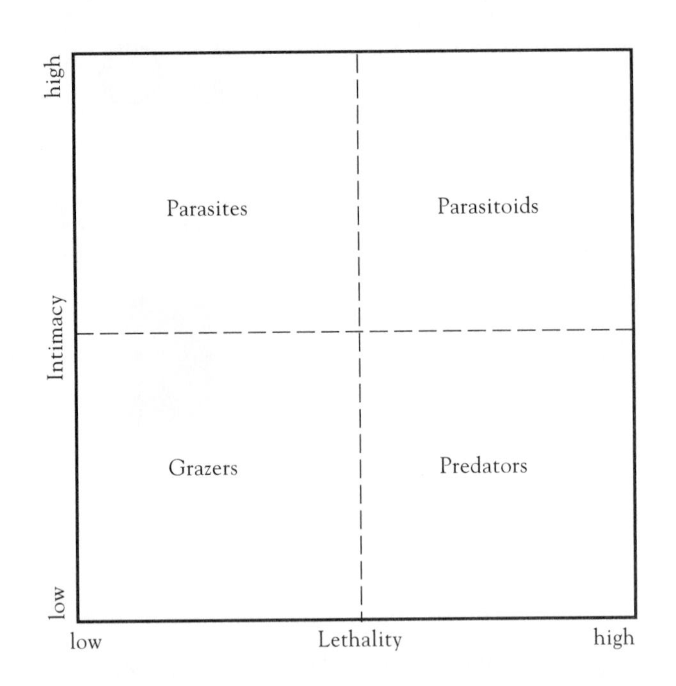

FIGURE 10.1 Possible interactions between populations. Lethality represents the probability that a trophic interaction results in the death of the organism being consumed. Intimacy represents the closeness and duration of the relationship between the individual consumer and the organism it consumes. (*After Pollard 1992.*)

10.1 The variety of strategies that organisms have evolved to avoid being eaten suggests that predation may be a strong selective force.

In response to predation, many prey species have developed strategies to avoid being eaten (see color insert). Edmunds (1974), Owen (1980), and Waldbauer (1988) have reviewed many case studies, which include the following:

- *Aposematic, or warning, coloration:* Aposematic coloration advertises a distasteful nature. Lincoln Brower (1970) and coworkers (Brower et al. 1968) showed how inexperienced blue jays took monarch butterflies, suffered a violent vomiting reaction, and learned to associate forever afterward the striking orange-and-black barred appearance of a monarch with a noxious reaction. The caterpillar of this butterfly gleans the poison from its poisonous host plant, a milkweed. Many other species of animals, especially invertebrates, are also warningly colored, for example ladybird beetles. Caterpillars of many Lepidoptera are bright and conspicuous too, because being noxious is their main line of defense—being soft-bodied, they would otherwise be very vulnerable to predators (Chap. 4).
- *Crypsis and catalepsis:* **Crypsis** and **catalepsis** are the development of a frozen posture with appendages often retracted. This is another common method of avoidance of detection by invertebrates. For example, many grasshoppers are green and blend in perfectly with the foliage on which they feed. Often, even leaf veins are mimicked on grasshopper wings. Stick insects mimic branches and twigs with their long slender bodies. In most cases, these animals stay perfectly still when threatened because movement alerts a predator. Crypsis is prevalent in the vertebrate world too. A zebra's stripes supposedly make it blend in with its grassy back-

ground, and the sargassum fish even adopts a body shape to mimic the sargassum weed in which it is found. Perhaps first prize should go to the chameleon, whose skin tones can be adjusted to match the background on which it is resting.

- *Mimicry:* Though many organisms may try to blend into their background, mimicking the foliage or background around them, some animals mimic other animals instead. For example, some hoverflies mimic wasps. Several types of mimicry can be defined. The three discussed here are illustrated in Photo 10.3.
- *Mullerian:* This refers to the convergence of many unpalatable species to look the same, thus reinforcing the basic distasteful design, as, for example, with wasps and some butterflies (Muller 1879).
- *Batesian:* (after the English naturalist Henry Bates [1862], who first described it): Mimicry of an unpalatable species by a palatable one or of dangerous species (coral snakes) by innocuous ones (pseudocorals). Wickler (1968) has documented many cases in which flies, especially hoverflies of the family Syrphidae, are striped black and yellow to resemble stinging bees and wasps. Also, some butterflies painted to mimic distasteful models were recaptured more frequently than those painted to mimic palatable forms, suggesting the great survival value of mimicry (Sternberg, Waldbauer, and Jeffords 1977). Until very recently, monarch and queen butterflies, aposematically colored themselves, were considered models for mimic viceroy butterflies. However, David Ritland (Ritland and Brower 1991; Ritland 1994) showed that all three species were unpalatable and thus coexisted as a Mullerian mimicry complex.
- *Aggressive*: In this case, the body coloration permits individuals not so much to escape predation as to be better predators themselves. Many praying mantises mimic flowers so as to entrap insect prey. This strategy is shared by the crab spiders. Certain bottom-dwelling ocean fish also mimic the substrate so as to get in closer proximity to their prey. In extreme cases, such as anglerfish, parts of the body are modified to act as lures for prey. Some species also use light-emitting organs to attract prey. On land, aggressive mimicry is practiced by female fireflies of the genus *Photuris* (Lloyd 1975). Normally males respond to the species-specific light flashes of their females and move toward them. *Photuris* females mimic the flashing patterns of females of other species (*Photinus* sp.) to lure the males of those species and eat them.
- *Intimidation displays:* An example of intimidation display is a toad swallowing air to make itself appear larger. Frilled lizards extend their collars when intimidated to create the same effect.
- *Polymorphisms:* **Polymorphism** is the occurrence together in the same population of two or more discrete forms of a species in proportions greater than can be maintained by recurrent **mutation** alone. Often this phenomenon takes the form of a color polymorphism; if a predator has a preference or **search image** for one color form, usually the commoner (Tinbergen 1960), then the prey can proliferate in the rarer form until this form itself becomes the more common (Cain and Sheppard 1954*b*) (so-called

apostatic selection, Clark 1962). Stiling (1980) advocated just this type of choice of prey, by a visually searching **parasitoid**, to maintain the difference between two distinct color morphs, orange and black, in some leafhopper nymphs of the genus *Eupteryx*, though Stewart (1986*a*) also implicated thermal **melanism** as an important agent of selection in some species. That is, black morphs occur in cooler climates because they heat up faster in the sun. However, Losey et al. (1997) showed how a balanced polymorphism between "green" and "red" color morphs of the pea aphid *Acyrthosiphon pisum* was maintained by the action of natural enemies. Green morphs suffered higher rates of parasitism than red morphs, whereas red morphs were more likely to be attacked by ladybird beetle predators than green morphs. Therefore, when parasitism rates in the field were high relative to predation rates, the proportion of red morphs increased relative to green morphs whereas the converse was true when predation rates were relatively high. Owen and Whiteley (1986) have pointed out that in many species the form of the polymorphism is such that *every* individual is slightly different from all others. This is true in brittlestars, butterflies, moths, echinoderms, and gastropods. They suggest that such a staggering variety of form thwarts predators' learning processes, and they suggest the term *reflexive selection* for this type of phenomenon.

- *Phenological separation of prey from predator:* Fruit bats, normally nocturnal foragers, are active by day and at night on some small species-poor Pacific islands such as Fiji. Wiens et al. (1986) suggest the fruit bats are constrained elsewhere to fly only at night by the presence of predatory diurnal eagles.
- *Chemical defenses:* One of the classic defenses involves the bombardier beetle (*Bradinus crepitans*) as studied by Tom Eisner and coworkers (Eisner and Meinwald 1966; Eisner and Aneshansley 1982). These beetles possess a reservoir of hydroquinone and hydrogen peroxide in their abdomens. When threatened, they eject these chemicals into an explosion chamber where they mix with a peroxidase enzyme. The resultant release of oxygen causes the whole mixture to be violently ejected as a spray that can be directed at the beetle's attackers. Many other arthropods have chemical defenses too, such as millipedes. This phenomenon is also found in vertebrates, as people who have had a close encounter with a skunk can testify.
- *Masting:* Masting is the synchronous production of many progeny by all individuals in a population to satiate predators and allow some progeny to survive (Silvertown 1980). It is commonly documented in trees, which tend to have years of unusually high seed production. A similar phenomenon is exhibited by the emergence of seventeen-year and thirteen-year periodical cicadas (*Magicada* sp.). They are termed *periodical* because the emergence of adults is highly synchronized to once every thirteen or seventeen years. Adult cicadas live for only a few weeks, during which time females mate and oviposit on the twigs of trees. The eggs hatch six to ten weeks later, and the nymphs drop to the ground and begin a long subterranean development feeding on the contents of the xylem tissue of roots. Because xylem is such a dilute medium, nymphal development is very

> slow, but there seems to be no reason why adults should not emerge say after ten, eleven, or twelve years. It is thought that predator satiation is maximized by exact synchrony of emergence. It is worth noting in this context that both thirteen and seventeen are prime numbers, so no predator on a shorter multiannual cycle could repeatedly use this resource. Williams, Smith, and Stephen (1993) studied the morality of populations of a thirteen-year periodical cicada that emerged in northwestern Arkansas in 1985. Birds consumed almost all of the standing crop of cicadas when the density was low (i.e., when the cicadas make a "mistake" and emerged in a nonmast year), but this became only 15 to 40 percent when the cicadas reached peak density in a mast year. Predation then rose to near 100 percent as the cicada density fell again. Karban (1997) has also shown how the four extra years' growth of seventeen-year cicadas over thirteen-year ones can produce females with much heavier ovaries and more eggs. This increased fecundity more than makes up for the delay in reproduction.

How common is each of these defense types? Brian Witz (1989) surveyed 354 papers that documented antipredator mechanisms in arthropods, mainly insects (Table 10.1), of 555 predator/prey interactions. By far the most common antipredator mechanisms were chemical defense, noted in at least 46 percent of the examples. I say at least, because many categories were not mutually exclusive, for example, aposematic coloration is also usually coupled with noxious chemicals.

It is obvious that predation constitutes a great selective pressure on plant and animal populations. Despite the impressive array of defenses, predators still manage to survive by eating individuals of their chosen prey, often by circumventing the defenses in some way. The coevolution of defense and attack can be seen as an ongoing evolutionary arms race. According to Dawkins and Krebs (1979), the prey are always likely to be one step ahead. The reason is what they termed the *life-dinner principle*. In a race between a fox and a rabbit, the rabbit is usually faster because it is running for its life whereas the fox is running "merely" for its dinner. A fox can still reproduce even if it does not catch the rabbit. The rabbit never reproduces again if it loses. It has been argued that this arms race is run not only between predators and prey but also between parasite and host and between plant and herbivore. In the latter case, the race often proceeds by the production of toxins by the host and detoxifying mechanisms by the predator. Usually, however, at least for plants and herbivorous insects, phylogenetic relationships among herbivores do not correspond well to those of their host plants (Futuyma 1983; Mitter and Brooks 1983). The result is that at each defensive turn by a plant (or other host), it is as likely to run into a new set of enemies as it is to escape the old ones.

10.2 The effects of predation on the dynamics of predator and prey populations can be described using mathematical models.

What effect has the predator on its prey population? The answer depends on many things, including prey and predator density and predator efficiency. Rosenzweig and MacArthur (1963) modeled predator-prey dynamics using a graphical method. First, they assumed that, in the absence of predators, the prey population increases exponentially according to the formula

ECOLOGY IN PRACTICE

Rick Karban, University California, Davis

I was an undergraduate at Haverford College, which was an excellent place to start. Students were encouraged to participate in their education, to ask questions, and to develop important skills. Haverford offered only cell and molecular biology, and a very brief exposure to this "small stuff" convinced me that biology was not for me. Fortunately, I had the opportunity to work for the National Park Service during my summers in college, and this experience convinced me that ecology might be more to my liking. I went to Cornell for my junior year and took ecology and other whole organism courses, reinforcing my suspicion that this "bigger stuff" was pretty interesting. Dan Janzen had recently moved to the University of Pennsylvania, which was a train ride away from Haverford. I attended an informal research seminar that he gave and was impressed. Dan was one of the most charismatic personalities I had ever witnessed. He was full of stimulating ideas and stories about how nature worked.

I decided to go to graduate school in ecology at the University of Pennsylvania. Starting grad school was exhilarating and intense. Unfortunately, my feeling of excitement of academia was short lived. My first surprise came following a seminar presented to the biology department by an invited speaker. I was told that my questions were naive and reflected poorly on the department. Over the next few years, it became clear to me that people around me were desperately afraid of being wrong, or worse, not as smart as the fellow down the hall. Perceptions were given a lot of thought, and there was no room for naïveté. Questions in seminars were about showing others about how smart you were, not about satisfying curiosity.

Because I was fascinated by population-level events that were synchronous, Dan suggested that I read papers about periodical cicadas that Monte Lloyd and his associates, JoAnn White and Chris Simon, had written. A personal visit to Monte in Chicago gave me the encouragement and advice I needed to start a field project that turned into my thesis.

Grad school continued to be an ordeal. I was told I was not a thinker and that I would not have a chance of getting an academic job. At one point, I left graduate school without telling anyone at the university. I went home, talked to my parents, spent time with my best friend, and thought a lot about potential careers. I decided that I really did want to be a professional biologist in some capacity and that I would not let the poor opinions that others had of my potential defeat me. I felt determined to make my own decisions. I felt that the hardest part would be returning and explaining why I had been gone for a week. I returned, fortified by the encouragement and support of being home, to find that no one had even noticed I had been gone.

Although my personal interactions in grad school were not very rewarding, field work was. Field work has always been the most enjoyable part of being an ecologist for me. Field seasons during my years in graduate school were fantastic, the kind of hard work that I most enjoy.

I applied for many jobs, both in academics and elsewhere. The first interview I was granted came from the entomology department at the University of California at Davis. A job in an entomology department seemed like a long shot since I had no formal training in that subject. I had the feeling that this was an opportunity that would not come twice. In any case, I was in the right place at the right time. Dan was, well, surprised at Davis's decision. At my thesis defense just prior to moving to Davis, I was advised not to spend much time looking back, advice that I generally try to heed.

I learned some important lessons in graduate school though I wouldn't choose to repeat the experience. I learned that working with large, soil-dwelling insects that take seventeen years to complete a generation was a lot of fun but not a good vehicle for answering the questions I was most interested in. I haven't abandoned cicadas but I have worked more recently with mites and thrips, small arthropods that complete a generation in only a few weeks. Most importantly, I learned that I won't let anybody sell me short again. Being a successful scientist is about daring to be naive, asking questions that other people think are stupid, doing the projects that are personally interesting whether or not your peers agree, and finding the confidence to keep going when you don't really possess it.

TABLE 10.1 Antipredator mechanisms in arthropods. (*After Witz 1989*)

MECHANISM AND RANK	EXAMPLE OF MECHANISM	% FREQUENCY
1. Chemical	Reflexive bleeding, toxic chemicals (especially beetles)	46
2. Fighting	Stinging (especially wasps), biting, kicking	11
3. Crypsis	Camouflage, especially caterpillars	9
4. Escape	Running away, flying	8
5. Mimicry	Batesian and Mullerian	5
6. Aposematism	Warning coloration	5
7. Intimidation display	Posturing	4
8. Dilution	Masting, satiation	4
9. Mutualism	Defense by other organism	3
10. Armor	Spines, thorns	2
11. Acoustic	Loud noise, e.g., grasshopper	2
12. Feigning death		1

$$\frac{dN}{dt} = rN$$

The rate of removal of prey increases with an increase in encounter rate by predators dependent upon the number of predators (C) and the number of prey (N). The rate of removal is also dependent upon a "searching efficiency" or attack rate of predators, termed a'. The consumption of prey by predators is then $a'CN$, so

$$\frac{dN}{dt} = rN - a'CN$$

The rate of growth of the predators is given by

$$\frac{dC}{dt} = fa'CN - qC$$

where q is a mortality rate based on starvation in the absence of prey and $fa'CN$ is predator birth rate based on f, the predator's efficiency at turning food into offspring.

The properties of this model can be investigated by locating zero isoclines (regions of zero growth, neither positive nor negative). In the case of the prey, when

$$\frac{dN}{dt} = 0, rN = a'CN$$

or

$$C = \frac{r}{a'}$$

Because r and a' are constants, the prey zero isocline is a line for which C, the number of predators, is itself a constant (Fig. 10.2a). Along this line, prey neither increase nor decrease in abundance.

For the predators, when

$$\frac{dC}{dt} = 0, fa'CN = qC$$

or

$$N = \frac{q}{fa'}$$

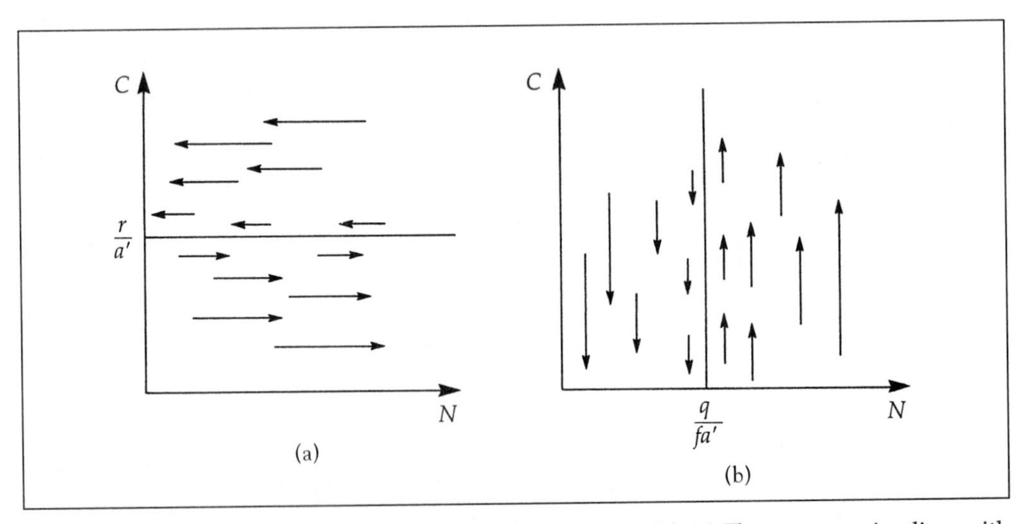

■ **FIGURE 10.2** A Lotka-Volterra type predator-prey model. (**a**) The prey zero isocline, with prey (N) increasing in abundance at lower predator densities (low C) and decreasing at higher predator densities. (**b**) The predator zero isocline, with predators increasing in abundance at higher prey densities and decreasing at lower prey densities. *(Redrawn from Begon, Harper, and Townsend 1986.)*

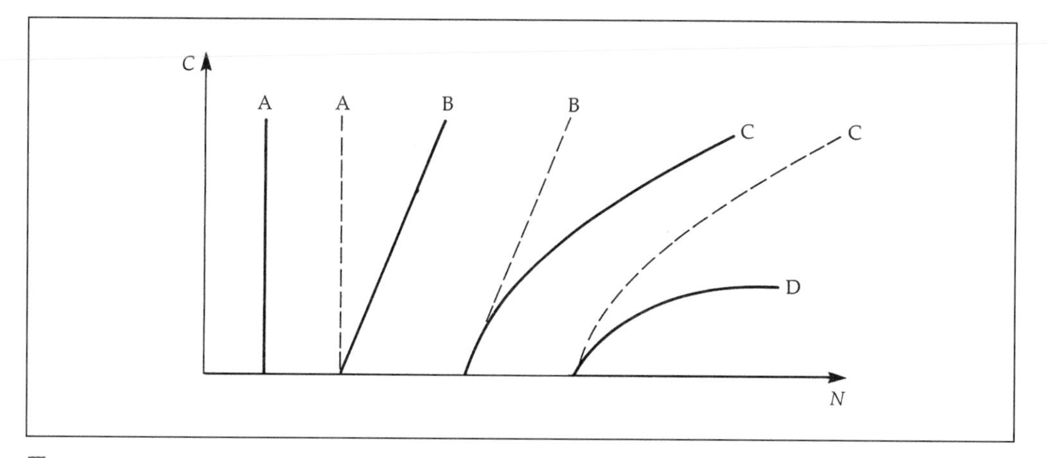

FIGURE 10.3 Predator zero isoclines of increasing complexity, A to D. A is the Lotka-Volterra isocline. B shows that more predators require more prey. C shows that the consumption rate is progressively reduced by mutual interference among predators. D shows that predators are limited by something other than their food. *(Redrawn from Begon, Harper, and Townsend 1986.)*

The predator zero isocline is also a straight line, one along which N, the number of prey, is constant (Fig. 10.2b). An assumption is that, if there are enough prey, a population of predators will increase and that if there are not enough prey they will starve. This is obviously a gross oversimplification. First, larger populations of predators require larger populations of prey to maintain them, so the zero isocline for predator growth should slant to the right (line B of Fig. 10.3). Second, as predator density increases, so will mutual interference between predators, who will spend more time fighting one another and less time tackling prey. For zero growth, then, more realistically, there must be even more prey (line C of Fig. 10.3). Finally, at the highest densities of prey, it seems that the rate of growth of predators will be limited by something other than prey availability—say, social constraints on territory size or the availability of burrows or nest sites, so the zero isocline will appear as in line D in Fig. 10.3. Remember that at predator-prey combinations to the left of this line, predator numbers decrease, whereas to the right of it they increase.

The refinement of the prey zero isocline is dependent on two concepts. The first is that the recruitment rate of prey into the population is low when N is low (because there simply aren't many reproducing adults) or when N is high and near the carrying capacity of the environment (because resources are limiting). The logic is that strong competition for resources at levels of N near K severely reduces the recruitment rate. At intermediate prey densities, recruitment is highest. Hence, the recruitment curve is semicircular (Fig. 10.4, top). The rate of growth of the prey population is also dependent on the level of predation, and different predation rates are represented by the steepness of the dashed lines in the top of Fig. 10.4. At each of the points where the consumption curve crosses the recruitment curve, the rate of population growth of the prey is zero, and these pairs of densities can be directly transferred to give the zero isocline for growth of the prey (Fig. 10.4, bottom).

The effect of different densities of predators can now be assessed by combining the two isoclines (Fig. 10.5). The highest levels of predation translate into large oscillations

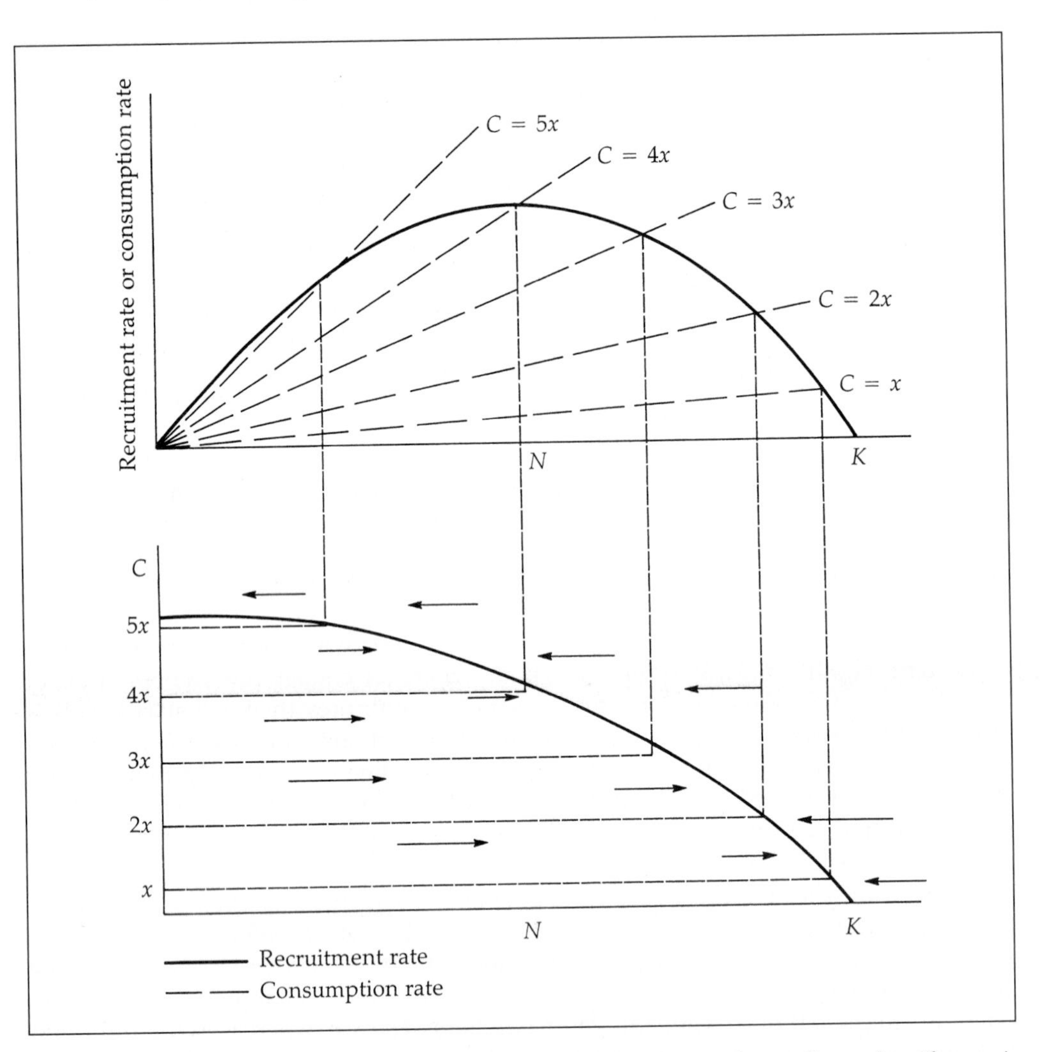

■ **FIGURE 10.4** Refinement of prey zero isocline. The solid line in the top figure describes variation in prey recruitment rate with density. The dashed lines in the upper figure describe the removal or consumption of prey by predators. There is a family of dashed curves because the total rate of consumption depends on predator density: increasingly steep dashed curves reflect these increasing densities. At the points where a consumption curve crosses the recruitment curve, the net rate of prey increase is zero (consumption equals recruitment). Each of these points is characterized by a prey density and a predator density, and these pairs of densities therefore represent joint populations lying on the prey zero isocline in the bottom figure. The arrows in the lower figure show the direction of change in prey abundance. *(Redrawn from Begon, Harper, and Townsend 1986.)*

of predator and prey (i), sometimes so violent that one or both species goes extinct. Predators in this case are relatively efficient and abundant. Less efficient predators (ii) give rise to small and often damped oscillations in the system. At the lowest levels of predation, perhaps when predators are territorial, the system is highly stable, but only low levels of predators are supported (iii). The problem with this analysis is that it is

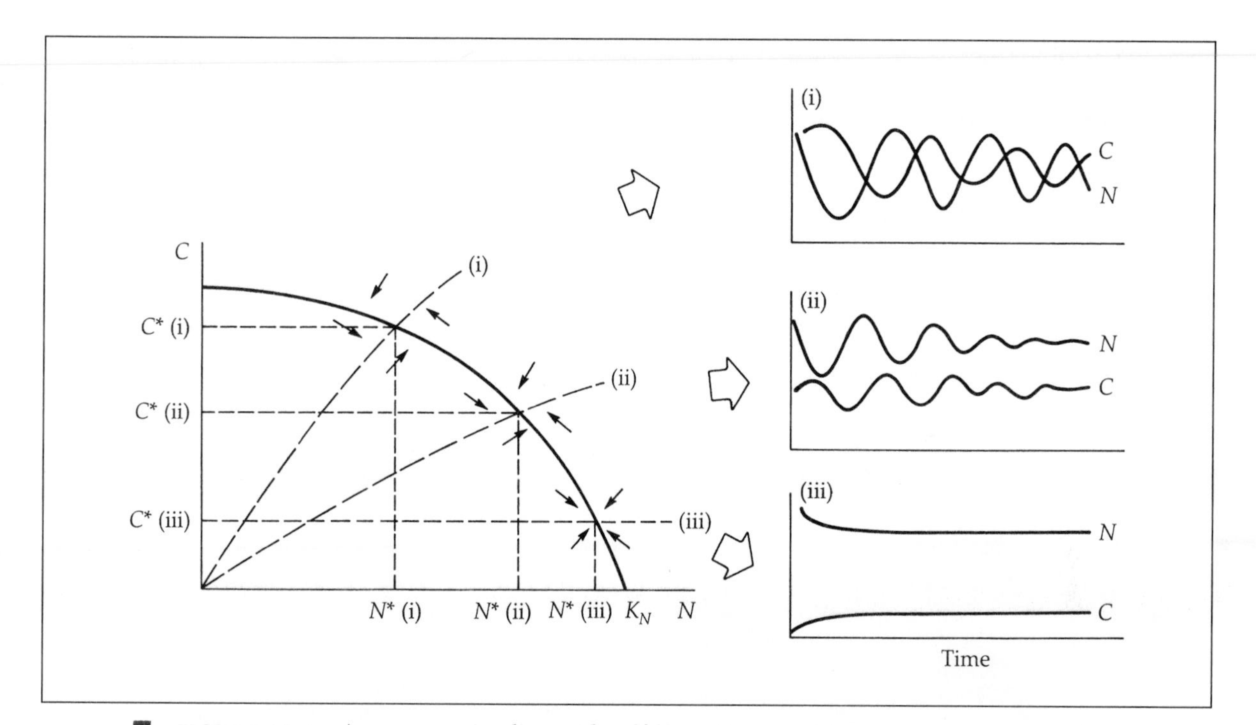

FIGURE 10.5 A prey zero isocline with self-limitation, combined with predator zero isoclines with increasing levels of self-limitation: (i), (ii), and (iii). C^* is the equilibrium abundance of predators, and N^* is the equilibrium abundance of prey. Combination (i) is least stable (most persistent oscillations) and generally has most predators and least prey: the predators are relatively efficient. Less efficient predators (ii) give rise to a lowered predator abundance, an increased prey abundance, and less persistent oscillations. Strong predator self-limitation (iii) can eliminate oscillations altogether, but C^* is low and N^* is close to K_N. *(Redrawn from Begon, Harper, and Townsend 1986.)*

almost impossible to distinguish this type of predator-prey cycling from population fluctuations resulting from environmental factors such as weather. The predictive value of these models seems to decrease as their complexity increases. However, the models can be a starting point for determining safe harvest levels for such applied endeavors as fishing, hunting, and logging (Applied Ecology: The Optimal Yield Problem). Moreover, they do show how, in theory, the interactions of predators and prey alone can have many different outcomes.

Finally, predator-prey isoclines can change dramatically with the incorporation of such factors as refuges from predators, the ability of predators to switch back and forth between multiple species, and other factors of prey. For example, some modification of the graphical model discussed so far is needed to reflect the instances in which prey populations have a refuge from predators. For example, many fossorial mammals escape from predators into underground burrows, and certain passerines exist in well-defined territories with abundant cover. Mammals outside their burrows or birds in suboptimal habitats are often exposed to predators, whereas those in their refuges are not. This situation can be represented graphically by a practically vertical prey zero isocline at low

APPLIED ECOLOGY

The Optimal Yield Problem

How many deer can be shot before a marked effect on population numbers occurs? How can people harvest a large part of a population without causing long-term changes in its equilibrium numbers? How can commercial fishermen prevent the catastrophic decimation of fish and the lowered economic yield of overfishing (Photo 1)?

PHOTO 1 Humans as the ultimate predators. Trap fishing in Rhode Island removes a staggering amount of fish from the sea, as does commercial fishing everywhere. (*George Bellerose, Stock Boston, GXB 1258B.*)

For a harvested population the important measurement is the *yield* expressed in terms of either weight or numbers over a particular time period, to give a catch per unit effort. Catch per unit effort can then be compared year after year to determine how well a particular managed resource is doing. Remember that the maximum yield of a population is related to the maximum population increase. Greatest population increase, dN/dt, occurs according to the equation

$$\frac{dN}{dt} = rN\frac{(K - N)}{K}$$

at the midpoint of the logistic curve as shown in Fig. 6.6 of Chapter 6. Thus, maximum yield is obtained from populations at less than maximum density, when they are constantly trying to expand their own population densities into unutilized resource areas. Adjusted for fishing losses, dN/dt becomes

$$\frac{dN}{dt} = rN\frac{(K - N)}{K} - qXN$$

where q is a constant and X equals the amount of fishing effort, so qX equals fishing mortality rate. As fishing and hence fishing losses increase, qXN can begin to affect the total catch (yield) severely. This relationship is shown in Figure 1, which details the relationship between fishing effort and total catch for the Peruvian anchovy. The Peruvian anchovy fishery was the largest fishery in the world until 1972, when it collapsed. In 1972 12.3 million metric tons were harvested, and this one species alone comprised 18 percent of the world's total harvest of fish. People may not have been the only cause of population collapse. In 1972, the phenomenon known as El Niño, a major climatic change, a warming of the oceans occurred that permitted warm tropical water to move into the normally cool, nutrient-rich upwellings near the Peruvian coast. The anchovies failed to spawn, and adult fish moved south to cooler waters. Whether or not the fishery might have been saved if fishing had ceased or its intensity had been reduced to allow recovery is a matter for speculation.

We can calculate the maximum sustainable yield (MSY), the largest number of individuals that can be removed without causing long-term changes. This is like creaming off the interest from a bank account but leaving the principal intact to ensure the same future interest payment in subsequent years. MSY can be estimated as

$$\text{MSY} = Nr$$

where r is the rate of natural increase and N is the average number of animals present throughout the year. The maximum value of MSY will generally be at the mid-point of the logistic curve (see Fig. 6.6), where the overall population growth rate is highest. Thus, if $N = 10{,}000$ and $r = 0.14$ then MSY = 1400. However, this approach assumes a constant rate of cropping throughout the year. If there is one "season," then MSY is best estimated with reference to the stochastic equation (Section 6.3):

$$\text{MSY} = N(1 - e^{-r})$$

which is likely to be a lower value. If $r = 0.14$ and $N = 10{,}000$ then

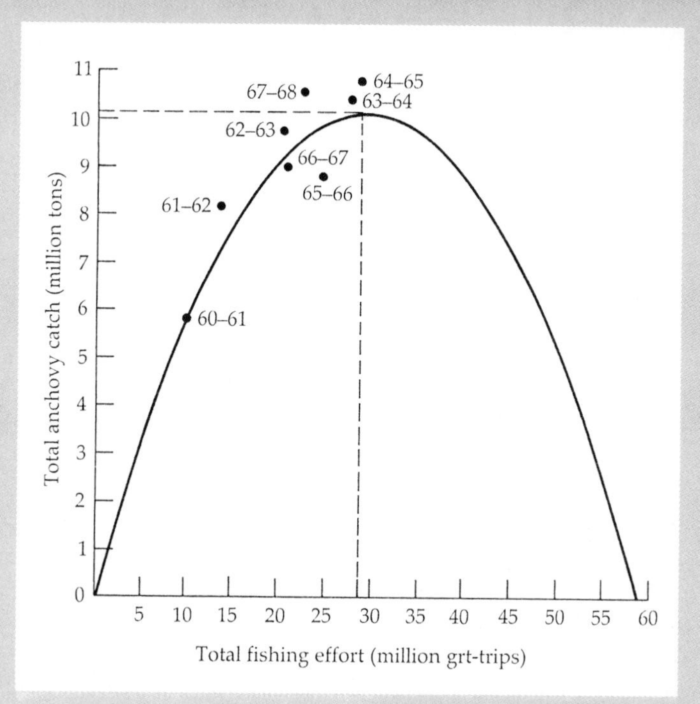

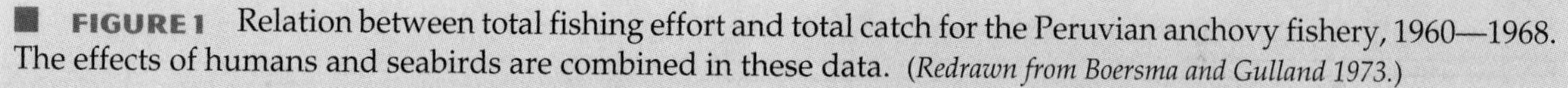

■ **FIGURE 1** Relation between total fishing effort and total catch for the Peruvian anchovy fishery, 1960—1968. The effects of humans and seabirds are combined in these data. *(Redrawn from Boersma and Gulland 1973.)*

$$\text{MSY} = 10{,}000(1 - e^{-0.14}) = 1{,}306$$

We can also argue that if the harvest is spread out over the year, N is likely to be reduced by natural deaths. N might be reduced by, for example, 10 percent so that $N = 9{,}000$ Then

$$\text{MSY} = 9{,}000 \times 0.14 = 1{,}260$$

In practice, cropping is often limited to seasons if only to give animals some respite during breeding.

Despite the existence of good scientific methods to identify sustainable yields and prevent overhavesting, there is case upon case of mismanagement of wildlife populations, forest trees, and fish stocks. It is the fisheries industry that provides some of the best data on overharvesting.

densities (Fig. 10.6). In other words, at low prey densities, prey can increase irrespective of predator densities—there are enough refuges for all individuals. At higher prey densities many prey exist outside the refuges and are available to predators. With respect to the relationship between predator and prey abundances, the refuge stabilizes the interaction and a cyclic rise and fall of predator and prey occur. This is because predator numbers increase when prey numbers increase beyond the refuges. Then predator numbers

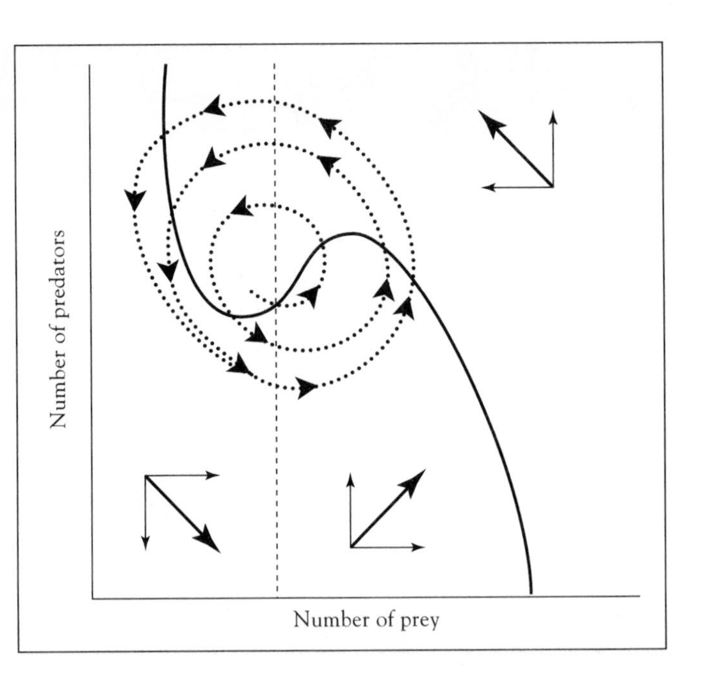

■ **FIGURE 10.6** Cycling of predator and prey populations because of prey refuges. If there are spatial refuges from predation, the prey isocline becomes vertical at low densities. In this case, the predator begins to starve once all the prey outside of the refuges have been consumed. After the predator population declines below a certain point, the population begins to increase again, repeating the cycle. (*After Gotelli 1995.*)

increase and reduce prey numbers. As prey are reduced so predators starve. Finally, prey numbers recover in the refuges and begin to expand, and the whole cycle starts again.

How do the data on predator and prey abundances in the field compare with these types of graphical models? Predators of the Serengeti plains of eastern Africa (lions, cheetahs, leopards, wild dogs, and spotted hyenas) seem to have little impact on their large-mammal prey (Bertram 1979). Most of the prey taken are injured or senile and are likely to contribute little to future generations. In addition, most of the prey sources are migratory, and the predators, residents, are more likely to be limited in numbers by prey also resident in the dry season when migratory ungulates are elsewhere. Pimm (1979 and 1980) has argued that the importance of predation is dependent on whether the system is "donor-controlled" or "predator-controlled." In a donor-controlled system, prey supply is determined by factors other than predation, so removal of predators has no effect. Examples include consumers of fruit and seeds, consumers of dead animals and plants, and intertidal communities in which space is limiting. In a predator-controlled system, the action of predator feeding eventually reduces the supply of prey and their reproductive ability. The removal of predators in a donor-controlled system is obviously likely to have little effect, whereas in a predator-controlled system such an action would probably result in large changes in abundance.

Many predator-prey systems appear stable; others appear to fluctuate dramatically. Determining which mechanism works in which situation is not easy. For example, in the Arctic there are two groups of primarily herbivorous rodents—the microtine varieties, lemmings and voles, and the ground squirrels. Ground squirrels exhibit the strongly self-limiting behavior of aggressive territorial defense of burrows (Batzli 1983). Their populations are remarkably consistent from year to year. On the other hand, the microtines are renowned for their dramatic population fluctuations. Thus, even in the same habitat, results for different species vary dramatically. One series of population fluctuations analyzed in great detail is that of the Canada lynx and snowshoe hare.

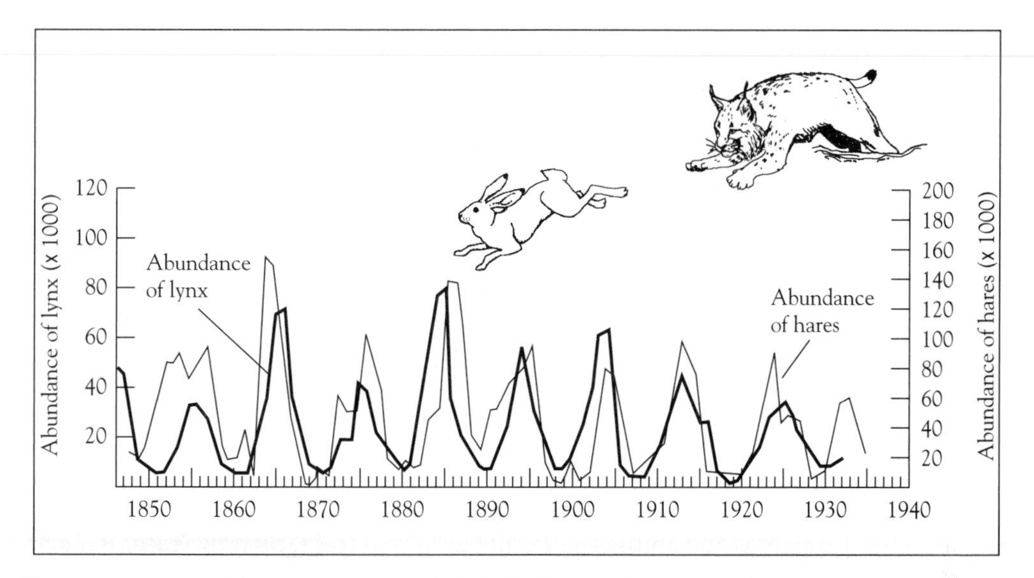

■ **FIGURE 10.7** The apparently coupled oscillations in abundance of the snowshoe hare (*Lepus americanus*) and Canada lynx (*Lynx canadensis*) as determined from numbers of pelts lodged with the Hudson's Bay Company.

The Canada lynx (*Lynx canadensis*) eats snowshoe hares (*Lepus americanus*) and shows dramatic cyclic oscillations every nine-to-eleven years (Fig. 10.7). Charles Elton analyzed the records of furs traded by trappers to the Hudson's Bay Company in Canada over a two hundred-year period and showed that a cycle has existed for as long as records have been kept (Elton and Nicholson 1942). This cycle has been interpreted as an example of an intrinsically stable predator-prey relationship (Trostel et al. 1987), but Keith (1983) has argued that it is winter food shortage and not predation that precipitates hare decline. He showed that heavily grazed plants produce shoots with high levels of toxins, making them unpalatable to hares. Such chemical protection remains in effect for two to three years, precipitating further hare decline. Predators, he argued, simply exacerbate population reduction. Thus, although lynx cycles depend on snowshoe hare numbers (donor controlled), hares fluctuate in response to their host plants. Subsequently, Smith et al. (1988) showed that, although food quality greatly affects hare **biomass**, most hares die of predation, not starvation. However, death due to predation is greatly exacerbated by poor quality of hares, which is of course greatly affected by food quality; thus, there seems to be a good deal of common ground between the predation and starvation camps. Experiments showed that three-trophic-level interactions among vegetation, hare, and predator populations seemed capable of explaining the cyclicity of snowshoe hare populations.

Sinclair et al. (1993) suggested a correlation between the frequency of sunspots, the level of herbivory by snowshoe hares and hare fur records stretching back over two hundred years. They were able to show a link between the frequency of feeding marks on trees and hare density over one hare cycle in Canada. The logic was that sunspots change the climate, which affected the spruce, which in turn affected herbivory and hence predator populations. Quite a bottom-up effect. However, Ranta et al. (1997) argued that if sunspots were influencing hare numbers than the pattern should be evident in other areas of the globe also, but they could show no link between snowshoe hare fluctuations and sunspots in Finnish data, strengthening the idea that tri-trophic interactions were driving the patterns.

10.3 Experimental studies and accidental introductions provide insight into the degree to which predators determine the abundance of their prey.

Perhaps the best way to find out whether predators determine the abundance of their prey is to remove predators from the system and to examine the response. One of the best examples involves dingo predation on kangaroos in Australia (Caughley et al. 1980). The dingo, *Canis familiaris dingo,* an introduced species, is the largest carnivore in Australia and an important predator of imported sheep. Dingoes have been intensively hunted and poisoned in sheep country, southern and eastern Australia, and the world's longest fence extends 9,600 km to prevent them from recolonizing areas from which they have been shot or poisoned, providing a classic experiment in predator control (Fig. 10.8). Because dingoes prey on native animals, as well as imported sheep, we can examine their impact on things like kangaroos and emus. The result has been a spectacular increase, 166-fold, of red kangaroos where the dingoes have been eliminated in New South Wales (Fig. 10.9), over their density in south Australia, where dingoes have not been molested.

Emus (*Dromaius novaehollandiae*) are also over twenty times more abundant in dingo-free areas. Dingoes are also frequent predators of feral pigs in tropical Australia (Newsome 1990). In Cape York, northern Queensland, there is a gross shortage of young pigs less than two years old on the mainland where there are dingoes. On neighboring Prince of Wales Island where dingoes are absent recruitment is considerable (Fig. 10.10). Also, the closest marsupial competitors to the dingo, the Tasmanian wolf and the Tasmanian devil, long ago disappeared from mainland Australia, possibly coincident with the dingo's arrival with Asian seafarers, thirty-five hundred years ago.

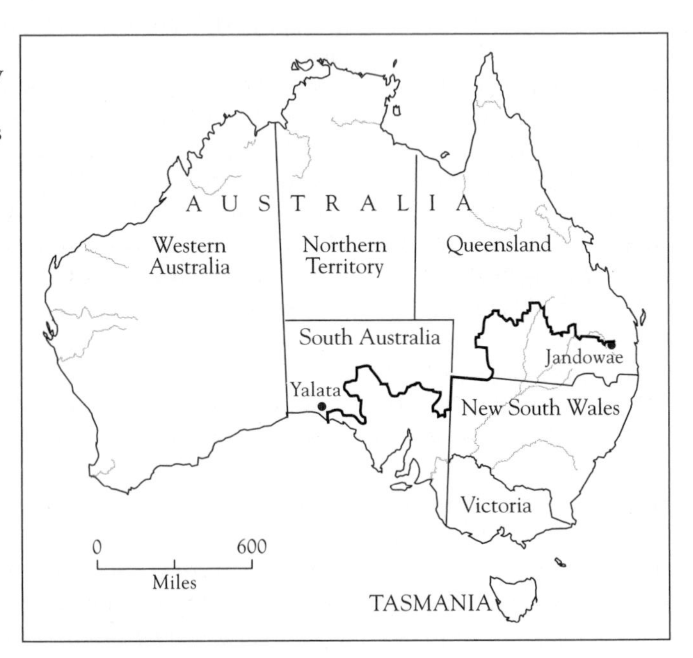

FIGURE 10.8 The world's longest fence—nearly a thousand miles longer than China's Great Wall—marches over sand hills on the South Australia-New South Wales border. It separates dingoes, to the north, from sheep, to the south.

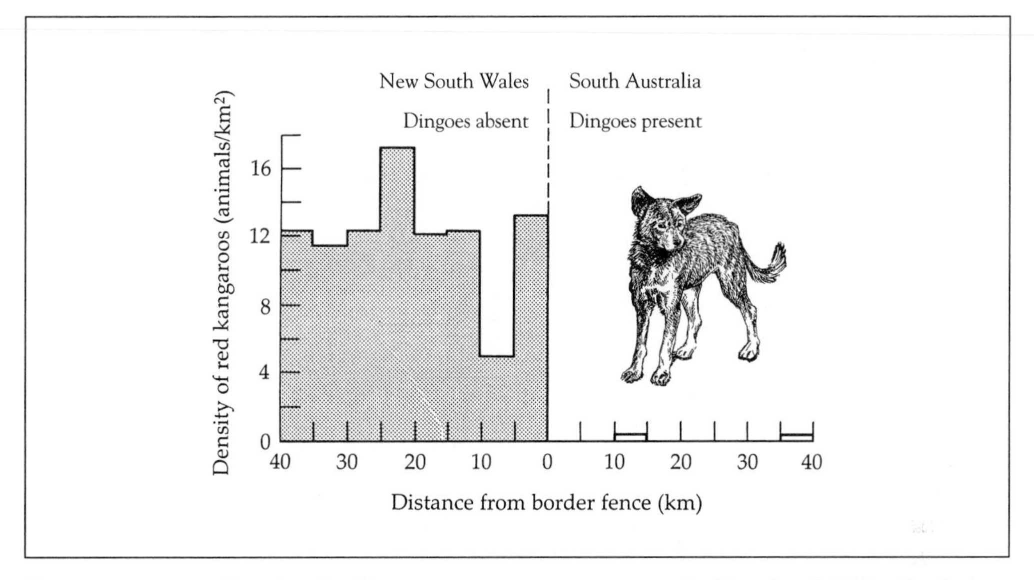

■ **FIGURE 10.9** Density of red kangaroos on a transect across the New South Wales-South Australia border in 1976. The border is coincident with a dingo fence that prevents dingoes from moving from South Australia into the sheep country of New South Wales. *(Redrawn from Caughley et al. 1980.)*

Other important exotic animals in Australia are European foxes and feral cats. Both can do damage to domestic livestock and are subject to eradication by shooting. In areas where these predators are shot, numbers of rabbits, also exotic in Australia, increased (Fig. 10.11). Where rabbits increase, valuable rangeland may become overgrazed. Also, where dingoes are absent, kangaroos can severely reduce forage quality for sheep and other organisms. Removal of exotic predators is a complex issue in Australia where so many exotic herbivores also exist.

Another striking example of predation pressure has been provided by an inadvertent introduction by humans. Marine sea lampreys (*Petromyzon marinus*) live on the Atlantic coast of North America and migrate into fresh water to spawn. Adult lampreys feed by attaching themselves to other fish, then rasping a hole and sucking out the body fluids. The passage of the lamprey to the upper Great Lakes was presumably blocked by Niagara Falls before the Welland Canal was built in 1829. The first sea lamprey was found in Lake Erie in 1921, in Lake Michigan in 1936, in Lake Huron in 1937, and in Lake Superior in 1945 (Applegate 1950). Lake trout catches decreased to virtually zero within about twenty years of lamprey invasion (Fig. 10.12). After 1951, control efforts were made to reduce the lamprey population, and lamprey became rarer. As a result, in the 1970s and 1980s lake trout numbers rebounded to near pre-lamprey levels in all lakes, underscoring the dramatic results of predation.

10.4 Reviews and natural systems.

Most of the examples discussed so far have focused on introduced species. The effects of such introductions are often very strong, but what of totally natural predator-prey systems where predator and prey have evolved together for long periods? In 1903, lions

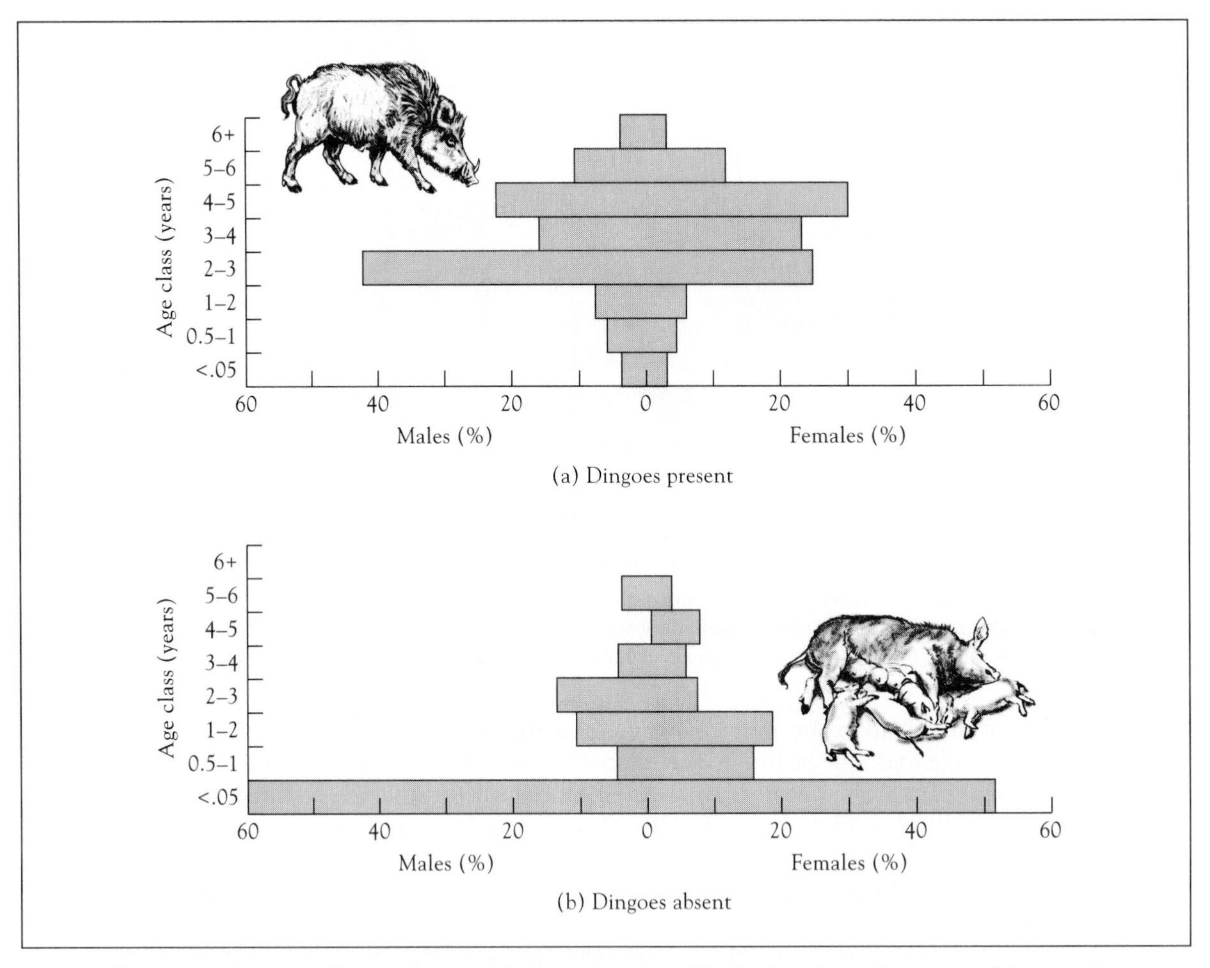

■ **FIGURE 10.10** Contrasting population structures of feral pigs where dingoes are (**a**) present and (**b**) absent in tropical northern Australia. *(After Newsome 1990.)*

were shot in Kruger National Park, South Africa, to allow numbers of large prey to increase. Shooting ceased in 1960, by which time wildebeest (*Connochaeter taurinus*) had increased so much that human culling was instigated from 1965 to 1972.

For many years the moose population on Michigan's Isle Royale, a forty-five-mile-long island enjoyed a wolf-free existence. Then, in 1949, during a particularly hard winter, a pair of Canadian wolves were able to colonize the island, walking across frozen Lake Superior. In 1958 wildlife biologist Durwood Allen of Purdue University began tracking wolf and moose numbers. The wolf population peaked at fifty individuals in 1980, and then in 1981 it took a severe nosedive (Fig. 10.13). Wildlife ecologist Rolf Peterson of Michigan Technological University has followed this population in recent years. The wolf population has continued to decline. Only four pups were born in 1992 and 1993, all to the same female in one wolfpack. The other two packs on the island are down to just a pair of wolves each. Many observers feel the total wolf population is on its way to extinction. As for the moose population, it has reached a record level of about nine-

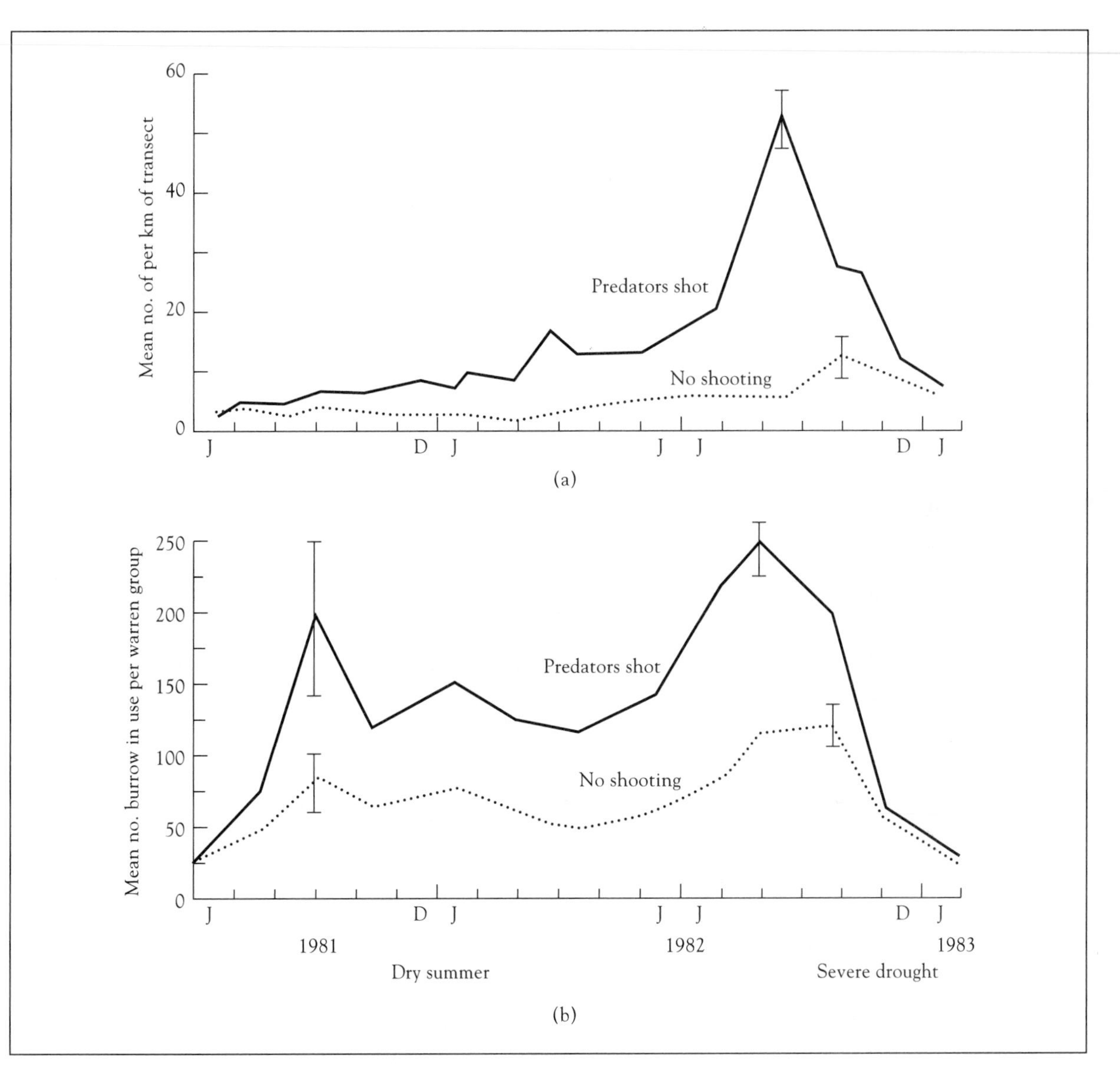

■ **FIGURE 10.11** Accelerated increase in rabbits with removal of European foxes and feral cats in an Australian field experiment. (**a**) Comparison of counts of rabbits per kilometer along transects where predator populations were continually shot (solid line) or left intact (dotted line). (**b**) Comparison of burrow use in warren groups where predator populations were continually shot (solid line) or left intact (dotted line). The error bars represent one standard error on either side of the mean. *(After Newsome et al. 1989.)*

teen hundred moose in 1993. This seems good evidence that predation does have a strong effect on natural populations.

Why did the wolves decrease in numbers? First of all, there was a narrow genetic base to begin with. Restriction enzyme analysis of the wolves' mitochondrial DNA turned up just a single pattern, indicating the wolves' were all descended from a single

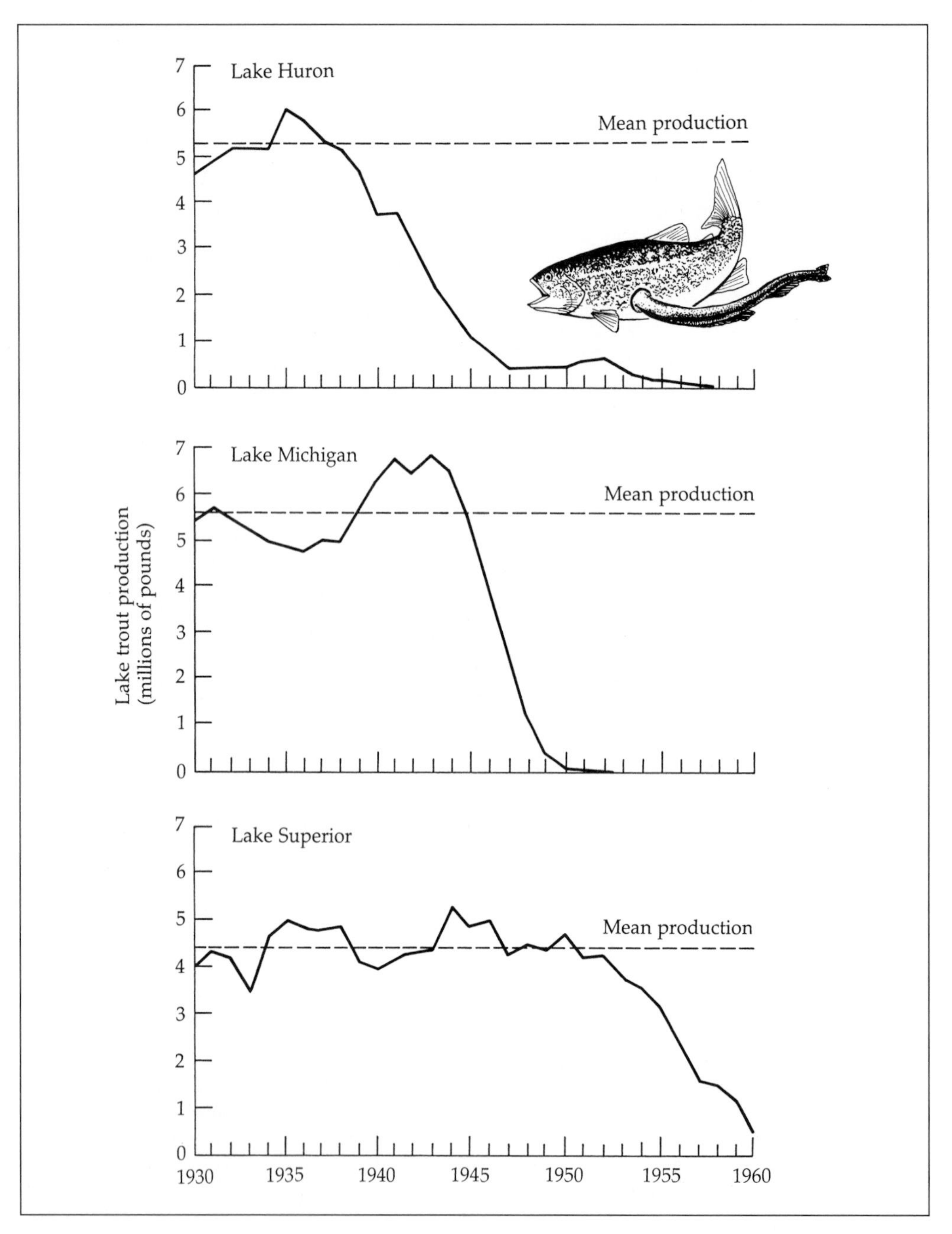

■ **FIGURE 10.12** Effect of sea lamprey introduction on the lake trout fishery of the upper Great Lakes. Lampreys were first seen in Lake Huron and Lake Michigan in the 1930s and in Lake Superior in the 1940s. *(Redrawn from Baldwin 1964.)*

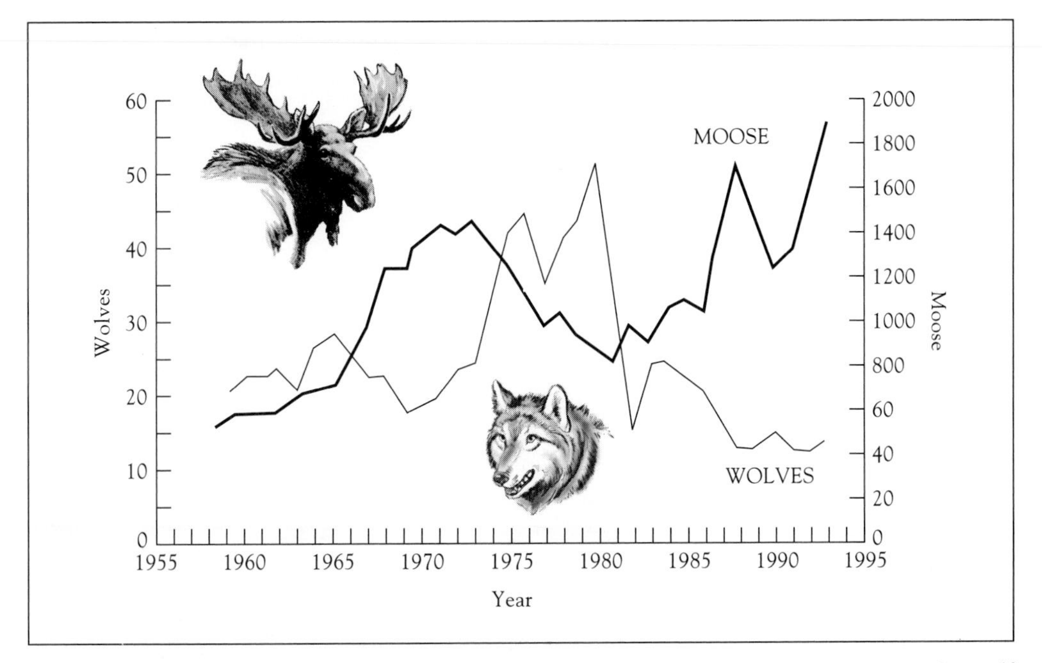

■ **FIGURE 10.13** The effects of wolf predation on moose numbers in Isle Royale. As the wolf population declined, the number of moose went up. (*After Mlot 1993.*)

female. They had only about half the genetic variability of the mainland wolves. Second, there was evidence of a deadly canine virus in 1981. This was probably a result of a parvovirus outbreak in nearby Houghton, Michigan, which was carried to the island on the hiking boots of visitors (Mlot 1993).

Insects are small enough that whole populations can be maintained inside cages. Ted Floyd (1996) performed predator-exclusion experiments on the insects feeding on creosote bushes (*Larrea tridentata*) at the Jornada long-term ecological research site in the Chihuahuan desert of southwest New Mexico. Bird predators were excluded using nylon mesh cages and arthropod predators were removed by hand. In each of two years when these experiments were performed herbivore densities became significantly higher on experimental bushes than on control bushes.

For game animals, managers are continually concerned that natural predators may take individuals that might otherwise be available to hunters. The grey partridge, *Perdix perdix,* is a widespread game bird in Europe with over twenty million shot annually in Britain in the 1930s (Photo 10.1). By the mid-1980s there was a reduction in the bag to 3.8 million (Potts and Aebischer 1995). Chick mortality caused by reduced insect abundance following the introduction of herbicides in the 1950s was suspected, and trials with reduced herbicide use proved this to be the case. However, it was also noted that populations were smaller in areas where there was no predator control by gamekeepers. Tapper, Potts, and Brockless (1996) described the result of a six-year predator-removal experiment designed to test the effect of predation in the breeding season. Foxes, crows, magpies, and jackdaws were shot with a high-powered rifle and stoats, weasels, and rats

■ **PHOTO 10.1** A grey partridge cock crowing in Germany. A six year predator removal experiment showed that predation significantly affected breeding success and brood size. (*Robert Maier, Animals Animals/Earth Scenes BIRI70MAR01001.*)

were trapped. After the nesting period these predators reestablished themselves. Predation control significantly increased the proportion of partridges that bred successfully, and the average size of their broods so that August numbers of partridges increased by 75 percent. Incorporating the effects on breeding stocks in subsequent years, this led to an overall 3.5-fold difference between autumn populations with and without predation control.

Other studies also suggest bird mortality due to predation can be high. For example, O'Connor (1991) reviewed seventy-four studies of nesting success of various bird species and found that one in three nests failed due to predation. Similar estimates were also found by Martin (1993, 41.4% of nests lost to predation in fifty-five species) and Côté and Sutherland (1995, 38.4% of nests lost to predation in ninty-eight species). In a meta-analysis of twenty published studies of predator-removal programs, Côté and Sutherland (1997) showed an average of 75 percent higher hatching success compared to control areas.

In 1985 Andrew Sih and colleagues (Sih et al. 1985) surveyed twenty years (1965–1984) of seven ecological journals for field experiments concerned with predation. Their survey yielded 139 papers involving 1,412 comparisons. Virtually every report, 132 of 139 or 95 percent showed some significant effects. In 85.5 percent of the reports there was at least one comparison where prey showed a large response to predator manipulations. Two-thirds of the studies showed depression of prey density by predators. Can we conclude that in most cases predators influence the abundance of their prey in the field? Probably, although the same

ECOLOGY IN PRACTICE

Andrew Sih,
University of Kentucky

The theme of my career as an evolutionary ecologist is integration and synthesis. That is, I try to understand nature by using a blend of approaches in evolution, behavior, and ecology. Most of my career has focused on predator-prey interactions. As a community ecologist, I am interested in why predators have major impacts on some prey while other prey are left relatively untouched. As a behavioral ecologist, I address this community-level issue by looking at the behaviors of predators and prey. As an ethologist, I try to understand predator and prey behaviors by looking at sensory mechanisms underlying behavior. Finally, as an evolutionary biologist, I try to explain both the behaviors and their sensory basis by studying evolutionary mechanisms including natural selection, genetics, and evolutionary histories. This integrative blend is, I think, both powerful and fun.

My integrative view of life is derived perhaps from my mixture of family influences. As a Chinese-American, I grew up immersed in a blend of Eastern and Western cultures. In addition, my parents lent me very different skills. My father is a retired engineer who taught math at the local college, while my mother's strengths were in music and art. In college, I dabbled in both math and art. Math made me comfortable with logic, theory, and modeling, while art helped to train my eyes to look at nature carefully. I joined these together when I became an ecologist whose research includes theory, field studies, and laboratory experiments.

In graduate school, I developed my lifelong interest in predator-prey interactions. My inspirations included some of the best and brightest ecologists of our time. Bill Murdoch, my major advisor, started me studying the interface between predator-prey behavior and population dynamics. Bob Warner got me excited about using the optimality framework to understand behavior. Joe Connell taught me community ecology and the value of learning by using carefully controlled experiments, and Peter Abrams and Wayne Sousa showed me that twenty-five-year-olds can write papers that shape an entire field. I drew all these influences together to mold my own scientific identity.

Perhaps my most influential work is a review paper that I wrote in 1985 with colleagues at the University of Kentucky. Our review of over a hundred experimental studies showed that both predation and competition are very often important factors in nature. More surprisingly, we found that in roughly one-third of all comparisons, prey did better in the presence of predators than in their absence. This finding helped to catalyze an interest in "indirect effects" (e.g., keystone predator effects, trophic cascades) that can help to explain this "unexpected effect".

I retain a strong optimism about the value of using an integrative approach to study nature. If anything, it is more important than ever for humans to understand how organisms cope (or do not cope) with environmental changes, often driven by human disturbances. The integrative science of ecology is critical to guiding our, hopefully, effective responses to these challenges.

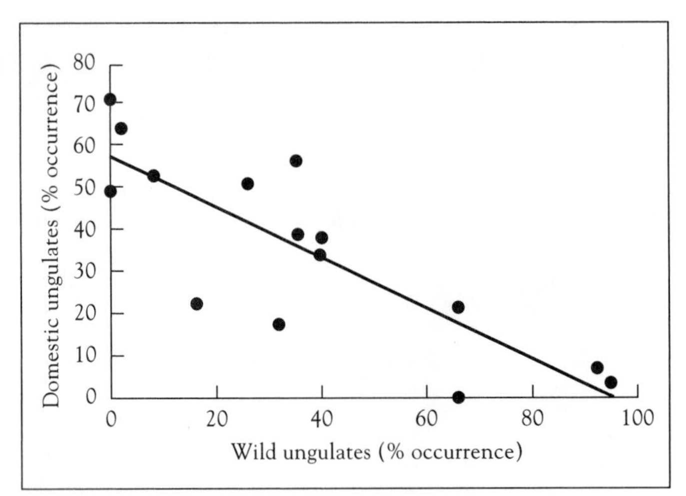

FIGURE 10.14 Linear regression suggesting that the presence of livestock in the wolf diet may be influenced by that of wild ungulates. Restoration of wolf populations may therefore be enhanced by restoring areas with wild game. (*After Meriggi and Lovari 1996.*)

caveats remain as in the reviews of field experiments investigating competition, that is, underreporting of negative results, biased views of investigating scientists and the like.

Finally, it is worth noting that studies on the effects of predators are particularly timely now given the desire of certain conservation groups to reintroduce large predators into certain areas. The U.S. Fish and Wildlife Service would like to reintroduce the wolf into Yellowstone National Park and to stabilize its numbers in Montana and Minnesota, the only states other than Alaska to possess viable populations (Mitchell 1994). Cattle ranchers are fearful that wolves would decimate their herds. In southern Europe the main conservation problem with the wolf lies with predation on domestic ungulates, which leads to extensive killing of wolves (Meriggi and Lovari 1996). The reintroduction of wild large herbivores has been advocated as a means of reducing attacks on livestock because the percentage of domestic stock eaten is reduced as the amount of wild game increases (Fig. 10.14). However, predation on the latter may remain high if domestic ungulates are locally abundant.

SUMMARY

1. Is predation a strong selective force in nature? The existence of the following phenomena suggest that it is: aposematic coloration, crypsis and catatepsis, Batesian and Mullerian mimicry, intimidation displays, polymorphisms, and chemical defenses.

2. The existence of mutual interference between predators, specific territory sizes, and the ability of predators to feed on more than one prey type make it very difficult to accurately predict or model how populations of predators and prey interact.

3. Accidental and deliberate introductions of predators in different parts of the world have often had profound effects on populations of native prey. Although this suggests that predators can have important regulatory effects on prey, this data is not from "natural" systems and should be treated with caution.

4. Evidence from natural systems, where both predators and prey are native, and from a 1985 review by Andrew Sih suggest that most studies that have looked for significant effects of predators on prey have found it. This means that predation is not a casual or unimportant force in nature but is frequent, with often strong effects.

DISCUSSION QUESTION

Should ranchers be concerned about the reintroduction of large predators like wolves or panthers? Do sea lions, otters, or dolphins decrease the stock of fish available for people who fish? Would the number of deer available for hunters be the same in the presence of large predators as it would in the absence of them? Do predators control herbivore populations or do they merely take weak and sickly individuals?

CHAPTER

11

Herbivory

Plants appear to present a luscious green world of food to any organism versatile enough to attack and use it. Why couldn't more of this food source be exploited? After all, plants cannot even move to escape being eaten.

There are three possible reasons that more plant material is not eaten. First, natural enemies, predators and parasites, might keep **herbivores** below levels at which they could make full use of their resources. Second, herbivores may have evolved mechanisms of **self-regulation** to prevent the destruction of the host plant, perhaps ensuring food for future generations. But this argument relies on group selection, and, as we saw in Chapter 4, this is unlikely to be the case. Third, the plant world is not as helpless as it appears—the sea of green is in fact tinted with shades of noxious chemicals and armed with defensive spines and tough cuticles (see color insert).

11.1 Plants have evolved a variety of defenses against herbivory, including both chemical and physical strategies.

An array of unusual and powerful chemicals is present in plants, such as alkaloids (nicotine, morphine, and caffeine), mustard oils, terpenoids (in peppermint and catnip), phenylpropanes (in cinnamon and cloves), and many others. A teaspoon of mustard should be enough to convince anyone of the potency of these chemicals. Such compounds are not part of the primary metabolic pathways of plants. They are therefore referred to as **secondary chemicals** and are now thought to be synthesized mainly as a deterrent to herbivores. However, for a long time secondary plant substances were thought merely to be the waste products of plant metabolism. Excretion is a necessary part of metabolism even for plants, but it must take a fundamentally different form (that is, storage) than it does in animals. Some authors (e.g., Muller 1970) argued that it is only coincidental that these waste metabolites have an effect on herbivores. Interestingly, this idea has recently resurfaced under a new term "redundancy" (Romeo, Saunders, and Barbosa 1996). Is there unecessary duplication of chemical defense in some systems, and if so, why? It is being increasingly recognized that functional diversity and multiplicity of function of secondary metabolites, now more commonly called "natural products" is the norm rather than the exception.

Ehrlich and Raven (1964) were the first to suggest the notion that secondary plant compounds evolved specifically to thwart herbivores. They proposed that most secondary compounds are produced only at a metabolic cost to the plant, not as energy-free by-products. This is the prevalent view among plant-insect ecologists today. Rhoades (1979) has formulated a general defense theory based on the idea that such plant compounds are costly to produce:

1. Higher herbivory levels lead to more defenses.
2. Higher costs of defense lead to fewer defenses.
3. More defenses are allocated to the most valuable tissues.
4. Environmental stress may lessen the availability of energy for defensive mechanisms. Alternatively, in nutrient-stressed plants because the carbon-nutrient balance (CNB) is changed plants may accumulate easier to make carbon-based secondary metabolites (e.g., phenolics) instead of nitrogen-based ones (Bryant, Chapin, and Klein 1983; Ruohomäḳi et al. 1996).
5. Defense mechanisms are reduced when enemies are absent and increased when plants are attacked. This is known as the *theory of induced defense*.

Plant defenses can be classified as quantitative or as qualitative, depending on the effects the defenses have on the herbivore.

Defensive reactions in plants can be classified into several main types, of which the most commonly used are quantitative and qualitative.

Quantitative defenses are substances that are eaten in large amounts by the herbivore as it eats and that prevent the digestion of food. Examples are tannins and resins in leaves, which may occupy 60 percent of the dry weight of the leaf (Feeny 1976). These compounds are not toxic in small doses, but they have cumulative effects. Tannins (the compounds in many leaves, like tea, that give water a brown color) act by binding with proteins in insect herbivore guts. The more leaf the herbivores ingest, the more difficult it is for them to digest it. Feeny (1970) was the first to document such a defense by testing oaks against externally feeding caterpillars. Zucker (1983) proposed that there are two main classes of tannins, each with differing biological functions. *Hydrolyzable tannins* inactivate the digestive enzymes of herbivores, especially insects, whereas *condensed tannins* are attached to the cellulose and fiber-bound proteins of cell walls, thereby defending plants against microbial and fungal attack. Some authors have found that insect herbivores are not much affected by quantitative defenses (Coley 1983; Karban and Ricklefs 1984; Faeth 1985; Mauffette and Oechel 1989), but the evidence that tannins affect vertebrate herbivory is stronger. For grey squirrels, acorns with higher concentrations of tannins are less preferred than acorns with lower concentrations (Smallwood and Peters 1986). Cooper and Owen-Smith (1985) showed that for browsing ruminants in Africa, kudus, impalas, and goats, palatability of fourteen species of woody plants was clearly related to the leaf contents of condensed tannins. The effect showed a distinct threshold; the browsers rejected all plants containing more than 5 percent condensed tannins. The reason is that ruminants depend on the microbial fermentation of plant cell walls for part of their energy needs. Most quantitative defenses are carbon rich, so they are relatively easy to produce, even in nitrogen-limited systems. In nutrient-poor systems, where conifers often grow, such defenses are common.

Qualitative defenses are, essentially, highly toxic substances, very small doses of which can kill herbivores. These compounds are present in leaves at low concentrations, like 1 to 2 percent of dry weight. Examples include cyanogenic compounds in leaves. Atropine, produced by European deadly nightshade, *Atropa belladonna*, is a most potent poison. Of course, the plant must store some of these poisons in discrete glands or vacuoles or in latex or resin systems in order not to poison itself. Others, however, like alkaloids, terpenes, and quinones can be moved around the plant's vascular tissues. Some compounds are stored as precursors and only become toxic when they are metabolized by the herbivore. For example, fluoroacetate found in certain Dichapetalaceae is metabolized by herbivores to fluorocitrate, a potent inhibitor of Krebs-cycle reactions (McKey 1979). Most qualitative defenses are nitrogen rich, so are "expensive" to make. For this reason, they are more common in nutrient-rich systems.

A good review of theory and pattern in plant defense allocation is given by Zangerl and Bazzaz (1992). The two defense strategies, qualitative and quantitative, are correlated with plant "apparency" (Feeny 1976; Rhoades and Cates 1976; Chew and Courtney 1991). Apparent plants are long lived and always apparent to the herbivores (for example, oak trees). Their defenses are thought to be mainly of the quantitative kind, effective against all herbivores, specialist and generalist, with a long history of association with these *K*-selected plants. Unapparent plants are weeds, which are ephemeral and unavailable to herbivores for long periods. Their defenses are thought to be mainly qualitative, guarding against generalist enemies which would find them only by chance. Thus, trees nearly all contain digestibility-reducing compounds, and weeds contain toxins. These terms are of dubious value, however. An "unapparent" plant is not likely to be unapparent to the herbivores that specialize in finding it by chemical cues and eating it. Apparency probably best reflects the ability of human searchers to find plants. Nevertheless, the terms have spawned a great deal of research and Table 11.1 illustrates some of the phenomena associated with apparent and unapparent plants.

As well as chemical compounds within the leaf there exist a variety of other plant defenses:

> *Defensive associations:* Hjalten and Price (1997) noted that palatable plants can gain protection from association with unpalatable neighbors. They positioned susceptible potted clones of the willow *Salix casiolepsis* among a matrix of resistant clones and found a positive correlation between sawfly density on the matrix clones and sawfly density on the potted clones.
>
> *Mechanical defenses:* Plant thorns and spines deter vertebrate herbivores, if not invertebrate ones. In Africa, in the presence of a large guild of vertebrate herbivores, much of the vegetation is thorny and spinose (Cooper and Owen-Smith 1986). Many neotropical plants are also armed in this fashion, for, although now absent, large browsers were abundant in this area until recently (Janzen and Martin 1982).

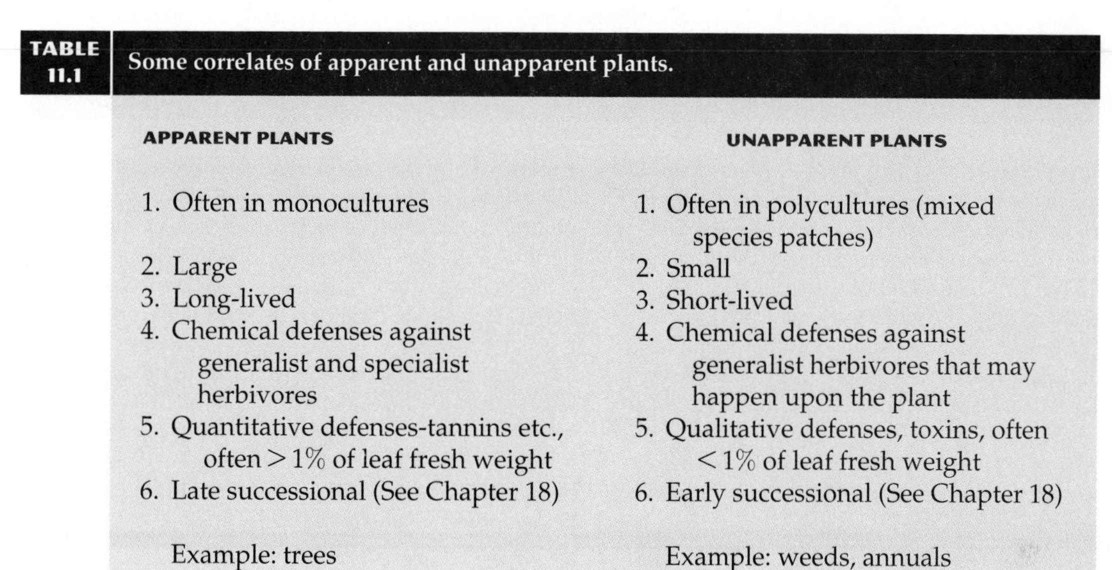

TABLE 11.1 Some correlates of apparent and unapparent plants.

APPARENT PLANTS	UNAPPARENT PLANTS
1. Often in monocultures	1. Often in polycultures (mixed species patches)
2. Large	2. Small
3. Long-lived	3. Short-lived
4. Chemical defenses against generalist and specialist herbivores	4. Chemical defenses against generalist herbivores that may happen upon the plant
5. Quantitative defenses-tannins etc., often > 1% of leaf fresh weight	5. Qualitative defenses, toxins, often < 1% of leaf fresh weight
6. Late successional (See Chapter 18)	6. Early successional (See Chapter 18)
Example: trees	Example: weeds, annuals

Failure to attract: Some plants may stop herbivory by failing to attract herbivores. They do so by lacking a certain chemical attractant that the herbivore uses as a cue.

Reproductive inhibition: Some plants, for example firs (*Abies* sp.), contain insect hormone derivatives that, if digested, prevent successful metamorphosis of insect juveniles into adults (Slama 1969). In this way, herbivory in the future is diminished by a decrease in the herbivore's reproductive output. Some insects may bypass such defenses by feeding on the sap, phloem, or xylem, which does not contain the range of defensive chemicals that appear in the foliage.

Masting (see also Chap. 10): The synchronous production of progeny, seeds, in some years satiates herbivores, permitting some seeds to survive. Nilsson and Wastljung (1987) compared seed predation on beeches *(Fagus sylvatica)* in mast and non-mast years. In mast years, 3.1 percent of seeds were destroyed by a boring moth; in nonmast years, this figure was 38 percent. Vertebrate predation of seeds was 5.7 percent in mast years but 12 percent in normal years.

An understanding of plant defenses is of great use to agriculturalists since the better crops are defended against pests, the higher their yields. This line of defense is known as host plant resistance (Applied Ecology: Host Plant Resistance). Some plants defend themselves against herbivores by enlisting the help of other animals. Such a relationship can, of course, be seen as mutualism. A very common example is that plants attract ants by providing sugary nectar secreted from extrafloral nectary glands (Barton 1986; Smiley 1986). African *Barteria* and neotropical *Cecropia* trees have hollow stems where ants maintain populations of scale insects. The ants are obligate occupiers of the trees; in return for food and shelter, they protect the trees from other herbivores and from encroaching vines, both of which they bite to death (Janzen 1979b). Schupp (1986) demonstrated the benefits to juvenile *Cecropia* of removing ants. The experimental plants suffered more damage

APPLIED ECOLOGY

Host Plant Resistance

It is has long been known that some varieties of crop plants are more resistant to pest attack than others (Photo 1). Crossing resistant with high-yielding varieties can produce a crop that is both pest resistant and high yielding. This usually involves genetic crosses of two potential types to produce the genetically desired offspring. Some authors have included grafting under the banner of host plant resistance. For instance, in the late nineteenth century the root aphid, grape phylloxera, was accidentally introduced in Europe. The European grape was so susceptible that the entire wine industry was threatened with ruin. The vineyards were saved by the discovery of pest-resistant American grapes and by the development of grafting techniques that enabled resistant roots to be joined onto popular European varieties. While grafting does not actually involve cross-breeding of different parents, it nonetheless is a technique, based on host plant resistance, that is very valuable to agriculturalists.

■ **PHOTO 1** Soybean field in Abilene, Kansas. Seventy-five percent of U.S. crops utilize pest-resistant varieties so understanding what makes plants resistant to herbivores, or diseases, is of vital importance. *(Inga Spence, Tom Stack, 3180-233.3.)*

Host plant resistance may be due to physiological factors (e.g., toxic compounds within plant tissues that inhibit the pest) or mechanical factors (e.g., a cuticle that is too tough for the pest to penetrate), or the plant may be highly tolerant, that is, it may continue to support pest populations and remain tolerant of pest damage.

The most serious problems associated with development of host resistance are as follows:

1. It may take a long time to develop—between ten and fifteen years.
2. Sometimes resistance to one pest is obtained at the cost of increasing susceptibility to other pests. Some plants can only be resistant to one pest, not to all.
3. Pest strains can appear that are able to overcome the plant's mechanism of resistance. Circumvention of resistance by pests develops in much the same way as resistance to pesticides.

Despite these problems, host resistance is a good tactic for the farmer. After the initial development of resistant varieties, the cost is minimal to the grower. Perhaps more importantly, host resistance is environmentally benign, generally having few side effects on the managed ecosystem. It is estimated that about 75% of U.S. cropland utilizes pest-resistant plant varieties, most of these being resistant to plant pathogens.

from nocturnal herbivorous Coleoptera than did unmanipulated controls. The *Acacias* of Central America have hollow thorns, which provide homes for extremely aggressive *Pseudomyrmex* ants, which also kill other herbivores and chop away encroaching vegetation (Janzen 1966). In numerical terms, the effects of ants can be quite substantial. Schemske (1980) found that seed production of *Costus woodsonii* in Panama was reduced 66 percent where ants were excluded. In England, Skinner and Whittaker (1981) used exclusion techniques to show that the wood ant, *Formica rufa*, was able to reduce herbivory from 8 percent of the leaves to 1 percent.

Not all ant attendance is beneficial to trees, however, because many ants simply farm sap-sucking Homoptera and effectively protect their populations, a behavior that is of little use to the plant.

Some defensive schemes backfire on the plants. Certain chemicals that are toxic to generalist insects actually increase the growth rates of adapted specialist insects, which can circumvent the defense or actually put it to good use in their own metabolic pathways. Danaid (monarch) butterflies are attracted to milkweed plants that contain cardiac glycosides. These substances are vertebrate heart poisons, and cattle will not eat the milkweeds, but monarch butterflies can assimilate these poisons and use them in their own bodies as a defense against their own predators, advertising their distastefulness with bright colors (Brower 1969). With decreased rates of predation, monarch caterpillars may actually cause more herbivory than they would if they were acceptable to predators.

Plant defenses can also have wide-ranging effects on nontarget organisms. Grime et al. (1996) showed that interspecific differences in rates of leaf litter decomposition between plants arise as a consequence of differences in the antiherbivore defenses of living leaves (Fig. 11.1). This suggests a critical role between leaf palatability, litter decomposition rates, and nutrient cycling in ecosystems.

There is some evidence that plant defenses may be induced by herbivore attack.

Plants do not necessarily keep their tissues permanently suffused with defensive and deadly chemicals. There is much evidence that chemicals are produced only as they are needed. The initiation of herbivore attack is usually sufficient to start the metabolic pathways of defense grinding. This defensive tactic is known an *induced defense* (Schultz and Baldwin 1982; Karban and Carey 1984; Edwards and Wratten 1985; Fowler and Lawton 1985; Faeth 1988; Rausher et al. 1993). For example, Rhoades (1979) removed 50 percent of the leaves of ragwort, *Senecio jacobaea*, and detected a 45 percent increase in leaf alkaloids and N-oxidases in the undamaged leaves. Some of these effects can persist for long

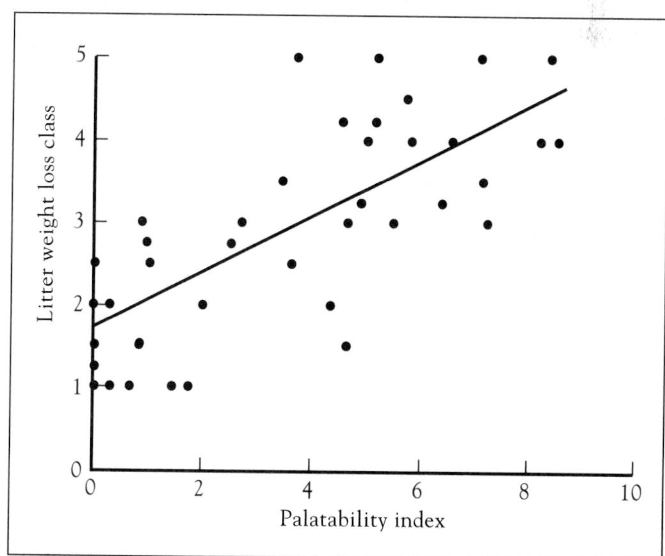

FIGURE 11.1 The relationship between leaf palatability to two invertebrate herbivores and litter decomposition in 43 plant species. (*After Grime et al. 1996.*)

periods. Haukioja (1980) found that leaves of a birch tree were still poor-quality food for a moth three years after the tree had been damaged. This phenomenon may help to explain the severe oscillations of many forest insects. Sheep fertility was reduced six weeks after an aphid attack on the alfalfa the sheep fed on had increased the production of its estrogen mimic, coumestrol (Schutt 1976).

Facultative defenses are not confined to chemical mechanisms. For example, the prickles on cattle-grazed *Rubus* plants were longer and sharper than those on ungrazed individuals nearby (Abrahamson 1975). Browsed *Acacia depranolobium* trees in Kenya have longer thorns than unbrowsed ones (Young 1997). Stinging nettles, *Urtica dioica*, exhibit increased density of stinging trichomes after herbivore damage (Pullin and Gilbert 1989). On holly trees, *Ilex aquifolium*, in Britain, lower leaves, subject to grazing, are heavily armed with spines; upper leaves, free from herbivory, are not (Crawley 1983). Interestingly enough, though the same phenomenon is apparent on American holly, *Ilex opaca*, Potter and Kimmerer (1988) regard it as an ontogenetic phenomenon rather than a defense against browsers. They suggest that the poor nutritional quality of holly and high concentrations of saponins are more important as deterrents of vertebrate herbivores and that fibrous leaf edges deter invertebrate herbivores, which are unlikely to be affected by spines. Williams and Whitham (1986) have even argued that leaves infested with sessile insects, for example, gall makers and leaf miners, are abscised prematurely as a defense against herbivory; the insects are killed as the leaf senescences on the forest floor. Stiling and Simberloff (1989) examined such a situation on oak trees in northern Florida. Although it is true that leaves infested with leaf miners do abscise earlier than noninfested leaves and that premature abscission kills miners inside the leaf, leaf abscission is not likely to be a complex induced defense because of the many trees that support whole complexes of herbivores, each with multiple overlapping generations with which it would be hard to synchronize leaf fall. It is more likely that premature abscission is a simple wound response on the part of the tree to rid itself of damaged leaves, possibly to avoid infection (Stiling and Simberloff 1989).

Though induced defenses undoubtedly occur in nature, their effectiveness as a deterrent is open to speculation and is probably less than that of permanent chemical defenses. Fowler and Lawton (1985) reviewed much of the work on induced defenses and concluded that as a result of such chemicals only small changes, generally less than 10 percent, occurred in such things as larval development time or pupal weights. Other studies have even shown certain insects even benefit by feeding on previously damaged plants (Myers and Williams 1984 and 1987; Niemelä et al. 1984).

11.2 Plant-herbivore interactions have been modeled by assuming that plant population growth rate is determined by the balance between gains from reproduction and losses from herbivory.

Crawley (1983) has provided a series of models of plant-herbivore interactions. The most simple of these assumes that there is an upper limit, a carrying capacity, K, for a population of plants. The rate of change of a population of plants is given by

$$\frac{dV}{dt} = A - B$$

where A is the gains and B the losses and V is plant abundance. Similarly, for the herbivores

$$\frac{dN}{dt} = C - D$$

where C and D are gains and losses, and N is herbivore numbers.

It is assumed that, in the absence of herbivores, plant populations increase exponentially such that gains, $A = rV$, where r is the plant's intrinsic rate of increase. Losses for plants, B, $= bNV$ where b is the feeding rate of herbivores (sometimes called the functional response). There are three recognized functional response types, I, II, and III (Fig. 11.2). In Type I the herbivore consumes more plant as plant density increases. In Type II the herbivores can eventually become satiated (stuffed) and stop feeding. Alternatively, they may be limited by the "handling time" needed to locate and eat the plants. In either case, there is an upper level or asymptote at which herbivores can process plants. The Type III response also supposes an asymptote, but here the curve has a sigmoid shape, similar to the logistic curve. Here the feeding rate is low at low plant density but increases quickly at high density. This sometimes happens if herbivores switch feeding between plants species and develop a search image for particular plant species once it reaches a certain threshold density. In any event, the functional response can have important consequences for the ability of herbivores to control plant populations because the proportion of herbivore population that is consumed by an individual herbivore changes as the functional response type changes (Fig. 11.3).

For the herbivores, $C = cNV$ where c describes how efficient herbivores are at turning food into progeny (sometimes called the numerical response), and $D = dN$ where d is herbivore death rate. The assumption is that, in the absence of plants, herbivores starve and their numbers decline exponentially. Assuming an upper carrying capacity, K, and following a basic Lotka-Volterra type of logistic model, then $A = rV(K - V)/K$. The other elements, B, C, and D, remain the same. Therefore,

$$\frac{dV}{dt} = \frac{rV(K - V)}{K} - bNV$$

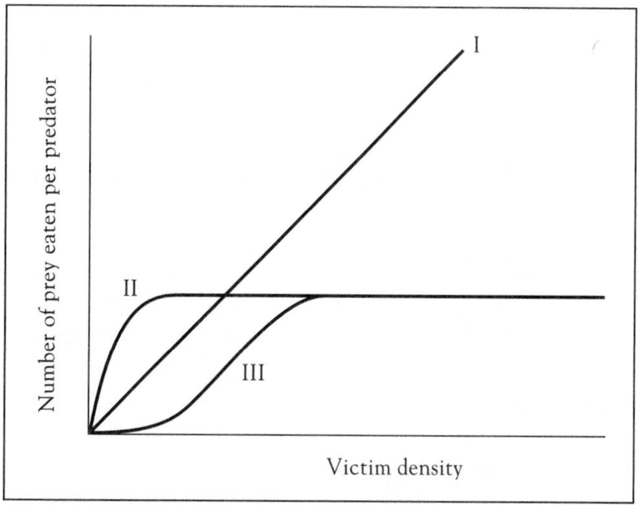

■ **FIGURE 11.2** Type I, Type II, and Type III functional responses. *(From Gotelli 1995.)*

FIGURE 11.3 The proportion of the prey population consumed by an individual predator as a function of victim density. (*From Gotelli 1995.*)

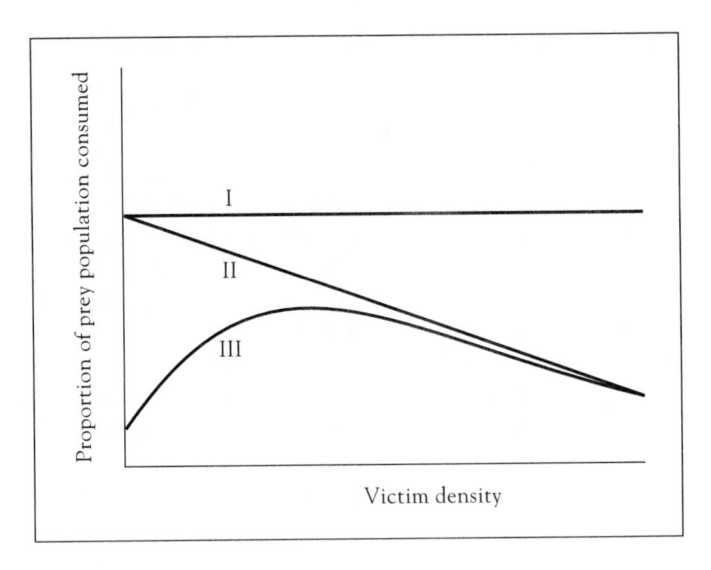

and

$$\frac{dN}{dt} = cNV - dN$$

At equilibrium both dV/dt and dN/dt are zero; there is no population change. The equilibrium plant density V^* and equilibriumn herbivore abundance N^* can then be solved for because $A = B$ and $C = D$. Therefore,

$$V^* = \frac{d}{c}$$

and

$$N^* = \frac{r(K - (d/c))}{bK}$$

The effect of each of the parameters a, b, c, and d can be assessed by graphical techniques. Each parameter can be made to vary while the others remain fixed. The effect of increasing the plant's intrinsic rate of growth, a, stabilizes and increases herbivore equilibrium density but has no effect on equilibrium plant abundance (Fig. 11.4): essentially, the faster the plants grow, the faster the herbivores can eat them up and increased herbivore densities are supported. Exactly the opposite happens when b, the feeding rate of herbivores, is increased. There is a lower herbivore equilibrium, but the size of the equilibrium plant population is unchanged. The more individual herbivores feed, the rarer the herbivores are because fewer herbivores can be supported by a given level of plant production. When we increase c, the efficiency with which herbivores turn food into progeny, we lower the equilibrium population of plants because an efficient herbivore turns all production into herbivores (Fig. 11.5). Perhaps more important is the effect of c on stability. When plant equilibrium density is low, much below K the carrying capacity, the population is under lax control and tends to increase exponentially when herbivore numbers decrease yet be dramatically decreased as herbivore numbers rise. The effects of herbivore death rate, d, are exactly opposite to those of c. Increasing herbivore death rate increases **stability** because it leads to an increase in equilibrium plant numbers.

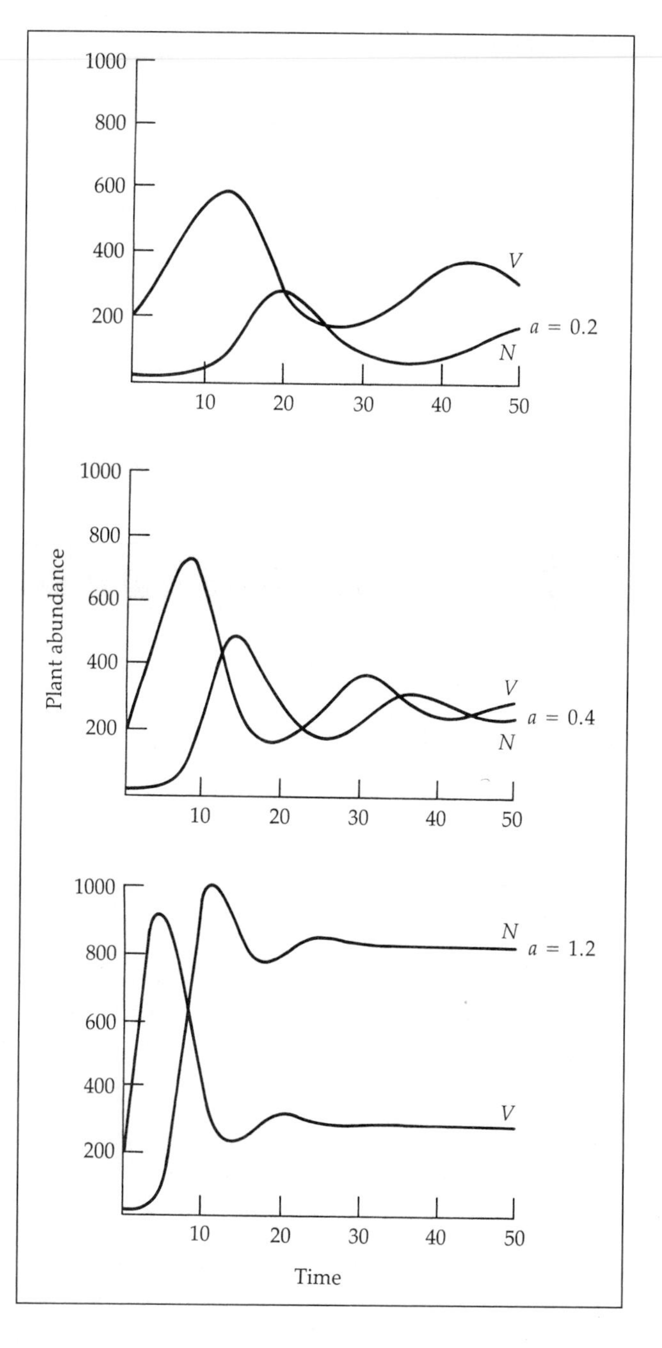

■ **FIGURE 11.4** The effect of the plant's intrinsic growth rate a on plant and herbivore abundance. Increasing a from 0.2 to 1.2 increases stability and increases herbivore equilibrium density but has no effect on equilibrium plant abundance. Other parameters are $b = 0.01$, $c = 0.001$, $d = 0.3$, $K = 1{,}000$. As in all these types of models, the curve V represents plant abundance, and the curve N represents herbivore numbers (in arbitrary units). *(Redrawn from Crawley 1983.)*

Crawley concluded that although plant and herbivore growth rates and herbivore efficiencies could dramatically affect population densities of both organisms in the field, the effects of plants on herbivores is stronger than the effects of herbivores on plants. However, the fact that C and b are independent parameters could be misleading because it suggests that herbivore birth rates are decoupled from the rate of con-

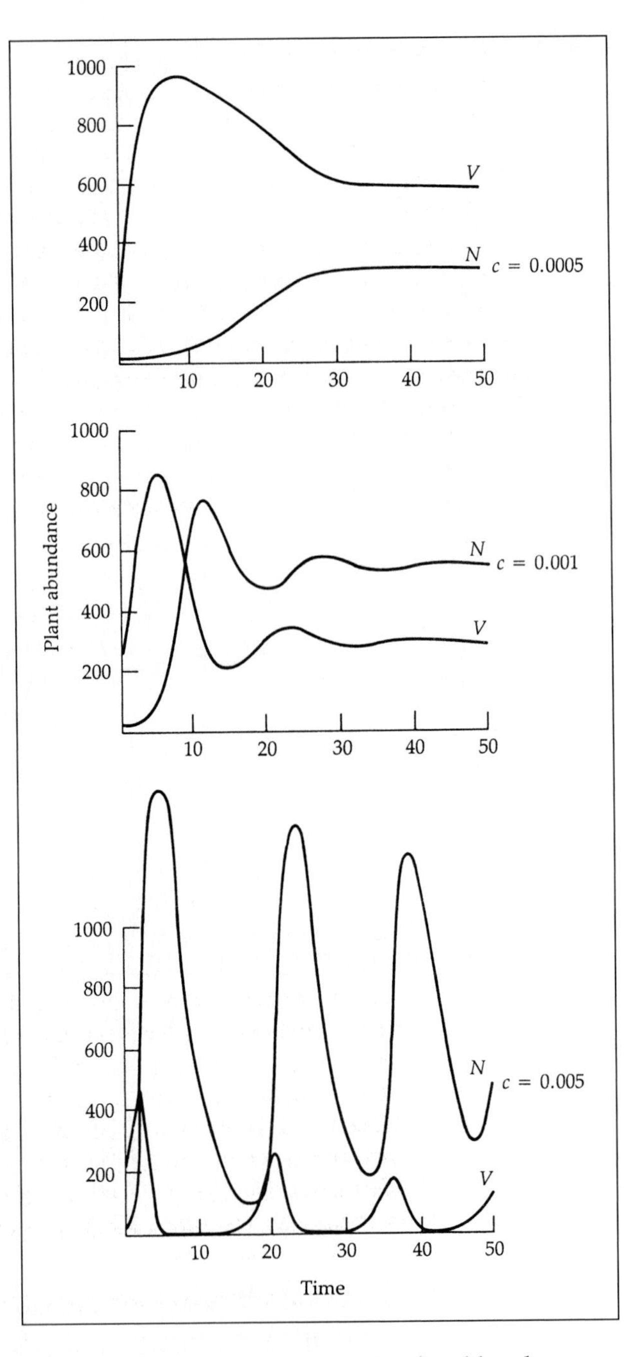

FIGURE 11.5 The effect of herbivore growth efficiency c. Increasing c from 0.0005 to 0.005 reduces equilibrium plant abundance and reduces stability but increases equilibrium herbivore numbers. Other parameters are $a = 0.5$, $b = 0.001$, $d = 0.3$, $K = 1{,}000$. *(Redrawn from Crawley 1983.)*

sumption. This could lead to unrealistic results. For example, if C is fixed but b goes to zero, i.e., low consumption, herbivore numbers become infinitely large (Robert Holt 1997). Further, there is much evidence from field exclosure studies that herbivores, especially large ones, do have a strong effect on plants (see next section, 11.3). Therefore, it is difficult to draw generalizations from these models other than that much variability is possible in nature.

11.3 Herbivores can influence the growth rate, reproduction rate, and stability of plant populations.

Despite the impressive array of defenses in their arsenals plants do not have things all their own way in the plant-herbivore interaction. Herbivores can detoxify many poisons by four chemical pathways: oxidation, reduction, hydrolysis, and conjugation (Smith 1962). Oxidation occurs in mammals in the liver and in insects in the midgut. It is brought about by a group of enzymes known as mixed-function oxidases (MFOs). Conjugation, often the critical step in detoxification, involves the uniting of two harmful elements into one inactive and readily excreted product. Given that herbivores can circumvent plant defenses in certain situations, what is their measured effect on plant populations in the field? There are numerous studies showing how herbivores reduce plant growth, flowering, reproduction, and survival (Fox and Morrow 1992; Wise and Sacchi 1996).

On average, no more than 10 percent of net primary productivity seems to be taken by herbivores and about 90 percent goes to decomposers, in most natural systems (Crawley 1983) (see also Chap. 21). In a review of ninety-three cases of leaf herbivory in terrestrial systems, an average of 7 percent of leaf area was found to be consumed (Pimentel 1988). It must be remembered, of course, that such figures mask large and important variations. For example, the larch budmoth may take less than 2 percent of the net production of forest trees in some years but 100 percent in others. Also workers interested in herbivory would probably not choose to study a plant that suffers very little herbivory, thus, even 7 percent could be an overestimate. A more recent review by Cyr and Pace (1993) showed much higher levels of primary production removed by herbivores than was previously thought; 18% in terrestrial systems ($n = 67$ studies) and 51% in aquatic systems ($n = 44$).

Sometimes damage from insect feeding facilitates increased losses to wilting or allows disease to reduce plant **biomass**. For example, grasshoppers feeding on needlerush, *Juncus roemerianus*, in salt marshes feed in the middles of the tall, narrow leaves, so even though only the middle of the leaf is actually digested, the top half is cut off and added to the litter layer (Parsons and de la Cruz 1980).

Ultimately, the best way to estimate the effects of herbivory on plant populations is to remove the herbivores and examine subsequent growth and reproductive output. Some of the best evidence on the impact of herbivores on plants comes from the biological control of weeds. Following its importation from the Americas in 1839, the prickly pear cactus, *Opuntia stricta*, became a serious pest in Australia, occupying by 1925 over 240,000 km^2 of once-valuable rangeland (Fig. 11.6). After some initial imports of insects that failed to control the growth of the cactus, *Cactoblastis cactorum*, a moth, was introduced from South America in 1925 (Osmond and Monro 1981). By 1932, the original stands of prickly pear had collapsed under the onslaught of the moth larvae. Despite a small resurgence of prickly pear in 1932–1933, *Cactoblastis* has devastated prickly pear

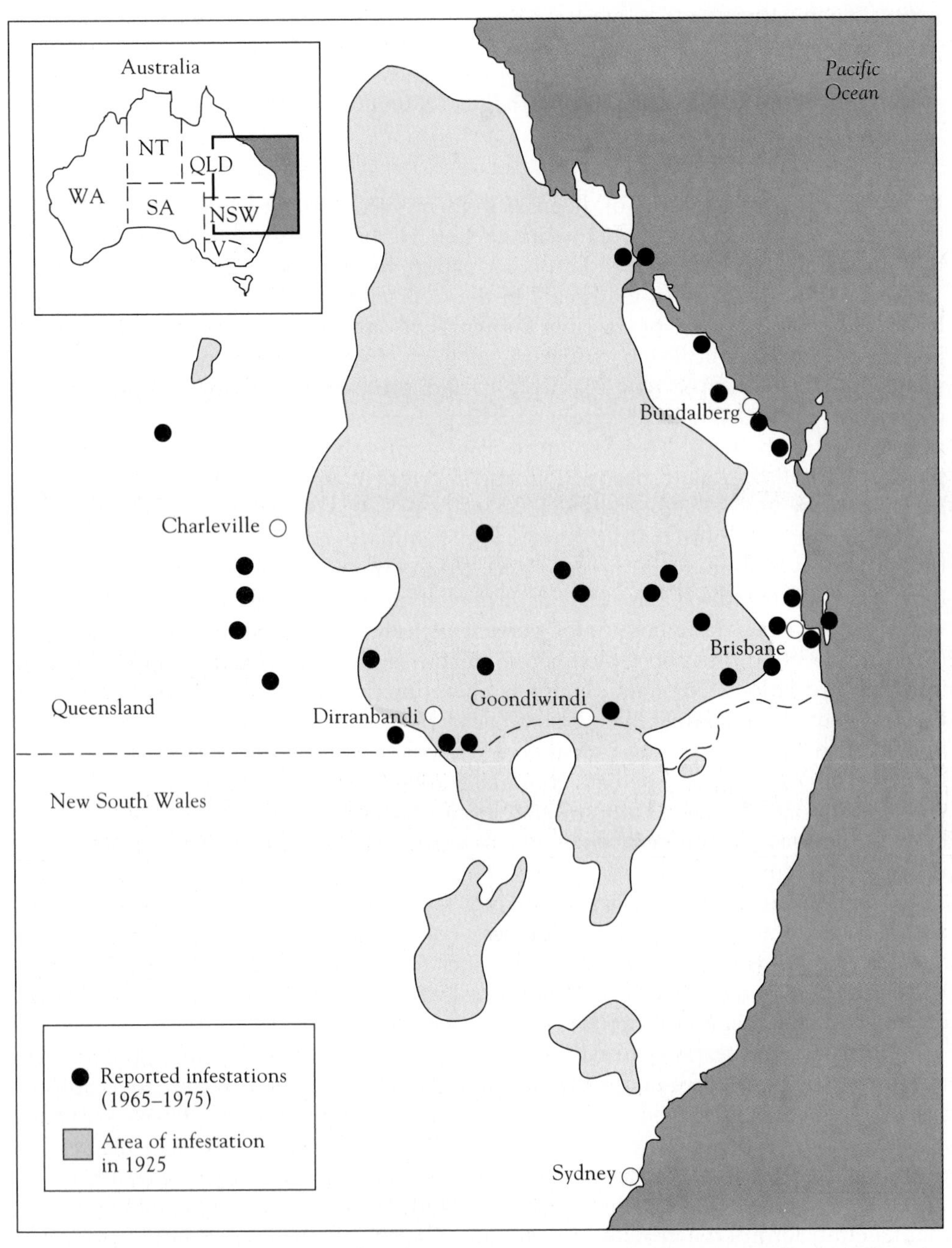

■ **FIGURE 11.6** Distribution of the prickly pear (*Opuntia*) in eastern Australia in 1925 at the peak of infestation and modern areas of local infestation. (*Redrawn from White 1981.*)

populations ever since, and the cactus is now confined to isolated areas. "Before" and "after" views appear in Photo 11.1a and Photo 11.1b. Similar success stories were reported in Hawaii and the Caribbean. Unfortunately, the moth has now invaded South Florida from the Caribbean and is damaging some rare and endemic cacti there (Johnson and Stiling 1996). This is important because it was thought to be one of the relatively few examples of biological control that has had unintended side effects, but others are now becoming apparent (Simberloff and Stiling 1996, Stiling and Simberloff 1999).

There have been many other successes in the biological control of weeds by natural enemies. Klamath weed (*Hypericum perforatum*), a pest of pastureland in California, was controlled by two French beetles (Huffaker and Kennett 1959). As illustrated in Photo 11.2a and Photo 11.2b, floating fern, *Salvinia molesta*, choked a lake in Australia and was controlled by the weevil *Cyrtobagus salvinae*, introduced from Brazil, where the fern is native (Room et al. 1981). Alligatorweed was controlled in Florida's rivers by the so-called alligatorweed beetle, *Agasicles hygrophila*, from South America, and hopes are high that water hyacinth can be controlled biologically, too (Buckingham 1987). On the other side of the coin, large numbers of insects have been introduced to control *Lantana camara*, an introduced weed in Hawaii, but few have had any impact on the growth of the plant, though its spread might have been slowed. Not all biological control campaigns result in success. Furthermore, even those that are successful involve plants and/or herbivores in exotic settings where one or both species may be changed by local environmental conditions. The relevance of such examples to native systems is therefore questionable.

▪ **PHOTO 11.1A** The strength of herbivory as illustrated by a terrestrial example of biological control. Rangeland in Australia infested with *Opuntia* cactus. (*Photo courtesy of Commonwealth Scientific and Industrial Research Organization of Australia.*)

■ **PHOTO 11.1B** The same site after introduction of a cactus-eating moth, *Cactoblastis*. (*Photo courtesy of Commonwealth Scientific and Industrial Research Organization of Australia.*)

■ **PHOTO 11.2A** An aquatic example of biological control. Lake Kabufwe, Papua New Guinea, choked with the floating fern *Salvina* in October 1983. (*Photo by P.M. Room, courtesy of Commonwealth Scientific and Industrial Research Organization of Australia.*)

■ **PHOTO 11.2B** The same lake clear of the weed in November 1984, after release of the herbivorous weevil *Cyrtobagus salvinae*. (*Photo by P.M. Room, courtesy of Commonwealth Scientific and Industrial Research Organization of Australia.*)

However, because biological control campaigns are often conducted in the absence of the herbivores' natural enemies, they provide strong circumstantial evidence that herbivores can provide strong control of plant populations when predation of herbivores is low.

In a natural setting, removal of herbivores from their host plants has been done less commonly.. Australian *Eucalyptus* trees with insects removed were almost 100 percent taller than control trees after three years of this treatment (Fox and Morrow 1992). Increased growth in herbivore removal experiments has also been shown for sanddune willow trees (Bach 1994). One of the strongest effects was shown by Gibbens et al. (1993) who excluded rabbits from browsing rangeland plants in New Mexico for over fifty years. By the end of this period the basal area of some plants were thirty-fold greater in the exclusion treatments than in the controls.

Perhaps the most serious effects of herbivory may be on reproductive output. Crawley (1985) removed herbivores from oaks in Britain by spraying techniques. Though unsprayed trees lost only 8 to 12 percent of their leaf area, the sprayed trees consistently produced from 2.5 to 4.5 times the number of seeds produced by unsprayed plants. Waloff and Richards (1977) found almost three times more seed on broom bushes sprayed for insect control than on unsprayed ones.

One of the most thorough experiments on herbivore removal has been done by Root (1996) who provided an elegant demonstration of the effect of native herbivores on goldenrod, *Solidago altissima*, in New York, by using the pyrethroid insecticide, fenvalerate, to remove insects that fed on aboveground parts. The main effects of normal abundances of herbivores were to reduce seed production by decreasing the proportion of stems that bloomed and the size of the inflorescenes. Stem density and length of stems was not much affected. Herbivore loads were never sufficient to affect plant performance beyond the current year, and there was no evidence that insects were attracted to lush plots that had escaped herbivory the previous year. However, infrequently, on the order of 5 percent of the time, one or other of several common insect species underwent population outbreaks, and such loads completely inhibited flowering and severely stunted vegetative growth. These outbreaks usually waned before threatening goldenrod's persistence in the community but were clearly sufficient to act as potent agents for selecting between goldenrod genotypes. Meyer (1993) was able to show, by selective removals, that xylem-feeding spittlebugs had more of an effect on goldenrod growth than either leaf-chewing beetles or phloem-feeding aphids. However, we are not yet in a position to know which type of herbivores have the greatest effects on which type of plants.

Hulme (1996) not only examined plant loss to herbivores but also compared the effects of three major herbivore taxa—rodents, molluscs, and insects—by the use of cages, molluscides, and insecticides, or all three treatments together, to remove all herbivores. He compared losses among twenty-one plant species sown experimentally in a resource-abundant grassland and resource-limited meadow (Fig. 11.7).

Rodents exerted the greatest influence on plant performance, reducing plant numbers by as much as 50 percent in the grassland and substantially decreasing plant biomass in the meadow. Molluscs and insects had similar but smaller effects. Among the

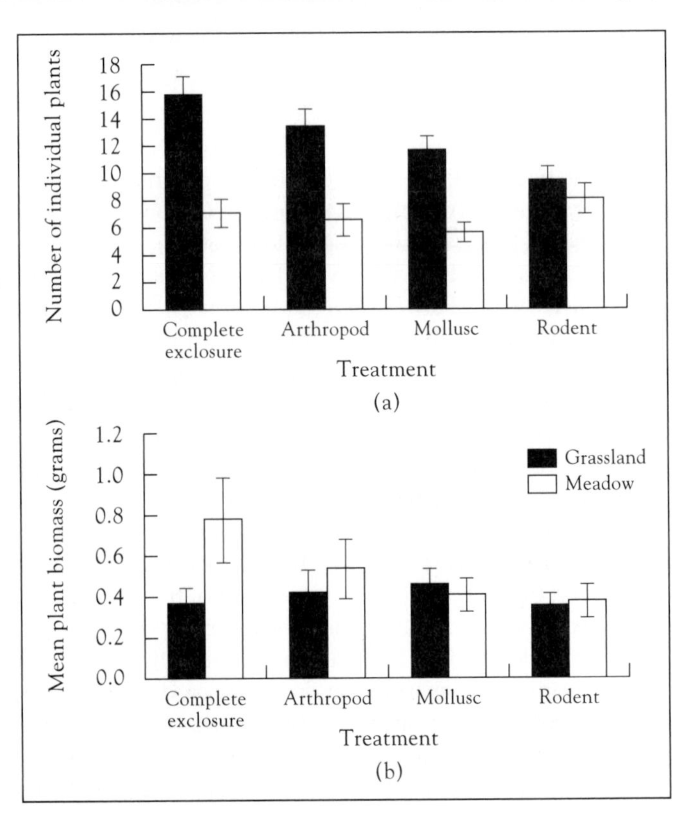

■ **FIGURE 11.7** (**a**) Mean number of plants harvested and (**b**) mean aboveground biomass in four treatments in (■) grassland and (□) meadow sites. Error bars represent one standard error of the mean. (*After Hulme 1996.*)

plants, legumes were more susceptible to herbivory than grasses, exhibiting both lower survival and greater loss of biomass. Loss of plants was less in the meadow where resources were more limiting and plant survival lower than in the grassland. In contrast, the influence of herbivores on plant growth in the meadow was considerable because plants were less able to compensate for tissue loss via regrowth. Thus, the effect of herbivores differed according to habitat. For some plants, whose growth and numbers were both affected by herbivory, numbers became very rare, while for others, who were less affected, numbers become abundant, suggesting herbivory has the power to dramatically alter plant community composition in grasslands (see also Chap. 18).

Hulme (1996) concluded that in most natural settings herbivores cause subtle alterations of growth rates of stems and roots rather than outright death of the plant. Their main effect may then be to reduce the competitiveness of grazed individuals rather than to cause outright mortality. Flower, seed, and fruit production can also be influenced, though it is likely that predators of fallen seed and fruit are more important in a scheme of plant fitness. In this respect, herbivores can be seen as successful **parasites** because they do not kill their hosts—they merely reduce the growth rate. In an agricultural setting, however, there are many examples of huge losses of crops to herbivores. This may be because of a lack of coevolutionary history between plant and herbivore. Most crops are exotics. Damage is often severe enough not only to justify pest control but to cause economic hardship to agriculturalists (May 1977; Pimentel et al. 1980; Barrons 1981). Even though plants are still rarely killed outright, the effects of even minor damage on crop yields can be economically substantial.

Beneficial herbivory?

Some authors have argued that herbivory can actually be beneficial to plants (McNaughton 1986; Crawley 1987; Owen and Weigert 1987). The rationale is that because plants are stimulated to regrow after damage, they may end up overcompensating, growing even more than they would have had they not been damaged. Valentine et al. (1997) manipulated sea urchin densities inside cages over patches of turtle grass. With higher urchin densities, turtle grass compensated for herbivory by increased recruitment of shoots. This led to a 40 percent increase in net aboveground primary production. Simberloff, Brown, and Lowrie (1978) noted that the action of isopod and other invertebrate root borers of mangroves tended to initiate new prop roots at the point of attack (Fig. 11.8). More prop roots meant greater stability of mangroves against wave and storm action, so root herbivory could in fact be beneficial. However, in a review of the twenty papers most commonly cited as evidence for beneficial herbivory, Belsky (1986) found fault with the logic, experimental design, or statistics of nearly all of them. Even newer papers that purported to support the beneficial-herbivory theory are usually fraught with methodological or technical errors (Belsky 1987). In addition, Strauss (1988) has pointed out that very carefully designed experiments involving measurements of plant size before and after herbivory are needed to address the issue of beneficial herbivory because herbivores themselves naturally choose larger plants, which might be expected to show more growth than would stunted plants, even after herbivory. One of the more recent studies has focused on the regrowth of scarlet gilia, *Ipomopsis aggregata*, in the San Francisco peaks area near Flagstaff, Arizona, following grazing by elk and mule deer (Paige 1992 and 1994, but see Bergelson and Crawley

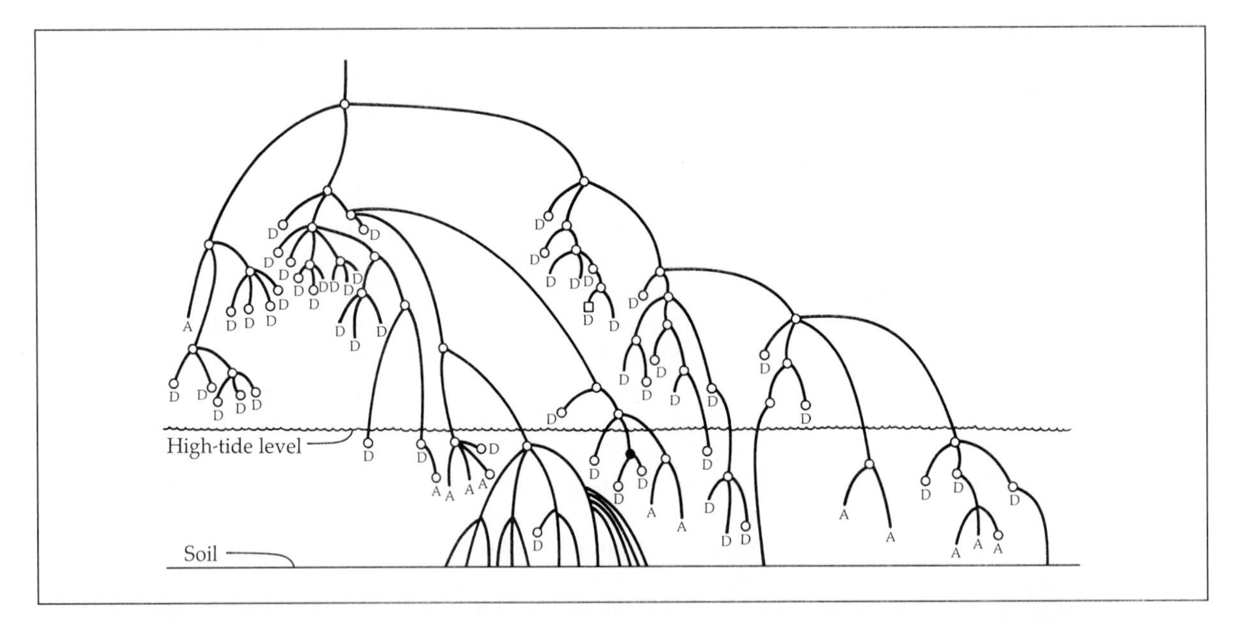

■ **FIGURE 11.8** Beneficial herbivory? Branching pattern for a single *Rhizophora* mangrove root from Clam Key, Florida, following attack by herbivores. A = alive; D = dead; open square = bored by *Ecdytolopha sp.*; open circle (above water) = bored by unknown insect; open circle (below water) = bored by *Sphaeroma terebrans*; shaded circle = bored by *Teredo sp.* (*Redrawn from Simberloff, Brown, and Lowrie 1978.*)

1992a, b). Ken Paige and his associates (Gronemeyer et al. 1997) actually showed that not only did grazing stimulate regrowth, but more flowers and fruits appeared on browsed plants than on nonbrowsed control plants (Fig. 11.9).

Because of the economic damage associated with some herbivores, especially insects, it is not trivial to assess whether or not defoliation could be beneficial. Mattson and Addy (1975) tried to model forest growth with and without insect herbivores. They examined two situations. In the first, aspen was defoliated by forest tent caterpillars. Forest tent caterpillars begin to infest a forest slowly, reach a peak in numbers, remain there for three or four years, and then subside. Stemwood production in the years of peak infestation is much reduced, but foliage production increases to compensate for insect defoliation. Within roughly ten years of the infestation, the biomass production was identical to that of unaffected forests. In the short term, the caterpillars reduced wood production, but in the long term they had no major effect.

In a second example, balsam fir was defoliated by spruce budworms. These larvae actually kill mature trees aged fifty-five to sixty years, but they leave young trees largely alone. The saplings grow quickly after their parents are killed, and a resurgence of the forest, from its juveniles or saplings, occurs. The end result is the same as in the previous case: in the short term there is a considerable effect on wood production, but there is not in the long term. Production rates in the young forest remain elevated above that of the mature forest for fifteen years because in a mature woodland most trees have passed their rapid-growth phase. The role of foresters in this cycle is not at all clear (Holling 1978).

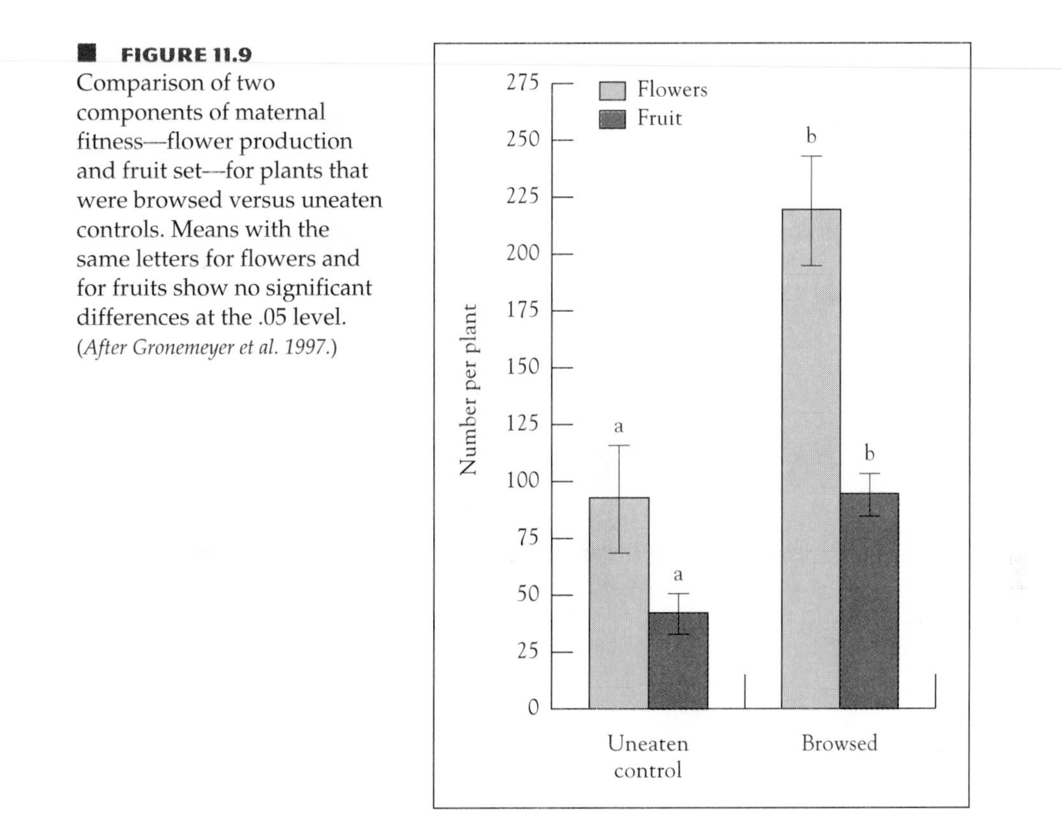

■ **FIGURE 11.9**
Comparison of two components of maternal fitness—flower production and fruit set—for plants that were browsed versus uneaten controls. Means with the same letters for flowers and for fruits show no significant differences at the .05 level. *(After Gronemeyer et al. 1997.)*

11.4 Herbivores are strongly influenced by plant quality and chemical defenses.

There is much evidence that herbivores themselves select the plants that are the most nutritionally adequate in terms of nitrogen content of the tissue (Mattson 1980; Scriber and Slansky 1981; White 1984) or amino-acid concentration of the sap (Brodbeck and Strong 1987). Iason, Duck, and Clutton-Brock (1986) showed how red deer fed preferentially on grasses defecated upon by herring gulls (*Larus argentatus*). Where the number of gull droppings increased, so did the vegetation nitrogen content. For birds, Watson, Moss, and Parr (1984) showed how food enrichment affects numbers and spacing behavior of red grouse. Waring and Cobb (1992) systemically reviewed the effects of nutrients and other stresses on insect performance. Plant fertilization, specifically nitrogen enhancement, had strong positive effects on herbivore population sizes, on survivorship, growth, or fecundity of individuals for most all herbivores including chewing organisms, sucking insects, galling insects, and phytophagous mites (Fig. 11.10). Nearly 60 percent of 186 fertilization studies reported positive responses by herbivores. The proportion of positive responses was greater in cultivated versus wild plants and in herbaceous and broadleaf trees versus conifers (Fig. 11.11). Although the addition of other nutrients such as phosphorous and potassium can increase herbivore densities, the overall responses to these nutrients were much more variable, and positive responses were generally not as common as no responses or negative responses.

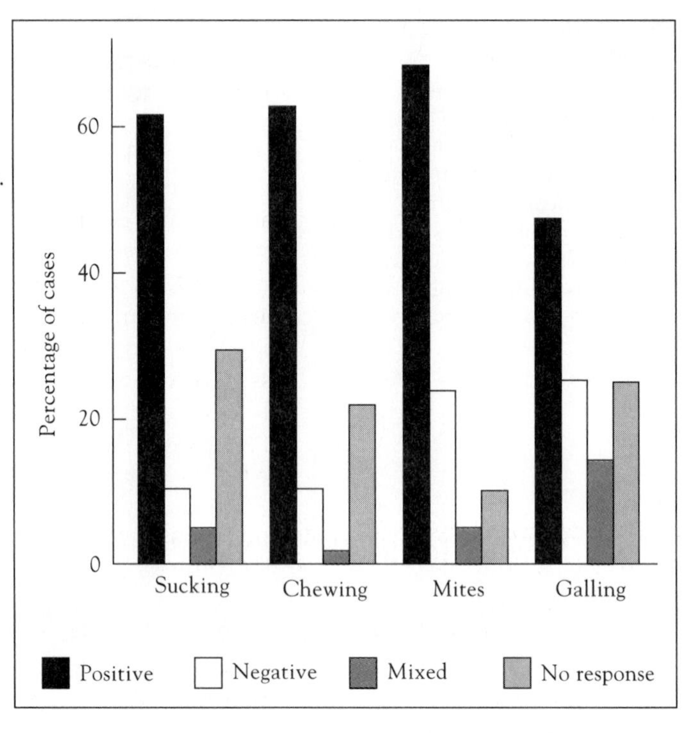

■ **FIGURE 11.10**
Responses of herbivores to nitrogen fertilization, measured as the percentage of studies for various herbivore feeding techniques. *(After Waring and Cobb 1992.)*

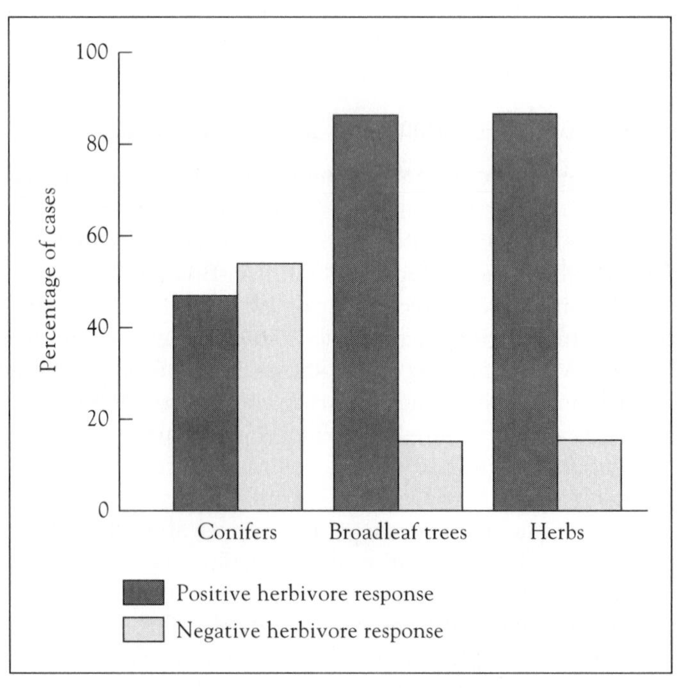

■ **FIGURE 11.11**
Percentage of studies in which herbivores responded positively or negatively to nitrogen fertilization in conifers, broadleaf trees, and herbaceous plants. *(After Waring and Cobb 1992.)*

In some cases, correlations between host-plant quality and herbivore density are present, but observed population patterns of herbivores are more dependent on other phenomena, such as predation or parasitism. In Florida salt marshes, leaf mining flies appear to feed primarily on the grass richest in nitrogen, but their population densities do not increase when the grasses are fertilized. Densities are actually governed by high parasitism rates, and the flies are primarily choosing plants not likely to be searched by parasites—coincidentally, these are also the richest in nitrogen (Stiling, Brodbeck, and Strong 1982). There seems to be no easy way to predict when herbivore densities are controlled by the quality of their hosts and when they are primarily influenced by other mortalities. However, other mortalities are often mediated by plant quality. When food quality declines, many herbivores, especially vertebrates, respond simply by feeding at a higher rate or for a longer time. This can result in increased exposure time to predators and parasites and higher enemy-induced mortality rates (Loaker and Damman 1991, Stiling et al. 1998). Although deaths of herbivores due to depletion of food plants are witnessed infrequently, perhaps because herbivores can leave an area of poor food availability, insect performance and subsequent fecundity of females is strongly affected by their host plants. The relatively few examples of mass starvation due to overexploitation of plants come mainly from studies of insects that habitually undergo periodic outbreaks or from cycles of Arctic rodents.

Herbivores are not only influenced by plant quality, they are affected by host defenses, many of which are genetically inherited. Sometimes there is an interaction between host genotype and environment such that at some sites plants are resistant to herbivores, but at other sites they are susceptible. Presumably, the abiotic environment, e.g., soil type, water levels, or nutrient levels influence whether or not plants can maintain their resistance to herbivores. Which of these two factors is most important: plant quality or plant defenses? Rick Karban (1992), in a review of the effect of plant variation on herbivorous insects, showed that over 80 percent of studies that had used experiments to test for the presence of plant genetic variation on insect herbivore densities found it. This is a high percentage. However, Karban also reviewed studies that compared the effects of plant genotype to other factors and found that in 17/30 cases (56.7 percent) plant genotype explained variation in herbivore numbers less than other factors, for instance, yearly variability. For example, Stiling and Rossi (1995, 1996) showed that environmental conditions affected the population densities of eight major herbivores on coastal plants in Florida more than did plant genotype. This result reminds us of the ideas of Davidson, Andrewartha, and Birch (Davidson and Andrewartha 1948a, b; Andrewartha and Birch 1954) who suggested that abiotic factors may be of paramount importance in determining herbivore abundance.

SUMMARY

1. A variety of plant defenses are testament to the strength and frequency of herbivory in nature. There are chemical defenses such as nicotine in tobacco and caffeine in tea and mechanical defenses such as spines and stinging hairs. Plants may also contain insect hormone mimics that disrupt insect molts. Other plants enter into mutualisms with ants that attack and remove herbivores in return for shelter and food (extrafloral nectaries).

2. Chemical defenses can be subdivided into qualitative and quantitative defenses. Quantitative defenses gradually build up inside herbivore guts and prevent food digestion. The more foliage that is eaten, the worse the situation becomes for the herbivore. Examples are tannins and resins in leaves. Qualitative defenses are toxic compounds, such as cyanogenic compounds in leaves, which are lethal in small doses. A good example is atropine, produced by deadly nightshade.

3. Because chemical defenses are energetically costly to produce, more are allocated only to the most valuable tissues. Also, some defense mechanisms are only initiated following herbivory. These are known as induced defenses.

4. Mathematical models have led some ecologists to conclude that plants have a much more important impact on herbivores than herbivores have on the dynamics of plants. Reviews suggest that, on average, between 7 and 10 percent of plant tissue is consumed by herbivores. Of course, this masks much important variation, and there are many systems in which there are periodic outbreaks of herbivores, such as locusts in the tropics or moth larvae on conifers in boreal zones.

5. Biological control projects have shown that many exotic weeds that have undergone population explosions in the absence of their native herbivores in foreign countries can be brought under control when the native herbivore is reunited with its host. Some experiments that have removed native herbivores from native plants have also shown the dramatic effects of herbivory. Thus, in some systems, the effects of herbivores are known to be substantial.

6. Population densities of herbivores are strongly influenced by plant quality and chemical defenses.

DISCUSSION QUESTION

Are grazing mammals or insects likely to be the most important herbivores? Which types of plants might suffer more herbivory? Are chemical defenses more likely to be found in temperate or tropical plants, in desert species or wetland species? Would your conclusion hold true for other types of defenses?

CHAPTER

12

Parasitism

When one organism feeds off another but does not normally kill it outright, the predatory organism is termed a parasite and the prey a host. Some parasites remain attached to their hosts for most of their lives, like tapeworms, which remain inside the host's alimentary canal. Others, such as ticks and leeches, drop off after prolonged periods of feeding. Mosquitoes remain attached for relatively short periods. By this definition, many species of phytophagous insects are parasitic upon their "host" plants. Still, there remain many problems of definition. Should organisms that feed off more than one individual, without killing them, be known as parasites or predators? For example, saber-tooth blennies (*Plagiotremus*) on the Great Barrier Reef dash out and bite chunks out of fish hosts/prey that swim by. Should the large ungulates of the Serengeti plains, wildebeest, zebra, and the like, be known as parasites? Although they feed off more than one individual host grass, the grass is not killed and will grow back later.

Should we retain the term *parasite* for organisms that remain in intimate contact with their hosts? Mosquitoes develop as larvae in a nonparasitic manner in pools of water, and the adults only come into contact with "hosts" for short periods. Rhinoceroses live on top of their food supply for their entire lives. And what about parasites of insects? Many of these develop as internal parasites of caterpillars or other immature stages. In these cases, the host almost never survives, and the term *parasitoid* is used to refer to these "parasites," each of which uses only one host but invariably kills it. Even in this case, further gradation between parasitoid/parasite and predator is evident, as when an egg parasitoid hatches from a host egg and has to devour several more in the clutch before it is mature (Askew 1971). May and Anderson (1979) have tried to distinguish two types of parasites—microparasites, which multiply within their hosts, usually within the cells (bacteria and viruses), and macroparasites, which live in the host but release infective juvenile stages outside the host's body. For most microparasite infections, the host has a strong immunological response. For macroparasitic infections, the response is short lived, the infections tend to be persistent, and hosts are subject to continual reinfection.

Despite these problems of definition, the biology of host-parasite relationships has a rich history of interesting, coevolved, and complex life history patterns. Parasites on animals include those of interest to the conventional parasitologist—viruses, bacteria, protozoa, flatworms (flukes and tapeworms), thorny-headed worms (Acanthocephala), nematodes, and various arthropods (ticks, mites, and so on). Parasitoids from the parasitic Hymenoptera and Diptera are of more interest to the entomologist and biological-control specialist. Such parasitoids are often deployed against other insects and may contribute over 70 percent of the insect fauna (Price et al. 1980). Because about 75 percent of the known global fauna consist of insects, then at least 50 percent of the animals on Earth might be considered parasitic. When the other large groups of parasites are considered—nematodes, fungi, viruses, and bacteria—it is clear that parasitism is a very common way of life. A free-living organism that does not harbor several parasitic individuals of a number of species is a rarity. The frequency of human infection by parasites is staggering. There are 250 million cases of elephantiasis in the world, over 200 million of bilharzia, and the list goes on and on.

12.1 Hosts have developed a variety of defenses against parasite attack, including immune responses, defensive displays, and grooming behavior.

The defensive reactions developed by hosts to resist parasites are impressive:

Cellular defense reactions: These reactions particularly are found in insect larvae as a defense against parasitoids, where eggs of the parasitoid are "encapsulated" or enclosed in a tough case rendering them inviable (Salt 1970).

Immune responses in vertebrates (Cox 1982): These responses are the vertebrate body's defense against the parasitic microbes that cause disease in humans and animals. Phagocytes may engulf and digest small alien bodies and encapsulate and isolate larger ones. For microparasites, the host may develop a "memory" that may make it immune to reinfection.

Defensive displays or maneuvers: These actions are intended to deter parasites or to carry organisms away from them. For example, gypsy moth pupae spin violently within their cocoons to deter pupal parasites (Rotheray and Barbosa 1984), and syrphid larvae often drop to the ground from the foliage they forage on to escape parasites (Rotheray 1981, 1986).

Grooming and preening behavior: This behavior is found in mammals and birds, respectively, to remove ectoparasites (Struhsaker 1967; Kethley and Johnston 1975).

12.2 The rate at which a parasite spreads among hosts is a function of the density of susceptible hosts, the transmission rate of the disease, and the fate of infected hosts.

The spread of parasitic diseases can be described by modeling the population dynamics of the parasitic organism (May 1981; Anderson 1982). Models of parasite population dynamics generally describe population growth rate by the term R_p, the average number of new cases of a disease that arise from each infected host, rather than R_o, the net reproductive output of the parasite. The reason is that in epidemiology, the study of the spread of disease, the number of infected hosts is the most important factor, not the number of parasites. The transmission threshold, which must be crossed if a disease is to spread, is therefore given by the condition $R_p = 1$. For a disease to spread, R_p must be greater than 1, and for a disease to die out it must be less than 1. The term R_p is influenced by

S, the density of susceptibles in the population.

B, the transmission rate of the disease (a quantity correlated with frequency of host contact and infectiousness of the disease).

L, the average period of time over which the infected host remains infectious.

The value R_p is related to these factors by the equation

$$R_p = SBL$$

Two generalizations can thus be made.

1. As L, the period of the host's life when it is infectious, increases, R_p increases. An efficient parasite therefore keeps its host alive. Some hosts remain infectious long after they are dead. This is especially true in plant parasites, which leave a residue of resting spores.
2. If diseases are highly infectious (have large Bs), R_p increases.

By rearranging the preceding equation, we can obtain the critical threshold density, N_T, where $R_p = 1$ and N_T is an estimate of the number of susceptible hosts needed to maintain the parasite population at constant size:

$$N_T = \frac{1}{BL}$$

Now, if B or L is large, N_T is small. Conversely, if B, or L is small, the disease can only persist in a large population of infected hosts. Cockburn (1971) has provided some interesting medical and anthropological evidence to back up these ideas, at least for humans. Measles, rubella, smallpox, mumps, cholera, and chicken pox, for example, probably did not exist in ancient times because the hunter-gatherer populations were small, bands of two hundred to three hundred persons at most. These bands were too small to constitute reservoirs for the maintenance of infectious diseases of the types described. In a small population there should be no infections like measles, which spreads rapidly and immunizes a majority of the population in one epidemic. Measles only occurs endemically in human populations larger than five hundred thousand. Instead, typhoid, amoebic dysentery, pinta, trachoma, or leprosy were probably the common afflictions, diseases for which the host remained infective for long periods of time. Malaria and schistosomiasis would still have been very prevalent because of the presence of outside vectors to serve as additional reservoirs. Paradoxically, civilization has increased the kinds and frequencies of diseases suffered by humans by enlarging the source pools and by domesticating certain animals. Most modern diseases have arisen because of intimate association with animals and their viruses (Foster and Anderson 1979). Smallpox, for example, is very similar to the cowpox virus, measles belongs to the group containing dog distemper and cattle rinderpest, and human influenza viruses are closely related to those found in hogs. AIDS is similar to a virus found in monkeys in Africa.

For parasites that are spread from one host to another by a vector, for example, an insect, the life cycle characteristics of both host and vectors become important in the calculation of R_p. For a disease to establish itself and spread, the ratio of vectors to hosts must exceed a critical level—hence, disease control measures usually aim directly at reducing the numbers of vectors and are aimed indirectly at the parasite. Insecticides are used to kill aphids and mosquitos, which transmit virus diseases of crops and malaria rather than directing chemicals directly at the parasite. Of course, this is not always true, for example, yellow fever was eradicated in the United States by inoculation rather than eradication of all mosquitos.

Many parasites have a complex life cycle involving several hosts. They are therefore faced with the problem of transmission from one host to the next. Many induce changes in behavior or in the color of one host, making them more susceptible to being eaten by a second host, thus facilitating parasite transmission (Moore 1995). For example, the ancanthocephalan parasite *Pompharhynchus laevis* includes color and behavioral changes in its crustacean host *Gammarus pulex* that make them more susceptible to fish predation. Mosquitoes aggressively seek more blood meals when infected with malaria (Morrell 1997).

One of the diseases causing the most concern in the United States recently is Lyme disease, a disease caused by a spirochete that is transferred from animals to humans by the bite of hard-bodied ticks. The disease came to public attention in the 1980s and has been increasing ever since. The disease is not new; it has been known from Europe for over a hundred years, but it is only in the 1980s and 1990s that it has become prevalent in the United States. More than forty thousand cases were reported from 1982 to 1991. There are two common species of tick implicated in the transmission of the disease, one in the eastern United States and one in the western United States, and both have many possible hosts, consisting of over a hundred species of mammals, birds, and lizards. Adults of both species of tick feed preferentially on large and to a lesser extent on medium-sized mammals. One of the problems in the increase of the disease appears to be related to land use patterns and limited hunting programs that have allowed one of the main hosts, white-tailed deer, to increase. Most evidence suggests that disease will become more prevalent and the ticks will continue to expand their range and become more numerous. A vaccine for Lyme disease has been tested and is going through the FDA approval process.

12.3 Parasites can dramatically decrease the size of host populations.

Once again, the best way to determine the effect of parasites on the population abundances of their hosts is to remove the parasites and to reexamine the system. This has rarely been done, probably because of the small size and unusual life histories of many parasites, which makes them difficult to exclude. Furthermore, dead hosts are often difficult to find. Also, as pointed out earlier, parasites may not always kill; they may merely impair the health of their hosts. This makes their effects even more difficult to gauge. However, inoculations, which combat disease, are known to save many lives in humans and in domestic animals. Agricultural sprays reduce crop losses to disease.

Evidence from the natural populations suggests that introduced parasites have substantial impacts on their hosts. In North America, chestnut blight, a fungus, has virtually eliminated chestnut trees (Photo 12.1). In Europe and North America, Dutch elm disease has devastated elms. Twenty-five million of Britain's original thirty million elm trees have been wiped out by the disease since the 1960s. In Italy, canker has had severe effects on cypress.

Rinderpest, caused by a virus, has at least forty-seven natural artiodactyl hosts (Scott 1970), most of which occur in Africa. It belongs to a class known as morbilliviruses,

▪ **PHOTO 12.1** American chestnut. This species of tree was common in eastern United States deciduous forests but was essentially eliminated by an introduced fungal disease, chestnut blight. Only a few individuals remain. *(Photo from National Archives, no. 95-G-250527.)*

ECOLOGY IN PRACTICE

Janice Moore, Colorado State University

I never intended for my graduate career to resemble a travelogue, but then again, there was a time when a graduate career was not exactly an intention, and maybe I should begin there. As an undergraduate at Rice University I thought I might go to medical school, the aspiration du jour of many of my fellow students. As a consequence, I took Clark Read's parasitology course; it was rumored among the undergraduates that he was a heavyweight and good to have among your referees for medical school. (It turns out he was one of the great parasitologists of the twentieth century, a fact pretty much lost on the premed contingent.)

Read taught a parasitology course that left me awestruck. He presented us with the puzzles that emerge when two organisms live in symbiosis and showed how these questions were nested in one another, from the dimension of the cell to the way predators find their prey. We studied animals that had no guts and were covered with microvilli, which depended on their hosts to digest nutrients before taking those nutrients themselves. We studied animals that could live in water, in snail innards, in cow livers, all in sequence, and that somehow knew where they were and what to do when they got there. Toward the end of the course, I told Dr. Read that I was especially interested in some stories he had told about parasites that made their hosts behave strangely, even to the point of getting eaten, thus transmitting the parasite. What a perfect amalgam of animal behavior, parasitology, and a touch of science fiction!

Read said that there weren't that many people in the United States doing this kind of work, and my best bet was to do a master's degree in behavior, a Ph.D. in parasitology, and create my study as I went along. Ever an optimist, I rushed in, surrounded by a host of angels beside themselves with terror. I studied insect behavior at the University of Texas, Austin, under the direction of Dr. Robert Barth and then entered a Ph.D. program at the Johns Hopkins University a bastion of parasitological research. An inner-city university was something of a shock to a kid from Texas, and I was becoming tired of the fact that so few people seemed to care about the behavioral or ecological aspects of parasitism. I decided to leave Hopkins and work for awhile.

Eventually, I returned to my graduate education—at the University of New Mexico. My graduate advisors, Don Duszynski and Rex Cates, were truly supportive, and in my case that allowed me to do what I wanted. For once, no one was trying to get me to do "mainstream" ecology. Moreover, they literally put their money where their mouths were, with in-house grants that enabled graduate students to pursue independent lines of research. So I asked if a parasitic worm could alter the behavior of its isopod host in a way that would get it transmitted (eaten). I found that parasitized isopods behave strangely, in ways that probably increase encounters with birds, and that starlings feed their offspring more parasitized isopods than you would expect from random foraging. Talks with Dan Simberloff about the potential that parasite communities have for replication and experimental manipulation resulted in a postdoctoral collaboration that led me into the world of community ecology, with every host harboring a complete community.

My work since then has focused on various aspects of parasite-induced behavioral alterations and parasite community ecology. Looking at parasites through the eyes of an ecologist presents no small problem, especially if one focuses on the parasites themselves and not on parasites as agents of doom for what some folks see as the main event, hosts. I don't fit most job descriptions. I was happy to get a faculty position as an "invertebrate zoologist" at Colorado State University—that's a reasonable statement of what I do, if anything is. And funding can be a challenge, as it is preferred that one fit neat categories in that realm too. Given these and other frustrations, why not simply find a neat category and fit in?

I imagine that most folks who work in the interstices of biology might have an answer similar to mine. No one is forcing us to do this. We like it. We like it a whole bunch. And it allows us to fill in gaps that should never have been there in the first place but for accidents of history and curricula. Working at the margins of ecology, behavior, neurophysiology, epidemiology . . . this gives me a world of colleagues, from vector biologists to theoreticians, that redefines the notion of intellectual diversity. Even my graduate school peregrinations were hardly the misfortune they may initially seem. The experiences and the people they brought into my life are without price.

which includes measles and distemper and is spread by food or water contaminated by the dung of sick animals. Wildlife get the disease from cattle. The disease is usually fatal in buffalo, eland, kudu, and warthog and less fatal in bushpig, giraffe, and wildebeest. Other species, such as impala, gazelle, and hippopotamus, appear to suffer little. A major epidemic swept through Africa in 1896 (from cattle imported from India or Italy), leaving vast areas uninhabited by certain species. More than 80 percent of the hoof stock died over the entire continent. The disease was brought under control in the 1960s largely through cattle vaccination programs, but even in the 1970s distribution patterns reflected the impact (May 1983; McCollum and Dobson 1995). For example, zebra were exterminated in an area of the Elizabeth National Park in Uganda, and they still had not recolonized the area by 1954 (Pearsall 1954). Because wildlife was eliminated, other parasites were affected too. Tsetse flies became absent from large areas of Africa south of the Zambesi River (Stevenson-Hamilton 1957). Therefore, large areas became free from trypanosomiasis, sleeping sickness, a disease borne only by the tsetse fly. One parasite, rinderpest, thus had a severe impact on the pattern of life in an area. Furthermore, in the absence of tsetse flies, humans and cattle could move in, supplanting wildlife even further. Another outbreak has occurred in Africa in the 1990s (Packer 1997), and since 1993 some national parks in Kenya have lost 90 percent of their buffalo, antelopes, gazelles, and wildebeest.

Prins and Weyerhaeuser (1987) described two recent and major epidemics in wild mammals in a national park in Tanzania, east Africa. An anthrax outbreak lasted almost a year and killed more than 90 percent of the impala population, and a rinderpest outbreak of just a few weeks killed some 20 percent of the buffalo. Their conclusion was that epidemics have a more severe impact on these populations than does predation. This is a disconcerting finding for conservation biology, for it means that certain populations on small reserves could be wiped out by disease unless recolonization is encouraged.

In the early 1980s, a canine distemper virus caused a large decline (more than 70%) in the last remaining population of black-footed ferrets (Thorne and Williams 1988), and a few years later another exploded among seals and dolphins (Osterhaus et al. 1988). In 1994, one thousand lions, one-third of the resident Serengeti National Park population, were wiped out by canine distemper virus probably transmitted from domestic dogs (Roelke-Parker et al. 1996). Many other endangered animals are threatened by disease from domestic animals (Table 12.1) (MacDonald 1996). Guiler (1961) suggested that the demise of the thylacine (marsupial wolf) in Tasmania was because of distemperlike disease, brought about by close association with dogs. To prevent the threat of disease, some populations of endangered species have been vaccinated, for example, mountain gorillas against measles and African hunting dogs against rabies.

What of the effect of native parasites on native populations? The population dynamics of bighorn sheep in North America are dominated by a massive mortality resulting from infection by the lungworms *Protostrongylus stilesi* and *P. rushi*. This parasite predisposes the animals to pathogens causing pneumonia. A fetus can become infected through the mother's placenta, and mortality in lambs can be enormous (Hibler, Lange, and Metzger 1972). The lungworm-pneumonia complex is regarded as one of the most influential mortality factors in many sheep populations, with mortalities of up to 50 to 75 percent reported (Uhazy, Holmes, and Stelfox 1973). Plants too can be affected by parasitic plants, of which there are over five thousand species worldwide (Parker and Riches 1993). *Cuscuta salina* (marsh dodder) is a common and widespread plant parasite in saline conditions among marshes of North America. Pennings and Callaway (1996) showed how *Cuscuta* preferentially infected the most common plant in California marshes, *Salicornia virginica*, thus promoting the growth of two other plants, *Simonium* and *Frankenia*. Thus, parasitic plants can have strong effects on the relative abundance of species and thus on plant community structure.

TABLE 12.1 The threat of disease to endangered wildlife from domestic animals.

ORGANISM	DISEASE	FROM	EFFECT SIZE
Lions	Canine distemper virus	Dogs	30% killed
Black-footed ferret	Canine distemper virus	?	70% killed
Ethiopian wolf	Rabies	Dogs	> 50% killed
Blanford's fox	Rabies	Dogs	Eventual extinction a possibility
Cheetah	Feline infectious peritonitis	?	Strong
Florida panther	Feline immunodeficiency virus	Cats	None as yet but threat remains strong
Iriomole cats (Japan)	Feline immunodeficiency virus	Cats	None as yet but threat remains strong
Scottish wild cats	Feline immunodeficiency virus	Cats	None as yet but threat remains strong
Serengeti wild dogs	Rabies	Dogs	Extinction threatened
Mednyi Arctic Fox (Aleutian Islands)	Mange	Dogs	90% killed

Another interesting situation exists in North America. The usual host of the meningeal worm *Parelaphostrongylus tenuis* is the white-tailed deer, *Odocoileus virginianus*, which is tolerant to the infection. All other cervids and the pronghorn antelope are, however, potential hosts, and in these species the worm causes severe neurological damage, even when very small numbers of the nematode are present in the brain. This differential pathogenicity of *P. tenuis* makes the white-tailed deer a potential competitor with other cervids because they cannot survive in the same area with white-tails. This is known as apparent competition. The deleterious effects of the parasite probably include direct mortality, increased predation, and reduced resistance to other disease.

The activities of humans have altered the normal distribution pattern of the white-tailed deer. As northern forests were felled, the deer expanded their range from a stronghold in the eastern United States, eventually coming into contact with moose. In Maine and Nova Scotia white-tailed deer have replaced moose as the major cervid, and they have also replaced mule deer and woodland caribou in some parts of their ranges. Whether or not reintroduction of caribou into regions now occupied by white-tailed deer is possible because of the action of parasites is now debatable (Schmitz and Nudds 1994). So, apart from direct mortality from parasites, competitive interactions between populations can be mediated by the action of parasites. This phenomenon is similar to that discussed in Chapter 9, in which Park (1948) compared competing populations of flour beetles with and without the parasite *Adelina triboli*. It is clear that parasites can have direct effects on the species they parasitize and indirect effects on other species. The subject of indirect effects is addressed later, but another interesting example concerns the parasites of the red grouse in England (Hudson, Dobson, and Newborn 1992). More birds are killed by foxes as their burdens of the caecal nematode *Trichostrongylus tenuis* increase, suggesting that parasites increase the susceptibility of red grouse to predation.

Cornell (1974) has argued that distributional gaps between bird species, where apparently favorable habitat exists, are maintained by the capacity of vectors to travel between populations. The rationale is that each population has a pathogen to which it is adapted but to which the other species is not. This concept has led to the idea that populations might compete for parasite-free space. Because the same type of phenomenon might operate for predators, the idea of enemy-free space is more widely circulated. Crosby (1986) argues that the Old World diseases smallpox, measles, typhus, and chicken pox were so devastating to peoples of the New World that the Old World invaders, who had a limited measure of immunity, found subjugation of the people much easier. In return, Columbus took syphilis back to Europe.

While observational data on parasite loads and host fitness provide compelling evidence of the strong effects of parasites on hosts, experimental removal of parasites in field-controlled experiments remains the ideal way to examine negative effects on host fitness (Hudson, Dobson and Newborn, 1992; Lehman 1993; Moller 1993). Two recent removal experiments provide such proof.

Fuller and Blaustein (1996) were able to show decreased overwinter survivorship of deer mice experimentally infected with the protozoan parasite *Eimeria arizonensis* as compared to noninfected individuals. These parasites are transmitted through ingestion of contaminated feces, making inoculation with parasite oocysts relatively easy. The survivorship and body mass of infected and uninfected free-living mice were compared in large outdoor enclosures. Unfortunately, the enclosures prevent terrestrial predators from exerting their normal effects on the mice, and the body mass of infected mice was not different from controls so that the mechanism through which the parasite affected mouse survival was not clear.

Stiling and Rossi (1997a) were able to manipulate parasitism levels of a gall-making midge on a coastal plant, *Borrichia frutescens*, on isolated offshore islands in Florida. To get low parasitism populations they allowed potted plants on one island to be colonized first by gall flies. They then removed the plants before the parasitoids could find them (low parasitism treatment). To get high parasitism populations they allowed the parasites to colonize the galls. Using these plants, replicates of both high and low parasitism treatments were established on other islands. Where parasitism was high, numbers of new galls were significantly lower than where levels of parasitism were low (Fig. 12.1).

12.4 Parasites and parasitoids are being used increasingly in biological control of insect pests.

Not all parasites are seen as detrimental by humans. Many are used as an effective line of defense against insect pests of crops, although only about 16 percent of classical biological-control attempts qualify as economic successes (Hall, Ehler, and Bisabri-Ershadi 1980). Although we cannot abandon chemical control as yet, there are many deterimental effects of pesticides, as well as advantages (Applied Ecology: The Pesticide Treadmill). This has spurred the scientific community into the search for more effective biological control agents.

Huffaker and Kennett (1969) have suggested five necessary attributes of a good agent of biological control:

General adaptation to the environment and host.
High searching capacity.
High rate of increase relative to the host's.
General mobility adequate for dispersal.
Minimal lag effects in responding to changes in host numbers.

Although these attributes seem necessary for a good control agent, they are clearly not sufficient. So far, the application of biocontrol agents has been carried out by a hit-or-miss technique rather than by a sound biological method. Some authors consider that this trial-and-error method probably makes the best economic sense given the high cost of research into the biology of natural enemies (van Lenteren 1980). Others have recommended new techniques, for example, presenting novel parasite-host associations as the most likely avenue for control where hosts have not had the opportunity to evolve complex defenses against these parasites from foreign lands (Hokkanen and Pimentel 1984). Arguments still rage as to whether it is better to introduce one parasite at a time or many. The problem is that if more than one enemy is introduced, competition between parasites could ensue, lessening the overall level of control. Ehler and Hall (1982) provided evidence from a world review of 548 control projects to show that this phenomenon could be a problem; the more parasites were released, the lower the rate of establish-

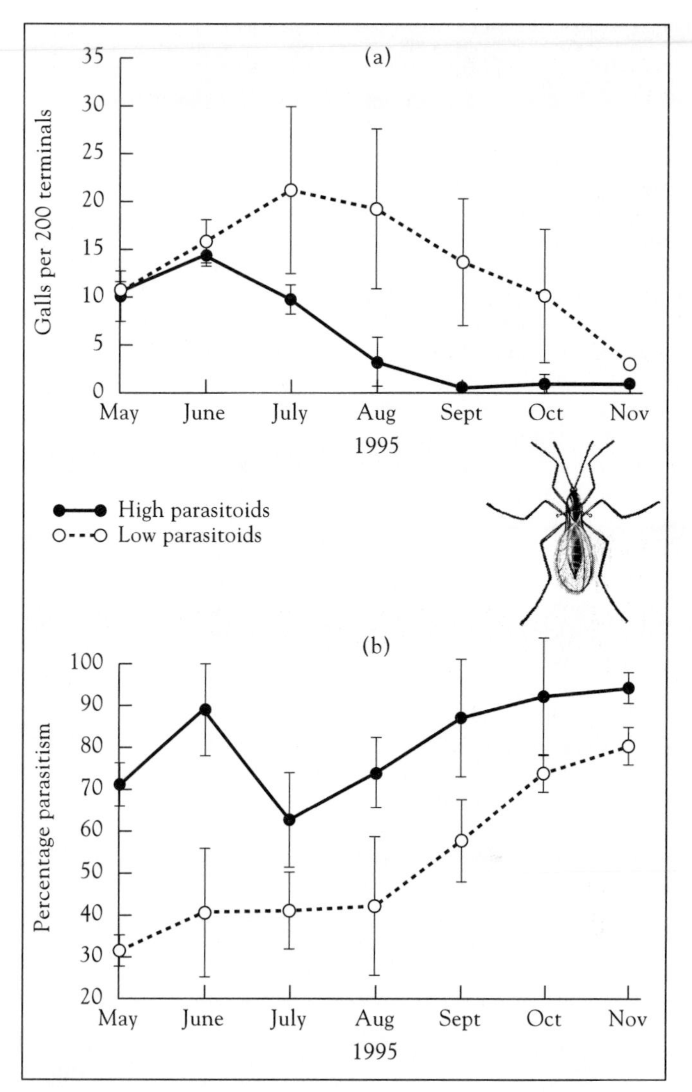

■ **FIGURE 12.1**
(**a**) Abundance of insect galls of *Asphondylia borrichiae* on experimental plots during 1995 in treatments with low densities of parasitoids and high densities of parasitoids. Abundance on both treatments tend to decrease dramatically in October/November as parasitism in both areas; (**b**), is high at this time. Data are means ±1.SE. (*After Stiling and Rossi 1997.*)

ment, although this analysis was disputed by Keller (1984). It is probably fairest to say that in this, as in so many ecological situations, the jury is still out.

Stiling (1990) reviewed the methods affecting success in biological control. The factor of greatest importance was the climatic match between the control agent's locality of origin and the region where it was to be released. This result stresses the value of studies in physiological ecology and that climatic variation is of vital importance in affecting biotic relations (Chap. 7). This was underscored by a separate analysis of reasons for biological failures (Stiling 1993) where reasons for failure related to climate (34.5 %) were more common than any other type of reason, including competition or parasitism by native insects.

It is important to note that biological control is not the risk-free alternative to chemical control it is often touted to be. As early as 1983 Frank Howarth (1983) lamented the

APPLIED ECOLOGY

The Pesticide Treadmill

The political controversy surrounding pesticides stems from their dual nature; they are valuable because they are lethal. Without the use of pesticides, preharvest crop losses in the United States would be about 18 percent from insects, 15 percent from diseases, and 9 percent from weeds. The benefits from pesticide use are valued at many billions of dollars. With the use of pesticides preharvest crop losses are estimated at 13 percent from insects, 12 percent from diseases, and 8 percent from weeds. Therefore, the use of pesticides causes a reduction in attack of 5 percent for insects, 3 percent for diseases, and 1 percent from weeds. This means that a roughly 10 percent additional harvest is reaped through the use of pesticides. Of course, these are overall figures, and in specific cases much higher yields are possible because of pesticides. The benefit-cost ratios are generally around 5:1. Pesticide use in agriculture can increase not only crop production but livestock production too and give greater longevity to food products on the shelf. In nonagricultural uses, pesticides have been important in reducing the incidence of malaria and other diseases, saving untold thousands of human lives. Also, many pesticides have rapid action and are the only alternative in an emergency situation.

A wide variety of pesticides are used in modern industrialized agriculture (Photo 1). Over forty-five thousand formulations (mixtures) and six hundred active ingredients are registered pesticides in the United States. Only about forty active ingredients, however, account for 75 percent of the use.

■ **PHOTO 1** The pesticide treadmill? Helicopter crop dusting a sugar beet field in California. More pesticides can mean more pesticide-resistant pests which therefore require newer and more lethal pesticides. (*Inga Spence, Tom Stack, CA 10493E.*)

Despite the value of pesticides, many factors give rise to environmental concern:

1. Pesticides generally have a broad spectrum of toxicity and have a lethal effect on a wide variety of organisms, not just insects but also fish, birds, and mammals. It has not generally been possible to develop a specific insecticide for a specific pest. Even if it were possible, it probably would be economically unfeasible for a company to make a profit on such a specialized product.
2. Some pesticides have a long life in the environment and, because of the broad spectrum of toxicity, can accumulate in food chains so that pesticide poisoning becomes serious in "higher" organisms.
3. There are manufacturing accidents and accidents when pesticides are applied to crops in the field.
4. Resistance can develop to the insecticide so that the pesticide essentially becomes ineffective.
5. Pesticides may kill the pest's natural enemies, allowing a resurgence of pests in the absence of their natural enemies. In both cases, this and number 4, more and different pesticides are often used to "get the job done." This is known as the pesticide treadmill.
6. People are exposed to pesticides by consuming residues in food and water. As long ago as the early 1970s, about 50 percent of food sampled by the U.S. FDA contained detectable levels of pesticide. This led to the so-called Delaney paradox—a ruling by Congress (the Delaney clause) that there be absolutely no traces of pesticide residue in certain foods. However, with such sensitive detection equipment as is now in use it is virtually impossible not to detect a miniscule fraction of pesticides. Dr. C. Everett Koop, former surgeon general of the United States, has commented that U.S. food supplies generally contain less than one-quarter of 1 percent of the allowable intake of pesticides and that there is not a food safety crisis. Nevertheless, to ensure that there are no traces of pesticide many people demand that their produce be "organically" grown without the use of chemicals.

reduction of native Hawaiian lepidopterans, partly due to wasp species introduced for biological control of lepidopteran crop pests. He called for a more narrowly focused release effort rather than a hit-or-miss campaign. Such a concern is even more important when the release of insect enemies to control weeds is considered. In this case, stringent host-specificity tests are performed to ensure that introduced insects will not turn to feed on valuable crops, even in times of starvation. Nevertheless, *Cactoblastis* moths, the agents that so successfully controlled the pest cactus *Opuntia* in Australia, have recently arrived in Florida and are decimating native cacti there (Johnson and Stiling 1996). Simberloff and Stiling (1996) have provided many more examples of the nontarget effects of biological control agents. Many biological control practioners are upset that they should be tarred with the same brush as the chemical control crowd, feeling that they have the environmental moral high ground. However, eventually chemicals will degrade, whereas biological control agents are essentially uncontrollable. Once released they are forever in the environment. That's why some people argue that biocontrol should be employed only as a last-ditch attempt at pest control.

The release of microparasites is a subject of even more concern. Australians use viruses to control exotic pest mammals. When the British colonized Australia they brought with them the English rabbit and the European red fox, both in the mid 1800s. The rabbits were for food, and the fox was to be hunted. Not long after these introductions people realized they had made a mistake; the rabbits began breeding like rabbits, and the predatory foxes turned their attention to the native animals. So far, they have been implicated in the extinction of twenty species of local marsupials (Morrell 1993). The Cooperative Center for Biological Control of Vertebrates (VBC) (a government and university consortium) released a rabbit calicivirus, in the fall of 1995. The results were staggering. Rabbit numbers dropped 95 percent, and huge increases were noted in kangaroos, other animals that competed for vegetation with rabbits, and in the vegetation itself, with a resurgence of many rare plants as well. Since the rabbits are the main item on the foxes' and feral cats' menu, a pleasant side effect was a population crash in the numbers of these species too! However, some scientists are concerned that these species might now be forced to seek alternative food—the likely choices being small endangered marsupials. In addition, fears are high in some circles that the calicivirus might attack other species or evolve to do so in the future (Drollette 1997).

SUMMARY

1. The true definition of a parasite is problematic. Parasites may include many species that feed on plants as well as more "traditional" parasites like tapeworms, leeches, bacteria, viruses, and parasitoids. Parasitism is undoubtedly an extremely common way of life, with perhaps 50 percent of all animals considered to be parasitic.

2. The presence of various defenses against parasites, such as the immune response in vertebrates, is testament to the importance of parasitism in nature.

3. Mathematical models suggest that efficient parasites are likely to keep their hosts alive as long as possible to facilitate the transmission to other hosts. Such diseases include leprosy, typhoid, and amoebic dysentery. Because human populations are now so large, diseases that used not to be common, such as measles and cholera, flourish with so many potential hosts. Finally, the intimate association of humans with animals has allowed the crossover of animal viruses to people. Smallpox in humans is derived from cowpox in cattle. AIDS originated in monkeys and spread to people.

4. The huge influence of introduced diseases, such as chestnut blight and dutch elm disease in America and rinderpest in Africa, are testament to the severe effects parasites can have on host populations, sometimes driving them close to extinction.

5. Parasites of insects can often be used as control mechanisms against agricultural and forestry pests. This is called biological control. Finding the attributes of successful biological control agents is a valuable ecological endeavor.

DISCUSSION QUESTION

Which types of plants and animals might be expected to have more species of parasites and suffer higher rates of parasitism? Why can we eradicate some diseases, such as yellow fever, through vaccinations, while we have not been able to eradicate other diseases, such as malaria? Can we ever expect chemical pesticides to be replaced entirely by biological control? If not, why not?

CHAPTER

13

Evaluating the Relative Strengths of Mortalities

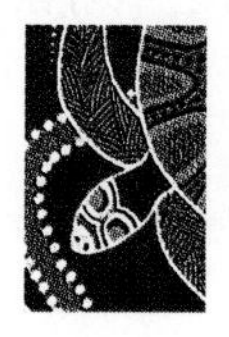

The density of any given population can be affected simultaneously by a number of factors, including parasitism, predation, herbivory, competition, and climate. Determining which of these many factors is an important influence on population density has proved to be a challenge to ecologists.

- First of all, it is necessary to determine the mean population density or equilibrium value, then determine which mortality factor causes the most change in population density. This sounds simpler than it actually is because population variability often makes it hard to determine equilibrium values.
- Second, while comparing the strengths of mortality factors is one of the most important pursuits in ecology, few scientific papers are broad enough in scope to be able to do this. Instead, most papers focus on one or two mortalities.
- Third, some authors have argued that species may exist not as a single population but as a series of mutually affected populations called **metapopulations**.

13.1 Many populations are so variable that it is difficult to know or to determine if they are at equilibrium.

Over one thousand years ago Chinese bureaucrats of the T'ang dynasty began collecting data on the severity and spatial extent of locust infestations in order to help forecast locust outbreaks (Photo 13.1 and Fig. 13.1). Population variability increased at longer timescales, a phenomenon referred to by some (Sugihara 1995) as a red shift, i.e., a shift to a longer power spectrum. Thus, two population values a century apart will on average differ more than two population values a decade apart, and so on. Similar patterns

PHOTO 13.1 Desert locust, (*Schistocerca gregaria*) oubreak in the grasslands of N. Africa. Chinese administrators have long collected information on locust outbreaks in China and the data are useful in population fluctuation analysis. (*Ken Lucas, Visuals Unlimited, AINS-1498.*)

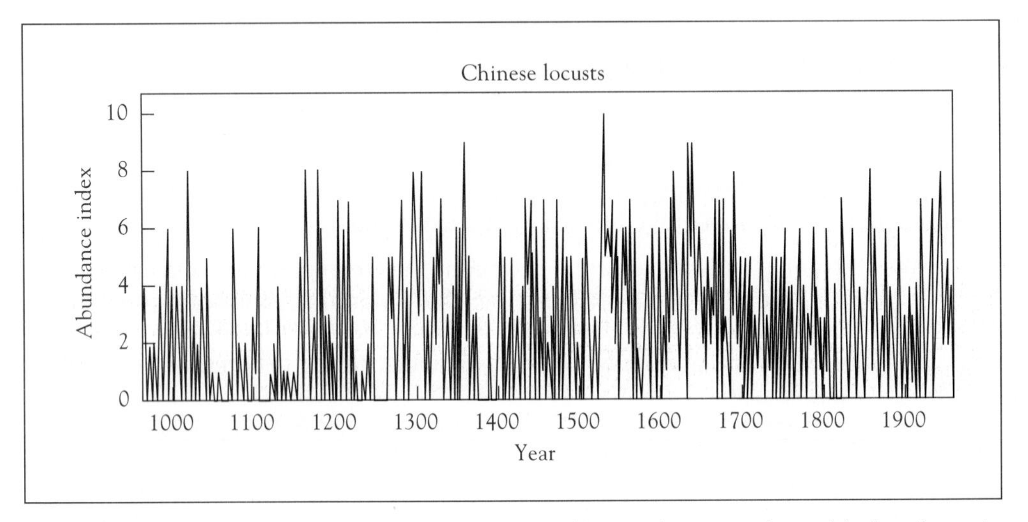

FIGURE 13.1 A 1,000-year record (957-1956) of locust (*Locusta migratoria*) abundance in China. (*After Sugihara 1995.*)

have also been noted for stocks of wild animals (Diamond and May 1977). Thus, the longer you observe a population, the more variable it seems to be (for example, see Fig. 13.2). As such it is almost impossible to know what the equilibrium value is. Large variation in population density has been shown to exist for many species of farmland birds, woodland birds, insects, and mammals (Pimm and Redfearn 1988). The existence of such a general phenomenon casts doubt over whether many populations have an equilibrium. If they do, it would be hard to know what it was without decades, even centuries, of data.

What causes these "red shifts"? The only published analysis of the locust data suggested a correlation between locust outbreaks and climatic factors, especially wet peri-

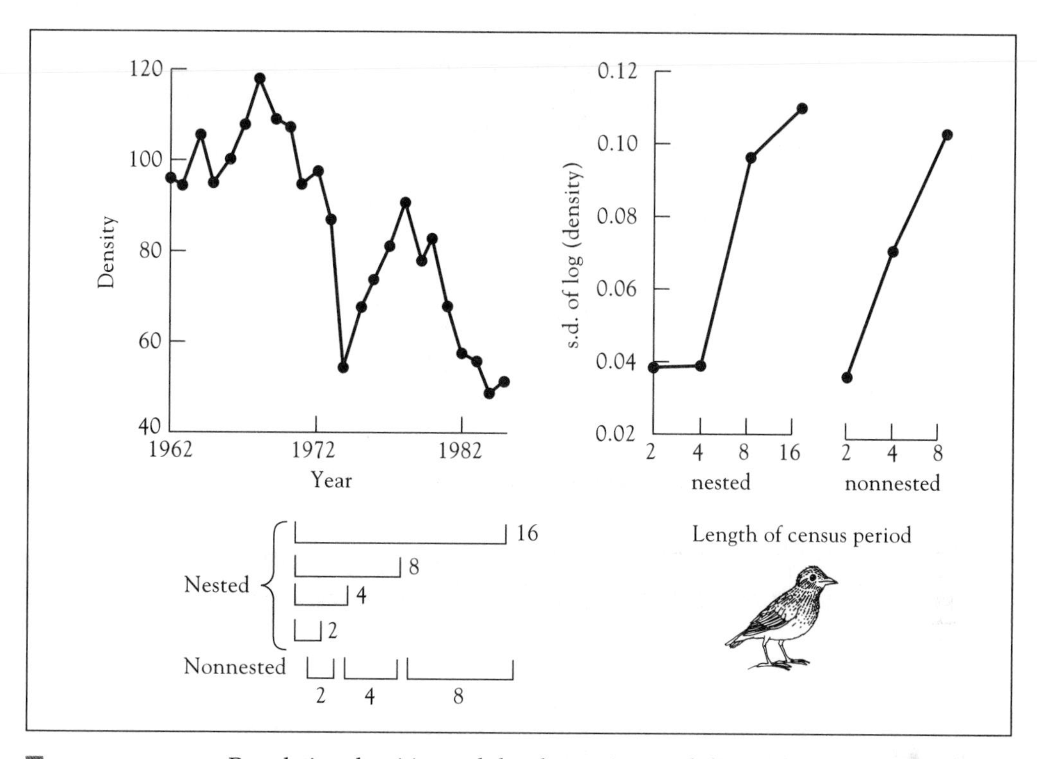

FIGURE 13.2 Population densities and the change in variability with the length of census. (**a**) Density of the skylark (*Alauda arvensis*) in English farmlands is plotted against year; the scale is relative and set to 100 in 1966. (**b**) For the same population, the standard deviations of the logarithms (SDL) of density are plotted against the period over which the calculation was made. For both nested years (2, 4, 6, and 16) or nonnested years (2, 4, and 8) SDL increases with period. *(After Pimm and Redfearn 1988.)*

ods, which promote nymphal survival. Indeed, climate variables themselves are often redshifted on this timescale (Sugihara 1995). It is for this reason that some entomologists subscribe to the view that external climatic factors control populations. However, other ecologists, including some that study longer-lived organisms like birds and mammals subscribe to the view that biotic mechanisms such as competition and predation are important. This debate permeates ecology and is known in various guises: density dependence versus independence, equilibrium versus nonequilibrium, biogeographical patterns versus null models, chaos versus noise, to name just a few (Sugihara 1995). This is an important issue because resource managers like fisheries biologists need to know if populations are behaving in an equilibrium or nonequilibrium manner in order to determine harvest levels. This is especially critical since many theoretical population models actually show a shift to the blue frequencies, i.e., adjacent generations tend to show more difference than nonadjacent ones (Cohen 1995). This may be because time-delayed birth/death responses are fast, causing the system to overshoot and undershoot a target equilibrium so that population fluctuations become periodic or chaotic. Low values tend to be followed by high values and vice versa.

This chasm between theory and data illustrates three points:

1. Natural population fluctuations are not chaotic (not blue shifted).
2. Existing population models are flawed (should be red shifted).
3. Climatic fluctuation or "environmental forcing" needs to be incorporated into population models to allow for the red shift. Recently, it has been argued that such forcing pushes simple models into a chaotic regime (Sugihara 1994).

13.2 Comparing the strengths of mortality factors may involve determination of the relative killing power of each of them.

There are a variety of factors that have the potential to influence populations. Competition, predation, parasitism, herbivory, and mutualisms are all common and powerful biotic forces. Climate can also influence populations. Which factor most commonly acts to cause structure in which communities and at what times? The argument is as old as the oldest writings on population ecology. Different reasons seem to go in and out of favor. In the 1970s, the prevalent view was that competition was of overriding importance (MacArthur 1972; Cody and Diamond 1975). In the 1980s this view was challenged (Strong, Lawton, and Southwood 1984), and more emphasis was given to nonequilibrial and stochastic factors such as disturbance (Chesson and Case 1986; Wiens 1986).

A comparison of mortality factors is also integral to community ecology theory. If we can determine which mechanism is most important then we may be able to tell whether or not communities are tightly structured entities. If biotic factors are of overriding importance then communities may really be tightly knit entities. If abiotic forces have the most influence in determining species abundance then structure may be loose and emphemeral.

Mortality factors can act in two ways: to disturb populations away from equilibrial levels or to return them toward equilibrium. Disturbing factors are often known as **key factors** and will be examined first in key-factor analysis. Regulatory factors act in a density dependent fashion and will be examined later.

Key-factor analysis is an approach for determining the relative importance of different mortality factors on population density.

One of the most well-known techniques for empirically comparing the importance of the effect of predators, parasites, or other factors on the size of field populations is key-factor analysis (Morris 1959). Key-factor analysis requires detailed information on the fate of a cohort of individuals: total mortality of a generation or cohort (K) is subdivided into its various causes, and the relative importance of these is compared. For example, in Table 6.4 the fate of a generation of spruce budworm is documented. At each stage, the researchers tracked the number of deaths and the cause of death (parasitism, predation, disease, etc.). The importance of each mortality factor (k) is estimated by calculating the amount that the factor reduced population size (or cumulative survivorship):

$$k = \log N_t - \log N_{(t+1)}$$

where N_t is the density of the population before it is subjected to the mortality factor and $N_{(t+1)}$ is the density afterward. Alternatively,

$$k = \log l_x - \log l_{x-1}$$

Total generational mortality, K, can then be defined as the sum of the individual mortality factors k_x:

$$K = k_1 + k_2 + k_3 + k_4 + \cdots + k_n$$

If this analysis is repeated for multiple generations, then correlations can be examined between the total generation mortality (K) and the individual sources of mortality. The source of mortality k_x that most closely mirrors overall generation mortality (K) is then termed the key factor.

More precisely, individual sources of mortality or k values can be plotted on the y-axis against total mortality, K, and the key factor is then the source of mortality with the biggest correlation coefficient, r, with K (Podoler and Rogers 1975). However, the visual method is preferred so that one can exclude a k that is correlated with K but makes only a small contribution to its variation. In oak winter moths in England (Fig. 13.3), a species probably subject to the most comprehensive key-factor analysis ever done (Varley, Gradwell, and Hassell 1973), the key factor is overwintering loss (Fig. 13.4). Young winter moth larvae emerge in the spring and feed on the newly developing foliage, which at this time has the lowest levels of antiherbivore chemicals like tannins. Often larvae emerge prior to bud burst and face starvation. To find fresh leaves, they disperse by ballooning away on silken threads to new trees. This is obviously a chancy business, and many larvae are lost; hence the high k value. Overwinter loss could be a key factor in many animal populations because of severe climatic stress. Key factors can usually be detected only by analysis of many generations; in **univoltine** animals this analysis may take many years.

For other animals, key factors are many and varied. There are few generalizations that can be made as to which type of key factors operate on which types of population (Table 13.1). Even for related species, like insects, there is no key factor of overriding importance (Stiling 1988b).

Although key-factor analysis has been an important tool for ecologists, it has not been without recent criticism (Royama 1996). First, key factors cannot always be precisely linked to specific mortality agents. Thus, overwintering mortality in the winter moth may be due to egg death, early larval death, or death by ballooning. The contribution of each of these to k_1, overwintering disappearance, is unknown, but it is probable that such a division would probably reduce the correlation between k and K so much that it would disappear. Often, census data are only good enough to determine key-factor phases in lifestyles, for example, juvenile mortality or adult mortality. Second, there may be intricate interactions between natural enemies, including hyperparasitoids (parasitoids of parasitoids), and such effects fail to show up in key-factor analysis. Third, populations can be hugely influenced by egg-bearing females that disperse into a population and that also never show up in key-factor analysis.

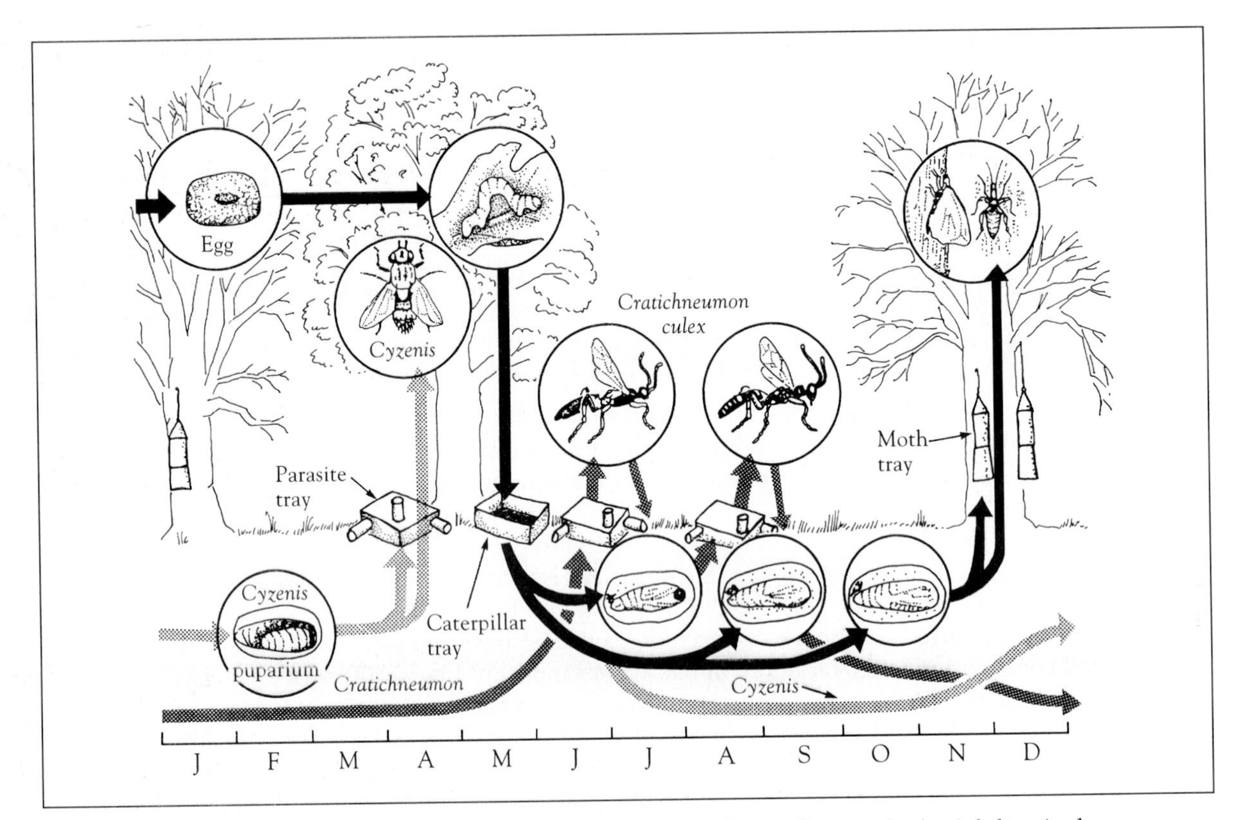

FIGURE 13.3 Life cycle of the oak winter moth and sampling methods. Adult, wingless, female winter moths, which climb up to lay their eggs, were counted in moth traps on the tree trunks; their larvae, which feed on the leaves, were counted on the foliage and in the caterpillar trays into which they feed when prepupal. Larvae of the parasite *Cyzenis*, which lay their eggs on the foliage also, were counted by dissection of the fallen caterpillars, and adult *Cyzenis* and *Cratichneumon*, which directly parasitize pupae, were counted upon emergence from the soil into the parasite traps. (*Redrawn from Varley 1971.*)

Southwood (1978) has outlined other ways at looking at life table data (see Table 13.2).

1. Real mortality d_x/n_o.
   This is mortality of the population compared to its density at the beginning of the generation. The real mortality row in Table 13.2 is the only percentage row that is additive and is useful for comparing the role of population factors within the same generation.
2. Indispensable (or irreplaceable) mortality.
   This is that part of the generation mortality that would not occur should the mortality factor in question be removed from the life system, after allowance is made for the action of subsequent mortality factors. For example, consider the egg stage mortality in Table 13.2: if there is no egg mortality, one thousand individuals enter the larval stage where a 40 percent mortality leaves six hundred survivors to pupate; in the pupal stage

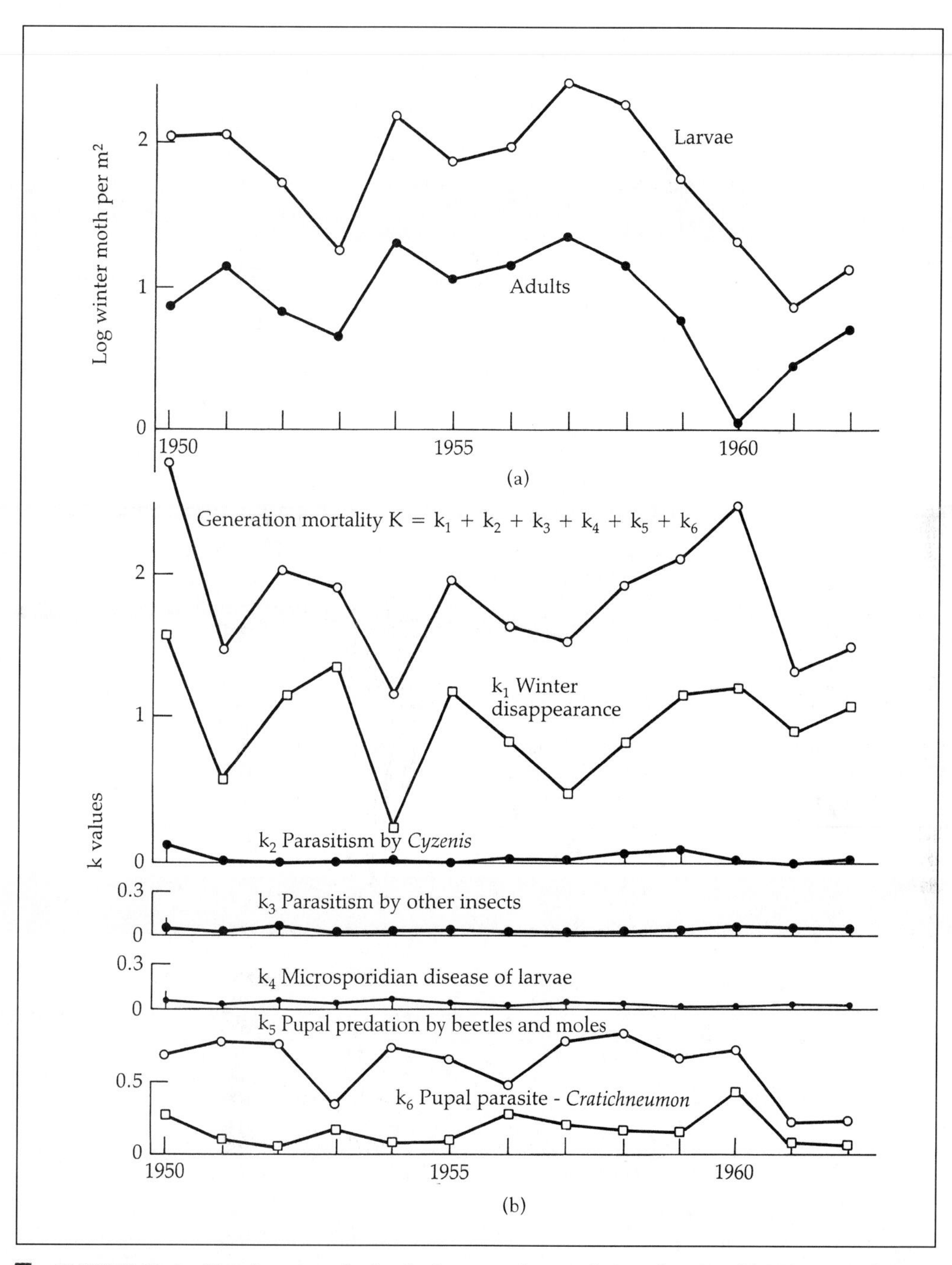

FIGURE 13.4 Key-factor analysis of winter moth population changes. (**a**) Winter moth population changes expressed as generation curves for larvae and for adults. (**b**) Changes in the mortality, expressed as k values, showing that the biggest contribution to changing the generation mortality K comes from changes in k_1, winter disappearance. *(Redrawn from Varley, Gradwell, and Hassell 1973.)*

a 90 percent mortality leaves sixty survivors, that is, thirty more than when egg mortality occurs, and thus its indispensable mortality.

$$= 30/1000 \times 100 = 3\%$$

TABLE 13.1 Key factors and density dependent factors for a variety of plants and animals. (*From Podler and Rogers 1975; Stubbs 1977; and Stiling 1988*)

ORGANISM	KEY FACTOR	DENSITY DEPENDENT FACTOR
Sand-dune annual plant	Seed mortality in soil	Seedling germination
Colorado potato beetle	Emigration of adults	Larval starvation
Tawny owl	Reduction in egg clutch size from maximum	Losses of birds outside the nesting season
African buffalo	Juvenile mortality	Adult mortality
Partridge	Chick mortality	Shooting of adults!
Great tit	Loss of birds outside the breeding season	a. Variation in clutch size b. Hatching success
Broom beetle	Larval mortality on foliage	a. Larval mortality in ground b. Survival of adults
Grass-mirid insect	No obvious key factor	No obvious density-dependent factor
Cabbage-root fly	Reduction in egg production	Pupal parasitism and predation

Note: Some species have no obvious density dependent or key factor; other species may have more than one density dependent factor. Different species tend to have different types of density dependent or key factors.

TABLE 13.2 Various ways of comparing mortality factors from life tables. (*After Southwood 1978*)

	STAGE						
MEASURE	EGGS		LARVAE		PUPAE		ADULTS
n_x	1000		500		300		30
d_x	500		200		270		
q_x	0.5		0.4		0.9		
% real mortality	50		20		27		
% indispensible mortality	3		2		27		
Mortality/survivor ratio	1.00		0.66		9.00		
log population	3.00		2.70		2.48		1.48
k-values		0.30		0.22		1.00	

The indispensable mortality of a factor may be useful for assessing its value in pest control programs.

3. Mortality-survivor ratio d_x/n_{x+1}
   This measure represents the increase in population that would have occurred if the factor in question had been absent. If the final population is multiplied by this ratio then the resulting value represents, in individuals, the indispensable mortality due to that factor.

13.3 Although population density changes over time in many populations, density dependent processes may regulate population size around some mean value.

If parasitism, predation, competition, and abiotic factors can all perturb the population densities of living organisms, which effect is most important in returning populations to equilibrial levels? This is again a very difficult question to answer for many reasons. First of all, it is necessary to compare the strength of different mortality factors. Second, it is appropriate to determine which of them act in a density dependent manner. Even abiotic factors could act in a density dependent manner if, for example, there were limited refuges, like burrows, to get away from inclement weather.

Density dependence can be detected where adverse effect, expressed as percentage, is plotted against population density (Fig. 13.5) or if k values are plotted against generational mortality k. If a positive slope results and mortality increases with density, then the mortality factor tends to have a lesser affect on sparse populations than on dense ones and is clearly acting in a density dependent manner. In the winter moth example, pupal predation of overwintering moth pupae in the ground (by beetles, shrews, and moles) is the density dependent factor (Fig. 13.6). This is also true in Canada where the winter moth was introduced in the 1930s (Roland 1988). In such plots, factors that appear not to change with density, and thus do not contribute to population regulation, are termed **density independent**. Those sources of mortality that decrease with increasing population size are called **inversely density dependent**.

Which factors tend to act in a density dependent manner? We can examine this question with reference to insects, the most speciose group on Earth and the group with perhaps the best available data to address this question, perhaps in part due to the pest status of many species and the large number of studies done on them. Stiling (1987) and

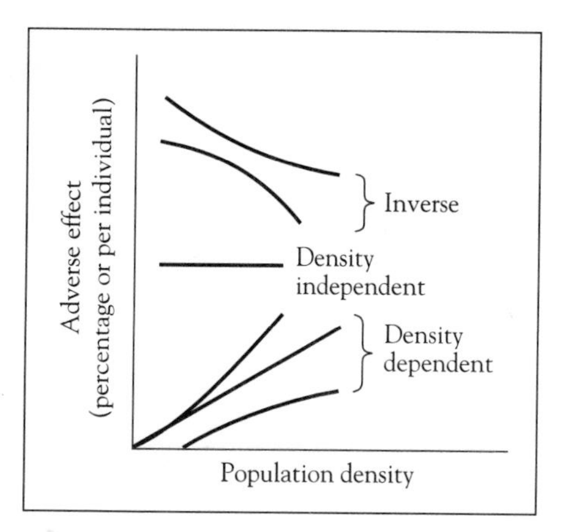

FIGURE 13.5 Types of response to changes in population density. Expressed as a percentage response to increasing population.

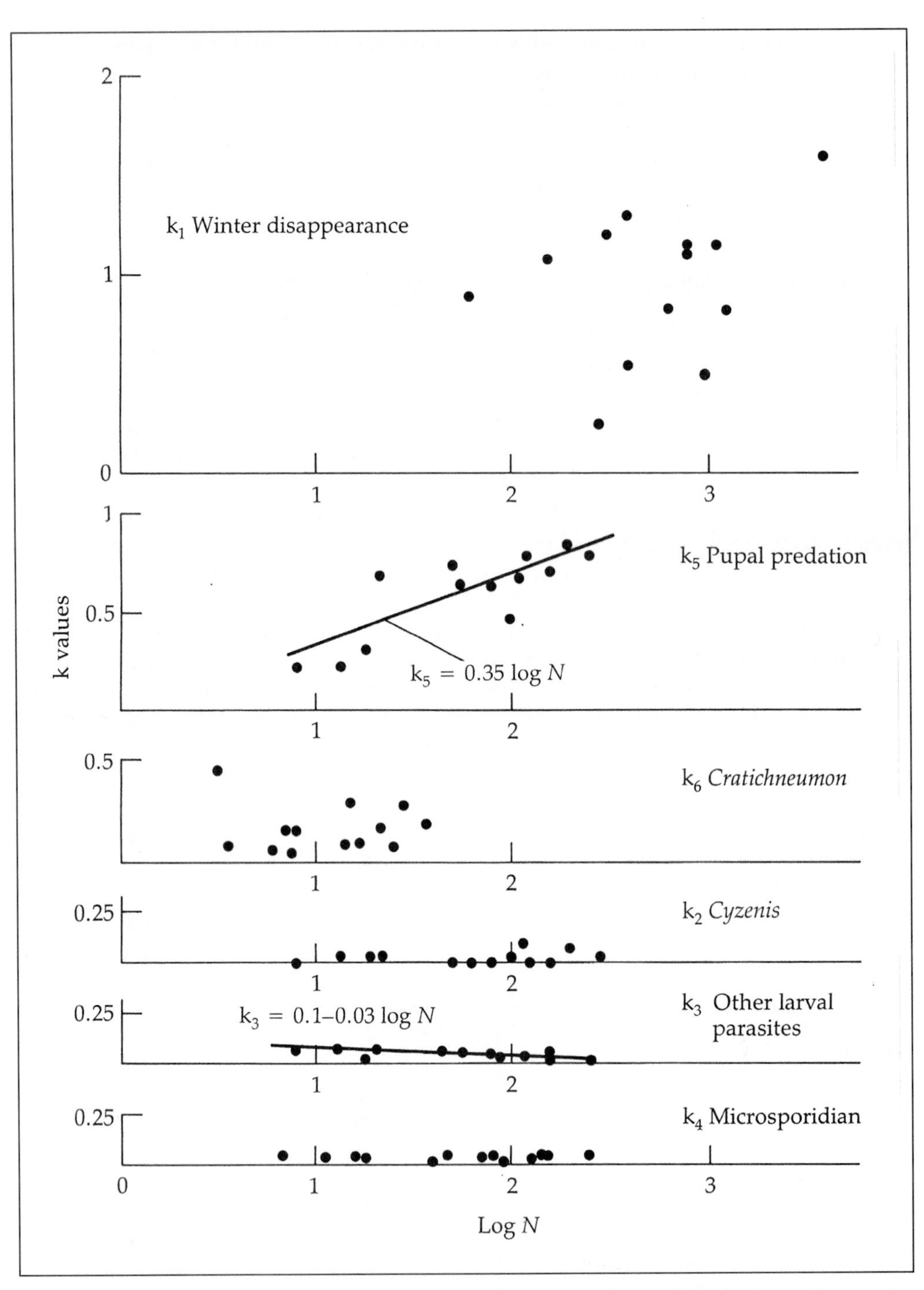

■ **FIGURE 13.6** *K* values for the different winter moth mortalities plotted against the population densities on which they acted. k_1 and k_6 are density independent and vary quite a lot; k_2 and k_4 are density independent but are relatively constant; k_3 is weakly inversely density dependent; and k_5 is quite strongly density dependent. (*Redrawn from Varley, Gradwell, and Hassell 1973.*)

Walde and Murdoch (1988) showed that the percentage of field studies on insects that detected density dependent parasitism is quite low, on the order of 25 percent.

Stiling (1988) went on to examine density dependence by any mortality factor in insect populations, using data sets for fifty-eight species. The density dependent factor that was most important was different for populations of different species at different times. No single process could be regarded as a regulatory factor of overriding importance. By way of illustration, Karban (1989a) examined the effects of different sources of mortality on three herbivores of the seaside daisy, *Erigeron glaucus*. For spittlebugs, *Philaneus spumarius*, damage caused by caterpillars increased rates of desiccation and mortality of nymphs. Thus, competitive effects were strongest for this species. For the caterpillars, *Platyphilia williamsii*, predation by savannah sparrows, *Passerculus sandwichensis*, was the strongest effect, and protecting caterpillars from sparrows by enclosing colonies under chicken-wire cages greatly increased caterpillar survival. Finally, for the third herbivore, the thrips *Apterothrips secticornis*, host-plant chemical effects were of greatest importance in determining survival (Karban 1987). The factor that was most important for each herbivore species did not interact with other biotic factors and was different for each herbivore. This is disconcerting because it means generalizations about which factors are likely to act in a density dependent manner are not easily made or are likely to be different, even for different consumers on the same plant.

Sinclair (1989) presented a broader review of the literature, reporting on the cause of density dependence in insects, fish, large mammals, small mammals, and birds (Table 13.3), though his literature base was not as current, at least for the insects. Sinclair found that there were few generalities to be drawn. However, there were some differences in density dependent factors between the taxa: food was more important for large mammals, and space and social interactions were important for smaller mammals and birds. Sinclair noted, however, that the effects of disease and parasitism had probably been grossly understudied and were therefore likely to be underrepresented in his table.

Stubbs (1977) suggested that for *r*-selected animals most density dependent factors acted on juvenile stages whereas for *K*-selected animals reduced fecundity of adults seemed to act more frequently. Sinclair (1989) revisited this issue. He found that species with very high reproductive rates (insects and fish) have early juvenile (eggs, larvae) density dependent mortality; those with intermediate reproductive rates (birds and

TABLE 13.3 **Number (%) of reports of separate populations recording cause of density dependence.** (*After various sources in Sinclair 1989*)

GROUP	SPACE *N* (%)	FOOD *N* (%)	PREDATORS *N* (%)	PARASITES *N* (%)	DISEASE *N* (%)	SOCIAL *N* (%)	TOTAL NO. POPULATIONS
Insects	0	23 (45)	20 (39)	19 (37)	5 (10)	4 (8)	51
Large mammals	1 (1)	77 (94)	4 (5)	0 (0)	2 (2)	0 (0)	82
Small mammals and birds	19 (53)	13 (36)	4 (11)	1 (3)	0 (0)	21 (58)	36

Note: These values do not normally add up to 100%, indicating more than one density dependent factor can operate on a population.

small mammals) have late juvenile and prebreeding regulation, while large mammals with low reproductive rates are at least partly regulated through change in fertility (Table 13.4).

Questions of which mortality factor operates in a density dependent manner and which do not notwithstanding, Stiling (1988b) noted that the overall detection of density dependence in insects was quite low—it was detected in only about half of the studies. One implication is that without density dependence communities may not commonly exist in tightly knit communities. However, Mike Hassell and colleagues (Hassell, Latto, and May 1989) showed that many of the data sets Stiling used may have been of insufficient duration to detect density dependence, even if it was present. They noted that the more generations were studied, the more frequently density dependence was detected (Fig. 13.7). Woiwood and Hanski (1992) found strong evidence for density dependence in moths and aphids when series of more than twenty years were used. Solow and Steele (1990) theorized that up to thirty generations would need to be studied before density independence could be rejected with a high probability. This has led to much debate about the number of generations needed and the type of test necessary to detect density dependence (Vickery and Nudds 1991; Holyoak and Crowley 1993; Wolda and Dennis 1993; Fox and Ridsill-Smith 1995; Reddinguis 1996).

While valuable, Hassell's study is disconcerting because it means only long-term studies are useful in the search for density dependence. Such studies are likely to be time consuming, difficult to perform, and probably difficult to get funded. Furthermore, how many scientists would perform a study for thirty years and admit to finding no density dependence? Remember the results of Csada, James, and Espie (1996) that showed that studies with nonsignificant results constituted less than 10 percent of published biological papers.

Ray and Hastings (1996) suggested that an even bigger reason that density dependence has not been detected more frequently is that ecologists have searched at the wrong spatial scale. Using seventy-nine insect population studies they showed that only

TABLE 13.4 Number (%) of reports of separate populations demonstrating density dependence at different life stages. (*After various sources in Sinclair 1989*)

GROUP	FERTILITY/ EGG PRODN. *N* (%)	EARLY JUVENILE MORTALITY *N* (%)	LATE JUVENILE MORTALITY *N* (%)	ADULT MORTALITY *N* (%)	TOTAL NO. POPULATIONS
Insects	14 (30)	19 (40)	13 (28)	6 (13)	47
Fish	2 (6)	33 (94)	0	0	35
Birds and small mammals	5 (16)	6 (19)	26 (81)	5 (16)	32
Large mammals	83 (73)	45 (40)	1 (1)	13 (11)	113

Note: Density dependence can occur at more than one life stage; hence, the total number of populations is not the total of the rows.

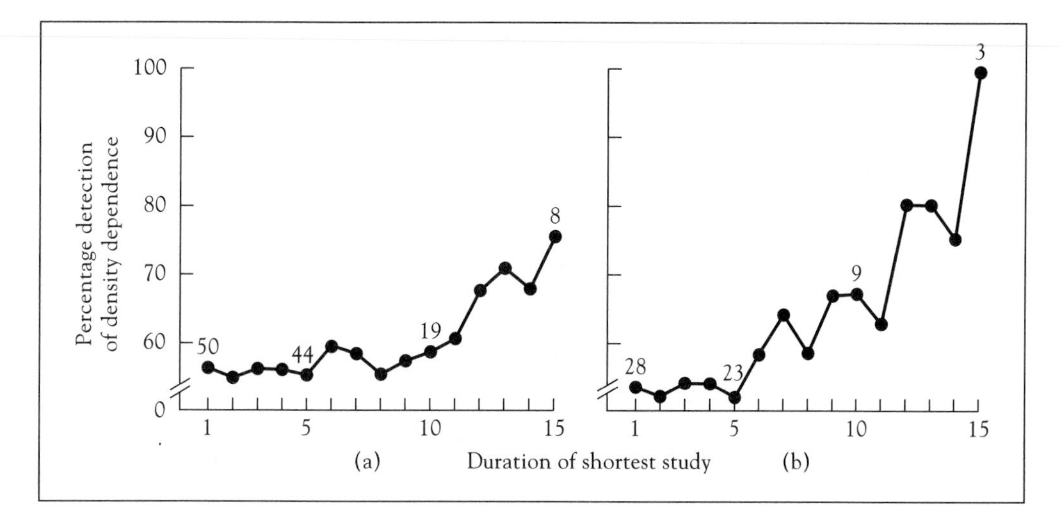

FIGURE 13.7 The effect of the duration of the studies (in generations) on the probability of detecting density dependence using fifty examples from Stiling (1988). The relationship rises as studies of shorter duration than the abscissa values are shed from the analysis. The falling sample sizes of the first, fifth, tenth, and fifteenth points are marked above each line. (*After Hassell et al. 1989.*)

a minority (23 to 35%) of studies with relatively large data-collecting areas found density dependence while the converse was true for relatively smaller-scale studies (69 to 73%). Ray and Hastings (1996) found a much higher overall rate of density dependence (75%) than either Stiling (1988) (46%) or Hassell, Latto, and May (1989) (54%). However, they also noted the possibility of failure to report negative findings in the literature and estimated a possible overall rate of density dependence of 26 to 47 percent if this were taken into consideration.

Addicott et al. (1987) pointed out that the correct scale for investigation depends on the movement patterns of the organisms involved. A parasite that moves only tens of meters is unlikely to be able to respond to differences in host densities over hundred of meters. But when Stiling et al. (1991) examined density dependence by two parasites of widely differing dispersal abilities on a salt marsh insect, no density dependence was found at any scale, small or large, by either parasitoid. In some systems, density dependence might not occur for other reasons. In the salt marsh this may be because continual tidal inundation disrupts parasitoid searching behavior (Stiling and Strong 1982), preventing density dependent attack by parasitoids.

Another reason for a lack of detection of density dependence is that in some situations the effect of density dependent factors on a population is delayed by one or two generations. In such cases, when percentage mortality is plotted against host density and the points are joined in a time series, the time lag results in a counterclockwise spiral (Fig. 13.8). While most studies have searched for density dependent regulation, very few have examined the frequency of time-lagged density dependence. Turchin (1990) evaluated the evidence for delayed density dependence in fourteen forest insects and found eight showed evidence of delayed density dependence while only three showed direct

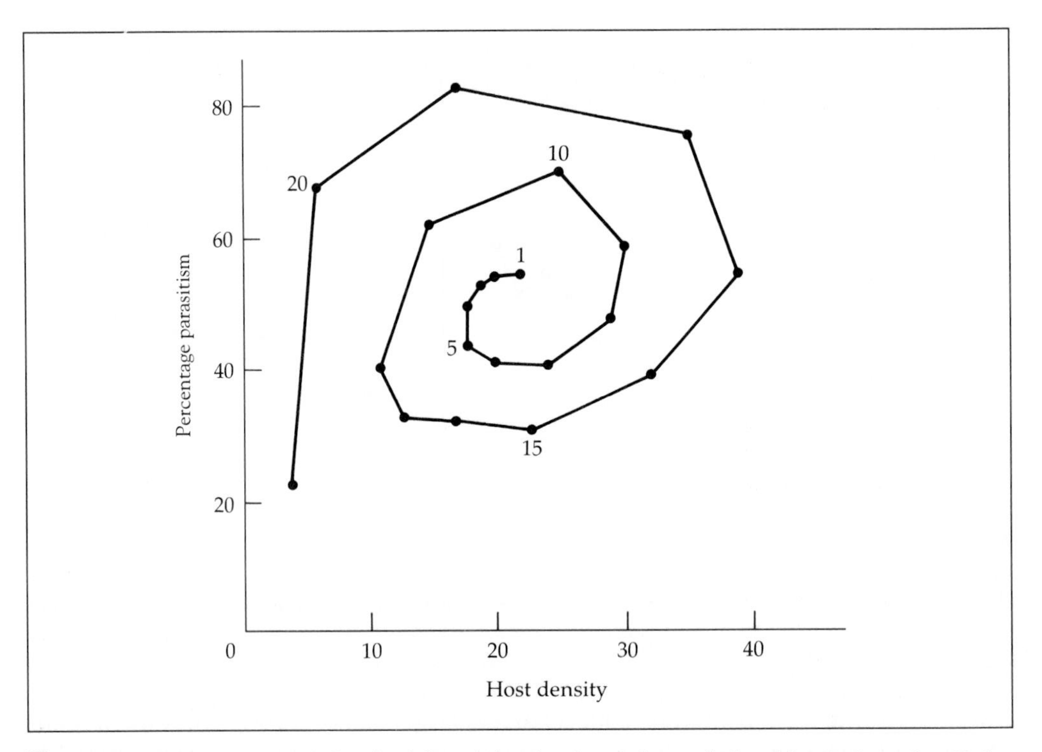

FIGURE 13.8 An example of a delayed density dependent relationship. *(Redrawn from Varley, Gradwell, and Hassell 1973.)*

density dependence. Such studies could have been prominent in the reviews of Stiling (1988); Hassell, Latto, and May (1989); and Ray and Hastings (1996) yet went unreported. However, Williams and Leibhold (1995) argued that Turchin's (1990) tests for delayed density dependence were themselves ambiguous and could give rise to spurious diagnosis of feedback effects. Berryman and Turchin (1997) and Williams and Leibold (1995) continue to debate this issue.

Because of the limitations of duration and spatial scale, Harrison and Cappuccino (1995) argued that the best way to test for density dependence was to do it experimentally, using "density perturbation" experiments that changed species density either upward or downward and examine subsequent effects of potential regulating agents like natural enemies. Harrison and Cappuccino (1995) scoured the ecological literature for such studies and found sixty studies on eighty-four regulatory forces (sometimes a study looked at more than one force) (Table 13.5). The most striking result was that 79 percent of studies showed direct density dependence, suggesting to the authors that density dependence is common in nature. However, there remains the possibility that the treatment levels may not always have represented natural variation in the field, clouding the issue of whether the results of such tests were always meaningful. Many of the perturbations involved reducing population densities by removals because this is often much easier to do than elevating densities, although an equal mix of the two would have been

ECOLOGY IN PRACTICE

Peter Turchin,
University of Connecticutt

I was born in 1957 in Obninsk, a small town 100 km from Moscow. My father was a theoretical physicist turned cyberneticist, so I never had a fear of math. Additionally, I attended high school for math and physics (sort of like the schools for math and sciences they have in many southern states; they accept only a portion of applicants, which is really a lot of fun because everybody is interested in studying some sort of science). I did not get into college on my first attempt (1974) and almost got drafted, but I managed to squeak in on the second try (1975). I studied for two years at Moscow University in the biological faculty and was going to specialize in ecology, but fate intervened. My father was one of the dissidents and a close friend of Sakharov. He organized the first group of Amnesty International in the USSR and was active in other dissident activities. He was fired from his job and was arrested several times, and our apartment was searched and the telephone was cut. This went on for four years. He almost ended up in prison, but for some reason the authorities decided to force him to emigrate, which we did in 1978. I finished at New York University in 1980.

I had lousy luck applying for academic jobs. Basically, it took me nine years after my Ph.D. to land one. I ascribe this to my specialization in combining theory with data—while it's a wonderful way to do science (at least I think so), it does take time to build a name for yourself, or maybe I was just plain unlucky. Anyway, I applied to literally thousands of jobs but had only two interviews. In one, they did not hire anybody but redefined the position; the other was the University of Connecticut. But I am not complaining, because in the interim I worked for the Forest Service, and I had a lot of time to do basic research. Working on Southern Pine Beetles (SPB) is what was responsible for me starting with delayed density dependence. There was a thirty-year-long time series for SPB oscillations, and it looked cyclic but not terribly so. The prevailing opinion was that SPB outbreaks result from fluctuations in weather, and there were several analyses that tried to relate SPB numbers to weather. To cut a long story short, I tried many ways of analyzing the data, some of them pretty stupid, and eventually hit on a better way of doing that. The indications were that delayed density dependence (DD) was very important, but direct DD was not statistically significant. I had recently read papers trashing the idea of DD, even one by the author of this book, and I thought that these did not consider delayed DD. I got pretty excited about it, rushed to the library in Baton Rouge, tried to get as many time series for forest insects that I could (since that's what I knew about). The analyses came out beautifully. In retrospect, I may have been lucky in selecting forest insect pests—the incidence of complex population dynamics is much higher among these creatures than among other groups of organisms.

One periodically hears that ecology has no general theory, or even that no general theory of ecology is possible. I completely disagree. Ecology has been a theoretical science at least since the days of Lotka, Volterra, Gause, and Park. It is no less "hard science" than most areas of physics. Unlike physicists, however, most ecologists are not well trained in mathematics, and a substantial proportion still have little use for theory. These attitudes are changing.

TABLE 13.5 Number of density perturbation studies (*N*) that investigated bottom-up, top-down, and lateral regulatory forces, and the percentage of these studies documenting direct density dependence. *(After Harrison and Cappucino 1995)*

	BOTTOM-UP		TOP-DOWN		LATERAL		TOTAL	
	N	(%)	*N*	(%)	*N*	(%)	*N*	(%)
Invertebrates	25	88.0	10	30.0	9	77.7	44	72.7
Herbivorous insects	9	66.7	6	33.3	5	80	20	60
Fish, amphibians, reptiles	18	94.4	1	100	9	77.8	28	89.3
Birds, mammals	4	75	2	50	6	83.3	12	75
Total	47	89.4	13	38.5	24	79.2	84	78.6

desirable. Nevertheless, some interesting comparisons were obvious. Bottom-up forces were more likely to produce density dependence than top-down ones: 89 percent of studies looking for resource-based density dependence found it, compared with only 38.5 percent of those examining enemy-based effects.

Population regulation that is not strictly density dependent has been described as "density vague."

An alternative notion of population regulation has sprung from the work of Pacala, Hassell, and May (1990). Strong (1986) had suggested that even if a trend of density dependence was observed in nature, the actual scatter of the points around the line of best fit was often substantial—there was much variance in real biological data. He termed such relationships "density-vague" and argued that real relationships were more often "density-vague" than density dependent (Fig. 13.9). Such a relationship was often taken to mean that nothing interesting was going on. Pacala and colleagues suggested that this "disorder" was actually a source of order. Consider a population of parasitoids attack-

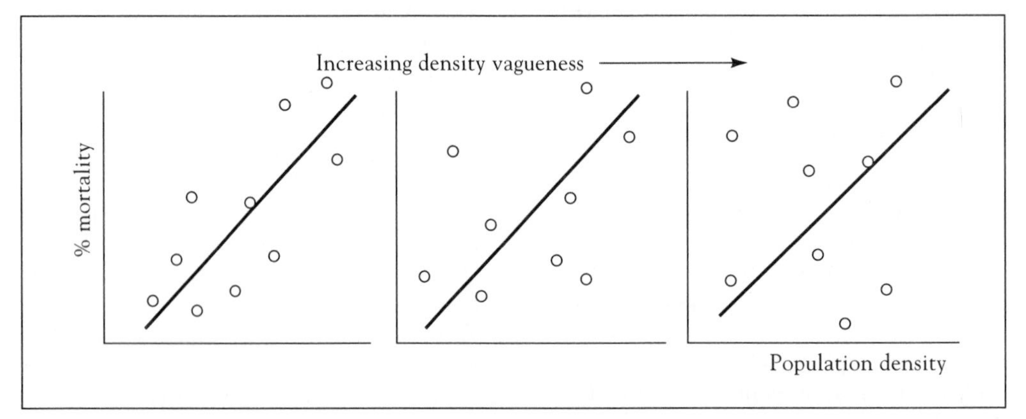

■ **FIGURE 13.9** Types of density dependent relationships likely to be found in real-world populations. An increasing scatter of points makes it difficult to determine if there is a relationship or what it is. *(After Strong 1984.)*

ing different populations of host insects. If rates of parasitism vary sufficiently from one patch of hosts to another, some patches of hosts can be counted on to escape attack and thereby sustain the host population into the next generation. Pacala and colleagues proposed what they called the "$CV^2 > 1$" rule. This rule states that, for a broad suite of simple models, if the square of the coefficient of variation (the scatter in the figures) among host patches is larger than one, there is enough variability to stabilize otherwise unstable host-parasitoid interactions. Data from thirty-four studies (twenty-one different interactions) showed that CV^2 was greater than one in about a third of the interactions.

Finally, there may be specific biological reasons why mortalities do not operate in a density dependent fashion. For example, a searching parasitoid may not always oviposit in a density dependent fashion, laying more eggs in dense concentrations of caterpillars; she may save some of her eggs to oviposit elsewhere, even in a seemingly suboptimal place. The reason may be that some local catastrophe could occur in the best area, wiping out all her progeny. For example, a flock of birds could also be attracted to a dense congregation of caterpillars and eat every one. To avoid losing all her eggs in such a catastrophe, the wasp oviposits in a few solitary caterpillars in out-of-the-way places. Ferguson and Stiling (1996) showed that parasitism of coastal aphids by wasps was not density dependent and that predatory ladybird beetles often devastated local aphid populations. This phenomenon has become known as **spreading the risk** (den Boer 1968, 1981).

Spreading the risk may provide a good explanation for the apparently random patterns that are so often observed in the field. Furthermore, even if they wanted to, animals may not always behave in a density dependent fashion in the face of a huge array of conflicting pressures. Their life history patterns may be adaptations to survival, not maximization of fitness. Searching parasitoid wasps may not oviposit in a density dependent manner because of the presence of a concentration of predatory spiders that preys on them. Many organisms and their enemies are sensitive to environmental change, and this can greatly affect correlations between predator and prey density (Levins and Schultz 1996). Sinclair (1989) noted that the appeal of the density dependent argument is that it predicts the existence of a process (density dependence) that can be measured in nature. The philosophical weakness of the argument is that it is very hard to reject; absence of density dependence can too easily be explained away on ad hoc methodological grounds. There is still no simple answer as to which factors, competition, predation, parasitism, mutualism, or abiotic factors, are most likely to affect population densities. As Strong (1988) has summed it up, no single factor along the gamut from plant chemistry to abiotic influences to natural enemies can be ruled out even for a minority of cases. A complex of influences participate in the coactions and regulations of herbs and insect herbivores, and probably other organisms too. Jim Brown (1997), writing in an essay on the Internet, is convinced that trying to make predictions about what factors will influence the future abundance of a plant or animal population is akin to trying to predict when and from what cause a human will die. For example, we know that age, smoking history, car accidents, and cholesterol can cause death but, because the human body is incredibly complex, we don't know how these things all interact. Because communities and indeed populations are at least as complex as the human body, it is difficult to predict future abundance of species and individuals.

Such questions are important to game managers as well as theoretical ecologists. The population of Thompson's gazelles in the Serengeti National Park, Tanzania, declined by almost two-thirds over a thirteen-year period, from 660,000 in the early 1970s to 250,000 in 1985. Predation, interspecific competition, and disease have all been implicated (Borner et al. 1987). Ecology, it might seem, is entering a period of pluralism in which simplistic, one-dimensional explanations of population phenomena are seldom sufficient.

Schoener (1986) has attempted to provide some generalities about sources of mortality by ranking communities along certain environmental and biotic axes. For example, he has listed six environmental factors: severity of physical stress, trophic position in food chain, resource input (from closed, a lake, to open), spatial fragmentation (fragmented to continuous), long-term climatic variation (high to low), and partitionability of resources (low, bare rock, to high; such as a complex leaf or prey of different sizes). To characterize organisms, Schoener has developed another set of axes: body size (small to large), recruitment, generation time (short to long), individual motility (sessile to mobile), and number of life stages (low to high, for example, holometabolous insects). He has proposed that, at some time in the future, it will be possible to describe communities in terms of such axes, with the result that, for "similia-communities" (his definition), say short-lived ephemeral marine algae, similar processes would shape the population densities. In this case, the algae are less affected by competition than by predation (herbivory).

13.4 A metapopulation is a group of separate populations that are united by the migration of individuals among them.

A metapopulation is a series of small, separate populations that mutually affect one another. In this scenario, even if one individual population goes extinct, other populations survive, and they supply dispersing individuals who recolonize the "extinct" patches. Even if the population that supplied the colonists itself becomes extinct, it will be recolonized again later. Metapopulations are viewed as sets of populations persisting in a balance between local extinction and colonization. The relevance of metapopulation theory to population biology is clear: populations could be maintained by a balance or colonization and extinction of small habitat patches. In this scenario, there is no mean or equilibrium level; extinctions happen at any time and density dependence is irrelevant. Persistence depends on factors affecting extinction and colonization rates such as interpatch distances, species dispersal abilities, and number of patches.

Harrison (1991) reviewed the empirical literature and found few situations that fit the classical metapopulation description. More common were three related situations (Fig. 13.10):

1. Core/satellite or source/sink or mainland/island metapopulations in which persistence depended on the existence of one or more extinction-resistant populations, usually large patches, which constantly supplied colonists to small peripheral patches that often went extinct.
2. Patchy populations in which dispersal between patches or populations was so high that colonists always "rescued" populations from extinction. The system was then effectively a single extinction-resistant population.

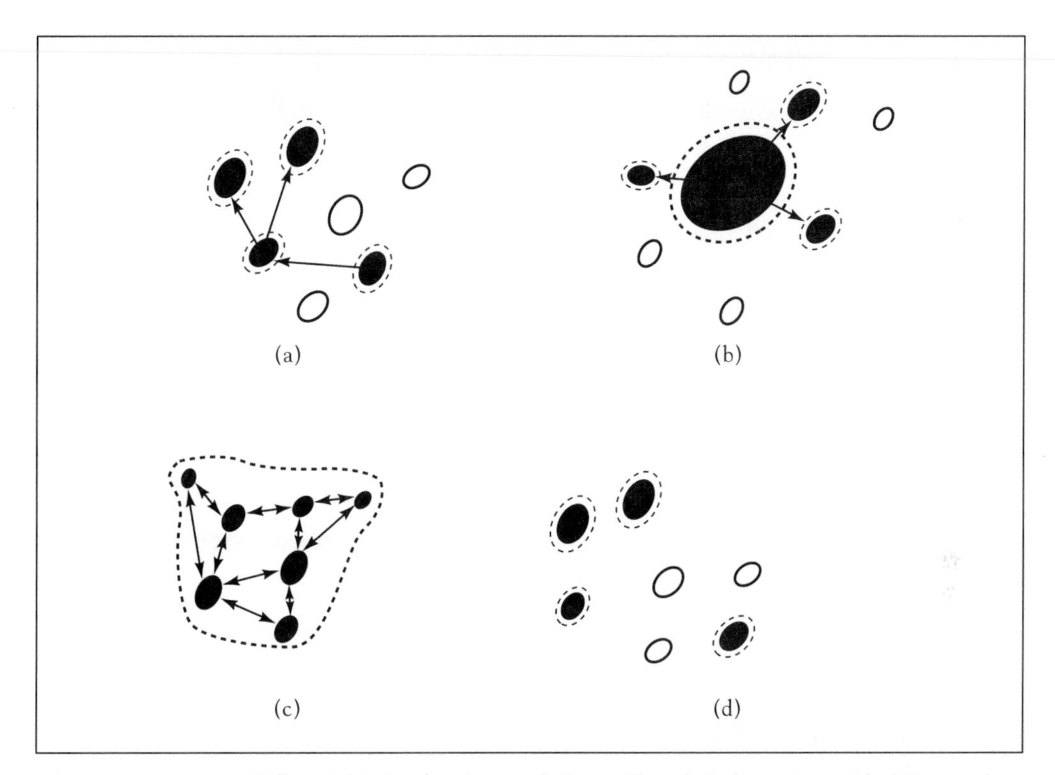

■ **FIGURE 13.10** Different kinds of metapopulations. Closed circles represent habitat patches; filled = occupied; unfilled = vacant. Dashed lines indicate the boundaries of "populations." Arrows indicate migration (colonization). (**a**) Classic metapopulation. (**b**) Core-satellite metapopulation (common). (**c**) Patchy population. (**d**) Nonequilibrium metapopulation (differs from *a* in that there is no recolonization) often happens as part of a general regional decline. (*After Harrison 1991.*)

3. Nonequilibrium metapopulations, in which local extinctions occurred in the course of species' overall regional decline. Many rare species conform to this scenario, with habitat fragmentation reducing population density. Lack of dispersal between populations effectively eliminates a true metapopulation scenario. There are many cases of regional declines of species abundance caused by human impacts on natural habitats (Wilcove, McLellan, and Dobson 1986; Rolstad, 1991) that may be halted by providing enough habitats to establish metapopulations. Conservation of species in fragmented habitats is an important area for the application of metapopulation models (Gilpin 1987; Hanski 1989, 1991). (Applied Ecology: Metapopulations and Endangered Species.)

Because of the lack of empirical support for classical metapopulation dynamics, there has been a call for a modified view of metapopulations in which local extinction is not a central feature (Hastings and Harrison 1994). It could be argued that it is unrealistic to expect real metapopulations to show enough subdivision so that populations can fluctuate and go extinct independently, yet enough connectedness that the metapopulation can persist. A broader, different view of metapopulations could presume neither independently functioning populations nor a balance between immigration and extinction.

APPLIED ECOLOGY

Metapopulations and Endangered Species

Many species survive for a long time even if there are no particular local populations that could be expected to persist for long time periods. In such cases, the maintenance of many habitat patches, allowing formation of a metapopulation, may be critical. This is especially true of butterflies living in successional habitat (Thomas 1995). In any one place, the right kind of habitat is present only for a limited period of time, and local extinctions are inevitable. However, species may survive regionally if they are able to establish new local populations elsewhere, where the right kind of habitat has appeared. An assemblage of the extinction-proof local populations constitutes a more robust metapopulation.

Many rare butterflies on nature reserves in Great Britain declined in the 1960s and 1970s probably because of small population sizes (Table 1). Ilkka Hanski and his colleagues (1995) have documented the existence of a metapopulation in the rare Glanville fritillary butterfly, *Melitaea cinxia*, in Finland. This butterfly has become extinct or very rare in many parts of Europe, and in Finland exists only on a 50 km × 70 m network of small islands. Here there are 1,502 suitable patches of habitat containing the host plants for the caterpillars, *Plantago* and *Veronica*, but only 536 of these were occupied. Smaller and more distant patches tended not to have such high occupancy rates, but butterflies were able to disperse to all islands, though most stayed on their natural patch. The conservation of this species, and many others, may be dependent on the preservation of a whole network of suitable habitat patches.

TABLE 1 Extinctions of rare butterflies on nature reserves in Britain between 1960 and 1982. (*Modified from Warren 1992*)

SPECIES	TOTAL NUMBER OF POPULATIONS IN BRITAIN	NUMBER OF POPULATIONS ON RESERVES	PERCENTAGE OF POPULATIONS ON RESERVES THAT WENT EXTINCT, 1960–81
Lycaena dispar	1	1	100%
Maculinea arion	4	4	100%
Carterocephalus palaemon	16	4	100%
Melitaea cinxia	18	7	0%
Melitaea athalia	29	7	29%
Satyrium pruni	30	11	0%
Hesperia comma	52	33[1]	27%
Lysandra bellargus	*c.* 75	19	21%

[1] *Note:* Nine populations on reserves went extinct, but four new reserves were colonized during the same time period.

In this way both core-satellite populations and patchy populations (numbers 1 and 2 in the preceding list) would fall under the umbrella of metapopulations.

Under these conditions, one of the most well-known examples of a metapopulation in nature was the metapopulation of the Bay checkerspot butterfly (*Euphydryas editha*) (Photo 13.2 and Fig. 13.11) (Harrison, Murphy and Ehrlich 1988). It consisted, in 1987, of a population on the order of 106 adult butterflies on a 2,000 ha habitat patch called Jasper

ECOLOGY IN PRACTICE

Susan Harrison, Univeristy of California, Davis

When I graduated from U. C. Davis with a bachelor's degree in 1983, I'd been interested enough in nature to major in zoology and to spend my junior year in Kenya taking ecology classes. Yet it never occurred to me that I could become a research biologist, and no professor or TA ever took me aside and said "Hey, you should consider graduate school." So I did what everyone else was doing—applied to medical school.

Then I got a summer job as Rick Karban's field and lab assistant, and Rick was great; within a week or two he was just about letting me design his experiments. We'd go out to do field work at Bodega Bay, and it was a series of revelations to me: collecting larvae and finding out what parasitoids were; using glue and Twist-Ties to imitate the way moths damage plants; the whole idea of asking nature questions. It stretched into a year of working for Rick, Jim Quinn (who was collaborating with Rick on a study of habitat fragmentation), and Rick Grosberg, all of whom were great mentors. I went to seminars, hung out with grad students, and started some research projects. So by the time I started grad school at Davis in 1984 I already had a sense of direction. I took the Organization for Tropical Studies (OTS) course in Costa Rica, did a side project at Barro Colorado Island, and finished a master's degree in 1986. My interests really formed at this time, and I've been trying to emulate Rick's experimental finesse and Jim's grasp of theory ever since.

It seems strange by present standards that in 1986, of the various places I was interested in going for a Ph.D., Paul Ehrlich's group at Stanford was the only one where conservation biology was respectable! Everywhere else, prospective advisors said that you should wait to get involved in conservation until you get your Ph.D (and maybe tenure). Paul said I could work on the conservation biology of an endangered butterfly with a fragmented habitat, so of course I went there, and he also involved me in other conservation-related projects. I finished in 1989 and did a postdoc with John Lawton in England, in 1990-1991, which was great for getting a European perspective on ecology.

My present faculty position at Davis came about through the "Targets of Opportunity for Diversity" program, which allows U.C. campuses to create extra positions for women and minorities who fill particular needs. This is supposed to lead to more role models for female and minority students, without taking away from other people's chances of getting jobs. Programs like this seem essential to making ecology less of a European-American dominated field.

In terms of significant obstacles, one that comes to mind is private land ownership. Metapopulation studies, and studies of habitat fragmentation in general, require that you survey all the patches of habitat in a large area—maybe ten by ten miles or more. This obviously can't be done within a single research reserve. I've had both wonderful and terrible experiences dealing with landowners and am terribly jealous of the Scandinavians, whose laws say that the only thing you can't do on someone's land is camp within sight of their house. This must be one reason so many excellent metapopulation studies come from Sweden and Finland.

As far as future directions, one very significant trend I see is that nearly all the top-notch applicants to graduate school in ecology are seriously interested in conservation biology and other applied areas. Thus, I expect an increasing amount of high-quality work in these fields. Another trend is that conservation biology itself is maturing, moving away from promoting simplistic rules and principles, toward a more scientific approach that includes experiments, modeling, and considering alternative hypotheses.

PHOTO 13.2 The Edith Checkerspot butterfly, *Euphydryas editha*, was much studied for its importance in understanding metapopulation theory. *(LeRoy Simon, Visual Unlimited AINS-3270.)*

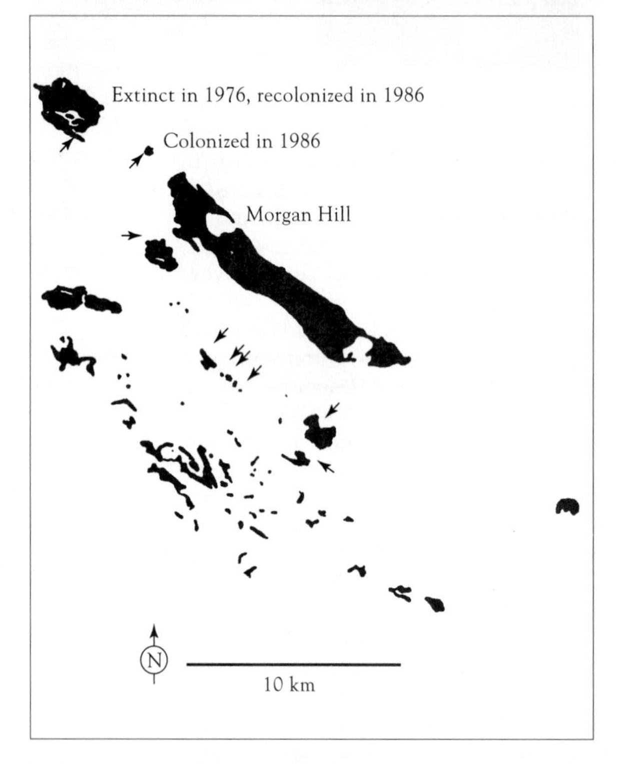

FIGURE 13.11 Metapopulation of the Bay checkerspot butterfly, *Euphydryas editha* (from Harrison et al. 1988). The black areas represent patches of the butterfly's serpentine grassland habitat. The 2000-ha patch labeled "Morgan Hill" supported a population in the order of 10^6 adult butterflies in 1987 and acted as the source population. The nine smaller patches labeled with arrows supported populations in the order of $10^1 - 10^2$ butterflies in that year. Eighteen other small patches were found to be suitable but unoccupied. *(Redrawn from Harrison 1991.)*

Ridge near Stanford University and nine populations of 10 to 350 adult butterflies on patches of 1 to 250 ha. Of twenty-seven small habitat patches in the region found to be suitable, only those closest to the large patch were occupied. This pattern of patch occupancy could not be explained by differences in habitat quality. Instead, the distance effect appeared to indicate that the butterfly's capacity for dispersal is limited and that the large population acts as the dominant source of colonists to the small patches. From this and other evidence, it appeared that persistence in this metapopulation is relatively unaffected by population turnover on the small patches, which act as sink populations. In 1996, the population disappeared from Jasper Ridge, having been studied there since at least 1934. Whether it will ever return is uncertain.

What of other systems? Data are just now being accumulated. Bengtsson (1993) showed how *Daphnia* may exclude one another from experimental pools but coexist in a complex of rockpools. Nieminen (1996) showed a metapopulation scenario for moths on small islands off the Finnish coast and Hermann and Ankerims (1996) invoked a metapopulation model to explain the presence of bush crickets on habitat patches in Norway. There has also been much theoretical interest in the possibility that the interaction between competitors, predators, and prey or hosts and parasitoids may promote a metapopulation scenario (Reeve 1988).

As long ago as the 1950s, Carl Huffaker was interested in the interaction of predators and prey over a fragmented landscape (Huffaker 1958; Huffaker, Shea, and Herman 1963). Huffaker studied a laboratory system of a phytophagous mite, *Eotetranychus sexmaculatus*, as prey and a predatory mite, *Typhlodromus occidentalis*, as predator. The prey mite infests oranges, so Huffaker used these for his experiments. When the predator was introduced onto a single prey-infested orange, it completely eliminated the prey and died of starvation. Huffaker gradually introduced more and more spatial heterogeneity into his experiments. He placed forty oranges on rectangular trays like egg cartons, but still the system eventually resulted in the extermination of the populations. Finally, Huffaker used a 252-orange universe. He partially isolated each one by placing a complex arrangement of vaseline barriers in the tray, which the mites could not cross. But he facilitated the dispersal of *Eotetrancyhus* by inserting a number of upright sticks from which they could launch themselves on silken strands carried by air currents. The prey were thus able to keep one step (or one orange) ahead of the predators. Overall, at any one time, there was a mosaic of unoccupied patches, prey-predator patches heading for extinction and thriving prey patches; and this mosaic was capable of maintaining persistent populations of both predators and prey (Fig. 13.12).

More recently, similar experiments were done by Holyoak and Lawler (1996) using a protist predator-prey pair, the bacterivorous ciliate *Colpidium*, and the predaceous ciliate *Didinium* in a microcosm of linked bottles. They again showed how critical dispersal rate was to the maintenance of stable systems. Dispersal rate was changed by providing single containers where dispersal between all areas was high and more complex linked containers where dispersal was slowed by linking tubes between containers. At the lowest dispersal rates, long-term persistence of the pair was impossible because recolonization did not balance local extinctions. At the highest dispersal rates, the entire system behaved as a single large population, in which the pair could not exist. The pair persisted as predator and prey metapopulations between these extremes.

13.5 Ecologists have developed conceptual models to describe the mortality factors that should be most important in different systems.

Many different theoretical models have been proposed to describe which types of mortality factors should be most important in which systems. Some stress the importance of "bottom-up" factors, like food, some "top-down factors," like natural enemies, and some theories incorporate components of both these ideas.

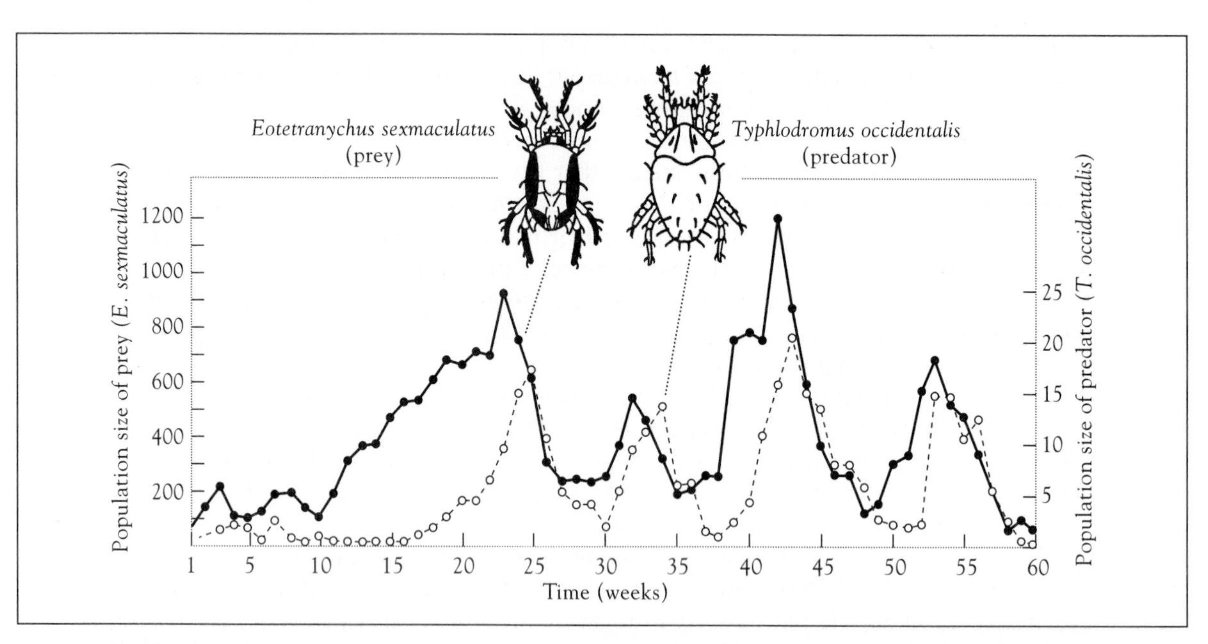

■ **FIGURE 13.12** Predator-prey interaction between predator and prey mites in a complex laboratory environment with a 252-orange system with sticks present to promote dispersal. (*After Huffaker et al. 1963.*)

Population regulation can either occur from the top-down or the bottom-up.

Bottom-up or trophodynamics.

Among the earliest contenders for a general theory to explain population processes was Lindeman's (1942) trophic-level concept, or **trophodynamics**. This theory explained the height of the trophic pyramid by reference to a progressive attenuation of energy passing up trophic levels from plants through herbivores, primary carnivores and secondary carnivores. Since about 10 percent of energy was transferred between levels, there was precious little left for trophic levels higher than about four. This was a true "bottom-up" theory based on thermodynamic properties of energy transfer, with little use for top-down control by natural enemies. This 1942 article was the last of a series of articles by Lindeman on the ecology of a senescent lake, Cedar Creek Bog in Minnesota, which was the subject of his dissertation. Sadly, Lindeman never saw the 1942 article in print because he died that year after a long illness; he was twenty-seven years old. The paper had been held up for publication because two eminent limnologists, Chancey Juday and Paul Welch, recommended that it not be published because it was too theoretical and went beyond the data available at the time. However, the editor of the journal *Ecology*, Thomas Park, decided to publish it anyway, despite the negative reviews.

Top-down effects.

Since Lindeman's work, many different models have been proposed to incorporate the effect of natural enemies in communities. Among the first was Hairston, Smith, and Slobodkin's (1960) idea (often called HSS) that since the earth appears "green" herbivores therefore have little impact on plant abundance. They suggested that this is because herbivores are ordinarily limited by their predators, not their food supply. The implication was that plants, being abundant, endure severe competition for resources, but herbivores, suffering high rates of mortality from natural enemies, do not compete. Natural enemies themselves, being limited only by the availability of their prey, also compete. These authors later (Slodbodkin et al. 1967) reformulated their arguments to include only consumers of producing tissue (i.e., not granivores, nectarivores, or frugivores), only nonintroduced species and only terrestrial systems. Perhaps because of these restrictive caveats it is the more general HSS hypothesis that has become entrenched in the literature.

Oksanen, Fretwell, and colleagues (1981) proposed that the strength of various types of mortalities varies with the type of system involved—particularly as a function of primary productivity. They termed this idea the exploitation ecosystem hypothesis or EEH. Thus, for very simple systems with low primary productivity, like Arctic tundra, productivity is so low that few plants exist. As productivity increases a little plants become resource limited (competition). As productivity increases still more, some herbivores can be supported, but there are too few herbivores to support carnivores. In the absence of carnivores, levels of herbivory can be quite high. Plant abundance becomes limited by herbivory, not competition. The abundance of herbivores, in the absence of carnivores, is limited only by competition for limiting plant resources. As primary productivity increases still more, carnivores can be supported, and there are three trophic levels—this is the HSS scenario. Finally, as productivity increases still further, secondary carnivores might be supported, and these in turn would depress the numbers of carnivores, which in turn would increase levels of herbivory and lessen competition between plants. Sound confusing? The scheme is outlined in Fig. 13.13 and Table 13.6. The importance of different mortalities in this scheme is linked to the number of links in the food chain or, more precisely, to primary productivity.

Support for the EEH hypothesis comes from freshwater systems (Wooton and Power 1993), low productivity terrestrial systems like tundra communities (Oksanen et al. 1995), and studies on the Bahama Keys (Schoener 1989). In the latter case, strengths of mortalities at different trophic levels were found to differ according to the number of trophic levels present. On some islands with few higher predators (lizards and spiders), levels of herbivory were greater than on islands where these organisms were present.

Perhaps the biggest support for EEH has come from freshwater studies in lake and rivers where, Strong (1992) argued, the communities are simple and species poor relative to terrestrial communities. In these systems, carnivores greatly reduce their prey populations, and this has a dramatic effect on the producer organisms one or even two links down the chain (Carpenter and Kitchell 1993; Vanni et al. 1990). Such effects are known as trophic cascades. Why is this? Winemiller (1990) and Power, Parker, and Wooton (1995) suggested that river food webs are as speciose as food webs in any other ecosystem, but Polis and Strong (1996) noted that a critical element is the exceptionally high edibility, nutritiousness, and vulnerability to herbivores of algae that form the base of the food chains. Such algae lack the sophisticated chemical and even physical defenses of higher plants on land. As a consequence, the link between primary productivity and secondary productivity is very tight. Also, detritivores, common in most systems, are relatively unimportant in these aquatic systems. Hairston and Hairston (1997), however,

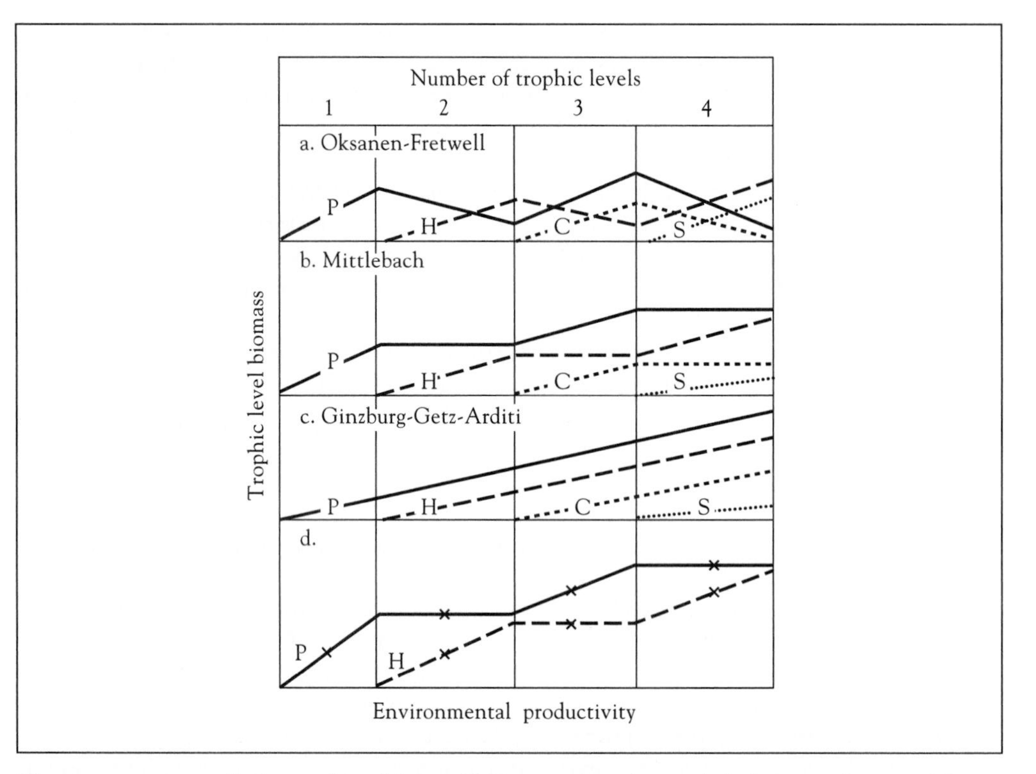

■ **FIGURE 13.13** Patterns of trophic level biomass accrual expected along environmental productivity gradients under pure top-down models (Oksanen-Fretwell and Mittlebach) (**a**, **b**) and joint control by predators and resources (Ginzburg-Getz-Arditi models) (**c**). P represents primary producers, H herbivores, C primary consumers that eat herbivores, and S secondary carnivores that eat primary carnivores. In (**d**), crosses represent positions along the gradient of hypothetical biomass samples taken for producers and herbivores. The positive covariance of consumers and resources predicted by the third model could be mistakenly inferred from a pattern that in reality was stepped if regions where population plateaus are undersampled, and if transitions between n and $n + 1$ trophic levels are undetected. (*After Powers 1992.*)

TABLE 13.6 Effects of number of trophic levels on major mortality factors in natural systems.

TAXA	PLANTS ONLY	PLANTS AND HERBIVORES	PLANTS, HERBIVORES, AND CARNIVORES	PLANTS, HERBIVORES, CARNIVORES, AND SECONDARY CARNIVORES
Plants	Competition	Herbivory	Competition	Herbivory
Herbivores		Competition	Predation	Competition
Carnivores			Competition	Predation
Secondary carnivores				Competition

took exception to Polis and Strong's critique of EEH theory, and they argued that many algae possess demonstrably effective grazer-inhibiting compounds while many terrestrial plants are quite edible, as evidenced by the many herbivores they support. Since then Moran, Rooney, and Hurd (1996) provided the first example of a top-down cascade by a generalist arthropod predator on a nonagricultural ecosystem. They added predatory preying mantids to old fields in Delaware. The most common other predators, spiders, migrated from the areas of mantid addition, but mantid predation alone was still sufficient to decrease the biomass of herbivorous insects. This, in turn, increased the biomass of plants.

There has been other criticism of Oksanen, Fretwell, and colleagues' ideas. Polis and Strong (1996) pointed out numerous inconsistencies with available data, such as the following:

1. Consumer species rarely aggregate into discrete homogeneous trophic levels. Omnivory and ontogenetic diet shifts are common and blur trophic levels (see also Chap. 20).
2. Most energy fixed by plants passes into the detrital food chain, not the traditional plant-grazer-carnivore chain envisioned in HSS or EEH. Detrital energy reenters the food web when detritivores are eaten by predators that also eat herbivores (Porter 1995; Vanni 1995).
3. All links must be "recipient controlled," that is, consumers substantially depress populations of "donors" (resources) for EEH to work. If a link is donor controlled, i.e., the number of prey limits the number of herbivores, predation effects will not cascade down the food web. Many systems appear to be donor controlled (Hawkins 1992).
4. Diseases and parasitism are very common (Burdon and Leather 1990); yet virtually no biomass of disease-causing organisms results even after heavy mortality (Loye and Zuk 1991). Furthermore, neither pathogens nor parasites are known to be controlled by their own natural enemies (Anderson and May 1982; Burdon 1987).

Middle ground.

The HSS and EEH models can both be categorized under the banner "top-down effects" since the effects of predators are important at least some of the time. This contrasts with the "bottom-up" scenario where energy is argued to control plant numbers, which in turn control herbivore densities, which in turn control carnivore numbers and so on. Hunter and Price (1992) argued that "bottom-up" forces must logically be the most important. The removal of higher trophic levels leaves lower trophic levels present—although greatly modified. However, the removal of primary producers leaves no system at all!

Needless to say, there is a middle ground. Some authors argue that both bottom-up and top-down effects are consistently important. In their nutrient/productivity/omnivory (NPO) model, Menge, Daley, and Wheeler (1995) argued that although influences on webs vary with productivity, competition and predation do not alternate in importance at longer food chain lengths as envisioned by EEH. Rather, at low food chain lengths food supply or nutrients are always most important. As food chain length

ECOLOGY IN PRACTICE

Mark Hunter, University of Georgia

Most ecologists that I talk to describe one part of the job that they really love: the thing that helps them get up in the morning. It might be field work, it might be writing papers, it might be designing experiments or teaching students. For me, it's the analysis of data. I love playing with numbers and searching for the patterns that those numbers (hopefully) reveal. It's the best mix of science and detective work that I can imagine. The data might be from an experiment we've done in the field or the laboratory, or they might be long-term data collected and passed on by a kind-hearted colleague. Either way, the best part of my work day is spent with numbers.

Growing up in the 1960s and 1970s in Scotland, there was still some gender bias in the subjects taken by high school students: boys did physics and math, and girls did biology. Thankfully that's changed, but it never really occurred to me to be a biologist until it was about time to go to university. I'd planned to study physics, but then a friend of mine lent me *King Solomon's Ring* by Konrad Lorenz. The enthusiasm for the study of animal behavior in that book was simply infectious, and I changed my mind almost overnight. I ended up at the University of Oxford to study zoology. One of the lecturers I met at Oxford was Martin Speight, who taught courses in the ecology of insects. He was an inspiring teacher and later became a good friend, and he's probably responsible for developing my interests in both ecology and entomology.

I stayed at Oxford for my doctoral studies, working with Dick Southwood and Willy Wint on the ecology of oak-feeding insects. I began to realize that my early interest in numbers and recent interest in ecology were naturally combined in the field of population ecology. Really, I've been a population ecologist ever since. I'm fascinated by the causes and consequences of population change in ecological systems. I want to understand the mechanisms that underlie the distribution and abundance of organisms in space and time. Insects are great organisms of study for population ecologists because they're relatively simple to manipulate experimentally: it's much easier to fit a hundred insects in a growth chamber that a hundred caribou! So I've continued to work with insects that feed on plants, hoping that some of the results of the experimental work can be applied to many more kinds of organism.

Right now, my main interest is in the relative roles of top-down, bottom-up, and lateral forces on the population ecology of herbivores. Top-down forces are the effects of natural enemies, bottom-up forces are variation in the quality and quantity of food, and lateral forces are interactions between competitors at the same trophic level. I think that all of these forces can be important to herbivores; they interact with one another and are modified by the environment in which the organisms live. It's fascinating to me to try to understand how the relative importance of these forces varies in space, varies over time, and varies from species to species. It's not an easy process and seems to require a mixture of field experiments, lab experiments, and long-term sampling data. Luckily, there are quite a few other ecologists who share this interest and are carrying out studies in a variety of terrestrial and aquatic habitats. I'm optimistic that within the next few years we'll begin to be able to make some generalizations about these processes in a variety of environments. We are slowly learning to deal with the complexities of this kind of research, such as interactions among ecological forces, variation in the timescales on which they operate, and lagged effects on the dynamics of herbivore populations. But we'll get there.

A few years ago, I moved to the Institute of Ecology at the University of Georgia, and Dac Crossley (now retired) has helped me to understand the importance of the ecosystem in providing a template for population-level studies. Dac has nudged me toward exploring the links between populations and ecosystems. In fact, the role of species in ecosystems may be a focus of ecological research over the next few years. Many ecologists are interested in the effects of biodiversity on ecosystem function for both theoretical and practical reasons. For example, some herbivore species appear to influence the rates of nutrient cycling and energy flow in systems. We need to understand the extent to which changes, natural and human induced, in the populations of these herbivore species will change the environment. I hope that collaboration between population ecologists and ecosystem ecologists will help to answer such questions.

increases, competition for such nutrients becomes important, and finally predation kicks in and the strength of this control increases with higher productivity. This is similar to the Ginzburg-Getz-Arditi model of Fig. 13.13.

Interestingly, Menge, Daley, and Wheeler (1995) noted that some components of the NPO hypothesis were consistent with an earlier hypothesis, known as the MS hypothesis (Menge and Sutherland 1987), which postulated that community structure was governed by environmental stress. The MS hypothesis envisioned that in stressful habitats, higher trophic levels have little effect because they are rare or absent, and plants are affected mainly by environmental stress. In habitats of moderate stress there is a little herbivory, but plant densities are affected by competition. In more benign environments there are many herbivores, and herbivory controls plant abundance, not competition or environmental stress. As a good example of the MS hypothesis in action, consider salt marshes. These are among the most productive ecosystems in the world, equivalent in productivity to tropical forests or coral reefs, but they are stressful, being inundated constantly by the tide. Here levels of herbivory are low, as predicted by the MS hypothesis. How can we tell the difference between the MS hypothesis and the NPO hypothesis? Menge, Daley, and Wheeler (1995) noted that model predictions were similar but that in the NPO hypothesis web structure varies with productivity, whereas in the MS hypothesis it does not.

What empirical support is there for the HSS, EEH, NPO, or MS models? The NPO model is so new as to be relatively untested. However, Menge, Daley, and Wheeler (1995) suggested that the prediction that predation increases in importance in marine benthic communities with increased productivity may apply broadly. In a comparison of the MS and HSS hypotheses, Sih et al. (1985) examined twenty-five papers on terrestrial plants and herbivores and found only two gave support to HSS. Perhaps less surprisingly, Menge and Farrell (1989) reviewed experimental studies in marine ecosystems and found that the MS model was well supported. It is interesting that Hairston (1991) has raised objections to the validity of many of the studies reviewed by Sih et al. (1985), and Sih (1991) has replied to these criticisms. It is educational to read first hand the arguments of both authors.

13.6 Ecologists have increasingly recognized that population dynamics may be strongly influenced by indirect interactions with other species.

It is becoming increasingly apparent that to study mortality factors like herbivory or predation in isolation is to study them out of context. For example, herbivory may only be important on plants in the absence of natural enemies (Price et al. 1980; Stiling and Rossi 1997a). In many cases, there are complex interactions between natural enemies, herbivores, and plants. In order for us to fully evaluate top-down and bottom-up models, we need to know how commonly species from one trophic level can affect species more than one level away. When one species affects another in this manner it does so indirectly. How common and potent a force are indirect effects compared to direct effects?

For HSS or EEH to work, indirect effects in communities must be frequent and strong. Indirect effects embrace a wide variety of phenomena, many of which are similar but bear different labels: apparent competition, facilitation, some mutualisms, cascading effects, tri-trophic level interactions, higher-order interactions, and nonadditive effects. Indirect effects are often defined as "how one species alters the effect that another species has on a third." Thus, three is often considered the minimum number of species required for indirect effects. However, species interactions can be mediated by a nonliving resource just as well as through a living one, so that in some instances indirect effects can be detected with just two species. Alteration of water quality in temporary ponds by early colonizing frogs may affect the subsequent success of other species after the first species is no longer present (Alford and Wilbur 1985). Some authors prefer to call such phenomena "priority effects" or "historical effects" rather than indirect effects.

If we accept that there are five major classes of interaction (competition, predation, herbivory, parasitism, and mutualism) then there are at least $5 \times 4/2 = 10$ classes of indirect interaction. I say at least because this does not include "within mechanism" effects, i.e., competition *x* competition. For example, the "competitor of my competitor is my friend." Examples of these follow:

1. **Competition *x* Predation.**
   Predators may reduce the abundance of competitors below the level where competition is important (Paine 1966). Alternatively, the risk of predation in some habitats compels prey species to feed and compete more often in the same habitat (true of coral reef fish, which find refuge in the coral from predators).
2. **Competition *x* Herbivory.**
   i. Resource competition. If two herbivorous animals compete for a common resource, the plant, then host-plant herbivory might be less than it would be if the two species did not compete. Resource competition is a common form of competition (Schoener 1985). Although a primary focus of ecology for years, resource competition was not appreciated as an indirect effect until relatively recently (Harper 1977; Strauss 1991).
   ii. Darwin's own classic work showed that grazing animals can increase plant species diversity by reducing dominant competitors. Under this scenario one grass would never be common enough to outcompete

the others. However, some herbivores don't always eat the competitive dominants, and their effects on plant communities are much more complex.

3. **Competition *x* Parasitism.**
   Individuals weakened by parasites may lose in competitive interactions (Park's [1954] work on *Tribolium* beetles and the parasite *Adelina* and Schmitz and Nudds (1994) work on white-tailed deer, caribou, and meningeal worm parasites Chapter 12). Also, two species might mistakenly be construed as competitors when their negative interactions are actually the consequence of an (unconsidered) shared predator or parasite. This is what Holt (1977) termed *apparent competition*. In the absence of the predator/parasite, no competition would be detected.
4. **Competition *x* Mutualism.**
   It may be possible that competition between two plants may be altered because of the presence of mycorrhizae, a mutualistic association of root and fungi.
5. **Predation *x* Herbivory.**
   Tree growth on Isle Royale in Michigan increased when wolf predation on herbivores reduced herbivory (McLaren and Peterson 1994). Also, insectivorous birds increased the growth of white oak trees in Missouri through the consumption of leaf-chewing insects (Marquis and Whelan 1994). This is often called a tri-trophic effect because it involves three trophic levels, the plant, its herbivore, and the herbivore's natural enemy (Price et al. 1980; Marquis and Whelan 1996).
6. **Predation *x* Parasitism.**
   i. Parasitized or sick individuals are often more subject to predation.
   ii. Red wax scale insects, *Ceroplastes rubens*, are parasitized less in the presence of honeydew-foraging ants, *Lasius niger*, which deter the scale's natural enemies (Itioka and Lowe 1996).
7. **Predation *x* Mutualism.**
   Pheidole megacephala ants on the plant *Pluchea indica* in Hawaii tend scale insects, *Coccus viridis* (Bach 1991). They remove predatory insects, like coccinellid larvae, which feed on the scales. Predation on the scale insects is then reduced because of their mutualism with the ants. In this case, the plant suffers from higher rates of herbivory, so we could also call this a tri-trophic effect.
8. **Herbivory *x* Parasitism.**
   i. Many plants suffer attack by gall insects, wasps, or flies who initiate tumorlike growths on plants and whose larvae feed inside. Galls of *Asphondylia borrichae* on coastal *Borrichia* plants are less frequent where parasitism levels of galls by hymneoptrous parasites are high (Stiling and Rossi 1997a). This is potentially important in agriculture where crop varieties may be bred to encourage or facilitate attack of insect pests by natural enemies such as parasites or predators (Price et al. 1980, Barbosa and Saunders 1983). Sometimes host defenses against insect pests can actually backfire and result in reduced rates of attack of

the pests by natural enemies. For example, the number of parasitic *Cotesia* wasps surviving to adulthood on tobacco horwarm caterpillars, *manduca sexta*, was greatest where the host fed on a low-nicotine diet not a high-nicotine diet (Thorpe and Barbosa 1986).

ii. Outbreaks of disease on herbivores may lead to reductions in grazing pressure and increases in plant abundance (Dobson and Crawley 1994).

iii. Herbivory on a plant parasite, *Casteilleja wightii*, by an aphid was lower where the plant parasite attacked lower quality plant hosts (Marvier 1996).

9. **Herbivory *x* Mutualism.**
   The mutualism of acacia plants and ants occurs because of herbivory. Ants reduce levels of herbivores (see Chap. 8 for a fuller discussion).
10. **Parasitism *x* Mutualism.**
    The mutualism of cleaner fishes and the fishes they clean occurs because of parasitism. The cleaner fish actually rid the customer fish of parasites. Menge (1995) has provided a different classification of nine major "types " of indirect effects (Fig. 13.14, Table 13.7) and suggests that several of them, such as keystone predation and apparent competition are common, at least in intertidal situations.

The relative importance of direct and indirect effects on population dynamics have not been widely compared.

Theory exists (Levine 1976; Vandermeer 1980) that shows how indirect effects can outweigh direct ones. However, Schoener (1993) noted that it is tempting to argue that the more indirect an effects pathway is, the weaker it should be on average. If an indirect effect is unlikely, an indirect effect involving a direct effect and perforce others is more unlikely still. Consider, for example, a predator that eats three kinds of prey. If one prey is perturbed, this will immediately affect, on average, one-third of its diet. A predator of that predator, that also eats three kinds of prey, will have one-third of one-third of its diet reduced so the initial perturbation is reduced, very roughly, to one-ninth.

The indirect effects of one species on others may be detected by removing the focus species for several generations.

There are very few empirical studies that attempt to compare the frequency and strengths of direct and indirect effects.

Bender, Case, and Gilpin (1984) proposed experiments designed to distinguish direct effects from indirect ones. In "pulse" experiments, a species is removed only once, and any change in population density of the remaining species reflects the direct effects of the removed species. In "press" experiments, removals are maintained over several generations of the target species. Any resultant effects are argued to be the result of both direct and indirect effects. This assumes that indirect effects take more time to appear. The differences between the two types of treatment reflect the extent of indirect effects. However, Menge (1997) studied the effect of experiment duration in a survey of marine intertidal interactions and found, contrary to expectation, that indirect effects appeared

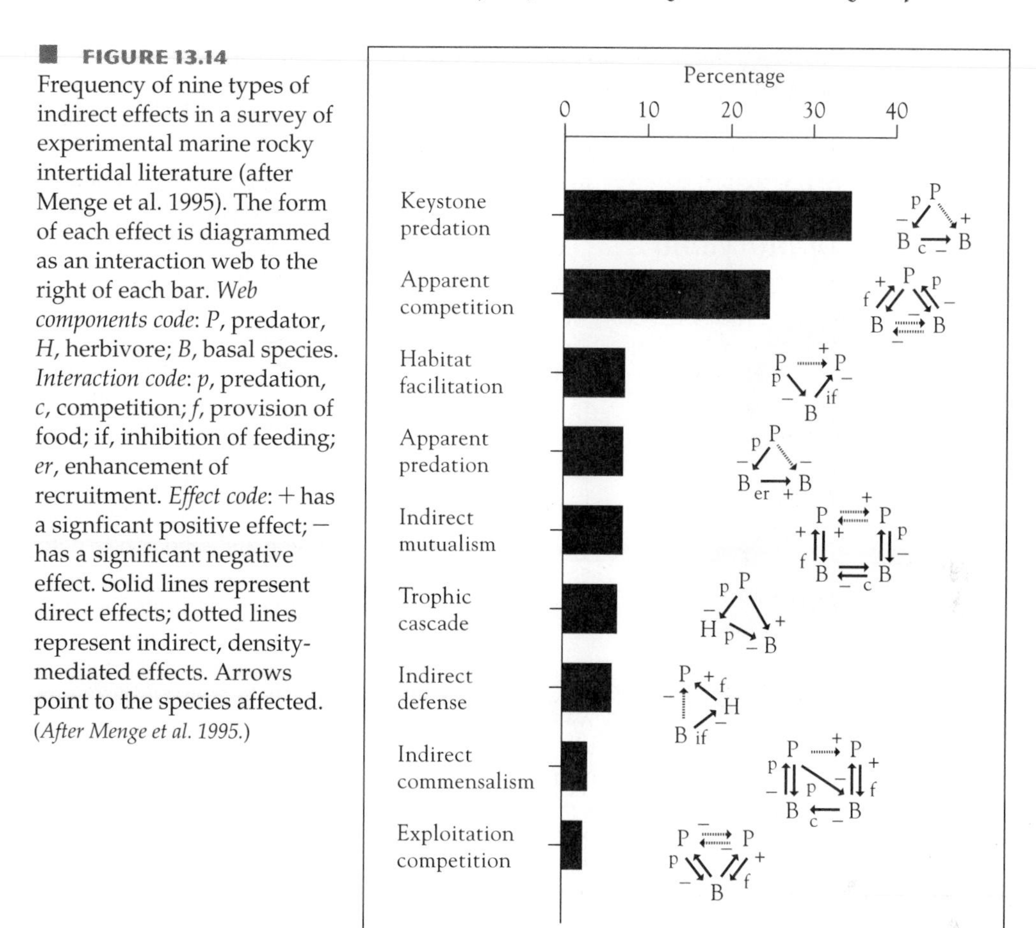

FIGURE 13.14 Frequency of nine types of indirect effects in a survey of experimental marine rocky intertidal literature (after Menge et al. 1995). The form of each effect is diagrammed as an interaction web to the right of each bar. *Web components code*: *P*, predator, *H*, herbivore; *B*, basal species. *Interaction code*: *p*, predation, *c*, competition; *f*, provision of food; if, inhibition of feeding; *er*, enhancement of recruitment. *Effect code*: + has a signficant positive effect; − has a significant negative effect. Solid lines represent direct effects; dotted lines represent indirect, density-mediated effects. Arrows point to the species affected. (*After Menge et al. 1995.*)

either simultaneously with direct effects or shortly after indirect effects were evident, casting doubt on the pulse-versus-press dichotomy.

Paine (1992) made one of the first comparisons of direct and indirect effects by painstakingly removing individual species in a marine intertidal food web in the Pacific Northwest. He found very little evidence for strong indirect effects and more evidence for strong direct effects of predation. On the other hand, Menge (1995) compared direct and indirect effects in twenty-three rocky intertidal habitats and found that indirect effects accounted for about 40 percent of the change in community structure resulting from manipulations, with a range of 24 to 64 percent. Indirect effects increased in frequency as food web diversity increased. However, the strength of indirect effects remained unchanged by web species richness. Menge (1995) compared the magnitudes of direct and indirect effects from his twenty-three food webs (Fig. 13.15) and found change in density due to all direct effects was greater than the mean change due to all indirect effects in eight of ten cases. Direct effects seemed particularly important in communities containing high species numbers. Of all indirect effects, keystone predation was the most common (35%), followed by apparent competition (25%), with all the rest

ECOLOGY IN PRACTICE

Pedro Barbosa, University of Maryland

Although I was born in Puerto Rico, my family moved to New York City when I was three. I was raised in Spanish Harlem and I attended and graduated from City College of New York. I went on to receive my M.S. and Ph. D. from the University of Massachusetts, becoming the first Puerto Rican to be granted a Ph. D. from that institution. Although I was not a very good student as an undergraduate, my newfound love of science motivated me to conduct enough research as a graduate student to publish about thirteen research publications by the time I was awarded a Ph. D. in 1971, in addition to the three publications that resulted from my dissertation research. Because of background and early experiences, I have always had an interest in increasing and encouraging the participation of minorities in science. Even though minorities are underrepresented in science, I have tried to let students know that as an entomologist and ecologist my life has been filled with years of the joy of exploration and discovery, as well as just plain fun. It is my hope that my career as an insect ecologist serves to highlight that, although historically minorities have not selected ecology as a career, the field is nevertheless open to students of color and offers exciting challenges and opportunities.

As an insect ecologist I not only have been fortunate enough to teach and do research in insect ecology but have visited and worked in a number of countries throughout the world, including Venezuela, Guatemala, Mexico, the People's Republic of China, New Guinea, Puerto Rico, and Jamaica. Having been raised in a large city, I know that for many students in urban areas, whether or not they are minorities, it is often difficult to imagine becoming an ecologist. Living in a city provides few opportunities to experience the variety of habitats in which ecologists work, and thus it's difficult to imagine exploring the multitude of strange and fascinating interactions that occur among plants, animals, and their environment. This impression may be reinforced when students are exposed, on television and school videos, to ecological research and explorations in strange and exotic locales. I fear that too many students conclude that doing the type of research they see on television is not relevant to their world. Unfortunately, students sometimes are led to believe that ecologists only work in exotic locations, even though most ecologists work in much less glamorous but equally exciting and interesting locations, which often are only a few hundred miles from the city. Having been raised in the city, I know that exposure to green habitats is not a major requirement for having a successful and enjoyable career in ecology.

For the past twenty-six years my research has reflected a blend of interests in theoretical and applied aspects of the ecology of parasitoid-host interactions and three trophic-level interactions. The latter are interactions involving natural enemies (parasitoids, predators, pathogens), plants, and herbivores. During the time of my graduate training, leading ecologists and entomologists debated whether insects were significant regulators of plant population abundance and distribution. The key to that debate was the issue of whether insects were food limited given the apparent abundance and availability of plant biomass. Because of the overwhelming abundance of apparently suitable plant biomass relative to insect biomass, the possibility that insects could regulate plants appeared unlikely. My own career has been strongly influenced by this debate as well as by subsequent attempts to resolve this debate. With the support provided by several grants from the National Science Foundation and the U.S. Department of Agriculture Competitive Grants Program and the able assistance of my students and colleagues, my research has helped to provide some of the data that has demonstrated the multifaceted and essential role of plants in insect herbivore ecology.

One of the most exciting and rewarding aspects of my career has been the opportunity to participate in the development of the discipline of ecology by conducting research that incorporates entomological perspectives and insights. Working in applied systems (i.e., with species of economic importance) I have been able to generate data that may help mitigate the negative impact of certain insect species, while at the same time help to provide evolutionary and ecological explanations that are harmonious with the patterns observed in unmanaged ("natural") environments. I have tried to serve as a bridge between two perspectives (or groups) that have often failed to communicate: basic and applied researchers. This schism was particularly apparent in the beginning of my career when researchers in these groups often attended different meetings, published in different journals, and were professionally organized in relatively self-contained and separate societies. I have made a concerted effort in my career to ignore, if not remove, these barriers by publishing in journals such as *Ecology*, *American Naturalist*, and *Science* as well as in journal such as the *Journal of Economic Entomology*, *Entomophaga*, and the *Journal of Entomological Science*. Similarly, I have purposefully given presentations and participated in symposia at the meetings of diverse professional societies as well as to biology, zoology, ecology, and entomology departments. I believe these interactions have been mutually advantageous. I am happy to see that today the unproductive division of the past is significantly less common.

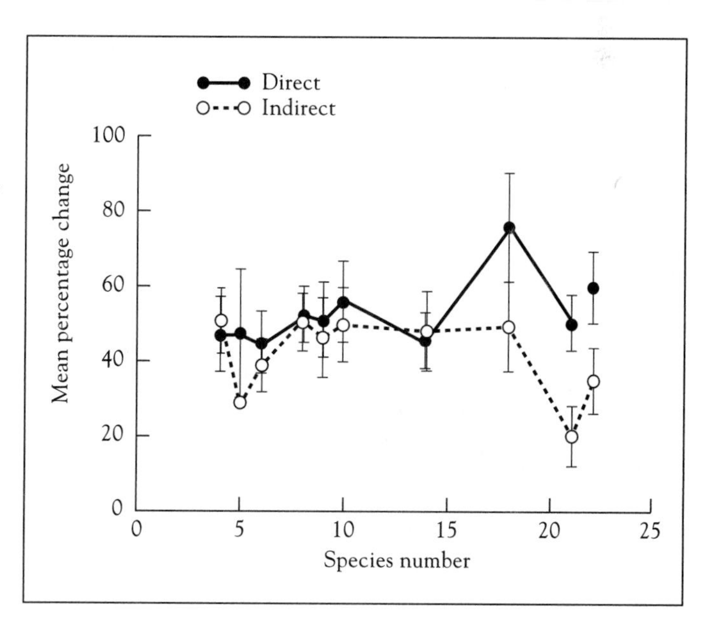

FIGURE 13.15 Mean percentage of total change (*absolute value*) caused by direct and indirect effects in experimental manipulations surveyed in Menge (*1995*) by number of species in each web (*webs of equal S were lumped*).

TABLE 13.7 **Different types of indirect effects as proposed by Menge (1997), based on his analyses of intertidal communities.**

TYPE OF INDIRECT INTERACTION	PROPOSED MECHANISM
1. *Keystone predation*	Predator indirectly increases the abundance of competitors of its prey via consumption of the prey.
2. *Tri-trophic interactions*	Increase in plant abundance caused by the control of herbivores by predators.
3. *Exploitation competition*	A reduction in a consumer or producer resulting from the reduction of its prey or resources by another consumer species.
4. *Apparent competition*	Reduction of a species resulting from increases in a second species that enhances predation by a shared enemy.
5. *Indirect mutualism*	Positively correlated changes in two species resulting from predation by each on the competitor of the other's main prey.
6. *Indirect commensalism*	Like indirect mutualism but one potential indirect mutualist is more generalized in diet and also feeds on the main prey of the other indirect mutualist.
7. *Habitat facilitation*	One organism indirectly improves the habitat of a second by altering the abundance of a third interactor.
8. *Apparent predation*	An indirect decrease in a nonprey produced by a predator or herbivore, e.g., when whelks remove barnacles, they eliminate shelters for littorine sails, and the abundance of snails declines.
9. *Indirect defense*	The indirect reduction of a predator or herbivore by a nonprey, e.g., competition by a nonprey can reduce the abundance of a prey and thus its predator.

occurring at between 6 to 7 percent frequency, except for indirect commensalism (3.2%) and exploitation competition (2.8%). Menge (1995) suggests these patterns may apply to other aquatic habitats, especially benthic marine and freshwater habitats. However, their applicability to terrestrial studies is unknown and may not hold.

Schoener (1993) surveyed six studies in detail, out of a total of forty-eight he found in the literature. The six studies involved one study in a desert, one on a tropical island, one marine intertidal study, one in a river, and two in lakes. The magnitude of direct effects was greater than indirect effects in six of eight experiments. In a larger survey, direct effects were stronger than indirect in 75 percent of the experiments. In the absence of more data, the reviews of both Schoener (1993) and Menge (1995) suggest indirect effects are less frequent than direct effects. The implications for HSS and EHH are clear. Trophic cascades probably do not get everyone wet. The environmental stress hypothesis (MS) of Menge and Sutherland (1987) and Menge, Daley, and Wheeler's (1995) nutrient/productivity/omnivory (NPO) model now look even more appealing.

SUMMARY

1. Do population densities commonly vary around mean or equilibrium values or do they fluctuate widely? Often, it is hard to tell. The longer a population is observed, the more it seems to vary (red shift hypothesis).

2. Which factors affect population densities the most? There can be two kinds of effects. Those that perturb populations away from mean levels can be thought of as key factors and identified by a technique known as key-factor analysis. The key factors for plants and animals are many and varied, and there seems to be no generalization as to which key factors are important for which types of organism.

3. Factors that act so as to return populations to equilibrium levels are called density dependent factors. Although there are many statistical problems inherent in the detection of density dependence, once again there are few generalizations as to which factors (such as predators, parasites, or disease) act most frequently in a density dependent fashion.

4. There have been several different types of models that have been proposed to describe the types of mortality factors that should be most important. The Hairston, Smith, and Slobodkin model (HSS) was among the first. It suggested that because the Earth is "green" with plants, herbivores must have little impact on plant abundance. This is so because predators keep herbivore numbers down. Thus, predators undergo severe competition for prey (herbivores); herbivores are limited by predators, and plants, which do not suffer much herbivory, undergo strong competition for resources.

5. The HSS model was the jumping-off point for many other models, the most well known of which is the ecosystem-exploitation hypothesis (EEH) models developed by Oksanen, Fretwell, and their associates. The EEH model suggests that the frequency of competition and predation in a food web varies with productivity and the number of trophic levels in that web.

6. An alternative to both HSS and EEH was suggested by Menge and Sutherland (MS) who postulated that biotic complexity decreases with increasing stress. Thus, in polar habitats the climate is so severe that predators and herbivores are absent, and the few plants that survive are limited by environmental stress. In benign environments such as tropical forests, herbivory and predation are important. Another alternative is the NPO model of Menge, Daley, and Wheeler (1995), which suggests that the strength of competition and predation all increase as food chain length increases.

7. The HSS and EEH models can be categorized under the banner of "top-down" models, with the effects of predators being potentially important. In contrast, "bottom-up" models suggest that populations of all species are ultimately affected by plant populations. Large plant populations permit the growth of herbivore populations and natural enemies. The removal of predators or herbivores still leaves plant populations intact. The removal of plant populations eliminates all other populations.

8. It is most likely that a variety of interactions exist in nature such that, for example, herbivory is less frequent on plants where carnivores depress herbivore numbers. These interactions are termed indirect effects and are very common in nature. Empirical evidence suggests, however, that the strength of indirect effects is outweighed by direct effects.

DISCUSSION QUESTION

There are clearly a plethora of ecological effects on populations, from direct effects such as predation, competition, parasitism, herbivory, and mutualism, to various types of indirect effects. Can we ever hope to erect a framework whereby we can predict which effects are important in which systems? How could such theories best be tested?

SECTION

5 Community Ecology

More rain in the tropics means more plants. But why are there more species in the tropics and not just more individuals of one species? Contrast the vegetation of the desert island of San Pedro Martir, Baja, with just one species of cactus, with the species-rich jungle vegetation of Southeast Asia. Community ecology attempts to address questions like these. *(Photo by E.R. Degginger.)*

Why are there many more species in tropical forests than in temperate forests? Why are there generally a few common species and lots of rare species in a given area? Why do we see weeds and herbs in an old field gradually being replaced by shrubs and trees? Community ecology attempts to answer these questions. Thus, community ecology deals with populations not in isolation but together as a whole (community) in the natural areas where they interact. It concerns why we see the numbers of species in a community rather than the numbers of individuals in a population. Community ecology is about biodiversity.

Communities can occur on a wide range of scales and can be nested—the tropical forest community encompasses the community living in the water-filled recesses of bromeliads, which in turn encompasses the microfaunal communities of cellulose-digesting insects' guts. Once a community has been identified, we can describe the basic type of community present, determine the trophic structure (who eats whom), and determine the relative biomass of individual components. We can also count the number of species present and the abundance of each species and try to come up with an index of diversity.

The understanding of communities is particularly important in a wide variety of disciplines, for example, modern agriculture, where the emphasis today is on the integration of pasture, trees, and livestock (so-called agroforestry), especially in tropical regions. Problems associated with particular facets of the community may develop; for example, in tropical Asia cattle can cause damage to young trees and, more importantly, their dung can serve as a breeding place for rhinoceros beetles, one of the major pests of coconut (Reynolds 1988). Adjusting the species mix for optimum yields is a complex problem (Young 1988). In conservation, land managers often strive to maximize biodiversity, so a thorough understanding of this concept is necessary. And in restoration ecology, practitioners are keen to know if, following the replanting of areas with natural vegetation, the restored area will recruit animals and attain the diversity of undisturbed areas.

Some ecologists view community ecology as a poor excuse for a science (Schrader–Frechette and McCoy 1993), arguing that communities are no more than the sum of their individual components. They claim that a grassland prairie community is simply a collection of populations of several species of grasses, each of which has the same environmental requirements. These grasses are fed upon by an array of insect and mammalian herbivores that also happen to have similar environmental requirements and requirements for similar host plants. Others view communities as unique integrations of different populations with special properties, in much the same way that salt has unique characteristics (taste, for example) and does not simply combine the attributes of sodium and chlorine. Such unforeseen characteristics of a community are often termed **emergent properties**. Life itself is an emergent property. In this view, some members of a community are thought to change their habitat subtly and fractionally, enabling other species to exist. For example, in a log community, the first invaders attack and weaken the wood slightly, facilitating the entry of other organisms, though both can be found at the same time and can be thought of as part of the same community. The existence of emergent properties may be important in applied situations because human interventions may alter not only the population biology of certain species but also disrupt these higher-order interactions.

CHAPTER 14

The Main Types of Communities

14.1 The early history of community ecology focused on the argument of "individualistic associations" of species versus real communities.

Much of the early work in community ecology focused on descriptions of groups of plants. Some of the scientists in the field considered communities to be equivalent to a "superorganism" in much the same way that a collection of organs are gathered together in the body of animal. Indeed, the champion of this group, Frederic Clements, suggested that ecology was to the study of communities what physiology was to the study of individual organisms. In the United States, it was the limnologist Stephen Forbes, in 1883, who first applied the analogy of organism to community: "a group or association of animals or plants is like a single organism" (Forbes 1883).

In 1887, Forbes espoused these viewpoints in a widely cited paper, "The lake as a microcosm" and described the lake as functioning as an organic unit (Forbes 1887). He urged community ecologists to go further than compiling species lists to understand the laws of community activity. At the same time, the community concept was being developed in Europe. In an 1870 study of oysters, Mobius (1877) had proposed the term *biocoenosis* for a community of species inhabiting a definite territory.

Although some early plant ecologists were concerned mainly with purely descriptive studies of plant communities, others emphasized the dynamic or changing nature of communities. Eugene Warming (1895) stressed that many plant communities were not static or at equilibrium, but that changes in physical conditions or changes induced by animals caused changes in the plant communities. By the turn of the century, Henry Chandler Cowles, whose research concentrated on sand dune plants in the Chicago region, developed a dynamic perspective that he called *physiographic ecology*. Cowles trained as a geologist and constantly stressed the ever-changing interactions between plant formations and underlying geological formations. His stroke of genius was to recognize that because the youngest sand dunes lay closest to the shore, a walk back from the shore would in effect constitute a walk back through time. This is sometimes known as the *space-for-time substitution*. Thus, a comparison of the vegetation of the sand dunes of Lake Michigan with the vegetation inland could reveal the temporal development of plant communities termed **succession** (Cowles 1899). Although Cowles believed succession tended toward an equilibrium or end state, he did not believe this state was ever reached. Abiotic or biotic changes would alter the path of succession.

While Cowles studied the changing flora of sand dunes, Frederic Clements studied the more stable plant communities of Nebraska grasslands. Perhaps as a result, his theories of plant communities differed from those of Cowles. Clements believed that succession reached an end or stable point, which he termed the "climax." In the absence of climate instability and human intervention, the community could remain stable for millions of years. Clements's (1916) monumental study Plant Succession, formally presented these ideas and treated plant communities as organisms that underwent a life cycle in which communities grew and matured, like individual organisms.

Clements believed the most important process directing change or "succession" was competition between plants, and he felt succession reduced competition within the community. He also thought that the influence of animals on communities was trivial—communities were structured mainly by plant formations. One of Clements's strengths was his emphasis on quantitative methods to build a rigorous discipline. The adoption of the quadrat to sample vegetation by Pound and Clements (1899) was seen by some as the leap to numerical quantification in ecology. Until then, impressions of vegetation or floristic lists had sufficed. Clements's (1905) book, *Research Methods in Ecology* emphasized quantification and was, arguably, the original ecology text.

There is no doubt that Clements's ideas on succession and the organismic analogy permeated many textbooks of the day. McIntosh (1985) commented that Clementsian ecology had an orderly neatness that made it pedagogically useful and appealing. Clements's ideas on climax were, and still are, a subject of debate among community ecologists. For example, perhaps the ultimate extension of Clements's ideas is to regard the whole Earth, or at least the biosphere, as behaving as one giant single living organism. In his "Gaia hypothesis," British scientist James Lovelock suggested this very thing. He also (Lovelock 1991) suggested that medical science could help save the planet.

Clements's ideas were challenged mostly by Henry Allen Gleason (1882–1975) who proposed an "individualistic" concept of plant association in place of Clements's organismic metaphor (Gleason 1926). Gleason and others suggested that distinct ecological communities did not exist. Instead, species of plants were viewed as being distributed independently along gradients, so that communities could not be assigned boundaries in a nonarbitrary way. While acknowledging that some communities were fairly uniform and stable over a given region, Gleason argued that not all vegetation could be segregated into such communities. Short-term environmental changes could profoundly affect community composition. To Gleason, and many other mid-twentieth-century ecologists, it was not possible to create a precisely logical classification of communities, as Clements had tried to do. By the 1950s plant ecologists had abandoned many of Clements' principles. Robert Whittaker's studies in particular (1953, 1970), on the vegetational communities along elevational gradients on mountains, asserted the "principle of species individuality," which stated that most communities intergrade continuously and that competition does not create distinct, separate vegetational zones.

Despite the studies of Gleason, Whittaker, and others, the Clements-Gleason conflict in community ecology never quite seems to be resolved (Underwood 1986; Wilson, Ullmann, and Bannister 1996). In the 1970s, some animal ecologists, notably ornithologists, adopted the concept of the community as an entity composed of assemblages of species at equilibrium with competition being a structuring force. These ideas were often attributed to the brand of theoretical ecology espoused by Robert MacArthur and his colleagues (MacArthur 1955; Cody and Diamond 1975). For example, Diamond (1978) claimed that the idea that processes other than competition contributed to community patterns "strained one's credulity." Many others, however, questioned whether or not communities even showed a pattern (Strong, Lawton, and Southwood 1984). For example, Wiens (1985) asked whether communities are real and strongly influenced by competition or whether they are casual associations of species that are individually reacting to abiotic features in the environment. If they are the latter, then communities have no real structure and no structuring mechanism such as competition. In this view, communities are but a descriptive convention without emergent properties (Gilbert and Owen 1990). Connell and Sousa (1983) suggested that an important question would be to decide whether a community was stable (at equilibrium) or not. This is not a trivial question to answer (see Chap. 17).

If the structure of communities depends very much on species interactions (Lack 1971) then the number of species present in a community is critical to determining structure. The concern of community ecology with number of species can perhaps be traced to Gilbert White, an eighteenth-century British pastor/naturalist who had commented that the more an area is examined, the more species are found (White 1789). This relationship was formalized mathematically by Arrhenius (1921), developed further by Preston (1948) and fully expanded upon by MacArthur and Wilson (1967) as a key to their book *The theory of island biogeography* (see review by Connor and McCoy [1979]; also Chap. 19). Diversity gradients from the poles to the tropics were perhaps first formally noted by Wallace (1876) who attributed this to the constancy of the tropical climate. But it wasn't until the 1960s that interest in diversity gradients revived (Fischer 1960). At about this time, Robert MacArthur focused on species packing along resource axes as a major determinant of species richness (number of species) (MacArthur and MacArthur 1961). He later recognized that while local diversity might be constrained by ecological interactions, the species richness of a large geographical region would be influenced by historical processes and events (MacArthur 1969). But in addition to species richness, studies on diversity were now beginning to recognize the proportionate distribution of individuals among species. Participants at a 1969 Brookhaven symposium addressed the link between diversity and stability, although the issue was never resolved. The link between diversity and community ecology was now firmly established. In the 1980s and 1990s the value of biodiversity both in its own right (Wilson 1992) and for its effects on ecosystem function began to be addressed.

14.2 A study of paleocommunities may give us insights into what structures modern-day communities.

It is possible that the community patterns that we see today may be influenced as much by historical events as by contemporary factors such as competition or rainfall. Does an understanding of why communities are organized or distributed as they are require a

ECOLOGY IN PRACTICE

Robert P. McIntosh, University of Notre Dame

A product of the Milwaukee, Wisconsin, school system, I gravitated by chance to Lawrence College in Appleton, Wisconsin, graduating in 1942 with a B.S. degree just in time for World War II. The salient biological event of my collegiate years was a summer collecting plants under the direction of Albert Fuller, curator of botany at the Milwaukee Public Museum. Following my military service, it was Albert Fuller who, in 1946, guided me to interim teaching at the University of Wisconsin in Milwaukee, whence I moved to the University at Madison. I ended up in plant ecology with John T. Curtis and received a M.S. degree in 1948 and a Ph.D. in 1950.

This was a most fortunate circumstance as Curtis was just beginning his extremely productive career in the study of the plant communities of Wisconsin, and the postwar era in plant ecology at Madison was enriched by numerous students who carried on his tradition of quantitative community ecology.

My own part in this tradition was stimulated by Curtis who secured support from the Wisconsin Alumni Research Foundation (this was in the era before the National Science Foundation). I worked on relic pine stands in the Driftless Area for my M.S. thesis and the hardwood forests of southern Wisconsin for the Ph.D. thesis. I also worked with Curtis on one of the earliest studies of an artificial population in the precomputer era. The highlight of my work at Wisconsin was the appearance of the "vegetational continuum" concept in an article with Curtis in *Ecology* in 1951. This marked the substantiation of H. A. Gleason's famous "Individualistic Concept" with extensive quantitative data that permeated plant ecology at Wisconsin. The apparently perpetual debate between proponents of Frederic Clements's superorganismic concept of community and Gleason's individualistic concept continues in various venues to the present, although Gleason's idea is now widely accepted.

It is widely recognized that Gleason's individualistic concept was a primary stimulus to the development, in the 1950s, of the ideas of continuum and gradient applied to community in the vegetation studies of J. T. Curtis, Robert Whittaker, and their students and associates.Their studies brought Gleason's ideas out of limbo and added a new dimension (actually multidimensional) to the discussion of community. Gleason's individualistic concept and the issues it addressed were introduced into general textbooks of ecology in the 1950s, but they were not immediately widely incorporated into animal community ecology until later.

Leaving Madison in 1950, I spent eight very pleasant years teaching at Middlebury College and Vassar College, moving to the University of Notre Dame in 1958. Research in these years was subordinated to teaching, but I laid the groundwork for later studies of the forests of the Catskill Mountains. My career as a field ecologist was generally fortunate although, in the pre Geographic Information Systems (GIS) era, I was sometimes not sure just where I was.

The watershed of my career as a field ecologist came in 1970 when I became editor of the *American Midland Naturalist* and found it difficult to do the fund-raising and field work necessary for ecological studies. I turned to the history of ecology in a series of papers and, finally, in a volume, *The Background of Ecology: Concept and Theory.* This interest, in what may be called transecology, persists to the present in the era of social ecology, ecological philosophy, and ecological theology, among a host of other hybrids.

I have been most fortunate in my life in ecology in having many stimulating and helpful associations—some immediate colleagues, others professional associates and friends. The writings of H. A. Gleason, Grant Cottam, Robert H. Whittaker, and Peter Greig-Smith were particularly significant to my thinking about plant ecology. The transition to history of ecology was aided by Frank N. Egerton, one of the first historians to devote his work to ecology and its origins.

The volume and diversity of current ecological work are overwhelming and its quality impressive. The impact of ecology since the 1970s on philosophy, sociology, economics, and even theology greatly complicates an already inherently complex discipline.

knowledge of events that occurred thousands or even millions of years ago? Paleoecology is the study of the ecological relationships of organisms in the ecological past. Examining fossil records of preexisting communities may give us valuable clues as to how modern-day communities are organized. Unfortunately, our knowledge of paleocommunities is hampered on several counts:

1. Many organisms have soft bodies that rarely if ever fossilize, so invertebrates are underrepresented in paleocommunities, thus distorting the true picture of community structure.
2. Terrestrial habitats have very few situations in which fossilizing conditions occur, whereas others, like those derived from deposits in shallow water of lakes or seas, are relatively overrepresented. However, the latter sites may contain plants and animals washed down from the upland habitats of the region, complicating the analysis of community patterns.
3. It is often difficult to sort out fossils into well-delimited time units. Any one collection of fossils can represent thousands of years of accumulation (Knoll 1986). The actual community present at any given time might have been different. We could, for example, overestimate diversity if species came and went with range contractions and expansions.

Despite the caveats, some tantalizing information has been revealed about paleocommunities. There are indications that the diversity of many groups has held constant for long periods of time (Webb 1987). For example, the total variety of amphibians and reptiles remained essentially level from Permian to Cretaceous times (Simpson 1969). This indicates that there is some structure to species diversity—as one group declines in diversity another becomes more diverse. There seems to be a large-scale relationship between area and diversity throughout the ages (see Chap. 19 for more modern examples). Thus, the most devastating mass extinction in the history of marine organisms occurred at a time of worldwide reduction in shallow shelf area in the late Permian

(Schoff 1974), as Pangaea coalesced. The fact that terrestrial vertebrate diversity did not experience a mass extinction at this time supports this hypothesis.

What factors other than area influenced paleocommunity species richness? One recurring theme is that environmental conditions, especially climate were of paramount importance. An example is the Ladds quarry fauna of northwestern Georgia, dated as ten thousand to eleven thousand years old (Holman 1985). The known fauna is very rich and includes about ninety vertebrate species, of which twenty do not occur in the region today. Of these twenty, ten are northern and ten are southern. The northern species include the spruce grouse, today occurring no closer than Ontario; southern species include the round-tailed muskrat that does not now occur north of central Florida as well as extinct species of tropical affinities. Climate is argued to have a big effect in richness because of climatic equability; that is how much the seasons differ. The "equability" argument is as follows: If certain northern species have their southern range limits set by hot summers, and if certain southern species have their northern range limits set by cold winters, then a more equable climate, with milder winters and cooler summers, will allow these species to expand and, possibly, overlap. This situation may have prevailed at the time of the Ladds quarry fauna. With a subsequent decrease in equability—colder winters and hotter summers—the ranges have separated. This gives us a valuable clue as to how abiotic factors like temperature can affect species richness.

14.3 Communities may be classified in various ways, according to climatic conditions and vegetational types.

Much of the early work in community ecology focused on classifying groups of plants into recognizable communities. The world encompasses an enormous range of terrestrial and aquatic environments, from polar ice caps to forests to coral reefs. The classification of this immense range of variation into a manageable system is of fundamental importance in the management and conservation of the biosphere. However, the classification of natural communities is problematic. This is because these classifications are ultimately based on the assumption that the natural environment can be divided into a series of discrete, discontinuous units rather than representing different parts of a highly variable natural continuum. As we shall see, the latter could be a more accurate description of the world.

In general, attempts to classify communities are based on an identification of the plant species that occur in them along with a description of the physical characteristics of the area. The more rigidly a community is defined, the more site-specific it becomes and hence the more limited its use in analysis and planning.

At the extreme, very general habitat classifications ("forest," "grasslands," "wetlands") are based on the physical characteristics and appearance of an area, independent of species composition. They cover such a wide range of possible conditions that they have little heuristic use: the term *forest* applies both to highly diverse lowland tropical rainforest and northern coniferous monocultures, two systems that may have no species in common. Furthermore, these general terms are often difficult to define and delimit in a universally applicable way. Thus, for example, the density of tree cover necessary before an area can be called a woodland is arbitrary. Similarly, it is impossible to determine for how long and how intensely an area must be flooded before it can be classified as a wetland rather than a terrestrial ecosystem. This naturally makes any mapping of communities a problematic task.

Most global community classification systems have attempted to steer a middle course between the complex communities and the oversimplified ones just discussed. Generally, these communities will use a combination of a general definition of habitat type with a climatic descriptor (e.g., "tropical forest," "temperate grassland," and "warm desert"). Community mapping may also take into account human activity to attempt to produce a realistic contemporary map of land-cover types or may create a potential vegetation map from an analysis of climatic or other environmental variables. Potential vegetation maps are independent of human activities on the landscape.

One of the most notable major global classification systems is presented here. The Holdridge Life Zone Classification map (Fig. 14.1) depicts potential vegetation using the life zone classification system developed by Holdridge (1967) (Fig. 14.2). Holdridge considered temperature and rainfall to prevail over other environmental factors, hence, his focus on these parameters in his classification scheme. A similar scheme was used by Whittaker (1975), who superimposed vegetation structural classifications on to Holdridge's scheme. The Classification of Biogeographical Biomes of the World by

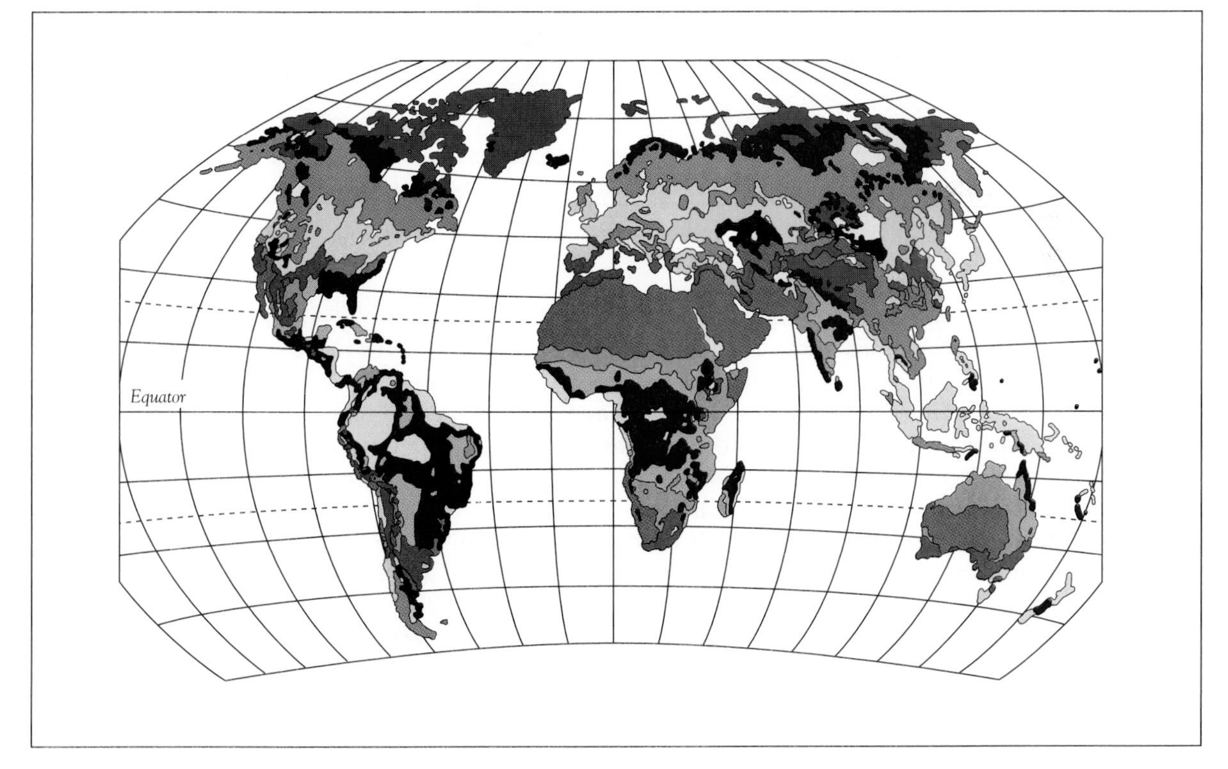

■ **FIGURE 14.1** Holdridge Life Zone Classification for current climate.

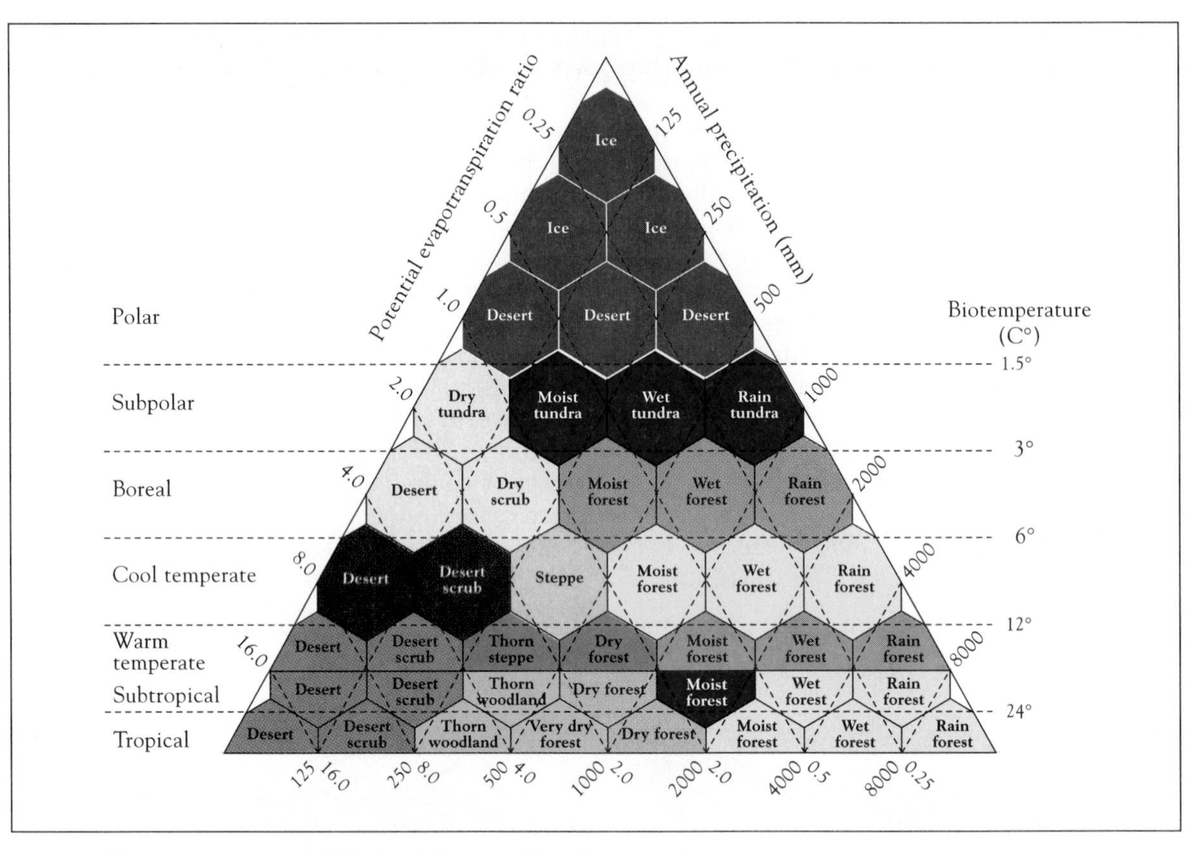

■ **FIGURE 14.2** Holdridge Life Zone Classification scheme.

(Udvardy 1975), provides a classification based largely on geography and potential vegetation. The Ecoregions of the Continents map (Bailey 1989) is produced from a combination of potential vegetation, hydrology, and actual land use. Table 14.1 presents the area contained in each region and its percentage of the global land area. There are many other classification schemes that rely on different criteria—for example, soil types, faunal species composition, and ecosystem properties. Each of these is equally justifiable so we should not expect there to be a single "best" classification for all different classifications to correspond (Hengeveld 1990; Heywood 1995). On a large scale, ecological classifications are based on types of ecosystem whereas on a small scale they may be replaced by classifications based on species present.

Given the difficulties of community definition outlined earlier, it is, unsurprisingly, extremely difficult to measure existing areas of any given community and even more problematic to estimate loss. Just as it is impossible to define rigidly the limits of any given community, so it is impossible to determine how much a given community has to change before it is considered destroyed or converted. Nevertheless, it is probably true that grasslands, forests, and woodlands have been greatly reduced because of conversion to crop-

TABLE 14.1 Terrestrial community types. (*After Bailey 1989*)

ECOREGION DOMAINS, DIVISIONS, AND PROVINCES	KM²	PERCENT
100 Polar domain	38,038,000	26.00%
110 Icecap Division	12,823,000	8.77%
M110 Icecap Regime Mountains	1,345,000	0.92%
120 Tundra Division	4,123,000	2.82%
121 Polar deserts	283,000	0.19%
122 Arctic tundras	1,231,000	0.84%
123 Oceanic moss–and–grass tundra	184,000	0.13%
124 Continental moss–and–lichen (typical) tundra	1,981,000	1.35%
125 Continental bush–and–shrub tundra	445,000	0.30%
M120 Tundra Regime Mountains	1,675,000	1.14%
M120 Tundra regime mountains (Antarctica)	60,000	0.04%
M121 Tundra–polar desert	795,000	0.54%
M122 Polar desert	820,000	0.56%
130 Subarctic Division	12,259,000	8.38%
131 Continental dark evergreen needleleaf open forest	2,285,000	1.58%
132 Continental light deciduous needleleaf open forest	1,286,000	0.88%
133 Eastern oceanic tayga	918,000	0.63%
134 Moderate continental dark evergreen needleleaf taiga	2,692,000	1.84%
135 Continental dark evergreen needleleaf taiga	1,880,000	1.29%
136 Continental and extreme continental light deciduous taiga	2,237,000	1.53%
137 Moderate continental small–leafed forest	251,000	0.17%
138 Continental mixed coniferous and small–leafed forest	710,000	0.49%
M130 Subarctic Regime Mountains	5,812,000	3.97%
M131 Open woodland–tundra	1,750,000	1.20%
M132 Open woodland–creeping trees–tundra	1,806,000	1.23%
M133 Forest–tundra of moderately and continental climate	686,000	0.47%
M134 Forest–creeping trees–tundra of extreme continental climate	1,203,000	0.82%
M135 Oceanic forest—tundra	367,000	0.25%
200 Humid temperate domain	22,455,000	15.35%
210 Warm Continental Division	2,187,000	1.49%
211 Eastern oceanic mixed monsoon forest	65,000	0.04%
212 Moderate continental mixed forests	2,122,000	1.45%
M210 Warm Continental Regime Mountains	*1,135,000*	0.78%
M211 Oceanic forest—tundra	67,000	0.05%
M212 Oceanic forest—creeping trees	331,000	0.23%
M213 Forest—tundra of moderately continental and continental climate	736,000	0.50%

(*continued*)

TABLE 14.1 (continued) Terrestrial community types. (*After Bailey 1989*)

ECOREGION DOMAINS, DIVISIONS, AND PROVINCES	KM²	PERCENT
220 Hot Continental Division	1,670,000	1.14%
221 Permanently humid eastern oceanic broadleaf forests	788,000	0.54%
222 Moderately humid broadleaf forest in moderately continental climate	882,000	0.80%
M220 Hot Continental Regime Mountains	485,000	0.33%
M221 Forest—alpine meadows	485,000	0.33%
230 Subtropical Division	3,568,000	2.44%
231 Oceanic mixed constantly humid forests	3,568,000	2.44%
M230 Subtropical Regime Mountains	1,543,000	1.05%
M231 Forest–meadow of eastern oceanic (monsoon climate)	1,264,000	0.86%
M232 Oceanic constantly humid forest–alpine meadows	278,000	0.19%
240 Marine Division	1,347,000	0.92%
241 Oceanic meadow	92,000	0.06%
242 Western oceanic coniferous and mixed forests	210,000	0.14%
243 Permanently humid western oceanic broadleaf forests	951,000	0.65%
244 Western oceanic taiga	95,000	0.07%
M240 Marine Regime Mountains	2,194,000	1.50%
M241 Oceanic meadow–tundra	21,000	0.01%
M242 Oceanic forest–tundra	1,068,000	0.73%
M243 Forest—alpine meadows	1,105,000	0.76%
250 Prairie Division	4,419,000	3.02%
251 Temperate prairies (humid steppes and wooded steppes) of eastern parts of continents	752,000	0.51%
252 Broadleaf–wooded steppes and meadow steppes of moderately continental climate	1,172,000	0.80%
253 Small–leafed and coniferous wooded steppes of continental climate	787,000	0.54%
254 Open woodland, savannas, and shrub of eastern parts of continents	925,000	0.63%
255 Subtropical prairies (humid steppes and wooded steppes) of eastern parts of continents	783,000	0.54%
M250 Prairie Regime Mountains	1,256,000	0.88%
M251 Continental steppe–forest–tundra and steppe–forest–meadow	690,000	0.47%
M252 Forest–alpine meadows	566,000	0.39%
260 Mediterranean Division	1,090,000	0.75%
261 Western oceanic mixed sclerophyll forest and shrub	927,000	0.63%
262 Dry steppes and shrub of moderate continental climate	163,000	0.11%

(*continued*)

TABLE 14.1 (continued) Terrestrial community types. (*After Bailey 1989*)

ECOREGION DOMAINS, DIVISIONS, AND PROVINCES	KM²	PERCENT
M260 Mediterranean regime Mountain	1,561,000	1.07%
M261 Forest–alpine meadows of western oceanic (mediterranean) climate	567,000	0.39%
M262 Shrub–forest–meadow of mediterranean climate	995,000	0.68%
300 Dry domain	46,806,000	32.00%
310 Tropical/subtropical Steppe Division	9,838,000	6.73%
311 Steppes and shrub of moderate continental climate	364,000	0.25%
312 Dry steppes, open woodland, and shrub of continental climate	846,000	0.58%
313 Shrub and semishrub semideserts of continental climate	1,392,000	0.95%
314 Desertlike savannas, open woodland, and shrub	5,807,000	3.97%
315 Dry steppes and shrub of moderate continental climate	1,429,000	0.98%
M310 Tropical/subtropical Steppe Regime Mountains	4,555,000	3.11%
M312 Forest–meadow–steppe of continental climate	670,000	0.46%
M313 Open woodland–steppe of continental climate	2,714,000	1.86%
M314 Open woodland–shrub–desert	770,000	0.53%
M315 Open woodland–steppe	400,000	0.27%
320 Tropical/subtropical Desert Division	17,267,000	11.80%
321 Shrub and semishrub semideserts and deserts of continental climate	1,321,000	0.90%
322 Semideserts and deserts	665,000	0.45%
323 Inner continental shrub semidesert	3,674,000	2.51%
324 Inner continental deserts of continental climate	7,921,000	5.42%
325 Western oceanic semideserts and deserts with high relative humidity	958,000	0.65%
326 Inner continental semideserts and deserts of extreme continental climate	2,727,000	1.86%
M320 Tropical/subtropical Desert Regime Regime Mountains	3,199,000	2.19%
M321 Desert–steppe and desert–steppe—desert of continental climate	1,193,000	0.82%
M322 Extreme continental desert	899,000	0.61%
M323 Desert–steppe	471,000	0.32%
M324 Desert	636,000	0.44%
330 Temperate Steppe Division	4,780,000	3.27%
331 Dry steppes of continental climate	1,790,000	1.22%
332 Steppes of moderately continental climate	1,581,000	1.08%
333 Dry steppes of extreme continental climate	1,409,000	0.96%

(continued)

TABLE 14.1 (continued) Terrestrial community types. (*After Bailey 1989*)

ECOREGION DOMAINS, DIVISIONS, AND PROVINCES	KM^2	PERCENT
M330 Temperate Steppe Regime Mountains	1,066,000	0.73%
M331 Forest–alpine meadows	893,000	0.61%
M332 Continental open woodland–steppe	173,000	0.12%
340 Temperate Desert Division	5,488,000	3.75%
341 Semideserts and deserts of continental climate	922,000	0.63%
342 Semideserts of continental climate	1,213,000	0.83%
343 Deserts of continental climate	1,647,000	1.13%
344 Semideserts of extreme continental climate	399,000	0.27%
345 Deserts of extreme continental climate	1,306,000	0.89%
M340 Temperate Desert Regime Mountains	613,000	0.42%
M341 Extreme continental desert–steppe	613,000	0.42%
400 Humid tropical domain	38,973,000	26.64%
410 Savanna Division	20,641,000	14.11%
411 Seasonally humid mixed (deciduous and evergreen) forest	1,346,000	0.92%
412 Savannas, open woodland and shrub with seasonal moisture supply	2,496,000	1.71%
413 Seasonally humid, predominantly deciduous forests	4,951,000	3.38%
414 Humid tall–grass savannas and savanna forests	3,699,000	2.53%
415 Moderately humid grassy savannas	4,771,000	3.26%
416 Dry savannas and open woodland	3,379,000	2.31%
M410 Savanna Regime Division	4,488,000	3.07%
M411 Forest—steppe and forest—meadow of seasonally humid type	1,102,000	0.75%
M412 Forest—meadow, seasonally humid	1,220,000	0.83%
M413 Forest—steppe, inner continental and leeward slopes	2,167,000	1.48%
420 Rainforest Division	10,403,000	7.11%
421 Eastern oceanic constantly humid forests	1,843,000	1.26%
422 Mixed forests with short dry season	2,893,000	1.98%
423 Constantly humid evergreen forests	4,280,000	2.93%
424 Humid forests with short dry season	1,387,000	0.95%
M420 Rainforest Regime Mountains	3,440,00	2.35%
M421 Forest—meadow of constantly humid eastern oceanic type	728,000	0.50%
M422 Forest–paramo and forest–meadow of constantly humid ocenic (and widward–slope type	1,013,000	0.69%
M423 Forest–paramo and forest–meadow	1,700,000	1.16%

lands. In addition, consideration of pure percentage changes may mask other changes that are deleterious to natural communities. In Europe, for example, forest area has actually increased during the twentieth century, but this is the result of the large–scale planting of species-poor coniferous monocultures; the area of species-rich natural and seminatural woodland has continued to decrease. Similarly, the area of grassland in Europe has remained static or nearly so over this period, but there has been wholesale conversion from low nutrient-input, species-rich grassland, to high-input, intensively cultivated, species-poor pasture. It is extremely difficult to map these changes and to measure their effects.

Finally, the increasing levels of the so-called greenhouse gases, (e.g., carbon dioxide, and methane) in the atmosphere could also have a large impact on worldwide community patterns if increases in these gases continue unabated. Shifts of community types would be most apparent in the mid- and high-latitude regions, with lesser changes in the tropics. The boreal and polar communities are likely to show the largest poleward shift, with a decrease in the extent of tundra and forested tundras. These communities currently form a continuous circumpolar band, but under a warmer climate only scattered patches would remain. Environmental change would favor an expansion of tropical forests. Yet human land-use patterns will probably mean an actual decrease in these forests.

Despite the difficulties in community classification some generalizations can be made about the larger types of communities such as tropical forests, temperate forests, marine communities, and so on.

14.4 Terrestrial communities include tropical and temperate forests, deserts, grasslands, tundra, and other communities.

Tropical forests contain an astonishing diversity of plant and animal species.

These forests are generally found in equatorial regions where annual rainfall exceeds 240 cm a year and the average temperature is more than 17°C. Thus, neither lack of water nor low temperature is a limiting factor. Surprisingly, soils in such areas can be fairly poor yet still support a luxuriant vegetation (Photo 14.1). Many of the nutrients are leached out by heavy rainfall. There is no rich humus layer as there is in temperate systems; fallen leaves are quickly broken down and nutrients returned to the vegetation, where most of the mineral reserves are locked up. Consequently, cleared tropical forest land does not support agricultural practices well for long.

Tropical forests cover much of northern South America, Central America, western and central equatorial Africa and some of Madagascar, southeast Asia, and various islands in the Indian and Pacific oceans. The total land area is about three billion ha, 23 percent of the world total (Bunting 1988). The human population in these areas is about 20 percent of the world total. Although concern is high that many biomes are being impacted by human activity, this concern is especially high for tropical forests (Applied Ecology: The Causes and Amounts of Tropical Deforestation). The diversity of species in tropical forests is staggering, often reaching more than fifty tree species per ha; indeed, the record for most tree species in an area alternates back and forth between southeast Asia and South America as different areas are censused. Gentry (1988) recorded 283 tree species in one ha of Peruvian

PHOTO 14.1 Tropical rain forest on the southwest coast of Costa Rica near Golfo Dulce. Tropical Forest has distinct vertical layers that provide niches for animals living on the top of the canopy, in the mid-layers, on epiphytes, and on the forest floor. Leaves and fruits exist year-round permitting a rich array of specialist feeders. *(Gregory G. Dimijian, Photo Researchers, Inc. 2R3493)*

rain forest. Sixty-three percent of the species in a one-ha plot were represented by a single tree, and there were only twice as many individuals as species.

Rain forest trees are often smooth barked and have large oval waxy leaves narrowing to "drip-tips" at the apex so that rainwater drains quickly. Many trees have shallow roots with large buttresses for support (Warren et al. 1988). The tallest trees reach heights of 60 m or more and emerge above the tops of lower trees, which interdigitate to form a closed canopy. Little light penetrates this canopy, and the ground cover is often sparse. Tropical rain forests are also characterized by epiphytes, plants that live perched on trees and are not rooted in the ground though some are rooted in accumulations of organic matter in the branches. Bromeliads are common epiphytes in New World forests. Climbing vines or lianas are also common.

Animal life in the tropical rain forests is also diverse; insects, reptiles, amphibians, and birds are well represented. Because many of the plant species are widely scattered in tropical forests, it is a more risky operation for plants to rely on wind to be pollinated or wind to disperse their seed. This means that animals are important in pollinating flowers and dispersing fruits and seeds. Many plants rely on mutualistic interactions with animals to deliver pollen. As many butterflies can be found in a small patch of rain forest as occur in the entire United States—five hundred to six hundred species. Tropical rain forests are the great reservoirs of diversity on the planet; as many as half the species of plants and animals on Earth live in them. Bright protective coloration and mimicry are rampant. Large mammals, however, are not common, though monkeys may be important herbivores. Though the genealogy of the major species in rain forests is different in different parts of the world, many species converge to a similar body form because they are adapted to a similar lifestyle (see Fig. 15.5).

Temperate forests occupy mid-latitude regions and are dominated by a combination of evergreen and deciduous tree species.

Temperate forest is the type of forest with which many people in the United States and Europe are most familiar (Photo 14.2). It occurs in regions where temperature falls below freezing each winter but not usually below −12°C, and annual rainfall is between 75 and 200 cm. Large tracts of such habitat are evident in the eastern United States, east Asia, and Western Europe. Commonly, leaves are shed in the fall and reappear in the spring, though there are exceptions. In the Southern Hemisphere, evergreen *Eucalyptus* forests occur in Australia and large stands of southern beech, *Nothofagus*, occur in southern South America, New Zealand, and Australia. Species diversity is much lower than in the tropics; several tree genera may be dominant or co-dominant in a given locality—for example, oaks, hickories, and maples are usually dominant in the eastern United States. Many herbaceous plants flower in spring before the trees leaf out and block the light (Heinrich 1976) though, even in the summer, the forest is usually not as dense as in trop-

PHOTO 14.2 Temperate forest, like this Douglas fir forest at Cathedral Grove, Vancouver Island, British Columbia, has sparser vegetation than its tropical counterpart, with fewer layers between the forest canopy and the ground. Lower temperatures mean a slower rate of decay so fallen leaves and branches remain on the ground longer, resulting in a richer humus layer in the oil. *(S. J. Krasemann, Peter Arnold, Inc. 36837.)*

The Causes and Amounts of Tropical Deforestation

Tropical rainforests exist primarily in three areas of the globe: Africa, Asia, and Latin America, with Latin America containing just over half the world's total:

REGION	AREA OF TROPICAL FOREST (1990) (10^6HA)	% WORLD'S	ANNUAL AVERAGE DEFORESTATION
Africa	527.6	30.0	−0.7
Latin America	918.1	52.3	−0.8
Asia	310.6	17.7	−1.2
Total	1756.3	100	0.8

Although there are different causes for topical deforestation in different areas, shifting cultivation has been identified as the prime cause of forest loss in Africa (70%), and it is also rampant in Asia (50%), and Latin America (35%). Although traditional shifting cultivators generally do not widely deforest, the recent rise in immigrant farmers in many countries has proved disastrous. For example, in 1995 as the Brazilian economy rebounded, many farmers began to expand, and there were more "queimadas" (burnings) than in the early 1990s. The burning mocked the message of the 1992 Earth Summit in Rio de Janeiro where Brazil promised sweeping treaties to protect the environment.

Traditional resident smallholders use selective felling, light burning, no tillage, and fallow periods. In contrast, immigrant farmers clear-cut the forest, burn it heavily, and employ soil tillage. In Brazil the problem has been compounded by large landowners moving in to buy up cleared land and convert it to cattle ranches. In Malaysia, oil palm and rubber tree plantations have sprung up.

Logging in excess of regrowth is also a significant cause of loss, particularly in Asian forests. Fuelwood collection can also be important, but generally fuelwood collection is more of a problem in woodlands, savannas, and scrub. Finally, the construction of mines, dams, and oil installations is a comparatively minor cause of direct deforestation, but indirectly the discharge of chemicals and silt into rivers can cause much damage. Worse still, the roads built into these regions open them up to further colonization.

Although a rate of loss of less than 1 percent of forest per year may not seem like much, the constant loss year after year soon mounts up. During the 1980s alone the world lost about 8 percent of its tropical forests. Some transformation of forests is necessary to generate badly needed capital and agricultural land. To deny countries the opportunity to do this would be to force them to keep living in the nineteenth century. On the other hand, the methods used should ensure the maintenance of the long-term productivity of the soil and the plants growing on it and should minimize the effects on native wildlife.

Will the tropical forests disappear forever, and if so how long will it take (Photo 1)? These questions cannot easily be answered. Rates of deforestation can decrease if the political will is there; they can also increase under economic hardship.

■ **PHOTO 1** Virgin rain forest cleared for transmigrant settlements near Pontianak, West Kalimantan, Indonesia. The relatively small annual loss of tropical forests mounts up to significant losses after just a few decades. *(Sean Sprague/Stock Boston SAS0100B.)*

ical situations, so there is often abundant ground cover. Epiphytes and lianas are few. Soils are richer because the annual leaf drop or detritus fall is not as quickly decomposed. With careful agricultural practices, soil richness can be conserved, and as a result agriculture can flourish. Like the plants, animals are adapted to the vagaries of the climate; for example, many mammals hibernate during the cold months. Birds migrate and insects enter diapause, a condition of dormancy passed usually as a pupa (though sometimes as an egg, larva, or adult instead). The reptile fauna, dependent on solar radiation for heat, is relatively impoverished compared to the tropics. Mammals include squirrels, wolves, bobcats, foxes, bears, and mountain lions.

Deserts occupy areas that experience regular water deficits and are dominated by species tolerant of extreme aridity.

Deserts are biomes suffering from water deficit. They are found generally around latitudes of 30°N and 30°S, between the latitudes of tropical forests and temperate forests or grasslands. About one-third of the Earth's terrestrial surface is occupied by these hot, dry regions. One reason for the locations of deserts is the movement of winds in the Earth's atmosphere. Warm, moist air rises over the equator; as it cools, rain is produced (causing the tropical rain forests), and the air moves north or south of the equator. This rising, raining, and movement continue to about 30° latitude, where the now dry and relatively cold air begins to sink groundward. It warms by compression and produces a downward flow of warm dry air at about 30° north and south of the equator. Such patterns of circulation, in which the drier air then filters back to the equator, are known as *tropical Hadley cells*. They produce deserts at characteristic latitudes, including the Sahara of north Africa, the Kalahari of southern Africa, the Atacama of Chile, the Sonoran of New Mexico and the southwest United States (shown in Photo 14.3), the Gobi of central Asia, and the Simpson of Australia.

Deserts are characterized by two main conditions, lack of water (less than 30 cm per year) and usually high daytime temperatures. However, cold deserts do exist and are found west of the Rocky Mountains, in eastern Argentina, and in much of central Asia. Lacking cloud cover, all deserts quickly radiate their heat at night and become cold. The degree of aridity is reflected in the ground cover. In true deserts, plants cover 10 percent or less of the soil surface; in semiarid deserts, like thorn woodlands, they cover 10 to 33 percent. Only rarely do deserts consist of ostensibly lifeless sand dunes, but such places do exist: in some places of the Atacama desert of western Chile no rainfall has ever been recorded.

Three forms of plant life are adapted to deserts: (1) the periodics or annuals, which circumvent drought by growing only when there is rain; (2) succulents, such as the saguaro cactus *(Carnegiea gigantea)*, and other barrel cacti of the southwestern deserts, which store water; and (3) the desert shrubs, such as the spraylike ocotillo *(Fouquieria splendens)*, which have short trunks, numerous branches, and small thick leaves that may be shed in prolonged dry periods. As a strategy against water-seeking herbivores, many plants have spines or an aromatic smell indicative of chemical defenses, although the physical structure of the desert plants—their few leaves and sharp spines—is probably also linked to water conservation and heat load. The aboveground parts of perennial desert plants are more widely spaced than those of their forest counterparts because their roots are longer and occupy greater areas to ensure maximum water-gathering

■ **PHOTO 14.3** Sonoran Desert, Arizona. The prominent plants include the tall, columnar saguaro cactus, *Canegiea gigantea*; the green spray-like ocotillo, *Fouquieria splendens*; and smaller cholla cacti, *Opuntia sp.* (*Photo by Peter Stiling.*)

potential (Fig. 14.3). Typical perennial plants of the desert are the succulent and thorny cacti of the Western Hemisphere and succulent thorny members of the milkweed (Asclepiadaceae) and spurge (Euphorbiaceae) families in African deserts. In North America the creosote bush *(Larrea)* is widespread over the southwestern hot desert, and sage brush *(Artemisia)* is more common in the cooler deserts of the Great Basin.

Seed-eating animals, such as ants, birds, and rodents, are common in deserts, feeding on the numerous small seeds. Reptiles are numerous because high temperatures permit these ecothermic animals to maintain their body temperature. Lizards and snakes are important predators of seed-eating animals. Like the plants, desert animals have also evolved many ways of conserving water, like dry excretion (uric acid and guanin), heavy wax "waterproofing" in insects, and generally crepuscular habits and the use of burrows. Some invertebrates such as brine shrimp (*Artemia* sp.) follow the strategy of annuals—grow and reproduce when it rains, leaving eggs during drought.

Irrigated deserts can, because of the large amount of sunlight, be extremely productive for agriculture, though large volumes of water must flow through the system or detrimental salts may accumulate in the soil because of the rapid evaporation rate. Desert civilizations that harnessed the flow of such rivers as the Tigris, Euphrates, Indus, and Nile dominated early human history. Unlike tropical forests, deserts seem to be expanding under human influence because overgrazing, faulty irrigation, and the removal of what little woody vegetation exists all speed up desertification. The Sahel region, a narrow low-rainfall band south of the Sahara whose name is derived from the Arabic word for border, is often argued to be a case in point. The acacia tree, ubiquitous in many arid

■ **FIGURE 14.3** Illustration of the regular spacing that allows desert plants to maximize their water uptake after rains.

zones and useful as firewood and forage, was common around the Sudan capital, Khartoum, as recently as 1955; by 1972 the nearest trees were 90 km south of the city.

Grasslands dominate regions that are too wet to be deserts and too dry to be forests.

Grasslands occur in the range between desert and temperate forest in which the rainfall, between 25 and 70 cm, is too low to support a forest but higher than necessary to support only desert life forms. But some ecologists (Bragg and Hurlbert 1976; Kucera 1981), feel the extensive grasslands of central North America, Russia, and parts of Africa are zones between forest and desert in which fire and grazing animals have together prevented the spread of trees (Photo 14.4). From east to west in North America and from north to south in Asia, grasslands show differentiation along moisture gradients. In Illinois, with about 80 cm annual rainfall, tall prairie grasses about 2 m high, such as big bluestem *(Andropogon)* and switchgrass *(Panicum)*, dominate, whereas along the eastern base of the Rockies, 1,300 km to the west, where rainfall is only 40 cm, shortgrass prairies exist, rarely exceeding 0.5 m in height and consisting of buffalo grass *(Buchloe)* and blue grama *(Bouteloua)*. Similar gradients occur in South Africa (the veldt) and in Argentina and Uruguay (the pampas). In some grasslands there may be just sufficient rainfall to support isolated trees. For example, African savannas contain isolated acacia trees, and the same is true in South America and Australia.

PHOTO 14.4 In between the areas of desert and temperate forest are the grasslands—vast areas of treeless plain such as this prairie pasture in Nebraska. Some of the largest mammalian herbivores and their predators exist on prairies and savannahs. *(Grant Heilman, Grant Heilman Photography, Inc. CNB-92B.)*

Nowadays, few original grasslands remain. Prairie soil is among the richest in the world, having twelve times the humus found in a typical forest soil. Historically and where the grasslands remain, large mammals are the most prominent members of the fauna; examples are bison (buffalo) and pronghorn (antelope) in North America; wild horses in Eurasia; large kangaroos in Australia; and a diversity of antelopes, zebras, and rhinoceroses in Africa, as well as their associated predators (lions, leopards, cheetahs, hyenas, and coyotes). Burrowing animals such as gophers (in North America) and mole rats (in Africa) are also common.

The taiga is the coniferous forest that dominates subarctic latitudes in North America and Eurasia.

North of the temperate-zone forests and grasslands lies the biome of coniferous forests, known commonly by its Russian name, taiga. Most of the trees are evergreens or conifers with tough, narrow leaves, needles that may persist three-to-five years. Spruces *(Picea)*, firs *(Abies* and *Pseudotsuga)*, and pines *(Pinus)* generally dominate, but some deciduous species such as aspens, alders, and willows occur in disturbed areas or along water courses. All these species are very freeze tolerant and can withstand temperatures of −60°C. Many of the conifers have conical shapes to reduce bough breakage from heavy loads of snow. As in tropical forests, the understory is thin because of the dense year-round canopies. Soils are poor and acidic because of the slow decay of fallen needles and organic matter that often builds up. Snakes are rare, and few amphibians exist. Insects are strongly periodic but may often reach outbreak proportions in times of climatic relaxation. Mammals such as bears, lynxes, moose, beavers, and squirrels are heavily furred. The taiga is famous for cyclic population patterns, of which the abundances of hares and lynxes are a well-known example. In the Southern Hemisphere, little land area occurs at latitudes at which one would expect extensive taiga to exist.

Regions that are too cold to support trees are occupied by treeless tundra communities.

The tundra is the last major terrestrial biome, occupying roughly 17 percent of the Earth's surface but existing only in the Northern Hemisphere, north of the taiga. Precipitation is generally less than 25 cm per year and is often "locked up" as snow and unavailable for plants. Deeper water can be locked away for a large part of the year in permafrost. With such little available water, trees cannot easily grow. Summer temperatures are only 5°C, and even in the long summer days the ground thaws to less than one meter in depth. Mid-winter temperatures average −32°C. Vegetation occurs in the form of fragile, slow-growing lichens, mosses, grasses, sedges, and occasional shrubs such as willow, which grow close to the ground. In some places desert conditions prevail because so little moisture falls. Because permafrost is impenetrable, water drainage is inhibited and surface water lies in shallow lakes and ponds on the surface of the earth in the summer. The anaerobic (oxygenless) conditions of the waterlogged soil and the low temperatures slow nutrient cycling. Organic matter cannot completely decompose, and it often accumulates as peat. Animals of the arctic tundra have adapted to the cold by having good insulation. Many birds, especially shorebirds and waterfowl, migrate. The fauna is much richer in summer than in winter. Many insects spend the winter at immature stages of growth, which are more resistant to cold than the adult forms. The larger animals include such herbivore as musk oxen and caribou in North America, and reindeer in Europe and Asia, as well as the smaller hares and lemmings. Common predators include arctic fox, wolves and snowy owls, and polar bears near the coast.

Tundra may occur not only in the far north but also in the higher elevations of mountains. Thus, alpine tundra can occur even in the tropics at the very highest mountain tops where nightly temperatures drop to below freezing. In tropical situations, of course, daylight varies little from the twelve hours per day throughout the year. So instead of an intense period of productivity, vegetation in the tropical alpine tundra exhibits slow but steady rates of photosynthesis and growth all year.

Other, smaller terrestrial biomes like temperate rain forests, tropical seasonal forests, and chapparal also exist.

Of course, not all communities fit neatly into these six major biome types. As with most things ecological, there exist characteristic regions where one biome type grades into another. For example, coniferous forests also occur in some temperate lowlands. A temperate rain forest extends along the coast all the way from Alaska into northern California. Again, most of the trees are conifers. That forest has some of the world's tallest trees—sitka spruce to the north and coastal redwoods to the south. On the eastern United States, most of New Jersey's coastal plain, which is sandy nutrient-poor soil, is dominated by pine barrens. This is a type of scrub forest with grasses and low shrubs growing among the open stands of pine and oak trees. Open stands of pine trees also occur in the coastal plains of North Carolina, South Carolina, Georgia, and Florida, and these forested regions are maintained by frequent fires. Tropical seasonal forests may be apparent where rainfall is heavy (between 125 and 250 cm a year) but occurs in a distinct wet season, as in India or Vietnam. In such monsoon forests leaves may be shed in the dry season. Another distinct biome type is chaparral, a Mediterranean scrub habitat adapted for fire, which is common along the coastlines of southern Europe, California, South Africa, and southwest Australia.

Mountain ranges must be treated still differently. Biome type relies predominantly on climate, and on mountains temperature decreases with increasing altitude. This decrease is a result of a process known as *adiabatic cooling*. Increasing elevation means a decrease in air pressure. When wind is blown across the Earth's surface and up over mountains, it expands because of the reduced pressure; as it expands it cools, at a rate of about 10°C for every 1,000 m, as long as no water vapor or cloud formation occurs. (Adiabatic cooling is also the principle behind the function of a refrigerator—freon gas cools as it expands coming out of the compressor.) Higher elevations are also cooler because the less dense air allows a higher rate of heat loss by radiation back through the atmosphere. A vertical ascent of 600 m is roughly equivalent to a trek north of 1,000 km. Precipitation changes with altitude, too, generally increasing in desert elevations but decreasing on the leeward side of slopes, which are in a rain shadow. Approaching clouds have usually dumped all their moisture on the windward side. Thus, biome type may change from temperate forest through taiga and into tundra on an elevation gradient in the Rocky Mountains, and even from tropical forest to tundra on the highest peaks of the Andes in tropical South America.

14.5 Aquatic communities.

Within aquatic environments, biome types can also be recognized, such as rivers, and freshwater lakes and, within saltwater oceans, the intertidal rocky shore, sandy shores, the neritic zone (encompassing shallow waters over continental shelves), coral reefs, seagrass beds, and the pelagic zone or open ocean. In each of these, the physical "climate" is different, varying in such parameters as salinity, oxygen content, current strength, and availability of light.

Like terrestrial regions, a variety of communities can be recognized in marine environments.

Marine environments are among the most extensive and uniform on Earth. Marine ecosystems are found over nearly three-quarters of the Earth's surface. Like those in freshwater areas, marine communities are affected by the depth at which they occur. The shallow zone where the land meets the water is called the intertidal zone. Beyond the intertidal zone is the neritic zone, the shallow regions over the continental shelves. Past the continental shelf is the open ocean or oceanic zone, which may reach very great depth. This is also often referred to as the pelagic zone. At the bottom is the sea floor or benthic zone. Just as in freshwater systems, we may also recognize the stratum of water near the surface where light penetrates as the photic zone and, below, the dark aphotic zone. Phytoplankton, zooplankton, and most fish species occur in the photic zone. In the aphotic zone production by plants is virtually zero and only a few invertebrates and luminescent fish live.

The area where a freshwater stream or river merges with the ocean is called an estuary. It is often bordered by extensive intertidal mudflats or salt marshes. The main plants of estuaries are, in temperate areas, including the United States, *Spartina* salt marsh grasses. In tropical areas of the world mangroves replace them along the mudflats. Most of the plant material is not eaten by herbivores but dies and rots on the mudflats. It is decomposed by bacteria and fungi, and particles of plant material are food for nematodes, snails, crabs, and some fish. Clams, oysters, barnacles, and other filter feeders may live in the estuaries too. The area is important as a nursery ground for shrimp and fish. Salt marshes are so productive that they are used as feeding grounds for ducks, geese, and other waterfowl or migratory birds.

The intertidal zone, where the land meets the sea, is alternately submerged and exposed by the daily cycle of tides (Photo 14.5). The resident organisms are subject to huge daily variation in the availability of seawater and in temperature. They are also battered by waves, especially during storms. In temperate areas they may be subject to freezing in the winter or very hot temperatures in the summer. At low tides they may be dry and subject to predation by a variety of animals including birds and mammals. High tides bring predatory fishes. There is commonly a vertical zonation. This is most evident on rocky shores, which may have three broad zones. The upper littoral is submerged only during the highest tides. The mid-littoral is submerged during the highest regular tide and exposed during the lowest tide each day. Life here may be quite rich and consist of green algae, sea anemones, snails, hermit crabs, and small fishes living in tide pools. Competition for space on rock faces may be quite intense. The lower littoral is exposed only during the lowest tide, and the diversity and richness of organisms is great. Along sandy and muddy shores, few large plants or other sessile organisms can grow

PHOTO 14.5 There is a distinct vertical zonation on rocky shores; upper littoral (submerged only during the highest tides), mid-littoral (submerged during normal tides), and lower littoral (exposed only in the lowest tides). Purple sea starts (*Pisaster ochraceus*) in the mid-littoral zone at the Olympic National Park. (*Jim Zipp, Photo Researchers, Inc. 2H2948.*)

■ **PHOTO 14.6** A coral reef in the Red Sea. Coral reefs have the most species-rich marine communities on Earth. (*Jeff Rotman, Peter Arnold, Inc. INV-02B04.*)

because the sand or mud is constantly shifted around by the tide. Instead, the ecosystem contains burrowing marine worms, crabs, and small isopods.

Coral reefs exist in warm tropical waters (Photo 14.6). This is a conspicuous and distinct biome. Currents and waves constantly renew nutrient supplies, and sunlight penetrates to the ocean floor allowing photosynthesis. Coral reefs are made of organisms that secrete hard external skeletons made of calcium carbonate. These skeletons vary in shape, forming a substrate that other corals and algae grow on. An immense variety of microorganisms, invertebrates, and fish live among the coral, making the reef one of the most interesting and richest biomes on earth. Probably 30 to 40 percent of all fish species on earth are found on coral reefs (Ehrlich 1975). Prominent herbivores include snails, sea urchins, and fish. These are in turn consumed by octopus, seastars, and carnivorous fish.

In the pelagic zone, nutrient concentrations are typically low, though the waters may be periodically enriched by upwellings of the ocean that carry mineral nutrients from the bottom waters to the surface. Pelagic waters are mostly cold, only warming near the surface. This is where many photosynthetic plankton grow and reproduce. Their activity counts for nearly half the photosynthetic activity on earth. Many scientists have suggested that if phytoplankton productivity is increased, much of the increased carbon dioxide from fossil fuel burning would be soaked up and global warming would be slowed. One of the limiting factors seems to be the availability of iron. Huge experimental additions of iron to the Pacific have increased phytoplankton production (Van Scoy and Coale 1994).

Zooplankton, including some worms, copepods (tiny shrimplike creatures), tiny jellyfish, and the small larvae of invertebrates and fish graze on the phytoplankton. The biome also includes free swimming animals, called nekton, which can move against the currents to locate food. The phytoplankton and zooplankton move with the current. The nekton include large squids, fish, sea turtles, and marine mammals that feed on either plankton or each other. Only a few of these live at great depth. Here the fish may have enlarged eyes, enabling them to see in the dim light. Others have luminescent organs that attract mates and prey. A number of marine animals are migratory, following seasonally available food sources or moving between summer breeding grounds and their winter feeding range.

Recently, other communities have been discovered that exist near the openings of deep-ocean volcanic vents (black smokers) in midocean ridges (Ballard 1977). The primary producers there are giant worms, which are nourished by symbiotic chemosynthetic bacteria that produce ATP by oxidizing sulfides and reducing carbon dioxide to organic compounds. In other areas, animals that harbor chemoautotrophic bacteria seem to be the most common. Hessler, Lonsdale, and Hawkins (1988) discuss the common occurrence of a "hairy snail" in the vents near the Philippines that contains bacteria in its gills that in turn oxidize sulfur to produce energy. Smith (1985) has studied the dense beds of mussels, *Bathymodiolus thermophilus*, that can be found along the Galapagos rift. Water in these areas is often 20°C warmer than in surrounding areas.

Freshwater habitats support a variety of plant and animal communities.

Freshwater habitats are traditionally divided into standing-water lentic habitats (from the Latin *lenis*, calm—lakes, ponds, and swamps) and running-water lotic habitats (*lotus*, washed—rivers and streams). Natural lakes are most common in regions that have been subject to geological change within the past twenty thousand years, such as the glaciated regions of northern Europe and North America. They are also common in regions of recent uplift from the sea, such as Florida, and in regions subject to volcanic activity. Volcanic lakes formed either in extinct craters or in valleys dammed by volcanic action are among the most beautiful in the world. Geographically ancient areas such as the Appalachian Mountains of the eastern United States contain few natural lakes.

The ecology of lentic habitats is largely governed by the unusual properties of water. First, water is at its least dense when frozen; ice floats. From a fish's point of view this property is advantageous because a frozen surface insulates the rest of the lake from freezing. If ice sank, all temperate lakes would freeze solid in winter, and no fish would exist in lakes outside the tropics. Water is at its densest at 4°C. Thus, as long as no water in the lake is colder than 4°C, the warmest water is at the surface, and temperature declines with depth, though not in a linear fashion. Normally, several layers are present (Fig. 14.4). There is an upper layer, called the epilimnion, that is warmed by the sun and mixed well by the wind. Below this lies the hypolimnion, a cool layer too far below the surface to be warmed or mixed. The transition zone between the two is known as the thermocline.

There are other divisions within a lake based on light availability. The upper layer, where light penetrates, is the autotrophic or photic zone. Below, in darkness, is the profundal or aphotic zone, where heterotrophs live, depending on the rain of material from above for their subsistence. The depth of the photic zone depends on light availability and water clarity. The level at which photosynthate production equals energy used up by respiration is the lower limit of the euphotic zone and is known as the compensation level or compensation point. In the summer, in temperate lakes, the compensation level is usually above the thermocline. Oxygen-needing organisms cannot usually live in the hypolimnion because it becomes oxygen depleted, a phenomenon known as summer stagnation.

The degree of summer stagnation in temperate lakes is partly determined by the degree of "productivity" of a lake. The least productive lakes are termed *oligotrophic*. Such lakes generally have low nutrient contents, largely as a result of their underlying substrate and young geologic age. Young lakes have not had a chance to accumulate as

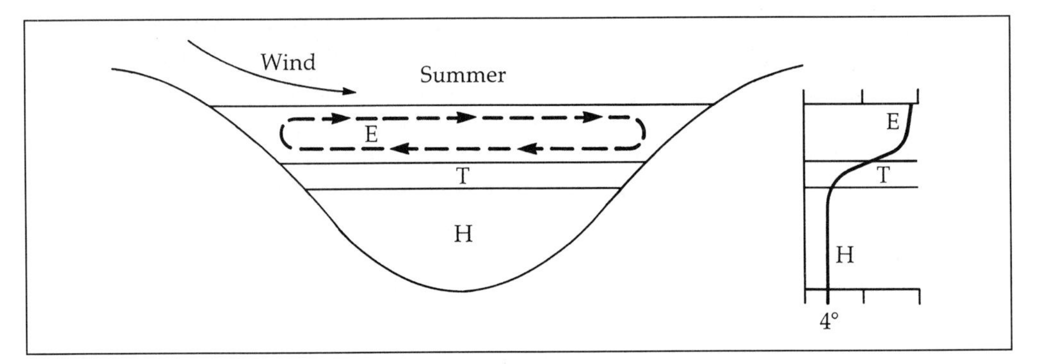

■ **FIGURE 14.4** Cross-section of lake stratification and profile of temperature with depth. E = epilimnion; T = thermocline, H = hypolimnion. Temperature scale on the right starts at 0 degrees C and increases to the right.

many dissolved nutrients as have older ones. Oligotrophic lakes are relatively clear, and their compensation levels may lie below the thermocline. In this situation, photosynthesis can take place in the hypolimnion, adding oxygen. Low nutrient concentrations keep the algae and rooted plants in the epilimnion sparse, and little debris rains down upon the inhabitants of the hypolimnion. As a result, oligotrophic lakes are clear and often contain human-desired fish such as trout. Even though few nutrients are present in oligotrophic lakes, eventually they do begin to accumulate; sediments are deposited and both algae and rooted vegetation begin to bloom. Organic matter accumulates on the lake bottom, respiration of bottom dwellers increases, the water becomes more turbid, and the oxygen levels of the water go down. Fish such as trout are excluded by bass and sunfish, which thrive in warm water and at low oxygen concentrations. This process of aging and degradation is natural and is termed eutrophication; its end result is a *eutrophic* lake. Eutrophication, however, can be greatly speeded up by human influences, which increase nutrient concentrations through the introduction of sewage and fertilizers from agricultural runoff. This is often termed cultural eutrophication. A measure of eutrophication is given by the dissolved oxygen concentration or biochemical oxygen demand (BOD). The BOD is the difference between the production of oxygen by plants and the amount of oxygen needed for the respiration of the organisms in the water. It is normally measured in the laboratory as the number of milligrams of oxygen consumed per liter of water in five days at 20°C.

The stratified nature of temperate-zone lakes in summer does not last all year (Fig. 14.5). In the fall, the upper layers cool and, as their density increases, sink, carrying oxygen to the bottom of the lake. The lake is thoroughly mixed by this action and by frontal storms, and the thermocline disappears. In winter the surface usually freezes, and no turnover of water occurs; once again a gradient is set up. Then, as spring returns, the ice melts and sinks, and water temperature rises producing another mixing, called the spring overturn. In contrast to temperate lakes, tropical lakes are often isothermal (that is, all at one temperature) or at most exhibit only a weak temperature gradient from top to bottom, with no seasonality. Little mixing occurs, and deep lakes are generally unproductive, with oxygen-poor, fishless lower depths, as the builders of tropical dams learn (to their chagrin). Worse still, most water from dams is drawn off from the base, meaning that the streams

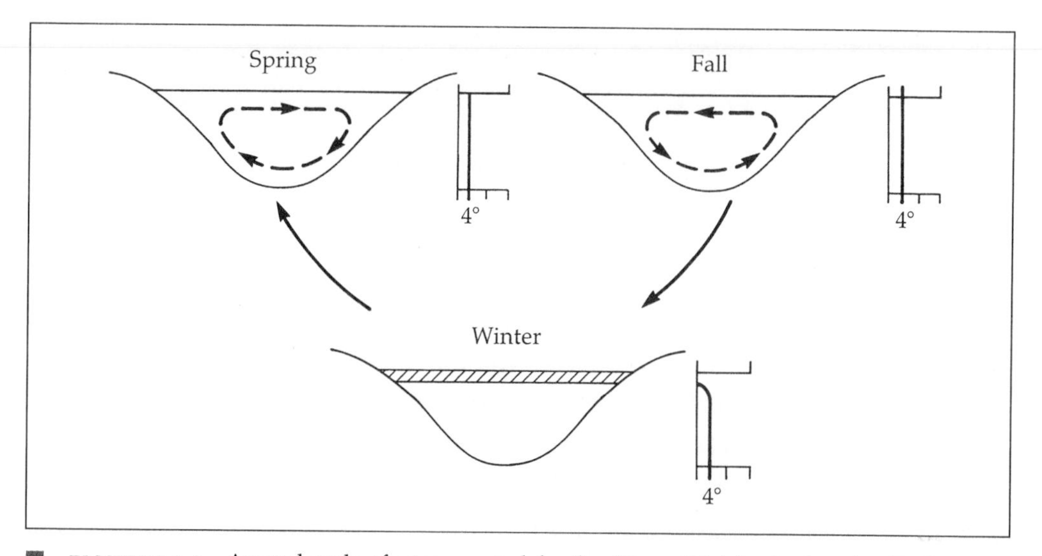

FIGURE 14.5 Annual cycle of a temperate lake. See Figure 14.4 for further details.

below dams are much less oxygen-rich than those above them. Shallow lakes and many slowly flowing reservoirs do not show thermal stratification in any region.

Abiotic phenomena other than nutrient content and dissolved oxygen content may be important to the lake communities also. Water pH is particularly relevant, as fish enthusiasts know. For example, some *Poecilia* species, such as live-bearing mollies, breed only in alkaline (high pH) waters, whereas *Hyphessobrycon*, neon tetras, breed only in low pH. Changes in the pH of lake water have been frequent over the past fifty years because of the impact of acid rain (Chap. 22). Fish exterminations have been recorded in over three hundred lakes in the Adirondack Mountains of the northeastern United States by acid rain. Suspended matter, often washed down into rivers from mining operations, is actually the most frequent freshwater pollutant, but others abound (Applied Ecology: Water Pollution).

Some ecologists also differentiate lakes into zones called littoral, limnetic, and profundal, which show the greatest differences in terms of organisms present. The littoral zone is a shallow well-lit zone extending all around the shore to the depth where rooted aquatic plants will not grow. Usually the greatest variety of organisms are found in the littoral zone, which is home to plants, snails, frogs, and many fish. The limnetic zone includes the open, sunlit waters down to the depth where energy fixed by photosynthesis no longer exceeds that used for respiration. Here we find plankton, communities of floating, weakly swimming, or passive organisms that are mostly microscopic. The phytoplankton include tiny plantlike organisms such as diatoms, green algae, and cyanobacteria. The zooplankton, or tiny animals, include rotifers and copepods. The final zone of lakes, the profundal zone, is the deep open water below the depth of effective light penetration. Shallow lakes may not have a profundal zone. Detritus, or rotting material, from the limnetic zone sinks to the profundal zone and into the bottom sediments, which contain communities of bacterial decomposers. The decomposers release nutrients to the water through their activities.

Lotic or running-water habitats (Photo 14.7) generally have a fauna and flora completely different from those of lentic waters. Plants and animals are adapted so as to

Water Pollution

Terrestrial fresh water, as part of the human environment, occurs in two main forms—as ground water and in surface freshwater bodies. Two types of pollution, point source pollution and nonpoint source pollution can be recognized. Things such as factories, power plants, and sewage treatment plants are classified

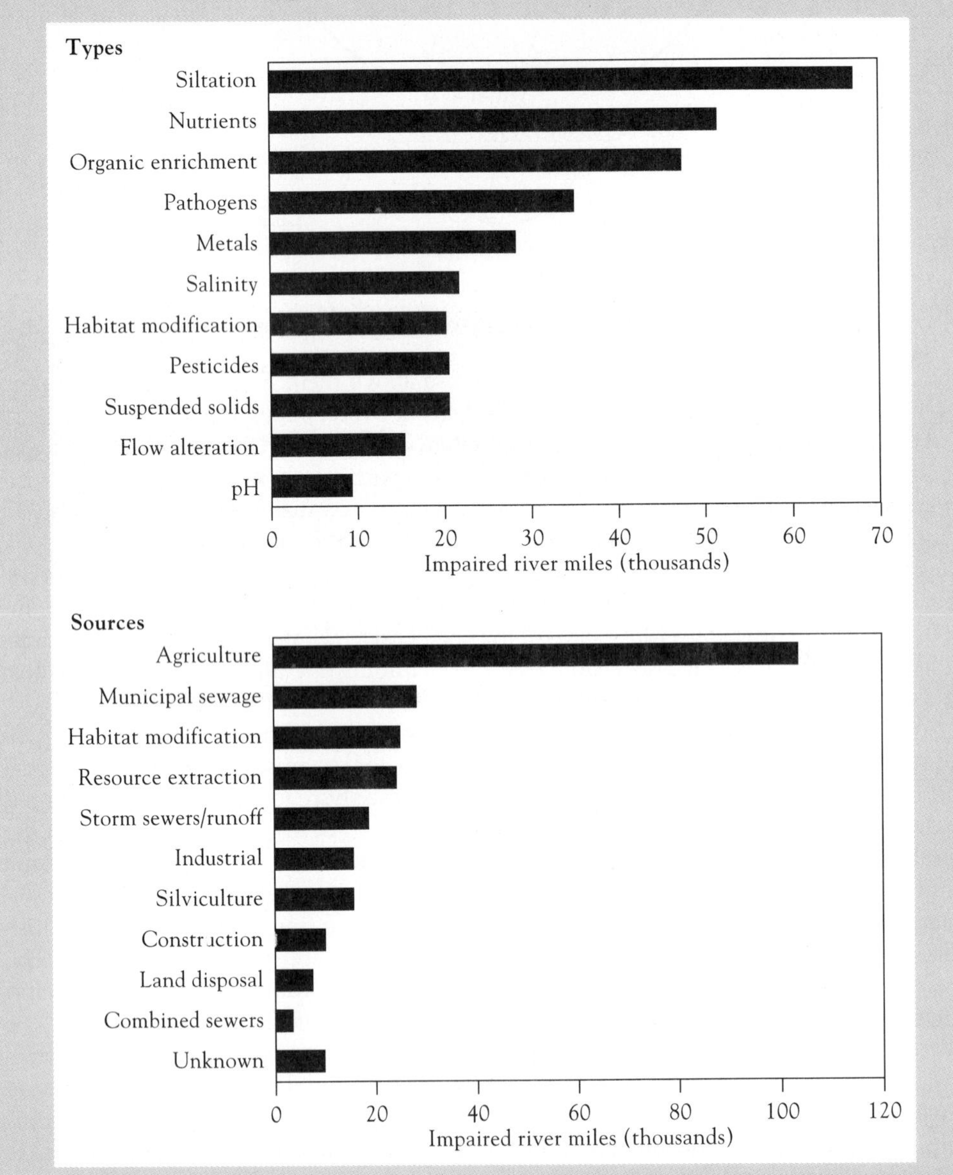

■ **FIGURE 1** The major types and sources of pollutants in U.S. rivers in 1990; data are in thousands of miles of impaired rivers.

as point sources since they discharge pollution from specific locations such as drain pipes, ditches, or sewer outflows. These sources are discrete and identifiable, so they are relatively easy to monitor and regulate. In contrast, nonpoint sources of water pollution are scattered and diffuse, having no specific location where they discharge into a particular body of water. Nonpoint sources include chemically laced runoff from farm fields, golf courses, lawns and gardens, construction sites, logging areas, roads, streets, and parking lots. Whereas point sources may be fairly uniform and predictable throughout the year, nonpoint sources are often highly episodic. The first heavy rainfall after a dry period may flush high concentrations of gasoline, lead, oil, and rubber residues off city streets, for instance, while subsequent runoff may have much less of these pollutants. The irregular timing of these events as well as their broad entry into a body of water makes them much more difficult to monitor, regulate, and treat than point sources. Unfortunately, nonpoint source pollution is responsible for over 75 percent of the pollution in both lakes and rivers.

The quality of natural waters varies enormously, both temporally and spatially, making broad comparisons between bodies of water difficult. For example, dissolved oxygen concentration, an important parameter for aquatic life, varies inversely with temperature. This makes comparisons of lakes in different seasons or between tropical and temperate lakes difficult. Also, specific bodies of water tend to be more or less susceptible to different types of problems. For example, lakes and reservoirs are more prone to acidification, sedimentation, and eutrophication than rivers. Ground water is often susceptible to agricultural chemicals yet is better protected from sewage.

In 1990, 70 percent of the U.S. rivers and 60 percent of the lakes met water quality standards and supported such uses as fishing and swimming; however, the other 30 percent and 40 percent respectively, were polluted to some extent. The main types of water pollution are illustrated in Fig. 1.

Siltation was the leading cause of pollution through the addition of suspended matter. The addition of nutrients like phosphorus and nitrogen was relatively high as a source of pollution, as was organic enrichment. Agriculture was by far the leading cause of water quality impairment, contributing to 60 percent of impaired stream miles and 57 percent of impaired lake acres. Municipal sewage treatment plants were next as a cause at 16 percent.

remain in place despite an often strong current. Nutrient accumulations and phytoplankton blooms do not occur because each would be quickly washed away. Current also mixes water thoroughly, providing a well-aerated regime. Animals of lotic systems are therefore not well adapted for low-oxygen environments and are particularly susceptible to high-BOD (oxygen-reducing) pollutants. Fish such as trout may be present in rivers with cool temperatures, high oxygen, and clear water. In warmer, murkier waters catfish and carp may be abundant.

SUMMARY

1. The early history of community ecology contrasted communities as "superorganisms" with emergent properties (Clements 1905) to individualistic associates (Gleason 1926).

2. Global classification schemes for communities, such as the Holdridge life zone classification map, are based mainly on vegetation in different temperature and rainfall regimes.

■ **PHOTO 14.7** North Fork of Payette River, central Idaho. Running-water habitats contain animals adapted to remain in place despite a strong current. *(Frazier, Photo Researchers, Inc. 2B9590.)*

3. Large-scale terrestrial communities are referred to as biomes and include tropical forests, temperate forests, deserts, grasslands, taiga,and tundra. Aquatic communities include: salt marsh, coral reef, open ocean, freshwater lakes, rivers and many other minor biome types that grade into these major biome types.

DISCUSSION QUESTION

Which do you think is the most meaningful scale on which to examine communities? For example, are the insects and other invertebrates that inhabit rotting logs a more tightly knit and therefore more biologically meaningful type of community than the temperate grassland community?

CHAPTER

15

Species Richness

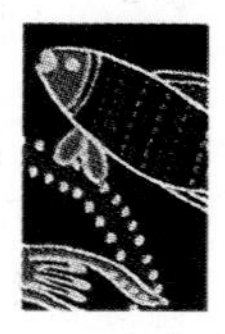

It has long been known that the number of species (species richness) of many plant and animal taxa is higher in the tropics than in more temperate regions. For example, the number of ant species in Alaska is 7, in Iowa 73, in Cuba 101, in Trinidad 134, and in Brazil 222. There are 293 species of snakes in Mexico, 126 in the United States, and only 22 in Canada. Over 1,000 species of fish have been found in the Amazon, whereas Central America has 456, and the Great Lakes of North America have only 142. The species richness of North American mammals increases from Arctic Canada to the Mexican border (Figs. 15.1a), and so does the richness of the birds (Fig. 15.1b).

Species richness is also affected locally by topographical variation, which increases diversity (hence, the increase of birds and mammals in the west) and by the peninsular effect, which reduces richness (hence, the drop of species toward Florida). It should also be noticed that there are more species of birds than mammals in any given region of the United States. Bird species diversity increases twelve-fold in the sixty degrees of latitude shown in Figure 15.1, whereas mammal diversity only increases eight times. Further examples of latitudinal gradients in species richness are given in Table 15.1. Sometimes a reverse pattern is found, as for sandpipers, family Scoloparidae, whose diversity increases toward the Arctic regions (Cook 1969); for aphids, whose diversity is highest in temperate realms (Dixon et al. 1987); and for helminth parasites of whales and sea lions (Rohde 1982). Richness of trees in North America is also not well linked to latitudinal gradients (Currie and Paquin 1987) (Fig. 15.2). Trees do not grow well in deserts of the U.S. Southwest, despite decreases in latitude and topographical variation. Some exceptions to the rule seem easy to explain. Aphids are highly host specific and often alternate between different plant species at different times in their life cycle. The tropics, with its extremely high diversity of trees, does not provide high densities of any one host-plant species. Thus, it would be difficult for aphids to find alternate hosts. Other exceptions, for example, the apparent trend of reduced diversity of parasitic insects in the tropics (Owen and Owen 1974) may be due to lack of sampling effort in the tropics

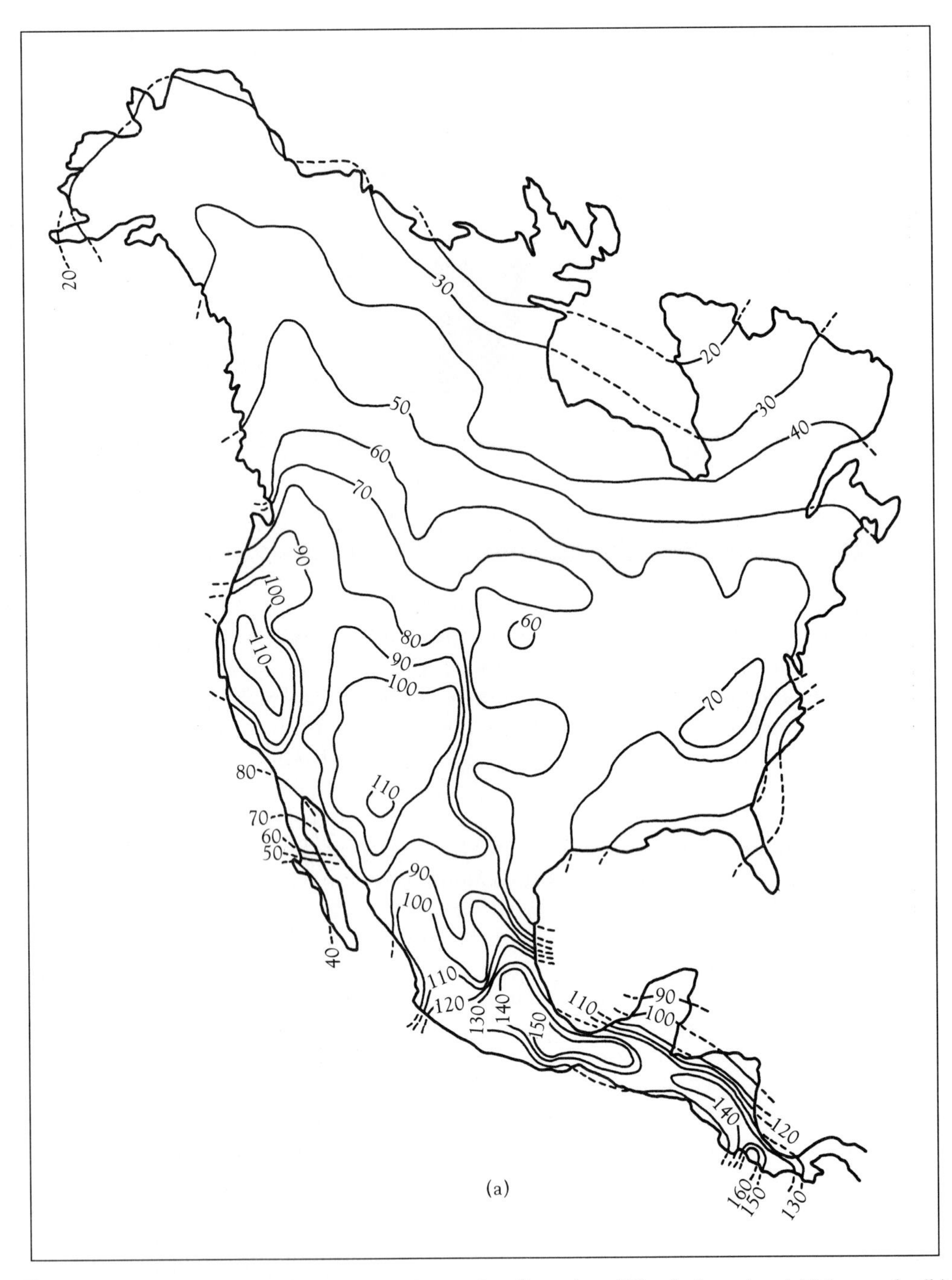

■ **FIGURE 15.1** Geographic variation in species diversity of North America. (**a**) Mammals. (**b**) Land birds. Note the pronounced latitudinal gradients in both groups and the high diversity in the southwestern United States and northern Mexico, a region of great topographic relief and habitat diversity. ([**a**] *Redrawn from Simpson 1964;* [**b**] *redrawn from Cook 1969.*)

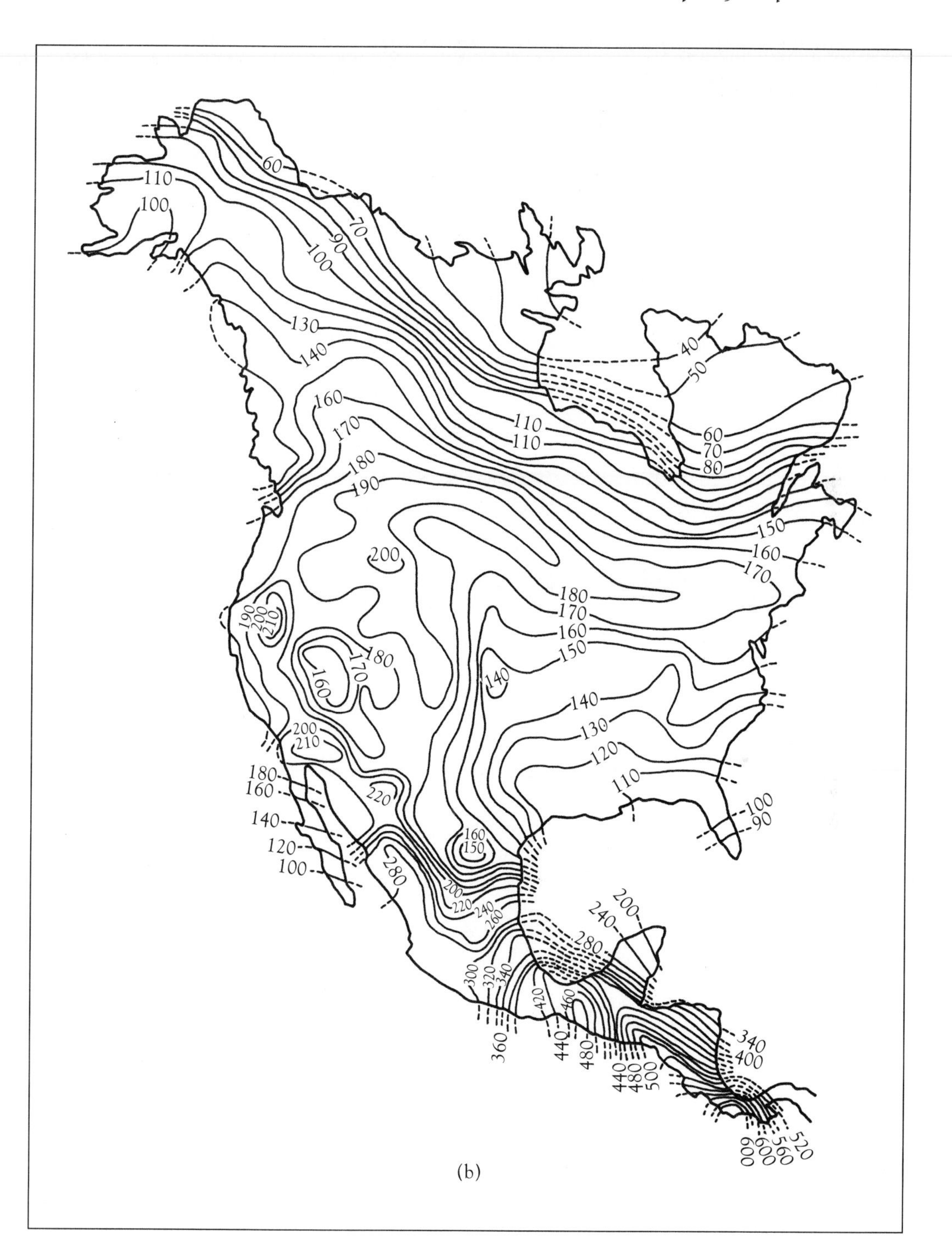
60
110
100
70
90
100
130
140
40
50
160
170
110
110
60
70
80
180
190
150
160
170
200
180
170
160
150
190
200
210
180
170
160
140
140
130
120
110
200
210
180
160
220
100
90
140
120
100
160
150
280
200
220
240
260
200
240
280
300
320
340
420
460
360
440
480
340
400
440
480
500
520
560
600
600

(b)

TABLE 15.1 Latitudinal gradients in species richness in various taxonomic groups. *(From Brown and Gibson 1983)*

TAXON	REGION	LATITUDINAL RANGE	RANGE OF SPECIES RICHNESS
Land mammals	North America	8°–66°N	160–20
Bats (Chiroptera)	North America	8°–66°N	80–1
Quadrupedal land mammals (all orders except Chitoptera)	North America	8°–66°N	80–20
Breeding land birds	North America	8°–66°N	600–50
Reptiles	United States	30°–45°N	60–10
Amphibians	United States	30°–45°N	40–10
Marine fishes	California coast	32°–42°N	229–119
Ants	South America	20°–55°N	220–2
Calanid Crustacea	North Pacific	0°–80°N	80–10
Gastropod mollusks	Atlantic coast of NA	25°–50°N	300–35
Bivalve mollusks	Atlantic coast of NA	25°–50°N	200–30
Planktonic Formainifera	World ocean	0°–70°N	16–2

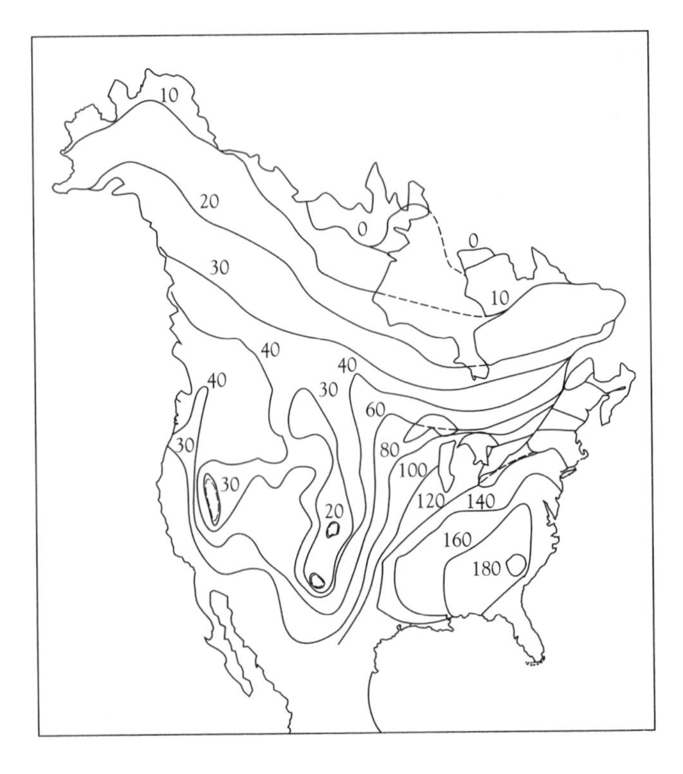

■ **FIGURE 15.2** Tree species richness in North America. (*After Currie and Paquin 1987.*)

(Morrison, Auerbach, and McCoy 1979). However, Hawkins (1994) has shown reduced numbers of insect parasitoid species per insect host species in colder climates than warmer ones, indicating that the pattern may be real. Certainly, the trends of increased diversity toward the equator seem more common.

15.1 Explanations for species richness gradients.

At least twenty-eight theories for temperate-to-tropical progressions have been advanced (Rohde 1992). Which one, if any, is correct, and why is this important? The importance lies in the current move to conserve biodiversity on earth—knowing what promotes diversity helps us preserve it. As for which theory is correct, the appropriate next step is to examine the most broadly proposed theories and examine the evidence in support of them. It is entirely possible that there are in fact multiple explanations operating for different taxa like birds, trees, or marine molluscs. Initially, however, it is useful to understand theories one at a time. We can divide the theories up into two main categories: biotic explanations and abiotic explanations.

Biotic explanations for diversity gradients argue that the increased diversity in the tropics reflects change in species interactions and habitat complexity.

Spatial-heterogeneity theory

Generally, there are more plant species in the tropics, which in turn support higher numbers of herbivorous animal species and hence carnivores. Richness of vegetation increases the numbers of herbivore species in two ways, by increasing the numbers of monophagous herbivores directly and also by creating a more diverse architectural complexity, providing more niches to occupy (Simpson 1964). MacArthur and MacArthur (1961) related bird species diversity to both plant species diversity and foliage-height richness (see also MacArthur 1972). Of course, the spatial-heterogeneity theory does not address the reason for the higher numbers of plant species themselves. Thus, this hypothesis remains incomplete.

Competition theory

It has been argued that, in temperate climates, natural selection operates mainly through harsh physical extremes and that species are generally *r* selected. In the more constant tropical temperatures, species are thought to become more *K* selected, to compete more keenly, and to interact more. This keen competition reduces species' niche breadth, allowing more species to pack along the resource axes (Dobzhansky 1950; Williams 1964). Again, no critical evaluation of this theory has been performed, and niche parameters have not been measured for a sufficient variety of species groups to determine how niche breadths are affected by tropical-to-polar gradients. Because competition has been found to be frequent and important in nature then the possibility that it could explain the pole-to-equator gradients in biodiversity warrants closer examination.

Predation theory

This theory, proposed by Paine (1966), runs contrary to the competition hypothesis and argues that there are more predators and parasites in the tropics and that these hold populations of their prey down to such low levels that more resources remain and competition is reduced, allowing more species to coexist. The increased richness in turn promotes more predators. Paine provided evidence to support the mechanism underlying this theory from studies on the intertidal communities of the U.S. Northwest coast at two wave-exposed sites: Mukkaw Bay and Tatoosh Island in Washington State, where the food web was fairly constant and the starfish *Pisaster* was the top predator:

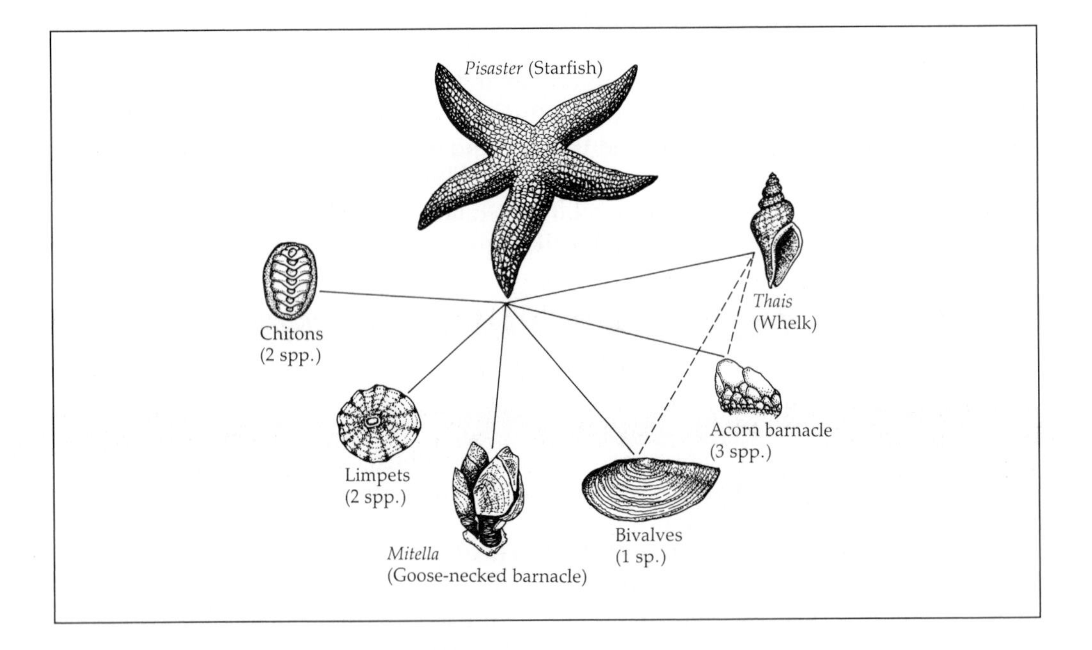

Thais preyed on bivalves and acorn barnacles. *Pisaster* preyed on those groups; on *Thais*; and on chitons, limpets, and *Mitella* as well. After removal of *Pisaster* from a section of the shore, diversity decreased from fifteen to eight species. A bivalve, *Mytilus*, increased, crowding out other species. In unmanipulated sections of shore, *Pisaster* tended to remove *Mytilus* and other species, preventing any one species from monopolizing space. Removal of any other single species from the system would not affect species diversity so drastically as *Pisaster* removal. For this reason, *Pisaster* was termed a *keystone species*, by analogy with the keystone that holds all the other stones in an arch in place (see Chap. 20). Top predators commonly specialize on the most abundant prey, developing a search image for it. Such a phenomenon, wherein predation allows the coexistence of more prey species, was noted even by Darwin (1859), who observed more grass species coexisting in areas grazed by sheep or rabbits than in ungrazed areas.

Of course, for such a system to explain tropical richness, the predation would have to be intense on the majority of species at all trophic levels, and few data are available as yet to test such an idea. We would have to explain why there are more predators in the tropics in the first place. Indeed, whether or not *Pisaster* acts as a keystone species even on most of the shores of the Pacific coast has been brought into question. Based on a survey of twenty sites in central and northern California, Foster (1990) argued that on some

Pacific rocky intertidal areas other processes may instead be operating. Mussels at some of these sites were never abundant enough to suggest that competitive displacement of other sessile organisms would occur. Paine (1991) noted that Foster did not advocate any other mechanisms as a causal factor in the observed distributions, but Foster (1991) replied that his concern was mainly to test the generality of Paine's original theories, not generate new theories. Their interchange makes fascinating reading. Once again, the frequency and strength of predation in nature needs to be examined more closely in order to more fully evaluate theories.

Animal-pollinators theory

In the tropics and other humid parts of the world, winds are less frequent and are of lower intensity than in temperate regions. This effect is accentuated by dense vegetative cover. Therefore, most plants are pollinated by animals: insects, birds, and bats. Even some grasses that are typically wind pollinated throughout most of the world are probably pollinated by insects in the tropics (Soderstrom and Calderon 1971). It is often thought that close associations built up between plants and specific pollinators, increasing the reproductive isolation between plant populations with a consequent increase in speciation rates. Coevolution of plants and pollinators then ensures high animal-pollinator speciation as well. If mutualisms like this are common in nature, they could explain increased biodiversity in certain areas. Although this theory is attractive for terrestrial systems, it cannot easily explain the similar diversity gradients that exist in aquatic systems. Furthermore, recent evidence suggests that many pollination systems are often more generalized than tradition suggests (Waser et al. 1996).

Abiotic explanations for diversity gradients emphasize the differences in environmental characteristics that parallel changes in diversity.

Ecological/evolutionary time theories

Proponents of these types of theories argue that communities diversify with time and that temperate regions have younger communities than tropical ones because of recent glaciations and severe climatic disruption. According to this view, species that could possibly live in temperate regions have not migrated back from the unglaciated areas into which the ice ages drove them (ecological theory) or resident species have not yet evolved new forms to exploit vacant niches (evolutionary theory) (Fischer 1960; Simpson 1964). In a useful analogy, tropical and temperate habitats can be compared to equal-sized libraries. The numbers of species, books in each library, is dependent on different things. In the tropics, the library is full, and size of books and available shelf space dictate the number of volumes held. In a temperate situation, there is plenty of available shelf space, and the number of books depends on their rate of purchase by the library and the length of time since the library opened.

Sanders (1968) provided evidence to support the evolutionary time theory by comparing diversity in glaciated and unglaciated northern hemisphere lakes that occur at similar latitudes. Lake Baikal in the former Soviet Union is an ancient unglaciated temperate lake and contains a very diverse fauna; for example, there are 580 species of benthic invertebrates (Kozhov 1963). A comparable lake in glaciated north Canada, Great Slave Lake, contains only four species in the same zone (Sanders 1968).

Another test of the time hypothesis has been provided by examining insect herbivore diversity on British trees. Since the end of the last Ice Age, trees have recolonized Britain, and in the past two thousand years, humans have introduced trees as well. Southwood (1961) was the first to examine the number of insect species associated with each tree species, and he found good correlations of insect diversity with length of tree tenure in Britain. Strong (1974a and 1974b), however, showed that insect species diversity was better correlated with the area over which a tree species could be found (see the section on area theory later in this chapter). Furthermore, Strong, McCoy, and Rey (1977) provided more detailed information on another system, sugarcane, its pest loads, and the dates of its introduction into at least seventy-five regions of the world over the past three thousand years. They found no support for the time hypothesis but good support for the area hypothesis. However, there have been several criticisms of Strong's papers based on the quality of data available on areas of sugarcane plantings in the tropics (for example, see Kuris, Blaustein, and Alio 1980) and for radiocarbon dating of tree pollen in Britain (Birks 1980). Still the time hypothesis remains weak. We might not expect terrestrial species to redistribute themselves quickly following a glaciation, especially if there is a barrier like the English Channel to overcome, but there seems no reason why marine organisms couldn't easily shift their distribution patterns in times of glaciations, yet the richness gradient still exists in marine habitats.

Productivity theory

This theory proposes that greater production results in greater richness; that is, a broader base to the energy pyramid permits more species in that pyramid (Brown 1981; Wright 1983). A common modification is that there is "room" for obligate fruit-eating birds (like parrots) or raptorial reptile eaters in the tropics but not in temperate regions (Orians 1969). Fruits appear year-round in the tropics, but a parrot would starve in a temperate winter. It is also argued that a longer growing season not only increases productivity but also allows component species to partition the environment temporally as well as spatially, thereby permitting the coexistence of more species. Thus, for species with annual life cycles, such as insects, some species could feed on leaves early in the year and others later. Currie and Paquin (1987) showed that the species diversity of trees in North America is best predicted by evapotranspiration rate (see Fig. 15.6). Realized annual evapotranspiration is correlated with primary production and is therefore a measure of available energy. Currie (1991) later expanded his arguments to discuss how diversity in North American birds, reptiles, amphibians, and mammals was also linked to energy. Turner, Gratehouse, and Carey (1987) demonstrated a correlation between the richness of British butterflies, exothermal species, and sunshine and temperature during the months they were on the wing, again suggesting a relationship between energy and species richness. Finally, a simple prediction from this theory is that the number of resident species in seasonal habitats should change according to the seasons. Turner, Lennon, and Lawrenson (1988) have shown that this is true for British birds. The number of birds present in Britain in the winter is less than that in summer, and this pattern is consistent with the amounts of energy present.

There are, of course, exceptions to this rule. Some tropical seas have low productivity but high richness. Eutrophic lakes have high productivity but low richness. So do coastal salt marshes, presumably because they represent a stressful environment for many organisms. Finally, Latham and Ricklefs (1993) have shown that while patterns of tree diversity in North America agree well with the evapotransporation theory, the pattern does not hold for broad comparisons between continents: the temperate forests of eastern Asia support substantially higher numbers of tree species (729) than climatically similar areas of North America (253) or Europe (124). These two areas have different histories and different access to source regions for new species as confounding factors.

Area theory

This idea is based on the notion that in larger areas the chances of isolation between populations increase, with corresponding increases in the chances of speciation (Terborgh 1973). It has also frequently been shown that larger areas support more species (see Chap. 19). Thus, large areas of climatic similarity will have greater species richness. On a worldwide scale, there is a symmetry of climates between polar regions and temperate areas, but only in the tropics do we see the symmetrically opposite climates adjacent, creating one large area. In a slight variation of this idea, Darlington (1959) argued that most dominant species evolved in the largest areas (which he also argued were the equatorial zones) and diffused out, creating a species-richness gradient. Neither theory, however, seems able to explain why, if richness is linked to area, there should not be more species in the vast contiguous land mass of Asia. North America has a larger area than Central America but has many fewer species of birds and mammals. Neither the surface nor the shelf areas of the northern Pacific are greater than those of the northern Atlantic but nevertheless the northern Pacific has a three-times-larger relative species richness of parasitic Monogenea (worms parasitic on the gills of fish). Rosenzweig and Sandlin (1997) have replied that while area is the primary determinant of species richness, it is also limited in unproductive environments (such as northern Asia), and the bleeding of species from one zone to another can obscure patterns. Their idea is that these three elements, area, productivity, and zonal bleeding cause most of the latitudinal gradients we see.

Rapoport's rule

Steven's (1989) compared the latitudinal ranges of North American trees, marine molluscs with hard body parts, freshwater and coastal fishes, reptiles and amphibians, and mammals between 25° and 70°N and found that high-latitude species had wider latitudinal ranges than low-latitude species (Fig. 15.3). He called the phenomenon Rapoport's Rule after Eduardo H. Rapoport who had referred to the phenomenon in his more sweeping study, *Areography* (Rapoport 1982).

This phenomenon suggests narrower environmental tolerances of tropical than of temperate species. If both groups dispersed to the same amount, a higher number of transient temperate species would occur in tropical habitats and be able to live there for a relatively short while. Their presence would inflate the species numbers in tropical zones. The establishment of species in sites where they cannot be self-maintaining has been shown to occur in plant communities on a local scale (Shmida and Wilson 1985). However, Rohde, Heap, and Heap (1993) showed that although the rule does apply to

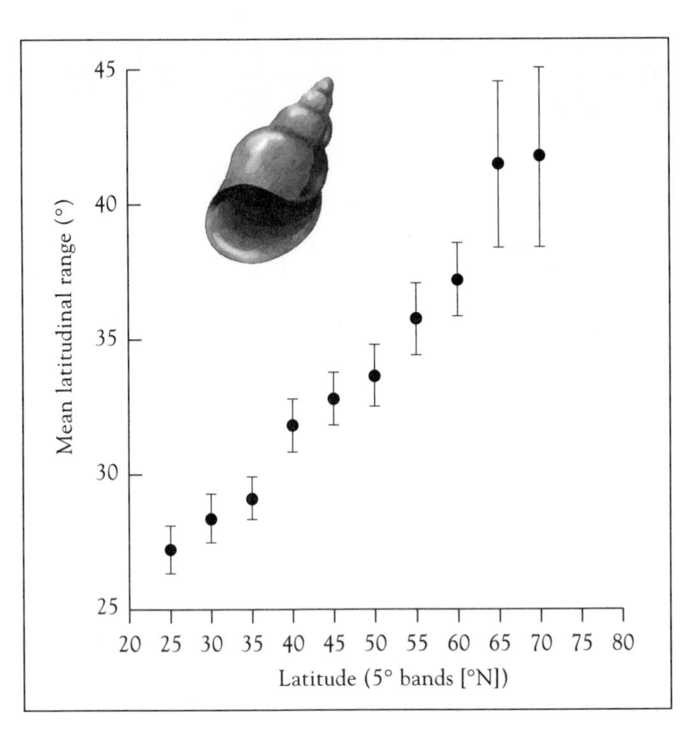

FIGURE 15.3 Mean latitudinal range of North American marine molluscs with hard body parts. For each latitudinal 5° band, all species occurring within that band irrespective of the midpoint of their distribution were considered, i.e., a species with a range of 50° appears in 10 or 11 bands. *(Modified from Stevens 1989.)*

North American freshwater and coastal fish above a latitude of 40°N it does not apply to marine teleosts in tropical waters.

Furthermore, Rohde, Heap, and Heap argued, the input of transients from adjacent habitats would increase local diversity but could not inflate total diversity because most times there just aren't adjacent habitats. For example, salt marshes occur in temperate coastal zones but not tropical ones; mangroves can be found only in the tropics, not in temperate areas. Thus, temperate salt marsh-living organisms would not be able to survive in the tropics. Rohde (1992) also noted the existence of Thorson's rule, which states that tropical benthic invertebrates are pelagic and dispersive, but cold-water benthic invertebrates are nonpelagic. This violates the equal dispersion assumption described earlier.

The evidence available to date suggests that the "rule" describes a local phenomenon, restricted to the northern land masses above a latitude of about 40 to 50°N. Support for the rule has come from studies of Palearctic mammals (Letcher and Harvey 1994), Nearctic mammals (Pagel, May, and Collie 1991), and Nearctic amphipods and crayfish (France 1992). So far, tropical animals have not been shown to have consistently narrower latitudinal ranges than temperate organisms (Smith, May, and Harvey 1994; MacPherson and Duarte 1994; Roy, Jablonski, and Valentine 1994, Colwell and Hautt 1994).

Of course, these eight theories are not exhaustive or mutually exclusive and can be combined in many permutations. Nevertheless, there is a strong tendency among ecologists to search for a common cause. Unfortunately, as pointed out by Rohde (1992) some "biotic" explanations are insufficient, that is, explanations that invoke increased competition or predation or disease are secondary explanations (Table 15.2). A primary

TABLE 15.2 Explanations of latitudinal gradients in species richness and their flaws. (*After explanations in Rohde 1992*)

TYPE OF FLAW	EXPLANATION
Secondary explanation—primary explanation still needed	Competition Predation Animal pollinators Spatial-heterogeneity
Insufficient evidence	Climatic stability Productivity Area Ecological/evolutionary time Rapoport's Rule
None as yet	Evolutionary speed

explanation is still needed to explain why these mechanisms might themselves be more or less important in certain areas. As for the "abiotic" explanations, there may be good correlations of various abiotic variables, like evapotranspiration rates with species diversity, but there is no reason why increased productivity promotes diversity and not simply higher population densities of just a few species. Furthermore, there often are numerous exceptions to these patterns.

Rohde himself concluded that the best explanation of the temperate-tropical richness gradient lies in terms of "evolutionary speed," which creates more species in the tropics. Higher energy levels in the tropics promote

a. Shorter generation times,
b. Higher mutation rates, and
c. Acceleration of selection leading to fixation of favorable mutants.

As yet, there is insufficient evidence to support or refute this theory (but see Jablonski 1993). Ricklefs and Schluter (1993) suggested that different processes may act on different scales. Biotic factors influence diversity more on a local scale and evolutionary processes on a provincial or global scale (Fig. 15.4).

15.2 Communities from different parts of the world may converge in structure.

Besides the generally recognized progression in species diversity from poles to the equator, there are some similarities in communities from different parts of the globe. For example, plant biologists have long noticed the similarity of vegetation formations in climatically similar areas around the globe. Cacti-like plants occur in deserts of all types. Systematists have also noted an astonishingly similar degree of convergence in morphology among distantly related organisms exploiting similar resources in different areas (Fig. 15.5). If species can converge in morphology, can communities converge in structure?

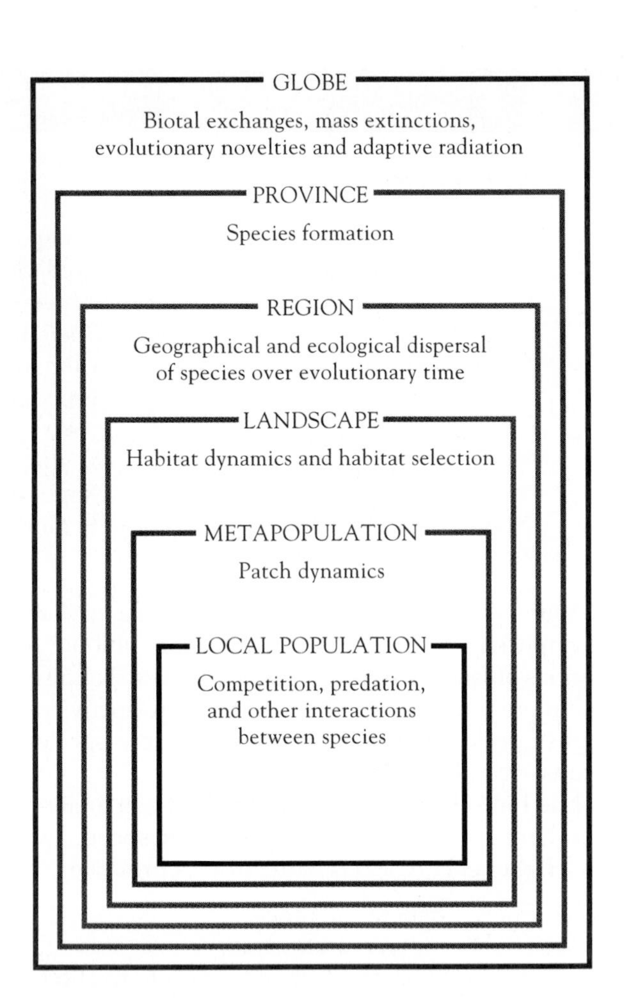

■ **FIGURE 15.4** A hierarchical viewpoint of processes influencing species diversity that have different temporal and spatial dimensions. Each level includes all lower ones and exists within all higher ones. Unique (chance) events may occur at any level of the hierarchy. (*After Ricklefs and Schluter 1993.*)

As is often the case, in ecology the evidence is equivocal. Schluter and Ricklefs (1993) provided a list of many examples in which there was nearly equal species diversity in similar habitats around the globe, then they provided another, practically equally sized list of dissimilar species diversities in similar habitats around the globe (Table 15.3a and 15.3b). They concluded that near identical richness does not always result when environmental conditions are judged to be the same.

One of the most comprehensive studies of convergence in species richness, or rather, lack of it, was done by Eric Pianka, who performed extensive field studies of desert lizards (summarized in Pianka 1986). Pianka was surprised to find quite different numbers of lizards in deserts around the world—sixty-one species in Australia and twenty-two in Southern Africa but only fourteen species in North America. Could it be that deserts are not functionally equivalent in different areas of the globe, or could there be historical evolutionary constraints also? Additional data from other habitats, like wetlands, show that while there are fewer species of lizards in wetlands globally, there still exists the same relative difference between Australia and South Africa (Fig. 15.6).

FIGURE 15.5 Convergence in mammals of tropical rainforests in South America and Africa. (**a**) Capybara and pigmy hippopotamus. (**b**) Paca and Africa chevrotain. (**c**) Agouti and royal antelope. (**d**) Three-toed sloth and Bosman's potto. (**e**) Giant armadillo and pangolin. The first three South American animals are rodents, and the first from Africa are ungulates, but convergence in shape and body size is still strong. (*Modified from Ehrlich and Roughgarden 1987.*)

TABLE 15.3A Examples of highly dissimilar species richness in similar habitats around the globe. (*After Schulter and Ricklefs 1993*)

ORGANISM	HABITAT TYPE	NO. SPECIES	NO. SPECIES
Mangroves	Mangal	Malaysia (40)	W. Africa (3)
Algae	Rocky shore	Washington (17)	S. Africa (3)
Chitons	Rocky shore	Washington (10)	S. Africa (3)
Insects	Streams	Australia (60)	N. America (26)
Insects	Bracken	England (21)	N. America (5)
Bees	Desert	Argentina (188)	Arizona (116)
Bees	Mediterranean scrub	California (171)	Chile (116)
Ants	Desert	Australia (37)	N. America (16)
Ants	Mediterranean scrub	California (23)	Chile (14)
Amphibians	Wetlands	Zambia (22)	Australia (14)
Lizards	Desert	Australia (27)	N. America (7)
Birds	Peatlands	Finland (33)	Minnesota (18)
Rodents	Desert	Arizona (16)	Argentina (5)

TABLE 15.3B Examples of nearly equal species richness in similar habitats around the globe. (*After Schlulter and Ricklefs 1993*)

GROUP	HABITAT	NO. SPECIES	NO. SPECIES
Plants	Desert	Arizona (250)	Argentina (250)
Plants	Semiarid	N. America (70)	Australia (65)
Sea anemones	Rocky shore	Washington (11)	S. Africa (11)
Ants	Desert	Arizona (25)	Argentina (25)
Sapr. insects	Tree hole	N. America (6)	Australia (6)
Fishes	Forest lakes	Wisconsin (4)	Finland (4)
Lizard	Mediterranean scrub	California (9)	Chile (8)
Birds	Desert	Arizona (57)	Argentina (61)
Birds	Mediterranean scrub	California (30)	S. Africa (28)
Birds	Shrub desert	Australia (5.5)	N. America (6.3)
Small mammals	Shrubland	California (7)	S. Africa (6)

Lizards are more speciose in Australia, whatever the habitat. Whittaker (1972) named the differences in diversity within habitats (in this case, deserts) alpha diversity and the difference between types of habitats beta diversity. The difference in diversity between two geographical regions, gamma diversity, would then be the product of alpha and beta diversity. Thus, the difference in species diversity of lizards in, say, Australia versus Southern Africa would be a combination of the numbers of different habitats in Australia and southern Africa (beta diversity) and the difference in diversity between identical habitats in each area (alpha diversity). However, as Rozenzweig (1995) notes,

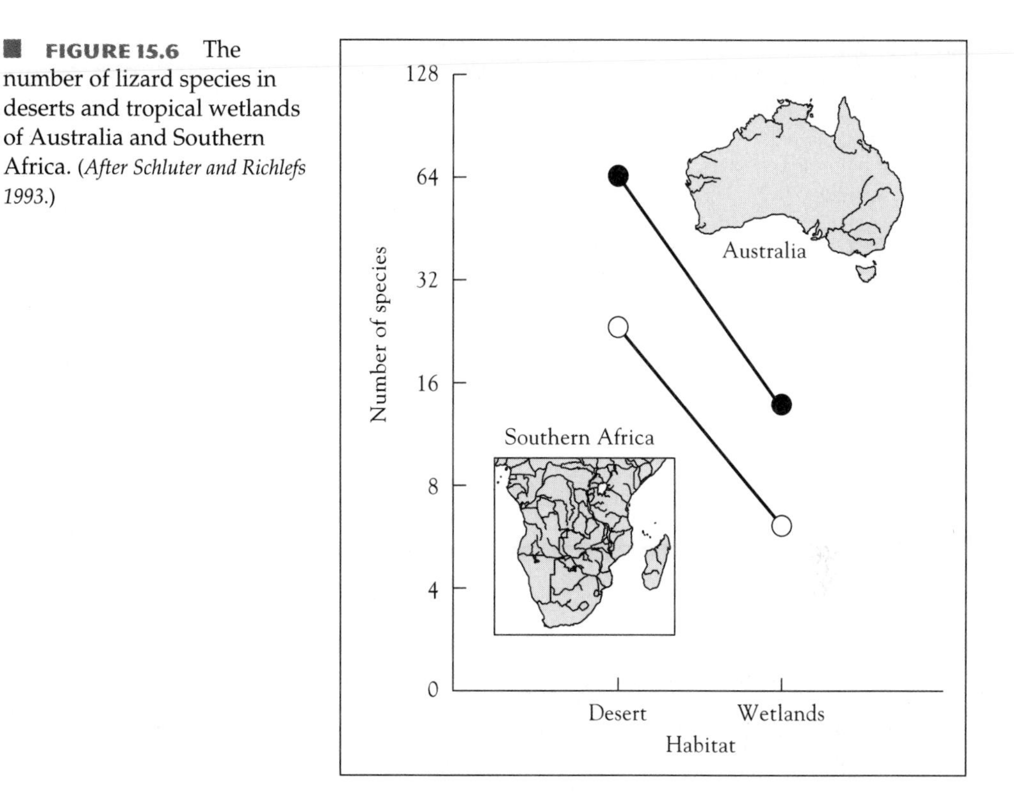

■ **FIGURE 15.6** The number of lizard species in deserts and tropical wetlands of Australia and Southern Africa. (*After Schluter and Richlefs 1993.*)

gamma diversity is no more than the number of species in a region, and having a jargon word for it doesn't help much. More than that, because gamma diversity is expressed as a number of species per unit area the units of alpha and beta diversity are not at all clear.

John Lawton and his colleagues (1993) have examined convergence in diversity in a different way, by examining convergence in guilds—the actual way species utilize their common resource. Bracken fern, *Pteridium aquilinum*, is a widespread and common native member of the flora of all the nonpolar continents. In many places it is rated as a serious weed, and its herbivorous insect fauna has been thoroughly studied, sometimes with a view to implementing biological control against it. Surveys of insects have been conducted in Hawaii, New Mexico, Great Britain, South Africa, Brazil, New Guinea, and Australia. The species assemblage varies remarkably, giving no evidence of taxonomic similarity in the fauna (Fig. 15.7). There are, for instance, no beetles on bracken in Britain and no hymenoptera in South Africa. Insects appear to have independently colonized bracken in different parts of its range over evolutionary time. Hawaii has no confirmed bracken herbivores, while New Guinea has the richest fauna with about thirty species. The variation in the total number of insect species exploiting bracken is partly a function of how common and widespread the plant is in each geographical region (Fig. 15.8). There is a strong species-area effect (see Chap. 19).

Lawton and colleagues have also asked if biotic interactions, in particular, competition, have shaped diversity. They argue that if competition were important one might

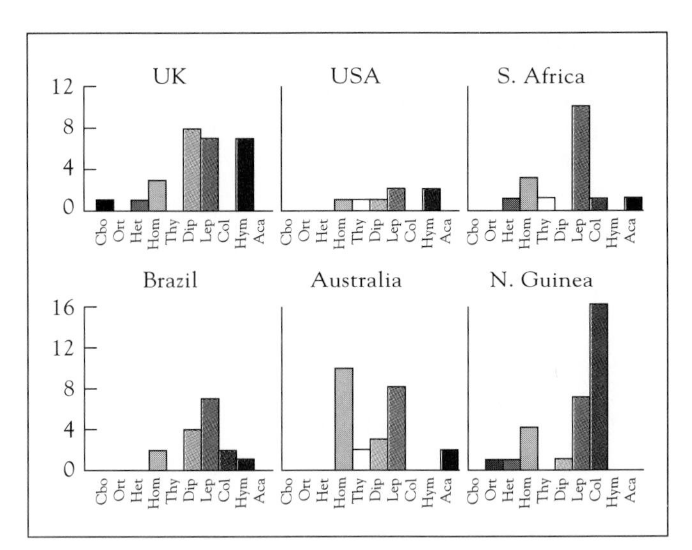

FIGURE 15.7 Taxonomic composition of the arthropod assemblages feeding on bracken in different parts of the world. Abbreviations: *CBO*, Collembola; *Ort*, Orthoptera; *Het*, Heteroptera; *Hom*, Homoptera; *Thy*, Thysanoptera; *Dip*, Diptera; *Lep*, Lepidoptera; *Col*, Coleoptera; *Hym*, Hymenoptera; *Aca*, Acarina. (*After Lawton et al. 1993.*)

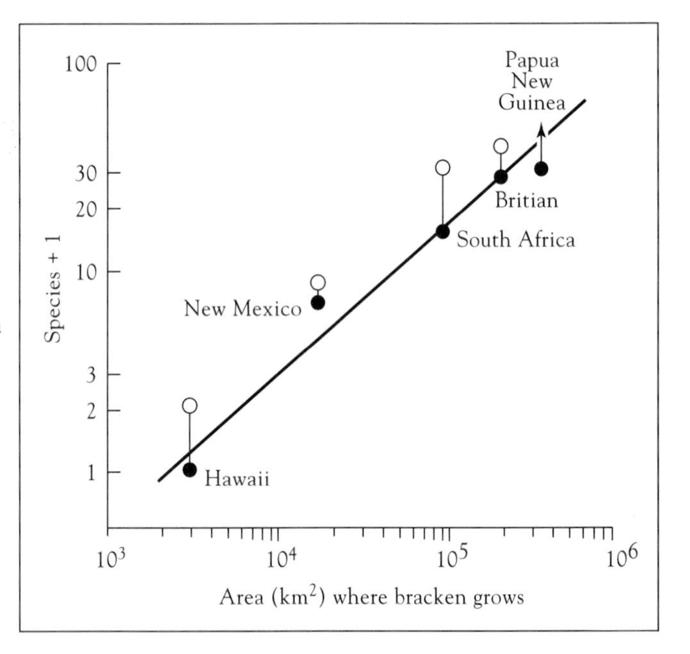

FIGURE 15.8 Species-area relationship for the number of species of herbivores definitely feeding on bracken (●) in different parts of the world. Also shown (○) are total numbers of species feeding on bracken including possible, occasional, and uncertain records. The arrow in the data for Papua New Guinea indicates that the number may be an underestimate. (*After Compton, Lawton, and Rashbrook 1989 and Lawton 1984.*)

expect it to produce convergence in the ways in which bracken is partitioned among herbivores, i.e., chewers, suckers, miners, and gallers, given that the plant is structurally similar everywhere, i.e., with rachis (stem), pinna (leaf), and costa (midrib). However, the distribution of species across resources on the plant is idiosyncratic from locality to locality with numerous vacant niches (Fig. 15.9). It does not look like there is conver-

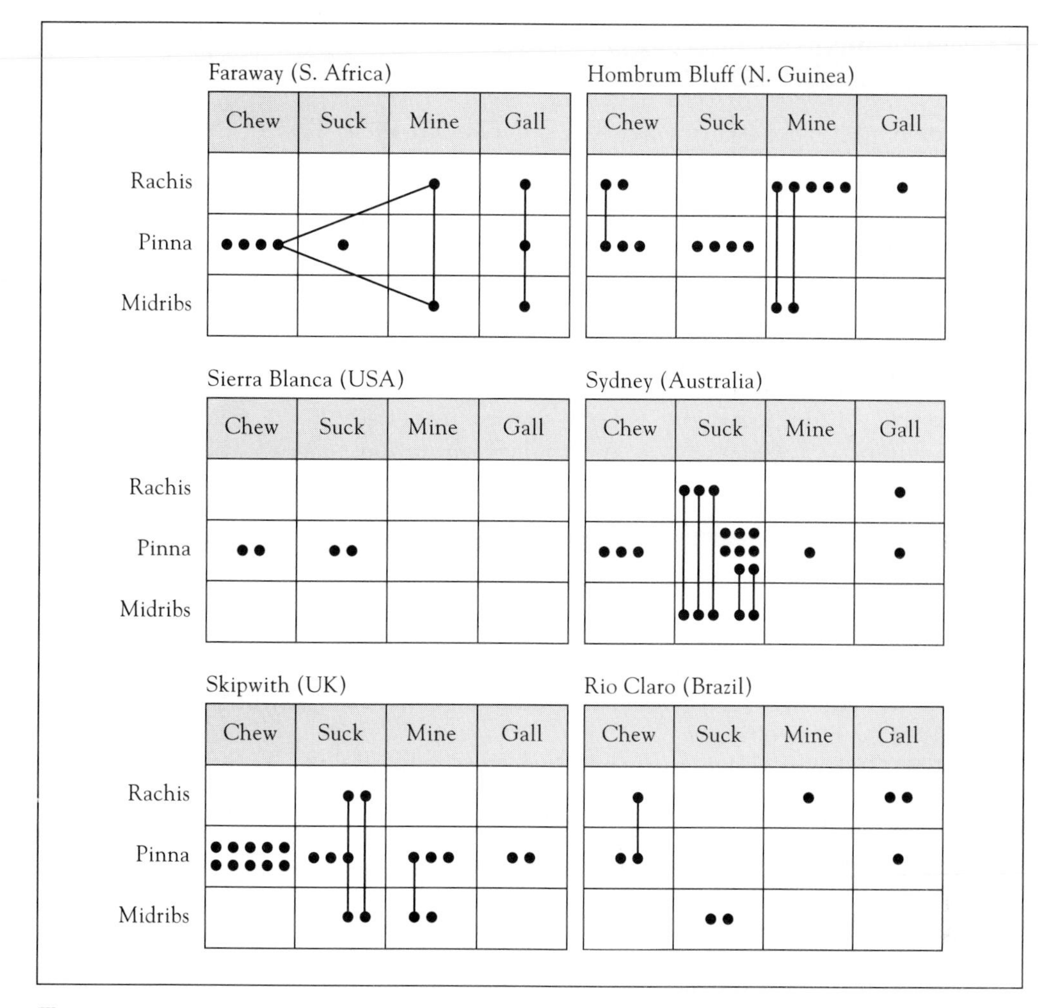

■ **FIGURE 15.9** Niche matrices defining the feeding sites and feeding methods of herbivorous arthropods on bracken at sites in different parts of world. Each dot represents one species, and feeding sites of species exploiting more than one part of the frond are joined by lines. Rachis = stem, pinna = leaf and cesta = midribs. *(After Lawton et al. 1993.)*

gence of feeding guilds across regions, nor does it look like competition is important. About the only pattern is that the pinna seems to be exploited more than the other parts, but this could be due to the fact that the pinna is the softest part of the plant. After twenty years of study on the system Lawton and colleagues (1993) have summarized the main rules that determine species diversity of insects on bracken (Table 15.4). The colonization of species in different parts of the world over evolutionary time has been idiosyncratic. Diversity is affected by latitudinal gradient and species-area relationships. The effects of species interactions look feeble.

TABLE 15.4 Factors that influence the diversity of insects feedings on bracken. (*After Lawton, Lewinsohn, and Compton 1993*)

FACTOR	SCALE	INDICATORS	IMPORTANCE
History	Continental	Faunal taxonomic composition	Strong
Plant range	Continental	Regional species-area effect	Strong
Latitude	Regional	Local assemblage size, turnover	Small but significant effects
Seasonality	Regional	Insect succession, dynamics	Variable
Habitat heterogeneity	Regional to local	Insect distribution	Variable
Patch size	Local	Local species-area effect	Weak
Community interactions	Local	Assemblage saturation, functional convergence	No detectable effects

15.3 As well as richness in species, communities also contain a richness of body size.

In addition to richness of species there are diversities in many other properties of organisms such as body size, reproductive rate, and longevity. Diversity of body sizes is perhaps one of the more obvious.

In 1959 G. Evelyn Hutchinson and his student Robert H. MacArthur (1959) wrote a paper on body-size pattern of species diversity. For most taxa they noted that there were many more intermediate-sized species than either very large or very small ones. Other authors have since found similar patterns (Bonner 1988; Dial and Marzluff 1988; Morse, Stork, and Lawton 1988; Brown and Nicoletto 1991; Brown, Margaret, and Taper 1993; and see Fig. 15.10). If one looks at the data on a logarithmic scale they are not normally distributed; they are right skewed. What is the explanation for this?

A recent theory (May 1986, 1988; Morse, Stork, and Lawton 1988) has employed the idea of fractal geometry. Fractals were developed by the mathematician Benoit Mandelbrot (1983) and concern the almost self-repeated patterns of similar structures over a wide range of spatial scales. Familiar examples are coastlines and snowflakes, which both appear to be equally irregular on whatever scale they are mapped. By a similar logic, the environment has fractal properties because it has many similar niches at a variety of different scales, each of which holds a specialized species. Brown (1995) gives the example of a Sierpinski gasket as an analogy (Fig. 15.11). As Brown points out, however, the fractal analogy has two drawbacks. First, the theory can account only for the decline in species size from the medium to large categories and actually predicts that the smallest species would be the most common. Second, what does it mean to say that the environment has a fractal structure and the breadth of niches is proportional to body sizes? How would one test whether such a pattern is different than that expected under any other scenario?

Rosenzweig (1995) has argued that there still is no satisfactory answer to the lognormal body-size question. Lawton (1991a) and Blackburn et al. (1993) have suggested

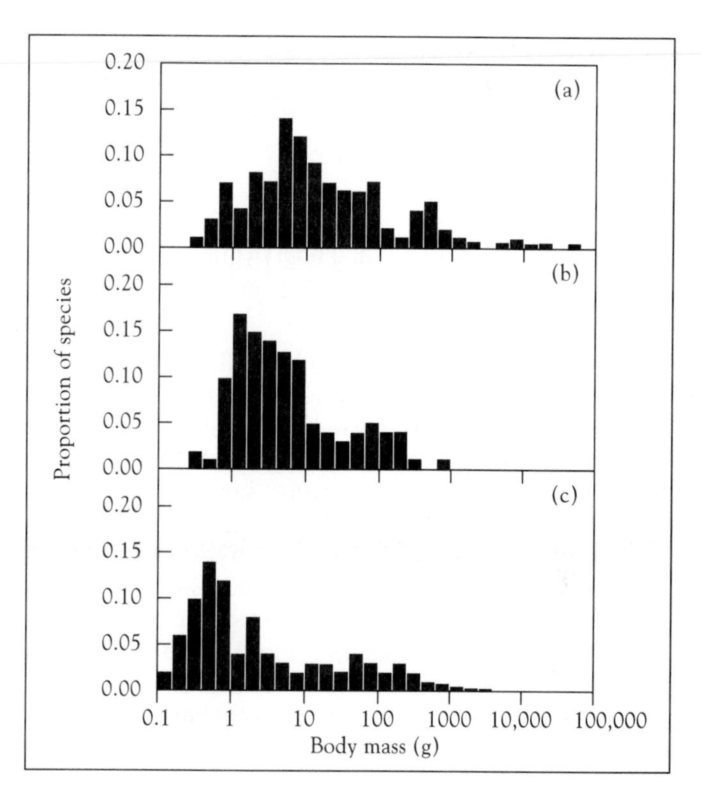

■ **FIGURE 15.10** Frequency distributions of numbers of species as a function of body mass (on a logarithmic scale) for three groups of vertebrates in North America; (**a**) Terrestrial mammals; (**b**) Land birds; (**c**) Freshwater fishes. All three distributions have a right-skewed shape. *(Adapted from Brown, Marquet, and Taper 1993.)*

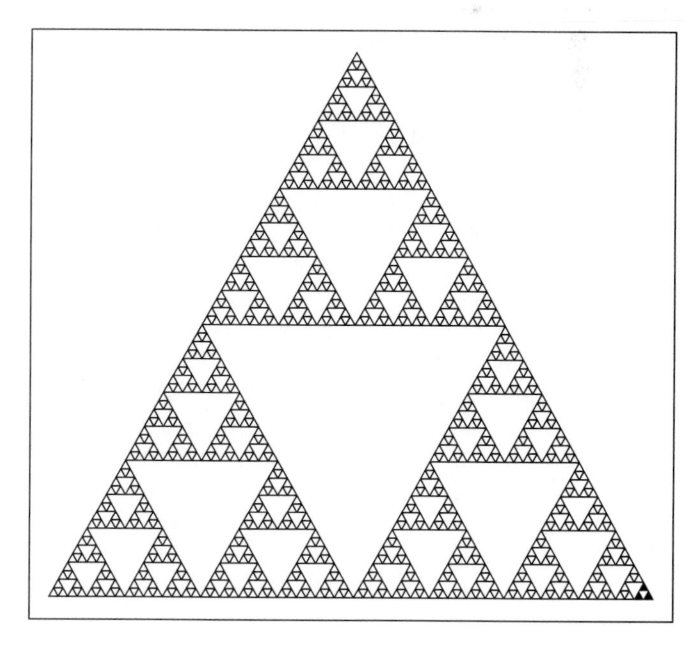

■ **FIGURE 15.11** The Sierpinski gasket, an example of a fractal pattern. Could this represent the packing of species of different body sizes (or the niches of those species), with there being more small species "inhabiting" the smaller triangles? *(From La Brecque 1992.)*

that species of intermediate body size are not only more diverse than others, they are also more abundant and that, ultimately, any explanation for the patterns of body size may be united with abundance. Brown (1995) suggests that, ultimately, patterns relating size and abundance will have to take into account energy use as well.

15.4 The estimate of the total global richness of species or biodiversity is about 10^7 species.

While ecologists strive to explain latitudinal gradients in species richness and diversity in body size, one even larger question remains: how many species are there on earth? Fewer than 2 million are currently classified (Table 2.1) while best estimates suggest about 12.25, million with maximum estimates of as much as 117.98 million (Table 15.5) . We have no idea why, theoretically, the global total is of the order of 10^7 rather than 10^4 or 10^{10} (May 1988). Insects are clearly the most diverse taxa on earth. Why? Remember that it is not the smallest or the biggest organisms that are the most abundant, but species intermediate between smallest and medium sized. A lognormal plot of the number of species on earth against body size, measured as length, gives the same right skew as most other graphs of this nature (Fig. 15.12). The body size of insects falls right where the number of species is highest (May 1978).

Even if one acknowledges the fact that insects are the most speciose group on earth, we still have to determine whether the total number of insect species on earth is closer

TABLE 15.5 Numbers of species in the groups of organisms likely to include in excess of 100,000 species (plus vertebrates). (*After World Conservation Monitoring Centre 1992*)

	ESTIMATED SPECIES	
	HIGHEST FIGURE	WORKING FIGURE
Viruses	500,000 +	500,000
Bacteria	3,000,000 +	400,000
Fungi	1,500,000 +	1,000,000
Protozoans	100,000 +	200,000
Algae	10,000,000 +	200,000
Plants	500,000 +	300,000
Vertebrates	50,000 +	50,000
Nematodes	1,000,000 +	500,000
Molluscs	200,000 +	180,000
Crustaceans	150,000 +	150,000
Arachnids	1,000,000 +	750,000
Insects	100,000,000 +	8,000,000
Total	118,000,000 +	12,230,000

to 2 million, 8 million, or 100 million. The estimate of the total number of species on earth will be influenced more by insects than by any other taxa. Besides this, estimates of total numbers of species are not likely to change much for the best-known groups like plants, birds, mammals, or indeed vertebrates as a whole. There is always, of course, the problem of lack of a uniform definition of what a species is (May 1990), but discussion of this topic is outside the scope of this book.

The basis of the estimate of high numbers of insects was the work of the entomologist Terry Erwin, who proposed, on the strength of samples of beetles taken from the canopies of nineteen specimens of one species of Panamanian tree, that there may be 30 million species of arthropods (principally insects) in tropical regions alone (Erwin 1982), or even 100 million (Erwin 1988). To arrive at this estimate Erwin had found 1,100 species of beetles on the canopy of one species of tree in Panama. He suggested that about 160 were specific to this tree and that beetles represent about 40 percent of all arthropods, so there would be about 400 canopy arthropods specific to this tree's canopy and about 600 on the canopy and the rest of the tree combined. If there are 50,000 species of tropical tree there are thus 30 million tropical arthropods. However, many problems with Erwin's assumptions were subsequently pointed out (Stork 1988; May 1988; Thomas 1990). For example, many tropical insects probably utilize more than one tree species—a necessary strategy because of the large intra-tree distances in the tropics. If each species utilized just two tree species then the total species estimate would be halved to fifteen million. Furthermore, Gaston (1991) surveyed insect taxonomists to obtain the input of

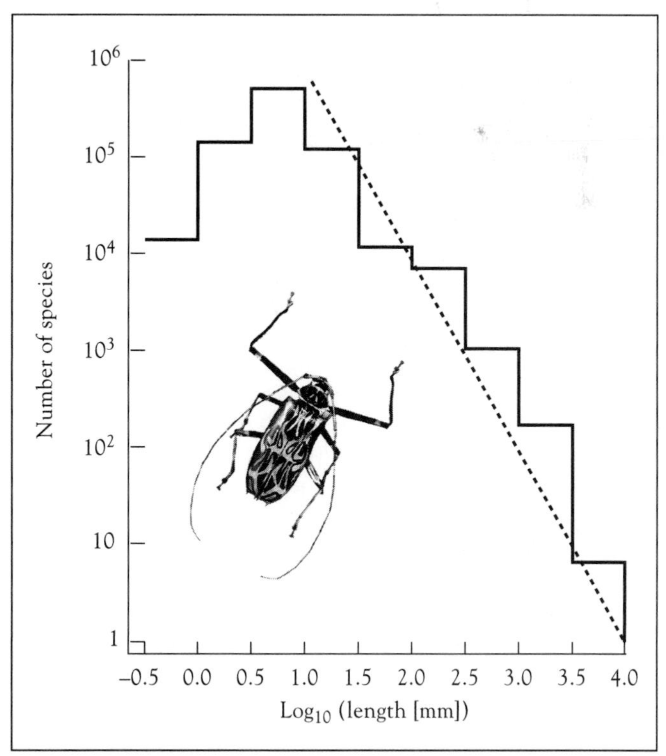

FIGURE 15.12 A crude estimate of the distribution of number of species of all terrestrial animals, categorized according to characteristic length L. The dashed line indicates the relation $S \sim L^{-2}$. (S is number of species.) *(After May 1998.)*

experts on likely insect diversity in different orders. Although new species are constantly being described, the rate of synonomy is about a third of this number, that is, experts finally recognize that a species they thought was new had actually been described by someone else earlier on. With such large numbers of insects this is quite a common problem—and one that may continue to grow given the ever-increasing numbers of species! Gaston concluded that a figure of between 5 and 10 million insects was probable.

15.5 Conservation of species richness emphasizes the preservation of areas of the world rich in total species.

A major goal of modern conservation biology is to ensure the maintenance of high species richness or biodiversity. In the past, most resources have been allocated to single, large species or "flagship" species like the Indian rhinoceros or the American buffalo, and preservation of these species has been ensured by trying to maintain their habitat. Conservation of biodiversity differs from this because the aim is to conserve the habitats that contain the most species. There are many reasons for conserving species (see Chap. 1), but why should we strive to preserve species richness? Because, at last, scientists are beginning to appreciate the link between richness and community function (Applied Ecology: Biodiversity and Ecosystem Processes) though this is not always the case. For example, a study by Wardle, Bonner, and Nicholson (1996) did not support the view that plant diversity affected in any way the degradation of plant litter.

Which habitats contain the most species? Generally tropical ones. A 13.7 km^2 area of La Selva Forest Reserve in Costa Rica contains almost fifteen hundred plant species, more than the total found in Britain, which has an area of 243,500 km^2. Ecuador contains more than thirteen hundred bird species, or almost twice the number as in the United States and Canada combined (Myers 1988). But there is much variation even between tropical countries. The South American and Asian tropics are much more diverse than the African tropics, and even here some areas are more diverse than others.

One method of targeting areas for conservation is to target countries with the greatest numbers of species, the so-called megadiversity countries (Mittermeier 1988; Mittermeier and Werner 1990). McNeely et al. (1990) used country species lists of vertebrates, plants, and butterflies to identify twelve such megadiversity countries: Mexico, Columbia, Ecuador, Peru, Brazil, Zaire, Madagascar, China, India, Malayasia, Indonesia, and Australia, which together hold up to 70 percent of the diversity in these groups. In this approach, bigger countries, because they hold more species (see also Chap. 19), fare better than smaller countries. Because conservation is most often managed at the national level, the proponents of this approach believe it works well. Perhaps its greatest drawback, however, is that although these areas contain the most species, they do not necessarily contain the most unique species (Williams et al. 1996). For example, the mammal species list for Peru is 344 and for Ecuador 271, with 208 species common to both. What is needed is some measure of the uniqueness of species—the endemics that are restricted to a country. Myers (1988, 1990) identified eighteen such hot spots for endemic tropical forest plants that, together contained 49,955 endemic species (20% of the world's total) in only 746,400 km^2 or 0.5 percent of the world's land area (Table 15.6). The next question, of course, is does the same pattern of endemism follow for other taxa, because if it does protection for these hot species will preserve a lot more than just plants.

TABLE 15.6 Numbers of endemic species present in 18 "hot spots." (*After World Conservation Monitoring Centre 1992*)

REGION	HIGHER PLANTS	MAMMALS	REPTILES	AMPHIBIANS	SWALLOWTAIL BUTTERFLIES
Cape Region (South Africa)	6,000	15	43	23	0
Upland western Amazonia	5,000	—	—	c.70	—
Atlantic coastal Brazil	5,000	40	92	168	7
Madagascar	4,900	86	234	142	11
Philippines	3,700	98	120	41	23
Borneo (north)	3,500	42	69	47	4
Eastern Himalaya	3,500	—	20	25	—
SW Australia	2,830	10	25	22	0
Western Ecuador	2,500	9	—	—	2
Colombian Choco	2,500	8	137	111	0
Peninsular Malaysia	2,400	4	25	7	0
Californian floristic province	2,140	15	15	16	0
Western Ghats (India)	1,600	7	91	84	5
Central Chile	1,450	—	—	—	—
New Caledonia	1,400	2	21	0	2
Eastern Arc. Mts (Tanzania)	535	20	—	49	3
SW Sri Lanka	500	4	—	—	2
SW Cote d'Ivoire	200	3	—	2	0
Total	49,955	375	892	737	59

Note: — = no data available yet. All areas are tropical forests, except Cape Region, California, SW Australia, and Central Chile, which have Mediterranean-type habitats.

Bibby et al. (1992) compared data for birds, mammals, reptiles, and amphibians and showed that, at least among larger vertebrates, areas rich in endemics of one taxa are often rich in endemics of another. An examination of Table 15.7, however, reminds us that such is not always the case. For example, while the Colombian Choco is rich in endemic reptiles and amphibians, it is poor in endemic mammals, and so is the Cape region of South Africa.

The downside of these types of arguments is that most areas rich in species or endemics are tropical rain forests. Scientists have recently argued that we also need to conserve representatives of all major habitats (Woinarski, Price, and Faith 1996). Thus, the Pampas region of South America, which is arguably the most threatened ecosystem on the continent, does not compare well in richness or endemics to the rain forests, but it needs to be preserved too. By selecting habitats that are most distinct from those already preserved, many areas that are not biologically rich but are threatened may be preserved in addition to the less threatened, but rich, tropical forest. The best strategy may then employ a "portfolio" of different habitat types.

APPLIED ECOLOGY

Biodiversity and Ecosystem Processes

Ehrlich and Wilson (1991) suggest that the loss of biodiversity should be of concern to everyone for at least three reasons. First, we have a moral responsibility to protect what are our only known living companions in the universe. The second reason is that humanity has already obtained enormous benefits from biodiversity in the form of foods, medicines, and industrial products like wood and rubber and has the potential to gain many more. But this second argument is couched more in terms of reasons for conserving individual species, not biodiversity. By this reasoning we should preserve varieties of wheat, rice, and corn and permit extinction of species with little value, like birds. We should promote the spread of trees whose timber we value over other species. This brings us to the third argument, which focuses on the array of essential services provided by natural communities. These services include maintenance of the "correct" gaseous composition of the atmosphere, which prevents global warming and the maintenance of soil biodiversity so as to maintain the ability of the soil to support forests and crops, decompose organic matter, recycle nutrients, and dispose of wastes. Other community functions include the maintenance of a reservoir of natural enemies to prevent pest outbreaks and of a reservoir of pollinators to pollinate crops and other plants. A skeptic might say that one grass or one tree can function as well as any other in helping control the hydrologic cycle, one predator will be as good as another in controlling a potential pest, and one tree species be as good, or better, than others in terms of lumber production. The truth is that organisms are generally adapted to specific physical and biotic environments—and substitutions are likely to prove unsatisfactory. But likely is the operative word here; only now are scientists starting to provide the data to back up the rhetoric. Two recent studies have provided data that diverse communities perform better than less diverse ones.

Shahid Naeem and colleagues (1994) used a series of fourteen environmental chambers, each 1 m^2, in a facility termed the Ecotron, at Silwood Park, England to replicate terrestrial communities that differed only in their biodiversity (Photo 1). The communities consisted of communities of nine, fifteen, and thirty-one species, spread across four trophic levels, with the species-poor communities being subsets of the more diverse communities (Fig. 1). Loss of richness was designed to be equivalent on all trophic levels: decomposers (earthworms and *Collembolla*), primary producers (annual plants), herbivores (aphids, whiteflies, snails, and slugs), and parasites attacking some of the herbivores. The experiment ran for just over six months, and species were added when the trophic level below them was established. For example, parasitoids were not added until herbivores were abundant. Researchers measured a whole host of variables such as community respiration, decomposition, nutrient retention rates, and productivity. The result was that community productivity increased two- to threefold as biodiversity increased two- to threefold (Fig. 2). Such experiments had never before been used to demonstrate that communities differing only in species richness differed in productivity. In the field experiments conducted by previous workers, many other environmental variables were altered as communities changed. The Ecotron experiments controlled environmental variables. The proposed

■ **PHOTO 1** Ecotron facility, Silwood Park, England, is a series of fourteen environmental chambers like this one, used to investigate the link between biodiversity and ecosystem function. (*Photo by Shahid Naeem, University of Minnesota.*)

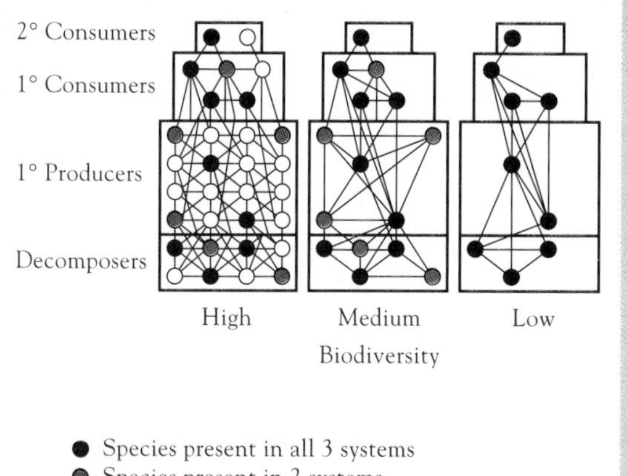

■ **FIGURE 1** Community diagrams of the three types of model terrestrial ecosystems developed in the Ecotron. Circles represent species, and lines connecting them represent biotic interactions among the species. Note that each lower-diversity community is a subset of its higher-diversity counterpart and that all community types have four trophic levels. *(After Naeem et al. 1994.)*

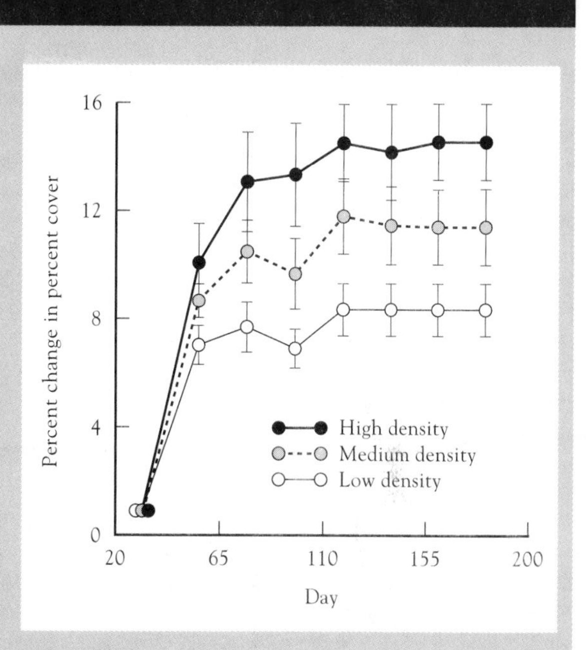

■ **FIGURE 2** Plant productivity is linked to community diversity as measured by the Naeem et al. (1994) Ecotron experiments. Percentage change in percentage vegetation cover (mean ± 1 S.E.) from initial conditions as determined by analysis of video images of canopies. Symbols: ●, low diversity; ▲, medium diversity; ■, high diversity. *(After Naeem et al. 1994.)*

mechanism for increased plant production was thought to be an increase in light interception associated with plant canopies that filled three-dimensional space more completely.

More recent experiments by Naeem and colleagues (1996) focused on a single trophic level (plants) and documented net aboveground primary productivity changes as a result of changes in plant diversity (See Ecology in Practice: Shahid Naeem that follows this section.) Species richness was used in replicate experiments in English weedy fields from one to sixteen species, and plant biomass was measured at the end of one growing season. On average, species-poor assemblages were less productive. Results also showed, however, that species-poor assemblages had a wider range of possible productivities than more diverse assemblages, reiterating the link between diversity and stability that Tilman (1996) had shown in plant assemblages in Minnesota (see Chap. 17).

David Tilman and colleagues (Tilman 1996; Tilman, Wedin, and Knops 1996) took such experiments a step further by performing them in the field, in a Minnesota prairie. Earlier, Tilman and Downing (1994) had suggested that species-rich grasslands are more resistant to the ravages of drought and recovered from drought more quickly than species-poor grasslands. That is, diversity increased both resistance and resilience. However, this older work was the result of a study of natural communities, not experimentally generated ones. The newer work directly manipulated plant diversity and measured plant production and nitrogen extraction rates from

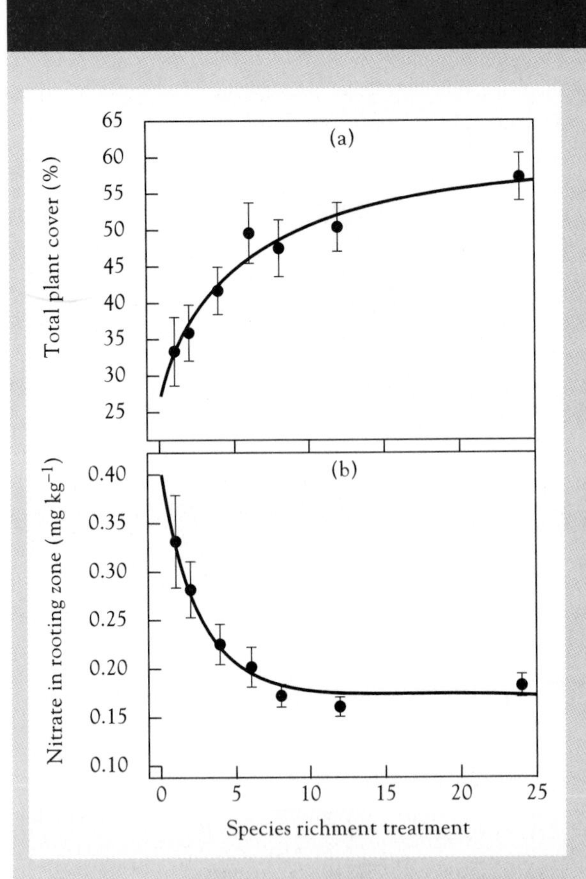

■ **FIGURE 3** The relationship between biodiversity and community biomass (**a**) and uptake of nitrogen (**b**) on experimental plots. (*After Tilman et al. 1996.*)

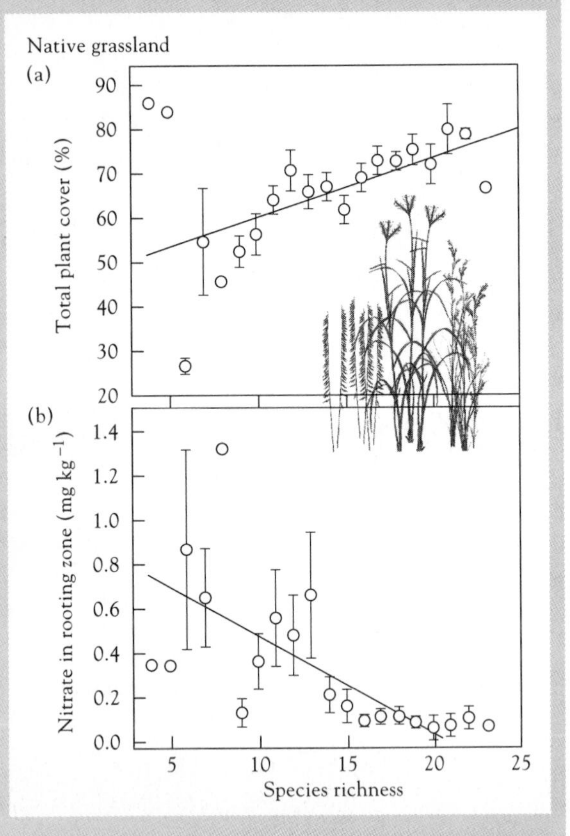

■ **FIGURE 4** The relationship between biodiversity and community biomass (**a**) and uptake of nitrogen (**b**) on native plots. (*After Tilman et al. 1996.*)

the soil. One hundred and forty seven plots, $3 \times 3\text{m}^2$, of comparable soils, were sown with seeds of one, two, four, six, eight, twelve, or twenty-four, species of prairie plants replicated twenty times for each treatment. Exactly which species were sown into each plot was a random draw from a pool of twenty-four native species. Thus, any differences in results between plots with one or two species and plots with twelve or twenty-four species would be due not to the identity of the species used, but to the number of species per se. The results were clear: more diverse plots used nutrients more efficiently than less diverse ones and had increased productivity (Fig. 3). Not only was this pattern evident in the experimental plots, it could also be found in a sample of thirty different unmanipulated native prairies (Fig. 4). The conclusion was that, in more diverse communities, interspecific differences in soil use allow fuller use of nitrogen, the main limiting nutrient.

Tilman argued that the relationship between productivity and diversity is, however, not linear but is strongest at lower levels of diversity and weaker at higher levels of diversity. Thus, diversity only affects community processes up to a point; after this it matters less.

Huston (1997) has urged caution in the interpretation of both Naeem's and Tilman's experiments. This is because, Huston suggested, when one varies species diversity in experiments it is very easy to, unintentionally, vary additional parameters at the same time. Thus, Naeem's experiments always used a particular subset of the high-diversity species group for the medium- and low-diversity treatments. As a result, the effect of the number of species present could not be distinguished from the particular set of species chosen for that treatment. This was compounded by the fact that the species chosen for the low-diversity treatment were plants that generally only grow to a small maximum size while the intermediate- and higher-diversity treatments included species of greater maximum size. Tilman's experiments avoided this effect but suffered from a more subtle problem. There is an increasing probability of selecting species with a specific property (e.g., large maximum height or nitrogen-fixation ability) in samples of increasing number that are randomly selected from any group of species. This is important because multispecies groups of plants are typically dominated by individuals of the largest species. Consequently, most of the biomass in each treatment was contributed by one or a few of the dominant species. Houston refers to these effects as hidden treatments of the experiments. Part of the problem is that many ecologists want to believe that increased diversity is "useful" to communities, and as diversity is under seige in so many areas of the world, such experiments are great ammunition in the fight to show why we should be preserving biodiversity. Indeed, more authors are beginning to perform other experiments to show how diversity and ecosystem functions are linked (Hooper and Vitousek 1997; Wardle et al. 1997; Tilman et al. 1997).

TABLE 15.7 Countries rich in endemic land vertebrates. (*After World Conservation Monitoring Centre 1992*)

COUNTRY RANK ORDER	ENDEMIC TAXON							
	MAMMALS		BIRDS		REPTILES		AMPHIBIANS	
1	Australia	210	Indonesia	356	Australia	605	Brazil	293
2	Indonesia	165	Australia	349	Mexico	368	Mexico	169
3	Mexico	136	Brazil	176	Madagascar	231	Australia	160
4	USA	93	Philippines	172	Brazil	178	Madagascar	142
5	Philippines	90	Peru	106	India	156	Ecuador	136
6	Brazil	70	Madagascar	97	Indonesia	150	Colombia	130
7	Madagascar	67	Mexico	88	Philippines	131	India	110
8	China	62	New Zealand	74	Colombia	106	Indonesia	100
9	USSR	55	Solomon Islands	72	Ecuador	100	Peru	87
10	PNG	49	India	69	Peru	95	Venezuela	76
11	Argentina	47	Colombia	58	Cuba	79	Cameroon	65
12	Peru	46	Venezuela	45	South Africa	76	Zaire	53

ECOLOGY IN PRACTICE

Shahid Naeem, University of Minnesota

There are two things that are great about being an ecologist. First, just about anything you learn can be useful in ecology. Second, it never gets boring. As a resident of New York City, I was exposed to a tremendous diversity of people and ideas, none of which had anything to do with ecology or nature. I went to the City College of New York largely because I was bored with my work as an inventory controller for a publishing firm in Manhattan. I moved to California and transferred to the University of California at Berkeley primarily because I ran out of courses to take at the City College of San Francisco. I majored in biology only because it allowed me to take a wide variety of courses, not because I had any plans to be a biologist. By my second year at Berkeley, I found the tertiary structure of the sodium ionophore in membranes of neurons to be the epitome of what the biological sciences stood for—focused, high-tech, abstract, aesthetically appealing, and ultimately boring. So I decided I'd be an illustrator.

My interests changed dramatically, however, when, quite by chance, I began working for Dr. Robert K. Colwell who was then a professor at Berkeley. Rob's research centered on understanding the significance of complexity in biological communities. This struck me as totally wacky, but I found the question way cool. Even more wacky was the fact that Rob's way of approaching the question was by studying mites that rode from flower to flower on the noses of hummingbirds, an approach that required endless trips to the tropics and many adventures, including nearly losing his life to a snake bite. As a city slicker, I had no particular desire to go through similar experiences, but I was infected with the desire to understand the role of complexity in nature. So I switched my major from an emphasis in cell-molecular biology to an emphasis in organismal population biology and later obtained both a master's and Ph.D. at Berkeley.

One might think that being a city person and trained primarily in cell biology would be a poor start. In fact, it turned out to be an ideal training for dealing with the logistical problems of international travel and field work as well as working in the laboratory whenever the need arose. I have worked in the lowland tropical rainforests of Central America and the Caribbean, in the low-elevation bogs of the Sierra Mountains, in the mud flats of Denmark, and now in the prairie grasslands of the North American Midwest. I have also worked on model ecosystems such as outdoor tanks stocked with plankton in Michigan; with assemblages of plants in greenhouses in Minnesota and England; with artificial terrestrial communities of plants, animals, and microbes in England; and currently with microbial microcosms in the laboratory here at Minnesota.

My research on artificial terrestrial ecosystems is a good example of where having a diversity of experiences helps a great deal. In collaboration with Professor John H. Lawton, Dr. Lindsey Thompson, Dr. Sharon Lawler, and Richard Woodfin, we attempted to answer our own personal wacky question: does biodiversity affect the functioning of ecosystems? This question may seem strange, but an international panel of ecologists who met in Germany in 1992 could provide no clear answer to it (Schulze and Mooney 1993). That is, aside from the obvious economic, moral, and aesthetic values of biodiversity, what benefits are there to having ten to one hundred million species on the planet? Does a rain forest need three hundred species of trees per hectare, or will a mono culture of bananas sequester carbon dioxide and cycle nutrients just as well? Does a prairie grassland need its two hundred species of plants, or can the depauperate corn fields that have replaced them do just as well in terms of nutrient cycling and energy flow?

Two books strongly influenced my approach in tackling this question; Schlesinger's *Biogeochemistry* (Schlesinger 1991) and De Angelis's *Dynamics of Nutrient Cycling and Food Webs* (De Angelis 1992). Together, these books suggested that the key to understanding the relationship between biodiversity and ecosystem functioning was understanding how variation in community structure affected biogeochemical processes. This meant that to test the idea, we had to build a series of replicate ecosystems identical in every way except one—community structure. Building and maintaining replicate model ecosystems was difficult work. It took a tremendous amount of planning and team effort to keep our fourteen model ecosystems going. Equipment failures, contamination, computer problems, and much more kept us up at all hours, often putting us in an ill humor—it made field work look easy. My experience in working with people; managing inventory; conducting research both in the laboratory and the field; and working with plants, animals, and microbes all paid off.

One particularly worrisome aspect of this research was fact that we carried out our experiment in the "Ecotron," a controversial system of controlled environmental chambers built at Imperial College of London's Centre for Population Biology under the directorship of Professor Lawton (Lawton et al. 1993). Nearly two million dollars went into building the Ecotron and the center that supported it. At a time when Thatcherism had ravaged the British economy and environmental research in the United Kindom was suffering from severe shortages of funds, the Ecotron was viewed with tremendous skepticism by the British ecological community. Was this the best way to spend limited funds? Was laboratory research even appropriate for ecology (compare, for example, Carpenter 1996; Drake et al. 1996)?

When I delivered our findings to the British Ecological Society meetings in 1994, the auditorium was packed and overflowing into the hallways with ecologists wanting to see and hear about the infamous Ecotron. What we had found was that even if ecosystems are intact, as biodiversity declines, ecosystems change their functioning (Naeem et al. 1994; Naeem et al. 1995). The audience was thrilled with our findings. Here was experimental evidence that the dramatic declines we are experiencing in global biodiversity could mean an irreversible change in the way our ecosystems function—an important reason to work toward the preservation of biodiversity. After we published the paper, the Ecological Society of America awarded the study its Mercer Award, and it became the fifth most-cited paper on global change research in 1996. Professor Lawton's vision of a large-scale laboratory facility for ecological experiments proved its value, and the Ecotron continues to do exciting research in global change ecology.

There has and continues to be much debate over the Ecotron's findings, but I am used to such debate. My research and that of my students focuses on addressing controversies in ecology by direct field or laboratory experimentation. This approach gets us in trouble, but I guess we feel that it is better to do something that is intellectually exciting even if it is risky than to do something technically flawless but boring.

As global change continues, the need for novel approaches will be critical. Ecology is fast becoming the premier science in environmental problem-solving, and these problems are topping both national and international agendas. The future of ecology will be one in which groups of investigators bring all their experiences and training together and take risks to tackle complex issues in ecology by any means possible.

SUMMARY

1. Species richness increases from the poles to the equator. Thus, tropical areas have more species than temperate areas, which in turn have more species than arctic areas. There have been a variety of explanations for this phenomenon. These explanations can be grouped under two headings: biotic explanations and abiotic explanations. However, as pointed out by Rohde (1992), most of these explanations seem secondary or incomplete. Perhaps the best explanation is that evolutionary speed is higher in the tropics, and this creates more species.

2. Besides the progression of species richness from poles to the equator there is convergence in species richness in different habitats, for example, in deserts or wetlands, although this is not always the case. There is also a richness in body sizes in different areas.

3. There are a variety of methods for managing richness or biodiversity on Earth. These include increased attention to so-called megadiversity countries, with very high species-richness, or to "hot spots" with high numbers of endemic species.

4. Evidence is emerging that richness can affect ecosystem function.

DISCUSSION QUESTION

If we are to preserve biodiversity on Earth should we focus primarily on species-rich areas, on "hot spots" of rare or endemic species, or on underrepresented habitats?

CHAPTER

16

Species Diversity

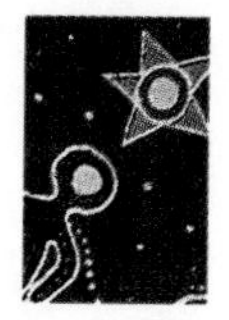

The simplest measure of diversity is to count the number of species (Photo 16.1) the result is termed *species richness* (Mcintosh 1967). Although most analyses of community diversity are compared this way, one major problem is that this approach does not take species abundance into account. For example, in a sample of one hundred individuals from a habitat, the species richness of a community of two species each of population size fifty would equal that of a community in which the population sizes of the same two species were one and ninety-nine. In actuality, the first community must be considered more diverse; one would be much more likely to encounter both species there than in the second community. To overcome this problem one can incorporate both abundance and species richness in measures of species diversity. Many diversity indices have been developed to do just that. An alternative is to rely on species richness but to control for the effects of sample size by a procedure called *rarefaction*.

16.1 The number of species found in a given sample is strongly dependent on the size of that sample.

Counts of species numbers are very dependent on sample size: the larger the sample the greater the expected number of species. One method of avoiding incompatibility of measurements resulting from samples of different sizes is called rarefaction. This involves calculating the number of species expected from each sample if all the samples were reduced to a standard size (such as one thousand individuals). The correct formula for doing this was derived independently by Hurlbert (1971) and Simberloff (1972):

$$E(S) = \sum_{i=1}^{S} \left\{1 - \left[\binom{N - N_i}{n} \Big/ \binom{N}{n}\right]\right\}$$

where E(S) is the expected number of species in the rarefied sample, n is the standardized sample size, N is the total number of individuals in the sample to be rarefied, and

■ **PHOTO 16.1** Diversity—a mixed herd at a water hole in Mkuze, Zululand. Most measures of diversity include number of individuals and number of species. There are 20 individual animals in this photo, and 5 species—warthog, zebra, nyala, impala, baboon, and an unknown number of plant individuals and species. (*Nigel Dennis, Photo Researchers, Inc. 7V7646.*)

N_i is the number of individuals in the *i*th species in the sample to be rarefied, summed over all species counted.

The term $\binom{N}{n}$ is a "combination" that is calculated as

$$\binom{N}{n} = \frac{N!}{n!\,(N - n)!}$$

where N! is a factorial, e.g., $5! = 5 \times 4 \times 3 \times 2 \times 1 = 120$

This combination allows us to calculate the possible numbers of unique species combinations. For example, if we have four species, A, B, C, and D, then we can have six species pairs: AB, AC, AD, BC, BD, and CD.

Using the equation

$$\binom{4}{2} = \frac{4!}{2!\,(4 - 2)!} = \frac{24}{4} = 6.$$

Thus, $\binom{N}{n}$ is the number of unique combinations of N taken n at a time, i.e., the number of different ways of picking species pairs from four different species.

Similarly,

$$\binom{4}{3} = \frac{4!}{3!\,(4-3)!} = \frac{24}{6} = 4,$$

Thus, there are four species triplets, which we can confirm by writing ABC, ABD, BCD, and ACD.

The following is a worked example (following Magurran 1988). Imagine two moth traps that have been operated for different lengths of time yield the following data:

	NUMBER OF INDIVIDUALS	
SPECIES	**TRAP A**	**TRAP B**
1	9	1
2	3	0
3	0	1
4	4	0
5	2	0
6	1	0
7	1	1
8	0	2
9	1	0
10	0	5
11	1	3
12	1	0
Number of species (S)	9	6
Number of individuals (N)	23	13

How many species would we have expected in Trap A if it too contained thirteen individuals? First, take the number of individuals of each species from Trap A and insert them into the formula. For species 1 in Trap A,

$$N = 23$$

$$n = 13$$

$$N_i = 9$$

$$N - N_i = 14$$

$$\binom{N}{n} = \frac{23!}{13!\,(23-13)!}$$

$$\binom{N - N_i}{n} = \frac{14!}{13!\,(14-13)!}$$

therefore:

$$\left\{1 - \left[\left(\frac{14!}{13! \times 1!}\right) \Big/ \left(\frac{23!}{13! \times 10!}\right)\right]\right\} = \{1 - [14/1144066]\} = 1 - 0.00 = 1.00$$

The results for each species are then summed:

N_I	EXPECTED
9	1.00
3	0.93
4	0.98
2	0.82
1	0.57
1	0.57
1	0.57
1	0.57
1	0.57
E(S)	6.58

Thus, if Trap A contained thirteen individuals we would expect it to contain 6.58 species—about the same as Trap B.

Rarefaction can be a bit difficult to perform by hand so ecologists have often preferred shortcuts. Thus, in addition to the rarefaction technique for measuring richness, several "richness indices," ratios of number of individuals per species, have been proposed:

1. Margalef (1969) $R_1 = (S - 1)/\ln N$
2. Menkinick (1964) $R_2 = S/\sqrt{N}$
3. Odum, Cantlon and Kornicher (1960) $R_3 = S/\text{Log } N$

where S is the number of species and N the number of individuals.

Using these indices, the richness values of the two moth traps listed earlier becomes

	TRAP A	TRAP B	DIFFERENCE IN INDEX VALUES
Margalef	2.55	1.95	30.8%
Menkinick	1.88	1.66	13.2%
Odum	6.61	5.39	22.2%

It is clear that the size of the difference in richness between Traps A and B depends on the richness index used. The degree to which an index can tell two samples apart is called the discriminant ability.

16.2 A variety of indices have been used to estimate diversity.

The diversity of a community is determined both by the number of species in the community and by the evenness with which individuals are distributed among species. A number of indices have been developed that measure diversity by incorporating the number of species and the degree of evenness.

Estimates of diversity from dominance indices are more strongly inflluenced by common species than by rare species.

This group of indices are referred to as dominance indices since they are weighted toward the abundance of the commonest species. They were first introduced into the ecological literature by Simpson (1949). Simpson's index gives the probability of any two individuals drawn at random from an infinitely large community belonging to different species. For example, the probability of two trees, picked at random from a tropical forest being the same species would be low, whereas in a boreal forest in Canada it would be relatively high.

$$D_s = \sum_{i=1}^{S} \frac{(n_i(n_i - 1))}{(N(N - 1))}$$

where n_i is the number of individuals in the ith species. Since D_s and species diversity are negatively related, Simpson's index is expressed as $1 - D$ so that increasing values mean increasing diversity. A worked example follows:

TREE SPECIES	NUMBER OF INDIVIDUALS
1	100
2	50
3	30
4	20
5	1
	201

For this hypothetical data set on tree abundance and richness in a community the calculation of D_s is

$[(100 \times 99)/(201 \times 200) + (50 \times 49)/(201 \times 200) \ldots + (1 \times 0)/(201 \times 200) = 0.338$

Then $1/D = 1/0.338 = 2.96$

The disadvantage of Simpson's index is that it is heavily weighted toward the most abundant species. Thus, the addition of a few rare species of trees with one individual will fail to change the index. This is obvious by examining the contribution of tree species #5 (in the preceding table) to the overall value of the index—it is zero. As a result, other indices are often used.

Berger and Parker (1970) proposed the index

$$D_{BP} = N_{max}/N$$

where N_{max} = the number of individuals in the most abundant species.

As with the Simpson index, the reciprocal form is usually adopted. Again the Berger-Parker index is still biased toward dominant species, and its discriminant ability is less than the Simpson index.

There is clearly no one "perfect" index. However, May (1975) concluded that the Berger-Parker index was one of the most satisfactory diversity measures available, based partly on its results and partly on its ease of computation. This seems strange because it only measures the relative abundance of the single most common species.

Information statistic indices have been developed that take into account both richness and evenness.

These indices are based on the rationale that diversity in a natural system can be measured in a way that is similar to the information contained in a code or message. The analogy is that if we can know the uncertainty of the next letter in a coded message, then use can also know the uncertainty of the next species to be found in a community. The uncertainty is measured as the Shannon Index, H. A message consisting of bbbbb has a low uncertainty and H=0. A high value of H indicates a high uncertainty of being able to tell the next letter in the sequence, or the next species in a community.

Shannon index

The Shannon index, H′ (Shannon and Weaver 1949) assumes that all species are represented in the sample and are randomly sampled:

$$H' = -\Sigma p_i \ln p_i$$

where p_i is the proportion of individuals found in the ith species and ln is the natural logarithm (although any base of logarithm may be taken). A worked example is given below:

	SPECIES	ABUNDANCE	P_I	P_I LN P_I
	1	50	0.5	−.347
	2	30	0.3	−.361
	3	10	0.1	−.230
	4	9	0.09	−.217
	5	1	0.01	−.046
Total	5	100	1.00	−1.201

Remember that the Shannon index has a minus sign in the calculation so the index actually becomes 1.201, not −1.201.

The most important source of error comes from the failure to include all species from the community in a sample, but this error decreases as the proportion of species represented in the sample increases and is minimal as it approaches the total actual number of species in the community. Values of the Shannon diversity index for real communities are often found to fall between 1.5 and 3.5.

The Shannon index is affected by both the number of species and their equitability or evenness. A higher number of species and a more even distribution both increase diversity as measured by the Shannon index. The maximum diversity of a sample, H_{max}, is found when all species are equally abundant, $H_{max} = \ln S$, where S is the total number of species. We can compare actual diversity value to the maximum possible diversity by using a measure called *evenness*. The evenness of the sample is obtained from the formula

$$\text{Evenness} = H'/H_{max} = H'/\ln S$$

E is constrained between 0 and 1.0. As with H′, this evenness measure assumes that all species in the community are present in the sample. It may even be possible to calculate replicated estimates of diversity from sites or communities so that t-tests or analyses of variance can be used to test for significant differences between the diversity of sites (Hutcheson

1970). Strictly speaking, the Shannon index should be used on random samples drawn from a large community in which the total number of species is known (Pielou 1966). Because this is not often the case, Pielou (1966) recommended using the Brillouin index.

Brillouin index

When the randomness of a sample cannot be guaranteed, as for example occurs in light trapping where different species of insects are differentially attracted to light, the Brillouin index H_B can be used:

$$H_B = \frac{\ln N! - \Sigma \ln n_i!}{N}$$

where N is the total number of individuals and n_i the number of individuals in the *i*th species. A worked example follows:

SPECIES	NUMBER OF INDIVIDUALS	LN N$_i$!
1	5	4.79
2	5	4.79
3	5	4.79
4	5	4.79
5	5	4.79
(S) = 5	N = 25	$\Sigma(\ln n_i!) = 23.95$

$$H_B = \frac{\ln 25! - 23.95}{25} = \frac{58.00 - 23.95}{25} = 1.362$$

Again, evenness is estimated from

$$E = \frac{H_B}{H_{Bmax}}$$

where H_{Bmax} represents the maximum possible Brillouin diversity, that is, a completely equitable distribution of individuals between species.

Generally, the Brillouin index gives a lower value for the same data than the Shannon index. One major difference is that the Shannon index does not change, providing the number of species and their proportional abundance remain constant, while the Brillouin index does. Whether this is desirable is open to debate. For example,

	INDIVIDUALS IN	
SPECIES	SAMPLE 1	SAMPLE 2
1	3	5
2	3	5
3	3	5
4	3	5
5	3	5
Shannon H'	1.609	1.609
Brillouin HB	1.263	1.362

From a conservationist's point of view, an area rich in individuals and in species (Sample 2) might be more valuable to preserve than an area that is species rich but with fewer individuals (Sample 1), so the Brillouin index may have some real advantage over the Shannon index after all.

A second major difference in the two indices is that the use of factorials in the equations quickly produces huge numbers that are unwieldy. This is why the Shannon index is often chosen—for its computational simplicity. Because values of diversity are often correlated (Magurran 1988) the choice of the "correct" diversity measure may not be as critical as might be feared. However, Hurlbert (1971) argued that in certain circumstances different indices can give different results.

A comparison of the performance and characteristics of most important indices is outlined in the Table 16.1. Some indices are more applicable in some situations than others. For conservation of biodiversity, indices sensitive to rare species might be useful; for comparing communities from very different sample sizes, species-richness indices would not be used.

Ideally, of course, the comparison of diversity indices should involve not only the ease of calculation or sensitivity to common or rare species but also their discrimnant ability, that is, how capable they are of detecting subtle differences between sites. Perhaps surprisingly, simple species richness has excellent discrimnant ability while the Shannon, Brillouin, and Berger-Parker indices fare less well (Magurran 1988). It is clear from an examination of Table 16.1 that there is no one index that is excellent across the board, having good discrimnant ability, low sensitivity to sample size, and ease of calculation. Thus, it is not possible to recommend a single index as superior to all others. The question being asked dictates the index to be used (May 1975; Peet 1974; Routledge 1979; Taylor 1978). There is now much interest in including weighting factors in indices so as to incorporate valuable biological information (Applied Ecology: Weighing Biodiversity Indices: The Use of Ordinal Indices).

16.3 Jackknifing.

Jackknifing is a technique that allows different estimates of virtually any statistic (Krebs 1989), including diversity estimates. The technique involves recalculating overall diversity

TABLE 16.1 A comparison of the effectiveness of different diversity indices. (*After Magurran 1988*)

INDEX	DISCRIMINANT ABILITY	SENSITIVITY TO SAMPLE SIZE	BIASED TOWARD RICHNESS (R) OR DOMINANCE OF A FEW SPECIES (D)	CALCULATION	WIDELY USED
S (species richness)	Good	High	R	Simple	Yes
Shannon	Moderate	Moderate	R	Intermediate	Yes
Brillouin	Moderate	Moderate	R	Complex	No
Simpson	Moderate	Low	D	Intermediate	Yes
Berger-Parker	Poor	Low	D	Simple	No

while omitting each sample in turn. The calculations are obviously tedious on large data sets, and a computer program is clearly desirable if calculations are to be done on a regular basis.

The jackknife technique was first suggested by Tukey (1958). Beginning with a set of n measurements, the jackknife is done as follows:

1. Omit one of the n replicates from the jackknife sample.
2. Calculate pseudovalues of the parameter of interest using the remaining data:

$$\varnothing_i = nS - (n - 1)S_t$$

where $\varnothing_i$ = Pseudovalue for jackknife estimate i

n = Original sample size

S = Original statistical estimate

S_t = Statistical estimate when original value i has been discarded from sample

i = Sample number (1, 2, 3, …, n)

3. Repeat for all the individual data points.
4. Estimate the mean and standard error of the parameter of interest from the resulting pseudovalues.

As an example, consider the diversity of trees in a forest measured at five different areas. The diversity could be calculated by summing all the data for all the trees from all the areas and generating one overall diversity statistic. In Jackknifing, five different diversities are created, with each area omitted once. A worked example follows. First, the diversity of all areas together, S, is calculated. Let's say this is 5.0. Next we calculate the five different estimates, S_t. Finally, each estimate is converted to a pseudovalue.

EXCLUDED AREA	DIVERSITY ESTIMATE	PSEUDOVALUE
1	4.9	5.2
2	5.2	3.6
3	4.9	5.1
4	5.5	2.7
5	4.6	6.3
Mean	–	4.6

The mean of the pseudovalues is 4.6, and this is the jackknife estimate of the diversity of the community. In this case, this value is different from the diversity calculated using all stations (5.0). Same authors use jackknifing to obtain a range of estimates or standard deviations of their diversity estimates by treating each pseudovalue as a data point.

APPLIED ECOLOGY

Weighing Biodiversity Indices: The Use of Ordinal Indices

It would be good to incorporate more biological information in a diversity measure, especially when trying to use these indices as measures of which types of community to conserve. All the indices discussed so far treat all species equally. Is this right, or should more weight be given to a rare species than to a common species? Should we give more weight to a large rare predator than a small rare nematode? Is this just a question of human preference, or would the larger, rare predator be more important in the community, using up more energy or acting as a keystone species (see Chap. 20)? Indices that attempt to rank species in importance can be thought of as ordinal indices. Indices that treat all species as equals are cardinal indices. All the indices discussed so far have been cardinal.

Counting the number of animal or plant species (S) in a weight class and the numbers of individuals (N) in a weight class and using cardinal indices within weight classes may be a useful next step up from pure cardinal indices and has been done for insects (Morse, Stork, and Lawton 1988). This allows us to at least tell at which size class most diversity is located. Some species, of course, could be counted in more than one weight class because they grow in size as they age. This further complicates the calculation of indices.

Vane-Wright, Humphries, and Williams (1991) have explored how to weight "taxonomically rare" species in diversity indices (Fig. 1). Species counts to be used in a diversity index are multiplied by a weighting factor. In a cladogram, which details the relatedness of species, species that have the most branches between the stem and the tip are set to a value of 1, then the sister groups are given a weight or score (W) equal to the sum of all the other branch values (in the case of Fig. 1, up to 8). However, this may overweight the value of the taxonomically distinct species (Photo 1). For example, for reptiles, May (1990) pointed out that the two living species of tuatara reptiles, which live only in New Zealand, would be weighted equally to the sum of all six thousand other species of snakes and lizards. A second approach is based on an information index (I), which is based on the number of branchings in the tree that include the species whose characters are being measured.

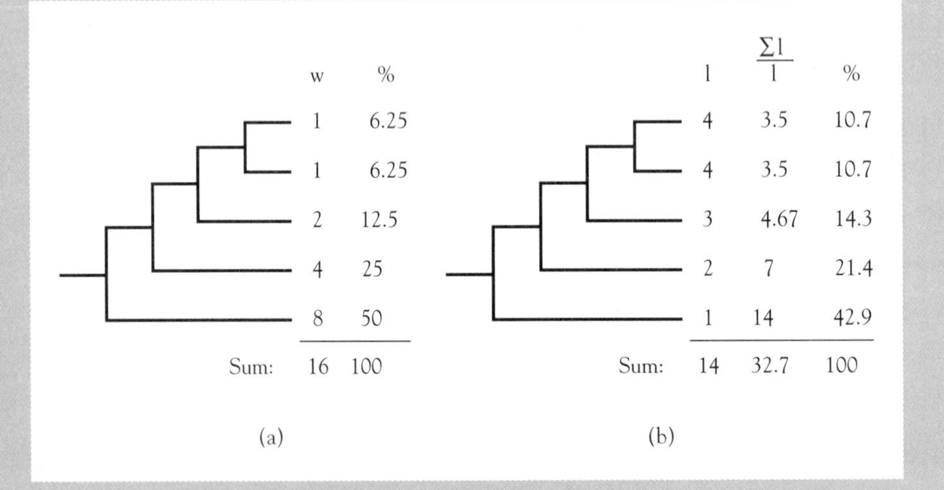

■ **FIGURE 1** Two ways of measuring taxonomic distinctiveness, (**a**) via common ancestry, where w represents a weighted score, (**b**) measures of genetic distance apart, where I represents an information index.

■ **PHOTO 1** *Sphenodon punctuatus*, the tuatara, from New Zealand, the most taxonomically isolated lizard in the world. How can we weigh rare or unusual species in diversity indices? (*John Cancalosi, Peter Arnold, Inc. 95776.*)

The sum of the l values is then divided by the value for the individual species itself. This contribution is then expressed as a percentage. Vane Wright, Humphries, and Williams (1991) used this technique on the *Bombus subrircus* group of bumblebee species, which is distributed worldwide. If a simple species count were used then maximum diversity occurred in Ecuador (10 species = 23% of world total), but if taxonomic distinctness was accounted for, maximum diversity occurred in the Gansu region of China (23% of world total).

Williams and Humphries (1994) and Williams, Gaston and Humphries (1994) discuss five different taxonomic diversity weights. However, these cladistic approaches are concerned mainly with conserving a representative genetic legacy of evolution, and there are other use values. For example, ecosystem-based methods for establishing priorities might include areas important in energy flow, such as coastal marshes or nutrient cycling.

The bootstrap technique was developed by B. Efron in 1977 (Efron 1982) and follows the same general procedure as the jackknife except that the original data of n measurements are placed in a pool and then n values are sampled with replacement. Thus, any measurement in the original data could be used once, twice, several times, or not at all in the bootstrap sample. Typically, one repeats this bootstrap sampling many times.

16.4 Rank abundance diagrams provide a graphical tool for describing how the individuals in a community are divided among species.

Descriptions of whole communities by one statistic of diversity run the risk of losing much valuable information. A more complete picture of the distribution of species abundances in a community is gained by plotting the proportional abundance (usually on a logscale) against rank of abundance. The abundance of the most common species appears on the extreme left and the rarest species on the extreme right (Fig. 16.1). A rank abundance diagram can be drawn for the number of individuals, biomass, ground area covered (for plants), and other variables. There are many different theoretical forms of ranked abundance diagrams, but we will consider the three best known: geometric, lognormal, and broken stick. The distribution of resources (and species abundance) is most equitable in the broken stick, less equitable in the lognormal, and less equitable still in the geometric series.

Some authors have argued that there is biological meaning behind these curves, especially if we can equate resource availability to species abundance. One biological explanation for the geometric series is that the first or most dominant species to colonize an area appropriates a fraction of the resource and by competitive interaction preempts that fraction. The second species preempts a similar fraction of the remaining resource and so on with further colonists (Fig. 16.2). Fits to this model have been found for plants from a subalpine fir forest community (Whittaker 1975) and for benthos in a polluted fjord (Gray 1981). The model has been argued to fit communities of relatively few species where a single environmental factor is of dominating importance. This distribution contrasts starkly with the more equitable distribution represented by the "broken stick" model of MacArthur (1957). Here the resources (the stick) are divided at the same time into segments over the whole stick length—implying instantaneous colonization by all

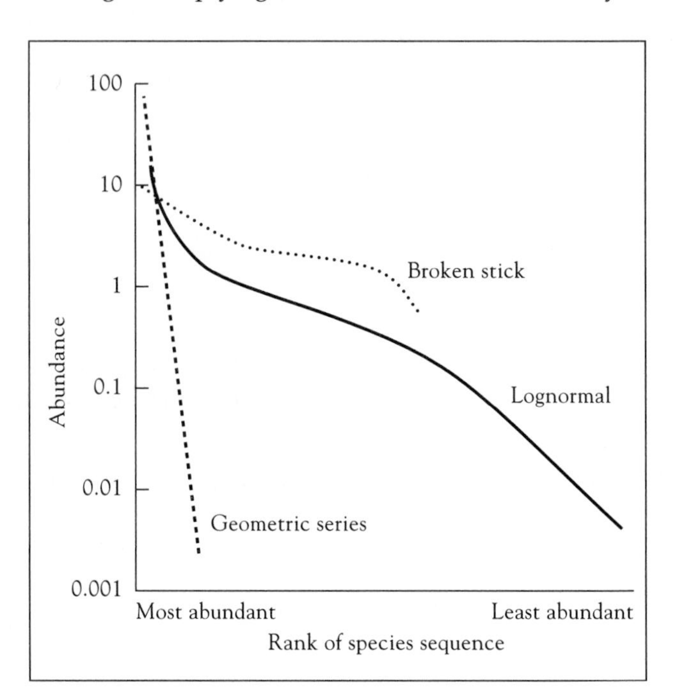

■ **FIGURE 16.1** Rank abundance plots illustrating the typical shape of three species abundance models: geometric series, lognormal, and broken stick. In these graphs the abundance of each species is plotted on a logarithmic scale against the species' rank, in order from the most abundant to least abundant species. Species abundances may in some instances be expressed as percentages to provide a more direct comparison between communities with different numbers of species.

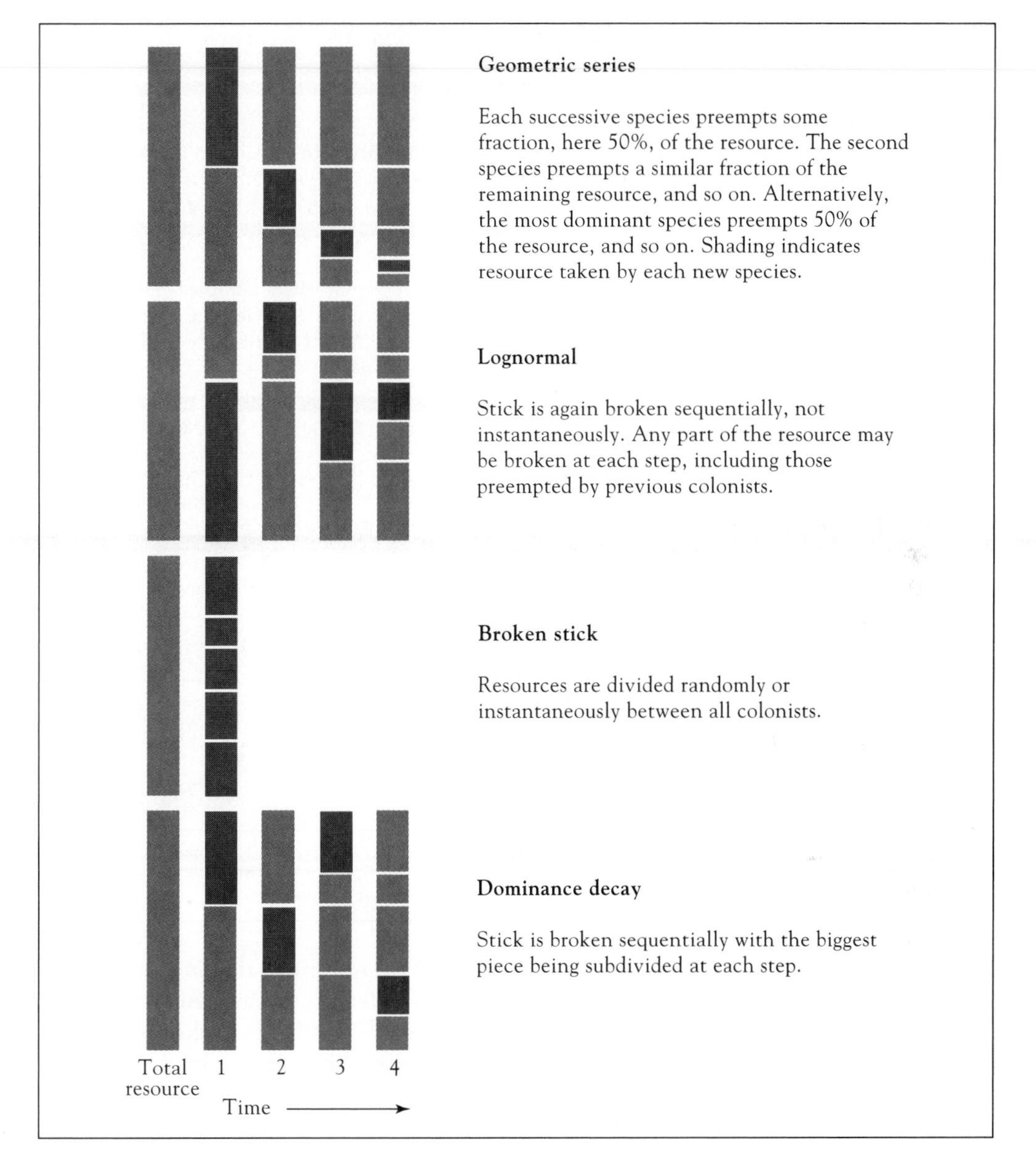

■ **FIGURE 16.2** Diagrammatic representation of species abundance curves and ecological explanations. Each community starts at time 1 and progresses toward time 4. The geometric series, lognormal, and broken stick models are commonly discussed in the ecological literature. However, there are a variety of other models in existence, like the dominance-decay model of Tokeshi (1990) (which gives a rank abundance curve much like the broken stick model).

species or random partitioning of resources. The segments are ranked into decreasing length order. The abundance of each species is assumed to be proportional to the length of the stick segment (Fig. 16.3). A few birds, fish, ophiuroid worms, and predatory gastropods were found to fit the broken stick model well (MacArthur 1960; King 1964), but the model is realistic on relatively few occasions (Hairston 1969).

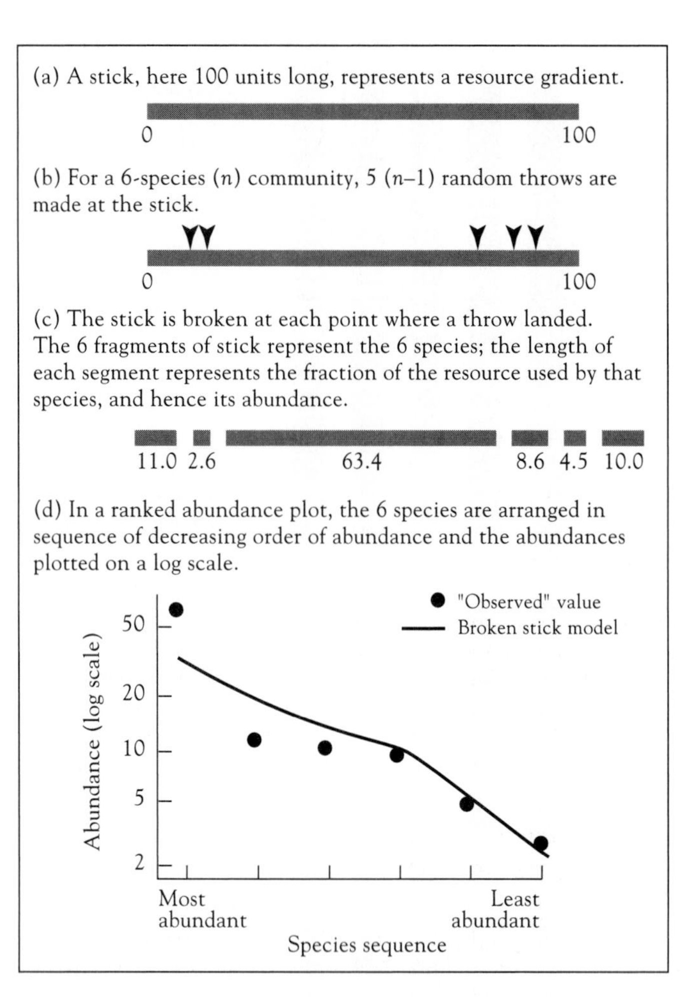

FIGURE 16.3 Simulation of MacArthur's (1957) broken stick model. *(From Wilson 1993.)*

The lognormal distribution is more common for communities rich in species. One ecological explanation for the lognormal distribution refers to the broken stick model, but the stick is broken sequentially not instantaneously. The stick is broken at random into two parts, then any one part is chosen at random and broken again, giving three parts. One of the three is chosen at random and broken again and so on. Various other ranked abundance patterns have been described (Tokeshi 1990 and see Fig. 16.2), and doubtless you may be able to think of more ways to break a stick.

The lognormal distribution owes its place in ecology to Frank Preston (1948) who obtained a normal or "bell"-shaped curve when plotting number of species (y-axis) against log species abundance (expressed as abundance classes [e.g., 1-10, 11-100, 101-1000, 1001-10,000]) on the x-axis (Fig. 16.4). Preston also recognized the existence of the truncated lognormal distribution in which plots of the numbers of species (y-axis) against individuals per species on a logarithmic scale, (x-axis), again followed a normal distribution but were truncated to the left of the mode (Fig. 16.5). The truncation was explained as being due to species that were present in the habitat but not in the sample. If larger samples were taken, more species would be obtained and the mode would move to the right. According to this

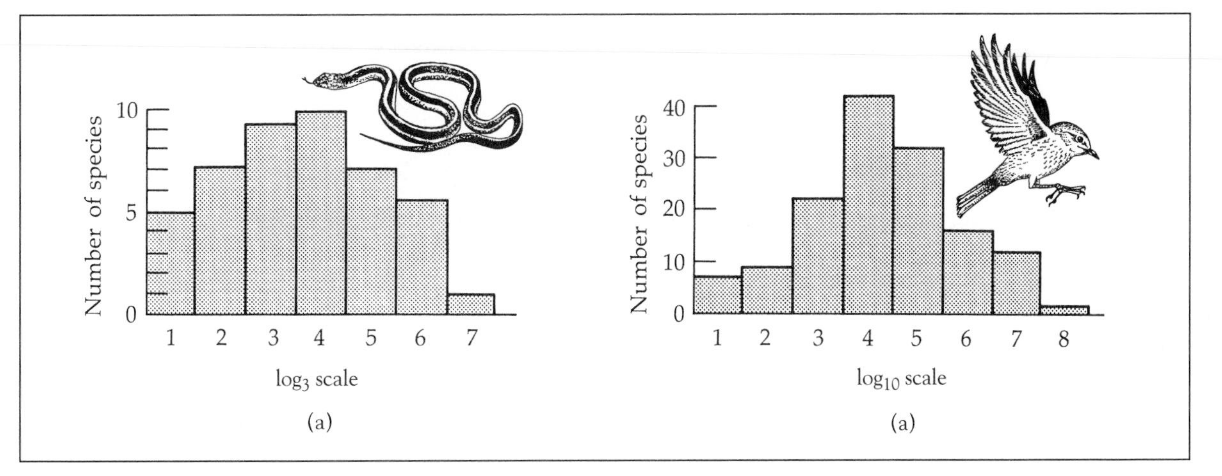

■ **FIGURE 16.4** The lognormal distribution. The "normal," symmetrical bell-shaped curve is achieved by logging the species abundances on the x-axis. A variety of log bases can be used. (**a**) log 3. Successive classes refer to treblings of numbers of individuals. Thus, in this example showing the diversity of snakes in Panama (Williams 1964) the upper bounds of the classes are 1, 4, 13, 40, 121, 364, and 1093 individuals. Although used widely by Williams (1964) log 3 is rarely employed today. (**b**) log 10. Classes in log 10 represent increases in order of magnitude of 1, 10, 100, 1000, 10000, 100000. Thus, the scale is 0-1, 2-10, 11-100, 101-1000, 1001-10,000, 10,001-100,000 and so on. The choice of log base is most appropriate for very large data sets, as for example in this case, the diversity of birds in Britain (data from Williams 1964). In all cases, the y-axis shows the number of species per class.

line of thinking, the size of the sample was seen as the main factor in determining which model fits the data best (Fig. 16.6), not how species partitioned resources.

Unfortunately, it is difficult to corroborate Preston's theory with data because it is extremely difficult to sample enough of a community to capture all the rare species. One exception is the British bird fauna, which is exceptionally well known because of all the amateur bird-watchers, many of whom flock to sites where rare species are thought to exist. Nee, Harvey, and May (1991) showed how the completely unveiled British bird community exhibited a left-skewedness (Fig. 16.7). Sugihara (1980) had actually earlier predicted that this would be the true shape of the unveiled curve. His model to explain left-skewedness was based on a sequential stick breaking. The stick would be broken into two pieces, three-quarters and one-quarter units long. One of these two pieces is then chosen at random and also broken into two pieces three-quarters and one-quarter units long. One of the three fragments is randomly chosen, and the process is repeated until there as many fragments as species in the community. The lengths of the fragments are taken to be proportional to the abundance of species. By starting out with a stick of unit length and breaking it to generate abundances, the model effectively imposes an upper limit on stick length or abundance but no lower limit because there can be an infinite number of tiny pieces—hence the left-skewedness.

While the original proponents of various ranked abundance diagrams argued that they captured real biological properties of communities, there are in fact "statistical" interpretations for all the major types of ranked abundance diagram (Gotelli and Graves 1996). Thus, it is hard to distinguish statistical artifact from real biological property.

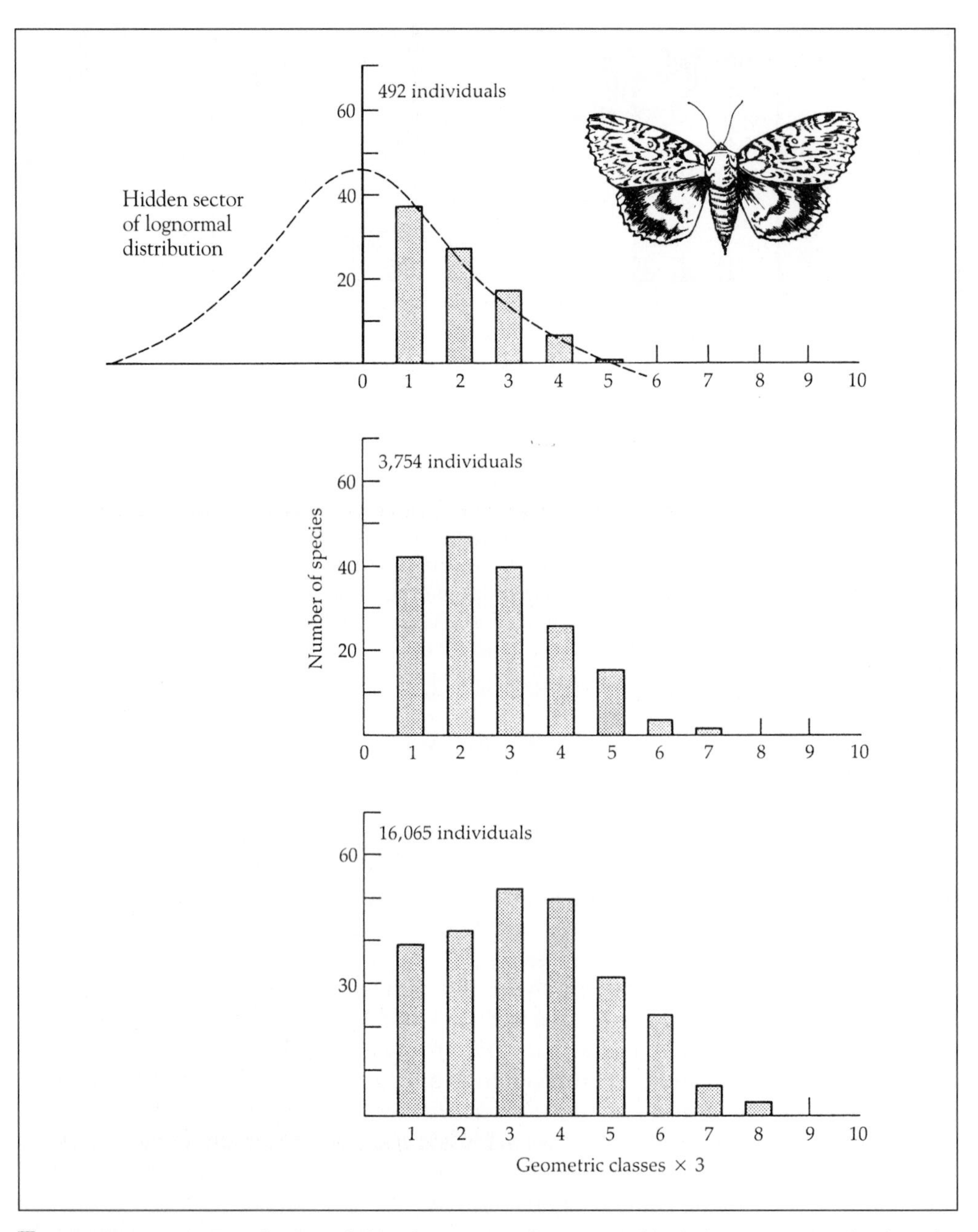

■ **FIGURE 16.5** Distribution of abundance in moths captured by light traps in England. In the top figure, the true distribution of abundance is hidden behind the veil of the *y*-axis. As sampling becomes more intense, the distribution pattern moves to the right to reveal the distribution of rare species. (*Modified from Williams 1964.*)

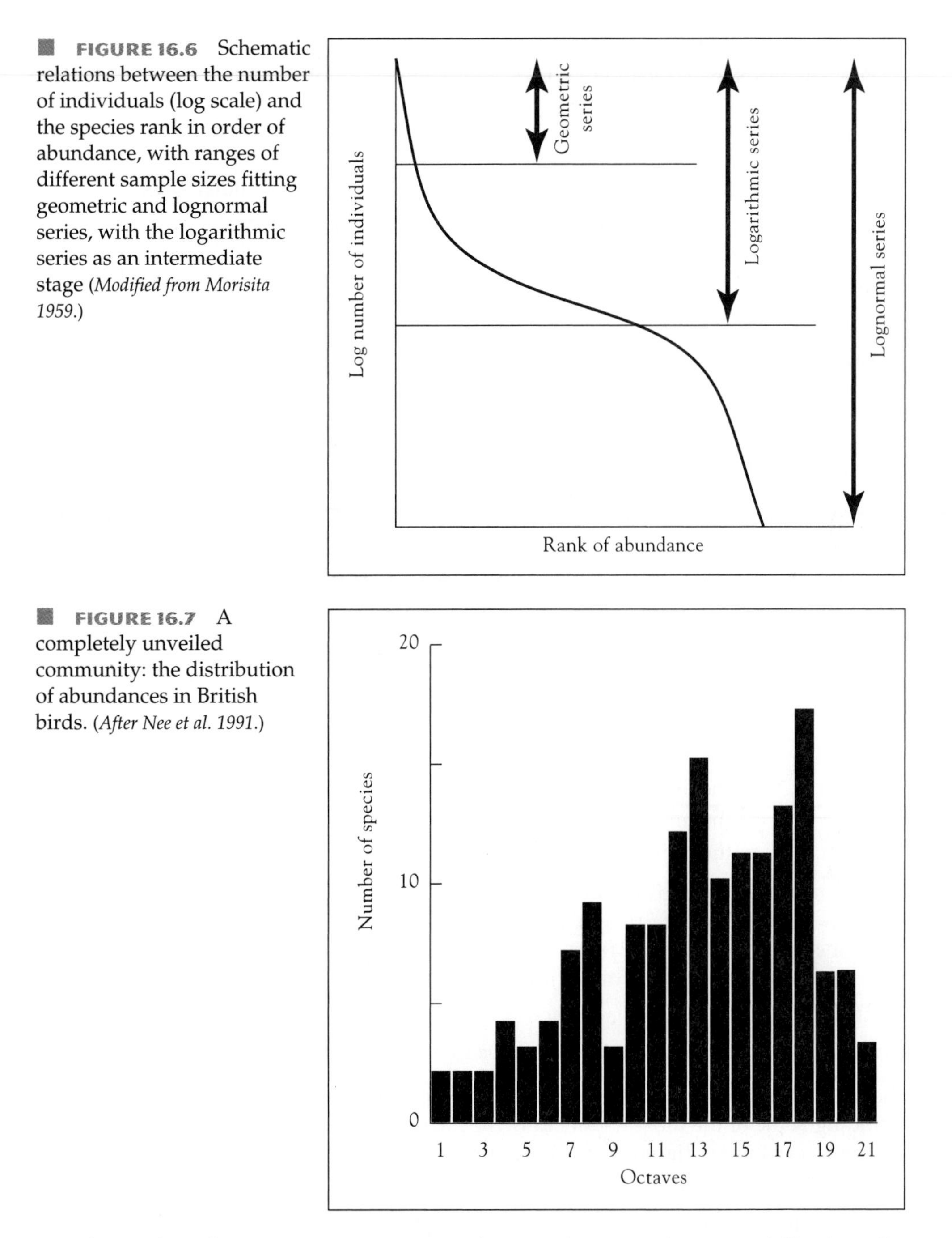

■ **FIGURE 16.6** Schematic relations between the number of individuals (log scale) and the species rank in order of abundance, with ranges of different sample sizes fitting geometric and lognormal series, with the logarithmic series as an intermediate stage (*Modified from Morisita 1959.*)

■ **FIGURE 16.7** A completely unveiled community: the distribution of abundances in British birds. (*After Nee et al. 1991.*)

Also, unless there are many species in the samples, it is often very difficult to distinguish a broken stick from a lognormal model (Wilson 1993). Finally, results are often sensitive to whether biomass or abundance are used (Tokeshi 1990). Resource use may not always be proportional to abundance. For example, large-bodied animals, which are usually rare, have high per-capita resource requirements compared to much smaller species, like birds (Harvey and Godfray 1987; Pagel, May, and Collie 1991). Thus, fitting ranked abundance curves to biological data may ultimately involve using some combination of species abundance, biomass, and energy use—quite a daunting task.

16.5 Community similarity is a measure of how many species, and individuals of those species, are common to two or more communities.

To compare diversity between areas one could simply compare diversity indices. Another method is to compare diversity of sites directly by use of indices called *similarity coefficients*. These coefficients compare the numbers (and sometimes abundances) of species common to all areas. Plant ecologists in particular have developed a wide array of statistical techniques to assist in the analysis of community patterns and to help investigate the influence of environmental variables on community patterns.

In discussing community similarity we will generally consider a simple matrix of presence/absence for two areas, A and B:

		AREA A	
		NO. OF SPECIES PRESENT	NO. OF SPECIES ABSENT
Area B	No. of species present	a	b
	No. of species absent	c	d

where

a = the number of species common to both sites
b = the number of species in site B, but not A
c = the number of species in site A, but not B
d = the number of species absent in both samples.

Is d biologically meaningful?—only if the pool of species is well known. If two areas both have few species despite a potentially rich pool of colonists then they may be more similar than some similarity coefficients would show. In this case, d, a measure of the negative matches, is potentially biologically meaningful. In reality, it is almost impossible to know d.

Similarity coefficients can determine the similarity between two communities.

There are many different types of similarity measures, and six will be discussed here including some of the most widely known. Two commonly used similarity coefficients are the Jaccard index (Jaccard 1912) and the Sorensen index (Sorensen 1948), probably for their ease of calculation:

1. Jaccard measure (*qualitative data, i.e., numbers of species not individuals*). This is calculated using the equation

$$C_J = a/(a + b + c)$$

Thus, for the data outlined in Table 16.2,

$$C_j = 12/(12 + 0 + 14) = 0.46$$

2. Sorensen measure (*qualitative data*).
   This measure is similar to the Jaccard index and uses identical variables.

$$C_s = 2a/(2a + b + c)$$

Thus

$$C_s = 24/(24 + 0 + 14) = 0.63$$

   The Sorensen measure weights matches in species composition between the two samples (which we know) more heavily than mismatches (which we are less sure of). Whether this is valuable or not has not yet been resolved.
3. The simple matching coefficient, C_{sm} (*qualitative data*).
   This index makes use of negative matches, d, mentioned earlier.

$$C_{sm} = \frac{a + d}{a + b + c + d}$$

   We cannot use the data in Table 16.2 without knowing the total number of species in the area. If we suppose d = 50 then

$$C_{sm} = \frac{12 + 50}{12 + 0 + 14 + 50} = \frac{62}{76} = 0.816$$

4. The Baroni-Urbani/Buser (1976) coefficient (*qualitative data*).
   This makes slightly more complex use of negative matches:

$$C_{BB} = \frac{\sqrt{ad} + a}{a + b + c + \sqrt{ad}}$$

   Again, if we assumed d = 50 then

$$C_{BB} = \frac{\sqrt{600} + 12}{12 + 0 + 14 + \sqrt{600}} = \frac{24.738}{50.494} = 0.490$$

   Although the range of these coefficients was thought to be 0 (no matching) to 1 (complete similarity), Wolda (1981) showed that often sample size and species richness affected the maximum value that could be obtained (just as these parameters affect values of different diversity indices). For this reason Krebs (1989) recommended rescaling the coefficients to a 0-1 scale of similarity. This could be done by dividing all coefficients by the maximum coefficient in any particular study.

   Another disadvantage of these first four similarity coefficients is that they take no account of the abundance of species. Rare and common species alike are weighted equally. Thus, the fifth index, a modified Sorensen index (Bray and Curtis 1957) attempts to use quantitative data.
5. Sorensen measure (*quantitative data, i.e., incorporates number of individuals of species*).
   This version of the Sorensen measure uses quantitative data and is calculated as

$$C_{SN} = 2aN/(aN + bN + cN)$$

where aN = the sum of the lower of the two abundances of species that occur in the two sites, bN = the number of individuals in site B but not A, and cN = the number of individuals in site A but not B. Using data from Table 16.2, aN is (2.9 + 10 + 5.7 + ... + 16.9) = 58.4. It is identical to the sum of the abundances in the managed area because abundances of the bird species are always lowest in this habitat. Thus

$$Cs_N = 2 \times 58.4/(58.4 + 58.4 + 146.1) = \frac{116.8}{262.9} = 0.44$$

TABLE 16.2 Abundance of birds in managed and unmanaged areas. *(From Edwards and Brooker 1982)*

	TERRITORIES PER 10 KM	
SPECIES	**UNMANAGED**	**MANAGED**
Great-crested grebe	1.4	0
Mallard	4.3	0
Mute swan	2.9	0
Moorhen	8.6	2.9
Coot	4.2	0
Common sandpiper	15.7	0
Kingfisher	2	0
Sandmartin	50	10
Dipper	1	0
Sedge warbler	11.4	0
Pied wagtail	11.4	5.7
Grey wagtail	4.3	2.5
Yellow wagtail	13	5.7
Reed bunting	14.3	8.6
Heron	8.6	5.7
Curlew	7.1	2.9
Lapwing	10	0
Redshank	1.4	0
Nuthatch	2.9	2.9
Tree-creeper	5.7	0
Whinchat	1.4	0
Blackcap	11.4	5.7
Garden warbler	2.9	0
Whitethroat	4.3	2.9
Lesser whitethroat	1.4	0
Spotted fly-catcher	2.9	2.9
Number of species (S)	26	12
Total number of Individuals (N)	204.5	58.4

However, Wolda (1981) noted that this index, in common with many others, was still strongly influenced by species richness and sample size. The only index that did not suffer from this problem was a modified version of the Morisita index (Wolda 1981).

6. Morisita-Horn measure (*quantitative data*).
This index is calculated from the equation

$$C_{MH} = \frac{2\Sigma(an_i \times bn_i)}{(da + db)aN \times bN}$$

where

aN = the number of individuals in site A,

bN = the number of individuals in site B,

an_i = the number of individuals in the ith species in site A,

bn_i = the number of individuals in the ith species in site B

$$da = \frac{\Sigma an_i^2}{aN^2} \quad \text{and} \quad db = \frac{\Sigma bn_i^2}{bN^2}$$

It was originally developed to measure niche overlap and is Horn's (1966) modification of an index of similarity proposed by Morisita (1959). In this example, following Magurran (1988),

$$\Sigma(an_i \times bn_i) = (1.4 \times 0 + 4.3 \times 0 + 2.9 \times 0 + 8.6 \times 2.9 + \cdots + 2.9 \times 2.9)$$

$$= 961.63$$

$$da = \frac{\Sigma an_i^2}{aN^2} = \frac{3960.27}{41820.25} = 0.0947$$

and

$$db = \frac{\Sigma bn_i^2}{bN^2} = \frac{352.22}{3410.56} = 0.1033$$

Thus

$$C_{MH} = \frac{2\Sigma(an_i \times bn_i)}{(da + db)aN \times bN} = \frac{2 \times 961.63}{(0.0947 + 0.1033)\ (204.5 \times 58.4)}$$

$$= \frac{1923.26}{2364.67} = 0.8133$$

16.6 Cluster analysis can determine the similarity of three of more communities.

When there are a number of sites in the investigation, not just two, one method of overall comparison is through cluster analysis. Cluster analysis starts with a matrix giving the similarity between each pair of sites as measured by the use of any similarity coefficient. The two most similar sites in this matrix are combined to form a single cluster. The analysis proceeds by successively clustering similar sites until all are combined in a sin-

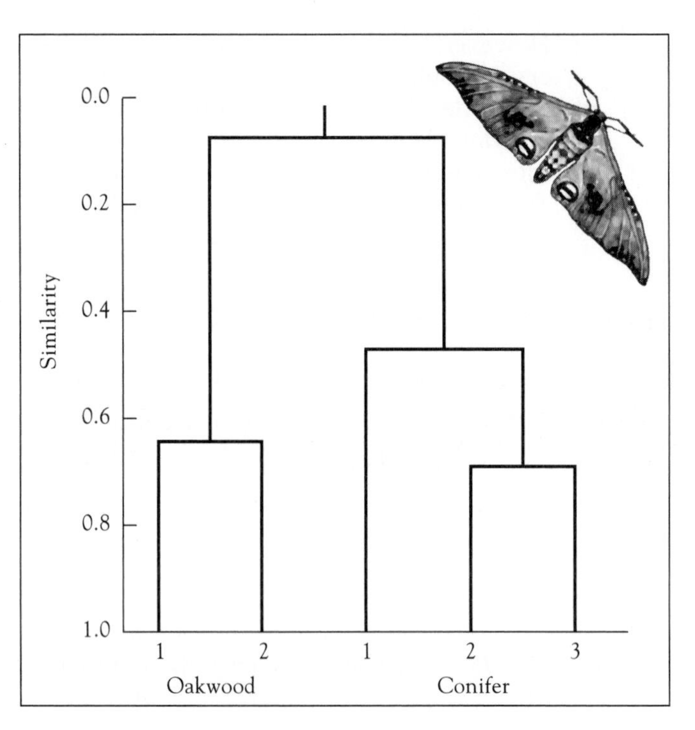

FIGURE 16.8 An example of a single-linkage cluster dendrogram showing the similarity between moths found at three light-trap sites in a conifer plantation and two light-trap sites in an oakwood. The dendrogram shows much greater similarity within the two woodland habitats than between them. (*After Magurran 1988 and based on data in Table 16.3.*)

gle figure called a dendrogram (Fig. 16.8). There are a variety of techniques for deciding how sites should be joined into clusters and how clusters should be combined with each other (Ludwig and Reynolds 1988; Kent and Coker 1992; Jongman, Ter Braak, and Van Tongeren (1995). Once sites have been fused into a cluster, the similarity between the clusters and remaining samples must be calculated. Three of the most widely used methods in ecology are single linkage clustering (nearest neighbor), complete linkage clustering (farthest neighbor), and group average clustering. Cluster analysis can be carried out using either presence and absence data or quantitative data.

Single linkage clustering.

As with all cluster analysis, we start out with a matrix of similarity coefficients (Table 16.3). In single linkage clustering we find the most similar pair of samples. In this case, the greatest similarity is between the conifer sites 2 and 3 = 0.70. Next, we find the second most similar pair of samples or the highest similarity between a sample and any of the samples in the first cluster. In our example, the next most similar pair of samples is between oakwood 1 and 2 = 0.65, whereas the highest similarity between a sample and the first cluster is 0.47. Thus, the two oakwood sites are linked together before conifer site 1 is linked with conifer 2 and 3, to form a three-community cluster. Finally, the last joining occurs between the oakwood 1 and 2 cluster and the conifer 1, 2, 3 clusters, at similarity 0.05. Fig. 16.8 shows the tree summarizing this cluster analysis.

Complete linkage clustering.

One of the problems with single linkage clustering is that it often produces long, drawn-out clusters that might appear like this:

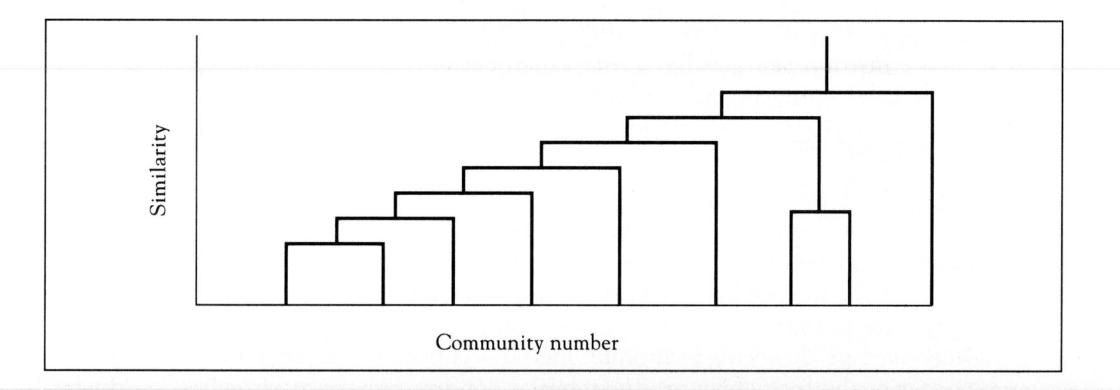

TABLE 16.3 Hypothetical matrix of similarity coefficients for the moth data shown in Figure 16.8.

		OAKWOOD 1	OAKWOOD 2	CONIFER 1	CONIFER 2	CONIFER 3
Oakwood	1	1.00	0.65	0.05	0.02	0.01
	2		1.00			
Conifer	1			1.00	0.47	0.4
	2				1.00	0.7
	3					1.00

STEP	ACTION	RESULT		ACTION	TREE
1	Find greatest similarity less than unity	0.7	(conifers 2-3)	Join conifer 2-3 at level 0.7	2 3
2	Find next highest similarity	0.65	(oakwood 1-2)	Join oakwood 1 to 2 at level 0.65	1 2 2 3
3	Find next similarity	0.47	(conifer 1-2)	Join conifer 1 to conifer 2 and 3 at 0.47 level	1 2 1 2 3
4	Find next highest similarity	0.4	(conifer 1-3)	1 already joined to cluster containing 3 therefore ignore	
5	Find next highest similarity	0.05	(oakwood 1 to conifer 1)	Join oak 1 and 2 cluster to conifer 1, 2, and 3 cluster at 0.05 level	1 2 1 2 3

The conceptual opposite is complete linkage clustering, which clusters together the most dissimilar pairs or a pair and the farthest member of that cluster. This technique tends to produce more tight, compact clusters.

Average linkage clustering.

Because both single linkage clustering and complete linkage clustering represent extremes of a continuum, average linkage clustering was developed to explore the middle ground. All types of average clustering require additional computational steps, steps that are normally done with a computer. Similar clusters are chosen by calculating the arithmetic mean of similarities between the sample and all members of the cluster.

Other average clustering techniques include centroid clustering and minimum variance clustering, both of which produce similar dendrograms as average linkage clustering (Jongman et al. 1995). Everitt (1980) also provides some nice examples.

16.7 Ordination.

Ordination means "to set in order" and is a term used to describe a set of techniques in which sampling units are arranged in relation to one or more axes such that their relative positions to the axes and to each other provide maximum information about their ecological similarities. It was first used by Goodall in 1954. By identifying those sampling units that are the most similar to one another based on coordinate position, we can then search for underlying factors that might be responsible for the observed patterns. Ordination is often used to simplify and condense massive data sets in the hope that ecological relationships will emerge. To illustrate how complex these types of analysis are we will consider the simplest method of ordination, polar ordination. Although polar ordination has largely been replaced by newer ordination techniques, a discussion of polar ordination is valuable because it introduces us to the logic behind many ordination techniques, and it is simpler to understand and work through than the other techniques, most of which require computer programs. Such newer methods include Principal Components Analysis (PCA), Detrended Principal Components (DPC), Correspondence Analysis (CA), Detrended Correspondence Analysis (DCA), and Nonmetric Multidimensional Scaling (NMDS). A detailed description of these techniques is well outside the scope of this book, and interested students should consult statistics books such as Ludwig and Reynolds (1988) or Jongman, Ter Braak, and Van Tongeren (1995), or vegetation analysis books such as Kent and Coker (1992).

Bray and Curtis (1957) first developed the method of polar ordination to analyze plant community data. A worked example, following Ludwig and Reynolds (1988), is shown here, which is based on the abundance of three species in five sampling units shown in Table 16.4.

First, the percentage similarity and dissimilarity between the pairs of sampling units are computed. There are various methods for determining similarity between areas of sampling units (see the previous section), but Bray and Curtis used the Czekanowski coefficient, which is very similar to the Jaccard coefficient,

$$\text{Percentage similarity beween sampling units j and k} = PS_{jk} = \left(\frac{2W}{A + B}\right)(100)$$

TABLE 16.4 Abundances of three species in five sampling units (SUs) used for ordination analysis in text.

			SUs		
SPP	**(1)**	**(2)**	**(3)**	**(4)**	**(5)**
(1)	2	5	5	3	0
(2)	0	3	4	2	1
(3)	2	0	1	0	2
Sums	4	8	10	5	3

where

$$W = \sum_{i=1}^{s} [\min(X_{ij}, X_{ik})]$$

$$A = \sum_{i=1}^{s} X_{ij}$$

and

$$B = \sum_{i=1}^{s} X_{ik}$$

X_{ij} represents the abundance of species i in SU_j, X_{ik} represents the abundance of species i in SU_k, and S = the total number of species.

Thus, PS between the jth SU and kth SU is a numerator of twice the sum of the minimum (min) of the paired observations X_{ij} and X_{ik} (the "shared" species abundance between each pair of SUs) divided by a denominator of the total of all the species abundance for the two SUs. For any pair of SUs with identical species abundances, their similarity is complete, that is, PS = 100%. Percent dissimilarity (PD), is computed as

$$PD = 100 - PS$$

PD may also be computed on a 0-1 scale as

$$PD = 1 - [2W/(A + B)]$$

Now we can use the data in Table 16.4 in a worked example: the percentage similarity and percentage dissimilarity between the (j, k) pairs of SUs are:

$$PS(1,2) = [(2)(2 + 0 + 0)/(4 + 8)](100) = (4/12)(100) = 33\%$$
$$PS(1,3) = [(2)(2 + 0 + 1)/(4 + 10)](100) = (6/14)(100) = 43\%$$

. . . .

. . . .

. . . .

$$PS(4,5) = [(2)(0 + 1 + 0)/(5 + 3)](100) = (2/8)(100) = 25\%$$

and

$$\begin{aligned} PD(1,2) &= 100 - 33 = 67\% \\ PD(1,3) &= 100 - 46 = 57\% \\ &\vdots \\ PD(4,5) &= 100 - 25 = 75\% \end{aligned}$$

Next the percentage similarity-dissimilarity matrix, along with the sum of the percentage dissimilarity for each SU, is drawn up (Table 16.5).

Next, endpoints are selected for the x-axis. Here, sampling unit 5 has the largest sum of dissimilarities (PD = 269) and is designated as endpoint Ax. SU2 has the greatest dissimilarity with SU5 (PD = 82%) and is thus selected as endoint Bx. The length (L) of the x-axis is thus 82. Now the remaining sampling units are positioned on the x-axis according to the formula

$$x(i) = \frac{L^2 + dA(i)^2 - dB(i)^2}{2L}$$

where $x(i)$ is the location of the ith SU along the x-axis, L is the length of the x-axis, $dA(i)$ is the PD of the ith SU to AX and $dB(i)$ is the PD of the ith SU to BX.

Using this equation with L = 82, the location of SU 1 on the x-axis is

$$\begin{aligned} x(1) &= [(82)^2 + (43)^2 - (67)^2]/(2)(82) \\ &= (6724 + 1849 - 4489)/164 = 4084/164 = 25 \end{aligned}$$

and, similarly, for SUs 3 and 4

$$x(3) = 69$$

$$x(4) = 72$$

TABLE 16.5 **Matrix of Proportional Similarity (PS) (above diagonal) and Proportional Dissimilarity (PD) (below diagonal) values between the five SUs. The sum of PD for each SU is also given.** *(After Ludwing and Reynolds 1988)*

		PERCENTAGE SIMILARITY (PS)				
	SUs	(1)	(2)	(3)	(4)	(5)
	(1)	—	33	43	44	57
Percentage	(2)	67	—	89	77	18
dissimilarity	(3)	57	11	—	67	31
(PD)	(4)	56	23	33	—	25
	(5)	43	82	69	75	—
	Sum PD	223	183	170	187	269

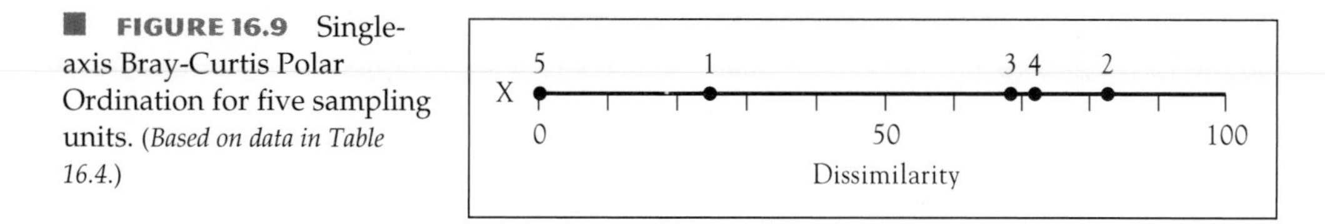

■ **FIGURE 16.9** Single-axis Bray-Curtis Polar Ordination for five sampling units. (*Based on data in Table 16.4.*)

The single-axis Bray-Curtis polar ordination for the table is shown in Fig. 16.9. Similar calculations can be used to perform ordination of another axis so that a two-dimensional ordination graph is produced (Fig. 16.10). We have succeeded in "reduced dimensionality" since the original data matrix of three species and five SUs has been transformed into a 2×5 matrix.

It is clear that ordination is a tedious procedure rather than being computationally difficult. Other ordination techniques like PCA, CA, and DCA are often even more computationally complex because the calculations for dissimilarity are sometimes even more involved, especially if species are weighted by size or importance. There are over forty different types of calculating dissimilarity coefficients included—those are due to Jacard, Sorensen, and Baroni-Urbani/Buser as discussed earlier, and others such as Euclidean, City block, Canberra, Penrose, and Czekanowski, as used by Bray and Curtis. Bray and Curtis's (1957) own work concerned ordination of the upland forest communities of southern Wisconsin. A more recent polar ordination of the same area was conducted by Peet and Loucks (1977) (Table 16.6). An ordination of the ten forest sites along the PO-axis (Fig. 16.11) shows that SUs 1, 2, and 4 are located near the endpoint SUs, indictating their close resemblance to SU3. As can be seen from Table 16.6, these SUs are dominated by bur oak and black oak. SUs 6, 7, 8, and 10 are located near the pole defined by SU9, and all are dominated by sugar maple and basswood.

Many ecologists believe ordination is useful for recognizing patterns in community data. Often ordination is combined with environmental information and classification techniques to gain a more complete description and understanding of the community. This may be done in a variety of ways. For example, we may equate the axes with envi-

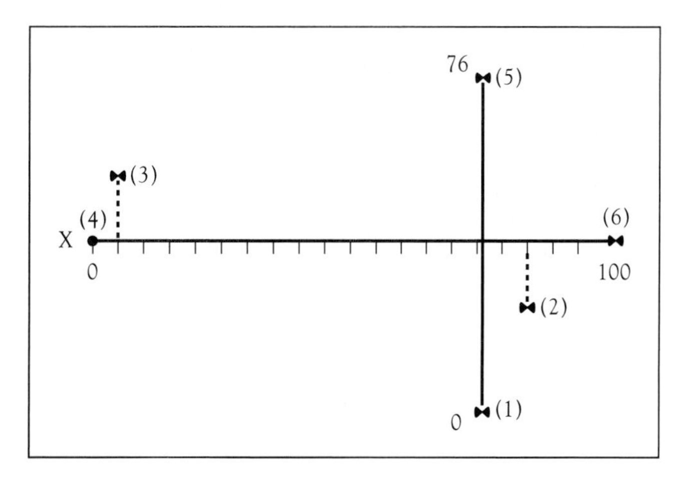

■ **FIGURE 16.10** Polar ordination of six localities based on abundances of five species. The x-and y-axes are drawn between their respective endpoint SUs to emphasize the concept of these SUs as "poles" for each axis.

TABLE 16.6 (1) Ecological data matrix of abundances for eight trees in ten upland forest sampling units in southern Wisconsin. (*Based on data from Peet and Loucks 1977*) (2) Polar ordination results: (a) PD matrix and sums and (b) X coordinates for each of the ten forest sites (SUs).

(1) ABUNDANCES

SPECIES		SAMPLING UNITS									
NAME	NO.	(1)	(2)	(3)	(4)	(5)	(6)	(7)	(8)	(9)	(10)
Bur oak	(1)	9	8	3	5	6	0	5	0	0	0
Black oak	(2)	8	9	8	7	0	0	0	0	0	0
White oak	(3)	5	4	9	9	7	7	4	6	0	2
Red oak	(4)	3	4	0	6	9	8	7	6	4	3
American elm	(5)	2	2	4	5	6	0	5	0	2	5
Basswood	(6)	0	0	0	0	2	7	6	6	7	6
Ironwood	(7)	0	0	0	0	0	0	7	4	6	5
Sugar maple	(8)	0	0	0	0	0	5	4	8	8	9

(2) (A) PERCENTAGE DISSIMILARITY MATRIX AND SUMS

SUs	(1)	(2)	(3)	(4)	(5)	(6)	(7)	(8)	(9)	(10)
(1)	—	7.4	29.4	25.4	43.9	70.4	56.9	71.9	81.5	75.4
(2)	7.4	—	33.3	25.4	43.9	70.4	53.8	71.9	77.8	75.4
(3)	29.4	33.3	—	17.9	48.1	72.5	64.5	77.8	92.2	77.8
(4)	25.4	25.4	17.9	—	25.8	55.9	42.9	61.3	79.7	67.7
(5)	43.9	43.9	48.1	25.8	—	40.4	32.4	53.3	71.9	60.0
(6)	70.4	70.4	72.4	55.9	40.4	—	35.4	19.3	40.7	43.9
(7)	56.9	53.8	64.5	42.9	32.4	35.4	—	29.4	32.3	26.5
(8)	71.9	71.9	77.8	61.3	53.5	19.3	29.4	—	22.8	23.3
(9)	81.5	77.8	92.2	79.7	71.9	40.9	32.3	22.8	—	15.8
(10)	75.4	75.4	77.8	67.7	60.0	43.9	26.5	23.3	15.8	—
Sum	462.2	459.4	513.5	402.0	419.6	448.9	374.1	431.1	514.6	465.8

(B) COORDINATES FOR SUs ON X-AXIS

SUs:	(1)	(2)	(3)	(4)	(5)	(6)	(7)	(8)	(9)	(10)
x(i):	77.4	72.9	92.2	78.8	61.6	26.5	29.2	16.1	0.0	14.6

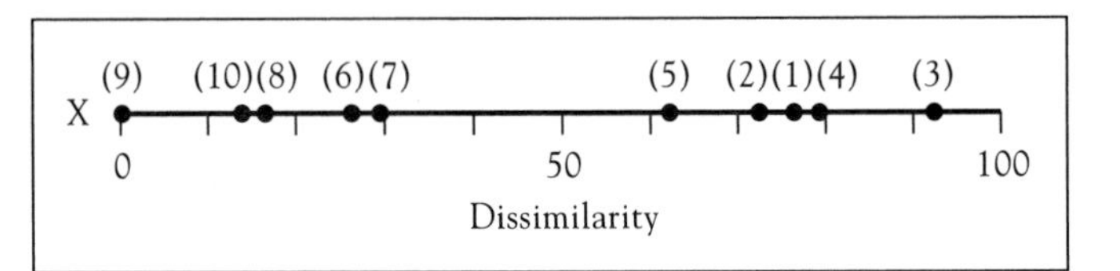

■ **FIGURE 16.11** Single-axis (x) polar ordination of 10 upland forest sites, Wisconsin. (*Based on data in Table 16.6.*)

ronmental variables (Fig. 16.12). One axis may be roughly equivalent to separation along a moisture gradient, another along a shade gradient. For added precision, we may correlate quantitative environmental variables with scores on ordination axes. Thus, if species separate out along a moisture gradient, we would expect a significant correlation between ordination score and moisture gradient. Even more precision is generated by so-called canonical ordination techniques, which are designed to detect patterns of variation in the species data that can be explained "best" by observed environmental variables or combinations thereof (e.g., Canonical Correspondence Analysis). Here the axes themselves are constructed by combinations of environmental data (See Fig. 16.13). This contrasts with normal polar ordination where the axes were defined by the species distributions themselves. The resulting ordination diagram expresses not only a pattern of variation in species composition but also the relationships between the species and each of the environmental variables. Canonical ordination thus combines aspects of regular ordination with aspects of correlation. But again there is no absolute agreement over which method of ordination is the best and which gives the least distortion to the data. A clear concensus as to which method should be recommended for general use has not yet emerged. A comparison of ordination techniques on simulated data (Gauch et al. 1981) indicated detrended correspondence analysis (DCA) was best, followed surpris-

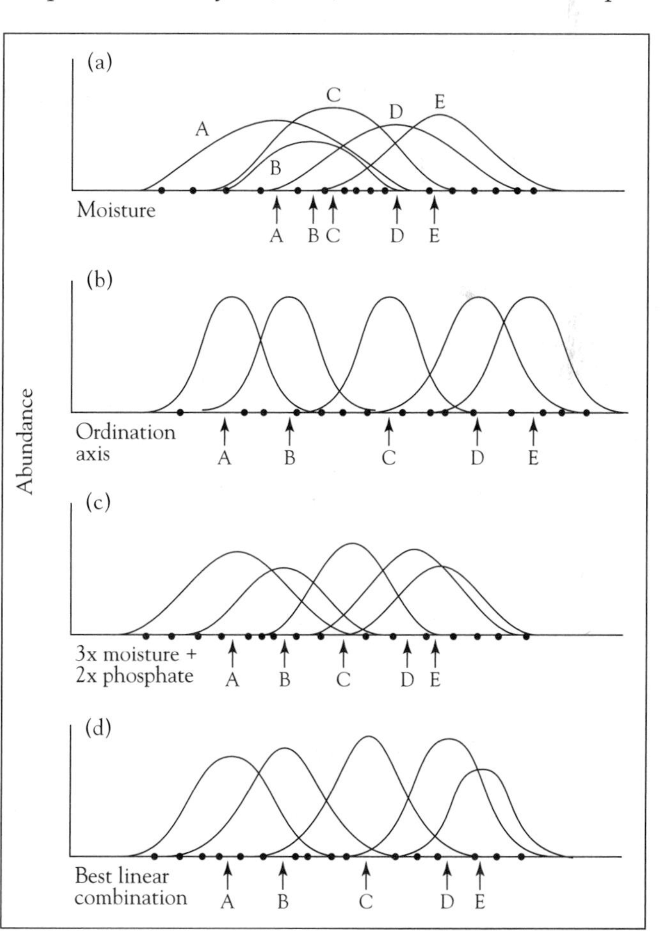

FIGURE 16.12 (**a**) Response curves of five species (A-E) with respect to moisture. Sites are shown as dots. (**b**) First axis of ordination constructs the theoretical variable that best explains the species data by choosing the best values for the sites, i.e., values that maximize the dispersion of the species scores. The variable shown gives a larger dispersion than moisture, and consequently the curves are narrower than in (**a**). (**c**) Linear combination of moisture and phosphate, chosen a priori. (**d**) Best linear combination of environmental variables, chosen by canonical ordination. (*After Jongman et al. 1995.*)

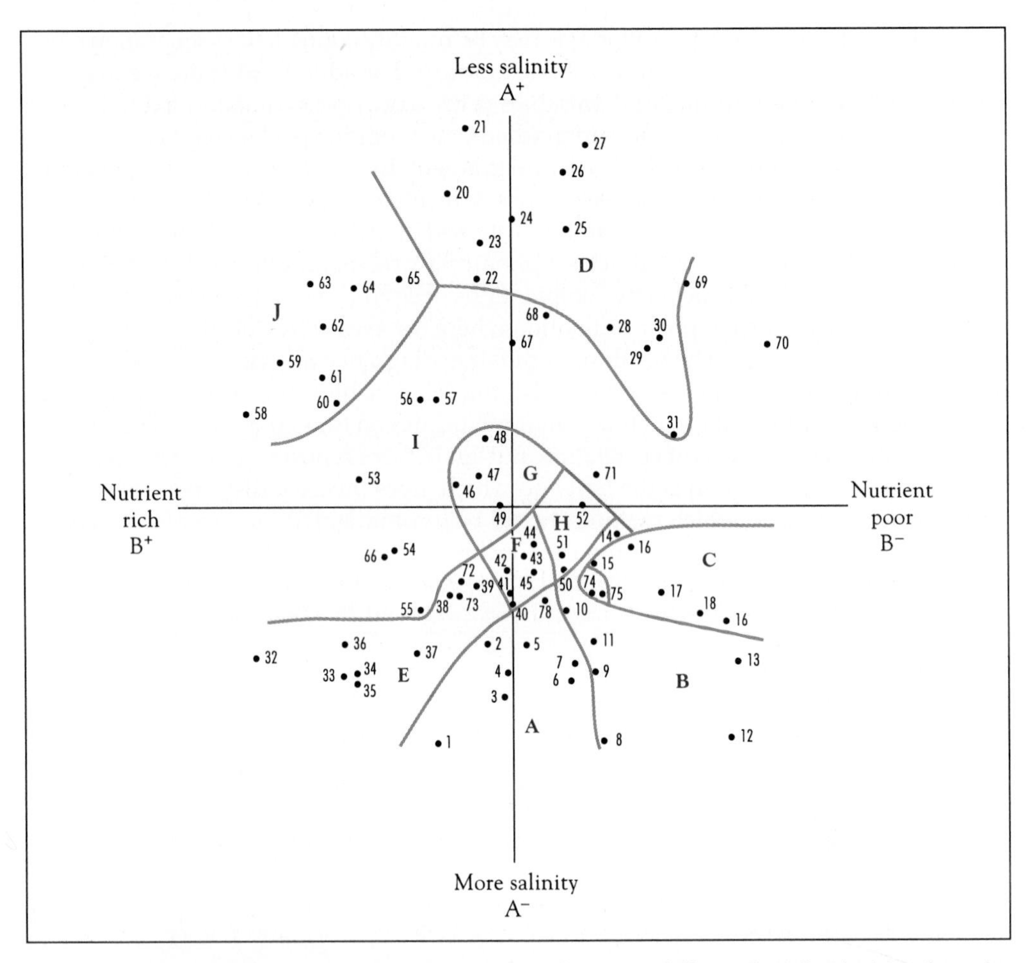

■ **FIGURE 16.13** Ordination of 76 plant species from sea cliffs on the island of Anglesey by principal components analysis. The species are arranged in an ecological space defined by two axes. The *A* axis is associated with salinity and the *B* axis with soil fertility. Species cluster into groups separated by dashed lines, defined by natural history observations. For example, species in group *A* are characteristic of bird colonies, those in group *D* characteristic of wet, acid heath. (*After Goldsmith 1973.*)

ingly by polar ordination. Where a good set of environmental data was available, CCA is most suitable (Kent and Coker 1992).

As we noted earlier, the management of biodiversity necessitates measurement, hence the interest in diversity indices and similarity coefficients. But there has always been another side to the importance of understanding diversity. It has been one of the hallowed tenets of community ecology that diversity begets stability. Stable systems are inherently easier to manage than unstable ones for land stewards, foresters, and wildlife managers. Hence, the added interest in understanding biodiversity. We shall now examine the link between diversity and stability. But before we do this we must first define what stability is.

SUMMARY

1. Biodiversity may be expressed in many ways: as species richness (numbers of species) or by various indices that take the numbers of species and the numbers of individuals of each species into account. Dominance indices include Simpson's index and the Berger-Park index, but both are biased toward dominant or common species. Information statistic indices include the Shannon index and Brillouin index. Each of these indices may give slightly different values from other indices, and each is based on certain assumptions. There is no one overall best index, and choosing the appropriate index for the appropriate situation depends on what sort of question is being asked.

2. Most diversity indices can be referred to as cardinal indices: they treat every species as equal. However, every species is not always equal. For example, in striving to preserve biodiversity, rare and unusual species might be viewed as more important than common or pedestrian species. Some ecologists give more weight to rare species in calculating diversity. Indices that attempt to weight rare species, or any other type species, are known as *ordinal indices*. The development of ordinal indices is undergoing much development the present time.

3. Diversity indices attempt to describe whole communities with just one statistic. A more complete description of a community can be obtained by plotting the proportional abundance of every species against its rank of abundance. The result is a rank-abundance diagram. The form of the rank-abundance diagram can be one of at least three slightly different shapes that are referred to as geometric, lognormal, and broken stick.

4. To compare diversity between areas one could simply compare diversity indices or one could use so-called similarity coefficients, which compare the numbers, and sometimes the abundances, of species common to all areas. Such indices include Jaccard, Sorenson, Baroni-Urbani/Buser, and Morisita-Horn. When two or more sites are compared, cluster analysis may be used.

5. Ordination is a sophisticated statistical technique used to condense large amounts of ecological data on communities into a one- or two-dimensional graph. Polar ordination is the most simple of these techniques though it has been superseded by others.

DISCUSSION QUESTION

In developing ordinal indices of species diversity, what factors would you use in weighting your indices: rarity (how would you measure it?), size, color, presence of fur or feathers, or any other features you can think of?

CHAPTER

17

Stability, Equilibrium, and Nonequilibrium

Perhaps because of the work of Clements and his followers, ecologists in the early twentieth century believed communities, like humans, functioned so as to maintain "homeostasis" (staying the same). In this scheme, the "balance of nature" was seen to return communities that underwent change to their "normal" state. In many respects, this view of nature has never died.

A community is often seen as stable when no change can be detected in the population sizes and numbers of species over a given time period. It may then be said to be at equilibrium. The frame of reference for detecting change may encompass a study of a few years or, preferably, a few decades. Long-term data on certain communities shows a "constancy" over time. For example, because bird watching is a popular pastime, long-term data exist for certain areas that show a stability over time (Fig. 17.1).

One of the most celebrated cases of the "stability" of bird numbers was in Selbourne, England, where as long ago as 1778 Gilbert White, pastor of the village, noted that there were eight pairs of swifts nesting in the village and that this number never varied (White 1789). John Lawton and Sir Robert May visited the village in 1983 and found twelve breeding pairs (Lawton and May 1984), which they argued was dramatic evidence for stability over such a long time span. (However, why they didn't argue that this actually represents a 50% change in population size is not clear.) Some long-term data actually obscure potential changes in the community—like the fact that some species go extinct or emigrate and others immigrate, even though the number of species remains constant. Thus, a definition of stability usually infers a stable equilibrium for each population in the community, not just a constant number of species. It is also worth pointing out that while a community in equilibrium for a long time is likely to be stable, a stable community need not be at equilibrium. For example, a cycle of predator and prey may exhibit a predictable periodicity over years, yet equilibrium in numbers is never reached (Williamson 1987).

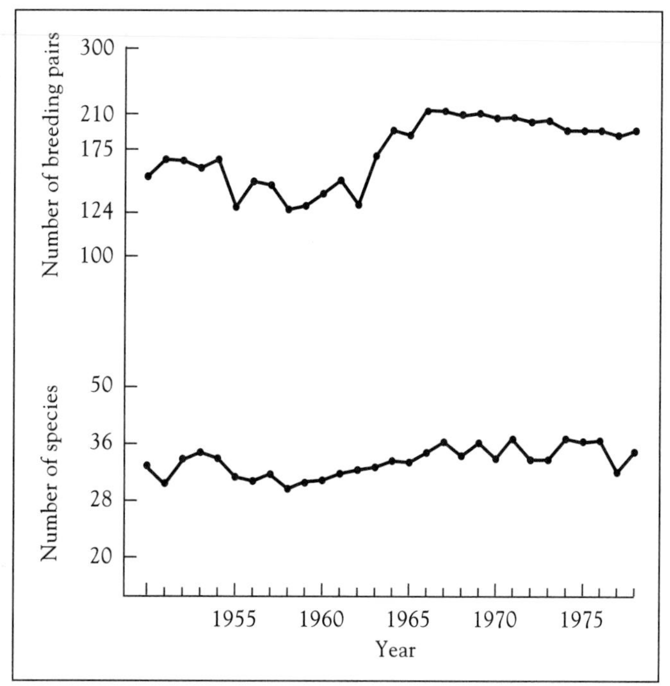

FIGURE 17.1 Time plots of the total number of breeding pairs, and the number of species breeding, on the same logarithmic scale, for the birds of Eastern Wood, Brookham Common. The relative lack of change is taken to represent stability in the community. (*After Williamson 1987.*)

Despite the apparent simplicity of the stability concept, stability is actually difficult to define. There are, in fact, many more different ways of thinking about stability (Orians 1975, Grimm and Wissel 1997) (Fig. 17.2):

a. Resistance to change.
b. Resilience: refers to the return to equilibrium after a perturbation.
   (i) elasticity: how quickly a community can return.
   (ii) amplitude: how much a disturbance it can return from.

Sometimes the ability to return from high-amplitude disturbances is viewed as global stability whereas the ability to return from only low-amplitude disturbances is seen as local stability; after high-amplitude disturbances communities with only local stability would never return to their previous state.

These independent concepts of resistance and resilience are sometimes correlated and sometimes not. Lakes are often only weakly resistant, because they concentrate pollutants from a variety of sources, and weakly resilient. Some truly barren deserts on the other hand are very resistant and resilient. Other deserts with limited vegetation may easily be changed by disturbance. Rivers are not particularly resistant but may be resilient because the fast-flowing water often cleanses them quickly. One could probably classify a set of biomes in this way as a class project.

17.1 What evidence is needed to establish stability?

How can we compare the stabilities of different communities? Two methods are useful depending upon what exactly we want to measure. To measure community resistance we could

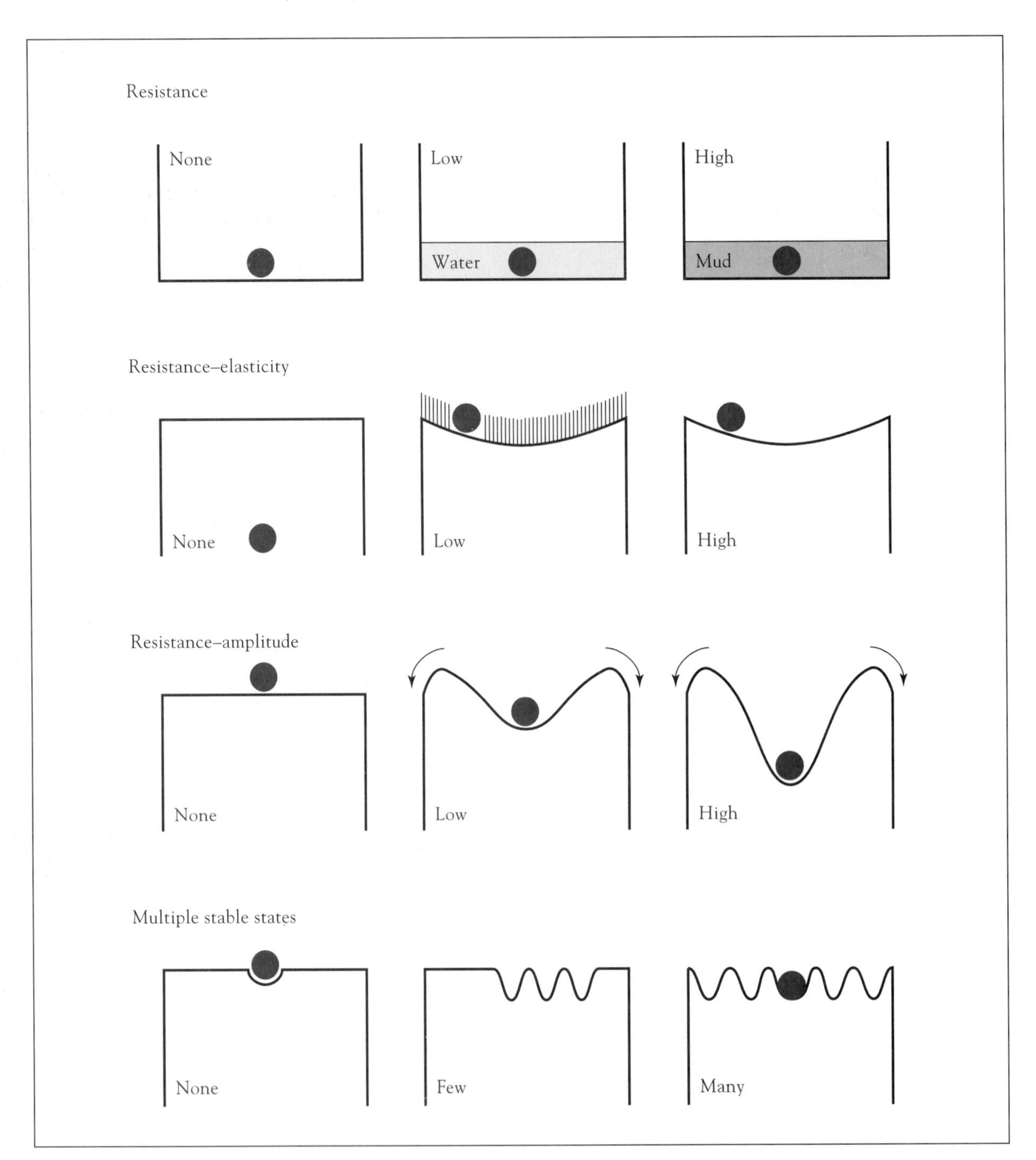

■ **FIGURE 17.2** Different ways of thinking about stability. Here the community is represented by a ball on an imaginary topographic "habitat" map. Once the ball goes "over the edge" the community goes extinct. Various ways of thinking about stability are presented diagramatically. You may be able to think of various ways to combine the concepts.

a. Determine the stable point of a community.
b. Apply a force.
c. See if the community changes.
d. Repeat for different communities.

To measure community resilience we could

a. Determine the stable point.
b. Apply a disturbance.
c. Measure the time the community takes to return to "normal."
d. Repeat for different communities.

Both of these procedures are difficult to do because

1. It is hard to know if the community is stable when measured without years of prior work.
2. There is often insufficient time between natural disturbances, e.g., storms or fires, for communities to return to stable points, even if these were known. This implies that communities are constantly reset by natural events.
3. Some communities, especially those present during various stages of succession, may never be "stable" in a conventional sense. By their very nature, they may vary in time.

Simberloff and Wilson (1969) defaunated mangrove islands in the Florida Keys and examined recolonization. Heatwole and Levins (1972) argued that the trophic structure of the community was remarkably constant following recolonization despite the fact that species identity and taxonomic composition varied widely on recolonized islands. They took this as strong evidence of a stable community. However, Simberloff (1976) himself noted that the recolonization patterns observed could not be shown to be different from random. Second, the concept of stability refers to the "bouncing back" of perturbed communities, not their recolonization patterns following extinction, when completely different rules (often called assembly rules, see Chap. 18) may apply. In fact, there are relatively few data on communities that are recovering from "mild" disturbances. Many disturbed communities have been obliterated by pollution or habitat destruction and distance from sources of colonists affects the recolonization process in such cases.

Bengtsson, Baillie, and Lawton (1997) examined the variability and stability of eighteen British woodland bird communities for twenty-two years, from 1971 to 1992. The analyses were made for the whole time period of twenty-two years and for shorter subsets to examine the effects of temporal scale on community variability. Community variability increased the longer the communities were observed, probably because of increased environmental and population variability with time, just as noted for populations (Chap. 13).

17.2 Can a community exist in multiple stable states?

Following a disturbance, must a community return exactly to how it was before in order for it to be labeled "stable"? If, following an oil spill, populations of ninety of one hundred intertidal species declined significantly, and only eighty-five of these populations recovered, would the community have recovered or would it be a different community? Also, if a "recovered" community had different abundances of species than the original pre-oil spill community, would we consider the community to have changed? In other words, can communities exist in multiple stable states? This was an important question following the Exxon Valdez oil spill (Applied Ecology: Community Recovery after Oil Spills). Connell and Sousa (1983) wrestled with these ideas and concluded that there are few multiple stable states in nature. Their three main arguments in support of this viewpoint were as follows:

1. Assemblages occupying an area before and after a disturbance are interpreted as alternative stable states when in fact the physical variables at the site have been changed, e.g., increased nutrient levels, agricultural runoff, water temperatures, sediment loads, hunting or fishing effort (Holling 1973).
2. Data are not taken for more than a generation or a year so as to establish longterm stability either before a disturbance or afterward. For example, Sutherland (1974, 1981) and Sutherland and Karlson (1977) studied marine fouling communities on panels submerged in the ocean. A community was considered stable if it occupied and held most of the space on a panel until it died and fell off. Often the vacated space was colonized and held by a different species, and the system was argued to have multiple stable states. In other systems, especially terrestrial ones, many other sets of organisms, like plants, hold onto space for their lifetimes and prevent others from invading. Should we regard all these as stable point? Connell and Sousa (1983) suggested that we need examples that show persistence of communities beyond the lifetime of an organism, but I am not so sure. How many generations are necessary to demonstrate multiple stable states?
3. One state may be artificially maintained by the constant addition or removal of predators or competitors by humans. This is true of some freshwater fish communities in England. For example, in Lake Windemere, perch were greatly reduced by fishing and did not recover even after the fishing ceased. Had the community moved to an alternative state? We are not sure because pike continued to be fished, especially large pike. Smaller pike became more numerous and attacked the young perch, probably keeping them from becoming more numerous.

17.3 Is there a link between diversity and stability?

The link between diversity and stability is intuitive to many but elusive to others. The intuitive part is that the effects of severe factors (catastrophes) will be cushioned by large numbers of interacting species and will not produce as drastic effects on species in a large community as they would on species in a small community. But the link between stability and diversity has never explicitly been proved. To quote McNaughton (1988):

> *There is something almost mythological about the hypothesis that greater species diversity in ecological communities is associated with greater community stability. Like Hydra, cutting off one of its heads leads to two sprouting back; like the Gorgons, looking directly at it turns researchers to stone. And, like Tantalus, ecologists who study diversity often seem to have resolution within their grasp, only to see it recede with the next data or theory paper.*

Given the difficulty of even defining what stability is, it seems premature to discuss the evidence for and against the diversity-stability concept. However, since the issue is potentially important, I will present some of the arguments in abbreviated form; many of which were discussed by Elton (1958), who was perhaps the first champion of the stability-diversity link.

Evidence for a link between diversity and stability.

1. Laboratory experiments by Gause (1934) confirmed the difficulty of achieving numerical stability in simple systems.
2. Small, faunistically simple islands are much more vulnerable to invading species than are continents. Most natural species on remote oceanic islands have been selected for high dispersal ability, not competitive dominance.
3. Outbreaks of pests are often found on cultivated land or land disturbed by humans, both of which are areas with few naturally occurring species.
4. Tropical rain forests do not often have insect outbreaks like those common in temperate forests.
5. Pesticides have caused pest outbreaks by eliminating predators and parasites, reducing the diversity of the insect community of crop plants.
6. In a review of forty food webs, the complexity of food webs in stable communities has been found to be greater than the complexity of food webs in fluctuating environments (Briand 1983).

Evidence against a link between diversity and stability.

1. The fluctuations of microtine rodents (lemmings, voles, and so on) are as pronounced in relatively complex temperate ecosystems as they are in simple Arctic environments.
2. Goodman (1975) has argued that the stability of tropical ecosystems is a myth and that there are reports of cases in which insects nearly completely defoliated Brazil nut trees and monkeys succumbed in large numbers to epidemics. Because tropical systems are not nearly as well known as temperate ones, we simply don't know as much about tropical instability as temperate instability. It is well known, however, that locusts in tropical countries undergo huge outbreaks.
3. Rain forests seem particularly susceptible to man-made perturbations (May 1979), yet these are very complex systems.
4. Agricultural systems may suffer from outbreaks, not because of their simple nature but because their individual components often have no coevolutionary history whatever, in complete contrast to the long associations evident in forest biomes (Murdoch 1975).
5. May (1973) argued that increasing complexity actually reduces stability in models. May argued that food webs are only likely to be stable if

   $$\beta\,(SC)^{1/2} < 1$$

   where β is interaction strength, S is number of species, and C is "connectance" (a measure of the number of links between species in a food web) (see Chap. 20). Any increase in S or C would clearly increase instability.

Community Recovery after Oil Spills

In marine systems a common contaminant is crude oil (Fig. 1). Oil can kill directly through coating and asphyxiation, which is especially acute for intertidal life, or by poisoning by contact or ingestion, as in plants and preening birds, respectively. Water-soluble fractions can be lethal to fish and invertebrates and may disrupt the body insulation of birds, resulting in their death from hypothermia.

In the United States a particularly severe oil spill resulted from the wreck of the *Exxon Valdez* near

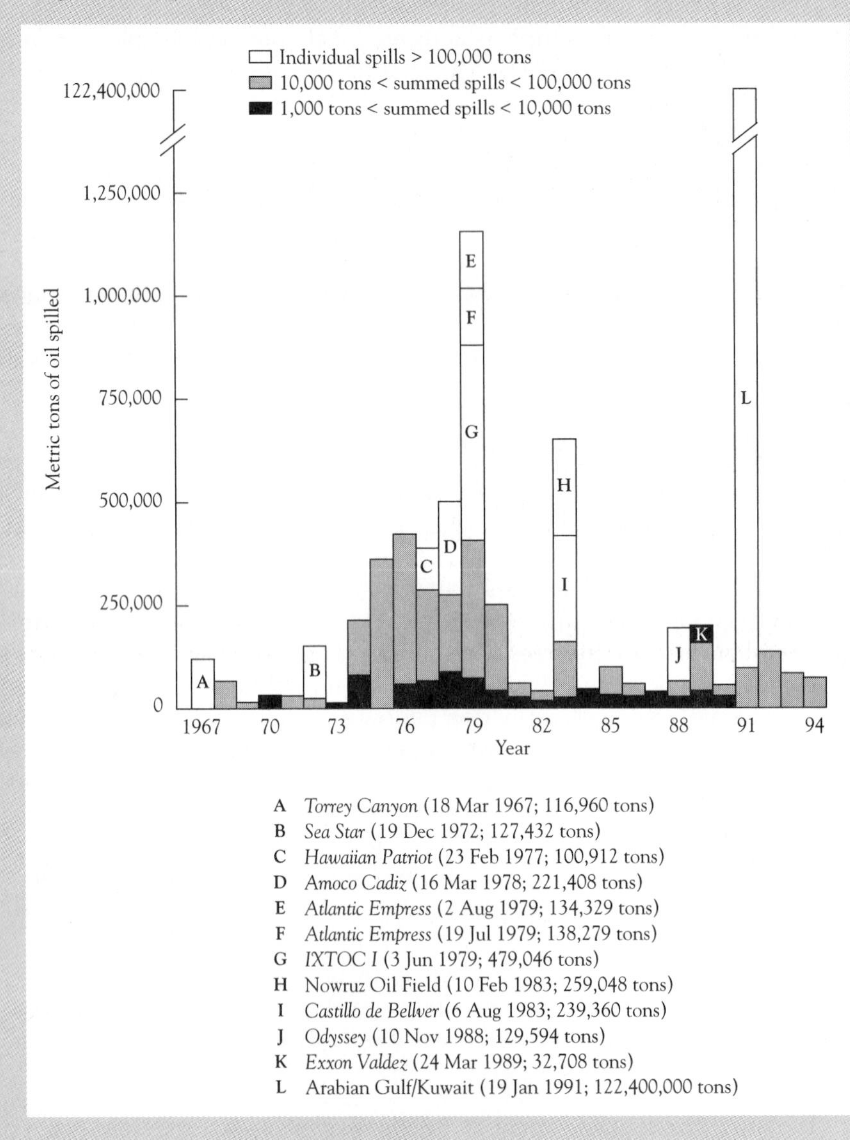

■ **FIGURE 1** Magnitudes of oil spilled into the ocean since 1967. (*After Paire et al. 1996.*)

■ **PHOTO 1** (**a**) Aftermath of a massive oil spill. The oil tanker Exxon Valdez, run aground off the Alaska coast, 1989, showing remaining oil being offloaded to another tanker and a boom placed in an attempt to minimize the amount of oil washed up on the shore. (*Photo by U. S. Coast Guard.*)

■ **PHOTO 1** (**b**) The adjacent shoreline, covered with oil despite containment efforts. Cleaning efforts were slow and laborious. Steam cleaning was sometimes employed. Although the hot water removed the oil more effectively than cold, it killed the animals that lived on the shore. (*Photo by U.S. Coast Guard.*)

Valdez, Alaska at 12:04 A.M. on Good Friday, March 24, 1989. The rocks at Bligh Island tore five huge gashes in the hull of the ship, one six-feet high by two hundred feet long. The result was one of the worst oil spills in U.S. waters; 10.8 million gallons spilled, most of it in the first twelve hours (Photo 1).

Exxon spent $3.2 billion on cleanup, an amount that would, at fiscal year 1995 rates, support the National Science Foundation's General Ecology program for 227 years (Paire et al. 1996). By international standards the spill was not exceptional. Between 1967 and 1994 thirty-nine other wrecks had spilled more oil, and several like the *Torrey Canyon, Amoco Cadiz*, and *Castillo de Bellver* spilled five to eight times as much (Fig. 1). Like many other spills, the *Exxon Valdez* accident took place in fine weather, clear visibility, and no traffic and was clearly the result of human error. A week after the spill, the resultant slick covered nearly nine hundred square miles. Hundreds of miles of shoreline were covered with oil, in places as much as six inches deep. Officially, 27,000 birds, 872 sea otters, and untold numbers of fishes died, although the true numbers are probably higher because many dead birds and otters probably sank and were not recovered.

The effects also carried over into the terrestrial ecosystem when bears, otters, and bald eagles feasted on the oily carrion washed up on the beach. Sitka black-tailed deer ate kelp on the beaches. Few of these animals were expected to be found dead on the beaches because they generally return to their normal habitats before the effects become apparent. Still, the Fish and Wildlife Service found over one hundred dead eagles, and most pairs in the area failed to produce young that year. Two billion dollars was spent in the rescue and rehabilitation of oiled wildlife including $18.2 million on sea otters alone.

The importance of a good definition of community recovery is seen in the contrasting goals of the two major agencies involved in the clean-up, the Exxon Corporation and the State of Alaska. The state viewed recovery as successful if it reached the point where it would have been had no oil spill occurred. The Exxon view was that recovery occurs by the reestablishment of a healthy biological community in which the plants and animals characteristic of that community are present and functioning normally. The first goal is probably unattainable because we can probably never know what the community was like before the spill. The second goal pays little attention to densities and age structures of the population.

Even if populations were shown not to be fully recovered we cannot be absolutely sure that the spill was to blame. Two months prior to the accident, in January 1989, Valdez suffered a record-setting freeze, with the coldest temperatures ever recorded. This could also have severely affected biological populations in the area.

The long-term consequences of the spill were examined by comparing nine oiled areas that were set aside and not cleaned to (a) areas that were oiled and cleaned and (b) areas that were not oiled. Percentage cover of rockweed, *Funcus gardneri,* which reaches 50 percent on unoiled areas, returned to normal values by 1991 on oiled, noncleaned sites but not until 1992 if those sites were cleaned. Cleaning also reduced the diversity of species found in soft-sediment cores for at least two years after spill while oiled sites that had not been cleaned showed reductions in species abundance but not diversity. This generally adds to the body of literature that says the best cleaning is to leave the beach alone.

When trophic links are assembled at random, the more diverse communities are more unstable than the simple communities. May cautioned ecologists that if diversity causes stability in nature it does so not as a direct consequence of the mathematics of the situation. However, some authors have noted that randomly assembled food webs often contain biologically unreasonable elements, like predators without prey. Analyses of food webs that are constrained to be reasonable show that they are more stable than their unrealistic counterparts (Lawlor 1978; Pimm 1979). But even here stability still declines with complexity. Another point of interest, however, is that May's analyses work only if the predators are affected by the food supply, and the food supply is affected by the predators. In nature, many food supplies are unaffected by their predators, and these systems are said to be "donor-controlled." In this type of food web, stability is unaffected by increases in complexity (DeAngelis 1975). This huge group of donor-controlled communities contains detritivores, nectar and seed feeders, and many phytophagous insects and is clearly too big a group to view as an annoying exception.

Experimental tests have often failed to show a link between diversity and stability.

What might help sort out the link between stability and diversity is an experimental approach (as well as a rigorous definition to begin with!). One of the first studies to do this was the laboratory study of Hairston et al. (1968) with microorganisms. *Paramecium,* when cultured with bacterial prey, showed less tendency to go extinct with only one species of prey present (extinction rates = 32% of cultures) than with two or three prey species available (extinction rates = 61% and 70% respectively). When the number of *Paramecium* species was increased to three, the results depended on which particular species was added to which other two. It was evident that species were not simply interchangeable numerical units. Finally, when a third trophic level was added, in the form of the predatory protozoa *Didinium* and *Woodruffia*, there was a further decrease in stability because *Paramecium* were usually forced to extinction regardless of how many species of *Paramecium* were present or whether one or two predator species were present. In these simple systems, diversity did not automatically lead to stability. Additional studies on small microcosms by Luckinbill (1973, 1974) and Salt (1967) showed very much the same thing: additions of a third trophic level (a predatory protist that fed on bacteriovores) either decreased the persistence of some or all protist species in the food chain or increased the temporal variability of the surviving species—clearly decreasing stability.

More recently, Sharon Lawler and Peter Morin at Rutgers University have used microcosm experiments with microorganisms to examine changes in population dynamics that occur as food chains increase in length from two to three trophic levels. In their experiments, bacteria were the prey, trophic level 1; bacteriovorous protists occupied trophic level 2; and predators that could either feed on trophic level 2 or trophic level 1 and 2 occurred at trophic level 3. Censuses focused on the variability of trophic level 2 as a measure of stability (Lawler 1993; Lawler and Morin 1993) and not on trophic level 1. Both species of protists, *Colpidium* and *Tetrahymena*, displayed remarkably constant abundances in shorter food chains where they occupied the top trophic level (i.e., food chains without predators on the third trophic level) (Fig. 17.3). The addition of predators significantly increased the variability of prey populations (on trophic level 2) in nine out of thir-

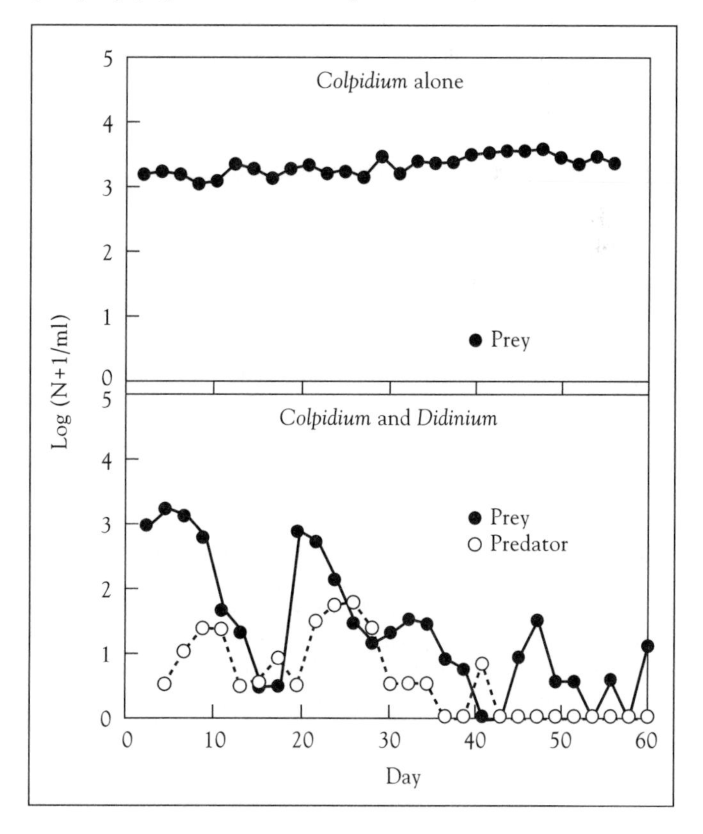

FIGURE 17.3 Population dynamics of *Colpidium* (trophic level 2) in the presence and absence of the predator *Didinium* (trophic level 3) in experimental microcosms. (*After Morin and Lawler 1995.*)

ECOLOGY IN PRACTICE

Sir Robert M. May, Oxford University

I went to Sydney University to get a degree in chemical engineering but ended up with a Ph.D. in theoretical physics. I had never even thought about a career as a researcher when I entered university, but once I became aware that it was possible to spend my life as a teacher/researcher, being paid (albeit not a lot!) for the pleasure of trying to understand how the natural world works and communicating that to others, I seized the chance.

After a postdoc at Harvard, I returned to Sydney University, becoming a full professor at age 33. Soon after, in the late 1960s, I became active in the movement for Social Responsibility in Science in Australia. This led to my becoming interested in environmental and ecological problems. As one result, and largely by accident, I came across the "stability–complexity" question. Around 1970, the conventional wisdom—following the work of two of my heroes, Elton and Hutchinson—was that "complex" communities (those with more species and richer webs of interaction among them) were more "stable" (better able to resist or recover from disturbance, human-created or natural). Comparing mathematical models for ecological communities with few species against the corresponding models with many species, I showed there could be no such simple and general rule; all things being equal, complex systems are likely to be more dynamically fragile. This work, I believe, has refocused the subsequent agenda in this area, as people have become more careful about distinguishing the productivity of a community as a whole from the fluctuations in individual populations and more broadly have sought to understand the intricate patterns that particular ecosystems have woven, over evolutionary time, in ways that undercut any glib generalizations.

Around this time I had arranged a sabbatical leave from Sydney at the Plasma Physics Laboratories near Oxford in the United Kingdom and at the Institute for Advanced Study in Princeton. But, meeting people like Southwood and the younger Hassell and Lawton in the United Kingdom and Robert MacArthur and others at Princeton University, I found my interests increasingly engaged by ecological problems. This was, I think, a romantic era for ecologists. In the hands of MacArthur (my number-one hero) and others, many key questions were being framed in physicslike ways that combined theory and field studies or experiments: what, if any, are the limits to similarity among coexisting competitors; how do the nonlinear feedbacks of intraspecific competition affect population dynamics; ultimately, what are the causes and consequences of biological diversity? By great good luck, I stumbled into this field at a time when a theoretical physicist could be useful. In 1973, following MacArthur's death of cancer at age 42, I moved to Princeton University as professor of biology. The ecology course I taught at Princeton was the first biology cause I had attended since the age of 12. This says a lot about Princeton University's willingness to take risks.

Many interesting projects, most of them collaborative with empirical researchers, followed. The 1973 book on *Stability and Complexity in Model Ecosystems* (with a second edition in 1974) drew some of this work together. Probably the most important thing I have done was, along with Jim Yorke, George Oster, and others, to bring "chaos" center stage through our studies of simple, deterministic, but nonlinear models for biological populations with discrete, nonoverlapping generations (first-order difference equations). The ideas thus generated in ecological research quickly spilled out into physics, chemistry, and other sciences. In the 1980s my interests turned (in collaboration with Roy Anderson) to how infectious diseases can often regulate the numerical abundance or geographical distribution of populations of plants and animals. This work preadapted us to contribute to early research on the AIDS epidemic in developed and developing countries. In 1988, feeling it was time for a change, I moved to Oxford as a Royal Society Research Professor. A narcissistic high point of my career was the award by the Royal Swedish Academy of Sciences of its 1996 Crafoord Prize for "pioneering ecological research in theoretical analysis of the dynamics of populations, communities, and ecosystems"; this award is intended to complement the Nobel Prizes, cycling on a three-year basis between mathematics, earth and space sciences, and the nonmedical areas of biology.

My current interests have mainly to do with conservation biology, and—rather differently—with theoretical immunology (in collaboration with Martin Nowak).

Through all the twists and turns of my research interests, I have had fairly consistent encouragement from funding agencies in Australia, the United States, and the United Kingdom. Entering unfamiliar fields of research, I have found many people enthusiastic about new ideas and new approaches, others interested only in ploughing their own furrow, and a few fiercely hostile. What else would you expect? Science is done by people! I hope this diversity of experience with jobs, people, and research areas has helped prepare me for my current position as chief scientific adviser to the U.K. Government and head of the U.K. Office of Science and Technology (which includes the Research Councils, roughly equivalent to the NSF plus the nonclinical part of the NIH).

teen different combinations of different species of predator on trophic level 3 and protists on trophic level 2 (Morin and Lawler 1995). The degree of stability reduction depended on the identity of both the prey and predator species. Once again stability in the systems was decreased by the addition of a third trophic level. The impact of a predator species on *Tetrahymena* was not a reliable predictor of its impact on *Colpidium*. Once again, species could not always be regarded as simply interchangeable. This again casts doubt on mathematical models of stability that treat species as simple equivalents.

Recently Tilman (1996) examined the relationships between biodiversity and stability for both population and ecosystem traits in a long-term study of 207 grassland plots in Minnesota. Results demonstrated that biodiversity stabilizes community and ecosystem processes but not population processes. What this means is that year-to-year variability in total aboveground plant community biomass was significantly lower in plots with greater plant species richness for the eleven-year period of the experiment (Fig. 17.4a and b).

In contrast, year-to-year variability in species abundances was not stabilized by plant species richness. Tilman (1996) suggested that the difference between species versus community biomass likely results from interspecific competition. When climatic variations harm some species, unharmed competitors increase. Such compensatory increases stabilize total community biomass but cause species abundances to be more variable. Tilman suggested that his results helped to reconcile the long-standing dispute over the diversity-stability relationships because they supported Robert May's theoret-

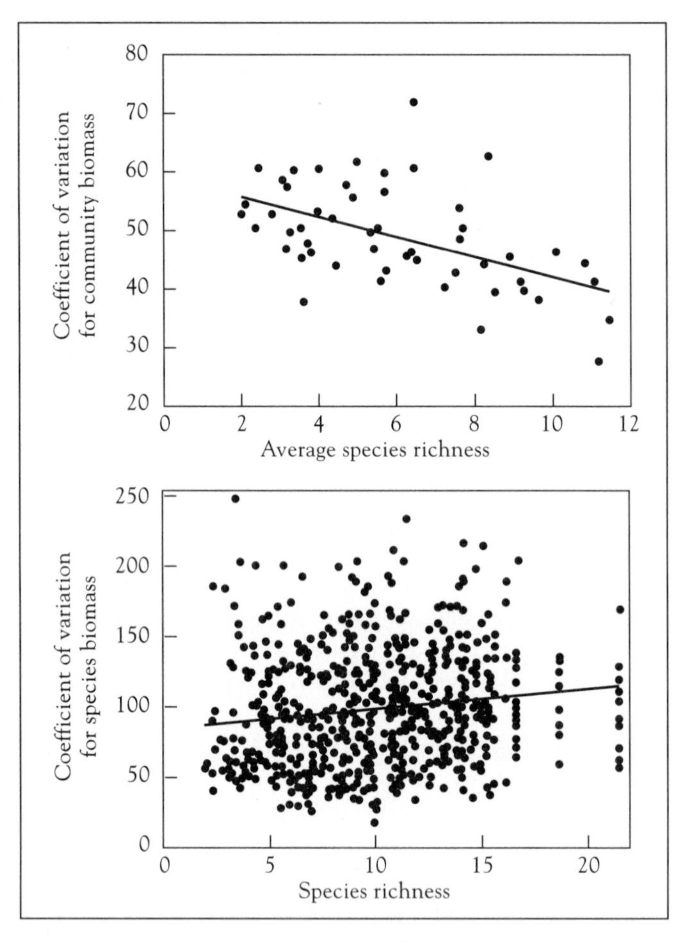

■ **FIGURE 17.4** (**a**) Coefficient of variation in community biomass for one of Tilman's (1996) experimental fields in Minnesota. Regression lines and Pearson correlation coefficients are shown. As plant communities become more diverse, the community biomass becomes less variable between the 11 years of the study. This contrasts with (**b**) where for any given species, biomass becomes more variable as richness increases. (*After Tilman 1996.*)

ical results concerning the effects of diversity on population stability while at the same time upholding Elton's original ideas of a diversity-stability link. Whether further work on other systems will substantiate Tilman's work remains to be seen.

17.4 Many communities do not exist at equilibrium, but instead are subject to multiple disturbance.

What then is the conclusion about how communities are organized around such concepts as stability and diversity? The older more conventional view can be termed the *equilibrium hypothesis*, in which local population sizes fluctuate little from equilibrium values, which are thought to be determined by predation, parasitism, and competition. Communities could also be stabilized by complementary changes in reproduction, survival, or movement when populations reach low densities. In this view, communities are stable and disturbances are damped out; species diversity is determined by the diversity of available niches. In the more modern nonequilibrium hypothesis, species composition is viewed as constantly changing and never in balance. Stability is elusive, and persistence (the length of time for which a community exists) and resilience are the most apt measures of community behavior (DeAngelis and Waterhouse 1987). Alternatively, local populations may be connected into so-called metapopulations at larger landscape levels (see Chap. 13). In this scenario, it may not matter that one patch is unstable and goes extinct. Species recolonize by dispersal between patches, and stability in metapopulations is shown at the landscape level, not at smaller, one-patch scales. As long as local patches are out of phase with one another, species will persist in the landscape.

The differences in organization of equilibrium and nonequilibrium communities are captured in Fig. 17.5, which depicts a theoretical gradient of nonequilibrium to equilibrium conditions. Nonequilibrium communities are often not saturated with species, and biotic interactions such as predation and competition are weak and density independent. Chance or stochastic events induced by abiotic factors such as weather are common. In contrast, equilibrium communities have many biotic interactions, and these would operate in a density-dependent manner to regulate population size. Because the community is saturated with species, invasions by new species are rare.

Chesson and Case (1986) reviewed the theoretical mechanisms that have been advanced to explain the existence of equilibrium and nonequilibrium communities, and these are summarized in Table 17.1. There have been three major equilibrium theories of community organization:

1. *Classical Competition Theory* (Hutchinson 1959)
   a. Competition is the major process controlling community structure, and limiting resources are required for coexistence of n species.
   b. Environmental influences are unimportant, and population growth rates can be determined by deterministic equations.
   c. The environment is spatially homogeneous, and there is little dispersal of organisms.
2. *Predation Extensions.* By adding predation to assumption (a) above, a new equilibrium model could be developed that allows n species to coexist on fewer than n resources (Levin 1970).

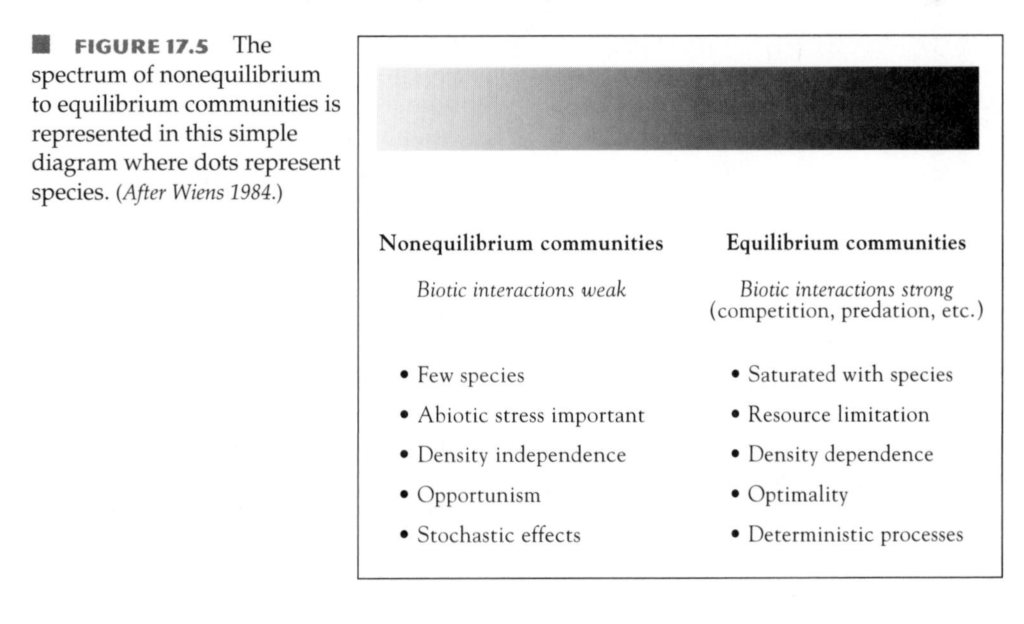

FIGURE 17.5 The spectrum of nonequilibrium to equilibrium communities is represented in this simple diagram where dots represent species. (*After Wiens 1984.*)

TABLE 17.1 Emphases of different community theories. (*From Chesson and Case 1986*)

	EQUILIBRIUM THEORIES			NONEQUILIBRIUM MODELS			
	CLASSICAL COMPETITION THEORY	PREDATION EXTENSION	SPATIAL VARIATION EXTENSION	(1) FLUCTUATING ENVIRONMENT	(2) DENSITY INDEPENDENCE	(3) CHANGING ENVIRONMENTAL MEAN	(4) SLOW COMPETITIVE DISPLACEMENT
Assumptions							
No environmental fluctuations	Yes	Yes	Yes	No	No	—	Yes/no
Constant mean environment	Yes	Yes	Yes	Yes	Yes	No	Yes
Spatial homogeneity	Yes	Yes	No	Yes/no	Yes/no	—	Yes
Life history traits adequately summarized by growth rates	Yes	Yes	Yes	No	No	No	Yes
Continuous competition	Yes	Yes	Yes	Yes	No	—	Yes/no
Predictions: Factors involved in coexistence of n species							
At least n resources, limiting factors, or patches	Yes	Yes	Yes	No	No	—	No
Limits to similarity of resource use	Yes	Yes/no	No	No	No	—	No
Differential responses to fluctuating environmental variables or resources	No	No	No	Yes	Yes	—	No
Particular kinds of life history traits	No	No	No	Yes	Yes	Yes	—
Shapes of functional responses	No	Yes	No	Yes/no	—	—	No
Stability	Yes	Yes	Yes	Yes	Yes	No	No
Overall similarity of species	No	No	No	No	No	—	Yes
History	No	No	No	No	No	Yes	Yes

Note: A dash indicates that no particular emphasis has yet emerged. Yes/no means that the theory suggests different answers in different circumstances.

3. *Spatial Variation Extension.* If species compete for a single resource but the environment favors different species in different patches, then it is possible for n species to coexist in a system of n patches. Each patch has a stable equilibrium (Levin 1974), and the resulting model is similar to a metapopulation model.

The four nonequilibrium theories can be summarized as follows:

4. *Fluctuating Environment.* Competition is still strong, but the environment changes seasonally or irregularly so that competitive rankings of species change temporally too and no one species can dominate (Armstrong and McGehee 1980).
5. *Density Independence.* The environment changes and population densities are not always high enough for competition or other biotic processes such as predation or parasitism to be important (Strong 1986).
6. *Changing Environmental Mean.* Instead of an environment fluctuating around a mean, as in no. 4, the mean actually changes, as it might if there were global climate change for example. Davis (1986) noted the importance of evolutionary changes in response to recent climate change that involved the retreat of glaciers in North America.
7. *Slow Competitive Displacement.* Competitive abilities are similar between species, and although competition is important random variation and chance play a big part in who wins. Hubbell and Foster (1986) argued that this was the case in species-rich tropical forests where many species are ecologically identical.

Which of these theories is best supported? Data from counts of wild game in East Africa indicated the existence of stable communities (Prins and Douglas-Hamilton 1990). Data from counts of thirteen species of large herbivores, such as buffalo, elephants, and other grazers or browsers from 1959 through 1984 with counts every three or four years showed little variability. This suggested an overall constancy of biomass in each of these herbivore groupings or guilds. The only exception was a gradual increase in buffalo numbers following an outbreak of rinderpest disease in 1959. Prins and Douglas-Hamilton (1990) implicated competition as a main mechanism contributing to stability.

On the other hand, in their studies of prairie bird communities in the United States, Wiens and Rotenberry (1980) could demonstrate no evidence of resource limitation, competition, species abundance, or change in species identities as the availability of resources changed. The abundance of each species changed independently as abiotic factors varied and the community could not really be seen as stable or cohesive. Wiens (1977, 1986) argued that most systems are not at equilibrium so that this is the rule rather than the exception.

17.5 The most diverse systems may exist at intermediate levels of disturbance.

We have so far focused on the question, "Are diverse communities more stable than species-depauparate communities?" Evidence from studies of microbial communities have failed to establish a link between diversity and stability. What if we rephrase the

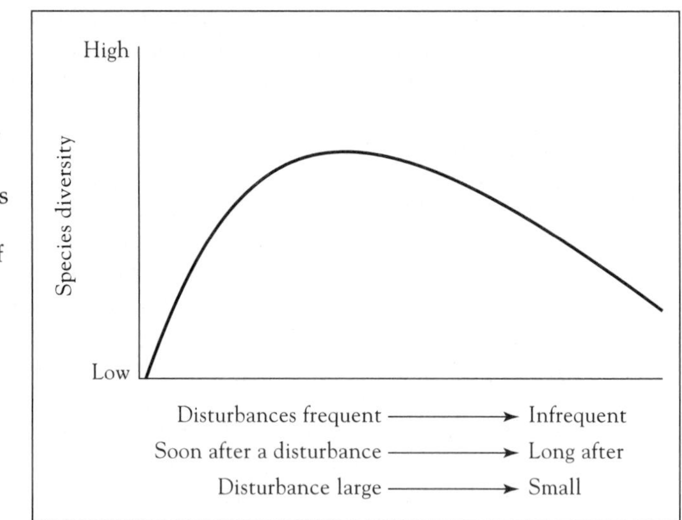

■ **FIGURE 17.6** The intermediate-disturbance hypothesis of community organization. The species composition of a community is never in a state of equilibrium, and high species diversity can be maintained only at intermediate levels of disturbances like fires or windstorms. (*Redrawn from Connell 1978.*)

question to, "Are stable systems more diverse than unstable systems?" Connell (1978) argued that the highest local diversities are actually maintained not in stable systems but in communities of intermediate levels of disturbance (Fig. 17.6). The rationale is that, at high levels of disturbance, for example, where environmental variables are extreme like frequent high winds or high frequency of fire, only good colonists, *r*-selected species, will survive, giving rise to low diversity. This theory also predicts that, at low rates of disturbance, competitively dominant species will outcompete all other species, and only a few *K*-selected species will persist, again giving low diversity. The most diverse communities lie somewhere in between. Natural communities seem to fit into this model fairly well. Tropical rain forests and coral reefs are both examples of communities with high species diversity, and both were used as evidence for the old equilibrium hypothesis. However, Connell (1978) has pointed out that coral reefs maintain their highest diversity in areas disturbed by hurricanes. He has also argued that the richest tropical forests occur where disturbance by storms is common. Some of the richest plant communities in the southeast United States occur on army bombing ranges. Reice (1994) argues that disturbance operates in habitats of all types, and naturally disturbed areas are nearly always more diverse than undisturbed areas. Although storms are frequent in many habitats, other common disturbances include fire, freezes in subtropical habitats, and floods along rivers.

Sousa (1979) provided an elegant experimental verification of the intermediate-disturbance hypothesis in a marine intertidal situation. He found small boulders, easily disturbed by waves, to carry a mean of 1.7 species of sessile plant and animal species; large boulders, rarely moved by waves, had a mean of 2.5 species; and intermediate-sized boulders 3.7 species. Sousa then cemented small boulders to the ocean floor and obtained an increase in species richness, showing that these results were a result of rock stability, not rock size.

What of other tests of the intermediate disturbance hypothesis—especially those done on a bigger scale? There have been very few. Scott Collins and his colleagues pointed out that the intermediate-disturbance hypothesis actually predicts two things:

highest species richness after moderate levels of disturbance and highest levels of richness at intermediate time spans following disturbance. They examined plant species composition from two long-term field experiments in North American tallgrass prairie vegetation where the community was disturbed by fire. In contrast to the first prediction of the intermediate-disturbance hypothesis, there was a significant monotonic decline in species richness with increasing disturbance frequency (fire) and with no evidence of an optimum. The average number of total plant species per quadrat and the number of grass, forb, and annual species were lowest on annually burned sites compared to unburned sites and sites burned once every four years. However, the number of species reached a maximum on field plots at intermediate time intervals since the last disturbance, supporting the second prediction of the hypothesis. They concluded that these two predictions of the intermediate-disturbance hypothesis are independent and uncoupled.

To evaluate whether the intermediate-disturbance hypothesis applies on regional scales, Hiura (1995) examined the relationship between species diversity and gap formation regime of beech forest in Japan over a scale of 10° latitude. The main gap size and the variation of gap sizes showed no correlation with species diversity. However, locations that sustained an intermediate frequency of disturbance in terms of the mean windstorm interval had the highest species diversity (Fig. 17.7).

However, when we examine the intermediate-disturbance hypothesis on a larger scale we run the risk of confounding latitudinal diversity gradients, which operate on large scales, with disturbance-related phenomena, which operate on smaller scales. Indeed, Hiura himself noted that the most predictable model for species diversity over a 10° latitude range in Japan was a multiple regression model composed of two factors, windstorm interval and cumulative temperature of the growing season.

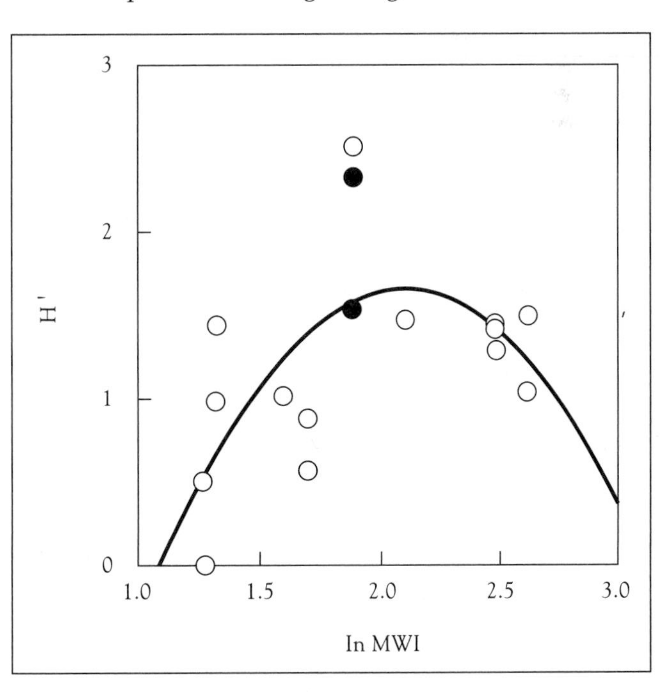

FIGURE 17.7 Relationship between disturbance interval (ln Mean Windstorm Interval) and species diversity, H'. Solid circles represent data from stands that have different flora from other stands (open circles). *(After Hiura 1995.)*

In summary, the relationship between diversity and stability is not clear-cut. Diversity may affect stability in some ways in some places and at certain times, but the relationship is by no means universal. Tilman's recent (1996) work on ecosystem stability suggests that more diverse systems may be more stable, but there is, as yet, little supporting data from other systems. On the other hand, disturbance regime, including frequency and intensity of disturbance, can affect community diversity and often promotes it.

SUMMARY

1. Stability can be thought of in different ways—as resistance to change or as resilience, which refers to return to equilibrium after perturbation. Resilience in turn can be divided into two concepts—elasticity, which measures how quickly a community can return, and amplitude, which measures how big of a disturbance it can return from.

2. A community may be able to exist in more than one form or stable state, though the evidence for this is not strong.

3. The circumstantial evidence for and against links of diversity with stability are many and varied, though experimental studies with communities of micrognisms do not show a strong link of diversity with stability.

4. Recently, Tilman (1996) has reconciled the diversity-stability debate by showing how communities of prairie plants show stability in biomass and ecosystem function but not species abundances.

5. The older more conventional view, called the equilibrium hypothesis, proposed that most communities are stable and predation, parasitism, and competition hold local population sizes at constant levels. The more modern nonequilibrium viewpoint argues that disturbances are frequent and species composition is constantly changing so that stability is elusive.

6. The intermediate disturbance theory suggests that the most diverse communities, like tropical forests and coral reefs, exist at intermediate levels of disturbance.

DISCUSSION QUESTION

How might we set about establishing a link between diversity and stability in nature?

C H A P T E R

18

Succession

Virtually all communities experience, from time to time, disturbances that remove all or some of the plant biomass and affect the abundance of animal species. Thus, diversity in most communities is drastically altered by fire, storms, grazing, or erosion. Succession is the process by which organisms recolonize an area following a disturbance (Glenn-Lewin, Peet and Veblen 1992). Species diversity gradually changes as the community returns to "normal," and early community ecologists suggested that this change was predictable and orderly. They termed such changes **succession**. Primary succession occurs when plants invade an area in which no plants have grown before, such as bare ground, or new lakes created by the retreat of glaciers. In primary succession on land (Photo 18.1), the plants must often build the soil; so a long time may be required for the process—hundreds to thousands of years. Only a tiny proportion of the Earth's surface is currently undergoing primary succession, though some argue that because of geological cycles and processes our climate is not stable, and primary succession never ends. Primary succession can have practical use in the reclamation of spoiled lands.

Secondary succession can be considered a modification of the longer-term primary succession. Clearing a natural forest and farming the land for several years is an example of severe forest disturbance that may lead to a distinct secondary succession following the cessation of farming. The ploughing and lack of plant cover cause substantial changes in the soil, particularly the loss of organic matter and nutrients. When the field is then abandoned, the succession that follows can be quite different from one that develops after a natural disturbance.

18.1 Facilitation assumes each invading species makes the habitat a little more favorable for succeeding species.

The concept of succession was first developed by botanists who monitored floristic changes along sharp environmental gradients, for example, from seashore and sand-dune halophytes back to scrub and finally mature woodland. Clements is often viewed

■ **PHOTO 18.1** Succession at Glacier Bay, Alaska. (**a**) First comes Dwarf fireweed then, (**b**) *Dryas drummondi* followed by (**c**) alder trees, *Alnus sinuata* which fix nitrogen and facilitate the entry of (**d**) spruce trees. (*Tom Bean, DRK Photo 262150, 898037, 898036, and 124561.*)

as the father of successional theory. His early work (Clements 1916) emphasized succession as a deterministic phenomenon, with a community proceeding to some distinct end point or climax community. A key assumption was that each invading species made the environment a little different—say, a little less salty, a little more shady, or a little more rich in soil nitrogen—so that it then became suitable for more *K*-selected species, which invaded and outcompeted the earlier residents (Clements 1936; Van Andel et al. 1993). This process, known as facilitation, supposedly continued until the most *K*-selected species had invaded, when the community was said to be at climax. Retrogression in this sequence was not possible unless another disturbance intervened. The climax community for any given region was thought to be determined by climate and soil conditions. What data fit this model?

Initially, succession following the retreat of Alaskan glaciers was argued to fit the Clementsian pattern of facilitation (Cooper 1923; Crocker and Major 1955). Over the past two hundred years, the glaciers in the Northern Hemisphere have undergone dramatic retreats, up to 100 km in some cases (Fig. 18.1), sure evidence of a climatic change. As glaciers retreat, they leave till and moraines, which are deposits of pulverized "soil" and stones brought forward by the ice, whose age has often been determined by direct observation. In Alaska the bare "soil" has low nitrogen content with little organic matter. In the early stages, it is first colonized by a "black crust" of blue-green algae, lichens, and liverworts. Next comes fireweed (*Epilobium latifolium*) and *Dryas drummondii*. In the intermediate stages, Alder trees, *Alnus sinuata*, invade the area and in the mature stage these are followed by spruce trees.

At first, ecologists suggested that each species facilitated the entry into the community of the next species. However, Chapin et al. (1994) showed that facilitation occurs only during part of the process. Because *Dryas* and alders can fix atmospheric nitrogen, the nitrogen content of the soil increases dramatically. This facilitates the invasion of spruce trees. On the other hand, competition is important because the taller alders shade out the shorter *Dryas*. Thus, after about 50 years, the dense stands of alder (*Alnus* sp.) begin to be shaded out by Sitka spruce, which after another 120 years may form a dense forest. Spruce trees cannot fix nitrogen directly; they take it from the soil, lessening the available nitrogen. Facilitation, originally thought to fuel the entire sequence of succession, was important only in the establishment of spruce. Competition was important in other phases of succession. This is probably true in other systems also (Callaway et al. 1996). It is important to remember that in classical facilitation only half of each interaction is facilitative: early facilitates late but late outcompetes early.

What other evidence is there of facilitation? The decomposition of plant material, such as logs, also incorporates elements of facilitation. A now classic experiment by Edwards and Heath (1963) demonstrated this phenomenon. They put oak and beech leaves in nylon bags in the soil and examined decomposition rates. By varying the mesh size of the bags, they could vary the sizes of the decomposers entering them. They obtained the following results:

BAG MESH SIZE (MM)	FAUNA THAT COULD ENTER BAGS	PERCENT OAK LEAVES GONE IN NINE MONTHS
7.0	All	93
0.5	Small invertebrates, microorganisms	38
0.003	Microorganisms only	0

Although microorganisms are very important in decay, they cannot begin their work until particle size is reduced by larger organisms. In the soil, earthworms are most important in the initial decay process. Thus, on a small scale, facilitation occurs in the decomposition of plant material.

Other examples of facilitation occur in terrestrial and marine systems. Nitrogen content of heathland soil in Western Europe increases from about 1 to 13g nitrogen per square meter (Nm^{-2}) over chronosequences of about fifty years (Berendse, Schmitz, and deVisse 1994). Such changes are accompanied by a replacement of the dwarf shrubs *Calluna vielgoris* and *Erica tetalix* by the grasses *Molinia caerulea* and *Deschampsia flexuosa*. Experi-

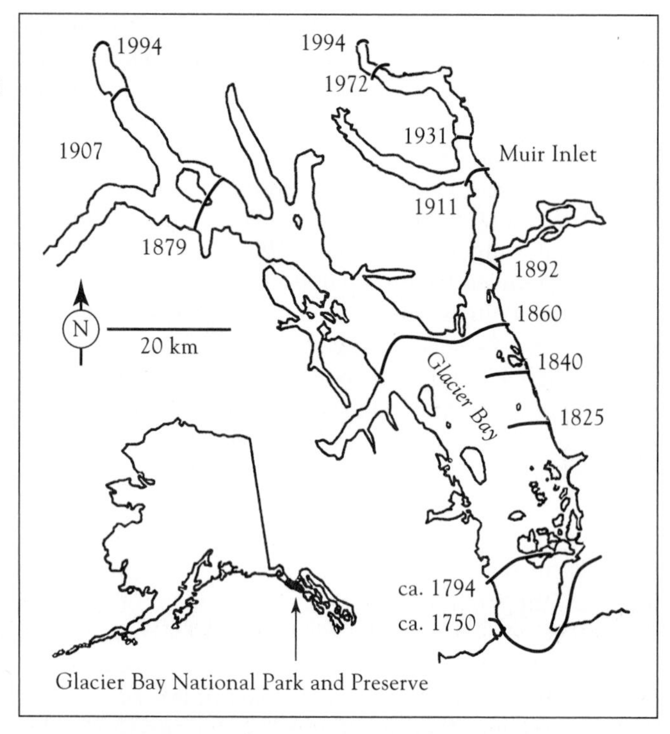

■ **FIGURE 18.1** Glacier Bay fjord complex of southeastern Alaska showing the rate of ice recession since 1760. (*After Fastie [1995] and others.*)

mental addition of *Calluna* litter, or nitrogen fertilizer, had the same result—an increase in the biomass of the grasses and a reduction in the biomass of *Calluna*. *Calluna* actually alters the environment, enriching the soil and hastening invasion by other species.

Although in marine systems soils do not develop, facilitation may still be encountered when one species enhances the quality of settling and establishment sites for another. Working with experimental panels placed subtidally in Delaware Bay, Dean and Hurd (1980) found that hydroids enhanced the settlement of tunicates, and both facilitated the settlement of mussels. Turner (1983) found that the establishment of the surfgrass *Phyllospadix scouleri* in rocky intertidal communities depended on the presence of certain successional algae to which its seeds cling and then germinate.

18.2 Enablement is said to occur if one species cannot invade unless another is already present.

The ultimate form of facilitation is enablement, where species *B* cannot invade unless species A has already invaded first. If this type of phenomenon occurs, the community is often thought to obey certain "assembly rules" (Drake 1991), which is analogous to the pieces of a jigsaw only fitting together in a certain order (Fig. 18.2).

Probably the first and best-known use of the term *assembly rule* for ecological communities was by J. M. Diamond (1975) who, while studying assemblages of birds on and around the islands of New Guinea, noted a checkerboard pattern to species' presence or absence on islands. He suggested that some species were precluded from colonizing certain islands by the presence of other species and implied that interspecific competition

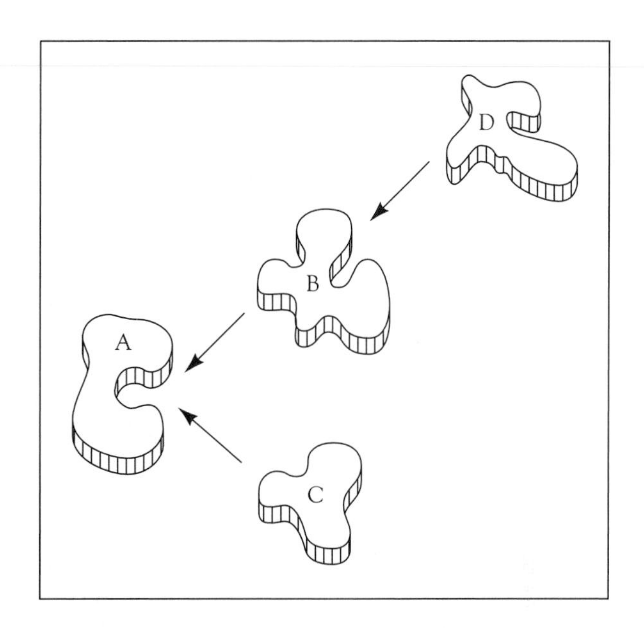

■ **FIGURE 18.2** Depiction of the "puzzle" analogy for community assembly. If species A colonizes first, then either species B or species C can colonize—competition is assumed. Depending on whether B or C is successful, species D is either a potential colonist or finds the community resistant to invasion. However, a real jigsaw puzzle has but one possible complete picture; a biological community may exist in alternative states.

for resources was the cause. In the context of assembly rules this would mean that during the colonization of the islands some bird species that arrived first could prevent other species from successful establishment and thus facilitate the establishment of others. Thus, in Diamond's original paper, the term *assembly rules* was loosely applied to demonstrating that a pattern exists in nature rather than to the mechanism that causes it. In contrast, Weiher and Keddy (1995) recommend reserving of the term *assembly rules* to describe those rules or constraints that govern the patterns.

In his original one-hundred-page paper, published as a chapter in the landmark book, *Community Ecology*, edited by Diamond and Ted Case, Diamond (1975) put together a set of seven "assembly rules" for the New Guinea island bird community:

a. If one considers all the combinations that can be formed from a group of related species, only certain of these combinations exist in nature.
b. Permissible combinations resist invaders that would transform them into forbidden combinations.
c. A combination that is stable on a large or species-rich island may be unstable on a small or species-poor island.
d. On a small or species-poor island, a combination may resist invaders that would be successful on a larger or more species-rich island.
e. Some pairs of species never coexist, either by themselves or as part of a larger combination.
f. Some pairs of species that form an unstable combination by themselves may form part of a stable larger combination.
g. Conversely, some combinations that are composed entirely of stable subcombinations are themselves unstable.

Connor and Simberloff (1979) criticized Diamond's assembly rules, calling them tautological and trivial and arguing that such rules would be suggested even if the species were distributed among islands by chance. To arrive at this conclusion, Connor and

Simberloff compared the observed distribution of birds on islands to a null model, a model where distributions were generated by chance. A series of rebuttals and rejoinders, based on the appropriate use of null models, followed (Gilpin and Diamond 1984; Connor and Simberloff 1984).

Gilpin and Diamond illustrated their point as follows: suppose one has *N* islands, on each of which either species *A* or species *B* is present but never both and never neither. If *N* is large, this "checkerboard" pattern could be regarded as evidence for competition between species *A* and *B*. For statistical rigor, it is desirable to test the pattern against some neutral model. The way Connor and Simberloff's neutral models were constructed was to randomly reshuffle the pool of species, subject to various constraints. The constraints were (1) the number of islands remains equal to the actual number; (2) the number of species remains equal to the actual number; (3) each species is present on exactly as many islands as in actuality; and (4) each island has as many species as does the real island. Although this procedure may appear reasonable, constraints (3) and (4) can have the effect of biasing the model. In fact, in our simple examples these constraints are so severe as to guarantee that the null model is identical with the observed pattern: each island in the hypothetical, "neutrally constructed" archipelago also must have one and only one of species *A* or *B*, and in the observed proportions.

Instead of rejecting the hypothesis that competition forged this pattern, one might more appropriately reject the construction of the null model. Colwell and Winkler (1984) called this the Narcissus effect: sampling from a postcompetition pool underestimates the role of competition since its effect is already reflected in the pool. To demonstrate the Narcissus effect, Colwell and Winkler used a computer program called GOD, which generated assemblies of species whose phylogenetic lineages obeyed specified rules. Subsets of this "mainland biota" then were made to colonize archipelagoes, as described by a program called WALLACE. In this colonization, competitive interactions were or were not important, depending on how WALLACE's rules were specified. Next, Colwell and Winkler took "field data" thus generated and examined to what extent the emergent community patterns stood out against various neutral models. An important conclusion was that the conventional neutral models, constructed by reshuffling the data, were often indistinguishable from the "observed" field data, even when competition had in fact been a strong force in WALLACE's colonization of the archipelago. Simberloff replied that in such circumstances we should not give up but rather should search for some more appropriate way of framing a neutral model.

The upshot of all this debate is that assembly rules may be difficult to uncover in natural communities because of our inability to view past events. Patterns may not have been generated during community assembly but instead may be due to contemporary ecological processes. Unfortunately, community assembly or whole community succession has rarely been documented in nature—except perhaps following volcanic eruptions. Despite this paucity of "hard evidence," Drake (1990) has argued that many scientists have in fact found circumstantial evidence for assembly mechanics, and he has cited as evidence studies on lizards (Roughgarden 1989), frogs (Wilbur and Alford 1985), plants (McCune and Allen 1985), and ants (Cole 1983).

An alternative to considering lists of species names in assembly rules is to use the traits they possess and seek patterns in them—so-called guild-based assembly rules or functional groups (see Chapter 20) (Wilson and Roxburgh 1994). There are several reasons why the value of assembly rules increases when they are based on guilds or traits. First, when rules are based on species names and a local species pool, they will not be easily comparable to other sites or habitat types. Assembly rules will be more generalizable if based upon guilds. Also, guild-based rules provide more information to readers outside any particular area of expertise. Lists of species names have little meaning to most ecologists outside each, often narrow, specialty. Finally, trait-based rules will usually be simpler to construct, whereas species-based rules will often ramify into a list of complex pairwise interactions. For example, the species form of a "rule" might be: if a community has species *A*, then it usually will also have species *C* unless species *E* is present, while if a community has species *B* then it will also have species *D* unless *F* is present. A more trait-based rule is clearer: the proportion of species from each functional group will tend to remain constant for each observation. Such rules have been argued to operate in desert rodent communities (Fox and Brown 1993, 1995; but see also Wilson 1995; Wilson and Roxburgh 1994), but, once again, another statistical reanalysis by Simberloff and colleagues (Stone, Dayan, and Simberloff 1996) casts doubt on this.

18.3 Inhibition implies early colonists prevent later arrivals from replacing them.

Many other types of succession do not show elements of facilitation or assembly rules at all. Another view is that possession of space is all important—who gets there first determines community structure. In such communities, the principal mechanism affecting successional change is inhibition of subsequent colonists. There are many examples of inhibition, indeed, it is often viewed as the main process determining the rate and direction of successional change (Gray 1987). For example, Facelli and Facelli (1993) removed the litter of *Setaria faberii*, an early successional species in New Jersey old fields and noted that the removal increased the biomass of a later species, *Erigeron annuis*. The release of phytotoxic compounds from decomposing *Setaria* litter or physical obstruction may contribute to the inhibition of *Erigeron*. Without the litter, *Erigeron* became dominant and reduced the biomass of *Setaria*.

If inhibition is the main process determining the rate of succession, what turns it off so that new species can invade? Van der Putten, Van Dijk, and Peters (1993) suggested that in coastal communities it might be species-specific soil pathogens. In Europe, sand dunes are built around marram grass, *Ammophila arenaria*, which is followed in sequence by fescue, *Festuca rubra*, and sand sedge, *Carex arenoria*, with sea couch, *Elymus athericus*, at the landward edge. Van der Putten and coworkers carried out reciprocal transplant experiments using sterilized and unsterilized dune sand. Two-week-old seedlings, raised from local seed, were planted into replicated series of pots containing sand collected from beneath each of the four main plant species. In half of the pots the soil was sterilized by gamma radiation. The results were striking. Biomass production of each plant species was reduced by the soil-borne diseases of its successors but not of its predecessors. Once a successor gains a small foothold, its associated pathogens help it outcompete its predecessor. Burial by new sand presumably provides marram grass with relatively sterile sand in which roots can develop and escape the ravages of pathogens (Little and Maun 1996). In addition, the presence or absence of mycorrhizal fungi may be critical because the presence of such fungi confers a degree of resistance to pathogen attacks.

18.4 Tolerance suggests that early colonists neither facilitate nor inhibit later colonists.

The huge differences in the two mechanisms of succession in nature, facilitation and inhibition, prompted Connell and Slatyer (1977) to view the different models as extremes on a continuum (Connell, Noble, and Slatyer 1987). Connell and Slatyer termed the classical view of succession proposed by Clements (1936), which supposes an orderly progression to a predictable climax community, the *facilitation model* because each species makes the environment more suitable for the next. For the type of succession in which competition between species was great, they formulated the term *inhibition model* because colonists tend to prevent subsequent colonization by other species. In this model, succession depends on who arrives first. Succession proceeds as colonists die, but it is not orderly. As a sort of an in-between model, Connell and Slatyer then proposed a third concept, which they termed the *tolerance model*. In this model, any species can start the succession, but the eventual climax community is reached in a somewhat orderly fashion. Connell and Slatyer could find little evidence to support the tolerance model. The best evidence for the tolerance model actually came from Egler's (1954) work on floral succession. Egler showed that in many flower communities there is a tendency in succession for most species to be present at the outset, as buried seeds or roots. Whichever species germinated first, or grew from roots, would start the succession sequence. Egler had termed this idea of succession "initial floristic composition sequence." But of course Egler's model works only for secondary succession. All three models predict that the most likely earlier colonists will be *r*-selected species, weeds in many cases. The key distinction between the models is in how succession proceeds. In the classical facilitation model, species replacement is facilitated by previous colonists, in the inhibition model it is inhibited by the action of previous colonists, and in the tolerance model, it is unaffected (Fig. 18.3). Understanding how succession proceeds is not merely of academic interest. Because of the emergence of restoration ecology as an important discipline, an understanding of succession on "repaired" communities is of vital importance (Applied Ecology: Restoration Ecology).

18.5 Biotic interactions may influence successional processes.

The facilitation and inhibition models of succession rely heavily on the presence of commensalism and competition, respectively. But there are a myriad of other biotic interactions that have the potential to affect how succession proceeds. Ten years after the facilitation, inhibition, and tolerance models were proposed, and many data later, they were criticized by Walker and Chapin (1987) who argued that not all studies fit neatly into them. Connell, Noble, and Slatyer (1987) replied that their models represented not strict alternatives but rather the ends of a continuum.

In the real world, components from each model may be prevalent in any one given study system. Thus, for the succession of alder to spruce on Alaskan floodplains, "stochastic, life history, facilitative, competitive and herbivory processes all affect the interaction between alder and spruce during succession and no single successional process or

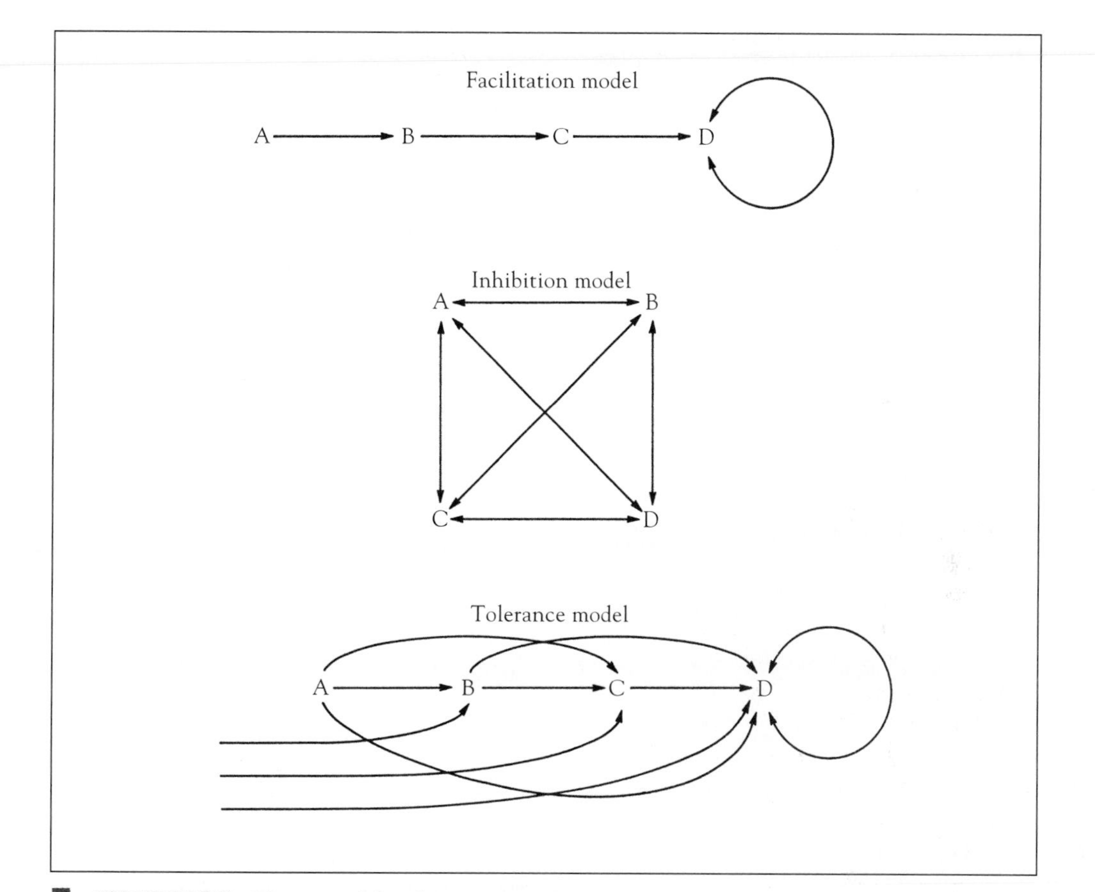

FIGURE 18.3 Three models of succession. Four species are represented by A, B, C, and D. An arrow indicates "is replaced by." The facilitation model is the classic model of succession. In the inhibition model, all replacements are possible, and much depends on who gets there first. The tolerance model is represented by a competitive hierarchy in which later species can outcompete earlier species but can also invade in their absence. *(Redrawn from Horn 1976.)*

model adequately describes successional change" (Walker and Chapin 1987, p. 131; see also Chapin et al. 1994). Walker and Chapin (1987) discussed the relative importance of each of a variety of major factors on communities of a different successional state and presented their results in a figure, which outlines how major processes determine change in plant-species composition (Fig. 18.4). Their conclusion was that succession is a complex process driven by many processes acting simultaneously in any given situation. Thus, the effects of such factors as competition, seed arrival time, insect and mammalian herbivory, and stochastic events vary in importance according to the stage of succession—early, middle, or late. For example, competition is not very important to colonizing species in the early stages of succession: there are very few competitors present. However, in mature communities, competition can be an important force. Similarly, stochastic events, like fire, can devastate early successional communities but may have less effect on species change in mature communities where species like large trees are able to withstand periodic burns.

APPLIED ECOLOGY

Restoration Ecology

Despite the best efforts of conservationists, some species or communities can go extinct in a particular area following habitat loss. If there are other populations of the species and other remnants of the habitat in other geographical areas, it may be possible to take seeds of plants and young animals from these areas and use them to restore the damaged area. How best to do this is the province of restoration ecology (Photo 1).

■ **PHOTO 1** Successful restoration of a strip mine in Fultan County, Illinois, reclaimed using sewage sludge. (*R. F. Ashley, Visuals Unlimited.*)

One might define restoration ecology as the full or partial replacement of biological populations and/or their habitats that have been extinguished or diminished and the substitution by the same or similar species that have social, economic, or ecological value. Restoration ecology can involve simply trying to return species to the wild such as, for example, returning peregrine falcons to suitable habitat in the eastern United States, returning Arabian oryx to Saudi Arabia, or returning the California condor to Southern California. Or it can focus on restoring complete habitats. For example, following open-cast mining, huge tracts of disturbed land have to have a large part of their original quota of biological species, such as grasses, shrubs, trees, and animals returned in order for the ecosystem to be anything like "normal." A knowledge of how succession proceeds can be vital if restored areas are expected to regain their full complement of species.

Some ecologists and conservationists are fearful of the restoration process. Their concern is that the admission of even the partial effectiveness of restoration will be viewed as a license for further ecological destruction in the name of progress and growth. They fear that some of the wilderness area, national parks, and other ecosystems that now have exceptional protection may have this protection reduced in light of the feasibility of repairing the damage caused by the exploitation of various resources in these systems. For example, there is much pressure to "open up" part of the Arctic Wildlife Refuge in Alaska because of the probability that there are huge oilfields in the area. However, large-scale restoration in such areas is likely to be prohibitively expensive. The costs of restoration are much greater than those of conservation, since in many cases whole communities of plants have to be replanted, soil and water conditions have to be altered, and keystone herbivores have to be brought into the area. This is generally much more expensive and demanding in terms of human time and money than simply conserving a habitat in the first place. Thus, only habitats and wildlife damaged as a result of small-scale oil spills could likely be restored using restoration ecology techniques. The whole process is analogous to the relationship between preventive medicine and surgery. Conservation can be viewed as preventive medicine and restoration ecology as surgery. Surgery is almost always more expensive.

Restoration ecology is in its infancy, with many of the techniques yet to be well worked out and refined. For example, following the wreck of the oil tanker *Torrey Canyon* in 1967, some cleanup methods appear to have caused more damage to the indigenous biota than the oil itself. A natural equilibrium was restored more rapidly to areas where no intervention occurred than to those areas where cleanup processes were used. The careless use of suction devices, scrapers, oil dispersants, and the like may cause more stress to the ecosystem, if improperly used, than the material of the spill itself.

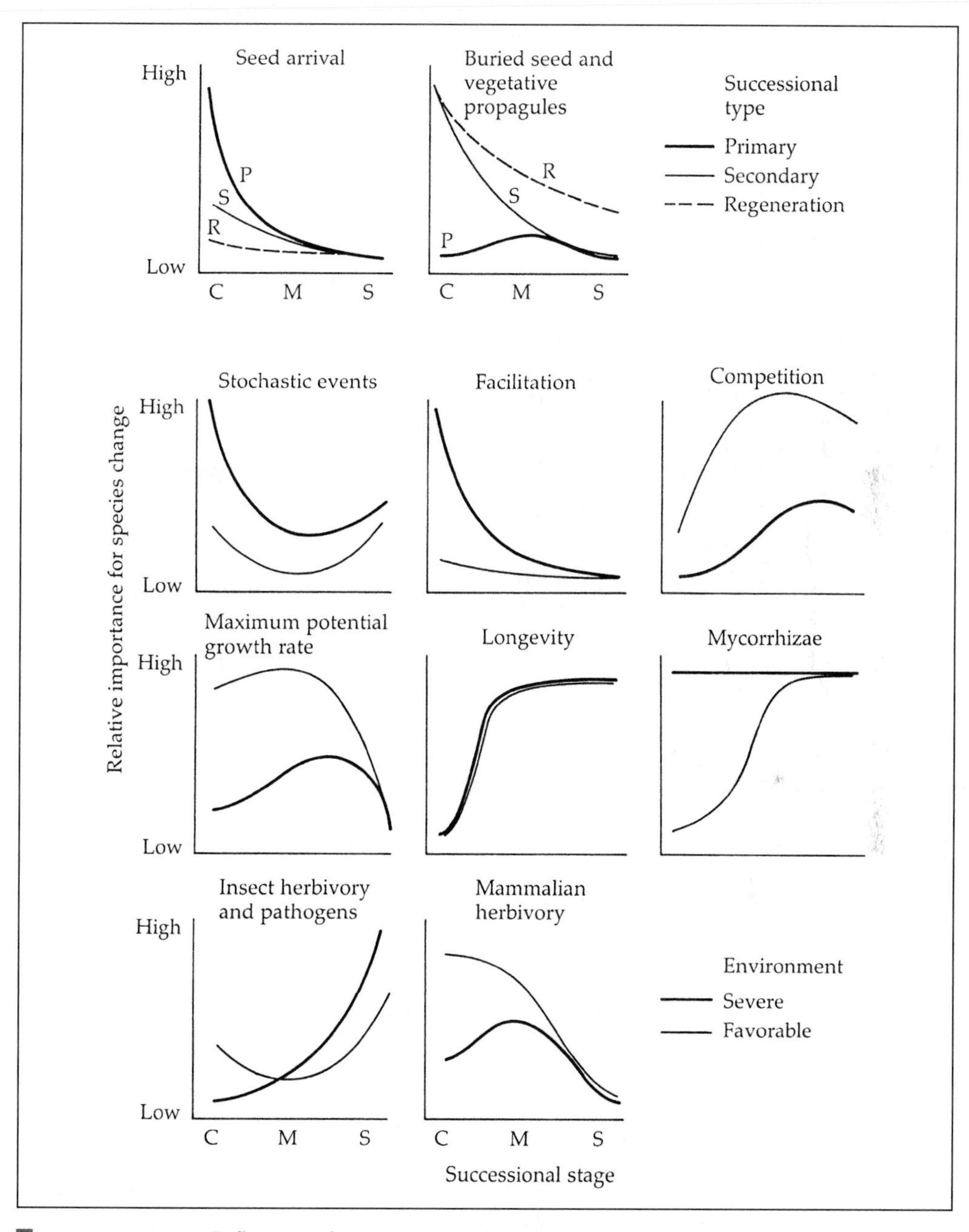

■ **FIGURE 18.4** Influence of succession and environmental severity on major successional processes that determine change in species composition during colonization (C), maturation (M), or senescence (S) stages of succession. *(Redrawn from Walker and Chapin 1987.)*

We can summarize Walker and Chapin's (1987) main conclusions as follows:

1. Dispersal mechanisms promoting seed dispersal are more important in primary succession, whereas buried seed and surviving vegetative propagules are more important in secondary succession.
2. Stochastic variation is more important in severe and low-resource environments. For example, fire is important in dry environments, flooding determines the colonization of river banks, and moisture levels are important in dune communities (see also Olff, Huisman, Van Tooren 1993).
3. Facilitation is more important in severe environments (like deserts), in primary succession, and in the early stages of community development.
4. Competition is probably more widespread than facilitation (Tilman 1985), especially in more favorable environments (Grime 1979).
5. Maximum potential growth is particularly important in favorable environments where resources are sufficient to promote rapid growth.
6. Differential longevity is more important in older communities.
7. Mycorrhizae may be especially important in severe environments (see also Little and Maun 1996).
8. Insect herbivory and pathogens are more important on more mature vegetation. Edwards and Gillman (1987) suggested that in the early stages of most primary successions herbivory is unimportant since it may take a relatively long time before vegetation is sufficiently developed to provide an adequate habitat for animals.
9. Mammalian herbivores are more important in early and mid-successional communities.

We can also add that

1. Symbiotic nitrogen fixers are often prominent in early succession (Chapin et al. 1994).
2. In primary succession early arrival is enhanced by the production of small wind-dispersed seeds (Fenner 1987).
3. Late successional species are more shade tolerant and often achieve greater height than early colonists. Indeed, Tilman (1985, 1988) discussed the effects of shade and resources in his resource-ratio hypothesis. This hypothesis predicted that when nutrients are in short supply in early succession, competition for nutrients is more important than competition for light. This favors high investment in roots. In contrast, in later levels of succession nutrient levels increase and so does plant biomass. Increased biomass leads to increased light interception and more intense competition for light, favoring tall species.

Interesting and valuable as these generalizations are, they should not necessarily be taken as uncontested. For example, Edwards-Jones and Brown (1993) argued that insects attacking ephemeral, early successional herbs will have high population densities because these have qualitative defenses (poisons). In contrast, insects on late successional plants will have lower population densities because these have quantitative defenses (tannins and resins) that act in a dosage-dependent fashion. This theory was tested for three separate insect taxa over a successional gradient in southern England

and found to be supported by the data. All taxa displayed greater population densities on short-lived plants of early succession than on trees of later succession. Davidson (1993) has also mustered evidence from a variety of studies in support of this position. She goes on to suggest that herbivory itself may actually hasten succession in earlier series and retard succession in later series.

In marine systems too, fish grazing may deflect the path of succession. Hixon and Brostoff (1996) performed some elegant manipulations of fish grazing in Hawaii. Where grazing was intense, all erect algae were removed and the path of succession was strongly deflected. The early successional stages were replaced by a low-biomass and low-diversity assemblage of crusts and prostrate blue-green mats. In the absence of grazing algal succession followed three stages over one year: early dominance by green and brown filaments, a midsuccessional stage of thin red filaments, and a late stage of coarsely branched thick filaments. Clearly, the facilitation, inhibition, and tolerance models are an oversimplification of what happens in nature where a myriad of things impact succession.

Brown and Southwood (1987) have taken a different tack in summarizing succession. They discussed general life history characteristics of the different organisms involved in the succession of a young field through an old field to a woodland. Because their studies were largely experimental and confined to a small spatial scale in Silwood Park, England, the chances of introducing confounding factors between the habitats was reduced. In general, they found organisms in early successional habitats, to have short generations, high reproductive effort, high dispersal ability, high niche breadth, and low morphological diversity and organisms in older series to have just the opposite (Fig. 18.5) with the exception of flight ability, which was about the same. This approach provides a valuable complement to Connell and Slatyer's original models.

18.6 Succession has been viewed as a Markov process.

Another contrast to Connell and Slatyer's models, which focus on the biological interactions of species, was the work of Horn (1976), who provided a simple model of succession that assumes succession is a Markovian replacement process.

A Markov chain is a stochastic process in which transitions between various states occur with characteristic probabilities that depend only on the current state and not on any previous state. A Markov chain is regular if any state can be reached from any other state in a finite number of steps and if it is not cyclic. A regular Markov chain can settle into a pattern in which the various states occur more or less randomly with characteristic frequencies that are independent of the initial state. If biological succession can be shown to operate in a similar manner we may be able to explain it purely in terms of chance changes.

To test his ideas, Horn gathered together information on a community, in this case the tree species present in a New Jersey forest. Next, he mapped the seedlings underneath the crowns of the forest trees; these were assumed to replace the mature trees in time. Horn assumed that the higher the number of seedlings of a particular species under a tree, the higher the likelihood that this species would replace the mature tree. Horn thus constructed a series of replacement probabilities like those shown in Table 18.1. For example, there were a total of 837 saplings under large grey birch trees (Horn 1975). Among these, there were no grey birch saplings. (Janzen [1970] has argued that this phe-

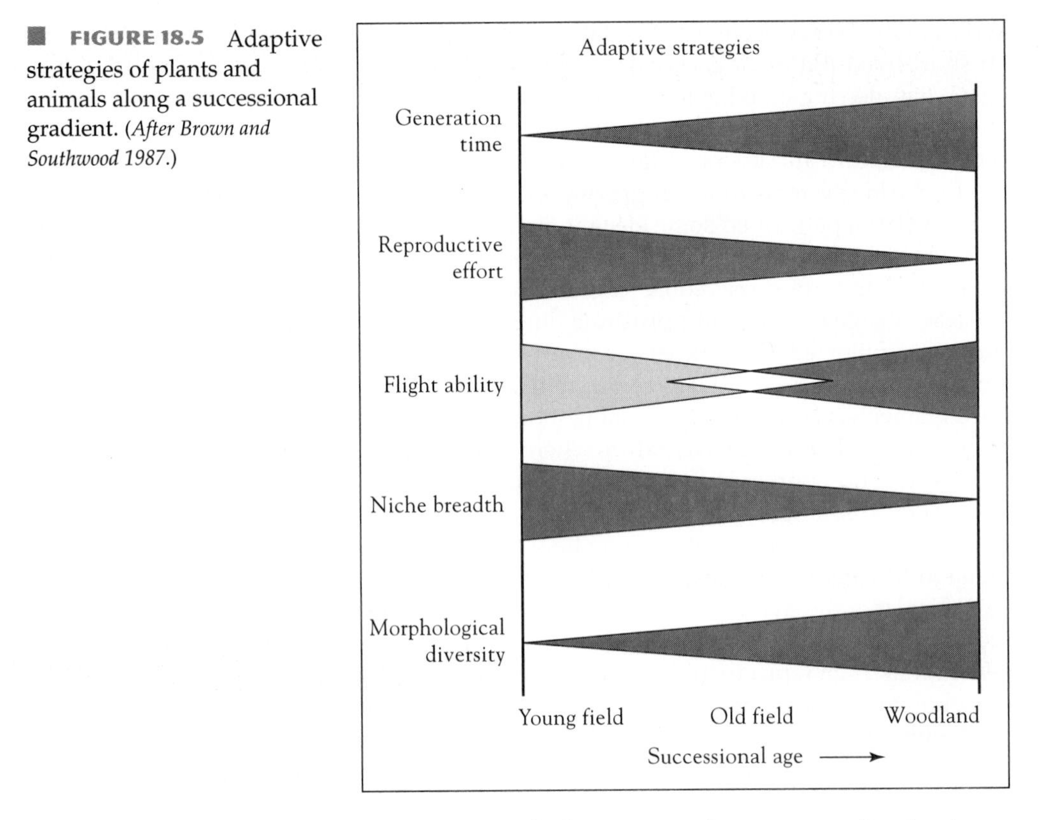

■ **FIGURE 18.5** Adaptive strategies of plants and animals along a successional gradient. (*After Brown and Southwood 1987.*)

nomenon, at least in tropical situations, results because seed eaters are often host specific and congregate under a parent tree, devouring seeds there. Only when seeds happen to fall under different species are they likely to survive. Thus, each tree casts a "seed shadow" in which survival of its own kind is reduced—hence, the high value of dispersive seeds.) There were 142 red maples, 25 beech seedlings, and a few other species. The replacement probability for grey birch under grey birch is then $0/837 = 0$.

TABLE 18.1 A 50-year tree-by-tree transition matrix showing the probability of replacement of one individual by another of the same or different species 50 years hence in a New Jersey forest. (*From Horn 1975*)

PRESENT OCCUPANT	OCCUPANT 50 YEARS HENCE: GREY BIRCH	BLACKGUM	RED MAPLE	BEECH
Grey birch	0+0.05	0.36	0.5	0.09
Blackgum	0.01	0.20+0.37	0.25	0.17
Red maple	0	0.14	0.18+0.37	0.31
Beech	0	0.01	0.03	0.35+0.61

Note: The diagonal is the percentage of trees replaced in 50 years by another of their own kind plus 5% of the original trees left standing (assuming a death rate of 95% for Grey Birch in 50 years, for Blackgum and Red Maple in 150 years, and for Beech in 300 years). Off-diagonal terms are percentage of trees replaced by another species in 50 years.

TABLE 18.2 The predicted percentage composition of a New Jersey forest consisting initially of 100% birch but subject to immigration of seeds of all species. (*From Horn 1975*)

AGE OF FOREST (YEARS):	0	50	100	150	200	\	DATA FROM OLD FOREST
Grey birch	100	5	1	0	0	0	0
Blackgum	0	36	29	23	18	5	3
Red maple	0	50	39	30	24	9	4
Beech	0	9	31	47	58	86	93

These replacement values can be summed to give the total predicted abundance of a species. Beginning with an observed distribution of the canopy species in a stand in New Jersey known to be twenty-five years old, Horn modeled the change in species composition over several centuries (Table 18.2). Theoretically, red maple is predicted

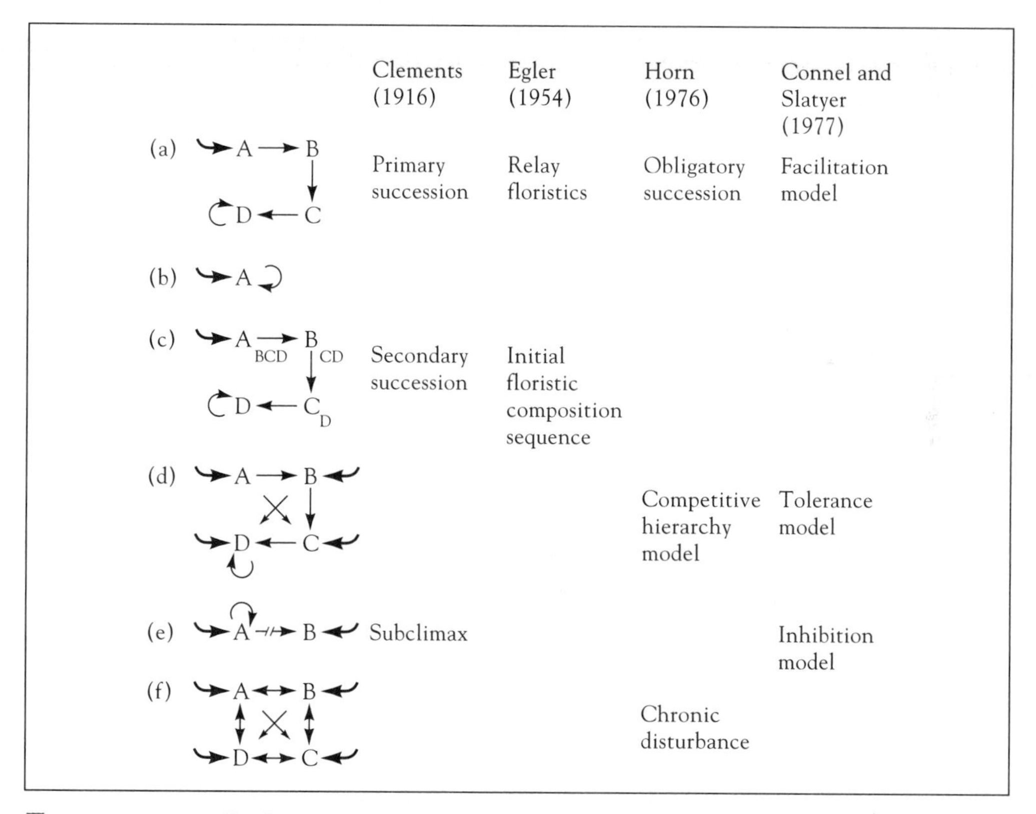

■ **FIGURE 18.6** Replacement sequences in succession proposed by different authors. The letters A-D in (a)-(f) represent hypothetical vegetation types or dominant species; subscript letters in (c) indicate that species are present as minor components or as propagules—for simplicity, they have been omitted from (d)-(f), however, in (f), all species are present as minor components at every stage; thin arrows represent species or vegetation sequences in time; bold arrows represent alternative starting points for succession after disturbance. (b) is usually referred to as direct succession. (*After Noble 1981.*)

to assume dominance quickly while grey birch disappears. Beech should slowly increase to predominate later. All these predictions agree with what happens in nature in forests of known age. Thus, Horn's model predicts a successional outcome independent of the initial forest composition—a prediction very similar to Connell and Slatyer's (1977) tolerance model. However, as Facelli and Pickett (1990) point out, such a result is not necessarily a good test of the Markovian nature of succession because such models were tested against the same data used for their construction. Subsequent tests of Markovian succession have been meager (Facelli and Pickett 1990; but see Hubbell and Foster 1987).

To sum up, succession may be affected by facilitation or inhibition, and it may be affected by a variety of other biotic processes. It may in some instances be adequately described by a Markov replacement process. Noble (1981) has provided a nice diagrammatic summary of succession as viewed by the proponents of these ideas: Clements, Egler, Horn, and Connell and Slatyer (Fig. 18.6). Included in this figure is direct succession (Fig. 18.6b). This is when, in simple communities like lakes, deserts, or tundra, the vegetation simply replaces itself. Chronic disturbance (Fig. 18.6f) was proposed by Lawton (1987) to provide a type of null model of succession. Here succession involves only the chance survival of different species and the random colonization by new species. There is no facilitation or competition. Noble notes that real case histories often show a mix of these succession types.

SUMMARY

1. Primary succession occurs when species invade an area in which no organisms have grown before, such as land unearthed after receding glaciers. Secondary succession is the change of species composition following a change in land usage, such as the return of forests to old fields after agriculture has stopped.

2. Early views of succession viewed the entire process as facilitative, where each species makes the environment more suitable for the next. The most severe form of facilitation is enablement, whereby a species can only occupy an area following the colonization of an earlier species. Such communities are said to obey certain assembly rules.

3. Later work revealed the existence of inhibition, where some colonists actually prevented colonization by other species. In 1977, Connell and Slatyer recognized the existence of a third type of succession, which they termed tolerance which in essence, was intermediate between the other two. In this model, any species can start the succession, but the eventual climax community is reached in a somewhat orderly fashion.

4. It has since been recognized that the three models represent not strict alternatives but different positions along a continuum of effects of earlier and later species.

5. More recently it has been recognized that the path of succession can be deviated by herbivory, disease, and other factors.

DISCUSSION QUESTION

If agriculture on once-virgin tropical forest areas were to stop, how could we speed up the process of secondary succession so that tropical forests returned?

Is mutualism a potent force which afffects the distribution and abundance of species? There are a variety of mutualisms in nature that are common and important.

▲ Cloudless sulfur on thistle, plant and pollinator. (Peter Stiling.)

▲ Beauty berry *Callicarpa americana*, seeds ready to be dispersed. (Peter Stiling.)

◀Coral *Millepora* sp. Dinoflagellates often enter into a mutualism with corals where they provide photsynthate to their hosts. (Roessler, Animals Animals, 620159.)

▶Cleaner fish and host. This moray looks as though it might at any moment cross the line from mutualist to predator. (Wu, Peter Arnold, AN-BE-70C.)

More mutualism.

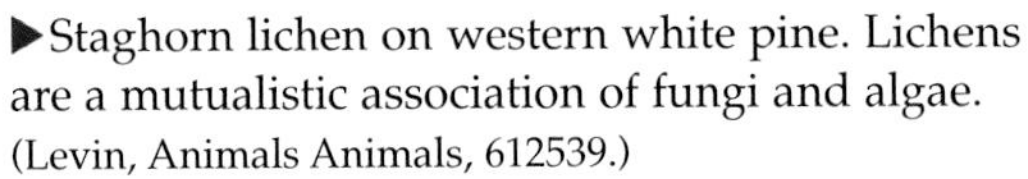

▶Staghorn lichen on western white pine. Lichens are a mutualistic association of fungi and algae. (Levin, Animals Animals, 612539.)

◀Commensalism. Black rhino and egrets at the Ngorongore crater, Tanzania. The egrets benefit from this association because the rhino dislodges insect prey for the egrets from the grass. The rhino is unaffected. (Murphy, Animals Animals, 423630-M.)

▶Mycorrhizae are a mutualistic association of fungus and root tissue which benefits both species. Here the root system of a lemon tree grown with and without mycorrhizae are shown at 4 1/2 months. (Runk/Schoenberger, Grant Heilman, CR2-80B.)

◀Ants tending aphids on a milkweed plant. The aphids secrete a sugar-rich substance called honey dew which the ants eat. In return, the ants protect the aphids from marauding natural enemies. (Robinson, Photo Researchers, Inc., 5J 5043.)

There are many lines of evidence to suggest competition is frequent and important in nature.

▲Competition may be intraspecific, between members of the same species, as between these California poppies on Portal Ridge, Antelope Valley, California. (Ulrich, DRK Photo, 164211.)

◀...or interspecific, as between Iceland poppies, baby's breath, and cornflower in a meadow garden. (Lefever/Grushow, Grant Heilman, CO4X-65B.)

Competition continued.

▲Consumptive competition over resources is common in nature. (Lynn, Photo Researchers, Inc., 2A1776.)

▲Plants commonly exhibit over-growth competition, where certain species are denied light as in this tropical forest. (Dimijian, Phot Researchers, Inc., 621642.)

◀Pre-emptive competition, over space, is common among intertidal organisms such as barnacles. Here barnacles compete for space on the body surface of a California gray whale. (Kolar, Animals Animals, 408301M.)

▶Allelopathy. Purple sage, *Salvia dorrii*, is surrounded by bare ground, although Indian Paintbrush can be seen growing among it. Very little work has been done on allelopathy, and chemicals which may be detrimental to one species may not have an effect on others. (Duffurrena, Grant Heilman, CO5SAD11A.)

CHAPTER

19

Island Biogeography

The 1980 eruption of Mount St. Helens in Washington State provided a good opportunity to study natural recolonization and succession. In the six years following this eruption, Wood and del Moral (1987) monitored vascular-plant invasion of the barren substrates. They found dispersal was limited in many species and that nurse plants played a key role in trapping seeds and promoting seedling establishment. They concluded that the path of early succession depended on spatial position of an area and dispersal abilities of species in the seed pool nearby. Chapin et al. (1994) also concluded that seed dispersal affected primary succession on Alaskan glaciers. Ultimately, who colonized what depended as much on the distance to the nearest floral or faunal source as on successional processes. The effects of distance and size of areas from source pools on successional processes is dealt with here under the heading of island biogeography.

19.1 The colonization of islands following disturbance has been described with the theory of island biogeography, which predicts that the equilibrium number of species on an island should be positively correlated with island size and inversely correlated with distance to mainland.

It was MacArthur and Wilson (1963, 1967) who first formally developed a comprehensive theory to explain the influence of distance (and size of island) on succession on islands in their "equilibrium theory of insular zoogeography." They suggested the following:

1. The number of species on an island, S, tended toward an equilibrium number, $\hat{S}$.
2. $\hat{S}$ is the result of a balance between the rate of immigration and the rate of extinction. The rate of immigration of new species is highest when no

species are present on the island so that each species that invades is a new species. As species accumulate, many subsequent immigrants no longer represent new species. The rate of extinction is low at the time of first colonization because few species are present. It increases with time as species accumulate. Thus, species may come and go extinct but $\hat{S}$ remains the same (Fig. 19.1). In actuality, MacArthur and Wilson (1963, 1967) reasoned that both the immigration and extinction lines would exhibit curvature (Fig. 19.2). First, species arrive on islands at different rates—some organisms are more mobile than others and will arrive quickly. Later, the

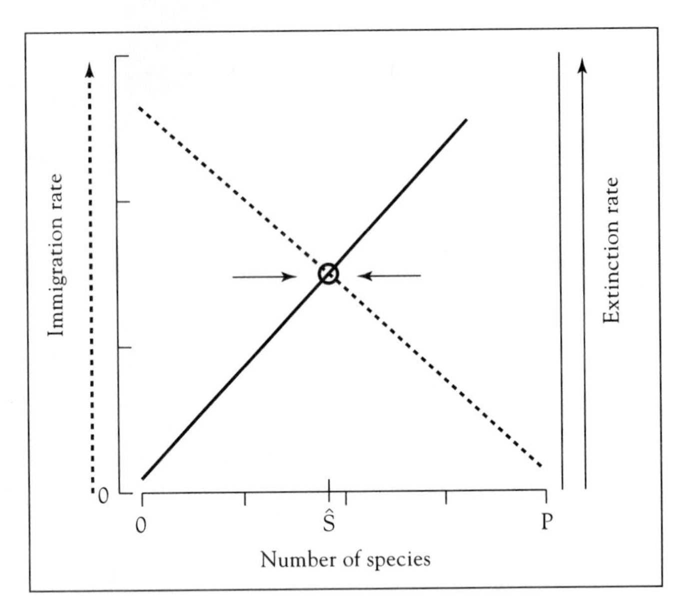

FIGURE 19.1 The interaction of immigration rate and emigration rate (extinction) to produce an equilibrium number of species on an island, $\hat{S}$. $\hat{S}$ varies from 0 species to P species, the total number in the species pool of colonists. The basis of the MacArthur-Wilson island biogeography model assumes species all have the same immigration and extinction rates and do not affect each other. Diversity on an island will then tend toward a steady state that is less than the diversity of the source pool.

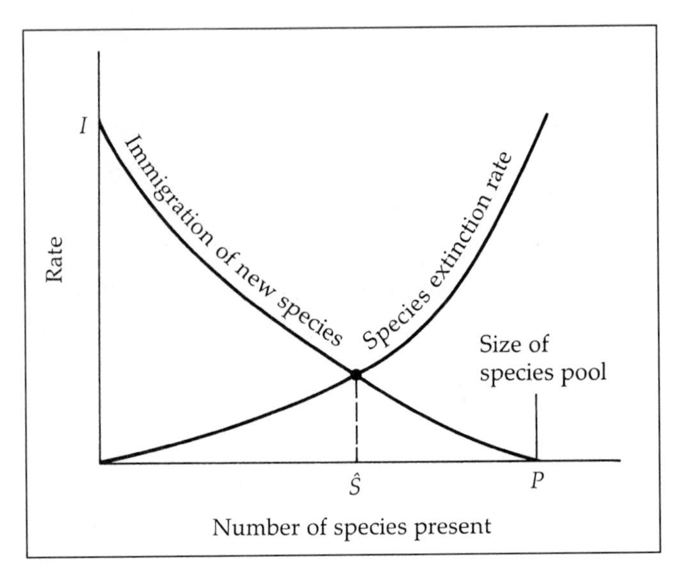

FIGURE 19.2 In MacArthur and Wilson's model for island biogeography the rate curves bend because species add to each other's extinction rates and because some species immigrate more readily than others.

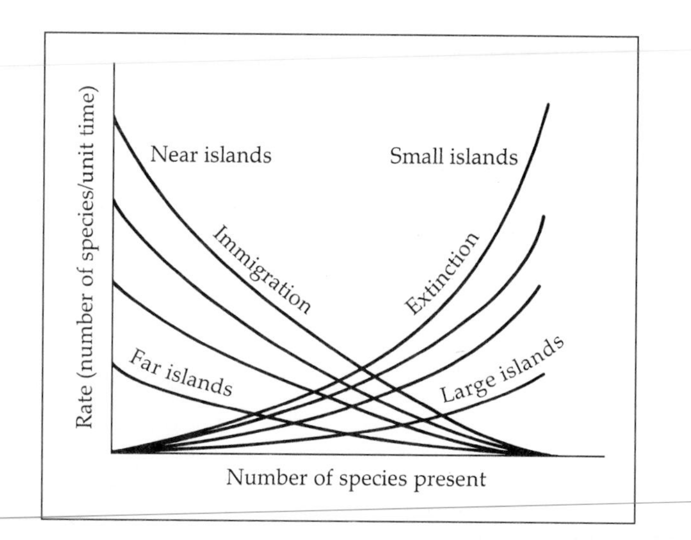

FIGURE 19.3 Equilibrium models of biotas of islands with varying distances from the principal source area and of varying size. An increase in distance (*near to far*) lowers the immigration curve; an increase in island area (*small to large*) lowers the extinction curve. (*Redrawn from MacArthur and Wilson 1967.*)

poorer dispersers arrive. This causes the immigration curve to start off steep but get progressively more shallow. For the extinction curve, extinctions rise at accelerating rates because, as later species arrive, competition between them increases and more species are likely to go extinct. Despite these modifications, the basic premises remain unchanged.

3. $\hat{S}$ is determined only by the island's area and position, which influence the rates of immigration and extinction. Extinction rates were argued to be greater on smaller islands and colonization rates greater on islands near source pools (Fig. 19.3). Extinction was more likely on small islands purely by stochastic processes, by chance alone, because population sizes would be smaller there.
4. Equilibrium is dynamic, and following colonization we should see the numbers of species stay constant through time but change in their identities. Thus, we will see turnover of species.

Since its original exposition, MacArthur and Wilson's (1963, 1967) theory of island biogeography has undergone some modifications:

1. *The "target effect"* (Whitehead and Jones 1969). The rate of colonization depends on an island's size as well as on its distance from a source pool of potential colonists because larger islands present larger "targets" for colonists than do smaller islands.
2. *The "rescue effect."* Brown and Kodric-Brown (1977) proposed that the distance from an island to a source pool of potential colonists affects the rate of extinction as well as the rate of colonization. This is because the immigration of individuals of taxa already resident on the island slows the rate of extinction of those taxa by keeping population sizes higher than they would be in the absence of immigration. Thus, although the basic MacArthur-Wilson model described only two mechanisms—the effect of area on extinction and the effect of distance on immigration—the target and rescue effects extend and effectively complete the model (Fig. 19.4).

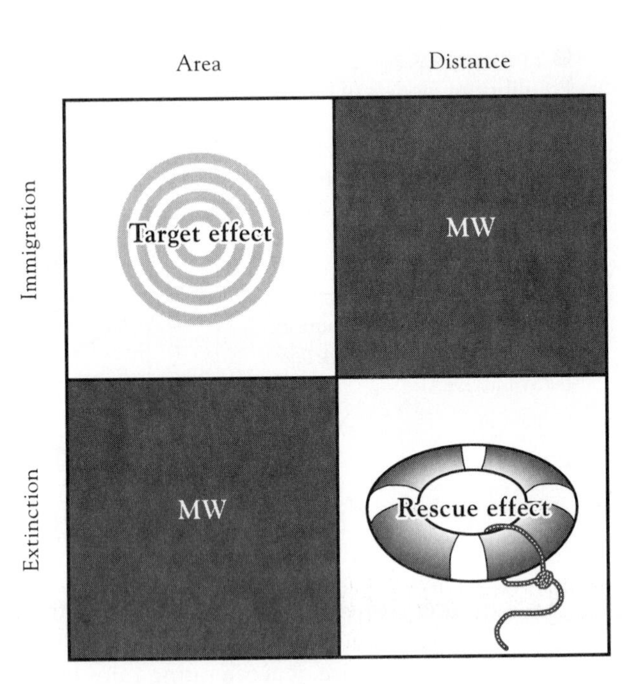

■ **FIGURE 19.4** The effects of area and distance on immigration and extinction of species on islands as originally proposed by MacArthur and Wilson (*MW*) and as modified by Whitehead and Jones (*Target Effect*) and Brown and Kodric-Brown (*Rescue Effect*). (*After Gotelli 1995.*)

3. The third modification concerned the nature of "islands." Patches of particular habitat on continents were viewed as "islands" in a sea of other, unsuitable habitat (Kilburn 1966; Vuilleumier 1970). Janzen (1968, 1973) extended the habitats-as-islands concept by proposing that individual host-plant species could be islands to their associated herbivore fauna, which was adapted only to feed on particular types of vegetation.

The strength of the MacArthur-Wilson theory was that it generated some falsifiable predictions:

a. Turnover of species should be considerable.
b. There should be relationships between island size and species number and distance from source pool and species number. Traditionally species-area relationships are plotted on a log-log plot, to give a straight-line fit. In an analysis of one hundred data sets Connor and McCoy (1979) found thirty-five to be best fit by an untransformed model and only thirty-six of the remainder to be best fit by a transformed, i.e., log-log model. They suggested that the appeal of a log-log plot rests on the fact that it turns most monotonic functions into a straight line. When the log-log transformation was used, seventy-five out of the one hundred data sets showed no lack of fit, that is, they could be fit according to statistical probabilities. Following this prediction,
    i. Log $\hat{S}$ should increase more rapidly with log A on distant islands than on islands near a source of potential immigrants.
    ii. Log $\hat{S}$ should increase more rapidly with the reciprocal of distance on smaller islands than on large.

Gilbert (1980) suggested that, in order to support the MacArthur-Wilson model, there should be data to suggest the following:

1. A close relationship exists between the area of different islands and the number of species they contain.
2. The number of species on an island remains constant but there is turnover of species in the system.

19.2 A number of studies support the prediction that the number of species should increase with island size.

Oceanic islands.

Preston (1962) provided many examples that showed a species-area relationship: birds, reptiles (Fig. 19.5), amphibians, and beetles on West Indian islands; ants in Melanesia; vertebrates on islands in Lake Michigan; and land plants in the Galapagos. At least nineteen data sets from islands were provided in a review by Quinn and Harrison (1988). However, there are many data sets that show no relationship between area and species richness (Gilbert 1980), and many more where the richness of island fauna is better predicted by variables other than area, for example, elevation or soil type. Gilbert (1980) suggested that MacArthur and Wilson (1963, 1967) were aware of this variation in the data and carefully selected the examples they used in their book to support their hypothesis.

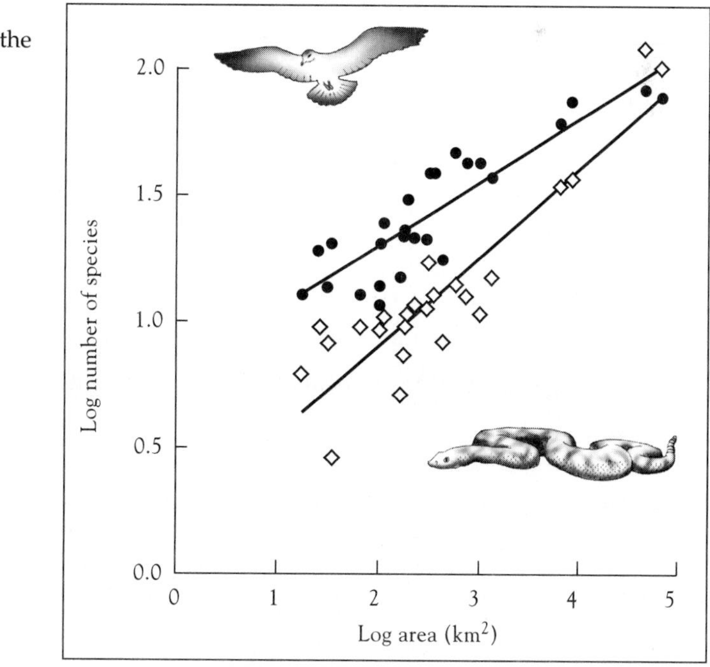

FIGURE 19.5 Two species-area curves from the Caribbean. Circles: birds. Diamonds: reptiles. *(Data from Wright 1981.)*

Among those studies for which the species-area relationship was viewed as proved, one of the early topics of discussion was the significance of variations in the value of z, the slope of the regression line relating species-richness and area in the equation

$$S = cA^z$$

or, in logarithmic form,

$$\log S = \log c + z \log A$$

where S = number of species, c = a constant measuring the number of species per unit area, i.e., per unit of forest, grassland or desert, and A = area.

Virtually no interest has been shown in comparing intercept parameters (Connor and McCoy 1979).

A high value of z indicates steep increases in species number as island size increases, whereas low values indicate much smaller differences in species numbers between islands. Preston (1962) calculated that "theoretically" the value of z should be 0.263. However, values of "around" 0.3 were often found in island studies. Such values were suggested to be permissible in "nonequilibrial" and "isolated situations" (Schoener 1976), but Gilbert (1980) suggested they were glaring examples of untestable ad hoc hypotheses erected to conveniently explain what amounts to pure speculation. There was rarely argument as to whether z values of 0.2 or 0.3 verified Preston's theory or not.

Is too much importance attached to z values? May (1975), Schoener (1976), and Connor and McCoy (1979) argued that z values of between 0.2 and 0.4 can be expected by chance alone given the variation in values of species richness and area. But this has been disputed by Sugihara (1981) (see reply by Connor, McCoy, and Cosby 1983). The crux of the argument rests on the equation relating the regression coefficient, or slope of the regression line z, to the correlation coefficient r:

$$z = r(s_y/s_x)$$

where s_y and s_x are the standard deviations of the dependent (species number) and independent (area) variables, respectively. Allowing that the value of r falls between 0 and 1 (as it must for a positive correlation) and that $s_y < s_x$ (because of the asymptotic behavior of species number), Connor and McCoy (1979) calculated expected values of the regression coefficient (shown in Table 19.1) by multiplying the marginal values of r and s_y/s_x to yield slope values. Even with these conservative assumptions, 30 percent of the slopes are expected to fall between 0.20 and 0.40. However, of one hundred actual species-area curves Connor and McCoy examined, 45 percent had slope values between 0.20 and 0.40. They argued that since the ranges of r and s_y/s_x of most relationships tend to be much smaller, slope values between 0.20 and 0.40 should be, and are, more frequently observed.

The question is, why do r and s_y/s_x have such narrow ranges? Connor and McCoy (1979) suggested that values of the correlation coefficient r are usually above 0.50 for logspecies/logarea regression, most likely because insignificant correlation coefficients are not published. The observed narrow range of s_y/s_x (usually between 0.20 and 0.60) is a consequence of the asymptotic behavior of species number; once species number

TABLE 19.1 In species-area relationships, z values of between 0.2 and 0.4 may be expected by chance alone given the variation in values of species richness and area. This table gives constructions of the expected values of the regression coefficent (z) with the constraints $0 < r < 1$ and $0 < sy/sx < 1$. (*After Connor and McCoy 1979*)

	SY/SX									
r	0.1	0.2	0.3	0.4	0.5	0.6	0.7	0.8	0.9	1
.1...	0.01	0.02	0.03	0.04	0.05	0.06	0.07	0.08	0.09	0.10
.2...	0.02	0.04	0.06	0.08	0.10	0.12	0.14	0.16	0.18	0.20
.3...	0.03	0.06	0.09	0.12	0.15	0.18	0.21	0.24	0.27	0.30
.4...	0.04	0.08	0.12	0.16	0.20	0.24	0.28	0.32	0.36	0.40
.5...	0.05	0.10	0.15	0.20	0.25	0.30	0.35	0.40	0.45	0.50
.6...	0.06	0.12	0.18	0.24	0.30	0.36	0.42	0.48	0.54	0.60
.7...	0.07	0.14	0.21	0.28	0.35	0.42	0.49	0.56	0.63	0.70
.8...	0.08	0.16	0.24	0.32	0.40	0.48	0.56	0.64	0.72	0.80
.9...	0.09	0.18	0.27	0.36	0.45	0.54	0.63	0.72	0.81	0.90
.10...	0.10	0.20	0.30	0.40	0.50	0.60	0.70	0.80	0.90	1.00

Note: Hatched area indicates values lying between 0.20 and 0.40.

becomes asymptotic, area can be increased virtually indefinitely, and concurrently s_y/s_x and the slope will decline. In other words, since species-area curves are characterized by inherently larger ranges of areas than species numbers, the numerator of the term s_y/s_x will always be smaller (usually much smaller) than the denominator. Hence, the small fractional values of s_y/s_x multiplied by r produce lower slopes the larger the area range. Connor and McCoy's conclusion was that the narrow range of observed slope values (0.20-0.40) is more parsimoniously explained to be a result of the characteristics of the regression system, and not the result of underlying biological reasons.

Habitat "islands."

Data from well over one hundred studies of islands or analogous habitat isolates are available for study (Quinn and Harrison 1988). Values of z obtained from continental areas are often lower than those from truly insular situations, 0.15-0.25 versus 0.20-0.40. This means that as larger areas are sampled, fewer new species are added on continents as on islands. Without wishing to attach too much significance to this variation, there may in this case be a sound biological explanation for this phenomenon. The rationale here is that each area in the continental studies, except the largest, contains some transient species from adjacent habitats, so the slope of the species-area curve is shallow. Islands are actual isolates with reduced migration rates because species cannot occur in the ocean between the islands, so the number of transients in an area is minimal, thus steepening the slope of the curve.

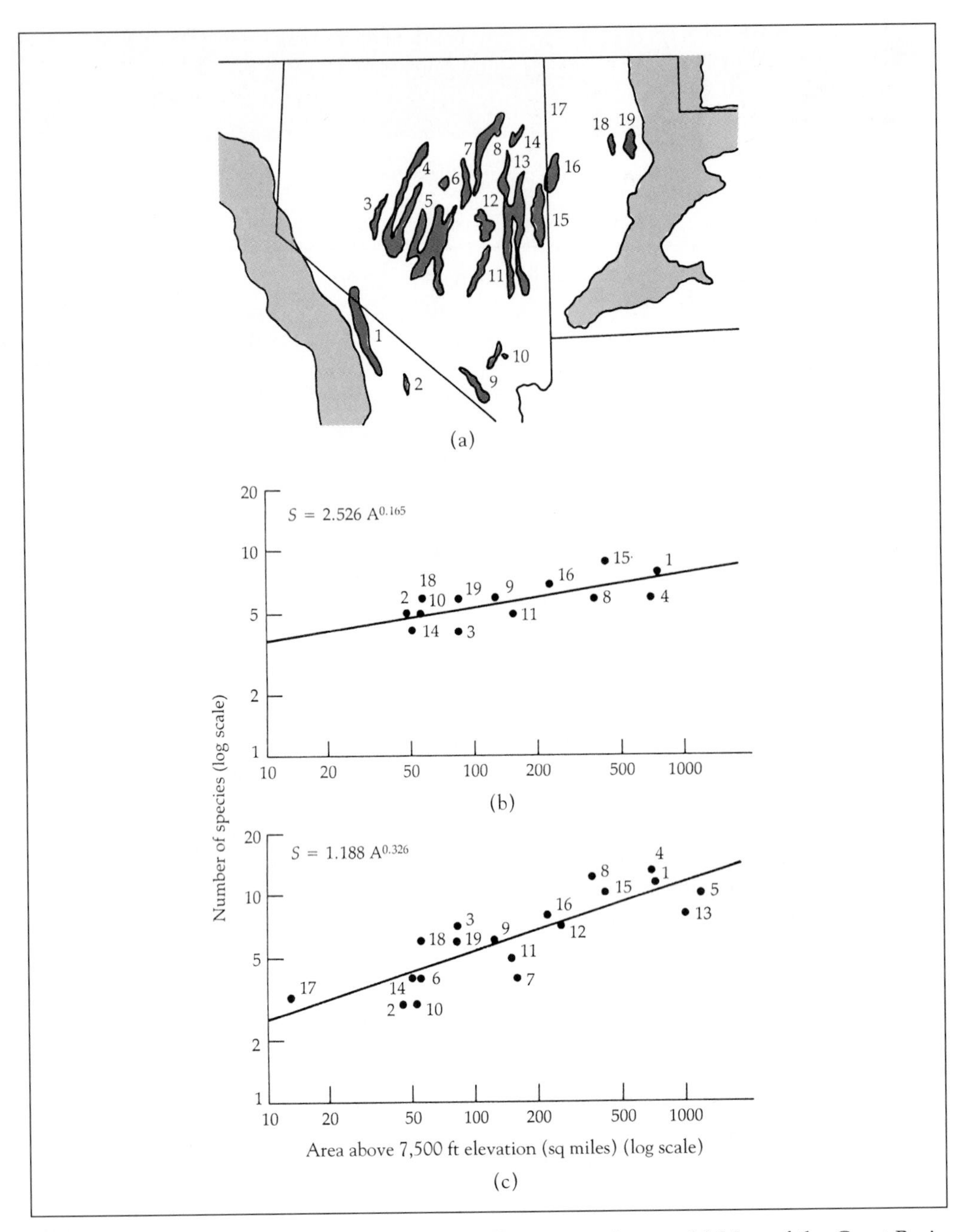

■ **FIGURE 19.6** Island biogeography applied to mountaintops. (**a**) Map of the Great Basin region of the western United States showing the isolated mountain ranges between the Rocky Mountains on the east (*right*) and the Sierra Nevada on the west (*left*). (**b**) Species-area relationship for the resident boreal birds of the mountaintops in the Great Basin. (**c**) Species-area relationship for the boreal mammal species. Numbers refer to sample areas on the map. *(Redrawn from Brown 1978.)*

Values of z can also be affected by the dispersal abilities of animals. Brown (1978) studied the distribution of boreal mammals and birds in the isolated ranges of the Great Basin in the United States. The mountain ranges are essentially isolated from one another, and the mammalian fauna is a relic community of a bygone age when rainfall was higher and this type of boreal habitat was contiguous. Each mountaintop is essentially a forest island in a sea of desert (Fig. 19.6). The species-area relationship for birds (Fig. 19.6b) had a slope of 0.165, that for mammals 0.326 (Fig. 19.6c). The slope of the line for mammals was more like that found on islands. The reason is that there is no mammalian migration between mountaintops because mammals would have to walk down the mountain across the valley and up the next mountain. In this situation, mountaintops behave as true islands. In contrast, birds disperse more than mammals because they can simply fly between mountaintops, and the z value in their case is much more in line with a mainland type of relationship.

Species as islands.

Janzen's ideas on species of host plant as islands for their herbivores were elaborated by Donald Strong (1974 a, b), who found a species-area relationship between the geographical area of distribution of British tree species and the number of insect herbivore species (Fig. 19.7). Some critics (e.g., Claridge and Wilson 1978; Birks 1980;

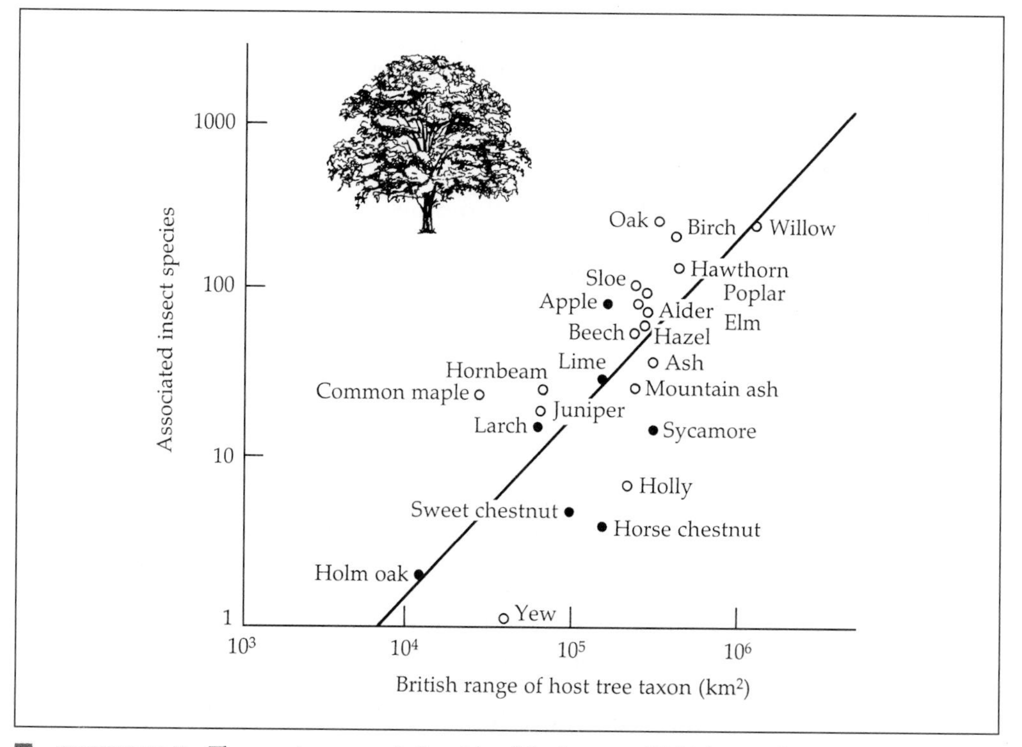

FIGURE 19.7 The species-area relationship of the insects of British trees ($r = 0.78$, $P < 0.001$). Closed dots indicate introduced taxa; open dots indicate British natives. *(Modified from Strong 1974a.)*

disputed this relationship because of supposed inadequacies in Strong's data. However, it is worth noting that since Strong's original work, many other workers have documented species-area relationships for insect herbivores on host-plant species. For example, Paul, Opler (1974) found a similar regression for California oaks and their leaf miners.

It is important to note that while many species-area relationships exist, the existence of species-area relationships does not necessarily validate the theory of island biogeography. Indeed, species-area relationships greatly antedate island biogeography theory. Hart and Horwitz (1991) provide at least five reasons for a species area relationship:

1. Extinction rates are greater on small islands—the original MacArthur-Wilson hypothesis.
2. Passive sampling. Bigger areas contain more individuals. The more time we sample these bigger areas, the greater our chances of collecting new species. The author has good experience of this from his experience rearing out the parasites of gallmakers and leaf miners—the more you rear out, the greater the number of species.
3. Speciation may be more likely in bigger areas.
4. Larger areas contain more "core" areas, which are less affected by disturbances, like wind.
5. Larger areas often contain a greater diversity of habitats—different soil conditions, slopes, elevations, salinities, pH, and the like. In fact, this is probably the biggest cause of species area relationships (see the discussion that follows).

Fox (1983) investigated the relationship between species and area in Australian mammals. He classified habitats into seven broad types and showed that larger areas do include more habitat types (Fig. 19.8). He also showed that the number of mammal species is even better predicted from the number of habitats than from area (Figs. 19.9 and 19.10). Moreover, area does not even help to explain the variance left over after habitat has done its predicting.

Tonn and Magnuson (1982) showed much the same thing. They recorded habitat variables and fish species from eighteen shallow lakes in Wisconsin. The area of a lake predicts its number of species $(r^2 = 0.525)$, but so does its habitat diversity $(r^2 = 0.423)$. The problem is that area and habitat correlate so closely that only one variable is useful in predicting diversity.

Of course, the best way to study the effect of area alone on species number is to choose a habitat that doesn't change as you sample bigger islands of it. Dan Simberloff (1976 a, b) did just that. We'll talk a little more about his experiments in the next section. Suffice it to say that Simberloff studied islands of pure mangroves, of varying size, in the Florida Keys. He experimentally created his own "reduced area" islands by taking a chain saw and felling trees to reduce island size. Although this treatment did nothing to habitat quality, the area reduction did cause a reduction in invertebrate species richness. Few other studies of this nature have been conducted (but see Rey 1981, 1983 and next section).

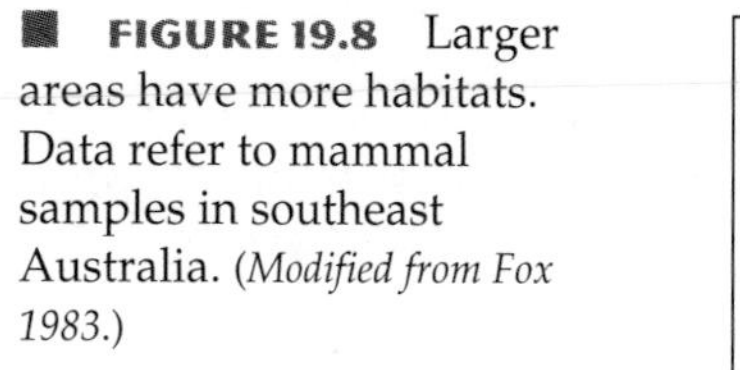

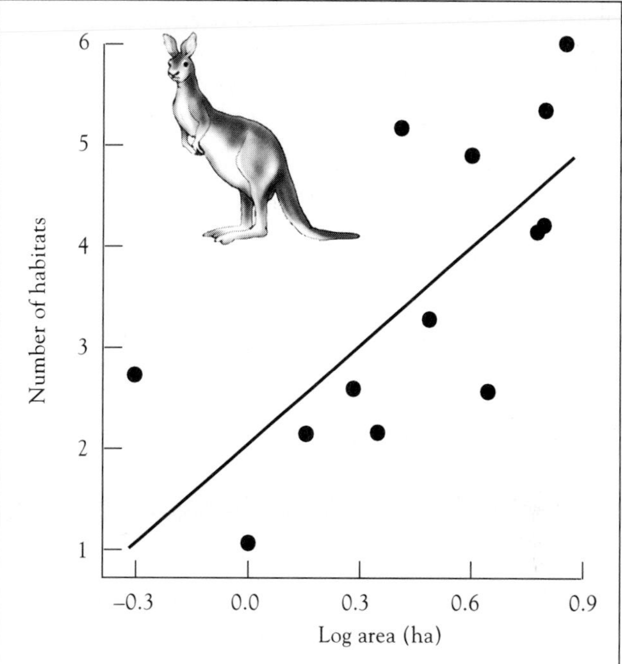

FIGURE 19.8 Larger areas have more habitats. Data refer to mammal samples in southeast Australia. (*Modified from Fox 1983.*)

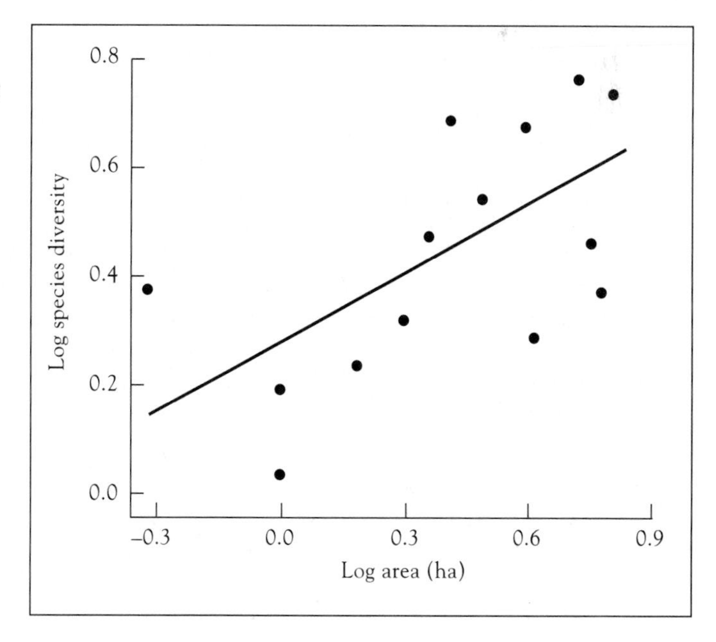

FIGURE 19.9 Species-area curve for mammals in southeast Australia. (*Modified from Fox 1983.*)

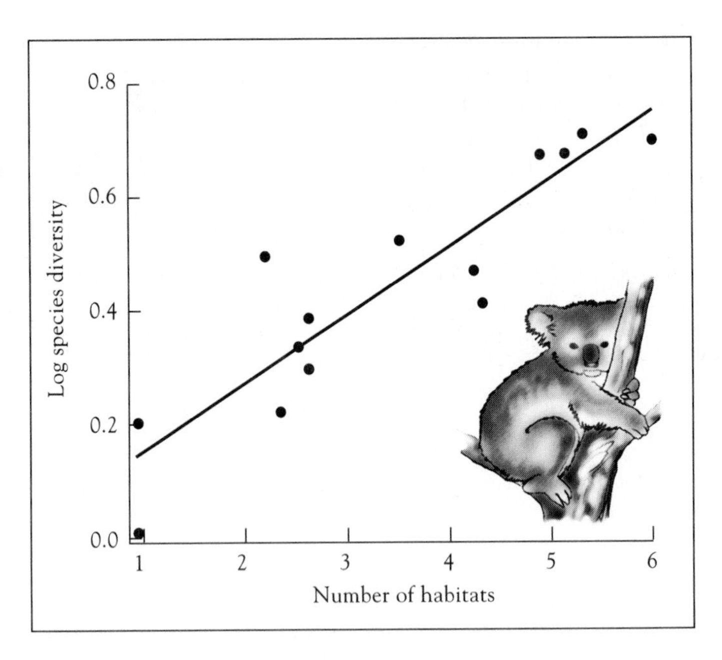

■ **FIGURE 19.10** The number of mammal species fits the number of habitats better than area. Area explains none of the residual variance. (*Modified from Fox 1983.*)

19.3 There are few estimates of the rates of colonization, extinction, and turnover on islands.

Gilbert (1980) managed to find twenty-five investigations carried out in order to demonstrate turnover. He dismissed virtually all of them as flawed in method, statistics, or quality of data. Only one, that of Simberloff and Wilson (1969, 1970), was seen as being of merit.

Simberloff and Wilson censused small (11 to 25m in diameter) red mangrove (*Rhizophora mangle*) islands in the Florida Keys for all terrestrial arthropods. They then fumigated some of them (experimental islands) with methyl bromide to kill all animals (Photo 19.1). Periodically thereafter they censused all islands for several years. After 250 days, most islands had similar numbers of arthropod species as they had to begin with (Fig. 19.11). Simberloff and Wilson (1969, 1970) observed colonization and extinction, but the infrequent nature of the sampling meant that it was difficult to determine the exact rates. The data did indicate that colonization rates during the first 150 days were higher on nearer islands than on far islands. Calculated rates of turnover were found to be very low—1.5 extinctions per year. However, again the length of time between censuses was so long that pseudoturnover could have occurred—colonization and subsequent extinction all occurring between censuses—masking true rates of turnover.

Simberloff (1974) initially seemed to view his mangrove experiments as good evidence in support of the MacArthur-Wilson theory, which he embraced, but later he seems to have had a change of heart (Simberloff 1976b) and suggested that the data were weak support for the MacArthur-Wilson theory and that very few other studies seemed to support it either. Part of this volte-face involves the reinterpretation of his

■ **PHOTO 19.1** Experimental defaunation of a mangrove islet in the Florida Keys. (**a**) Construction of a scaffold frame. (**b**) Installation of a large tent into which insecticide was introduced, killing all life. Commercial pest control operators from Miami were hired to perform the fumigation. Tent and scaffold were removed after defaunation, and recolonization was monitored. *(Photos by Daniel Simberloff, Florida State University.)*

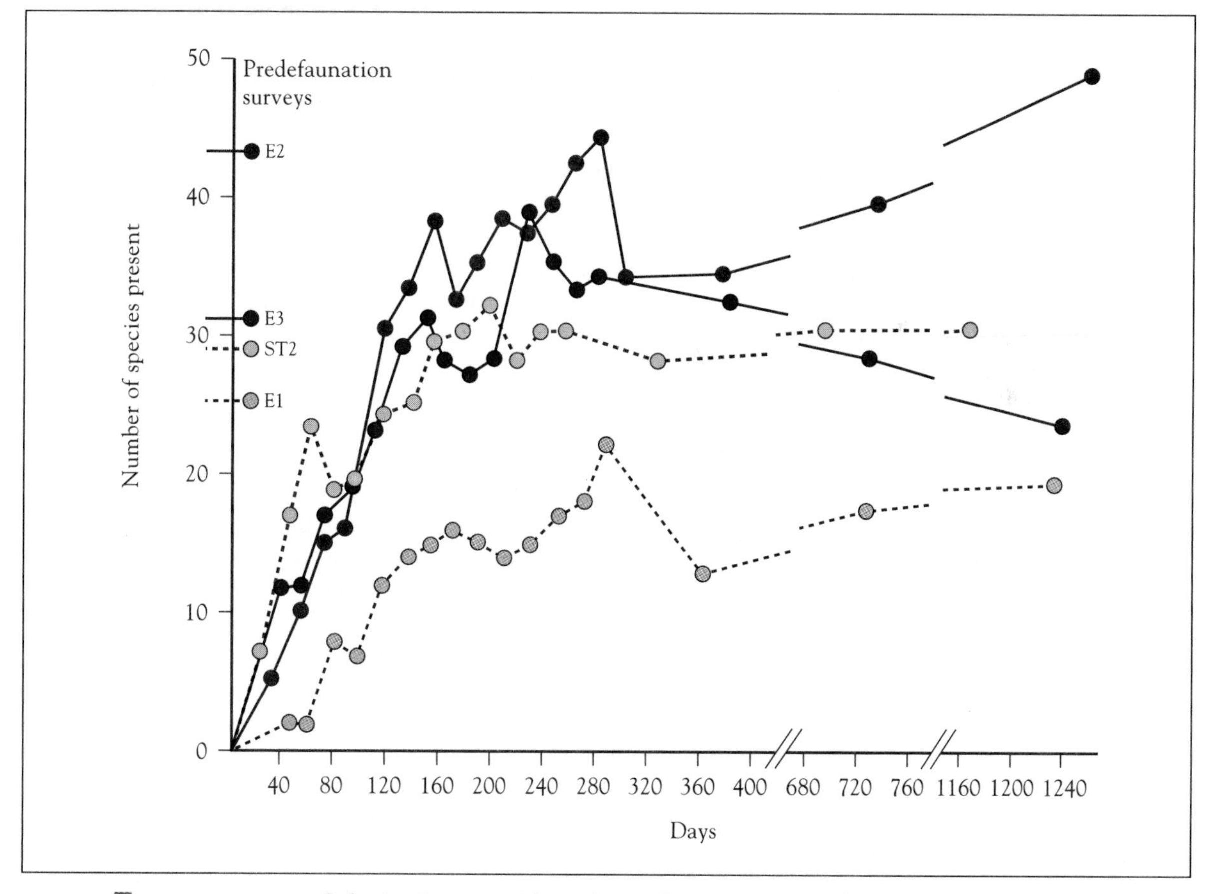

■ **FIGURE 19.11** Colonization curves for arthropods on mangrove islands near Sugarloaf Key, south Florida, after defaunation by fumigation. Predefaunation species numbers are on the ordinate. *(Redrawn from Simberloff 1978.)*

own data, which showed initial random colonizations of mangroves through a set of well-defined communities into a final assortative equilibrium with low turnover. This implies the existence of biological processes such as competition between colonists or predation and other biotic interactions that shaped final community structure. Under this scenario, extinctions and colonizations could not be regarded as having certain probabilities on islands of different sizes or distances from the source pool without a knowledge of the existing species present. This is contrary to the theory of island biogeography, which treats the dynamics of different colonizing species as essentially the same, with community properties essentially unimportant. As Simberloff (1978b) and Schoener and Spiller (1987) have concluded, turnover probably only involves a subset of transient or unimportant species with the more important species being permanent after colonization.

What of the more recent studies? Rey (1981, 1983) defaunated islands of pure salt marsh cordgrass, *Spartina alterniflora*, in north Florida. Turnover was estimated at about 0.14 species per island per week, but rates of colonization and extinction could not be shown to be related to the distances of islands from a source pool. In addition, later extensive and detailed studies of herbivory on *Spartina* revealed only about thirty herbivores and their parasites associated with *Spartina* (Stiling, Brodbeck, and Strong 1982; Stiling and Strong 1983), indicating that some of Rey's fifty-six recorded species may have been transients and not subject to the MacArthur-Wilson theory.

Each of sixteen offshore British islands were censused for birds at least six and as many as forty-nine times (Pimm, Jones, and Diamond 1988; Tracy and George 1992). Average immigration probability was about 0.0073 species per year and average extinction probability 0.0078 (Rosenzweig 1995). That the immigration and extinction rates were almost identical was good evidence in support of the MacAthur-Wilson theory. However, as Williamson (1983) pointed out, such turnover seems ecologically trivial (although Rosenzweig disputes this).

Abbott and Black (1980) measured the known extinctions and immigrations of plants on forty islands off the coast of western Australia. The forty islands were censused sporadically (one to four times between 1956 and 1978). Thus, as in Simberloff and Wilson's studies, some turnover could have occurred between censuses and never been recorded. Extinctions and immigrations tended to match each other, but again the majority of turnover was low, with most islands showing turnover rates of between zero and two species per census (or < 0.1 species per year). Morrison (1997) found low turnover for seventy-seven cays in the Exumas region of the Bahamas, which he surveyed annually over a four-year period. Most of the observed turnover (usually $< 1\%$ per year or less than one species per year) was due to immigrants that never became established. The take-home message from most of these studies is that recorded rates of turnover are low, which gives little support to the MacArthur-Wilson theory. Although actual rates could be higher (Rosenzweig 1995), what is needed is more accurate data.

Finally, we should note that island biogeography theory is not just an academic issue. It is important because, in the 1970s and 1980s, the theory came to be uncritically accepted by conservationists and park planners and was regularly incorporated into refuge design (Boecklen and Simberloff 1986). Island biogeography theory was thought to be particularly useful in the continuing debate over the design of wildlife preserves, particularly in the question of whether planners should design many small preserves or a few large ones (given the unlikely prospect of choice) (Applied Ecology: The Theory of National Park Design). The International Union for Conservation of Nature and Natural Resources (IUCN), a major worldwide conservation group, stated that refuge design criteria and management practice should be in accord with the equilibrium theory of island biogeography (IUCN 1980). Clearly, this idea is on shaky ground at best, and more consideration should probably really be given to the autoecology of target species. However, island biogeography can be relevant to conservation science. Ceballos and Brown (1995) showed how species richness of the world's mammals was strongly correlated with land area (Fig. 19.12), though the number of endemic species was only weakly correlated with area. Their analysis showed that certain countries, Australia, Madagascar, Indonesia, the Phillipines, and Mexico, had more species or more endemics than predicted by their land area and should thus receive the special concern of conservation biologists. Here, such factors as latitude, topography, habitat heterogeneity, and historical biogeography may promote especially rich mammalian faunas.

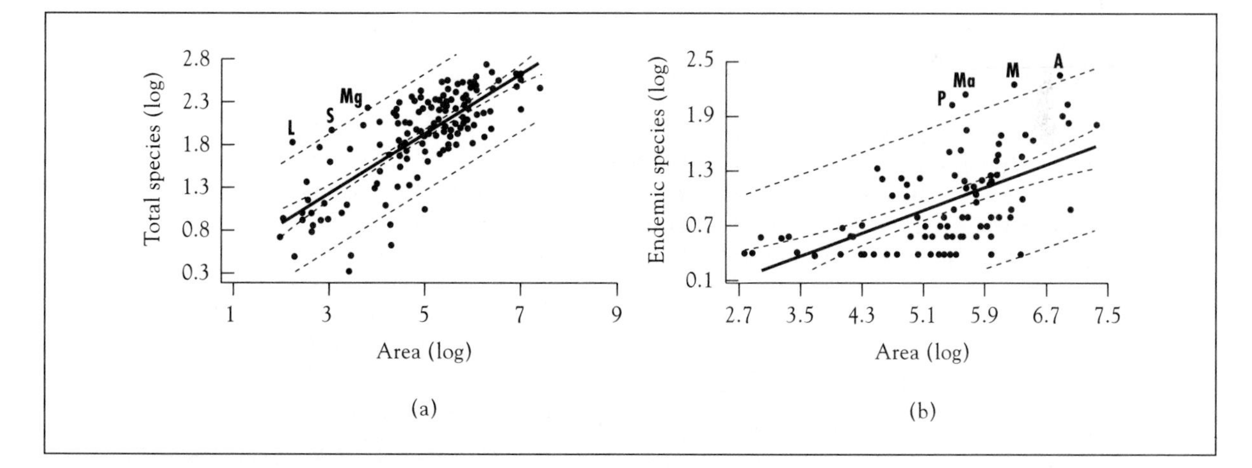

FIGURE 19.12 Global species richness-to-area relationships for terrestrial (**a**) and endemic (**b**) mammals. Data for 155 countries include bats and exclude cetaceans, pinnipeds, and sirenians. Outliers—countries with more species than expected by their land area—are Liechtenstein (L), Singapore (S), Martinique (Mq), the Philippines (P), Madagascar (Ma), Mexico (M), and Australia (A). (*After Ceballons and Brown 1995.*)

The Theory of National Park Design

There has been much discussion about the shape, design, and management of nature reserves since the early 1970s. The theories of park design are summarized in Figure 1. The debate on reserve design has centered on island bioeographic theory, which suggests that large parks would hold more species than small parks. It is widely agreed that the ideal strategy would be to have lots of large refuges. Of course, large refuges cost a lot of money, so many countries have to settle for preserving small areas of land rather than large ones. The debate then becomes, should single large areas be preserved or several small ones (the so-called SLOSS debate, Simberloff 1986a) (Photo 1). Single large preserves may buffer populations against catastrophe and possible extinctions, but many studies suggest that a few, dispersed small sites often contain at least as many species as does a single site of equal area. Usually multiple small sites contain more species. This is because a series of small sites is likely to have a broader variety of habitats contained within it than one large site. Hence, in practical terms, it may actually be better to have several small sites.

Quinn and Harrison (1988) have reviewed data from over thirty censuses for fifteen island groups where all data were reported island by island, species identities of individuals were confirmed, and at least six islands were included in each survey. The resultant species-area relationships for vertebrates, land plants, and insects show that collections of small islands generally harbored more species than do comparable areas composed of one or a few large islands. National park faunas were shown to be richer in collections of small parks than in the larger parks. This finding has an important bearing on future land purchases for conservation purposes. Furthermore, in twenty-nine out of the thirty cases, the small-islands curve saturated with species more rapidly than did the large-island curve, and in eighteen out of the thirty cases the two curves never intersected, indicating that smaller, more numerous islands always had more species than fewer large ones.

Of course, at some point a potential refuge is simply too small to conserve for long the species for which it is designed, and that minimum critical area

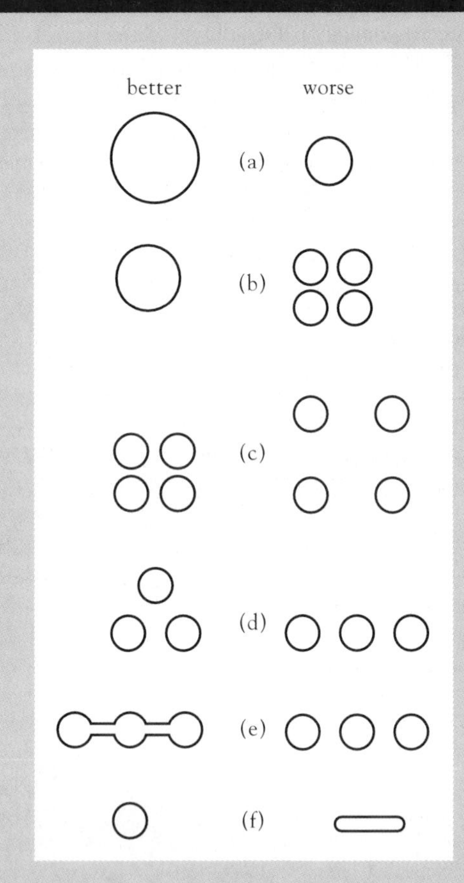

■ **FIGURE 1** The theoretical design of nature reserves based on the tenets of island biogeography. (**a**) A larger reserve will hold more species and have lower extinction rates. (**b**) Given a certain area available, it should be fragmented into as few pieces as possible because species incapable of dispersing readily among fragments may be confined to areas too small to avoid extinction. (**c**) If an area must be fragmented, the pieces should be as close as possible to minimize inhibition of dispersal. (**d**) A cluster of pieces in contrast to a linear arrangement will enhance dispersal among the fragments. (**e**) Maintaining or creating corridors of natural habitat between fragments may enhance dispersal. (**f**) A circular-shaped area will minimize the amount of "edge" habitat adjacent to nonreserve areas. (*After Wilson and Willis 1975.*)

■ **PHOTO 1** Debate over the best size and shape of a nature reserve seems out of place when in reality, protection of many species like this white rhino in Meru National Park, Kenya, is better linked to adequate protection by park rangers from poachers. (*Shane Moore, Animals Animals/Earth Scenes 616427M.*)

varies for different species. Detailed ecological study is probably the best means of telling just how large minimum critical areas are and where to locate them for given species. In this way, the habitat requirements for individual animals are documented, and the minimum area required for one individual can be multiplied by the number of individuals desired in the population to give a minimum area requirement.

Another area of contention in the design of nature reserves is how close to situate small reserves if these are chosen over large ones. Thus, should we have three or four small reserves close to each other, or should they be farther apart? In practice it is probably true that having small sites far apart would preserve more species than having them close together, for no other reason than distant sites are likely to incorporate slightly different habitats.

A variation of the argument as to whether small reserves should be situated close together or far apart is the suggestion that they should be linked together by biotic corridors and networks, thin strips of land that may permit the migration of species between small patches. This argument is based on the fact that corridors may facilitate movements of organisms with poor powers of dispersal between habitat fragments. If a disaster befalls a population in one small reserve, it could be recolonized by immigrants from neighboring populations. This obviates the need for humans to physically move new plants or animals into an area. Many types of habitat have been considered for corridors, like hedgerows in Europe, riparian habitat (along rivers) in the United States, and strips of tropical forest. Reserves with occasional interrefuge migrations may appear to be optimal for the genetic mixing of populations. However, there are many disadvantages associated with corridors (Simberloff et al. 1993). Corridors also facilitate the spread of disease, invasive species, and fire between small reserves. Having isolated populations greatly lessens the chances that all of them will succumb to the same catastrophe. For example, there are many small oceanic islands that have had their ground-nesting birds wiped out by introduced predators such as goats, rats, and even lighthouse-keepers' cats. It is not widely appreciated that even smaller oceanic islands have remained bastions for wildlife because they were not invaded by exotic species. Had these small oceanic islands been connected, the spread of exotic species would have been greatly facilitated, and there would have been many more human-induced extinctions than at present.

It has been said that the application of island biogeography theory to conservation was a worthwhile experiment, but experience and further deliberation have shown that it is not very helpful. Habitat considerations of individual species are ususally much more important. In reality, there is rarely any choice as to the size, shape, and location of nature reserves, and this constitutes a strong reason for being critical of adopting theories from biogeography for the establishment of nature reserves. Management practicalities, costs of acquisition and management, and also politics usually override ecological considerations, especially in developing countries, where management costs for large reserves may be relatively high. Economic and ecological considerations are both justified in choosing which areas to preserve. Typically, many countries protect areas in those regions that are the least economically valuable rather than choosing areas to ensure a balanced representation of the country's biota. For example, in the United States most national parks have been chosen for their scenic beauty, not because they preserve the richest habitat for wildlife.

While existing protected areas may be able to conserve populations of plants and animals for a very long period of time, it is unlikely that extinctions will be prevented forever. This is because virtually no national parks, even the largest, are large enough to support the process of speciation within them. Thus, as habitats change or climates modify, species will not have enough genetic variation to adapt to such changes, and populations may well become extinct (Soule and Wilcox 1980) (Fig. 2). The basis for this

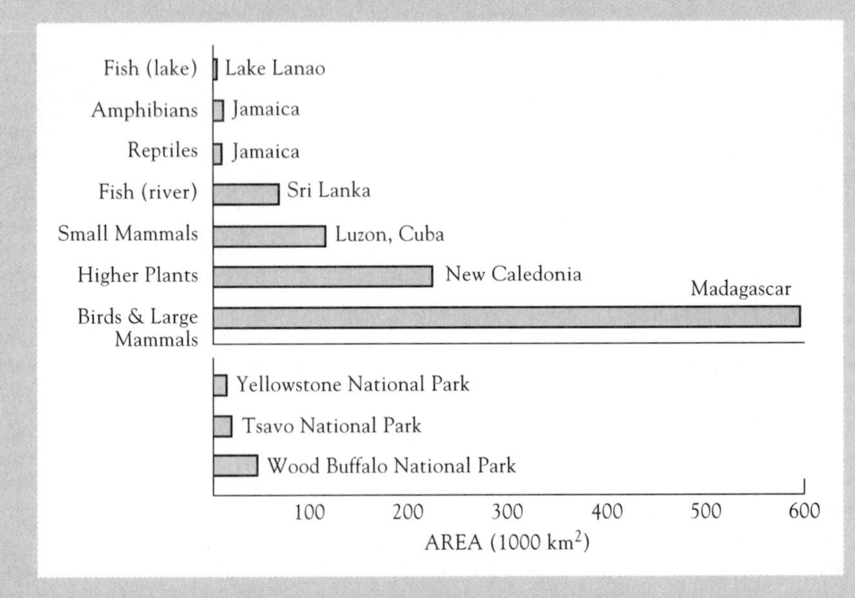

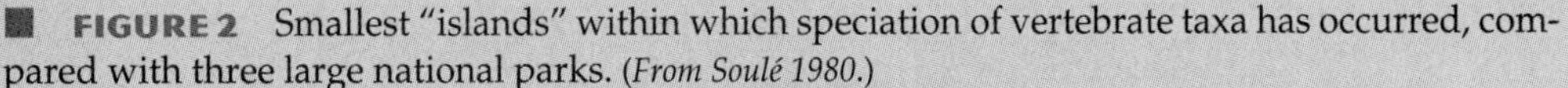
■ **FIGURE 2** Smallest "islands" within which speciation of vertebrate taxa has occurred, compared with three large national parks. (*From Soulé 1980.*)

conclusion was a survey of vertebrate taxa on islands, which looked for the minimum-sized island on which there was evidence for in situ speciation for a particular group. The results suggested that small mammals such as rodents have enough room for speciation on islands like Cuba and Luzon (about 110,000 km^2), but birds or large mammals as large or larger than jackals or vervet monkeys require an area the size of Madagascar (nearly 600,000 km^2). Higher plants appear to fall somewhere in between these extremes. For large organisms, natural speciation requires too much space, and probably too much time, for humans to be able to construct necessary nature reserves to allow such a process to continue. Perhaps the best we can hope for is that existing species will continue to survive for relatively long periods of time.

It has been argued that rather than concern ourselves with island biogeography theory when designing nature reserves, we would do better to consider how to finance their protection—this will dictate how big reserves can be. The amount of money spent to protect nature reserves may better determine species extinction rates than reserve area. According to island biogeography theory, large protected areas minimize the risk of extinctions because they contain sizable populations. In Africa, several parks, such as Serengeti, Tsavo, Selous, and Luangwa in Zambia, are large enough to fulfill theoretical ideals. However, in the 1980s, populations of black rhinos and elephants declined dramatically within these areas because of poaching, showing that there was a wide gap between theory and reality. In reality, poaching is a far more serious threat than the threat of genetic isolation.

In reality, the rates of decline of rhinos and elephants, largely a result of poaching, are related directly to conservation effort and spending (Milner-Gulland and Leader-Williams 1992; Leader-Williams, Dalban, and Berry 1990) (see Fig. 3). The conclusion is that nature reserves must be adequately funded if local extinctions are to be avoided. Instead of conservation biology concentrating on the theoretical implications of the size of protected areas, it should perhaps be more concerned with the direct level of funding for the protection of species. Indeed, the successes of conservation in Africa have resulted from situations in which resources were concentrated in particular areas to preserve certain species. For example, the few remaining black rhinos in Kenya, the lowland gorillas and pygmy chimpanzees in Africa, and the vicuna in South America have all shown the greatest stabilization of numbers in areas that have been heavily patrolled and where resources have been concentrated. If parks are too large they may not be adequately patrolled.

Also, it is not always remembered in theoretical debates that species are not simply numbers in an equation and cannot be treated as such. Some plants and animals are of more interest to conservationists than others, and island biogeographic theory fails because it does not draw any distinctions between desirable and undesirable species. Some areas such as oligotrophic lakes are species-poor yet are preserved because of their pristine appearance. Other areas like agriculturally improved grassland or even old railway sidings can be very rich in plant species, yet we often do not preserve these areas. Richness seems to be important only when it suits us; otherwise, we ignore it.

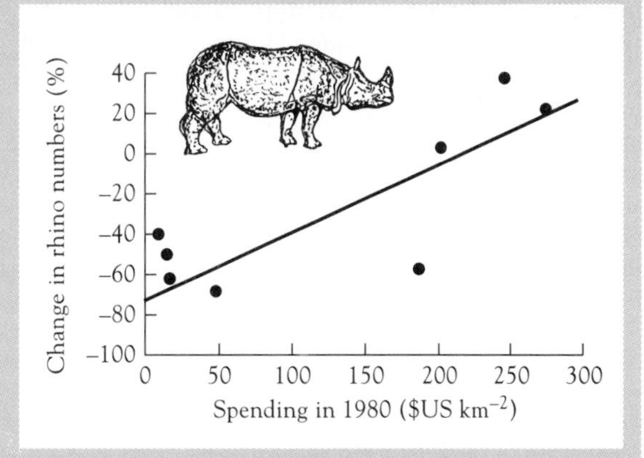

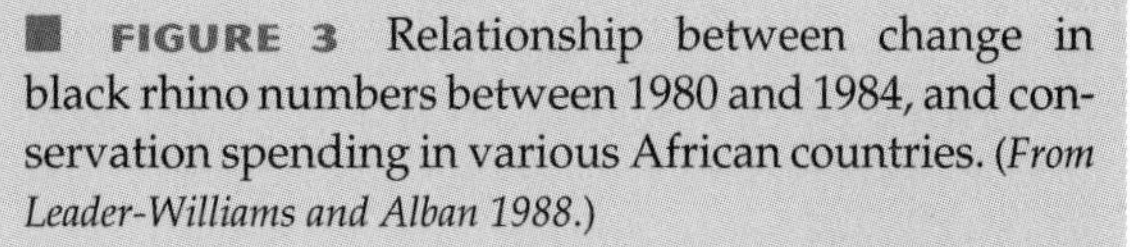
FIGURE 3 Relationship between change in black rhino numbers between 1980 and 1984, and conservation spending in various African countries. (*From Leader-Williams and Alban 1988.*)

Finally, Mabey (1980) and others have argued that conservation has catered principally to an elite minority and that government agencies and trusts select preserves on pseudoscientific grounds for ecological specialists. Can everyone afford to drive to national parks? What does the public really want? Goldsmith (1983) suggests that peoples' principal interest lies near to home, not at some distant reserve. They would like the "feeling" of nature in their immediate neighborhoods. They want to see pretty flowers, birds, and butterflies in the countryside. Swallowtail butterflies and bluebell flowers may do just as well as panthers and bobcats, which are very secretive anyway. It might be argued that the importance of different taxonomic groups could be assessed from the membership of the organizations that study them. For example, in Britain, the Royal Society for Protection of Birds has about 250,000 members and the Botanical Society of the British Isles has 2,300. But, of, course membership is affected by the marketing efficiency of the organization as well as the size, color, and attractiveness of the organisms concerned, and without sound habitat management for plants the suitability of an area for birds would certainly deteriorate.

SUMMARY

1. Succession may be affected by the distance to an area undergoing succession from source areas of colonists and by the size of that area. The influence of these variables on succession is often termed island biogeography.

2. Island-biogeographic theory predicts that the equilibrium number for species in an area S is determined by a balance between immigration and emigration (extinction). S is determined by an island's size and position. Extinction increases on small islands, and immigration decreases on far islands.

3. Island biogeography theory may be used for habitat islands as well as real islands.

4. There is much data to suggest that species richness increases with island size. Much discussion focuses on the slope of the line relating richness to area, which may be steeper for true islands than habitat islands, and steeper for poor dispersers like mammals than for good dispersers like birds.

5. There is little data to back up other predictions of the theory of island biogeography, such as the turnover of species on islands.

DISCUSSION QUESTION

Is island biogeography theory valuable for conservation biologists?

SECTION

6

Ecosystems Ecology

Three links in the food chain: seven-spot ladybird larva eating black bean aphids feeding on a plant leaf. How much energy is passed from one link to another? How many links are there in food chains? Ecosystems ecology focuses on the flow of energy and nutrients through food chains and attempts to answer questions like this. *(Nigel Cattlin, Photo Researchers, Inc. 2N0578.)*

The term **ecosystem** was coined by the British plant ecologist Tansley (1935) to include not only the **community** of organisms in an environment but also the whole complex of physical factors around them. Ecosystem ecology involves organisms and their **abiotic** environment and concerns the movement of energy and materials through communities. The concept can be applied at any scale: a drop of water inhabited by protozoa is an ecosystem and a lake and its biota constitute another. Lovelock (1979) took this idea to its extreme and regarded the whole Earth as one totally interlocked ecosystem, which he named Gaia after the ancient Greek Earth goddess. In this viewpoint, Gaia was reminiscent of one superorganism, forever regulating temperature, oxygen, and moisture levels to ensure the continuation of life. Lovelock pointed out that the levels of such things have not changed appreciably in hundreds of millions of years, whereas on the basis of physical changes alone such alterations would have been expected.

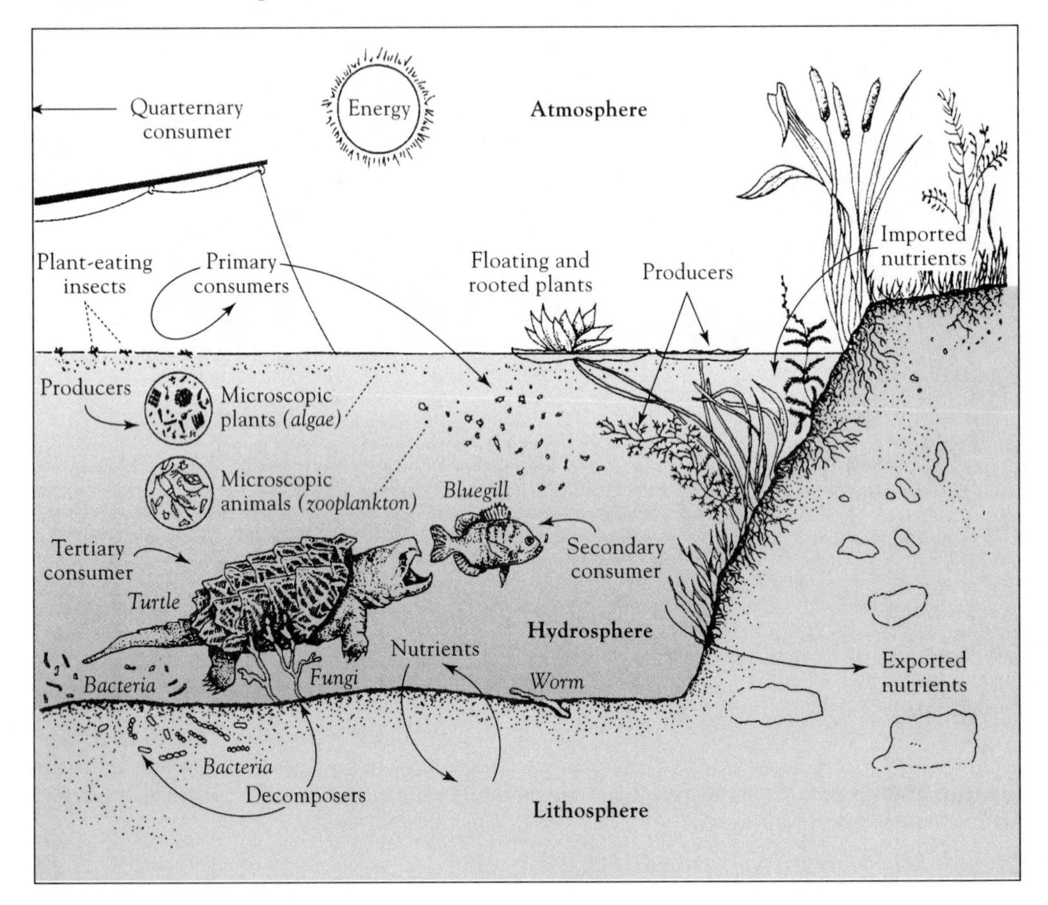

Producer, consumer, and decomposer in a pond ecosystem. Each of these roles is filled by a number of different organisms. For example, additional secondary or tertiary consumers might be water snakes, snapping turtles, and various birds of prey. Although ecosystems are often thought of as closed systems, none really is. Typically, both living and nonliving things are imported and exported.

Most ecosystems can never really be regarded as having definite boundaries. Reiners (1986) has argued that, for this reason and others, the ecosystem remains the least coherent of the organization levels of ecology. He suggests that it lacks a logical system of interconnected principles and a well-understood and widely accepted focus. Even in a clearly defined pond ecosystem (see figure on page 516) waterfowl may be moving in and out. The big advantage of ecosystems ecology, however, is the common currency of energy or nutrients, which allows the biology of communities and populations to be compared between and within trophic levels, something no other ecological discipline can boast.

In attempting to determine the importance of individual units within the scheme of an ecosystem, at least three major constituents can be measured. The first is **biomass**, the standing crop of an organism. Attaching too much importance to biomass, however, may lead to erroneous conclusions in some cases. In the applied world of timber technology, for example, a small standing crop may indicate small potential harvest or yield, but this is not necessarily true if the tree species in question has a high growth rate—new biomass will be produced rapidly and a higher rate of harvest could be sustained. In this situation, energy flow may be more critical. Then the community is regarded as an energy transformer; **energy flow** is the second constituent. Third, the ecosystem may be most limited by the availability of a **rare chemical or mineral**. In this case, the flow of limiting chemicals through ecosystems becomes the most important factor in understanding how systems work.

CHAPTER

20

Trophic Structure

Many authors suggest that food webs can be used to provide an insightful analysis of community organization (Cohen, Briand, and Newman 1990; DeAngelis 1992; Martinez and Lawton 1995; Raffaelli and Hall 1995; Polis and Winemiller 1995). Few ecosystems are so simple as to be characterized by a single unbranched food chain. Many types of primary consumers usually feed on the same plant species; for example, one may find tens to hundreds of insect species feeding on one tree species as well as several vertebrate grazers. Also, many species of primary consumer eat several different plants. Such branching of food chains occurs at other trophic levels as well. For instance, frogs eat many different types of insect species that also may be eaten by different types of birds. Owls may eat primary consumers such as field mice and also prey on predatory organisms like snakes. It is more correct, then, to draw relationships between these plants and animals, not as a simple chain but as a more elaborate interwoven food web, as illustrated in Fig. 20.1. The classification of organisms by trophic levels is one of function rather than species. For example, male horseflies are herbivores, feeding on nectar and plant juices, whereas females are blood-sucking ectoparasites, and the larvae are detritivores!

Most food webs are extremely complicated and imperfectly understood because they are so vast (Polis and Winemiller 1995; Polis and Strong 1996). Ecologists have recognized three kinds of food web:

a. *Source webs.* One or more kinds of organisms and the organisms that eat them, their predators, and so on.
b. *Sink webs.* One or more kinds of organism, the organisms they eat, their other prey, and so on.
c. *Community webs.* A group of species within a defined area or habitat (like the pitcher plant [Photo 20.1]).

The relative complexity of a food web can be denoted by a measure known as connectance, where

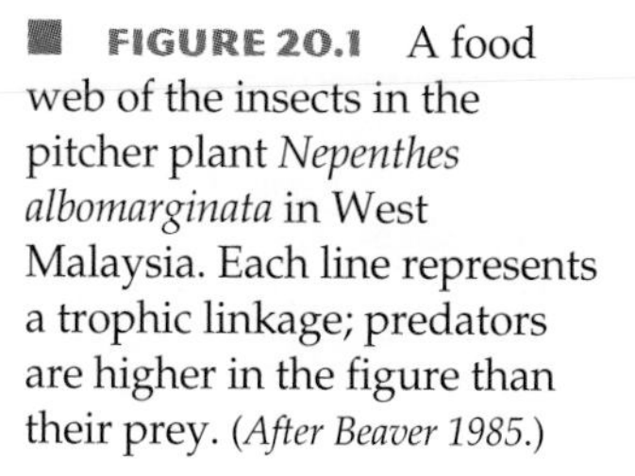

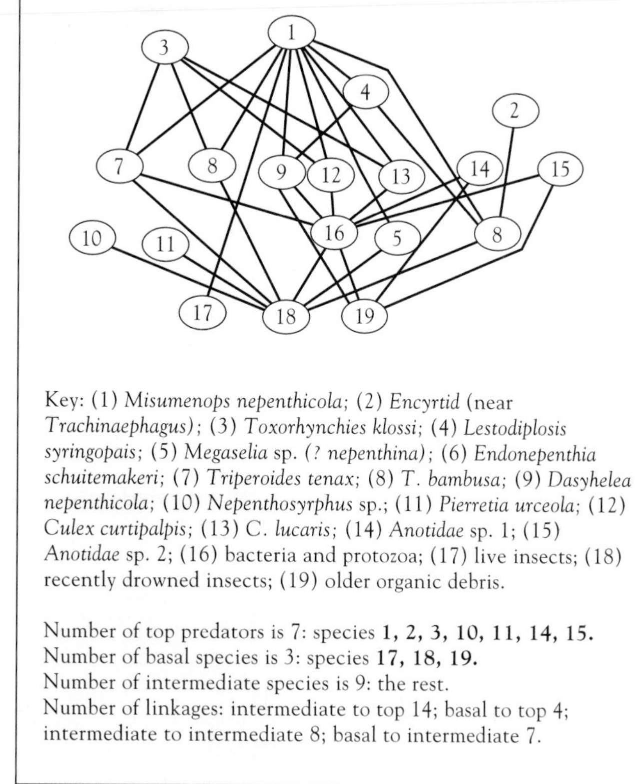

FIGURE 20.1 A food web of the insects in the pitcher plant *Nepenthes albomarginata* in West Malaysia. Each line represents a trophic linkage; predators are higher in the figure than their prey. (*After Beaver 1985.*)

$$\text{connectance} = \frac{\text{actual number of interspecific interactions}}{\text{potential number of interspecific interactions}}$$

so that for n species the number of potential interspecific interactions (assuming links cannot go in both directions)

$$= \frac{n(n-1)}{2}$$

The number of links per species is called linkage density, d. In Fig. 20.1, there are nineteen "species," thirty-three actual interactions, $(19 \times 18)/2 = 171$ potential interactions, and a connectance of 0.19. Linkage density $= 33/19 = 1.74$

20.1 Proposed patterns in food webs focus on connectance values, chain lengths, and linkage densities.

A number of generalizations have been made about food webs, and some of these were summarized by Pimm, Lawton, and Cohen (1991):

1. Cycles are rare. That is, species A eats species B, species B eats species C, and species C eats species A. Cannibalism is a cycle in which one species feeds on itself.

■ **PHOTO 20.1** Pitcher plant, *Nepenthes macfarlanei*, in Montane Forest at Bintang, Malaysia. Pitcher plants like this have their own self-contained ecosystems which lend themselves well to food web analysis. (*Fletcher & Baylis, Photo Researchers, Inc. 7Y8275.*)

2. The average proportion of top predators (nothing preys on them), intermediate species (with both predators above and prey below), and basal species (autotrophs) remains constant in webs regardless of the number of species.
3. The proportion of trophic links between top predators and intermediates, intermediates and intermediates, basal species and intermediates, and basal to top remains constant. However, in fifty food webs representing the pelagic communities of small lakes, the fraction of intermediate species increased with species number (Havens 1997), and this is probably true in large food webs in general (Martinez and Lawton 1995).
4. Linkage density is often constant, except for webs with large numbers of species. Links per species is scale invariant (does not change with the size of the food web) and is roughly equal to two. (Cohen, Briand, and Newman 1990).
5. Connectance remains constant as the number of species in the food web increases (Martinez 1992). This is the opposite of Cohen, Briand, and Newman's (1990) scale invariant law (see no. 4). Imagine an insectivorous

bird that feeds in two communities A and B, with A having twice the number of insect species that B does. Martinez (1992) argues that it is biologically more reasonable to suppose that a bird would eat twice as many species of insect in community A as in community B. Because this would likely apply to all species in the community, connectance remains unchanged as species richness increases.

6. Omnivory is rare (omnivores feed on organisms from at least two trophic levels).
7. Contrary to intuition, food chain lengths do not different greatly among ecosystems with different primary productivities. This was elegantly shown by Pimm and Kitching (1987), who fertilized tree holes, increasing primary productivity over a fourfold range but noted no increase in the numbers of trophic links. Spencer and Warren (1996) also increased the energy input into laboratory microcosms of protists and bacteria but got less complex food webs. However, Schoener (1989) presented evidence that food chain lengths are linked to the amount of productive space, a quantity that combined productivity with the area (or volume) occupied by the food web. Furthermore, experiments that increased tree hole productivity one hundredfold did link food chain length to productivity (Jenkins, Kitching, and Pimm 1992).
8. There are no significant differences in chain lengths where consumers are vertebrates compared to where they are all invertebrates.
9. Chain length is smaller on smaller islands. Food webs are also more complex in larger habitats. This was shown experimentally in laboratory microcosm communities of protists and bacteria by independently manipulating habitat size and productivity (Spencer and Warren 1996).
10. Chains are shorter in areas with frequent natural or experimental disturbances.
11. Chains are shorter in two-dimensional habitats, like grasslands, than in three-dimensional habitats like forests or reefs.
12. The ratio of the number of predator species to the number of prey species is constant. It was suggested to be about one to one by Briand and Cohen (1984), about 0.5 by Sugihara, Schoenley and Trombla (1989), and 2:1 by Havens (1997).
13. Top predators tend to be rather large and sparsely distributed, whereas herbivores are smaller and more common. This generalization is often termed the *pyramid of numbers*. In a small pond, the numbers of protozoa may run into millions and those of *Daphnia* and *Cyclops* (their predators) into hundreds of thousands, whereas there will be fewer beetle larvae and even fewer fish. One can think of several exceptions to this pyramid. The elm tree, one producer, supports many herbivorous beetles, caterpillars, and so on, which in turn support even more predators and parasites. The best way to reconcile this apparent exception is to weigh the organisms in each trophic level. The elm tree weighs twenty-seven metric tons, the herbivores 50 g, and the predators, say, 5 g. Then the elm tree is not a real exception. Inverted pyramids can still occur, even when biomass is used as

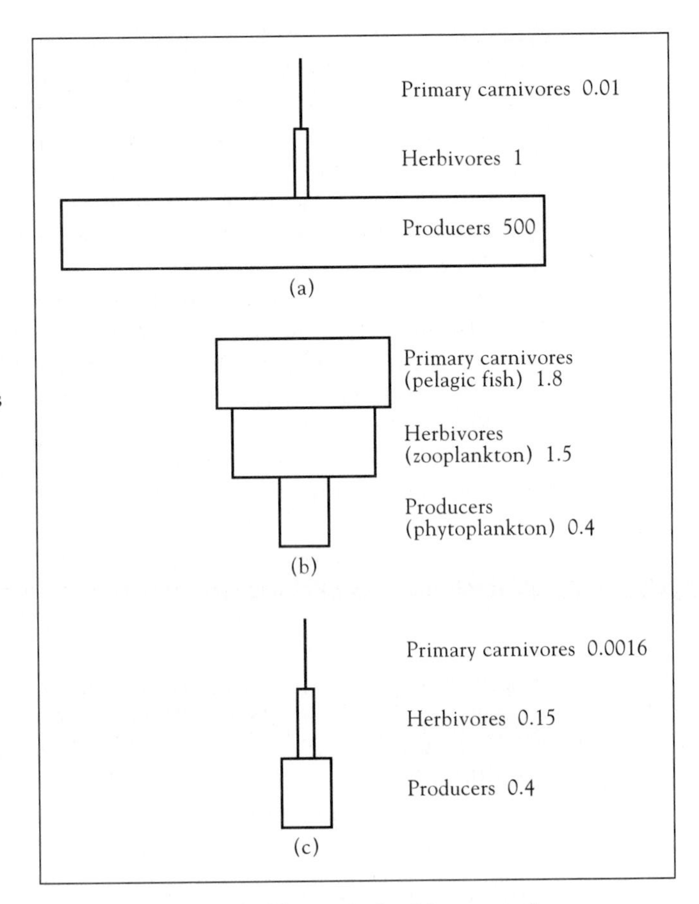

FIGURE 20.2 Biomass pyramids. (**a**) A pyramid of biomass seen in an old field in Georgia (g dry wt/m^2). (**b**) An inverted pyramid of biomass in the English Channel (g dry wt/m^2). Phytoplankton has a lower standing crop than the zooplankton, and the zooplankton biomass is less than that for the fish. (**c**) Production rates of organisms in the English Channel (g dry wt/m^2/day). The inversion in (b) is made possible by the high rate of production by producers in that system. *(Redrawn from Price 1975.)*

the measure (Fig. 20.2). In the English Channel, the biomass of phytoplankton supports a higher biomass of zooplankton. This is possible because the production rate of phytoplankton is much higher than that of zooplankton, and the small phytoplankton standing crop (biomass at any one point in time) processes large amounts of energy. The most realistic pyramid is thus the energy pyramid, which never becomes inverted.

20.2 Problems with food web theory focus on trophic aggregation and imperfectly known strengths of linkages.

While food web theory has undoubtedly led to an increase in the understanding of biological systems, many authors have pointed out numerous problems with the theory and the data:

1. Predation on "minor" species is often omitted; linkages are often informal and idiosyncratic. Often, authors are not sufficiently knowledgeable about organisms other than in their area of specialty so that some links are never drawn. For example, Polis (1991) noted that food webs in the real world are much more complex than reported in the food web literature. In a relatively

simple desert system in the Coachella Valley, California, Polis noted 174 species of plants, 138 vertebrates, 55 spiders and scorpions, an estimated 2,000 to 3,000 species of insects, and an unknown number of microorganisms and nematodes all in a single food web. Polis compared his web to catalogs of published webs by authors such as Joel Cohen and Keith Schoenly (Table 20.1). Chains were longer, omnivory and loops were common, and connectivity was greater. Lancaster and Robertson (1995) showed the food web parameters of an English stream as presented by Hildrew, Townsend, and Hasham (1985) were drastically altered by more intense sampling, which revealed an additional nine species in the web over the twenty-four noted by Hildrew. Goldwasser and Roughgarden (1997) suggested that most food webs that are published are too poorly sampled to be useful in developing theory. By using computer simulations they were able to randomly sample a fully known Caribbean food web. Web properties varied markedly with the intensity of sampling.

2. Data on quantities of food consumed, i.e., thicknesses of connecting links, are usually absent. In the first example in which interaction strengths have been calculated in a food web (Paine 1992), most connections were found to be

TABLE 20.1 Comparison of statistics from the Coachella Valley food web with means of the statistics from cataloged webs. (*After Polis 1991*)

	COHEN, BRIAND, AND NEWMAN AND BRIAND	SCHOENLY, BEAVER, AND HEUMIER	COACHELLA
Total number of "kinds" or species (S)	16.7	24.3	30
Total number of links per web (L)	31	43.1	289
Number of links per species (L/S)	1.99	2.2	9.6
Number of prey per predator	2.5	2.35	10.7
Number of predators per prey	3.2	2.88	9.6
Total prey taxa/total predator taxa	.88	.64	1.11
Minimum chain length	2.22	1	3
Maximum chain length	5.19	7	12
Mean chain length	2.71	2.89	7.34
Connectance ($C = L/S[S - 1]/2$)		.25	.49
Basal species (%)	19	16	10
Intermediate species (%)	52.5	38	90
Top predators (%)	28.5	46.5	0
Primarily or secondarily herbivorous (%)		14.6	60
Primarily or secondarily saprovorous (%)	21	35.5	37
Omnivorous (%)	27	22	78
Consumers with "self-loop" (%)	<1.0	<1.0	74
Consumers with mutual predation loop (%)	$\ll 1.0$	$\ll 1.0$	53

weak so the food web was essentially very simple. This is essentially the opposite of problem no. 1 just given—there may be many more links than we realize, but most of them are unimportant. Benke and Wallace (1997) showed how misleading a connectivity food web was for a riverine system in Alabama because it implied the equivalence of all food resources. An energetic analysis that looked at ingestion fluxes actually showed great variation in the strength of linkages (Fig. 20.3). Wootton (1997) also showed many weak per-capita interactions and a few strong ones when he examined and measured the impact of bird predation on different prey taxa in a rocky intertidal community. On the other hand, a superabundant species exhibiting numerous weak interactions could be important in food web dynamics.

3. There are few data on the importance of chemical nutrients. One apparently feeble link may be very important if it supplies a limiting chemical (Tilman 1982).
4. There are few data on temporal variation. Some species may constitute a large proportion of the web at one point in the year but not at other times (Closs and Lake 1994). The cumulative web, based on all the possible temporal webs, may be an accurate portrayal of all the possible interactions but an inaccurate predictor of what is going on at any one time (Thompson 1988).
5. Detritus is not often correctly positioned in food webs. Most authors put it at the bottom of the food chain. The correct procedure is to place material according to the number of acts of assimilation it has undergone since

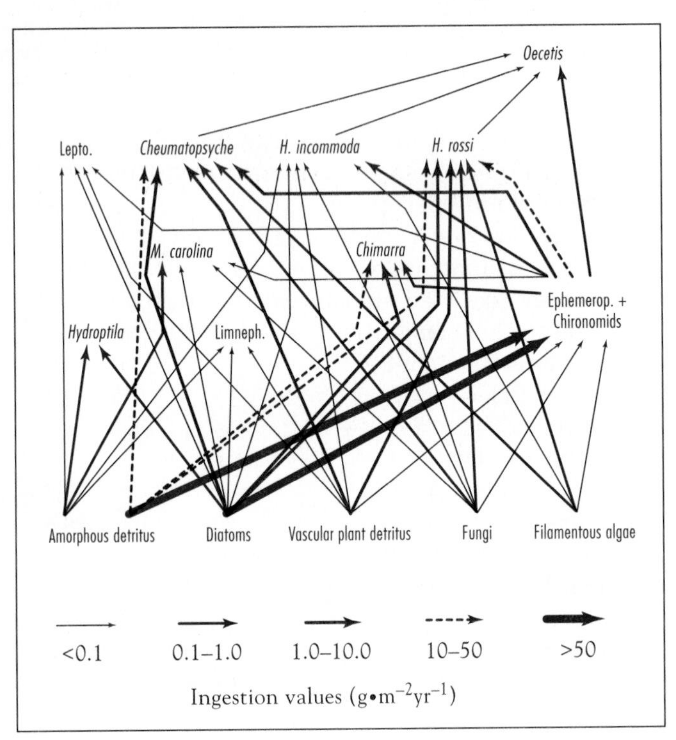

FIGURE 20.3 Quantitative food web for the major caddisfly taxa in an Alabama stream ecosystem. Line thickness indicates magnitude of ingestion fluxes. (Ephemerop. = Ephemeroptera) Lepto. = Leptoceridae, Limneph. = Limneophilidae (*After Benke and Wallace 1997.*)

being part of a green plant. Thus, dung, since it is unassimilated, is equivalent to green plants, but carcasses should be placed at the same level as when the bodies were alive. Treating carcasses in this way gives many food chains a higher number of trophic levels (Cousins 1987).

6. In food web theory species are often aggregated into "trophic species," e.g., insect larvae, plankton. For example, types of "insects" are often likely to be lumped together because of the difficulties of identifying them all. This disguises much important biology. Aggregation is rife in many published webs. The rejoinder to this criticism is that if aggregation has introduced any errors, patterns from aggregated webs should differ from nonaggregated webs where all species are known. Pimm, Lawton, and Cohen (1991) suggest that most patterns appear across all types of webs. A test is provided by taking completely known webs, aggregating them, and comparing patterns. For instance in Fig. 20.1, the three pairs of biological species (10 and 11, 12 and 13, 14 and 15) that have the same sets of predators and prey could be joined into three trophic species. The web could be further aggregated by joining 7, 8, 12, and 13, which share similar species of predators and prey. The aggregation of webs, in this case, using trophic criteria, affects webs' properties only slightly. In contrast, progressive aggregation by taxonomic affinities alters webs' properties more rapidly. However, Paine (1988) showed how connectance values of between 0.31 and 0.61 occur for the same rocky intertidal community in New England depending on which scientist studied the system and which methods they used for aggregation. The disconcerting point about this is that these different values for connectance in the same intertidal system span about half the range of 0.05-0.60 given in forty published webs presented in Briand (1983).
7. It is hard to define web boundaries. Some highly mobile species, like predatory gulls, may be very important predators in food webs, but they are almost always under represented in webs (but see Wootton 1997).
8. Many species like starfish or fish exhibit size- or age-related changes in diet and are not easy to assign a single position in a web.
9. Many individuals do not fit into discrete trophic levels. For example, a hawk feeding on a small bird that fed on an insect that fed on a plant would be feeding at trophic level 4. If we insert two extra trophic levels, for instance, the insect is attacked by a parasite that in turn is eaten by a spider before the small bird attacks it, then the hawk would be feeding at trophic level 6.
10. Constancy of the 1:1 ratio of predator-prey ratios may be an arithmetical artifact. Many taxa can be recorded as both predator and prey and, hence, be double counted. Closs, Watterson, and Donnelly (1993) argued that in many webs the proportion of species that are double counted is large, so the observed ratio of predator species to prey will inevitably be roughly equal to one. Furthermore, Wilson (1996) suggests that predator-prey ratios from communities constructed at random are not different from ratios in real communities.

11. Some links that appear negative may actually be positive. If pollinators are considered merely in the context of trophic interactions then they have a negative influence in terms of their consumption of energy that may otherwise have been used for other functions. Obviously, this is wrong; pollinators have a net positive effect on their hosts, but the point is that some links within the trophic network can portray a direct effect that is opposite in sign to the real effect. The same argument could be made for seed dispersers or other mutualists.
12. It is difficult to know which is the best way to analyze what the most important links in food webs are. One approach is to estimate the amount of energy flowing from a resource to a consumer. A second approach is to experimentally manipulate linkages (usually by removing predators) and assess subsequent changes in community structure. Unfortunately, these two approaches can yield different results. For example, Paine (1980) showed that, in the intertidal, several grazers consumed great amounts of resource species, yet when these grazers were removed there were no significant changes in community structure. In contrast, there was neglible energy transfer from a kelp to a sea urchin, but when the sea urchin was removed there were dramatic increases in kelp abundance, with effects cascading through the community. How can we incorporate these results into a food web diagram?

A similar picture emerges from studies of the community associated with the Ythan River estuary in Scotland where there are large populations of macroinvertebrates with 95 elements and 409 feeding links (Hall and Raffaelli 1991). The major elements are mudflat algae, mussels and other invertebrates, and predatory fish and shorebirds. The relative importance of linkages in terms of energy flow was compared to results from exclosure experiments designed to break the trophic linkages between fish, shorebirds, and invertebrates (Raffaelli and Hall 1992). If only energetically dominant linkages were considered, the web could be represented by only 31 of the original 409 linkages (Fig. 20.4).

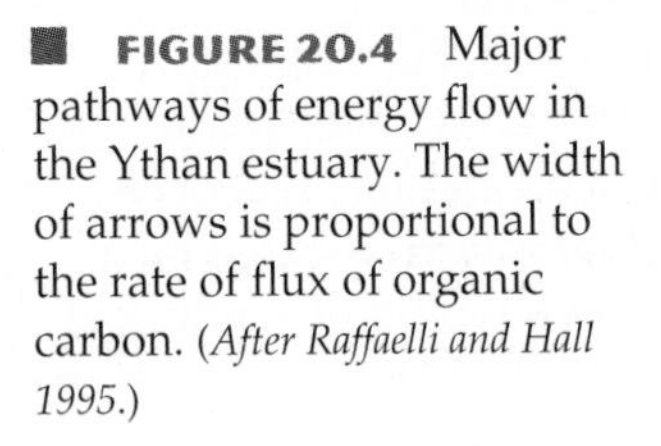

FIGURE 20.4 Major pathways of energy flow in the Ythan estuary. The width of arrows is proportional to the rate of flux of organic carbon. (*After Raffaelli and Hall 1995.*)

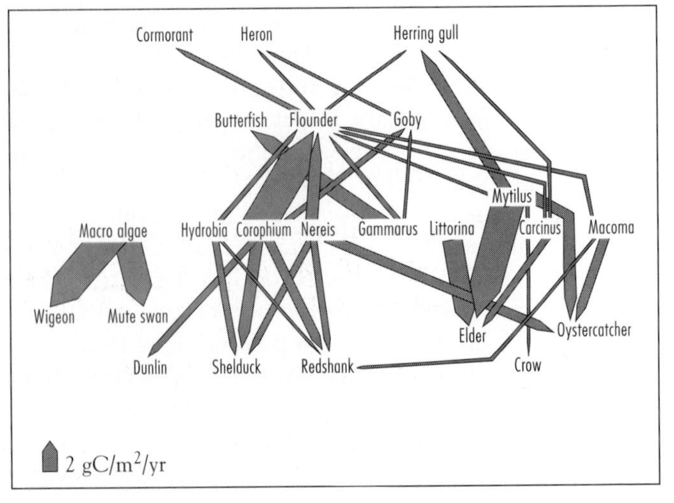

Three sets of interactions seemed especially important in terms of energy flow: wildfowl eating algae, flounders feeding on the amphipod *Corophium*, and eider ducks preying on the mussel *Mytilus*. In contrast, the predator manipulations revealed no statistically significant reductions in prey density for two of these three energetically dominant interactions. The manipulations did reveal a highly significant effect of eiders on mussels and smaller but statistically significant effects of shorebirds, flounders, and gobies on some polychaete and oligochaete species, linkages deemed too energetically trivial to include in Figure 20.4. Although the eider-mussel link was significant, and breaking it resulted in a twofold increase in mussel densities, such mussel increases did not affect the densities of other community members, so the link was not considered functionally important (Raffaelli and Hall 1995). This was in stark contrast to the cascading effects noted by Paine (1980) for rocky shores when *Mytilus* densities increased. Of 36 per-capita effects investigated by Hall and Raffaelli, 26 (76%) were very small: < 0.1. In fact, the per-capita effects were approximately normally distributed about zero (Fig. 20.5). The few large negative effects were between wading birds and their prey. The large positive effects were recorded for flounders feeding on annelids. Such effects are probably due to predators holding down potential competitors, thereby allowing nonpreferred resources to increase in density: that is, indirect effects (Paine 1992). Flounders disturb the mudflat while feeding, and this could maintain poor competitors or opportunists in the system (Hall, Rafaelli, and Thrush 1994).

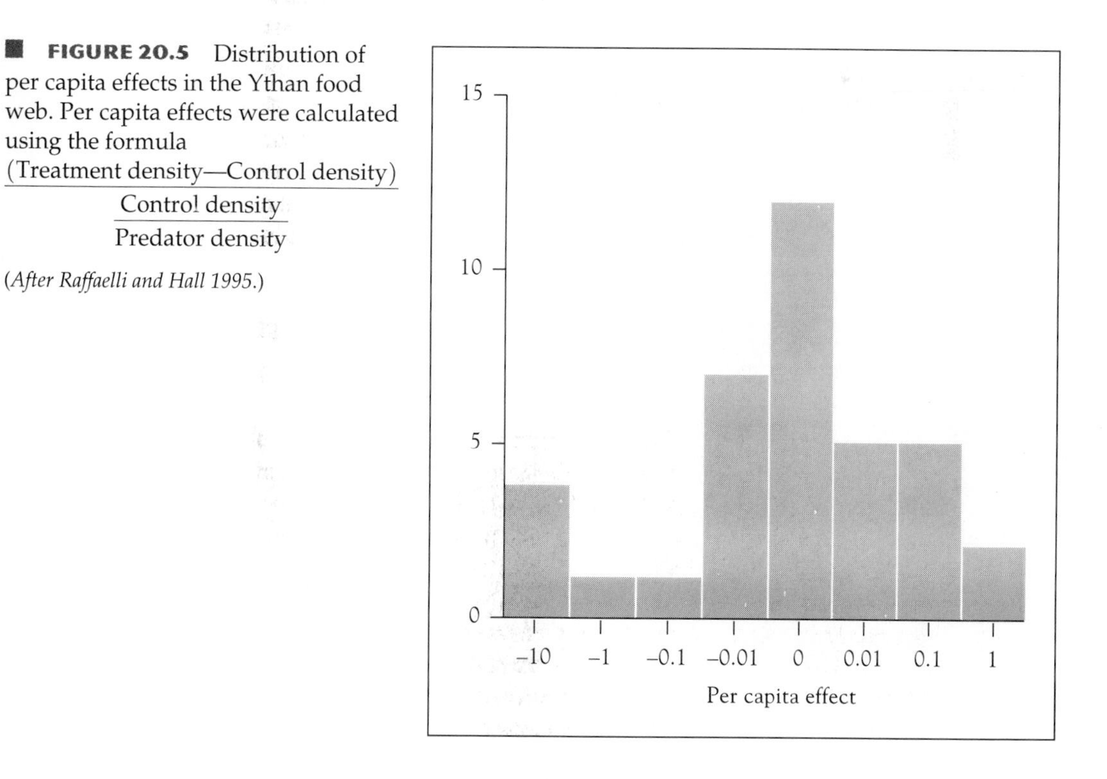

FIGURE 20.5 Distribution of per capita effects in the Ythan food web. Per capita effects were calculated using the formula

$$\frac{\dfrac{(\text{Treatment density}-\text{Control density})}{\text{Control density}}}{\text{Predator density}}$$

(*After Raffaelli and Hall 1995.*)

Food Webs and the Passage of Pesticides

DDT, dichloro-diphenyl-trichloroethane, was first synthesized in 1874. In 1939, its insecticide quality was recognized by the scientist Mueller in Switzerland, who won a Nobel Prize in 1948 for that discovery and subsequent research on the uses of DDT. The first important application of DDT was in human health programs during and after World War II, and at that time its use in agriculture also began. The global production of DDT peaked in 1970 when 175 million kilograms (kg) were manufactured. The peak of production in the United States was 90 million kg in 1964. Most industrialized countries banned the use of DDT after the early 1970s, and only a few less developed countries still use it today.

DDT has several chemical and physical properties that profoundly influence the nature of its ecological impact. First, DDT is persistent in the environment. It is not easily degraded to other, less toxic chemicals by microorganisms or by physical agents such as sun and heat. The typical persistence in soil of DDT is about ten years. This is two to three times as much as other organochlorine insecticides. The good news is that following the outlawing of DDT in the United States the DDT amounts in the soils are by now negligible. Another important characteristic of DDT is its low solubility in water (less than 0.1 parts per million [ppm]) and its high solubility in fats or lipids, a characteristic that is shared with other chlorinated hydrocarbons. In the environment, most lipids are present in living tissue. Therefore,

■ **PHOTO 1** Duck's nest with clutch of eggs illustrating effects of DDT, DDT accumulates in food chains and causes thin egg shells, which break under the weight of the parent birds. *(Photo from U.S. Department of Agriculture.)*

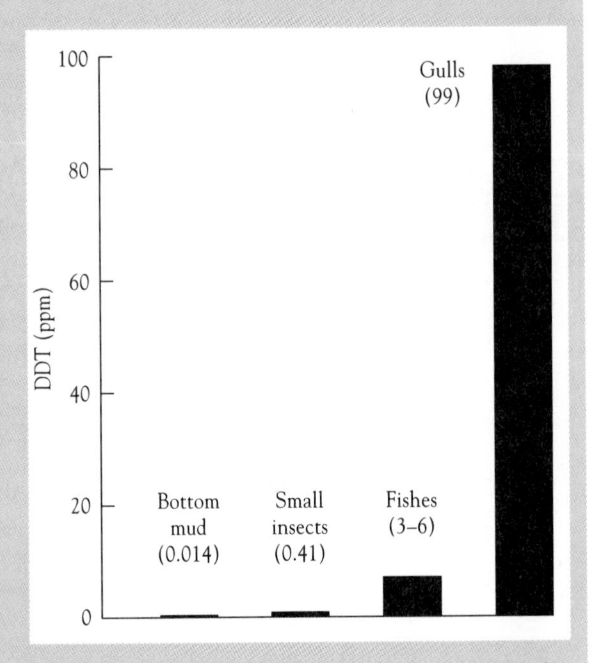

■ **FIGURE 1** DDT concentration in a Lake Michigan food chain. DDT load in gulls was about 240 times that of small insects.

because of its high lipid solubility, DDT has a great affinity for organisms, and it tends to bioconcentrate by a factor of several orders of magnitude. Furthermore, because organisms at the top of a food web are effective at accumulating DDT from their food they tend to have an especially large concentration of DDT in their lipids (Photo 1). A typical pattern of food web accumulation of DDT is illustrated in Fig. 1, which summarizes a pattern of DDT residues in a Lake Michigan food chain. The prime reason for the introduction of DDT there was to control mosquitoes. The largest concentration was in gulls, an opportunistic species that often fed on small fish. Large residues were also present in some game fish, which became unfit for human consumption. Had a more thorough knowledge of the relevant food webs been available, some of these side effects might have been predicted.

There are obviously many disadvantages, as well as many advantages, to looking at communities in terms of their food webs. However, a thorough knowledge of food webs would be valuable to allow us to be able to predict the indirect effects of toxic substances in communities. For example, the toxic effects of DDT proved disastrous for higher organisms like birds and fish, even though these were never the intended targets (Applied Ecology: Food Webs and the Passage of Pesticides.) Perhaps we could gain more realism and perhaps more insights by looking at finer subsets of organisms within food webs, sets of organisms that often feed in the same way and may be expected to interact closely. Such sets of species are often known as *guilds*.

20.3 A guild is a group of functionally similar organisms within a trophic level.

In human society, craft guilds connoted mutual aid and protection functions. There were guilds of goldsmiths or wheelwrights, each of whom made a living in the same way. In nature, the term *guild* seems to have departed subtly from this original meaning because although guild members feed in the same way, a guild often connotes groups of potentially competing species, not cooperating ones. The term was originally coined by Root (1967) in his studies of insectivorous tree-inhabiting birds:

> *A guild is defined as a group of species that exploit the same class of environmental resources in a similar way ... without regard to taxonomic position ... [This is] the most evocative and succinct term for groups of species having similar exploitation patterns.*

Thus, for example, in an insect community feeding on a plant, we may have the leaf-chewing guild, the sap suckers, leaf miners, stem borers, stem gallers, root feeders, and flower feeders (Fig. 20.6). Guilds have also been referred to as "functional groups" (Cummins 1974), particularly for aquatic invertebrates, but this type of classification focuses on the resource itself, i.e., the size of particle filtered from a stream and not on the techniques using for feeding. Hawkins and MacMahon (1989) noted that Root is cited more than Cummins in the literature and that the term *guild* has really "caught on," being cited over four hundred times between the time Root defined it and 1986, the last year for which they had data. In the 1980s in particular, guilds are mentioned many times, perhaps because interest in this branch of community ecology increased, or perhaps because science citation rates in community ecology in general went up.

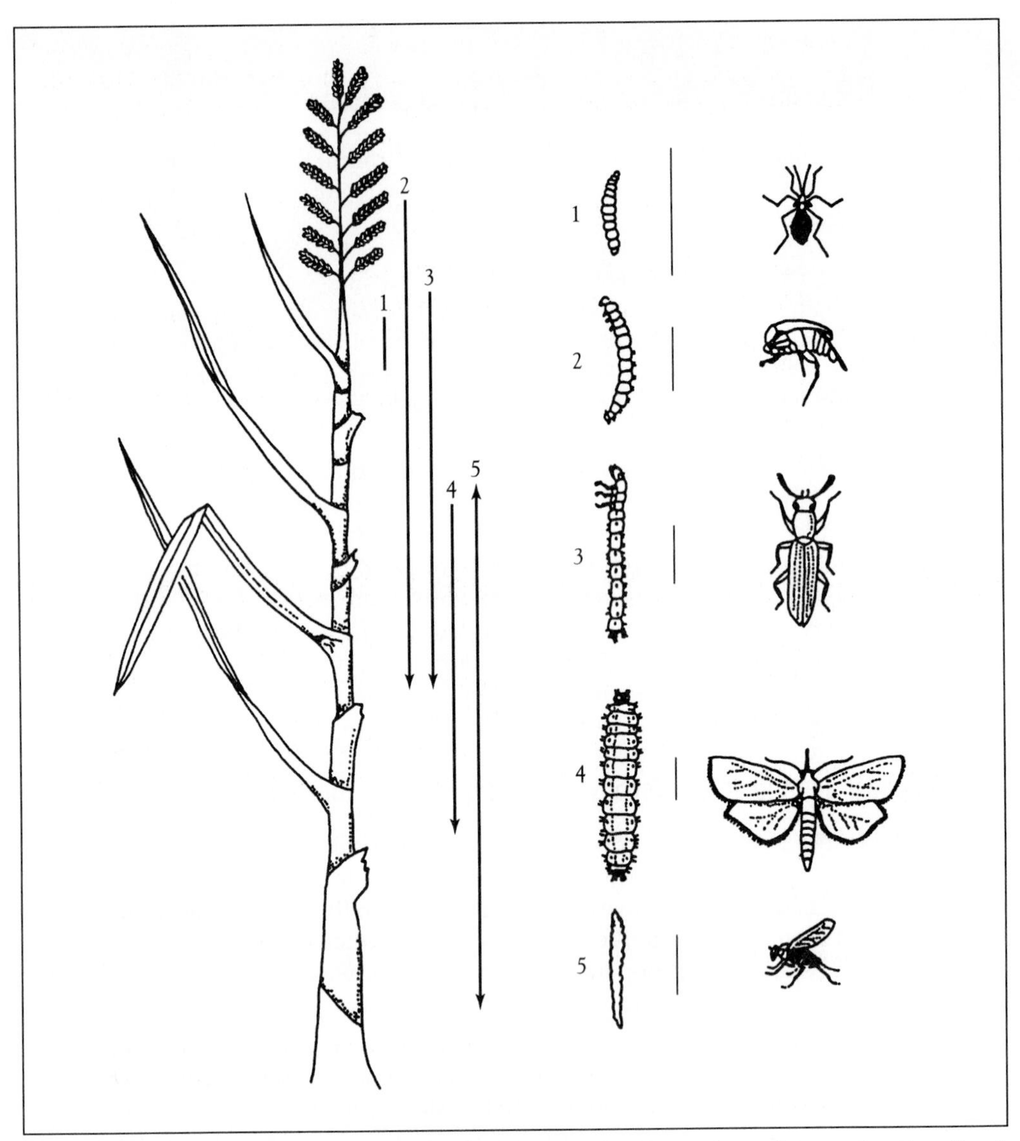

■ **FIGURE 20.6** The stem-boring guild associated with salt-marsh cord grass, *Spartina alterniflora*, on the Gulf Coast of North America. (1) *Calamomyia alterniflorae* (Diptera). (2) *Mordellistena splendens* (Coleoptera). (3) *Languria taedata* (Coleoptera). (4) *Chilo plejadellus* (Lepidoptera). (5) *Thrypticus violaceus* (Diptera). Arrows indicate where in the stem the larva of each species is found and the direction it bores. Adults, shown at extreme right, are free living. The length of each scale line between the larva and adult of each species represents 0.5 cm in the drawing.

Why study guilds?

Three reasons are usually given for studying guilds:

1. Attention is focused on all competing species, regardless of taxonomic relationship. For example, competition in a seed eating guild may involve

rodents, birds and ants. Thus for completeness sake, investigations on competition should involve all guild members. This forces researchers to cross taxonomic lines. For example, consider the following examples of guilds:

i. Frogs and insects feeding on pond periphyton (Morin, Lawler, and Johnson 1988)
ii. Lizards and spiders feeding on insects (Schoener and Spiller 1987)
iii. Ducks and fish feeding on insects (Hill, Wright, and Street 1987)
iv. Bees and finches feeding on nectar (Schluter 1986)
v. Lizards and birds feeding on insects (Wright 1981)
vi. Flamingos and fishes feeding on zooplankton (Hurlbert, Loayza and Moreno 1986)

2. Guilds might represent the basic building blocks of communities (Hawkins and McMahon 1989). Pianka (1988) envisioned a periodic table of niches, analogous to the periodic table of elements, with entries such as insectivorous bats, and so on.
3. Guilds might represent arenas of the most intense competition (Pianka 1980).

Are there different guilds for different resources?

Guilds have been established mainly in terms of the food items they use, but sometimes for other reasons. For example, some authors have suggested that a bird can be a member of the foliage-gleaning guild and/or the hole-nesting guild by virtue of its nest site requirements. Other guilds have been erected for biomass or mobility, but these seem to violate the original intent of guilds, which focused on organisms that feed on their resources in a similar manner, even though Root's original definition used the word *exploit*, not *feed*.

Problems with guild theory.

As with food web theory in general, the attraction of studying guilds is offset by some substantial problems:

1. Guilds must be correctly defined, not created out of an arbitrary subset of species from a community. Errors of omission are critical in that they may change conclusions as to the intensity, or lack thereof, of competition (Simberloff and Dayan 1991). For example, does the desert seed-feeding guild feed on all seeds, seeds of only a few plants, seeds of particular sizes, or seeds in particular habitats? Any of these might be appropriate depending on an investigator's frame of reference. A large raptor may overlap in the prey items it takes with a medium-sized raptor but not with a small raptor. Are all three raptors part of the same guild?
2. How much overlap in diet does there have to be for species to constitute the same guild? In his original definition, Root (1967) omitted some species of birds from the foliage-gleaning guild because they took prey from this resource habitat only occasionally. Jaksic and colleagues (Jaksic 1981; Jaksic and Delibes 1987) suggested a 50 percent overlap in diet as the minimum value for guild association. However, is the overlap 50 percent by prey

ECOLOGY IN PRACTICE

Tamar Dayan,
Tel Aviv University, Israel

I grew up in Israel and did the entire course of my academic studies in that country, at Tel Aviv University, where I am now a senior lecturer. This would seem an unusual course of events in a large country, but there are few universities in Israel, and Tel Aviv University has a strong program in zoology. My undergraduate degrees were in archaeology and the life sciences, and for my master's thesis I studied the faunal remains of a pre-pottery Neolithic B campsite, in an attempt to reconstruct ancient environments and food procurement methods.

When I started studying the Quaternary carnivores of Israel for my Ph.D. dissertation (with Yoram Yom-Tov and Eitan Tchernov as advisors), it was primarily with the desire to analyze morphological change through time and to understand the selective pressures involved. It was, therefore, from a morphological viewpoint that I approached my research on guild structure and the morphological relationships among carnivores rather than as an ecologist attempting to answer a specific question in community ecology.

Interestingly, in other countries other researchers were beginning to converge on the idea of looking at carnivore teeth for the same purpose at the same time. But I was across the ocean, and like many other students from faraway places I could not yet afford to attend scientific meetings, so was little exposed to what was going on and where until it was published.

In time I got swept up by the ecological issues that fascinate us all. I think I was fortunate to become an ecologist during a period when more than ever before ecological research involves a wide perspective, with evolutionary history, behavior, physiology, morphology, paleontology, and other disciplines becoming part of the large, and hopefully one day, coherent picture of life. In fact, this volume is an excellent example of this approach. I think that combining all these approaches gives us hope for the future, as well as being a source of constant frustration because it is so difficult to master all the disciplines. In the meantime, I think we all enjoy the excitement of working in a field where so much awaits discovery.

Science is made by people, and I was fortunate to interact with excellent people who have influenced my career. In particular, I should mention Daniel Simberloff who was receptive and generous to the thoughts of an unknown graduate student from a foreign country and from whom I try to learn to be critical yet open minded and to carry his spirit of generosity to my own graduate students. Also Steven J. Gould in whose lab I was exposed to many exciting and insightful ideas and whose provocative questions regarding the macroevolutionary role of microevolution have made studying microevolution for me an even more challenging venture.

Today, I teach ecology at Tel Aviv University, and with my group of graduate students I study various aspects of mammalian evolution within communities. This involves ecological field work, morphological research in museum collections, and the study of fossil mammals. We also still do anthropologically oriented research, in particular trying to understand the relationships between paleoecological changes and changes in human subsistence strategies and the role of humans in inducing micorevolutionary morphological changes. And like everyone else nowadays, we try to pitch in for conservation-related issues and research. I have discovered that interacting with my own graduate students and being exposed to their fresh perspectives adds another dimension to my work as well as to my enjoyment.

Doing research can be a frustrating proposition. Floods sweep enclosures open, animals do not always cooperate, fossils appear at unexpected places, unappreciative reviewers damage manuscripts, and ungenerous reviewers kill grant proposals. Trying to hold it all together while maintaining a family life with three children and a husband even busier than I am is not trivial, but the excitement and challenge are such that I do not even have the time to have regrets.

types or 50 percent by prey weight? For example, a red fox is likely to eat many insects, which would comprise 90 percent of its prey types, but the other 10 percent would be mammals, especially rabbits, which constitute 80 percent of the prey by weight. So is the fox an insectivore or not? Although quantitative statistical methods for guild assignment such as cluster analysis, principal components analysis, and canonical correlation exist, guilds can still not be unambiguously set because the investigator sets the levels for clustering. Also, these types of classification depend on selecting the "right" resources for analysis.

Patterns from guild analysis.

Despite the difficulties associated with assigning guilds, some authors have suggested that guild structure may be more predictable and stable than either the abundance of individual species or of species composition. Density compensation within guilds could maintain overall guild abundance at or near carrying capacity, while the fortunes of different species within a guild vary individually in response to factors other than resource availability, e.g., weather, predators, and the like. With this in mind, it is worthwhile to examine some case studies.

Cornell and Kahn (1989) examined the insects occurring on twenty-eight species of British trees, some rare, some common, and attributed them to one of four guilds: chewers, sap feeders, leaf miners, and gallers. The proportion of "chewers" increased on more common trees. However, guilds increased or decreased in richness independently of one another with only one exception: the sap-feeder and chewer guilds increased in parallel. Thus, in general, the system did not have a stable, predictable guild structure.

Mills (1992) divided up parasitoids attacking caterpillar hosts in the family Lepidoptera: Tortricoidea into three types of guilds:

1. Stage of host attacked (egg; early, mid, and late larva; pupa; adult)
2. Mode of parasitism (ecto or endoparasitism)
3. Form of parasitoid development (continuous or protracted)

These guilds are not mutually exclusive. There are thus $6 \times 2 \times 2 = 24$ possible guilds. Only fifteen of these are known to be utilized by parasitoids and only eleven by parasitoids of torticid hosts. Thus, tortricids were undersatured by parasitoid guilds, a factor also noted by Hawkins and Mills (1996) in a wider review of insects from many taxa where the occupancy rate of guilds was about 70 percent. Comparison of torticoid parasitoids from Nearctic hosts and Palearctic hosts showed a strong similarity, with larval ectoparasites with delayed development and endoparasitic pupal parasites that develop immediately being the most common types of parasitoid in both regions. Guild structure between Nearctic and Paleartic looked similar at this stage. Mills (1992) also noted that the range of host species attacked within each guild is often not significantly variable, regardless of parasite taxon. For example, both braconid wasps and ichneumonid wasps that attack larval stages attack the same number of species. Furthermore, for any one taxa, like ichneumonid wasps, the number of host species attacked varies according to guild so that ichneumonids in a larval guild attack 5.1 species of hosts whereas ichneumonids in the cocoon guild attack 10.5. Thus, Mills (1992) argued that guild type as well as taxonomic type are important in determining the biological patterns of host usage. One taxonomic type of parasitic wasp simply does not parasitize more species of caterpillar than another; guild of parasite is important too.

From Chapter 15, recall the studies of John Lawton and colleagues (1984, 1993) on the herbivore community on bracken fern *Pteridium aquilinum*, in different regions of the globe: England, Brazil, South Africa, New Mexico, the United States, Borneo, and Hawaii. A species-area relationship was evident: more species were found where the area covered by bracken was larger. In Hawaii, the area covered by bracken was very small, and only one leaf-tying lepidopteran caterpillar was found. Where the insects were more common, Lawton arranged them into four different guilds—chewers, suckers, miners, and gallers—and further divided them according to where they fed on the plant, rachis (main stem), costae (main stalks off the rachis), costules ("leaf veins"), and pinnae (leaves). There was no constancy in guilds between geographic areas. For example, there are no rachis miners in England but many in New Guinea. Lack of support for guild theory? Yes, but Ashbourne and Putnam (1987) did the same thing as Lawton using the insect herbivores feeding on red oak, *Quercus rubra*, and aspens, *Populus tremuloides*, in Canada and the United Kingdom. Their result was the opposite of Lawton's in that striking similarities were shown between the guild composition of the two regions (Fig. 20.7).

A different and very interesting example comes from the work of Nigel Stork (1987), who examined the arthropod (mainly insect) fauna feeding on five different species of rain forest trees in Borneo. Some 23,275 individuals were sorted into three thousand species and eight guilds, which not only included herbivores but predators and parasitoids as well. Thus, the total fauna was assigned to guilds, not just the herbivorous component. There was evidence of constancy of guilds between tree species for four guilds. That is, the proportion of species in four guilds was the same on the five tropical tree species. For the other four guilds, however, there was no constancy. Stork also noted that the way an investigator assigned species into guilds affected results. He assigned his insects into guilds previously erected by two other authors (Moran and Southwood 1982, who had argued for constancy in guilds when comparing the insect fauna of British and South African trees) and came up with slightly different results. Five guilds showed constancy and three did not. It follows that assigning insects into guilds from different trees could vary as much by the scientist summarizing the data as by geographic areas. This is similar to Paine's (1988) argument that measures of connectance in food webs varied with author as much as with system.

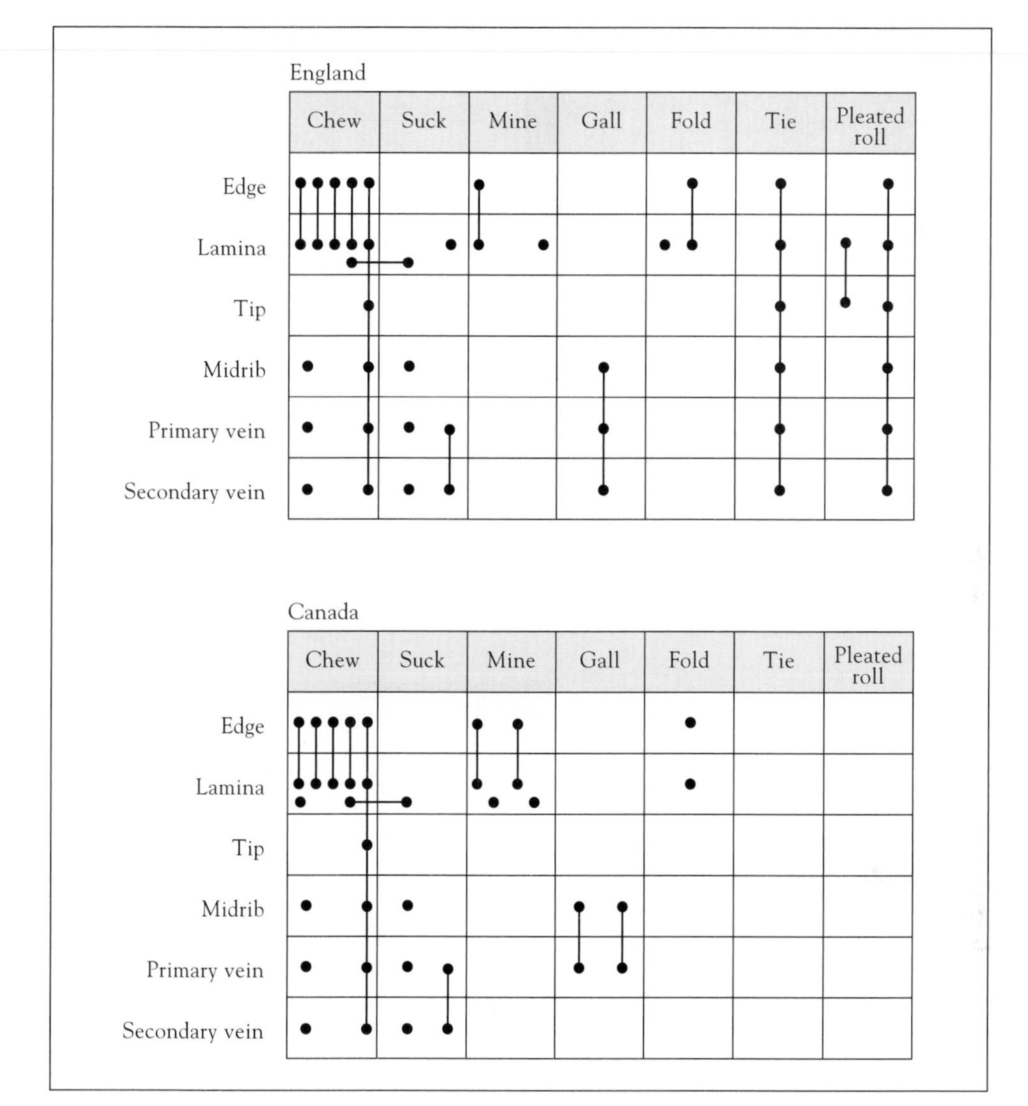

■ **FIGURE 20.7** Guilds of insects feeding on red oak (*Quercus rubra*) in Canada and the United Kingdom. Dots represent species and areas of oak leaves on that they feed. Lines connecting dots represent species that feed in more than one area. There are many similarities between the two countries in the area of the oak leaves where each guild feeds and in the number of species in each guild. (*After Ashbourne and Putnam, 1987.*)

20.4 Keystone species are members of a community which have an effect out of proportion to their commonness.

Within communities, food webs, and even guilds, some species may have an effect out of all proportion to their commonness or biomass (Fig. 20.8a). This is in contrast to a pattern in which the importance of component species in a community is normally distributed (Fig. 20.8b) or in which each species is equally important (Fig. 20.8c). The removal

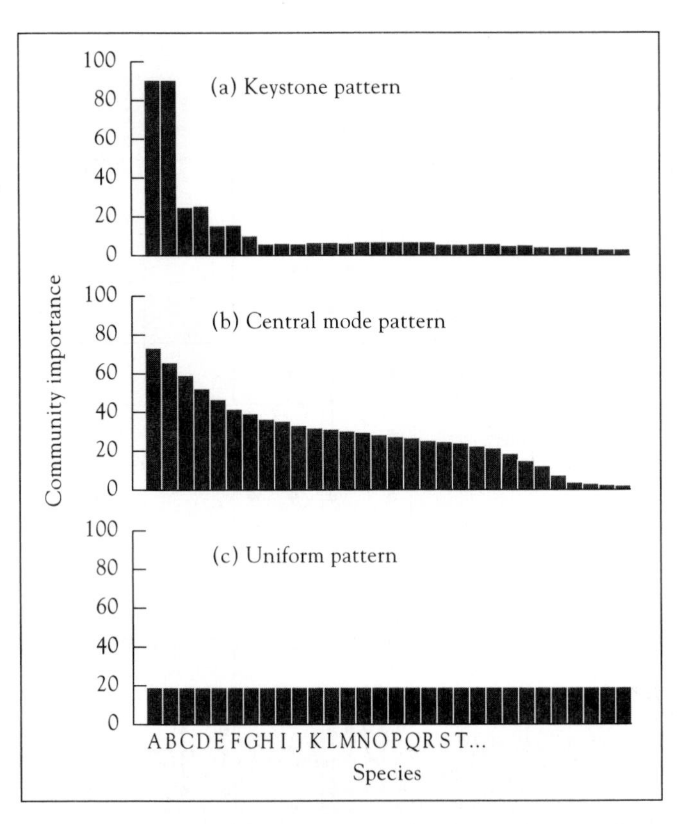

FIGURE 20.8 Community importance values (percentage of species lost from a community upon removal of a given species) for a hypothetical community based on the keystone-species model (**a**) and based on food-web theory (**b** and **c**). (*After Mills et al. 1993*).

of such species may have severe ramifications for a majority of community members. The term for such species, *keystone species*, has enjoyed an enduring popularity in the ecological literature (much like the term *guild*) since its introduction by Robert Paine in 1969. It is important at the outset to distinguish keystone species from dominant species, species that are so common in a community, like *Spartina* cordgrass in a salt marsh, that they are important purely because of their biomass and role in energy flow (Demaynadier and Hunter 1994). Other common dominant species include trees (T), prairie grass (G), corals (Cr), and giant kelp (K) (Fig. 20.9). A rare rhinovirus that makes a wildebeest sneeze, V_R, would not qualify as a keystone species because of its relatively low impact, but a rare distemper virus, V_O, that kills lions or wild dogs, would.

Several different types of keystone categories have been recognized: keystone predators, keystone prey, and keystone habitat modifiers (Mills, Soulé, and Doak 1993). Among the most famous keystone predators are *Pisaster* starfish (P) and predatory whelks, *Conchloepas*, (C) in the rocky intertidal (Paine 1966, 1969). Removal of these carnivores led to the nearly complete dominance of the substrate by one or two sessile species (mussels), which outcompeted the other consumer species, resulting in greatly reduced species diversity.

Navarrette and Menge (1996) further analyzed the role of the keystone species *Pisaster* on the northwest coast of the United States. The per-capita predation strength of seastars was two to three orders of magnitude greater than the next most important predator, the whelks, *Nucella*. The interesting thing about this study, however, was that in the absence of *Pisaster*, predation by *Nucella* was greatly elevated, indicating that after

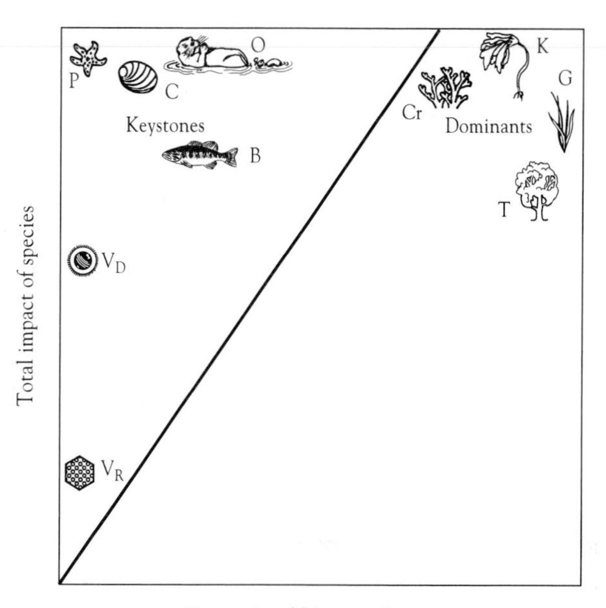

■ **FIGURE 20.9** Keystone species are those whose effects exceed their proportional abundances by some large factor and whose total effect exceeds some threshold. Dominants are species that dominate community biomass and whose ecosystem impacts are large, but not disproportionate to their biomass. Letters represent examples of particular species described in the text. (*After Power and Mills 1995.*)

■ **PHOTO 20.2** Sea otter, *Enhydra lutris*, have been labeled as keystone predators because they limit the densities of sea urchins, which, in turn, eat kelp and other macroalgae that form the basis of a different community in their absence. This one is biting down on a mussell shell in Prince Willian Sound, Alaska. (*Johnny Johnson, DRK Photo 265583.*)

the loss of a keystone species, species that might be considered "redundant" can partially compensate for the reduced predation and adopt a major role in the altered system.

Sea otters (Photo 20.2) have also been labeled as keystone predators (Estes and Duggins 1995; Menge, Daley, and Wheeler 1994) because they limit the densities of sea urchins, which, in turn, eat kelp and other macroalgae that form the basis of a different community in their absence (von Blaricom and Estes 1988). Mittlebach et al. (1995) noted a strong effect from removing the largemouth bass from a lake in Michigan. The elimination of the bass (B) via winter kills in 1976–1978 led to a dramatic increase in the density of planktivorous fish, the disappearance of large zooplankton, and the appearance of a suite of small-bodied zooplankton. Interestingly, following the reintroduction of the bass in 1986, the system returned to its previous state.

Terborgh (1986) considered palm nuts, figs, and nectar to be keystone prey/hosts because they are critical to tropical forest fruit-eating guilds including primates, rodents, and many birds. Together, these vertebrates account for as much as three-quarters of forest bird and animal biomass. The North American beaver (*Castor canadensis*) was argued to be a keystone species because its dams alter hydrology, biochemistry, and productivity on a wide scale (Naiman, Melillo, and Hobbie 1986). In the southeastern United States, gopher tortoises have been regarded as keystone species because their burrows provide homes for an array of mice, possums, frogs, snakes, and insects. Without tortoise burrows, many of these creatures would be unable to survive in the sandhill areas where they are found (Sunquist 1988). Both beavers and gopher tortoises have been called "ecosystem engineers" because of the ecological changes that result from the habitat modification they cause (Lawton and Jones 1995). Finally, African elephants may act as keystone species because of their browsing activity. Elephants destroy small trees and shrubs through their browsing and can change woodland habitats into grasslands. Ungulates that graze the grasses are favored by the elephant's activities.

Are keystone species present in most communities? It is too early to tell. Indeed, few studies have analyzed the community importance, or interaction strength, of species. In the rocky intertidal, Paine (1992) found that two of seven species of browser had strong effects on producer organisms, supporting the ideas of skewed interaction strengths. But this is a lone study, and clearly more data are needed. However, Paine (1988) suggests that humans, introduced pests, and diseases may commonly act as keystone species and that physical influences (which he calls critical processes) such as fire and disturbance are clearly keystone in their effects.

SUMMARY

1. An ecosystem includes not only a community of organisms in an environment but also the flow of energy and nutrients through that community.

2. There are two major ways in which organisms derive energy: autotrophs derive energy from the sun and are usually plants, and heterotrophs eat living or dead matter originally made by autotrophs.

3. Every step in the food chain is termed a trophic level, and herbivores and carnivores feed at different trophic levels. Omnivores feed on at least two trophic levels. Most food chains are aggregated into food webs.

4. Most food webs are imperfectly known because they are so vast. At least three different types of web are recognized: source webs, sink webs, and community webs.

5. Over a dozen generalizations about the properties of food webs have been made in the literature. Most of these are based on the numbers of species feeding at each trophic level and the number of connecting links between them.

6. There are numerous problems associated with food web theory based on the imperfect knowledge of food webs. Many of these focus on the lack of knowledge about the strength of connecting links between species and the taxonomic lumping of species into certain groups or guilds.

7. Some ecologists split up trophic levels into functional units called guilds. For example, among herbivores the leaf-chewing guild may be recognized as may the stem-boring guild and the root-feeding guild.

However, the precise limits of guilds are not easy to define: where would one place a species that feeds on leaves and stems?

8. Keystone species are species that have an effect out of all proportion to their commonness or biomass. Examples include the starfish *Pisaster* on the northwest U.S. coast, sea otters, beavers, gopher tortoises, and elephants.

DISCUSSION QUESTION

Would there be any meaningful way to delimit food webs more precisely—such as including only prey items that make up more than 5 percent of the diet (by some measure)?

CHAPTER

21

Energy Flow

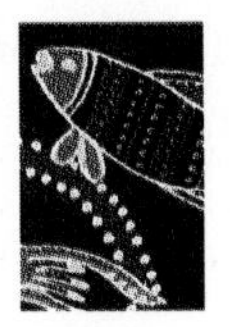

The organization of communities into trophic levels describes the fundamental pattern of energy flow from primary producers to primary consumers to secondary consumers and beyond. Viewing species and individuals as energy transformers in a community has a great strength, which is that the calorie is used as a lowest common denominator, and species and individuals can be reduced to caloric equivalents. In the search for common theories in communities, an ecosystems approach may provide great insights into how biological systems function by closely examining energy flow.

The first law of thermodynamics states that energy can be neither created nor destroyed. The second law states that, in every energy transformation, potential energy is reduced because heat energy is lost from the system in the process. Thus, as food passes from one organism to another, the potential energy contained in the system is dissipated as heat. Therefore, there is an almost unidirectional flow of energy through the system, with little possibility of recycling (Fig. 21.1). The process of energy transfer is inefficient, as a rule of thumb about 10 percent is transferred from one level to the next and 90 percent is lost—this may be why some food chains remain short. In contrast, chemicals are not dissipated and remain in the ecosystem indefinitely unless erosion occurs. Chemicals constantly circulate or recycle in the system, often becoming more concentrated in higher organisms (Fig. 21.2), leading to disastrous results in the case of chemical pollutants. This is especially true for the insecticide DDT (Applied Ecology: Food Webs and the Passage of Pesticides; see Chap. 20).

A relatively complete energy-flow diagram for a Georgia salt marsh is shown in Fig. 21.3. In the salt marsh, the **producers** are also the most important **consumers**; in other words, most primary production is used in plant respiration. The bacteria are next in importance; as **decomposers**, they degrade about one-seventh of the energy that plants use. As with many ecosystems, most **primary production** goes to the decomposers. Animal consumers are a poor third in importance, degrading only about one-seventh of the

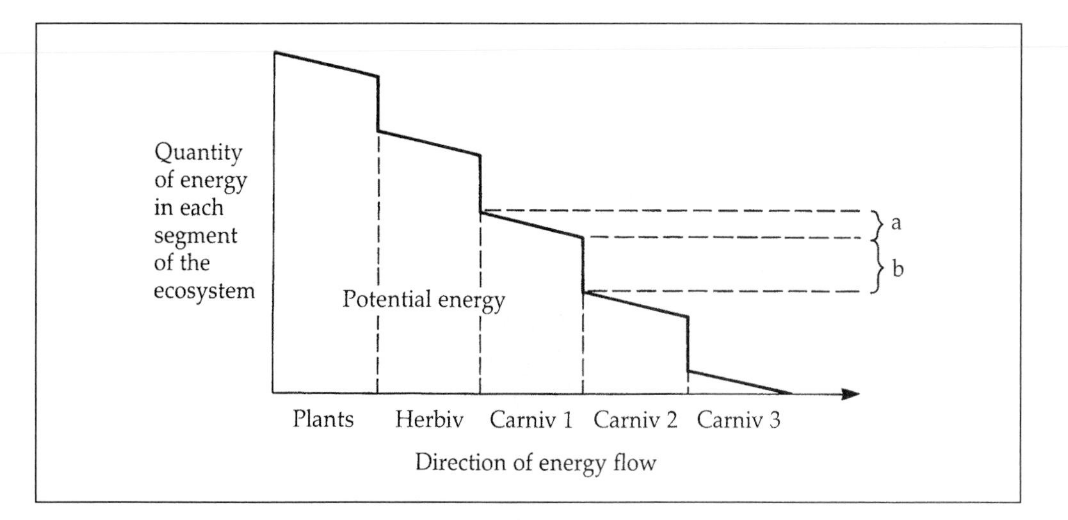

■ **FIGURE 21.1** Energy flow through the ecosystem. At each trophic level maintenance energy is lost as heat (a); energy is also lost as heat in each transformation from one trophic level to the next (b).

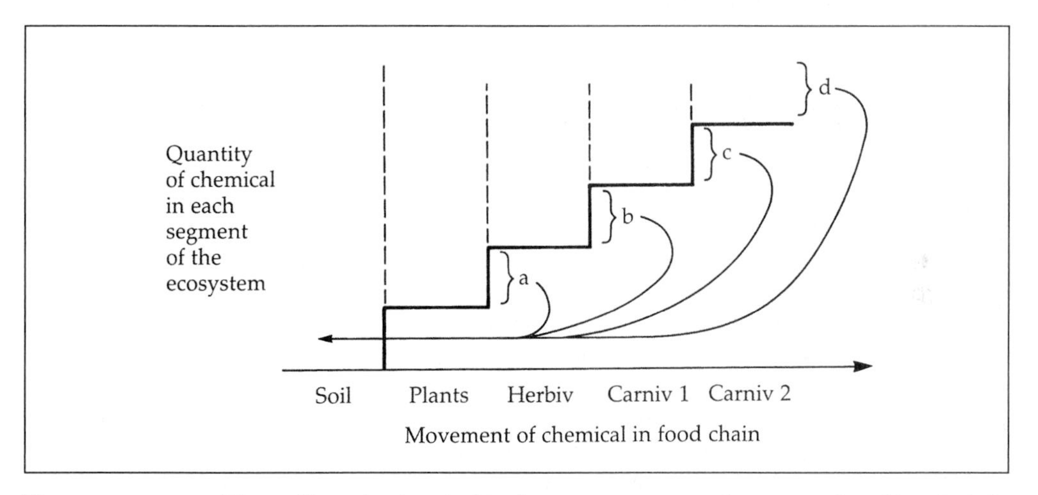

■ **FIGURE 21.2** The cycling of a chemical in the ecosystem assuming no erosion. Chemicals in organisms not consumed by the higher trophic level (a) to (d) slowly return to the soil as they are released through decay.

energy the bacteria use, though it must be stressed that, since this study was performed, many additional species of insects have been found to feed on *Spartina* in the U.S. Southeast (Rey 1981; McCoy and Rey 1987 and references therein), including leaf miners (Stiling, Brodbeck, and Strong 1982) and stem borers (Stiling and Strong 1983) not originally noticed in Teal's pioneer studies. The relative importance of these new herbivores, however, is probably less than that of the more abundant plant hoppers and grasshoppers that Teal (1962) recognized.

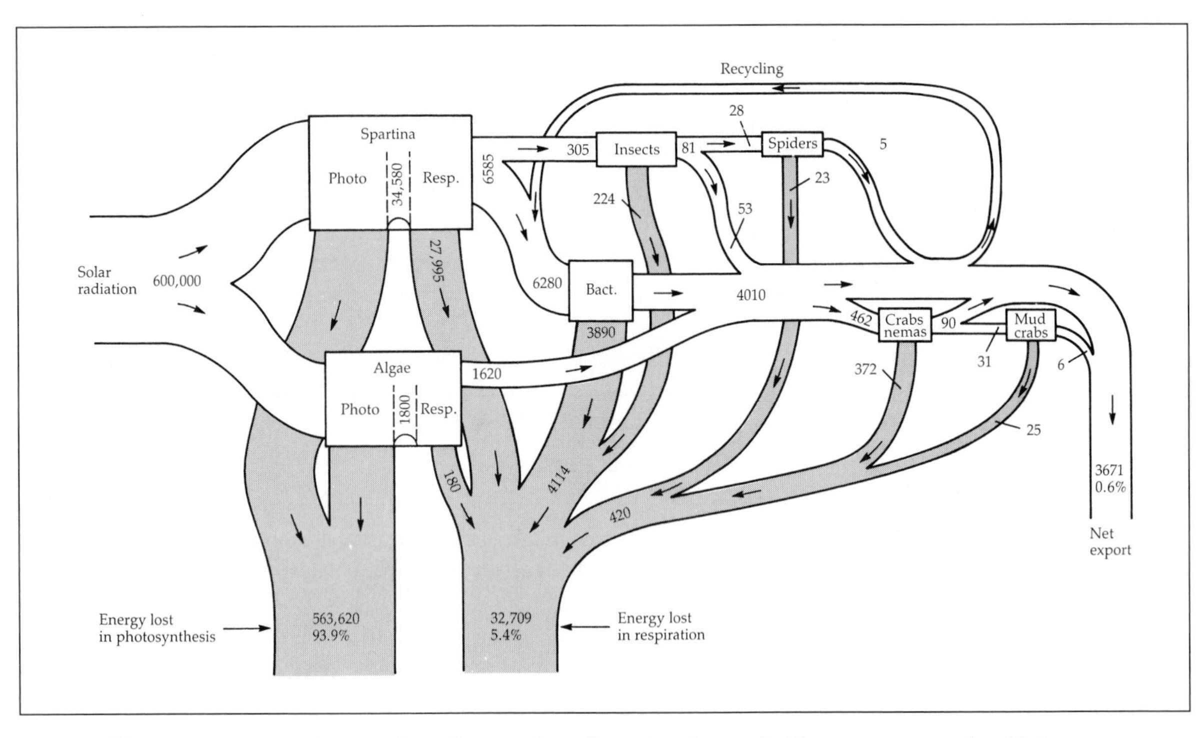

■ **FIGURE 21.3** Energy-flow diagram for a Georgia salt marsh. Nemas = nematodes. Units are kilocalories per square meter per year. (*Redrawn from Teal 1962.*)

21.1 Primary production, in which solar energy is converted into biomass by the process of photosynthesis, is the ultimate basis of all food chains.

The process of **photosynthesis** is the cornerstone of all life and the starting point for studies of community metabolism. The bulk of the Earth's living mantle is green plants (99.9% by weight) (Whittaker 1975); only a small fraction of life is animal. **Gross primary production** is equivalent to the energy fixed in photosynthesis, and **net primary production** is gross primary production minus energy lost by plant respiration.

The simplest method of measuring net primary production is the harvest method. The amount of plant material produced per unit of time, is equivalent to ΔB, the biomass change in the community between time 1 and time 2, $\Delta B_2 - \Delta B_1$. Two possible losses must be recognized: L = biomass lost by death of plants, and G = loss due to consumer organisms. Knowing these two elements one can estimate net primary production as follows:

$$\text{net primary production} = \Delta B + L + G$$

The energy equivalent of the production in biomass can then be obtained if one burns the biomass in a bomb calorimeter, but it must be remembered that this method cannot effectively gauge new root growth, which may be at least as much as aboveground productivity. In herbaceous communities, below-ground root biomass can be 40 percent of total biomass. Also, Wallace and O'Hop (1985) showed that in some cases where grazing by animals is high net productivity can be grossly underestimated when based solely on estimates of standing-crop biomass. In their study on beetle herbivory on water lilies, larval production of beetles alone—without adjustments for egestion, respiration, or adult feeding—surpassed plant biomass availability. Therefore, rapid macrophyte turnover was going on to support the beetle population.

A good estimate of global primary production is 110 to 120 $\times$ 10^9 metric tons dry weight yr^{-1} on land and 50 to 60 $\times$ 10^9 metric tons in the seas (Lieth 1975; Whittaker 1975). How does primary production vary over the different types of vegetation on the Earth? In general, primary production is highest in the tropical rain forest and decreases progressively toward the poles (McNaughton et al. 1991) (Table 21.1). Productivity of the open ocean is very low, approximately the same as that of the Arctic tundra, and is highest on coastal shelves, particularly in upwelling zones where the movement of water from the ocean bottom to the surface circulates nutrients. Because of the interest in biomass fuels as an alternative to fossil fuels, a more precise knowledge of where primary production is highest is important (Applied Ecology: Biomass Fuels).

TABLE 21.1 **Primary production and plant biomass for the Earth.** *(From Whittaker and Likens 1975, cited by Whittaker 1975)*

ECOSYSTEM TYPE	AREA (10^6 KM^2)	MEAN NET PRIMARY PRODUCTIVITY (G M^{-2} YR^{-1})	WORLD NET PRIMARY PRODUCTION (10^9 TONS YR^{-1})	MEAN BIOMASS (KG M^{-2})	WORLD BIOMASS (10^9 TONS)
Tropical rain forest	17.0	2,200	37.4	45	765
Tropical seasonal forest	7.5	1,600	12.0	35	260
Temperate evergreen forest	5.0	1,300	6.5	35	175
Temperate deciduous forest	7.0	1,200	8.4	30	210
Boreal forest	12.0	800	9.6	20	240
Woodlands and shrubland	8.5	700	6.0	6	50
Savanna	15.0	900	13.5	4	60
Temperate grassland	9.0	600	5.4	1.6	14
Tundra and alpine	8.0	140	1.1	0.6	5
Desert and semidesert scrub	18.0	90	1.6	0.7	13
Extreme desert, rock, sand, ice	24.0	3	0.07	0.02	0.5
Cultivated land	14.0	650	9.1	1	14
Swamp and marsh	2.0	2,000	4.0	15	30
Lake and stream	2.0	250	0.5	0.02	0.05
Total continental	149	773	115	12.3	1,837
Open ocean	332.0	125	41.5	0.003	1.0
Upwelling zones	0.4	500	0.2	0.02	0.008
Continental shelf	26.6	360	9.6	0.01	0.27
Algal beds and reefs	0.6	2,500	1.6	2	1.2
Estuaries	1.4	1,500	2.1	1	1.4
Total Marine	361	152	55.0	0.01	3.9
Grand Total	510	333	170	3.6	1,841

APPLIED ECOLOGY

Biomass Fuels

The energy content of plants can be recovered as "heat" energy in the combustion of biomass materials like wood, straw, plant matter, and other organic materials. Wood was the main source of energy for heating, transportation, and industrial processes in the United States until about 1850. It still is the main source of energy for many lesser-developed countries and features prominently as fuel for cooking. What are the prospects of harvesting solar energy by cropping plants and using these for fuel?

One important advantage is that photosynthesis does not require bright sunlight but continues at a slow and steady rate even if the intensity of sunlight is very low. This means that energy crops can be grown almost anywhere on the Earth's surface. There are, of course, some areas that would produce far greater yield than others (Fig. 1). Also, fuel crops

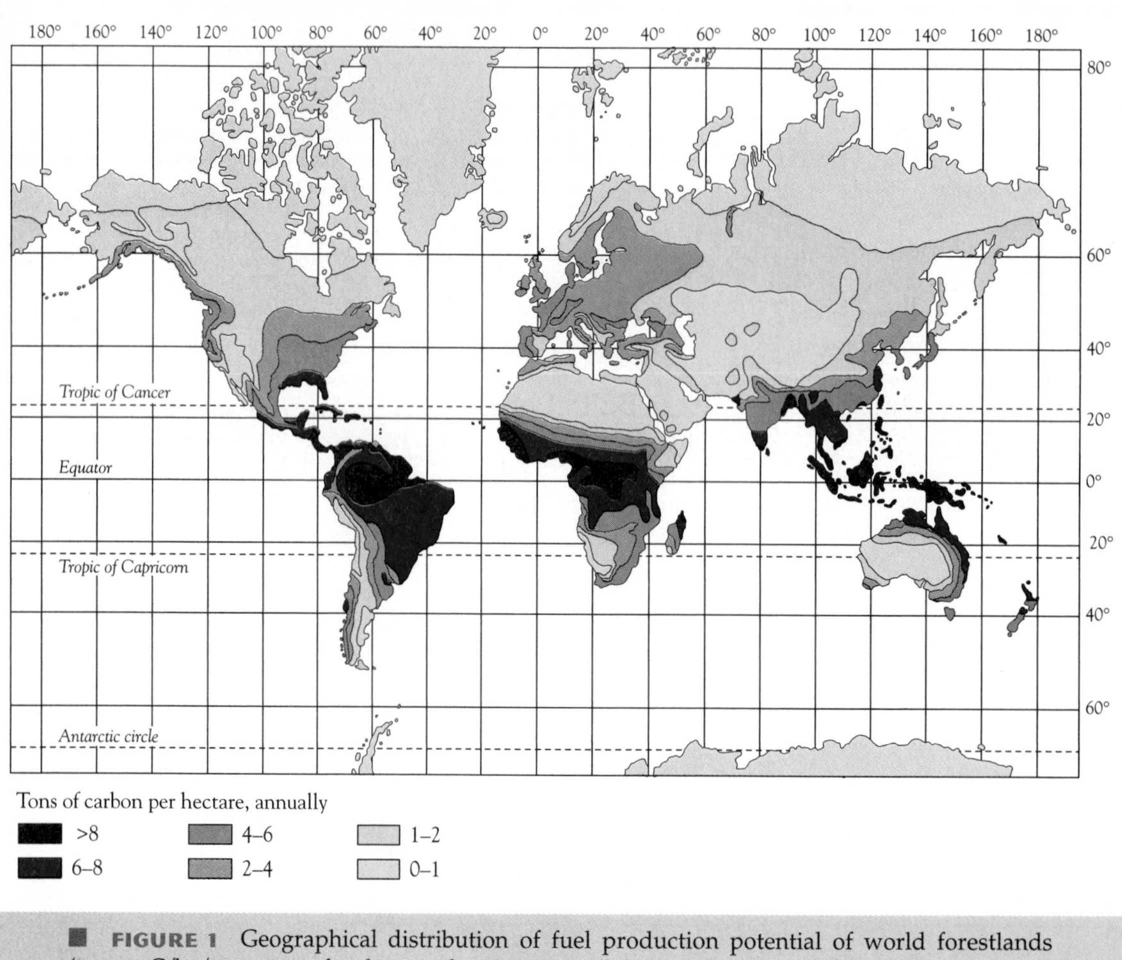

■ **FIGURE 1** Geographical distribution of fuel production potential of world forestlands (tonnes C/ha/an = tons of carbon per hectare, annually).

would be in severe competition with food crops in densely populated countries at high latitudes. So the potential for producing biomass is greatest in humid, densely vegetated, less densely populated tropical countries. The second advantage of biomass energy is that it can be converted into easily transported fuels. It can be used itself as a fuel, but plant material has several disadvantages as a fuel. First, most biomass is of low density compared with coal and oil, and because much of it is carbohydrate rather than hydrocarbon its energy density or calorific value is very much lower than coal or oil (crude oil $= 36 \times 10^9$ Joules per cubic meter [Jm^{-3}], coal $= 55 \times 10^9$ Jm^{-3}, compared with wood at 9×10^9 Jm^{-3}). Thus, large volumes of biomass will be required as fuel. Second, even assuming that biomass is dried before combustion—a process that itself may use energy—there will still be a large residual moisture content that will make it burn less efficiently than coal or oil. Third, because biomass breaks downs readily during storage (i.e., it rots), large volumes cannot be stored indefinitely—this is the main reason why biomass is converted into other fuels.

Any serious attempts to develop the potential of biomass energy should involve one of the following conversion processes:

1. Controlled burning to produce biogas, a mixture of methane and carbon dioxide.
2. Pyrolysis, in which organic molecules are strongly heated in the absence of oxygen yielding solid, liquid, and gaseous hydrocarbons. Solid fuels produced this way are known as "char" (as in charcoal), a carbon-rich substance that has a high calorific value.
3. Digestion by bacteria under anaerobic conditions (without oxygen), which again produces biogas.
4. Anaerobic glycolysis (fermentation), in which the organic polymers are broken down to produce liquid alcohols.

The main advantage of producing such concentrated fuels is that they can be stored indefinitely. The main potential of liquid fuels would be in transport. Since the late 1970s a large part of the Brazilian sugar crop has been used to manufacturer ethanol, which is added in a 1:3 ratio to gasoline for some cars and trucks. These developments require only relatively small modifications to engine design. The main practical disadvantages are that ethanol has a lower calorific value than petroleum (a little over half that of petroleum) and, at least in the early 1990s, was more expensive to produce. However, a main attraction of the alcohol fuels is that they are relatively clean burning as compared with conventional fuels, and, for this reason, much research has been carried out into their use in California, traditionally one of the worst smog areas of the world. Already one million pure ethanol-run cars are used in Brazil.

The efficiency of primary production can be calculated by comparing the energy produced by photosynthesis to the energy available in incident sunlight.

How efficient are the vegetation types of different communities as energy converters? One can determine the efficiency of utilization of sunlight from the ratio

$$\frac{\text{efficiency of gross}}{\text{primary production}} = \frac{\text{energy fixed by gross primary production}}{\text{energy in incident sunlight}}$$

Phytoplankton communities have very low efficiencies of usually less than 0.5 percent, herbaceous communities 1 to 2 percent, and crops generally less than 1.5 percent. The highest values occur in forests—2 to 3.5 percent (Cooper 1975), with conifers being more efficient than deciduous trees. Perhaps 50 to 70 percent of the energy fixed by photosynthesis is lost in **respiration**, so usually less than 1 percent of the sun's energy is actually converted into net primary production. In the temperate zone, this percentage is equivalent to 300 to 600 calories of primary production cm^{-2}.

Another useful way of looking at productivity is to determine productivity to biomass ratios. This is usually done as kilograms produced per year per kilogram of standing crop. Values of forests are traditionally very low, at around 0.04, probably because there is a huge volume of living support tissue, tree trunks, and branches, that is not photosynthetic. In grasslands P:B ratios increase to about 0.29 and in aquatic communities there is no need for support tissue or roots to absorb water, and P:B values may reach 17.0.

Primary productivity is limited by light, water, nutrients, and temperature.

What limits primary production? In terrestrial systems water is a major determinant, and production shows an almost linear increase with annual precipitation, at least in arid regions (Fig. 21.4). However, the temperature conditions are also important as there is a much greater range of temperatures over land than over water. Temperature affects production in two ways: directly by slowing or accelerating metabolic processes and indirectly by affecting evapotranspiration rates of water by plants. Raich, Russell, and Vitousek (1997) showed a direct, linear relationship between total annual net primary productivity (ANPP) and mean annual temperature on an elevational gradient in Hawaii (Fig. 21.5). For each 1°C increase in mean annual temperature, total ANPP increased by 54 $gm^{-2}\ yr^{-1}$. Decomposition rates of leaves also increased, releasing more nutrients under warmer conditions. Community-level ANPP was also related to rates of

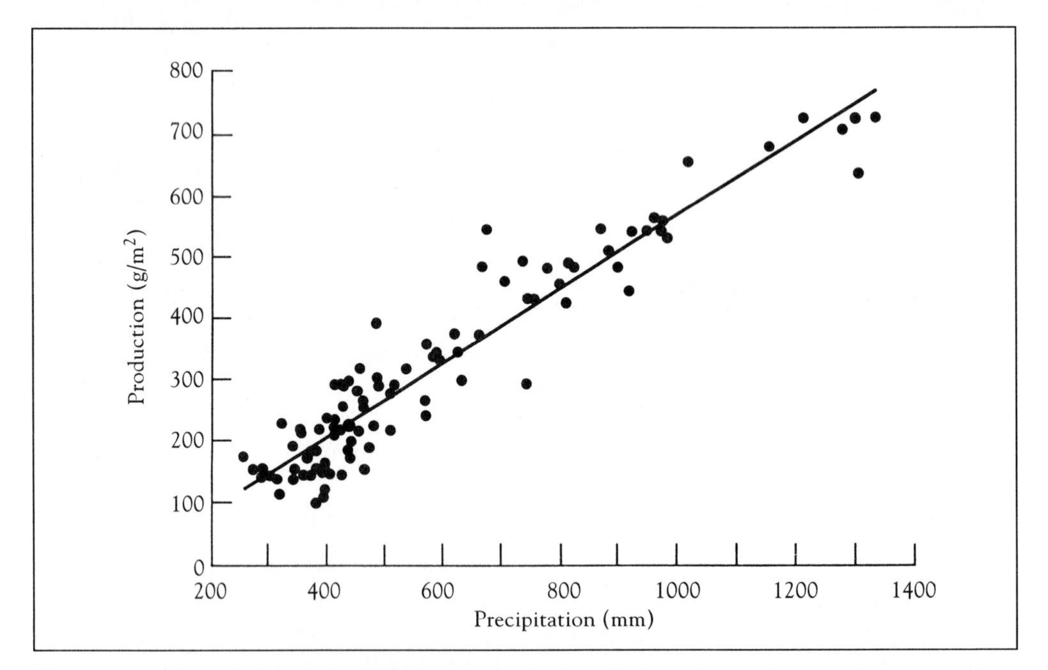

■ **FIGURE 21.4** Relationship between mean annual precipitation and mean aboveground net primary production for 100 major land resource areas across the central grassland region, Great Plains, of the United States. (*Redrawn from Sala et al. 1988.*)

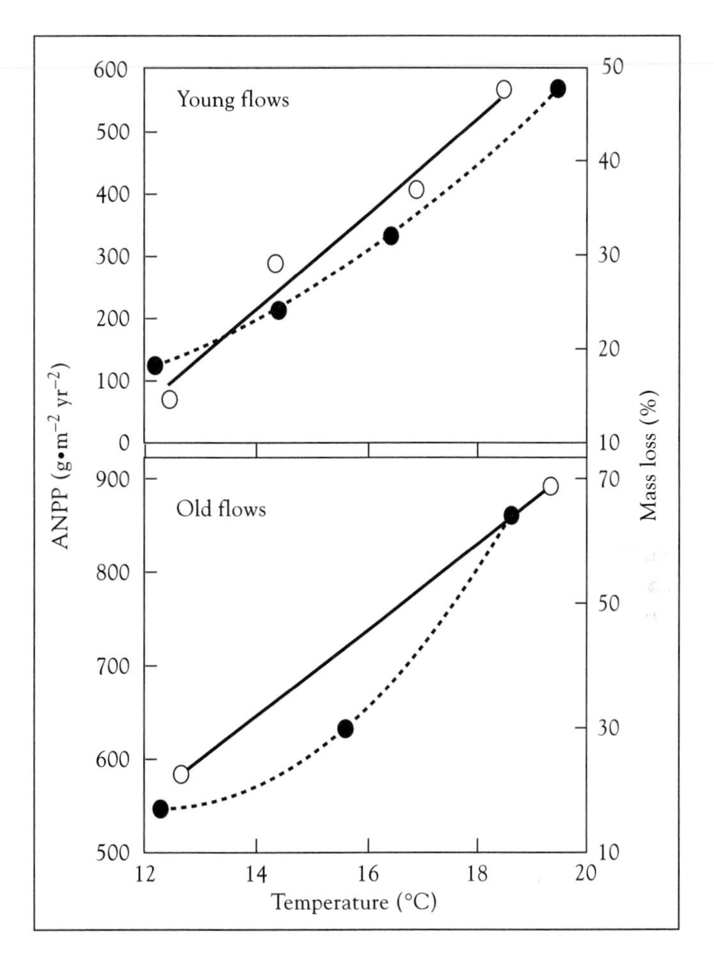

FIGURE 21.5 Aboveground net primary productivity (○) and decomposition rates (●) in relation to temperature on windward Mauna Loa, Hawaii. Mass loss refers to the percentage of, *Metrosideros* leaf mass lost during the first year of decomposition in situ. (*After Raich et al. 1997*).

nitrogen and phosphorus uptake by the vegetation. Rosenzweig (1968) noted that the combined effects of temperature and moisture, i.e., the actual evapotranspiration rate, could predict the aboveground production with good accuracy in North America.

Lieth (1975) showed that, in addition to evapotranspiration rate, length of growing season was well correlated with net primary production of forests, at least in North America (Fig. 21.6). This helps explain why tropical wet forests, with a very long growing season, are so productive and why conifers are so predominant in northern realms: they effectively extend the short growing season at high latitudes by retaining their leaves (needles) for long durations.

Nutrient deficiency, particularly of nitrogen and phosphorus, can limit primary productivity too, as agricultural practitioners know only too well. Fertilizers are commonly used to boost the productivity of annual crops. What is less commonly appreciated is that harvesting timber also removes large amounts of nutrients from the forest ecosystem. Foresters make good such losses by nutrient supplements. Rennie (1957) showed that pines removed fewer nutrients from soil than hardwoods. Some foresters have provided good evidence to show that fertilization increases timber yields. Among the most impressive studies are those involving fertilization of pines (*Pinus radiata*) with phosphate after a

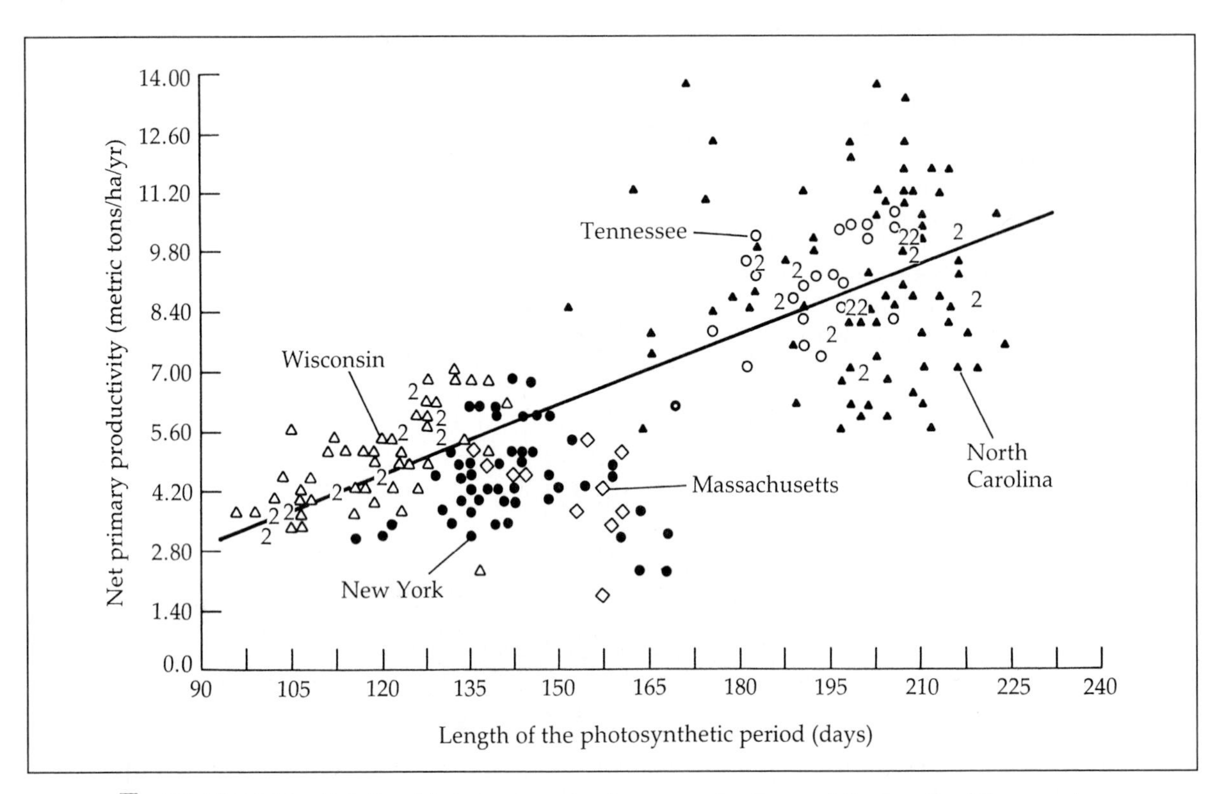

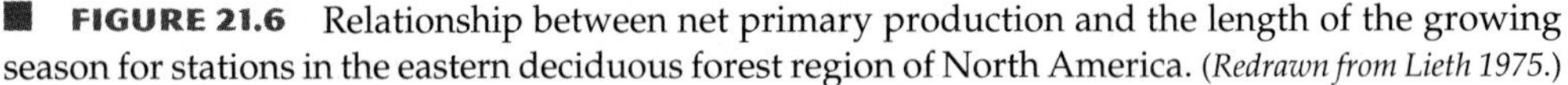
■ **FIGURE 21.6** Relationship between net primary production and the length of the growing season for stations in the eastern deciduous forest region of North America. *(Redrawn from Lieth 1975.)*

fire. Fertilized trees grew to more than twice the size of unfertilized ones after fifteen years (Gentle, Humphreys, and Lambert 1965). The basal area of pines on unfertilized plots was only 12 m^2/ha, compared to 27 m^2/ha in fertilized areas. Similar results are obtained in natural ecosystems, for example, fertilized *Metrosideros polymorpha* trees on Hawaii showed an annual rate of growth up to twice that of unfertilized controls (Gerrish, Mueller-Dombois, and Bridges 1988). In Britain, unfertilized grassland swards yield about 2.5 tons dry matter ha^{-1} yr^{-1}, and grass/legume swards about 6 tons ha^{-1} yr^{-1}. Adding 400 kg ha^{-1} nitrogen increases the yield of both types to about 10 tons ha^{-1} yr^{-1}.

If nutrient availability is so critical to plants, there should be an obvious relationship between soil nutrient content and plant production. Surprisingly, such relationships have not commonly been found (Gessel 1962), which tends to contradict results from fertilization studies. The probable explanation is that in many cases measured nutrients are locked up in a form unavailable to plants. An analogous problem exists in estimating what fraction of plant biomass is available to animals. In many cases, nitrogen availability is critical to animals but measures of total percentage nitrogen in plant tissues may not give a good indication of what is available to herbivores because much nitrogen is locked up in indigestible forms (Bernays 1983) or is unavailable because of the presence of digestibility-reducing substances (Mattson 1980; Wint 1983; Brodbeck and Strong 1987; see also Chap. 11).

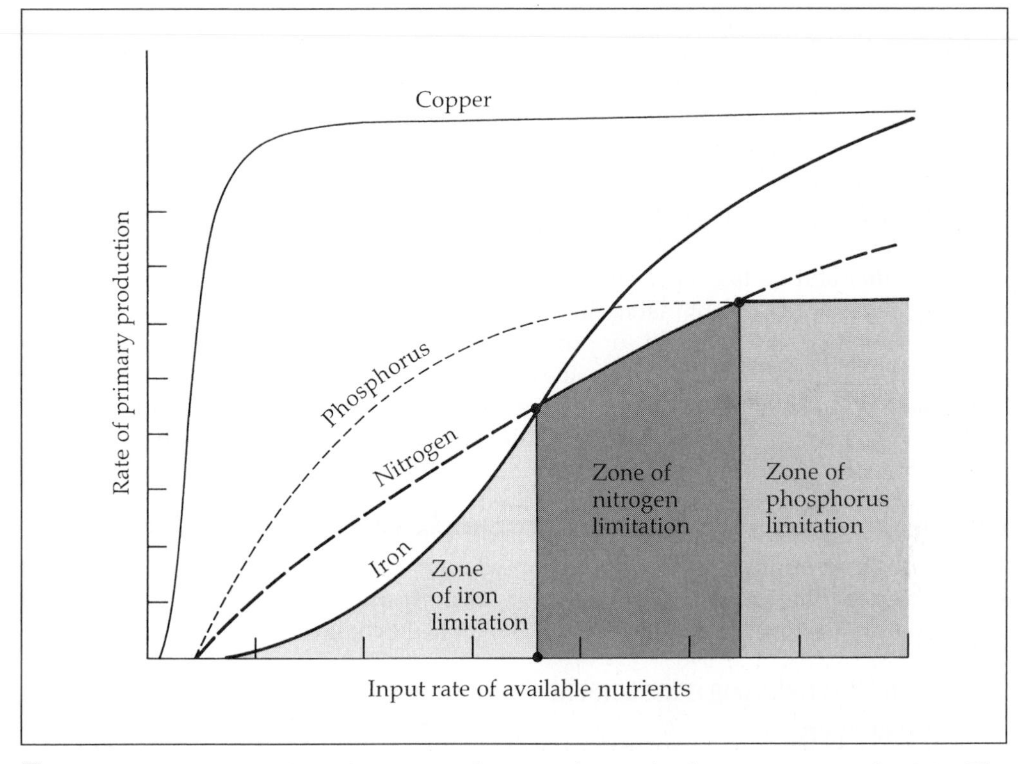

FIGURE 21.7 Hypothetical sequence of nutrient factors that limit primary productivity. The rate of primary production follows the line bordering the shaded areas and is limited first by iron, then (as more iron becomes available) by nitrogen, and finally by phosphorus. Some nutrients, such as copper, may always be present in superabundant amounts. (*Redrawn from Krebs 1985*a.)

In many systems, the productivity of an ecosystem will be limited by a succession of these factors. For instance, in a Kansas prairie, productivity would be limited by cold temperature in winter; by lack of water in summer; by herbivory, which reduces light gathering leaves; and perhaps by nitrogen mobilization at all times. Thus, if the level of one particular factor is raised, then another factor becomes limiting. This type of sequential progression of limiting factors is illustrated in Fig. 21.7. Such a phenomenon was demonstrated in nature by Menzel and Ryther (1961), who studied primary productivity of the Sargasso Sea. They found that iron levels were the most obvious limiting factor, but that nitrogen and phosphorus were likely to be limiting in the presence of sufficient iron. Increasing iron levels in the world's oceans may permit increased phytoplankton production, which will increase carbon dioxide uptake from the atmosphere and possibly reduce global warming.

Primary productivity in aquatic systems is limited primarily by light and nutrient availability.

Of the factors limiting primary production in aquatic ecosystems, among the most important are available light and available nutrients. Light is particularly likely to be in short supply because water absorbs light very readily. Even in "clear" water, only about 5 to 10 percent of the radiation may be present at depths of only 20 m. Too much light can also inhibit the growth of green plants by overheating them. Such a phenomenon can be found in tropical and subtropical surface waters throughout the year, where maximum primary production occurs several meters beneath the surface of the sea.

The most important nutrients affecting primary productivity in aquatic systems again are nitrogen and phosphorus. Important only locally in terrestrial systems, both often limit production in the oceans, where they occur in low concentrations. Few nutrients are tied up in the standing crop, in contrast to terrestrial systems, especially forests, where large amounts of nutrients occur in plants themselves. Rich, fertile soil contains about 5 percent organic matter and up to 0.5 percent nitrogen. One square meter of soil surface can support 1 kg dry weight of plant matter. In the ocean, the richest water only contains 0.00005 percent nitrogen, and 1 m^2 could support no more than 5 g dry weight of phytoplankton (Ryther 1963). Enrichment of the sea by the addition of nitrogen and phosphorus can result in substantial algal blooms. Such enrichment occurs naturally in areas of upwellings, such as the Antarctic or the coasts of Peru and California, where cold, nutrient-rich, deep water is brought to the surface by strong currents, resulting in very productive ecosystems.

Phosphorus is particularly important in limiting productivity in freshwater lakes. Some lakes in North America become polluted by runoff from the land of rainwater enriched with phosphorus from fertilizer application or from sewage. This new input into the lakes causes huge blooms of blue-green algae, which clog the lake, and increased turbidity, a process termed **eutrophication** (Schindler 1974, 1977) (Applied Ecology: Eutrophication).

21.2 Secondary production is the amount of new biomass produced by consumers.

The biomass of plants that accumulates in a community as a result of photosynthesis can eventually go in one of two directions: to herbivores or to detritus feeders. Most of the biomass dies in place and is available to detritivores. This is known as dead organic matter (DOM). Heal and Maclean (1975) calculated that detritivores carry out over 80 percent of the consumption of matter, often "working it over" on a number of occasions to extract the most energy from it. However, it is the herbivores that feed on the living plant biomass and, as a result, constitute the greatest selective pressure on living plants. Energy flow in such species will therefore be considered in more detail.

The efficiency of secondary production is estimated by comparing new tissue production to consumption or assimilation.

In viewing an herbivore as an energy transformer, three main types of efficiency can be measured in the animal world and used to compare species from different ecosystems:

$$\text{growth efficiency or production efficiency} = \frac{\text{net productivity}}{\text{assimilation}} \times 100$$

Growth efficiency in vertebrates is generally less than that in invertebrates because much more energy is used in respiration. Even within vertebrates, much variation occurs. In endotherms, over 98 percent of assimilation energy may be used in respiration, whereas in ecotherms, this figure is less about 90 percent, reflecting the energy cost of homeothermy. Thus, growth efficiencies of around 10 percent are common for ectotherms and 1 to 2 percent for endotherms. Invertebrates generally have higher efficiencies, of about 30 to 40 percent, losing much less energy as heat. Microorganisms also have very high production efficiencies. One consequence of these differences is that sparsely vegetated deserts can support healthy populations of snakes and lizards where mammals might easily starve. The large monitor lizard known as the Komodo dragon eats the equivalent of about a pig a month, its own weight every two months, whereas a cheetah consumes something like four times its own weight in the same period (Kruuk and Turner 1967). It is also interesting to note that growth efficiencies are higher in young animals than in old (35%, as opposed to 3 to 5%)—hence the practice of raising broiler chickens and calves for meat. Further, smaller species often have higher growth rates than larger species. For a given amount of hay, rabbits produce the same quantity of meat as beef cattle but do so four times as quickly.

Efficiency between trophic levels can be measured by two other indices:

Tropical level transfer efficiency or Lindeman's efficiency

$$= \frac{\text{assimilation at trophic level } n}{\text{assimilation at trophic level } n - 1} \times 100$$

This measure of **assimilation efficiency** was named after ecologist R. J. Lindeman (1942) in recognition of his classic work on ecological energetics. Lindeman's efficiency appears to average around 10 percent though there is much variation (Fig. 21.8), and some data on marine food chains shows that it can exceed 30 percent (Steele 1974).

$$\text{Consumption efficiency} = \frac{\text{intake at trophic level } n}{\text{net productivity at trophic level } n - 1} \times 100$$

Consumption efficiency measures the relative efficiency of one trophic level in converting the energy from the one beneath. Values seem generally to fall in the range of 0 to 15 percent for terrestrial herbivores (Table 21.2), meaning that 85 to 100 percent of the net terrestrial plant production goes into the decomposer chain. These low values arise because most plants are chemically well defended or herbivore densities are maintained at low levels by their natural enemies. In aquatic systems, zooplankton may be more efficient grazers, and

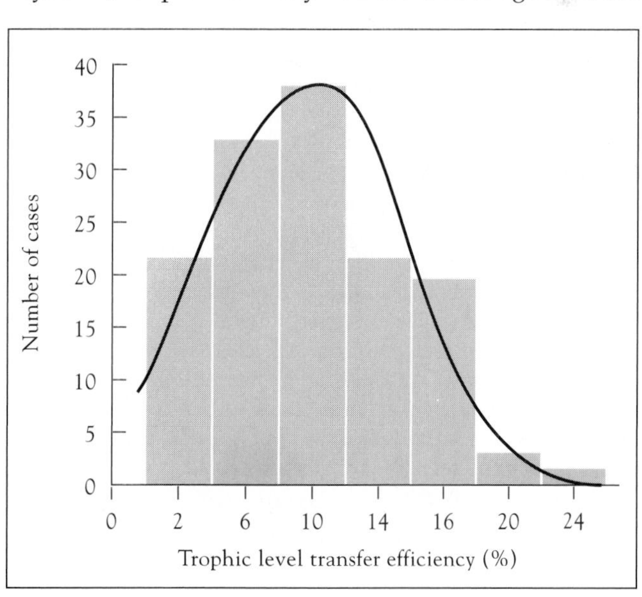

FIGURE 21.8
Frequency distribution of trophic level transfer efficiencies in 48 trophic studies of aquatic communities. There is considerable variation among studies and amongst trophic levels. The means is 10.13% (standard error = 0.49). (*After Pauly and Christensen 1995.*)

Eutrophication

APPLIED ECOLOGY

Eutrophication is the enrichment of waters with nutrients, primarily phosphorus and nitrogen. This usually leads to enhanced plant growth. These changes may occur as a result of anthropogenic changes (cultural eutrophication) or as a result of succession, the natural aging of a lake (natural eutrophication). As outlined in Chapter 14, bodies of water that are not rich in nutrients are called oligotrophic, and those rich in nutrients are called eutrophic. The most common attributes of a eutrophic lake are blooms of algae, make the water more turbid, more unattractive to swimmers, and less suitable to certain kinds of fish.

What is the state of eutrophication of lakes around the world? The data indicate that there is great disparity between countries and regions. Of the three-quarters of a million lakes in Canada, the great majority, 75 percent, are still oligotrophic, although the lakes of the southern, more densely populated regions are predominantly eutrophic, which reflects the increase in eutrophication associated with anthropogenic effects. Among the smaller U.S. lakes there are many that are surrounded by farmland. Their condition, therefore, reflects eutrophication from agriculture and animal husbandry. The result is that up to 70 percent of U.S. lakes may be eutrophic. In Europe, too, many lakes seem to be eutrophic.

However, the great majority of lakes and rivers in the eutrophic category are relatively small in both surface area and volume, averaging only 2.2 km^3 in volume compared to 67.6 km^3 for oligotrophic lakes. Smaller lakes are perhaps more susceptible to cultural eutrophication than bigger lakes. Since there are lots of small lakes and few big ones, the percentage of lakes that are eutrophic is high. However, the volume of fresh water in a eutrophic condition is small, only 12 percent of the volume compared to 52 percent that is classified as oligotrophic. The other 36 percent of lake water is in intermediate condition. It is clearly difficult to reach a consensus on the state of eutrophication of U.S. lakes. On the other hand, many of the smaller water bodies are important either for drinking water supply or for recreational purposes and should be cleaned up. There is clearly room for improvement.

The data for rivers are less definitive, but eutrophication effects are generally less acute than for standing waters because nutrient inputs are often quickly washed away. On the other hand, the data for human-constructed reservoirs show much higher rates of eutrophication than for natural lakes, although the volume of these waters is small in comparison to the area and volume of natural lakes.

How can we control cultural eutrophication? First, we have to know what the anthropogenic causes are. The degree of eutrophication for the Great Lakes shows a striking resemblance to the maps of human population density along the shores. Similarly, the areas most affected by eutrophication in the Mediterranean coincide with densely populated lands, most of which are either areas of intensive agriculture or high industrial development. High population densities lead to cultural eutrophication via three paths:

1. A strong tendency for urban waste to increase and be discharged directly into waterways. An added factor to this, since the 1940s, is the use of detergents containing polyphosphates.
2. Rapid industrialization with a corresponding increase in industrial wastes of all kinds.
3. Intensification of agriculture and the increased use of chemical fertilizers, especially those containing phosphorus; the concentration of livestock breeding; and the direct discharge of agricultural wastes, rich in nitrogen, into waterways.

The measures taken to control eutrophication fall into two main headings: preventive measures and corrective measures. Preventive measures include the following:

1. Treatment of waste waters (removal of phosphorus and nitrogen).
2. Diversion of wastewaters from lakes or rivers.
3. Primary sedimentation basins in wastewater streams to let phosphorus and nitrogen-rich materials settle to the bottom where it can be removed.
4. Watershed protection (reforestation, restriction of livestock, controlled fertilization/irrigation).
5. Substitution of phosphate detergents by other detergents not rich in phosphorus.

Corrective measures to try and bring eutrophic lakes back to an oligotrophic condition include the following:

1. Physical manipulation (withdrawal of water; aeration to increase levels of oxygen).
2. Chemical manipulation (application of herbicides to kill blooms of algae or large plants [macrophytes]).
3. Biological manipulation (mechanical harvesting of algae, macrophytes; direct manipulation of the food chain by adding exotic fish).

The treatment of waste waters is already underway in many areas. The reduction of polyphosphates in detergent has been imposed by law in Canada and some U.S. states bordering on the Great Lakes. In Sweden, 80 percent of treatment plants include a third stage for the elimination of phosphorus, and only 20 percent of waters discharged into the waterways receive no treatment. Strategies for control of nitrogen outputs, especially from agricultural activities, are in less well-developed stages, but there are some dramatic success stories in lake restoration.

TABLE 21.2 Consumption efficiencies of various herbivore groups for their plant food. (*After Brylinski and Mann 1973; Pimentel, 1975*)

HOST PLANT	TAXON OF FEEDING ANIMAL	PERCENTAGE OF PRODUCTIVITY CONSUMED
Beech trees	Invertebrates	8.0
Oak trees	Invertebrates	10.6
Maple-beech trees	Invertebrates	5.9-6.6
Tulip poplar trees	Invertebrates	5.6
Grass + forbs	Invertebrates	< 0.5-20
Alfalfa	Invertebrates	2.5
Grass	Invertebrates	9.6
Aquatic plants	Bivalves	11.0
Aquatic plants	Herbivorous animals	18.9
Algae	Zooplankton	25.0
Phytoplankton	Zooplankton	40.0
Phytoplankton	Herbivorous animals	21.2
Marsh grass	Invertebrates	4.6
Meadow plants	Invertebrates	14.0
Sedge grass	Invertebrates	8.0
Forests	All	5.0
Grasslands	All	25.0
Phytoplankton	All	50.0

consumption efficiency values of 10 to 40 percent have been reported (possibly because of the relatively low levels of chemical deterrents in phytoplankton and their higher digestibility because of a lack of complex structural cells; see Chap. 13). For carnivores, however, consumption efficiencies may reach 50 to 100 percent, showing that animals are much better equipped to handle meat than they are dead material or chemically protected plants.

Whittaker (1975) concluded that in the general scheme of things birds and mammals, the animals most keenly observed by humans, contribute almost nothing to secondary production and very little to consumption. Insects are a little more important, but soil animals, including nematodes, remain the most important. Thus, ecosystems ecology, unlike population ecology, does not stress the importance of higher trophic levels because they are relatively unimportant in energy flow.

Secondary production is limited by the availability and quality of primary production.

What controls secondary production is a complex question, but it is generally thought to be limited to a high degree by available primary production. McNaughton et al. (1989) and Moen and Oksanen (1991) have documented a tight correlation between primary productivity in a variety of ecosystems and the biomass of and consumption by herbivores (Fig. 21.9). More recent data from Cyr and Pace (1993) have confirmed this relationship.

The existence of this correlation is not as obvious as one might think because it

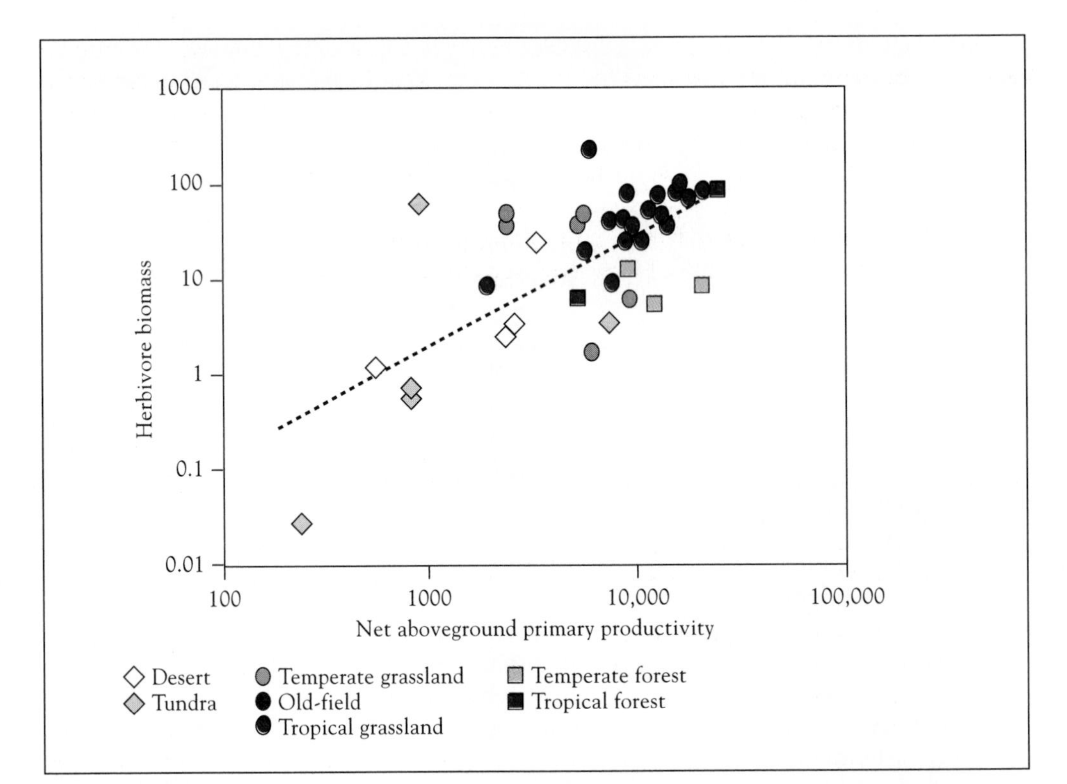

FIGURE 21.9 The relationship between herbivore biomass and net aboveground primary productivity. (*After Moen and Oksanen 1991*).

means that secondary chemicals that make plants poisonous or distasteful to herbivores are much less important influences on consumption at the ecosystem level than they are at the plant and population levels (see Chap. 11). When host quality and quantity are increased experimentally by an input of fertilizer, herbivore biomass often increases too. After reviewing eighteen laboratory and field nutrition-interaction studies, Onuf (1978) concluded that a general correlation could be found between nitrogen levels and the susceptibility of plants to insect attack. Leaf-feeding insects removed four times as much foliage from mangrove, *Rhizophora mangle*, on a high-fertility site enriched by the droppings from a large colony of egrets and pelicans as from a nearby low-fertility site (Onuf, Teal, and Valiela 1977). Nitrogen concentration in the phloem sap of grasses increases under fertilization, and sucking species like aphids and plant hoppers become much more successful and prolific (Prestidge and McNeill 1983). Vince, Valiela, and Teal (1981) found a dramatic increase in insect biomass on fertilized salt marshes of the sort where Teal (1962) had done his pioneering energy-flow studies (see also review by Waring and Cobb 1992 and Chap. 11). But fertilization does not always increase the growth rates of herbivores. Stark (1964), in a review of fifteen tree-insect interactions involving nitrogen fertilizers, showed that insect survival was reduced. Auerbach and Strong (1981) could find no increase in growth rates or nitrogen-accumulation rates of two species of tropical hispine beetles feeding on *Heliconia* spp. in Costa Rica. In such cases, natural enemies may also be involved in determining the numbers of herbivores in the field, but this type of argument belongs more in the realm of population biology than in ecosystems.

Finally, Odum (1969) suggested that the energy relations of communities can also be used as a measure of their maturity (Table 21.3). Gross and net primary productivity are very low in mature communities, high in developing ones. Food chains are shorter in the developing communities; those in mature communities are more complex. Odum viewed the ratio of production to biomass (P/B) as the single most important measure of maturity in an ecosystem; P/B ratios become lower as maturity increases. Of much interest to applied ecologists is that high P/B ratios mean high crop yields, so agricultural practices should be aimed specifically at exploiting immature ecosystems rather than mature ones.

SUMMARY

1. In ecosystems, energy is lost with each transfer up the food chain. In contrast, chemicals are not dissipated; they remain in the ecosystem and often concentrate at higher trophic levels (bioaccumulation).

2. In most ecosystems, plant material goes not to herbivores but to decomposers after the plant dies. Most primary production is used in plant respiration, so plants are the most important consumers. Bacteria as decomposers are next in importance and degrade about one-seventh of the energy that plants use. Herbivores degrade only about one-seventh of the energy the bacteria use.

3. In general, primary production is highest in tropical forests and decreases progressively toward the poles. There are exceptions to this trend. Temperate salt marshes are as productive as many tropical areas. Tropical deserts, because of a lack of moisture, are not as productive as temperate grasslands. Thus, temperature and rainfall both limit primary productivity.

4. Nutrient deficiency, particularly of nitrogen and phosphorus, can limit primary productivity too. In aquatic systems, light availability can also be important. In freshwater lakes, excess phosphorus can cause huge algal blooms and turbid water, a process known as eutrophication.

5. The limit to secondary productivity is available primary productivity. This means that, at a large scale, plant defenses do not effectively reduce consumption by herbivores.

TABLE 21.3 Model of ecosystem development: general trends in 18 variables during ecological succession. *(From Odum 1969 and 1997)*

ECOSYSTEM ATTRIBUTES	DEVELOPMENTAL STAGES	MATURE STAGES
Community Energetics		
1. Gross production/community respiration (P/R ratio)	Greater than 1	Approaches 1
2. Gross production/standing crop biomass (P/B ratio)	High	Low
3. Biomass supported/unit energy flow (B/E ratio)	Low	High
4. Net community production (yield)	High	Low
5. Food chains	Linear	Weblike
Community Structure		
6. Total biomass (B)	Low	High
7. Growth form	Rapid growth (r selection)	Feedback control (K selection)
8. Stratification and spatial heterogeneity (pattern diversity)	Poorly organized	Well organized
Life History		
9. Niche specialization	Broad	Narrow
10. Size of organism	Small	Large
11. Life cycles	Short, simple	Long, complex
Nutrient Cycling		
12. Mineral cycles	Open	Closed
13. Nutrient exchange rate, between organisms and environment	Rapid	Slow
14. Role of detritus in nutrient regeneration	Unimportant	Important
Overall Homeostasis		
15. Symbiosis	Less mutualistic	More mutualistic
16. Nutrient conservation	Poor	Good
17. Stability (resistance to external perturbations)	Poor	Good
18. Entropy	High	Low

DISCUSSION QUESTION

What advantages and disadvantages does the study of whole ecosystems, through the analysis of nutrient or energy flows, have over population or community ecology, which emphasizes biotic interactions?

C H A P T E R

22

Nutrient Cycles

Nutrients such as nitrogen and phosphorus often limit primary or secondary production. For example, McNaughton (1988) reported that the mineral content of foods is an important determinant of the spatial distribution of animals within the Serengeti National Park of Tanzania. Areas of grassland containing higher concentrations of magnesium, sodium, and phosphorus support higher densities of large herbivores than areas of low concentrations of these minerals. It is often argued that ecosystems can best be understood not from the path of energy through them but by the paths of nutrients, the **biogeochemical cycles**. Because chemicals are not dissipated, but remain in the ecosystem indefinitely, they tend to accumulate in individuals or species, which then act as "pools" of nutrients. The rate of nutrient movement between pools is called the **flux rate** and is measured as the quantity of nutrient passing from one pool to another per unit of time. Nutrients cycle between pools through meteorological, geological, or biological transport mechanisms. Meteorological inputs include dissolved matter in rain and snow, atmospheric gasses, and dust blown by the wind. This is how sulfur alters the pH of lakes. It is emitted through smokestacks and becomes incorporated in the rain, which then falls as "acid rain." Geological inputs include elements transported by surface and subsurface drainage, and biological inputs result from the movements of animals, or animal parts, between ecosystems.

Nutrient cycles have often been studied by introducing radioactive tracers into ecosystems. For example, Whittaker (1961) followed the introduction of ^{32}P-labeled phosphoric acid into an aquarium. He found a definite sequence of nutrient uptake (Fig. 22.1):

1. ^{32}P was rapidly taken up by phytoplankton and subsequently discharged.
2. Filamentous algae on the sides and bottom slowly picked up ^{32}P.
3. Crustaceans grazing on algae picked up ^{32}P even more slowly.
4. ^{32}P began to accumulate in bottom sediment and was tied up in less active forms.

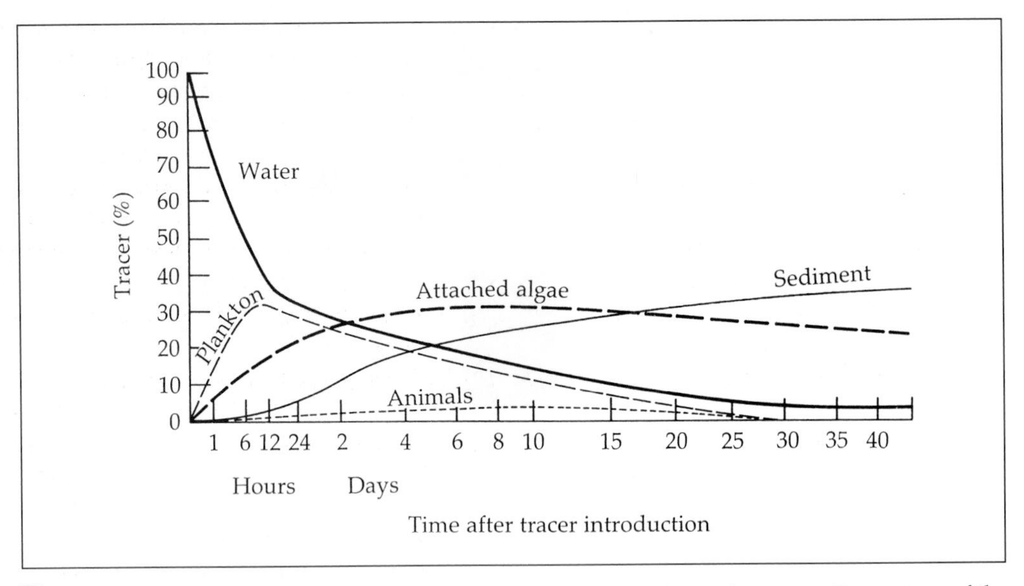

FIGURE 22.1 Movement of radiophosphorus in an aquarium microcosm. Percentage of the tracer present at a given time (after correction for radioactive decay) is on the vertical axis; time after tracer introduction (on a square-root scale) is on the horizontal. *(Redrawn from Whittaker 1961.)*

This type of cycle is broadly similar to that occurring in natural lakes; phosphorus and other nutrients accumulate in the bottom sediment, largely unavailable for use, which is why phosphorus is so often limiting in lake ecosystems.

Nutrient cycles can be divided into two broad types:

Local cycles, such as the phosphorus cycle just described, which involve elements with no mechanism for long-distance transfer.

Global cycles, which involve an interchange between the atmosphere and the ecosystem and are particularly applicable to elements such as nitrogen, carbon, oxygen, and water. Global nutrient cycles unite all the world's living organisms into one giant ecosystem called the **biosphere**, the whole Earth system.

22.1 Nutrient availability in ecosystems is determined by processes of nutrient cycling among biological and geological pools.

The turnover times of nutrients in ecosystems seem to get longer with increasing latitude because colder temperatures limit nutrient transfer rates. Jordan and Kline (1972) quote 10.5 years for the cycling time of nutrients in a tropical rain forest and 42.7 years for the taiga in the former Soviet Union. Turnover times are generally long because in almost every case the inputs and outputs of nutrients are small in comparison with the amounts held in biomass and recycled within the system. This was illustrated by a study of the annual nutrient budgets in a forest in the U.S. northeast called Hubbard Brook (Fig. 22.2). For example, nitrogen was added to the system only in precipitation $(6.5 \times 10^{-4}\ \text{kg m}^{-2}\ \text{year}^{-1})$ and atmospheric nitrogen fixation by microorganisms $(1.4 \times 10^{-3}\ \text{kg m}^{-2}\ \text{year}^{-1})$. The export, in streams, was only $4 \times 10^{-4}\ \text{kg m}^{-2}\ \text{year}^{-1}$, although dentrification by other

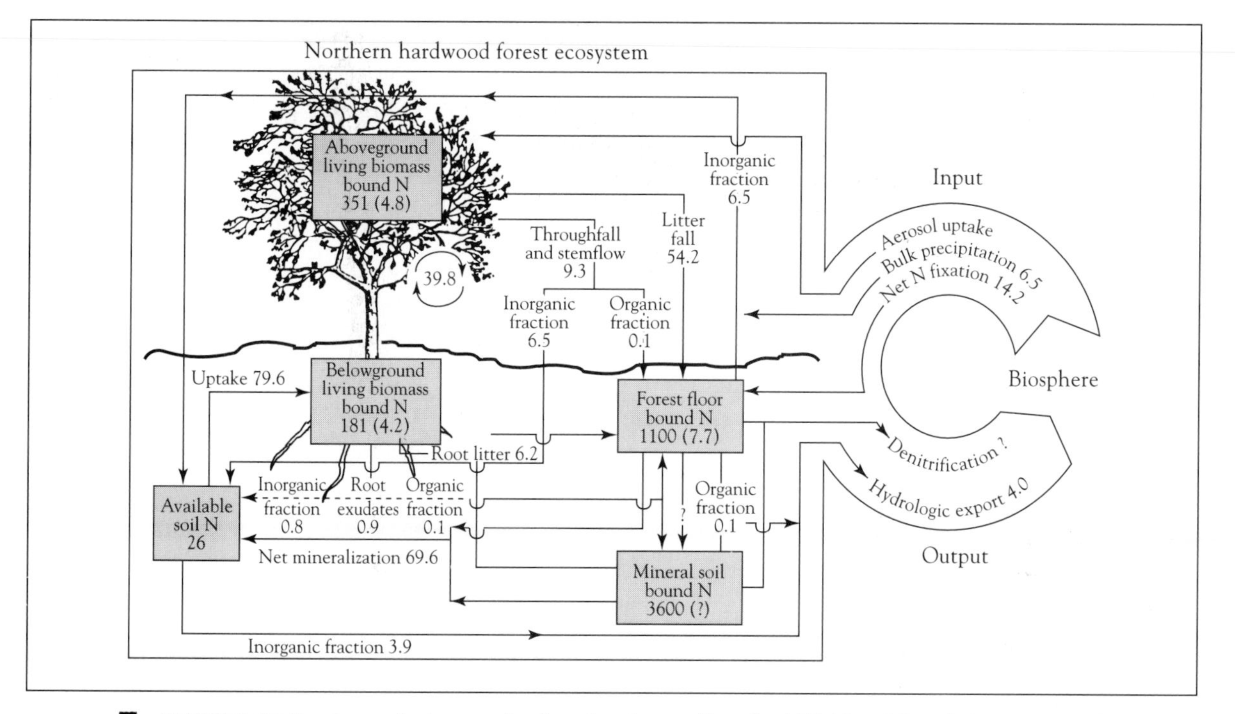

■ **FIGURE 22.2** Annual nitrogen budget for the undisturbed Hubbard Brook Experimental Forest. Values in boxes are the sizes of the various nitrogen pools in kilograms of nitrogen per hectare. The rate of accretion of each pool (in parentheses) and transfer rates are expressed in kilograms of nitrogen per hectare per years. (After Bormann et al., 1977.)

microorganisms, releasing nitrogen to the atmosphere, would also have been occurring but was not measured. This emphasizes how securely nitrogen is held and cycled within the forest biomass. Stream output represents only 0.1 percent of the total (organic) nitrogen standing crop held in living and dead forest organic matter.

The only major exception to this pattern is sulfur. The amount of sulfur leaving the system annually (about 24 kg ha^{-1} [2.4×10^{-3} kg m^{-2}]) was far in excess of the amount in annual litterfall (5.5 kg ha^{-1} [5.5×10^{-4} kg m^{-2}]), principally because of pollution resulting from the burning of fossil fuels.

These observational data were confirmed by a large-scale experiment in which all the trees were felled in one of the Hubbard Brook catchments. The overall export of dissolved inorganic nutrients from the disturbed catchment rose to thirteen times the normal rate (Fig. 22.3). Two phenomena were responsible. First, the enormous reduction in leaves led to 40 percent more precipitation passing through the groundwater to be discharged to the streams, and this increased outflow caused greater rates of chemicals leaching and rock and soil weathering. Second, and more significantly, deforestation effectively broke the within-system nutrient cycling by uncoupling the decomposition process from the plant-uptake process. In the absence of nutrient uptake in spring, when the deciduous trees would have started production, the inorganic nutrients released by decomposer activity were available to be leached in the drainage water. Once again, sulfur was exceptional. Its rate of loss actually decreased after deforestation. No satisfactory explanation for this is available. It seems likely that, in the majority of terrestrial

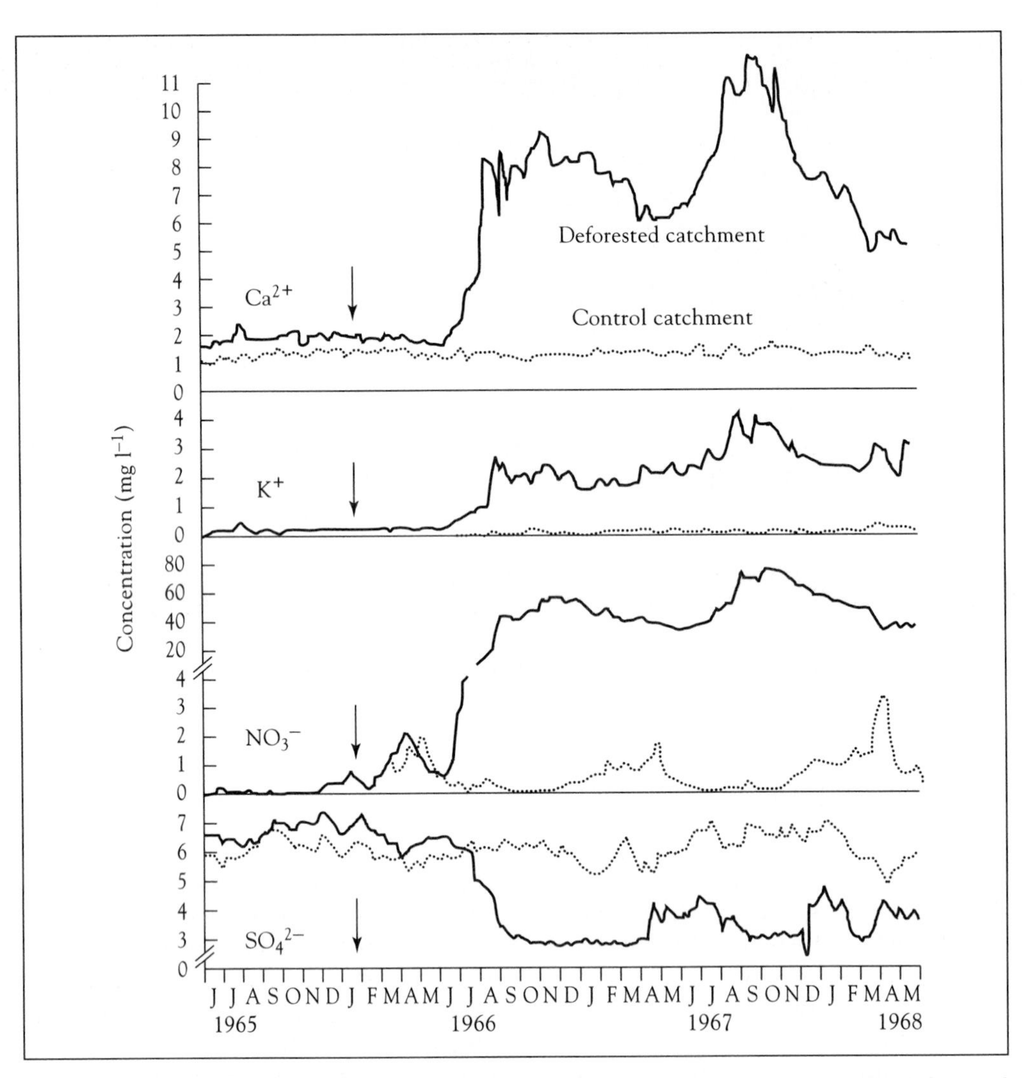

FIGURE 22.3 Concentrations of ions in stream water from the experimentally deforested catchment and a control catchment at Hubbard Brook. The timing of deforestation is indicated by arrows. Note that the "nitrate" axis has a break in it. (*After Likens et al. 1970.*)

environments, nutrient cycling within a community will be similar to that in Hubbard Brook with inputs and outputs relatively low; and this has indeed been shown in other forests (Waring 1989). However, nutrient-use efficiency does vary according to species and location. Thus, evergreen plants generally use nutrients more efficiently than deciduous species (Chapin 1980). This is partly due to the greater longevity of their leaves, with less nutrient being lost per unit time in leaf fall. This may explain why evergreen species are common in nutrient-poor environments. Some authors have also argued that the efficiency of nutrient use is greater in nutrient-poor environments (Shaver and Melillo 1984), though this is debated (Birk and Vitousek 1986).

In contrast to terrestrial systems, in freshwater streams and rivers only a small fraction of available nutrients are locked up in living biomass (Winterbourn and Townsend

TABLE 22.1 **Aboveground tree accumulation of organic matter and nitrogen and turnover time for different forest regions.** (*Data from Cole and Rapp 1981*)

FOREST REGION	% ORGANIC MATTER ABOVE GROUND	% NITROGEN ABOVE GROUND	MEAN TURNOVER TIME NITROGEN (YR)
Boreal coniferous	19	4	230
Boreal deciduous	20	6	27.1
Temperate coniferous	54	7	17.9
Temperate deciduous	40	8	5.5

1991). Thus, in freshwater systems nutrients may cycle more (Moss 1989). Lake Nakuru, Kenya, a saline lake, supports huge aggregations of plankton-filtering flamingos. This results in a high input of phosphorus-rich feces, which is once again taken up by phytoplankton. Thus, there is a tight phosphorus cycle here.

Nutrient cycling also varies by location. In the boreal forests of Alaska only about 20 percent of the organic matter is present in the trees above ground. Low decomposition rates in these forests cause most of the nitrogen to be tied up in the soil (Table 22.1). In temperature zones nutrient cycles operate more quickly, and nutrient turnover times are more rapid. Note again the quick nutrient turnover in deciduous forest compared to coniferous forest.

22.2 Phosphorus cycles between geological and biological pools.

The phosphorus cycle is a relatively simple cycle. This is because phosphorus essentially does not have an atmospheric component, that is, phosphorus is not moved around by the wind or the rain. Phosphorus tends to cycle only locally over short time periods. Exact rates of movement of phosphorus between different components of ecosystems vary, but the general patterns are outlined in Fig. 22.4. Any long-distance transfers of phosphorus involve movement from land to sediments in the sea and then back to land. The Earth's crust is the main storehouse for this particular mineral.

Generally, small losses from terrestrial systems caused by leaching through the action of rain are balanced by gains from the weathering of rocks. The geologic components from the phosphorus cycle take millions of years to release significant amounts of phosphorus from rocks into the living ecosystem. Compared to this, the ecosystem phase of the phosphorus cycle is much more rapid. All living organisms require phosphorus, which becomes incorporated into substances that help give plants and animals their energy. Plants have the metabolic means to absorb dissolved ionized forms of phosphorus. The most important form of phosphorus occurs as phosphate. Plants can take up phosphate rapidly and efficiently. In fact, they can do this so quickly that they often reduce soil concentrations of phosphorus to extremely low levels. Herbivores obtain their phosphorus only from eating plants; and carnivores obtain it by eating herbivores. Herbivores and carnivores excrete phosphorus as a waste product in urine and feces. Phosphorus is, of course, also released to the soil when plant or animal matter decomposes.

In aquatic systems, plants can take up phosphate even quicker than in terrestrial systems. In most aquatic systems, phosphorus is the limiting element. In other words, the

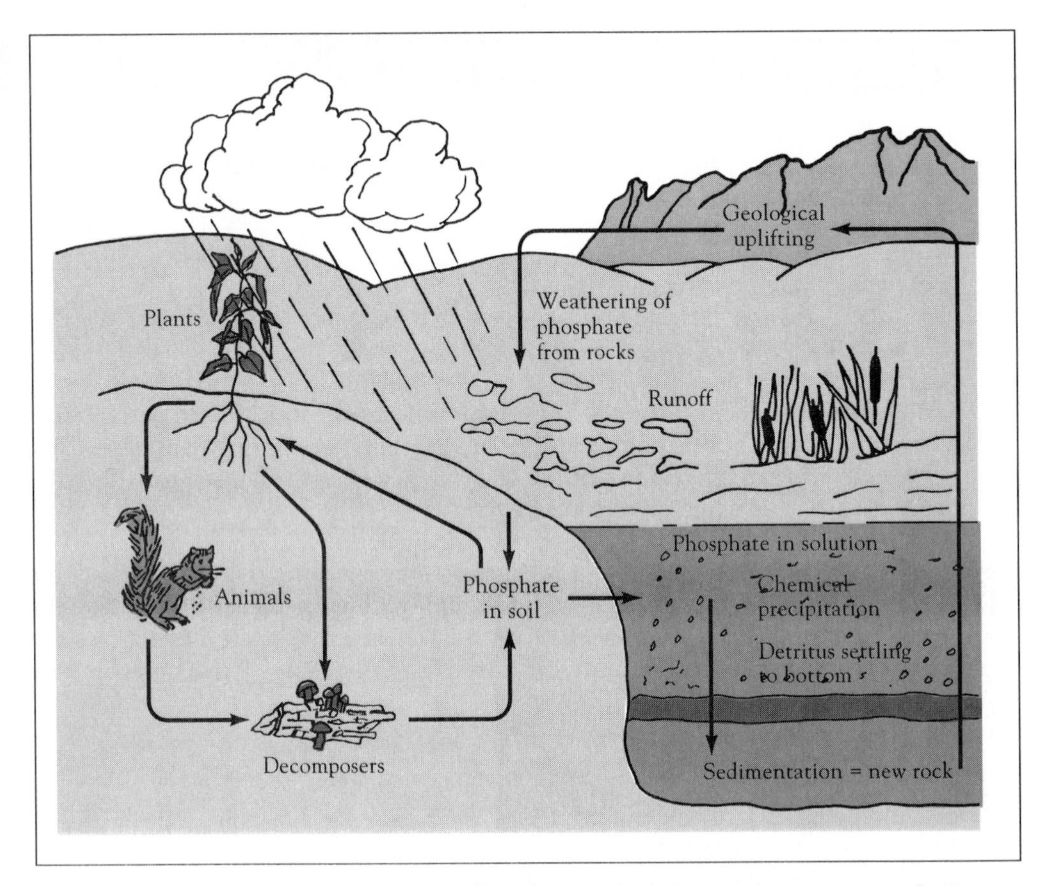

■ **FIGURE 22.4** The phosphorus cycle. Phosphorus, which does not have an atmospheric component, tends to cycle locally. Small losses from terrestrial systems caused by leaching are balanced by gains from the weathering of rocks. In aquatic systems, some phosphorus is lost from the ecosystem because of chemical processes that cause precipitation or through settling of detritus to the bottom, where sedimentation may lock away some of the nutrient before biotic processes can claim it. This may be why phosphorus is more often a limiting nutrient in aquatic systems. This general pattern of local cycling applies to many other nutrients, including trace elements.

more phosphorus that is added, the more aquatic productivity increases. Of course, an overabundance of phosphate can be damaging and can cause blooms of algae in lakes. This process is known as eutrophication. Cultural eutrophication results when humans induce eutrophication in a lake by releasing phosphorus-rich effluent into it.

22.3 Carbon cycles among biological, geological, and atmospheric pools.

A simplified picture of the carbon cycle is shown in Figure 22.5. Carbon dioxide is present in the atmosphere at the relatively low concentration of about 0.03 percent. Autotrophs (primarily plants) acquire carbon dioxide from the atmosphere and incor-

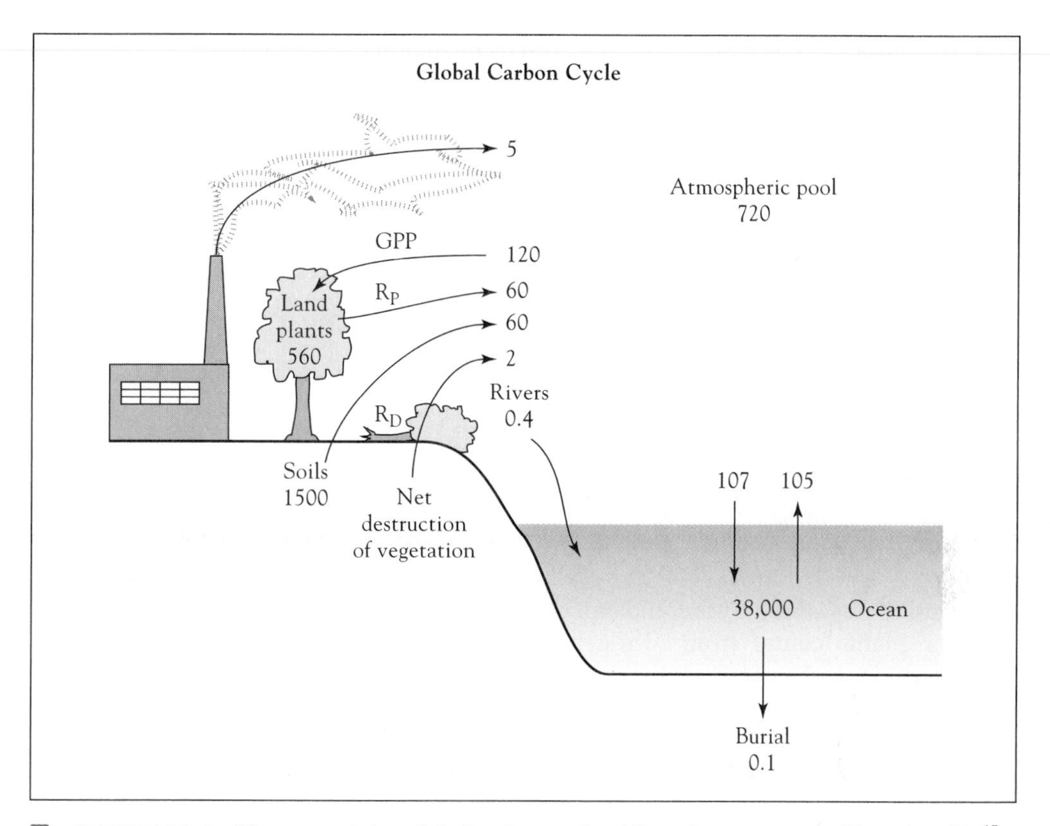

■ **FIGURE 22.5** The present-day global carbon cycle. All pools are expressed in units of 10^{15} g C and all annual fluxes in units of 10^{15} g C/yr. (*From Schlesinger 1991.*)

porate it into the organic matter of their own biomass via photosynthesis. Plant respiration returns some CO_2 to the atmosphere. Each year, plants remove approximately one-seventh of the CO_2 in the atmosphere. The decomposition of plants eventually recycles most of this carbon back into the atmosphere as CO_2, although fires can oxidize organic material to carbon dioxide much faster. Herbivores can also return some carbon dioxide to the atmosphere, eating plants and breathing out CO_2, but the amount flowing through this part of the cycle is probably minimal.

The amount of carbon dioxide in the atmosphere can increase considerably when volcanoes erupt. It can also vary with the seasons in temperate environments. Concentrations of carbon dioxide are lowest during the Northern Hemisphere's summer and highest during the winter. This is because there is more land in the Northern Hemisphere than in the Southern Hemisphere and therefore more vegetation. The vegetation has a maximum photosynthetic activity during the summer, reducing the global amount of carbon dioxide. During the winter, the vegetation respires more carbon dioxide than it uses for photosynthesis, causing a global increase in the gas. The combustion of fossil fuels is thought to have caused a far greater amount of CO_2 to enter the atmosphere than normal (see Applied Ecology: Global Warming, Chap. 7).

22.4 The nitrogen cycle is strongly influenced by biological processes that transform nitrogen into forms that are usable by primary producers.

The nitrogen cycle (Fig. 22.6) is a good example of a global cycle. There are five basic steps in the cycle:

1. Nitrogen fixation
2. Nitrification
3. Assimilation
4. Ammonification
5. Denitrification

Nitrogen fixation.

The quantity of nitrogen tied up in living organisms is very small compared with the total amount of nitrogen in the atmosphere. Despite the atmosphere being almost 80 percent nitrogen, plants cannot assimilate this form of nitrogen. Only certain bacteria, and a few algae, can fix nitrogen—that is, reduce atmospheric nitrogen to ammonia, which can be used to synthesize other biological compounds. In fact, almost all nitrogen available for plants comes from nitrogen-fixing bacteria or in aquatic environments, cyanobacteria, and there is severe competition between plants for this available nitrogen. This is why nitrogen is often a limiting factor for plant growth and why fertilizers contain so much nitrogen and have such a pronounced effect on plant growth. It is usually true that limiting nutrients such as nitrogen, and indeed phosphorus, are usually bound up tightly in the ecosystems in living components, whereas nonlimiting elements such as sodium do not accumulate in the food chain. Organisms that fix nitrogen, of course, are fulfilling their own metabolic requirements, but they release excess ammonia, and this is the nitrogen that becomes available to other organisms. The most important of these bacteria in the soil are called *Rhizobium*, which live in special swellings, called nodules, mostly in plants called legumes.

In terms of the global nitrogen budget, industrial fixation of nitrogen for the production of fertilizer makes a significant contribution to the pool of nitrogen-containing materials in the soils and waters of agricultural regions. Human alterations of the nitrogen cycle have aproximately doubled the rate of nitrogen input into the terrestrial nitrogen cycle (Vitousek et al. 1997a). Industrial fixation of nitrogen may occur in many other processes too.

Nitrification.

For most plants, the most useful form of nitrogen in the soil is nitrates, though many can use ammonia and some can use amino acids. Nitrates are formed from ammonia by other bacteria such as *Nitrosomonas*, *Nitrococcus*, and *Nitrobacter*.

Assimilation.

Plant roots assimilate nitrogen mainly in the form of nitrates, and animals assimilate their nitrogen by eating plants.

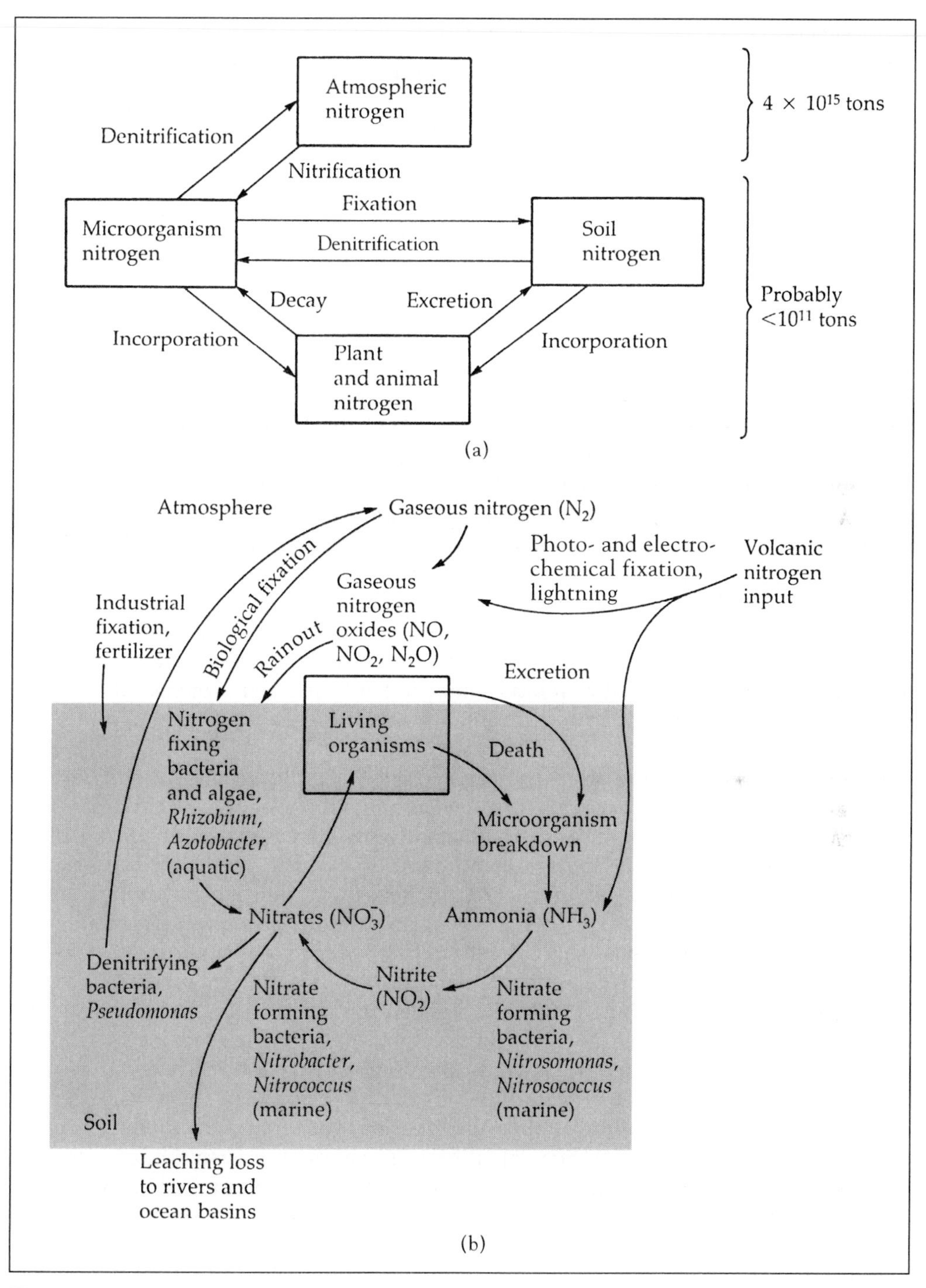

■ **FIGURE 22.6** The nitrogen cycle. (**a**) Major relationships between the very large atmosphere pool of gaseous nitrogen and the biosphere. (**b**) The complex interrelationships of the soil-based portion of the cycle.

Ammonification.

Ammonia can also be formed in the soil through the decomposition of plants and animals and the release of animal waste. This process is again carried out by many bacteria and fungi. Again, the ammonia in the soil is generally not used directly by plants.

Denitrification.

Denitrification is the reduction of nitrates to gaseous nitrogen, and denitrifying bacteria perform almost the reverse of their nitrogen-fixing counterparts.

The most biologically important components of the nitrogen cycle are the nitrogen compounds in the soil and the water, not in the atmosphere. Although nitrogen fixation from the air has been important in the gradual build-up of a pool of available nitrogen, in most systems it contributes only a fraction of the nitrogen assimilated by total vegetation. However, locally, direct nitrogen fixation by bacteria in the root nodules of legumes can be important. Legumes can grow in poor soils devoid of nitrogen and thus avoid competition with many other plants. Nitrogen-fixing plants can be very important in agriculture for this reason and because they replenish soil nitrogen. In agricultural systems, because so much nitrogen is removed when the plants are harvested, large quantities of nitrogen have to be added into the soil in the form of fertilizers.

Nitrogen availability can be critical in limiting individual species and population cycles. One of the most striking examples was often thought to be that of the brown lemming, which lives in the tundra areas of North America and Eurasia. The ecosystem of the tundra is simple compared with more temperate or tropical ecosystems, and the lemming is often the major herbivore. Every three to four years, numbers of these small rodents build up, only to crash again in a never-ending cycle (Fig. 22.7). This ecosystem has been studied in some detail on the Arctic coastal plain tundra near Point Barrow, Alaska. The traditional story is as follows. As lemming numbers increase in winter, their feces and urine stimulate plant growth, which in turn increases lemming production. Eventually, the lemming numbers become so great that the vascular plant cover is thinned out. More sunlight reaches the ground, thawing out the soil overlying the **permafrost**, and plant roots can penetrate a little farther. Permafrost lower down prevents water movement, so nutrients are not **leached** out but accumulate in peat, which is not available to the plants. The intense grazing and lack of root nutrients soon reduce forage quality. This trend, coupled with high predation rates, social strife, and dramatic emigration reduces lemming numbers drastically. Low nutrient quality also prevents breeding the following year. Gradually, during the next two to three years the vegetation slowly recovers, nutrients increase, and the cycle begins again. This whole description of the process that controls population variation was termed the *nutrient-recovery hypothesis* by Pitelka (1964) and Schultz (1964), and although it is an attractive idea, it must be stressed that not all the elements of the cycle have been proved beyond question. Much work remains to be done, and Batzli et al. (1980) have pointed out some inconsistencies between the data and the theory.

First, food supplementation in times of high densities should prevent population decline, but early experiments failed to produce this result. Second, "intrinsic" changes in the behavior of the rodents themselves may be causing variations in population num-

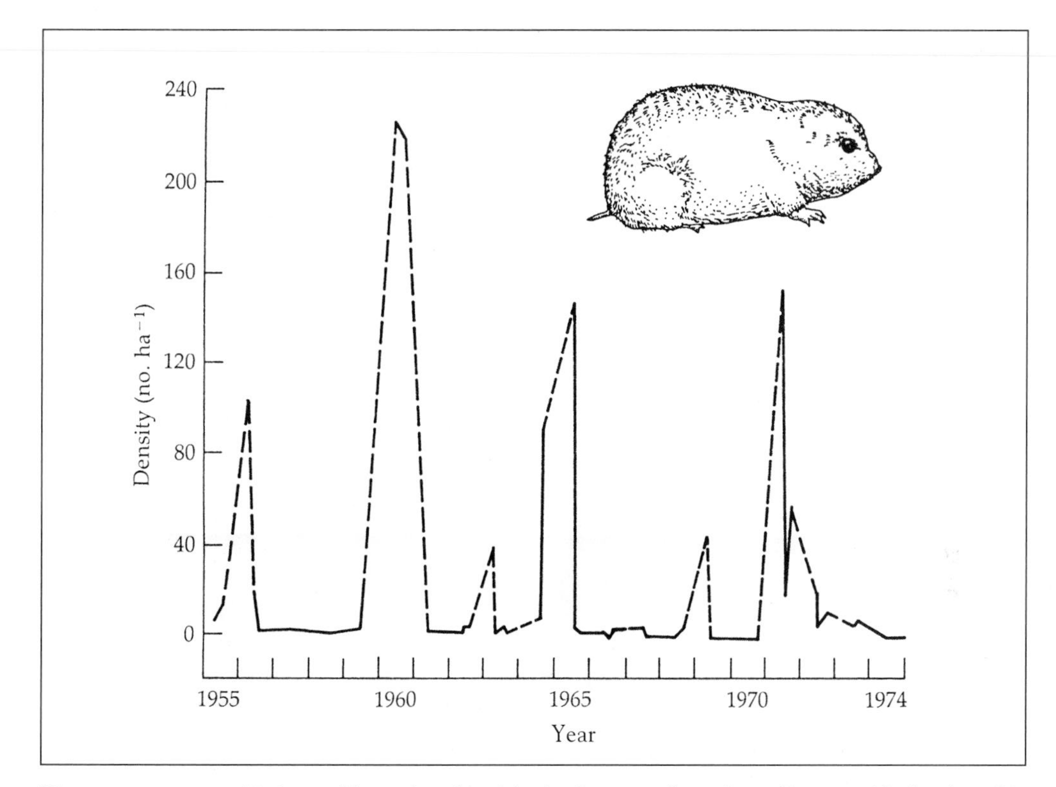

■ **FIGURE 22.7** Estimated lemming densities in the coastal tundra at Barrow, Alaska, for a 20-year period. (*Redrawn from Batzli et al. 1980.*)

bers. As densities of animals increase, aggressiveness between individuals goes up, affecting hormonal levels (Krebs 1985b). This process in turn promotes a huge dispersal of animals away from centers of high population density (Stenseth 1983), so the population cycles of lemmings can be much affected by the behavior of the animals themselves.

Similar processes probably occur in other ecosystems as well but are harder to unravel. The ecosystem of the Arctic is simple. There are about two hundred species of plant but only about ten are important, comprising 90 percent of the plant biomass. The lemming is the only major herbivore, and it has two main predators and six minor avian and mammalian enemies. Other ecosystems are much more complex. In addition, much carbon, nitrogen, and phosphorus is held in the soil in the Arctic; only about 2 percent is held in living material. This proportion is in sharp contrast to those in temperate and tropical forests, where between 20 and 70 percent of these nutrients are found in living plants (Fig. 22.8). Finally, it must be emphasized that Arctic systems are commonly not in a state of equilibrium; much organic matter is slowly accumulating as peat. This is not the case in many other ecosystems. Thus, although some of the peculiarities of other ecosystems could probably be explained by studies of nutrient cycling, the answers are likely to be different from those of the Arctic.

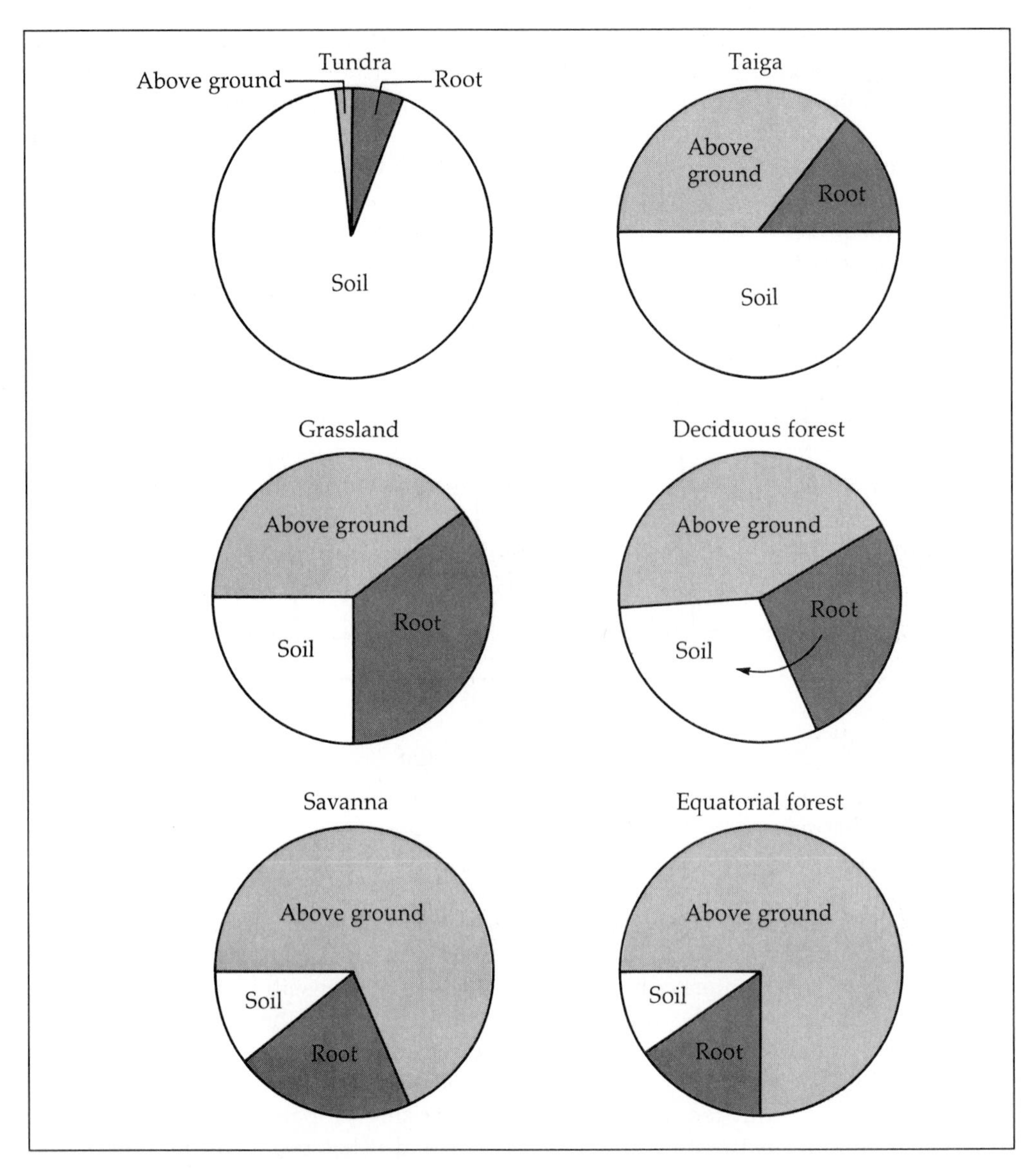

■ **FIGURE 22.8** Deposition of nitrogen in the three organic-matter compartments (above-ground, root, and soil) for each of six biome types. In arctic ecosystems most nitrogen is held in the soil. Soil nutrient availability may thus be critical in the population processes of arctic life. However, in most other ecosystems nitrogen is held more in living matter and biotic interactions may be more important in ecosystem processes. (*Redrawn from Swift, Heal, and Anderson 1979.*)

22.5 The sulfur cycle is heavily influenced by anthropogenic effects.

Sulfur has a long-term sedimentary phase in which it is bound up in organic matter such as coal, oil and peat, and inorganic matter like rocks with high sulfur deposits. The weathering of rocks and the decomposition of organic matter releases sulfur in a salt solution. However, most sulfur in the gaseous phase comes from the gas hydrogen sulfide (H_2S), which is released during volcanic eruptions and decomposition, especially in wetland environments (Fig. 22.9). The H_2S quickly oxidizes into sulfur dioxide, SO_2. Because SO_2 is soluble in water, it returns to earth as weak sulfuric acid, H_2SO_4. Here sulfate-reducing bacteria may release sulfur as H_2S, or the sulfate may be incorporated by plants into their tissue. Some bacteria, especially the purple bacteria found in salt marshes, can also use sulfur by oxidizing H_2S to sulfate. In the presence of iron sulfur can precipitate as ferrous sulfide, Fe S_2, and be incorporated in pyritic rocks. Because such rocks commonly overlay coal deposits, mining exposes them to the air and water, resulting in a discharge of sulfuric acid and other sulfur-containing compounds into aquatic ecosystems. Mining has polluted hundred of miles of streams and rivers in the eastern United States in this way.

Human activity through the combustion of fossil fuels has altered the sulfur cycle more than any of the other nutrient cycles. While human-produced emissions of CO_2 and N are only about 5 to 10 percent of the level of natural emissions, humans produce about 160 percent of the level of natural emissions of sulfur (Likens, Bormann and Johnson 1981). This leads to the widespread problem of acid rain (Applied Ecology: Acid Rain).

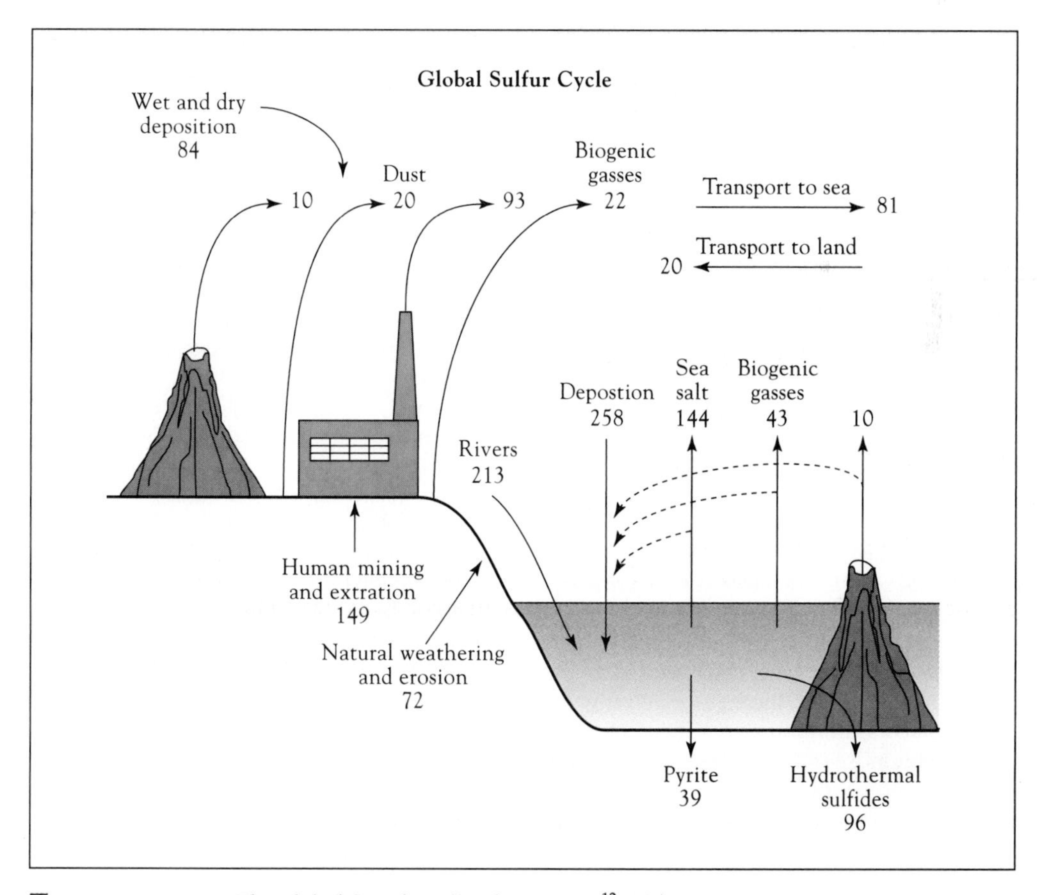

■ **FIGURE 22.9** The global S cycle. All values are 10^{12} g S/yr. *(From Brimblecombe et al. 1989.)*

Acid Rain

The current concern over acid rain is with its increased volume and strength, which are a result of modern industrial activity. For example, the burning of coal and oil, to provide energy for space heating or to fuel electric power stations, produces SO_2. As a by-product of the smelting process, particularly when nonferrous ores are involved, considerable amounts of SO_2 are also released into the atmosphere. Automobile engines contribute to acid rain through the release of nitrogen oxide into the atmosphere. Though both sulfur oxides, hydrogen sulfide and nitrogen oxides, contribute to acid rain, the contributions of sulfur compounds far outweigh the effect of nitrogen oxides, so attention is focused on sulfur, particularly the sulfur dioxides produced by human activity.

Initially, the effects of sulfur oxides were restricted to the local areas from which they originated and where their impact was often obvious. For example, the detrimental effects of SO_2 on vegetation around the smelters at Sudbury (Ontario), Trail (British Columbia), Anaconda (Montana), and Sheffield (England) have long been recognized. However, as smokestacks became taller and taller the problem became less of a regional issue, because the pollutants were whisked away in the atmosphere, and more of global issue, because these same pollutants were deposited hundreds or even thousands of miles away, often in different countries.

One of the main differences between human-produced acid rain and natural acid rain is its strength. Human-produced acid rain is more acidic than the natural variety. The acidity of a solution is measured by its hydrogen concentration or pH (potential hydrogen). The lower the pH, the higher acidity. A chemically neutral solution, such as distilled water, has a value of 7. Alkaline solutions have pHs ranging from 7 to 14 and acid solutions from 7 down to 0. The pH scale is logarithmic and, as a result, a change of one point represents a tenfold increase or decrease in the hydrogen ion concentration. A two-point change represents a one-hundredfold increase or decrease. Thus, a solution with a pH of 4 is ten times more acidic than one with a pH 5; a solution of pH 3.0 is one hundred times more acidic than one with a pH 5.0.

The difference between "normal" and "acid" rain is commonly on the order of 1.0 to 1.5 points. In North America, for example, naturally acid rain has a pH of about 5.6 while measurements of rain falling in southern Ontario, Canada, sometimes provide values in the range of 4.1 to 4.5. To put these values in perspective, it should be noted that vinegar has a pH of 2.7 and milk a pH of 6.6. Thus, Ontario rain is about one hundred times more acidic than milk but a hundred times less acidic than vinegar. Similar values for the background levels of acid rain are indicated by European studies, although the Central Electricity Generating Board (CEGB) in Britain has argued for a pH of 5.0 as a normal level for naturally acid rain.

The earliest concerns about acid rain were expressed as long ago as 1852 by the author Robert Smith in England, but modern interest in the problem dates only from the 1960s. Most attention has concentrated on the impact of acid rain on the aquatic environment, which can be particularly sensitive to even moderate increases in acidity. Harmful effects in lakes will begin to be felt by most bodies of water when the pH falls to 5.3, although this will vary from lake to lake (Photo 1). In some lakes, adverse effects on fish begin when pH reaches 6.0. For example, direct exposure to acid water damages brook trout and rainbow trout. The whole salmonid group of fish are much less tolerant than such fish as pike and perch. The stage of development is also important. Adult fish, for example, may be able to survive relatively low pH values, but newly hatched fry may be much less tolerant. As a result, the fish population in acid lakes is usually wiped out by low reproductive rates even before the pH reaches levels that would kill mature fish.

Fish in acid lakes may also succumb to toxic concentrations of metals such as aluminum, mercury, magnese, zinc, and lead. These are leached from the surrounding rock by acids. Thus, acid rain has direct and indirect effects. For example, many acid lakes have elevated concentrations of aluminum, which can cause the breakdown of salt regulation systems in fish. Aluminum may also be deposited on the gills where it inhibits breathing and eventually leads to suffocation.

The impact of acid rain on the environment depends not only on the level of acidity of rain but also on the nature of the environment itself. Areas underlain by granitic or quartzitic bedrock, for example, are particularly susceptible to damage since the soils and water are already acidic and lack the ability to "buffer" or neutralize the additional acidity from the precipitation. The areas greatest at risk here are the pre-Cambrian Shield areas of Canada and Scandinavia, where the acidity of the rocks and soils is already high. In contrast, areas that are geologically basic, underlain by limestone or chalk, for example, may even benefit from the additional acidity. The highly alkaline soils and water of these areas ensure that the acid added to the environment by the rain is very effectively neutralized.

Besides damage to aquatic systems, there is also evidence that terrestrial ecosystems show adverse affects from acid rain. Reduction in forest growth in Sweden, physical damage to trees in western Germany, and the death of sugar maples in Quebec and Vermont have all been blamed on the increased acidity of the precipitation in these areas together with increased ozone levels in the atmosphere. Again, the effects of this precipitation may be direct, brought about by the presence of acid particles on the leaves, for example, or indirect, associated with changes in the soil or the biological processes controlling plant growth. Vegetation growing at higher elevations, and therefore enveloped in clouds for longer periods, may display more symptoms of acid rain than forests in lower elevations since cloud moisture is often more acidic than rain itself.

The most obvious effect of acid rain on foliage has been what has been called "tree die-back," which describes the gradual wasting of the tree inward from the outer most tips and branches. Die-back has been likened to the premature arrival of autumn on deciduous trees. The leaves on the outermost branches begin to turn yellow or red in midsummer; they dry out and eventually fall, well ahead of schedule. These branches will fail to leaf in the spring. In succeeding years, the problem spreads from the crown until the entire tree is devoid of foliage and takes on a skeletal appearance, even in summer. Coniferous trees react in the same way. Needles turn yellow, dry up, and fall off branches. Also the weakened trees may be more likely to succumb to the effects of insect attack, disease, and the ravages of weather. Some authors believe that the maple groves in Ontario, Quebec, and Vermont have been suffering progressive die-back since 1980.

■ **PHOTO 1** Fish kills from acid rain can be substantial. (*Hank Andrews, Visuals Unlimited.*)

SUMMARY

1. Nutrients may cycle locally, like phosphorus in a lake, or globally, like carbon and nitrogen.

2. Nutrient turnover time is affected by species and location. Turnover times are faster in warmer than in colder environments and are faster in deciduous than in coniferous forests.

3. The phosphorus cycle is relatively simple because it does not have a large atmospheric component. The Earth's crust is the main storehouse for phosphorus.

4. In the carbon cycle, carbon dioxide is present in low concentrations in the atmosphere (0.03 %) and at higher levels in plants. Each year, plants remove about one-seventh of the CO_2 in the atmosphere. Decomposition eventually returns most of this CO_2 to the air. Forest fires can accelerate this process.

5. The nitrogen cycle is, like the carbon cycle, a global cycle. There are five basic steps to the nitrogen cycle:

a. Nitrogen fixation of atmospheric N_2 to ammonia by soil bacteria.
b. Nitrification—the formation of nitrates from ammonia by soil bacteria.
c. Assimilation of nitrates by plant roots.
d. Ammonification—the production of ammonia through plant decomposition and the production of animal wastes.
e. Denitrification—the reduction of nitrate to gaseous nitrogen. The limiting step is generally the first one, fixation of atmospheric nitrogen by bacteria.

6. The sulfur cycle, more than any other, is profoundly influenced by humans, with around 160 percent of the level of natural emissions of sulfur being produced anthropogenically.

7. Anthropogenic changes can affect all nutrient cycling dramatically.

DISCUSSION QUESTION

Will human-induced changes in the carbon cycle increase or decrease productivity? Is this a good or a bad thing?

Glossary

abiotic factors environmental influences produced other than by living organisms, for example, temperature, humidity, pH, and other physical and chemical influences; contrast with "biotic factors."
acclimation changes by an organism, often biochemical, subjected to new environmental conditions that enable it to withstand those conditions.
acid rain rainfall acidified by contact with sulfur dioxide (a by-product of the burning of fossil fuels) in the atmosphere.
acquired character a character not inherited but acquired by an individual organism during its lifetime.
adaptation a change in an organism's structure or habits that produces better adjustment to the environment; a genetically determined characteristic that enhances the ability of an organism to cope with its environment.
adaptive radiation evolutionary diversification of species derived from a common ancestor into a variety of ecological roles.
aerobic pertaining to organisms or processes that require the presence of oxygen.
age class individuals in a population of a particular age
aggressive mimicry resemblance of predators or parasites to harmless species that causes potential prey or hosts to ignore them.
agricultural pollution contamination of the environment with liquid and solid wastes from all types of farming, including pesticides, fertilizers, runoff from feedlots, erosion and dust from plowing, animal manure and carcasses, and crop residues and debris.
air pollution the presence of contaminants in the air in concentrations that overcome the normal dispersive ability of the air and that interfere directly or indirectly with human health, safety, or comfort or with the full use and enjoyment of property.
algal bloom a proliferation of living algae in a lake, stream, or pond.
allele one of two or more alternative forms of a gene located at a single point (locus) on a chromosome.
allelopathy the negative chemical influence of plants, exclusive of microorganisms, upon one another.
Allen's (1878) rule among homeotherms, the tendency for limbs and extremities to become shorter and more compact in colder climates than in warmer ones.
allochronic speciation separation of a population into two or more evolutionary units as a result of reproductive isolation arising from a difference in mating times.
allopatric speciation separation of a population into two or more evolutionary units as a result of reproductive isolation arising from geographic separation.
altruism in an evolutionary sense, enhancement of the fitness of an unrelated individual by acts that reduce the evolutionary fitness of the altruistic individual.
anadromous living in salt water but breeding in fresh water.
anaerobic pertaining to organisms or processes that occur in the absence of oxygen.
apatetic coloration coloration of an animal that causes it to resemble physical features of the habitat.
apomixis parthenogenetic reproduction in which offspring develop from unfertilized eggs or somatic cells.
aposematism conspicuous appearance of an organism warning that it is noxious or distasteful; warning coloration.
apostatic selection selective predation on the most abundant of two or more forms in a population, leading to balanced polymorphism (the stable occurrence of more than one form in a population).
apterous wingless.
aquaculture farming of aquatic or marine systems; rearing of organisms such as fish, algae, or shellfish under controlled conditions.
aquifer a layer of rock, sand, or gravel through which water can pass; an underground bed or stratum of earth, gravel, or porous stone that contains water; the place in the ground where groundwater is naturally stored.
assimilation efficiency the percentage of energy ingested in food that is assimilated into the protoplasm of an organism.
association a group of species occurring in the same place.
assortative mating nonrandom mating; the propensity to mate with others of like phenotypes.
autecology study of the individual in relation to environmental conditions; contrast "synecology."
autogamous able to produce offspring sexually by the fusion of gametes from the same individual, for example, by fusion of pollen and ovules from the same plant.
autosome a chromosome other than a sex chromosome.
autotroph an organism that obtains energy from the sun and materials from inorganic sources; contrast with "heterotroph."
average GNP the gross national product of a country divided by its total population.
balanced polymorphism the stable occurrence of more than one form in a population.
barrier island a narrow, elongated, sandy island paralleling the coast, separated from the mainland by a bay or lagoon.
Batesian mimicry resemblance of an edible (mimic) species to an unpalatable (model) species to deceive predators.
benthic pertaining to aquatic bottom or sediment habitats.
benthos bottom-dwelling aquatic organisms (for example, burrowing worms, molluscs, and sponges).
Bergmann's (1847) rule among homeotherms, the tendency for organisms in colder climates to have larger body size (and thus smaller surface-to-volume ratio) than those in warm climates.
bioassay the use of living organisms to measure the biological effect of some substance, factor, or condition.

biochemical oxygen demand (BOD) the amount of oxygen that would be consumed if all the organic substances in a given volume of water were oxidized by bacteria and other organisms; reported in milligrams per liter.
biodegradable capable of being decomposed quickly by the action of microorganisms.
biogeochemical cycle the passage of a chemical element (such as nitrogen, carbon, or sulfur) from the environment into organic substances and back into the environment.
biogeography the branch of biology that deals with the geographic distribution of plants and animals.
biological control use of natural enemies (diseases, parasites, predators) to regulate populations of pest species.
biological magnification the concentration of a substance as it "moves up" the food chain from consumer to consumer.
biomass dry weight of living material in all or part of an organism, population, or community; commonly expressed as weight per unit area, biomass density.
biome a major terrestrial climax community; a major ecological zone or region corresponding to a climatic zone or region; a major community of plants and animals associated with a stable environmental life zone or region (for example, northern coniferous forest, Great Plains, tundra).
biosphere the whole Earth ecosystem.
biota all the living organisms occurring within a certain area or region.
biotic factors environmental influences caused by living organisms; contrast with "abiotic factors."
boreal occurring in the temperate and subtemperate zones of the Northern Hemisphere.
breeder reactor a nuclear reactor that produces more fissionable material than it consumes.
burner reactor a nuclear reactor that consumes more fissionable material than it produces.
calcareous in soil terminology, rich in calcium carbonate and having a basic reaction.
canonical distribution a lognormal distribution of the numbers of individuals or species according to the mathematical formulation of Preston (1962).
carbon-14 a radioactive isotope of carbon (atomic weight 14) that can be used for dating organic materials.
carcinogen a chemical or physical agent capable of causing cancer.
cardinal index a cardinal number used in counting or to show how many, e.g., two, five, ten. A cardinal index, therefore, treats all things as equal in importance.
carnivore an animal (or plant) that eats other animals; contrast with "herbivore."
carrying capacity the amount of animal or plant life (or industry) that can be supported indefinitely on available resources; the number of individuals that the resources of a habitat can support.
catadromous living in fresh water but breeding in sea water.
character displacement divergence in the characteristics of two otherwise similar species where their ranges overlap; caused by selective effects of competition between the species in the area of overlap.
character divergence evolution of differences between similar species occurring in the same area.
chimera a piece of DNA incorporating genes from two different species; an organism whose cells are not all genetically alike.
China syndrome a popular term for the consequence of core meltdown in a nuclear reactor in which a molten mass of intensely radioactive material plummets through vessel and containment and into the Earth beneath, in the direction (from the Western Hemisphere) of China.
clade the set of species descended from a particular ancestral species.
clearcutting the practice of cutting all trees in an area, regardless of species, size, quality, or age.
cleistogamy self-pollination within a flower that does not open.
climax community the community capable of indefinite self-perpetuation under given climatic and edaphic conditions.
cline a gradient of change in population characteristics over a geographic area, usually related to a corresponding environmental gradient.
clone a lineage of individuals reproduced asexually.
coadaptation evolution of characters of two or more species to their mutual advantage.
coevolution development of genetically determined traits in two species to facilitate some interaction, usually mutually beneficial.
coexistence occurrence of two or more species in the same habitat; usually applied to potentially competing species.
cohort those members of a population that are of the same age, usually in years or generations.
coliform index a measure of the purity of water based on a count of the coliform bacteria it contains.
commensalism an association between two organisms in which one benefits and the other is not affected.
community a group of populations of plants and animals in a given place; used in a broad sense to refer to ecological units of various sizes and degrees of integration.
competition the interaction that occurs when organisms of the same or different species use a common resource that is in short supply ("exploitation" competition) or when they harm one another in seeking a common resource ("interference" competition).
competitive exclusion principle the hypothesis, based on theoretical considerations and laboratory experiments, that two or more species cannot coexist and use a single resource that is scarce relative to demand for it.
conspecific belonging to the same species.
consumer an organism that obtains its energy from the organic materials of other organisms, living or dead; contrast with "producer."
continental drift the movement of the continents, by tec-

tonic processes, from their original positions as parts of a common land mass to their present locations.
continental island an island that is near to and geologically part of a continent, for example, the British Isles or Trinidad.
continental shelf the shallow part of the sea floor immediately adjacent to a continent.
convergent evolution the development of similar adaptations by genetically unrelated species, usually under the influence of similar environmental conditions.
coprolite fossil excrement.
core the heart of a nuclear reactor where energy is released; the region of a reactor containing fuel (and moderator, if any) and within which the fission reaction occurs.
courtship any behavioral interaction between individuals of opposite sexes that facilitates mating.
crop rotation the farming practice of planting the same field with a different crop each year to prevent nutrient depletion.
crypsis coloration or appearance that tends to prevent detection of an organism by others, especially predators.
cultural change any modification of characteristics specific to a population that is transmitted by learning rather than by genetic mechanisms.
DDT 1,1,1-trichloro-2,2-bis(p-chloriphenyl) ethane; the first of the modern chlorinated-hydrocarbon insecticides.
decomposers consumers, especially microbial consumers, that change their organic food into mineral nutrients.
deforestation removal of trees from an area without adequate replanting.
deme a local population, usually small and panmictic.
denitrification enzymatic reduction by bacteria of nitrates to nitrogen gas.
density the number of individuals per unit area.
density dependent having an influence on individuals that varies with the number of individuals per unit area in the population.
density independent having an influence on individuals that does not vary with the number of individuals per unit area in the population.
desalinization the process of removing salt from water.
desert a region receiving very small amounts of precipitation or where (for example, ice caps) the moisture present is unavailable to vegetation.
deterministic model a mathematical model in which all relationships are fixed and stochastic processes play no part; contrast "stochastic model."
diapause a period of suspended growth or development and reduced metabolism in the life cycle of many insects, during which the organism is more resistant to ınfavorable environmental conditions than during other periods.
dimorphism the occurrence of two forms of individuals in a population.
dioecious characterized by individuals each of which has only male or only female reproductive organs.
diploid of cells or organisms, having two sets of chromosomes.
direct competition exclusion of individuals from resources, by other individuals, by aggressive behavior or use of toxins.
dispersal movement of organisms away from the place of birth or from centers of population density.
disruptive selection selection against individuals in a population that have intermediate values of a trait, leading to the divergence of subpopulations with extreme values of the trait.
distribution the area or areas (taken together) where a species lives and reproduces.
diversity the number of species in a community or region; alpha diversity is the diversity of a particular habitat; beta diversity is diversity of a region pooled across habitats.
dominance the influence or control exerted by one or more species in a community as a result of their greater number, coverage, or size.
ecologic pertaining to the living environment.
ecologic efficiency the percentage of energy in biomass produced by one trophic level that is incorporated into biomass by the next highest trophic level.
ecological impact the total effect of an environmental change, whether natural or man-made.
ecological release the expansion of habitat or food use by populations in regions of low species diversity, permitted by reduced interspecific competition.
ecology the branch of science dealing with the relationships of living things to one another and to their environment.
ecosystem a biotic community and its abiotic environment.
ecotone the transition zone between two diverse communities.
ecotype a subspecies or race that is specially adapted to a particular set of environmental conditions.
ecumene the portion of the Earth's surface occupied by permanent human settlement.
edaphic pertaining to soil.
emergent property a feature of a community or system not deducible from the features of single species or lower order processes.
emigration the movement of organisms out of a population.
endangered species a species with so few living members that it will soon become extinct unless measures are taken to slow its loss.
endemic an organism that is native to a particular region.
endogenous produced from within; originating from or due to internal causes; contrast "exogenous."
energy resource a natural supply of energy available for use, for example, the Earth's internal heat, fossil fuels, hydropower, nuclear energy, solar energy, and wind.
enrichment the addition of nutrients to an ecosystem, for example, the addition of nitrogen to waterways by agricultural runoff.
environment all the biotic and abiotic factors that affect an individual organism at any one point in its life cycle.
epidemiology the study of disease in populations or groups.
epilimnion the upper layer of water in a lake, usually warm and containing high levels of dissolved oxygen.

epiphyte a plant that lives on another plant but uses it only for support, drawing its water and nutrients from natural runoff and the air.
epistasis a synergistic effect whereby the effect of two or more gene loci is greater than the sum of their individual effects.
equilibrium a condition of balance, such as that between immigration and emigration or birth rates and death rates in a population of fixed size.
euphotic zone that part of the water column that receives sufficient sunlight to support photosynthesis; usually limited to the upper 60 m.
eutrophication the normally slow aging process by which a lake fills with organic matter, evolves into a bog or marsh, and ultimately disappears.
evapotranspiration the sum of the water lost from the land by evaporation and plant transpiration. Potential evapotranspiration is the evapotranspiration that would occur if water were unlimited.
evolution the gradual accumulation of genetic change that is thought to have given rise, beginning with common ancestors, to the diversity of life.
exogenous orginating outside an organism.
exponential growth the steepest phase in a growth curve, that in which the curve is described by an equation containing a mathematical exponent.
exponential rate of increase the rate at which a population is growing at a particular instant expressed as a proportional increase per unit time.
extant of a species; currently represented by living individuals.
extinct of a species; no longer represented by living individuals.
extrafloral nectaries nectar-secreting glands found on leaves and other vegetative parts of plants.
facilitation enhancement of a population of one species by another, often during succession, a type of one-way mutualism.
fecundity the potential of an organism to produce living offspring.
feral having reverted from domestication to the wild but remaining distinct from other wild species.
fission splitting or division; nuclear fission is the splitting of the nuclei of the atoms of certain elements into lighter nuclei and is accompanied by the release of relatively large amounts of energy.
fitness genetic contribution by an individual's offspring and heir offspring to future generations.
fixation attainment by an allele of a frequency of 1 (100 percent) in a population, which in effect becomes monomorphic for that allele.
food chain figure of speech describing the dependence for food of organisms upon others, in a series beginning with plants and ending with the largest carnivores.
forest a region that, because it receives sufficient average annual precipitation (usually 75 cm [30 inches] or more), supports trees and small vegetation.
fossil fuels coal, oil, and natural gas, so-called because they are derived from the fossil remains of ancient plant and animal life.
fossorial living in burrows.
founder effect the principle that a population started by a small number of colonists will contain only a fraction of the genetic variation of the parent population.
functional response a change in the rate of exploitation of prey by an individual predator resulting from a change in prey density (see also "numerical response").
fusion the combination of two atoms into a single atom as a result of a collision, usually accompanied by the release of energy.
gene a unit of genetic information.
gene flow the exchange of genetic traits between populations by movement of individuals, gametes, or spores.
generation time the time between the birth of a parent and the birth of its offsping.
genetic drift change in gene frequency caused solely by chance, usually unidirectional and more important in small populations.
genome the entire genetic complement of an individual.
genotype the genetic constitution of an organism or a species in contrast to its observable characteristics; contrast "phenotype."
genus the taxonomic category above the species and below the family; a group of species believed to have descended from a common direct ancestor.
geometric rate of increase the factor by which the size of a population changes over a specified period; contrast "exponential rate of increase."
geometric series a series in which each number is obtained by multiplying the previous one by one factor, e.g., 1, 3, 9, 27, etc.
glacial epoch the Pleistocene epoch, the earlier of the two epochs comprising the Quaternary period, characterized by the extensive glaciation of regions now free from ice.
global stability ability to withstand perturbations of a large magnitude without being affected; contrast "local stability." .
grassland a region with sufficient average annual precipitation (25-75 cm [10-30 inches]) to support grass but not trees.
green belts areas from which buildings and houses are excluded, often serving as buffers between pollution sources and concentrations of population.
greenhouse effect the heating effect of the atmosphere upon the Earth, particularly as CO_2 concentration rises, caused by its ready admission of light waves but its slower release of the heat they generate on striking the ground.
gross production production before respiration losses are subtracted; photosynthetic production for plants and metabolizable production for animals.
group selection elimination of a group of individuals with a detrimental genetic trait, caused by competition with other groups lacking the trait.
guyot a flat-topped submarine volcano.
habitat the sum of the environmental conditions in which an

organism, population, or community lives; the place where an organism normally lives; the environment in which the life needs of an organism are supplied.
half-life the time it takes for one-half the atoms of a radioactive isotope to decay into another isotope; the time it takes certain materials, such as persistent pesticides, to lose half their strength.
haplodiploidy the presence of haploid males and diploid females in the same species, for example, in the Hymenoptera.
haploid containing one set of chromosomes.
harem a group of females controlled by one male.
herbivore an organism that eats plants; contrast "carnivore."
heredity genetic transmission of traits from parents to offspring.
heterotroph an organism that obtains energy and materials from other organisms; contrast "autotroph."
hierarchy a rank order; the pecking order, leadership, or dominance patterns among the members of a population.
holotype the single specimen chosen by the original author of a species as the archetypical example of that species and which any revised description of the species must include; contrasts with "lectotype" a term for such an archetypical specimen when it is chosen, not by the original author, but by a later author in the absence of a holotype.
home range the area in which an individual member of a population roams and carries on all of its activities.
horizon in a soil, a major stratification or zone, having particular structural and chemical characteristics.
host the organism that furnishes food, shelter, or other benefits to an organism of another species.
hybridization breeding (crossing) of individuals from genetically different strains, populations, or, sometimes, species.
hypolimnion the layer of cold, dense water at the bottom of a lake.
immigration the movement of individuals into a population.
inbreeding a mating system in which adults mate with relatives more often than would be expected by chance.
inclusive fitness The total genetic contribution of an individual by way of its sons, daughters, and all other relatives combined.
independent assortment the separate inheritances, without mutual influence, of genes occurring on different chromosomes.
indirect competition exploitation of a resource by one individual that reduces the availability of that resource to others.
innate capacity for increase r measure of the rate of increase of a population under controlled conditions (also referred to as intrinsic rate of increase).
interdemic selection group selection of populations within a species.
interspecific between species; between individuals of different species.
intraspecific within species; between individuals of the same species.
isolating mechanism any condition, for example, a genetically determined difference or a mechanical or geographic separation, that prevents gene flow between two populations.
iterative evolution the repeated evolution of similar phenotypic characteristics at different times during the history of a clade.
kin selection a form of genie selection in which alleles differ in their rate of propagation because they influence the survival of kin who carry the same alleles.
LDC less-developed country, typically with low GNP, high population growth, low literacy, and low industrialization.
landfill a waste-disposal site in which layers of solid waste are laid down in alternation with layers of soil.
leaching the process by which soluble materials in the soil, such as nutrients, pesticides, or contaminants, are washed into a lower layer of soil or are dissolved and carried away by water.
lek a communal courtship area on which several males hold courtship territories to attract and mate with females; sometimes called an arena.
lentic pertaining to standing freshwater habitats (ponds and lakes).
life table tabulation presenting complete data on the mortality schedule of a population.
ligase an enzyme that joins DNA together.
limiting resource the nutrient or substance that is in shortest supply in relation to organisms' demand for it.
linkage occurrence of two loci on the same chromosome; functional linkage occurs when two loci do not segregate independently at meiosis.
local stability the tendency of a community to return to its original state when subjected to a small perturbation.
locus the site on a chromosome occupied by a specific gene.
logistic equation a model of population growth described by a symmetrical S-shaped curve with an upper asymptote.
lognormal distribution a frequency distribution of species abundance in which abundance, on the x-axis, is expressed on a logarithmic scale.
lotic pertaining to running freshwater habitats (streams and rivers).
MDC more-developed country, typically with high GNP, low population growth, high lit eracy, and a strong economy.
Malthusian theory of population the theory of English economist and religious leader Thomas Malthus that populations increase geometrically (2, 4, 8, 16) while food supply increases arithmetically (1, 2, 3, 4), leading to the conclusion that humans are doomed to overpopulation, misery, and poverty and that population levels will be reduced by disease, famine, and war.
meiotic drive a preponderance (generally a frequency greater than 50 percent) of one allele among the gametes produced by a heterozygote.
melanism occurrence of black pigment, usually melanin.
meltdown of a reactor core, the consequence of overheating that allows part or all of the solid fuel to reach the temperature at which cladding and possibly fuel and support structure liquify and collapse.

modifier gene a gene that alters the phenotypic expression of genes at one or more other loci.
monoculture cultivation of a single crop to the exclusion of all other species on a piece of land.
monoecious having separate male and female reproductive organs on the same individual, used mainly of plants.
morph a specific form, shape, or structure.
Mullerian mimicry mutual resemblance of two or more conspicuously marked, distasteful species to reinforce predator avoidance.
multivoltine having several generations during a single season; contrast "univoltine."
mutant an organism with a changed characteristic resulting from a genetic change.
mutation a change in the genetic makeup of an organism resulting from a chemical change in its DNA.
mutualism an interaction between two species in which both benefit from the association.
natural selection the natural process by which the organisms best adapted to their environment survive and those less well adapted are eliminated.
nektonic free swimming in the upper zone of ocean water and strong enough to swim against the current.
neotenic exhibiting neoteny, the ability of species to reproduce sexually when still exhibiting juvenile characteristics.
neritic pertaining to the shallow, coastal marine zone.
net production production after respiration losses are subtracted.
net production efficiency percentage of assimilated energy that is incorporated into growth and reproduction.
net reproductive rate *R* the number of offspring a female can be expected to bear during her lifetime; for species with clearly defined discrete generations.
niche the place of an organism in an ecosystem; all the components of the environment with which an organism or population interacts.
nitrate a salt of nitric acid; a compound containing the radical NO_3; biologically, the final form of nitrogen from the oxidation of organic nitrogen compounds.
nitrogen cycle the biogeochemical processes that move nitrogen from the atmosphere into and through its various organic chemical forms and back to the atmosphere.
nitrogen-fixing bacteria bacteria that can reduce atmospheric nitrogen to cell nitrogen.
nonrenewable resource a resource available in a fixed amount (such as minerals and oil); not replaceable after use.
norm of reaction the set of phenotypic expressions of a genotype under different environmental conditions.
numerical response change in the population size of a predatory species as a result of a change in the density of its prey.
oligotrophic low in nutrients and organisms; low in productivity.
omnivore an organism whose diet includes both plant and animal foods.
operational sex ratio ratio of sexually ready males to fertilizable females.
organic of biological origin; in chemistry, containing carbon.
palaeontology the science that deals with life of past geologic ages; treating fossil remains.
panmixis the condition in which mating in a population is entirely random.
parasite the organism that benefits in an interspecific interaction in which one organism benefits and the other is harmed.
parasitoid a specialized insect parasite that is usually fatal to its host and therefore might be considered a predator rather than a classical parasite.
particulate matter in air pollution, solid particles and liquid droplets, as opposed to material uniformly dispersed among the air molecules.
parthenogenesis reproduction without fertilization by male gametes, usually involving the formation of diploid eggs whose development is initiated spontaneously.
PCBs polychlorinated biphenyls, a family of chemicals similar in structure to DDT.
Pelagic pertaining to the upper layers of the open ocean.
per-capita rate of population growth *r* rate of population growth per individual; used for species with overlapping, nondiscrete generations.
permafrost a permanently frozen layer of soil underlying the Arctic tundra biome.
persistence of pesticides; the length of time they remain in the soil or on crops after being applied.
pH a measure of acidity or alkalinity.
phenology study of the periodic (seasonal) phenomena of animal and plant life (for example, flowering time in plants) and their relations to weather and climate.
phenotype the physical expression in an organism of the interaction between its genotype and the environment; the outward appearance of an organism.
phoresy the transport of one organism by another of a different species.
photic zone the surface zone of a body of water that is penetrated by sunlight.
photoperiodism seasonal response (for example, flowering, seed germination, reproduction, migration, or diapause) by organisms to change in the length of the daylight period.
photosynthesis synthesis, with the aid of chlorophyll and with light as the energy source, of carbohydrates from carbon dioxide and water, with oxygen as a by-product.
phyletic evolution genetic changes that occur within an evolutionary line.
phylogeny the line, or lines, of direct descent in a given group of organisms; also the study or the history of such relationships.
phylum one of the primary divisions of the animal and plant kingdoms; a group of closely related classes of animals or plants.
phytoplankton the plant community in marine and fresh-

water situations, containing many species of algae and diatoms, that floats free in the water.

plate tectonics the study of the global-scale movements of the Earth's crust that have resulted in continental drift and the creation of many geological formations.

pleiotropy the phenotypic effect of a gene on more than one characteristic.

Pleistocene a geological epoch, characterized by alternating glacial and interglacial stages, that ended about 10,000 years ago, lasted one to two million years, and is subdivided into four glacial stages and three interglacials.

point source an individual, stationary source of large-volume pollution, usually industrial in origin.

pollutant any natural or artificial substance that enters the ecosystem in such quantities that it does harm to the ecosystem; any introduced substance that makes a resource unfit for a specific purpose.

polygamy a mating system in which a male pairs with more than one female at a time (polygyny) or a female pairs with more than one male (polyandry).

polymorphism occurrence in a population of more than two different forms, independent of sexual differences.

population a group of individuals of a single species.

primary production production by autotrophs, normally green plants.

producer a green plant of chemosynthetic bacteria that converts light or chemical energy into organismal tissue.

production amount of energy (or material) formed by an individual, population or community in a specific time period; see also "primary production," "secondary production," "gross production," "net production."

protandry the condition of an individual that, during the course of its development, changes from male to female.

proximate factors the mechanisms responsible for an evolutionary adaptation with reference to its physiological and behavioral operation; the mechanics of how an adaptation operates; contrast "ultimate factors."

punctuated equilibrium a model that depicts macroevolution as taking place in the form of short periods of rapid speciation alternating with long periods of relative stasis.

***r* and *K* selection** alternative expressions of selection on traits that determine fecundity and survivorship to favor rapid population growth at low population density r or competitive ability at densities near the carrying capacity K_r.

recombinant DNA a single molecule combining DNA from two distinct sources.

recruitment addition, by reproduction, of new individuals to a population.

recycling the process by which waste materials are transformed into usable products.

Red queen hypothesis the idea, named for the character in Lewis Carroll's *Through the Looking Glass*, that a species must continually evolve just to keep pace with environmental change and with other species, let alone to get ahead in the coevolutionary struggle.

resource a substance or object required by an organism for normal maintenance, growth, or reproduction.

respiration the complex series of chemical reactions in all organisms by which stored energy is made available for use and that produces carbon dioxide and water as by-products.

restriction enzyme an enzyme that recognizes a specific base sequence on DNA and cuts DNA. Some restriction enzymes cut the DNA at a particular point, others at random.

riparian related to, living in, or located on the bank of a natural watercourse, usually a river, sometimes a lake or tidewater.

runoff water entering rivers, lakes, reservoirs, or the ocean from land surfaces.

saprophyte a plant that obtains food from dead or decaying organic matter.

secondary production production by herbivores, carnivores, or detritus feeders; contrast "primary production."

search image a behavioral selection mechanism that enables predators to increase searching efficiency prey that are abundant and worth capturing.

self-regulation a process of population regulation in which population increase is prevented by a deterioration in the quality of individuals that make up the population; population regulation by adjustments in behavior and physiology internal to the population rather than by external forces such as predators.

sere the series of successional communities leading from bare substrate to the climax community.

sessile of animals; attached to an object or fixed in place, as for example barnacles.

sewage the organic waste and waste water generated by residential and commercial establishments.

sewerage the entire system of sewage collection, treatment, and disposal; all effluent carried by sewers, whether sanitary sewage, industrial wastes, or storm-water runoff.

sex linked gene a gene carried on one of the sex chromosomes (expressible phenotypically in either sex).

sexual selection selection by one sex for specific characteristics in individuals of the opposite sex, usually exercised through courtship behavior.

sibling species species that are difficult or impossible to distinguish by morphological characters.

sigmoid curve an S-shaped curve, for example, the logistic curve.

slash-and burn cultivation a primarily tropical practice in which forest vegetation is cut, left to dry, and burned to add nutrients to the soil before the land is planted with crops, then abandoned after two to five years as a result of falling yields.

smog a term coined from "smoke" and "fog" to describe photochemical air pollution.

sociobiology the study of the biological reasons behind animal behavior and sociology.

species (both singular and plural) organisms forming a natural population or group of populations that transmit specific characteristics from parent to offspring; a group of organisms

reproductively isolated from similar organisms and usually producing infertile offspring when crossed with them.
stability absence of fluctuations in populations; ability to withstand perturbations without large changes in composition.
stochastic model a mathematical model incorporating factors determined by chance and providing not a single prediction but a range of predictions; contrast "deterministic model."
succession replacement of one kind of community by another; the progressive changes in vegetation and animal life that tend toward climax.
supergene a group of two or more loci between which recombination is so reduced that they are usually inherited together as a single entity.
symbiosis in a broad sense, the living together of two or more organisms of different species; in a narrow sense, synonymous with "mutualism."
sympatric occurring in the same place.
sympatric speciation formation of species without geographic isolation; reproductive isolation that arises between segments of a single population.
synecology the study (including population, community, and ecosystem ecology) of groups of organisms in relation to their environment; contrast "autecology."
synergism the situation in which the combined effect of two factors is greater than the sum of their separate effects.
taiga the northern boreal forest zone; a broad band of coniferous forest south of the Arctic tundra.
territory any area defined by one or more individuals and protected against intrusion by others of the same or different species.
thermocline the thin transitional zone in a lake that separates the epilimnion from the hypolimnion.
threatened species a species not yet endangered but whose population levels are low enough to cause concern.
timberline the uppermost attitudinal limit of forest vegetation.
time lag delay in response to a change.
topsoil the top few inches of soil; rich in organic matter and plant nutrients.
trophic level the functional classification of an organism in a community according to its feeding relationships.
tundra level or undulating treeless land, characteristic of Arctic regions and high altitudes, having permanently frozen subsoil.
turnover rate of replacement of resident species by new, immigrant species.
ultimate factors the evolutionary survival values of adaptations; the evolutionary reasons for an adaptation; contrast "proximate factors."
univoltine having only one generation per year; contrast "multivoltine."
upwelling the process whereby, as a result of wind patterns, nutrient-rich bottom waters rise to the surface of the ocean.
urban pertaining to city areas.
vector an organism (often an insect) that transmits a pathogen (for example, a virus, bacterium, protozoan, or fungus) acquired from one host to another.
vicarants disjunct species, closely related, assumed to have been created when the initial range of their common ancestor was split by some historical event.
vicariance biogeography the study of distribution patterns of organisms that attempts to reconstruct events through the study of shared characteristics (cladistics), often with little or no attention to dispersal capabilities or ecological properties.
watershed the land area that drains into a particular lake, river, or reservoir.
wilderness undisturbed area; as it was before human made changes.
wild type the allele, genotype, or phenotype that is most prevalent in wild populations.
zoogeography the study of the distributions of animals.
zooplankton the animal community, predominantly single-celled animals, that floats free in marine and freshwater environments, moving passively with the currents.

Literature Cited

Abbott, I., and R. Black. 1980. Changes in species compositions of floras on islets near Perth, Western Australia. *Journal of Biogeography* 7: 399–410.

Abele, L. G., D. S. Simberloff, D. R. Strong, and A. B. Thistle 1984. *Community ecology: Conceptual issues and the evidence*. Princeton University Press, Princeton, New Jersey.

Abrahamson, W. G. 1975. Reproductive strategies in dewberries. *Ecology* 56: 721–26.

Achiron, M., and R. Wilkinson 1986. Africa: The last safari? *Newsweek* 108(7): 40–42.

Ackery, R., and R. I. Vane-Wright. 1984. *Milkweed butterflies*. British Museum (Natural History), London.

Adam, P. 1990. *Saltmarsh ecology*. Cambridge University Press, Cambridge, U.K.

Adams, P. A., B. A. Menge, G. G. Mittlebach, D. A. Spiller and P. Yodzis. 1995. The role of indirect effects in food webs. In G. A. Polis and K. O. Winemiller (eds.) *Food webs: Integration of patterns and dynamics*, pp. 371–95. Chapman and Hall, NY.

Addicott, J. F., and H. I. Freedman. 1984. On the structure and stability of materialistic systems: analysis of predator-prey and competition models as modified by the action of a slow growing mutualist. *Theorethical Population Biology*, 26: 320-39.

Addicott, J. F. 1986. Variation in the costs and benefits of mutualism: The interaction between yuccas and yucca moths. *Oecologia* 70: 486–94.

Addicott, J. R., J. M. Aho, M. F. Antolin, D. K. Padilla, J. S. Richardson, and D. A. Soluk. 1987. Ecological neighborhoods: Scaling environmental patterns. *Oikos* 49: 340–46.

Alcock, J. 1979. Animal behavior: An evolutionary approach, 2d ed. Sinauer Associates, Sunderland, Mass.

Alexander, M. M. 1958. The place of aging in wildlife management. *American Scientist* 46: 123–31.

Alexander, R. D. 1974. The evolution of social behavior. *Annual Review of Ecology and Systematics* 5: 325–83.

Alexander, R. D., and P. W. Sherman. 1977. Local mate competition and parental investment in social insects. *Science* 196: 494–500.

Alford, R. A., and H. M. Wilbur. 1985. Priority effects in experimental pond communities: Competition between *Bufo* and *Rana*. Ecology 66: 1097–105.

Allee, W. C. 1931. *Animal Aggregations: A Study in General Sociology*. University of Chicago Press, Chicago.

Allee, W. C., A. E. Emerson, O. Park, and K. P. Schmidt. 1949. *Principles of Animal Ecology*, W. B. Saunders, Philadelphia.

Allen, M. F. (ed.) 1992. *Mycorrhizal functioning: An integrative plant-fungal process*. Routledge, Chapman and Hall, New York.

Allendorf, F. W. 1994. Genetically effective sizes of grizzly bear populations. pp. 155–56 in G. K. Meffe and C. R. Carrol (eds.). *Principles of Conservation Biology*. Sinauer Associates, Mass.

Allendorf, F. W., and R. F. Leary. 1986. Heterozygosity and fitness in natural populations of animals. pp. 57–76 in M. E. Soule (ed.), *Conservation biology: The Science of scarcity and diversity*. Sinauer Associates, Sunderland, Mass.

Ames, B. H. 1983. Dietary carcinogens and anticarcinogens. *Science* 221: 1256–64.

Anderson, D. R., and K. P. Burnham. 1976. *Population Ecology of the Mallard: VI. The Effect of Exploitation on Survival*. U.S. Fish and Wildlife Service Publication 128.

Anderson, J. M. 1981. *Ecology for environmental sciences: biosphere, ecosystems, and Man*. Wiley, New York.

Anderson, R. M. 1982. Epidemiology. In F. E. G. Cox (ed.), *Modern parasitology*, pp. 204–51 Blackwell Scientific Publications, Oxford.

Anderson, R. M., and R. M. May. 1982. Coevolution of hosts and their parasites. *Parasitology* 83: 411–26.

Andrewartha, H. G., and L. C. Birch. 1954. *The distribution and abundance of animals*. University of Chicago Press, Chicago.

Angermeier, P. L. 1995. Ecological attributes of extinction-prone species: Loss of freshwater fishes of Virginia. *Conservation Biology* 9: 143–58.

Antonovics, J., A. D. Bradshaw, and R. G. Turner. 1971. Heavy metal tolerance in plants. *Advances in Ecological Research* 7: 1–85

Applegate, V. C. 1950. *Natural History of the Sea Lamprey*, Petromyzon marinus, *in Michigan*. U.S. Fish and Wildlife Service, Special Scientific Report, Fisheries, No. 55, Washington, D.C.

Armstrong, R. A., and R. McGehee 1980. Competitive exclusions. *American Naturalist* 115: 151–70.

Arnold, M. L., and S.A. Hodges. 1995. Are natural hybrids fit or unfit relative to their parents? *Trends in Ecology and Evolution* 70: 67–71.

Arnquist, G., and D. Wooster. 1995. Meta-analysis: Synthesizing research findings in ecology and evolution. *Trends in Ecology and Evolution* 10: 236–40.

Aron, W. I., and S. H. Smith. 1971. Ship canals and aquatic ecosystems. *Science* 174: 13–20.

Arrhenius, O. 1921. Species and area. *Journal of Ecology* 9: 95–99.

Ashbourne, S. R. C., and Putnam. 1987. Competition, resource-partitioning, and species richness in the phytophagous insects of red oak and aspen in Canada and the U.K. *Acta Oecologia (Generalis)* 8: 43–56.

Askew, R. R. 1968. Considerations on speciation in Chalcidoidea (Hymenoptera). *Evolution* 22: 642–45.

Askew, R. R. 1971. *Parasitic Insects*. Heineman, London.

Auerbach, M. J., and D. R. Strong. 1981. Nutritional ecology of *Heliconia* herbivores: experiments with plant fertilization and alternative hosts. *Ecological Monographs* 51: 63–83.

Azim Abul-Atta, E. A. 1978. *Egypt and the Nile after the construction of the High Aswan Dam*. Department of Irrigation and Land Reclamation, Egypt.

Bach, C. E. 1991. Direct and indirect interactions between ants, scales and plants. *Oecologia* 82: 233–39.

Bach C. E. 1994. Effects of herbivory and genotype on growth and survivorship in sand-dune willow *(Salix cordata)*. *Ecological Entomology* 19: 303–309.

Bagla, P. 1997. Ivory trade seen as threat. *Science* 276: 1972.

Bailey, R. G. 1989. *Ecoregions of the continents*. U.S. Department of Agriculture, Forest Service, Washington, D.C.

Baker, S. J., and C. N. Clarke. 1988. Cage trapping coypus (*Myocastor coypus*) on baited rafts. *Journal of Applied Ecology* 25: 41–48.

Bakker, T. C. M., D. Mazzi, and S. Zala. 1997. Parasite-induced changes in behavior and color make *Gammarus pulex* more prone to fish predators. *Ecology* 78: 1098–104.

Baldwin, N. S. 1964. Sea lamprey in the Great Lakes. *Canadian Audubon Magazine*, November-December: 142–47.

Ballard, R. D. 1977. Notes on a major oceanographic fluid. *Oceanus* 20: 35–44.

Bambach, R. K. 1983. Ecospace utilization and guilds in marine communities through the Phanerozoic. In M. J. S. Tevesz and P. L. McCall (eds.), *Biotic interactions in recent and fossil benthic communities*, pp. 719–46. Plenum, New York.

Barber, R. T., and F. P. Chavez. 1983. Biological consequences of El Niño. *Science* 222: 1203–10.

Barbosa, P., and J.A. Saunders, 1985. Plant allelochemicals: Linkages between herbivores and their natural enemies. *Recent Advances in Phytochemistry* 19: 107–37.

Baroni-Urbani, C., and M. W. Buser 1976. Similarity of binary data. *Systematic Zoology* 28: 251–59.

Baross, J. A., and J. W. Deming. 1995. Growth at high temperatures: Isolation and taxonomy, physiology, ecology. In D. M. Karl (ed.) *Microbiology of deep-sea hydrothermal vent Habitats* CRC Press, New York.

Barrons, K. C. 1981. *Are pesticides really necessary?* Regnery Gateway, Chicago.

Barrowclough, G. F. 1980. Gene flow, effective population sizes, and genetic variance components in birds. *Evolution* 34: 789–98.

Bartholomew, G. A. 1986. The role of natural history in contemporary biology. *BioScience* 36: 324–29.

Barton, A. M. 1986. Spatial variation in the effect of ants on an extrafloral nectary plant. *Ecology* 67: 495–504.

Baskin, Y. 1994. Ecologists dare to ask: how much does diversity matter? *Science* 245: 202–03.

Bates, H. W. 1862. Contributions to an insect fauna of the Amazon Valley. *Transactions of the Linnaean Society of London* 23: 495–566.

Bateson, P. P. G., W. Lotwick, and D. K. Scott. 1980. Similarities between the faces of parents and offspring in Bewick's swans and the differences between mates. *Journal of the Zoological Society of London* 191: 61–74.

Batzli, G. O. 1983. Responses of arctic rodent populations to nutritional factors. *Oikos* 40: 396–406.

Batzli, G. O., R. G. White, S. F. Maclean Jr., F. A. Pitelka, and B. D. Collier. 1980. The herbivore-based trophic system. In J. Brown, P. C. Miller, L. L. Tiezen, and F. L. Bunnell (eds.), *An Arctic Ecosystem*, pp. 335–40. Dowden, Hutchinson and Ross, Stoudsburg, Pennsylvania.

Bawa, K. S., S. Menon, and L. R. Gorman. 1997. Cloning and conservation of biological diversity: Paradox, Panacea, or Pandora's Box? *Conservation Biology* 11: 829–30.

Bazely, D. R., M. Vicari, S. Emmerich, L. Filip, D. Lin, and A. Inman. 1997. Interactions betwen herbivores and endophyte-infected *Festuca rubra* from the Scottish Islands of St. Kilda, Benbecular and Rum. *Journal of Applied Ecology* 34: 847–60.

Bazzaz, F. A. 1991. Habitat selection in plants. *American Naturalist* 137: S116–S130.

Beattie, A. J. 1985. *The evolutionary ecology of ant-plant mutualisms*. Cambridge University Press, Cambridge, U.K.

Beattie, A. J., C. Turnbull, R. B. Knox, and E. G. Williams. 1984. Ant inhibition of pollen function: A possible reason why ant pollination is rare. *American Journal of Botany* 71: 421–26.

Beaufait, W. R. 1960. Some effects of high temperatures on the cones and seeds of jack pine. *Forest Science* 6: 194–99.

Beaver, R. A. 1985. Geographical variation in food web structure in *Nepenthes* pitcher plants. *Ecological Entomology* 10: 241–48.

Begon, M., J. L. Harper, and C. R. Townsend. 1990. *Ecology: Individuals, populations, and communities*. Blackwell Scientific Publications, Oxford.

Belovsky, G. E. 1981. Food plant selection by a generalist herbivore: The moose. *Ecology* 62: 1020–30.

Belovsky, G. E. 1987. Extinction models and mammalian persistence. In M. E. Soule (ed.) *Viable populations for conservation*, pp. 35–58. Cambridge University Press, Cambridge, U.K..

Belsky, A. J. 1986. Does herbivory benefit plants? A review of the evidence. *American Naturalist* 127: 870–92.

Belsky, A. J. 1987. The effects of grazing: Confounding of ecosystem, community, and organism scales. *American Naturalist* 129: 777–83.

Bender, E. A., T. J. Case, and M. E. Gilpin. 1984. Perturbation experiments in community ecology: Theory and practise. *Ecology* 65: 1–13.

Bengtsson, J. 1993. Interspecific competition and determinants of extinction in experimental populations of three rockpool *Daphnia* species. *Oikos* 67: 451–64.

Bengtsson, J., S. R. Baillie and J. Lawton. 1997. Community variability increases with time. *Oikos* 78: 249–56.

Benke, A.C., and J. Bruce Wallace. 1997. Trophic basis of production among riverine caddisflies: Implications for food web analysis. *Ecology* 78: 1132–45.

Berendse, F., M. Schmitz, and W. deVisser. 1994. Experimental manipulation of succession and heathland ecosystems. *Oecologia* 100: 38–44.

Bergelson, J., and M. J. Crawley. 1992a. The effects of grazers on the performance of individuals and populations of scarlet gilia, *Ipomopsis aggregata*. *Oecologia* 90: 435–44.

Bergelson, J., and M. J. Crawley. 1992b. Herbivory and *Ipomopsis aggregata*: The disadvantages of being eaten. *American Naturalist* 139: 870–82.

Berger, J. 1990. Persistence of diferent-sized populations: An empirical assessment of rapid extinctions in bighorn sheep. *Conservation Biology* 4: 91–98.

Berger, J. 1996. Animal behavior and plundered mammals: Is the study of mating systems a scientific luxury or a conservation necessity? *Oikos* 77: 207–16.

Berger, W. H. and F. L. Parker. 1970. Diversity of planktonic foraminifera in deep sea sediments. *Science* 168: 1345–47.

Bergerud, A. T. 1980. A review of the population dynamics of caribou and wild reindeer in North America. In *Second International Reindeer/Caribou Symposium*, pp. 556–81. Roros, Norway.

Bernays, E. A. 1983. Nitrogen in defence against insects. In J. A. Lee, S. McNeill, and I. H. Rorison (eds.), *Nitrogen as an Ecological Factor*, pp. 321–44. 22d Symposium of the British Ecological Society. Blackwell Scientific Publications, Oxford, U.K..

Bernays, E. A., and E. Graham. 1988. On the evolution of host specificity in phytophagous arthropods. *Ecology* 69: 886–92.

Berry, R. J. 1988. Natural history in the twenty-first century. *Archives of Natural History* 15: 1–14.

Berry, R. J. 1990. Industrial melanism and peppered moths (*Biston betularia* (L.)). *Biological journal of the Linnean Society* 39: 301–22.

Berryman, A., and P. Turchin. 1997. Detection of delayed density dependence: Comment. *Ecology* 78: 318–20.

Bertness, M. D. 1987. Competitive and facilitative interactions in acorn barnacle populations in a sheltered habitat. *Ecology* 70: 257–68.

Bertness, M. D., and S. D. Hacker. 1994. Physical stress and positive associations among marsh plants. *The American Naturalist* 144: 363–72.

Bertness, M. D., and G. H. Leonard. 1997. The role of positive interactions in communities: lessons from intertidal habitats. *Ecology* 78: 1979–89.

Bertram, B. C. R. 1975. Social factors influencing reproduction in wild lions. *Journal of the Zoological Society of London* 177: 463–82.

Bertram, B. C. R. 1976. Kin selection in lions and in evolution. In P. P. G. Bateson and R. A. Hinde (eds.), *Growing Points in Ethology*, pp. 281–301. Cambridge University Press, Cambridge, U.K.

Bertram, B. C. R. 1979. Serengeti predators and their social systems. In A. R. E. Sinclair and M. Morton-Griffiths (eds.), *Serengeti: Dynamics of an Ecosystem*, pp. 221–48. University of Chicago Press, Chicago.

Bibby, C. J., M. J. Crosby, M. F. Heath, T. H. Johnson, A. J. Lang, A. J. Sattersfield, and S. J. Thirgood. 1992. *Putting biodiversity on the map: global priorities for conservation*, ICBP, Cambridge, U.K.

Birch, L. C. 1953. Experimental background to the study of the distribution and abundance of insects. I. The influence of temperature, moisture and food on the innate capacity for increase of three grain beetles. *Ecology* 3: 698–711.

Birk, E. M., and P. M. Vitousek. 1986. Nitrogen availability and nitrogenase efficiency in loblolly pine stands. *Ecology* 67: 69–79.

Birks, H. J. B. 1980. British trees and insects: A test of the time hypothesis over the last 13,000 years. *American Naturalist* 115: 600–605.

Bishop, J. A. 1981. A neo-Darwinian approach to resistance: examples from mammals. In J. A. Bishop and L. M. Cook (eds.), *Genetic Consequences of Man-made Change*, pp. 37–51. Academic Press, London.

Bishop, J. A., and L. M. Cook. 1980. Industrial melanism and the urban environment. *Advances in Ecological Research* 11: 373–404.

Bishop, J. A., and L. M. Cook (eds.). 1981. *Genetic consequences of man-made change*. Academic Press, London.

Bjorkman, O. 1973. Comparative studies and photosynthesis in higher plants. In A.C. Grese (ed.) *Photophysiology*, pp. 1–63. vol. 8, Academic Press, New York.

Black, F. L. 1975. Infectious diseases in primitive societies. *Science* 187: 515–18.

Blackburn, T. M., V. K. Brown, B. M. Doube, J. J. D. Greenwood, J. H. Lawton, and N. E. Stork. 1993. The relationship between abundance and body size in natural animal assemblages. *Journal of Animal Ecology* 62: 519–28.

Boecklen, W. J., and D. Simberloff. 1986. Area-based extinction models in conservation. In D. K. Elliott (ed.), *Dynamics of Extinction*, pp. 247–76. John Wiley, New York.

Boersma, L. K., and J. A. Gulland. 1973. Stock assessment of the Peruvian anchovy (*Engraulis ringens*) and management of the fishery. *Journal of the Fisheries Research Board of Canada* 30: 2226–35.

Bohn, H., B. McNeal, and G. O'Connor. 1979. *Soil Chemistry*. John Wiley, New York.

Bonnell, M L., and R. K. Selander. 1974. Elephant seals: genetic variation and near extinction. *Science* 184: 908–09.

Bonner, J. 1994. Wildlife's roads to nowhere? *New Scientist* 20 August 1994, pp. 30–34.

Bonner, J. T. 1965. *Size and Cycle: An Essay on the Structure of Biology*. Princeton University Press, Princeton, N.J.

Bonner, J. T. 1988. *The evolution of complexity by means of natural selection*. Princeton Univ. Press, Princeton, NJ.

Booth, W. 1988. Animals of invention. *Science* 240: 718.

Bormann, F. H., G. E. Likens, and J. M. Melillo. 1977. Nitrogen budget for an aggrading northern hardwood forest ecosystem. *Science* 196: 981–83.

Borner, M., C. D. FitzGibbon, M. Borner, T. M. Caro, W. K. Lindsay, D. A. Collins, and M. E. Holt. 1987. The decline of Serengeti Thompson's gazelle population. *Oecologia* 73: 32–40.

Bowen, G.D. 1980. Mycorrhizal roles in tropical plants and ecosystems. In P. Mikola (ed.) *Tropical Mycorrhizal Research*, pp. 165–90. Clarendon Press, Oxford U.K.

Boyce, M.S. 1984. Restitution of r- and K- selection as a model of density-dependent natural selection. *Annual Review of Ecology and Systematics* 15: 427–47.

Bradbury, J. W., and R. M. Gibson. 1983. Leks and mate choice. In P. Bateson (ed.), *Mate Choice*, pp. 109–38. Cambridge University Press, Cambridge.

Bradbury, J. M., R. M. Gibson, C. E. McCarthy, and S.L. Vehrercam. 1989. Dispersion of displaying male sage grouse: The role of female dispersion. *Behavioral Ecology and Sociobiology* 24: 15–24.

Bragg, T. B., and L. C. Hurlbert. 1976. Woody plant invasion of unburned Kansas bluestem prairie. *Journal of Range Management* 29: 19–29.

Brakefield, P. M. 1990. A decline of melanism in the peppered moth *Biston betularia* in the Netherlands. *Biological Journal of the Linnean Society* 39: 327–34.

Bray, J. R., and J. T. Curtis. 1957. An evaluation of the upland forest communities of southern Wisconsin. *Ecological Monographs* 27: 325–49.

Breen, J. P. 1994. *Acremanium* endophyte interactions with enhanced plant resistance to insects. *Annual Review of Entomology* 39: 401–23.

Brett, J. R. 1959. Thermal requirements of fish: Three decades of study, 1940–1960. Transcript of the second seminar on Biological Problems in Water Pollution, April 1959. U. S. Public Health Service, Taft Center, Cincinnati, Ohio.

Briand, F. 1983. Environmental control of food web structure. *Ecology* 64: 253–63.

Briand, F., and J. E. Cohen. 1984. Community food webs have scale-invariant structure. *Nature* 307: 264–66.

Brimblecombe, P., C. Hammer, H. Rodhe, A. Ryaboshapko, and C. F. Boutron. 1989. Human influence on the sulphur cycle. In P. Brimblecombe and A. Y. Lein (eds.) Evolution of the Global Biogeochemical Sulphur Cycle, pp. 77–121. John Wiley, New York.

Brodbeck, B. V., and D. R. Strong. 1987. Amino acid nutrition of herbivorous insects and stress to host plants. pp. 347–64 in P. Barbosa and J. C. Schultz (eds.). *Insect Outbreaks*. Academic Press, San Diego.

Brody, S. 1945. *Bioenergetics and growth*. Van Nostrand Reinhold, New York.

Bronstein, J. L. 1991. Mutualism studies and the study of mutualism. *Bulletin of the Ecological Society of America* 72: 6–7.

Bronstein, J. L. 1994a. Our current understanding of mutualism. *Quarterly Review of Biology* 69: 31–51.

Bronstein, J. L. 1994b. Conditional outcomes in mutualistic interactions. *Trends in Ecology and Evolution* 9: 214–217.

Brookhaven Symposia in Biology 1969. *Diversity and stability in ecological systems*. No. 22. Brookhaven National Laboratory, Upton, New York.

Brower, L. P. 1969. Ecological chemistry. *Scientific American* 220: 22–29.

Brower, L. P. 1970. Plant poisons on a terrestrial food chain and implication for mimicry theory. In K. L. Chambers (ed.), *Biochemical Coevolution*, pp. 69–82. Proceedings of the 29th Annual Biological Colloquium. Oregon State University Press, Corvallis.

Brower, L. P., W. M. Ryerson, L. L. Coppinger, and S. C. Glazier. 1968. Ecological chemistry and the palatability spectrum. *Science* 161: 1349–1351.

Brown, A. A., and K. P. Davis. 1973. *Forest Fire Control and Its Use*, 2d ed. McGraw-Hill, New York.

Brown, C. R. 1988. Enhanced foraging efficiency through information transfer centers: A benefit of coloniality in cliff swallows. *Ecology* 59: 602–13.

Brown, C. R., and M. Bomberger Brown. 1986. Ectoparasitism as a cost of coloniality in cliff swallows (*Hirundo pyrrhonota*). *Ecology* 67: 1206–18.

Brown, J. H. 1978. The theory of insular biogeography and the distribution of boreal birds and mammals. *Great Basin Naturalist Memoirs* 2: 209–27.

Brown, J. H. 1981. Two decades of homage to Santa Rosalia: Toward a general theory of diversity. *American Zoologist* 21: 877–88.

Brown, J. H. 1989. Patterns, modes, and extents of invasions by vertebrates. In J. A. Drake, H. A. Mooney, F. di Castri, R. H. Groves, F. J. Kruger, M. Rejmanek, and M. Williamson (eds.), *Biological invasions: a global perspective*, pp. 85–109. Wiley, Chicester, England.

Brown, J.H. 1995. *Macroecology*. The University of Chicago Press, Chicago.

Brown, J. H. 1997. An ecological perspective on the challenge of complexity. Ecoessay series number 1. National Center for Ecological Analysis and Synthesis, Santa Barbara, CA.

Brown, J. H., D. W. Davidson, J. C. Munger, and R. S. Inouye. 1986. Experimental community ecology: The desert granivore system. In J. Diamond and T. J. Case (eds.), *Community Ecology*, pp. 41–61. Harper and Row, New York.

Brown, J. H., and A. C. Gibson. 1983. *Biogeography*. C.V. Mosby Company, St. Louis, Mo.

Brown, J. H., and A. Kodric-Brown. 1977. Turnover rates in insular biogeography: effect of immigration and extinction. *Ecology* 58: 445–49.

Brown, J. H., and M. V. Lomolino. 1989. Independent discovery of the equilibrium theory of island biogeography. *Ecology* 70: 1954–57.

Brown, J. H., P.A. Margret, and M. L. Taper. 1993. Evolution of body size: Consequences of an energetic definition of fitness. *American Naturalist* 142: 573–84.

Brown, J. H., and P. F. Nicoletto. 1991. Spatial scaling of species composition: body masses of the North American land mammals. *American Naturalist* 138: 1478–512.

Brown, J. L. 1969. The buffer effect and productivity in tit populations. *American Naturalist* 103: 354–74.

Brown, J. L., E. R. Brown, S. D. Brown, and D. D. Dow. 1982. Helpers: effects of experimental removal on reproductive success. *Science* 215: 421–22.

Brown, J. L., D. D. Dow, E. R. Brown, and S. D. Brown. 1978. Effects of helpers on feeding and nestlings in the grey-crowned babbler, *Pomatostomus temporalis*. *Behavioral Ecology and Sociobiology* 4: 43–60.

Brown, L. R. 1970. Human food production as a process in the biosphere. *Scientific American* 223(3): 161–70.

Brown, V. K., and T. R. E. Southwood. 1987. Secondary succession: patterns and strategies. In A.J. Gray, M.J. Crawley, and P.J. Edwards (eds.), *Colonization, succession and stability*, pp. 315–38. Blackwell Scientific Publishers, Oxford U.K..

Bruce, H. M. 1966. Smell as an exteroceptive factor. *Journal of Animal Science*, Supplement 25: 83–89.

Bryant, J. P., F. S. Chapin III, and D. R. Klein. 1983. Carbon/nutrient balance of boreal plants in relation to vertebrate herbivory. *Oikos* 40: 357–68.

Brylinski, M., and K. A. Mann. 1973. An analysis of factors governing productivity in lakes and reservoirs. *Limnology and Oceanography* 18: 1–14.

Buckingham, G. R. 1987. Florida's #1 weed: *Hydrilla* vs. biocontrol. In Research 87, pp. 22–25. IFAS Editorial Department, University of Florida, Gainesville.

Burdon, J. J. 1987. *Diseases and plant population biology*. Cambridge University Press, Cambridge, U.K.

Burdon, J. J., and S. R. Leather (eds.) 1990. *Pests, pathogens, and plant communities*. Blackwell, Oxford.

Burton, J. A. 1976. Illicit trade in rare animals. *New Scientist* 72: 168.

Bush, G. L. 1975a. Modes of animal speciation. *Annual Review of Ecology and Systematics* 6: 334–64.

Bush, G. L. 1975b. Sympatric speciation in phytophagous parasitic insects. In P. W. Price (ed.), *Evolutionary strategies of parasitic insects and mites*, pp. 187–206. Plenum, New York.

Bush, G. L. 1994. Sympatric speciation in animals: New wine in old bottles. *Trends in ecology and evolution* 9: 285–88.

Bygott, J. D., B. C. R. Bertram, and J. P. Hanby. 1979. Male lions in large coalitions gain reproductive advantage. *Nature* 282: 839–41.

Cain, A. J., and P. M. Sheppard. 1954a. Natural selection in *Cepaea*. *Genetics* 39: 89–116.

Cain, A. J., and P. M. Sheppard. 1954b. The theory of adaptive polymorphism. *American Naturalist* 88: 321–26.

Cairns, J., J. Overbaugh and S. Miller. 1988. The origin of mutants. *Nature* 335: 142–45.

Callaway, R., E. Delucia, D. Moore, R. Nowak, and W. H. Schesinger. 1996. Competition and facilitation: Contrasting effects of *Artemisia tridentata* on desert versus montaine pines. *Ecology* 77: 2130–141.

Callaway, R. M. 1997. Positive interactions in plant communities and the individualistic-continuum concept. *Oecologia* 112: 143–49.

Cameron, R. A. D. 1992. Change and stability on *Cepaea*, populations over 25 years: a case of climatic selection. *Proceedings of the Royal Society* B 248: 181–87.

Caraco, T., and L. L. Wolf. 1975. Ecological determinants of group sizes of foraging lions. *American Naturalist* 109: 343–52.

Carlton, J. T., and J. B. Geller. 1993. Ecological roulette: The global transport of nonindigenous marine organisms. *Science* 261: 78–82.

Caro, T. M., and M. K. Laurenson. 1994. Ecological and genetic factors in conservation: A cautionary tale. *Science* 263: 485–86.

Carothers, J. H. 1986. Homage to Huxley: On the conceptual origin of minimum size ratios among competing species. *The American Naturalist* 128: 440–42.

Carson, R. 1962. *Silent Spring*. Houghton Mifflin, Boston.

Carpenter, S. R. 1996. Microcosm experiments have limited relevance for community and ecosystem ecology. *Ecology* 77: 677–80.

Carpenter, S. R., and J. F. Kitchell. 1993. *The trophic cascades in lakes*. Cambridge University Press, New York.

Caswell, H. 1978. Predator mediated coexistence: A nonequilibrium model. *American Naturalist* 112: 127–54.

Caughley, G., G. C. Grigg, J. Caughley, and G. J. E. Hill. 1980. Does dingo predation control the densities of kangaroos and emus? *Australian Wildlife Research* 7: 1–12.

Ceballos, G., and J. H. Brown 1995. Global patterns of mamalian diversity, endemism and endangerment. *Conservation Biology* 9: 559–68.

Chalk, P. M., and C. J. Smith 1983. Chemodenitrification. In Gaseous loss of nitrogen from plant-soil systems. *Developments in plant and soil sciences*. Vol. 9, ed. J. R. Freney and J. R. Simpson, pp. 65–89. Martinus Nijhoff/Dr. W. Junk. The Hague.

Chapin, F. S. 1980. The mineral mutation of wild plants. *Annual Review of Ecology and Systematics*. 11: 233–60.

Chapin, F. S. III, L. R. Walker, C. L. Fastie, and L. C. Sharman. 1994. Mechanisms of primary succession following deglaciation at Glacier Bay, Alaska. *Ecological Monographs* 64: 149–75.

Charlesworth, B. 1984. The cost of phenotypic evolution. *Paleobiology* 10: 319–27.

Cherfas, J. 1986. What price whales? *New Scientist* 110(1511): 36–40.

Cherif, A. H. 1990. Mutualism: The forgotten concept in teaching science. *American Biology Teacher* 52: 206–8.

Cherrett, J. N. 1984. Key concepts, the results of a survey of our members opinions. Pages 1–16 in J. N. Cherrett, editor, *Ecological concepts: The Contribution of Ecology to an Understanding of the Natural World*. Blackwell Scientific, Oxford, England.

Chesson, P. L., and T. J. Case. 1986. Overview: Non-equilibrium community theories: chance, variability, history, and coexistence. In J.

Diamond and T. J. Case (eds.), *Community Ecology*, pp. 229–39. Harper and Row, New York.

Chew, F. S., and S. P. Courtney. 1991. Plant apparency and evolutionary escape from insect herbivory. *The American Naturalist* 138: 729–750.

Christensen, M. L. 1981. Fire regimes in southeastern ecosystems. In H. A. Mooney, T. M. Bonnicksen, M. L. Christensen, J. E. Lotan, and W. A. Reiners (eds.), *Fire Regimes and Ecosystem Properties*, pp. 117–36. U.S.D.A. Forest Service, General Technical Report WO-26, Washington, D.C.

Christie, W. J. 1974. Changes in the fish species composition of the Great Lakes. *Journal of the Fisheries Research Board of Canada* 31: 827–54.

Clapham, W. B., Jr. 1981. *Human Ecosystems*. Macmillan, New York.

Claridge, M. F., and M. R. Wilson. 1978. British insects and trees: A study in island biogeography or insect/plant coevolution? *American Naturalist* 112: 451–56.

Clark, B. C. 1962. Balanced polymorphism and the diversity of sympatric species. *Systematics Association Publication* 4: 47–70.

Clarke, C. A., F. M. M. Clarke, and H. C. Dawkins. 1990. *Biston betularia* (The peppered moth) in West Kirby, Wirral, 1959–1989: Updating the decline in *F. carbonaria*. *Biological Journal of the Linnean Society* 39: 323–26.

Clarke, C. A., G. S. Mani, and G. Wynne. 1985. Evolution in reverse: Clean air and the peppered moth. *Biological Journal of the Linnean Society* 26: 189–99.

Clay, K. 1990. Fungal endophytes of grasses. *Annual Review of Ecology and Systematics* 21: 275–97.

Clements, F. E. 1905. *Research methods in ecology*. University Publishing Company, Lincoln. Reprinted Arno Press, New York, 1977.

Clements, F. E. 1916. *Plant succession: Analysis of the development of vegetation*. Carnegie Institute of Washington Publication 242.

Clements, F. E. 1936. Nature and structure of the climax. *Journal of Ecology* 24: 252–84.

Clements, F. E., J. Weaver, and H. Hansson. 1926. *Plant competition: An analysis of the development of vegetation*. Carnegie Institute, Washington, D. C.

Closs, G., G. A. Watterson and P. J. Donnelly. 1993. Constant predatory-prey ratios: An arithmetical artifact? *Ecology* 74: 238–43.

Closs, G. P., and P. S. Lake. 1994. Spatial and temporal variation in the structure of an intermittent-stream food web. *Ecological Monographs* 64: 1–21.

Clutton-Brock, T. H., and P. H. Harvey. 1984. Comparative approaches to investigating adaptation. Pages 7–29 in J. R. Krebs and N. B. Davies (eds.), *Behavioural Ecology, an Evolutionary Approach*, 2nd edition. Blackwell Scientific Publications, Oxford.

Clutton-Brock, T. H., F. E. Guinness, and S. D. Albon. 1982. *Red Deer: Behavior and Ecology of Two Sexes*. University of Chicago Press, Chicago.

Cockburn, A. T. 1971. Infectious diseases in ancient populations. *Current Anthropology* 12: 45–62.

Cody, M. L. 1974. *Competition and the Structure of Bird Communities*. Princeton University Press, Princeton.

Cody, M. L., and J. M. Diamond (eds.). 1975. *Ecology and the evolution of communities*. Belknap Press, Cambridge, Mass.

Cohen, J.E. 1995. *How many people can the earth support?* W.W. Norton, New York.

Cohen, J. E., F. Briand, and C. M. Newman. 1990. Community food webs: Data and theory. *Biomathematics*. Vol. 20. Springer Verlag, Berlin.

Cole, B. J. 1983. Assembly of mangrove ant communities: Colonization abilities. *Journal of Animal Ecology* 52: 349–55.

Cole, D. W., and M. Rapp. 1981. Elemental cycling in forest ecosystems. In D. E. Reichle (ed.) *Dynamic properties of forest ecosystems*, pp. 341–409. Cambridge University Press, Cambridge, U.K.

Cole, L.C. 1954. The population consequences of life history phenomena. *Quarterly Review of Biology* 29: 103–37.

Coleman, D. C., C. F. Cooper, M. L. Rosenzweig, C. R. Tracey, P. A. Werner, and L. M. Miller 1982. ESA Publications Committee, subcommittee on journal content: final report, 10 January 1982. ESA Bulletin 63: 26–41.

Coley, P. D. 1983. Herbivory and defensive characteristics of tree species in a lowland tropical forest. *Ecological Monographs* 53: 209–33.

Coley, P. D. 1988. Effects of plant growth and leaf lifetime on the amount and type of anti-herbivore defense. *Oecologia* 74: 531–36.

Collins, N. M., and M. G. Morris. 1985. Threatened Swallowtail Butterflies of the World. The IUCN Red data book, IUCN, Cambridge, UK.

Collins, S.L., S.M. Glenn and D.J. Gibson. 1995. Experimental analysis of intermediate disturbance and initial floristic composition: Decoupling cause and effect. *Ecology* 76: 486–92.

Collins, S. L., and L. L. Wallace. 1990. *Fire in North American Tallgrass prairies*. University of Oklahoma Press, Norman.

Collinson, A. S. 1977. *An introduction to world vegetation*. George Allen and Unwin, London.

Collinson, A. S. 1988. *An introduction to world vegetation*. 2nd edition. Chapman and Hall, N.J.

Colwell, R. K. 1973. Competition and coexistence in a simple tropical community. *American Naturalist* 107: 737–60.

Colwell, R. K. 1984. What's new? Community ecology discovers biology. In P. W. Price, C. N. Slobodchikoff, and W. S. Gaud (eds.), *A New Ecology: Novel Approaches to Interactive Systems*, pp. 387–96. John Wiley, New York.

Colwell, R. K., and G. C. Hautt. 1994. Nonbiological gradients in species richness and a spurious Rapoport effect. *American Naturalist* 144: 570–95.

Colwell, R. K., and D. W. Winkler. 1984. A null model for null models in biogeography. In L. G., Abele, D. S. Simberloff, D.R. Strong, and A. B. Thistle, 1984. *Community ecology: Conceptual issues and the evidence*, pp. 344–59. Princeton University Press, Princeton, New Jersey.

Compton, S. G., J. H. Lawton, and V. K. Rashbrook. 1989. Regional diversity, local community structure, and vacant niches: The herbivorous insects of bracken in South Africa. *Ecological Entomology* 14: 365–73.

Compton, S. G., D. Newsome, and D. A. Jones. 1983. Selection for cyanogenesis in the leaves and petals of *Lotus corniculatus* L. at high latitudes. *Oecologia* 60: 353–58.

Connell, J. H. 1961. The influence of interspecific competition and other factors on the distribution of the barnacle *Chthamalus stellatus]*. *Ecology* 42: 710–23.

Connell, J. H. 1978. Diversity in tropical rain forests and coral reefs. *Science* 199: 1302–10.

Connell, J. H. 1980. Diversity and the coevolution of competitors, or the ghost of competition past. *Oikos* 35: 131–38.

Connell, J. H. 1983. On the prevalence and relative importance of interspecific competition: Evidence from field experiments. *American Naturalist* 122: 661–696.

Connell, J. H. 1990. Personal communication with author.

Connell, J. H., I. R. Noble, and R. O. Slatyer. 1987. On the mechanisms of producing successional change. *Oikos* 50: 136–37.

Connell, J. H., and R. O. Slatyer. 1977. Mechanisms of succession in natural communities and their role in community stability and organization. *American Naturalist* 111: 1119–144.

Connell, J. H., and W. P. Sousa. 1983. On the evidence needed to judge ecological stability or persistence. *American Naturalist* 121: 789–24.

Connor, E., and E. D. McCoy. 1979. The statistics and biology of the species-area realtionship. *American Naturalist* 113: 791–833.

Connor, E. F., E. D. McCoy, and B. J. Cosby. 1983. Model discrimination and expected slope values in species-area studies. *American Naturalist* 122: 789–96.

Connor, E., and D. Simberloff. 1979. The assembly of species communities: Chance or competition? *Ecology* 60: 1132–40.

Connor, E. F., and D. Simberloff. 1984. Neutral models of species cooccurrence patterns. In Abele, L. G., D. S. Simberloff, D. R. Strong, and A. B. Thistle, 1984. *Community ecology: conceptual issues and the evidence*, pp. 316–31. Princeton University Press, Princeton, New Jersey and see Rejoinders, pp. 332–43.

Conover, W. J. 1980. *Practical Nonparametric Statistics*, 2nd ed. Wiley, New York.

Constanza, R., R. d'Arge, R. de Groot, S. Farber, M. Grasso, B. Hannon, K. Limburg, S. Naeem, R. V. O'Neill, J. Paruelo, R. G. Raskin, P. Sutton and M. van den Belt. 1997. The value of the world's ecosystem services and natural capital. *Nature* 387: 253–60.

Cooke, E. 1975. Flow of energy through a technological society. in J. Lenihan and W. W. Fletcher (eds.), *Energy resources and the environment*, pp. 30–62. Blackie, Glasgow.

Cook, R. E. 1969. Variation in species diversity of North American birds. *Systematic Zoology* 18: 63–84.

Cooper, J. P. (ed.). 1975. *Photosynthesis and productivity in different environments*. Cambridge University Press, London.

Cooper, S. M., and N. Owen-Smith. 1985. Condensed tannins deter feeding by browsing ruminants on a South African savanna. *Oecologia* 67: 142–46.

Cooper, S. M., and N. Owen-Smith. 1986. Effects of plant spinescence on large mammalian herbivores. *Oecologia* 68: 446–55.

Cooper, W. S. 1923. The recent ecological history of Glacier Bay, Alaska. II. The present vegetation cycle. *Ecology* 4: 223–46.

Cornell, H. 1974. Parasitism and distributional gaps between allopatric species. *American Naturalist* 108: 880–83.

Cornell, H. V., and D. M. Kahn. 1989. Guild structure in the British arboreal arthropods: is it stable and predictable; *Journal of Animal Ecology* 58: 1003–20.

Costa, J.T. 1997. Caterpillars as social insects. *American Scientist* 85: 150–69.

Côté, I. M., and W. J. Sutherland. 1997. The effectiveness of removing predators to protect bird populations. *Conservation Biology* 11: 395–405.

Cousins, S. 1987. The decline of the trophic level concept. Trends in ecology and evolution 2: 312–215.

Cowles, H. C. 1899. The ecological relations of the vegetation on the sand dunes of Lake Michigan. *Botanical Gazette* 27: 95–117, 167–202, 361–91.

Cox, C. B., I. N. Healey, and P. B. Moore. 1976. *Biogeography, an Ecological and Evolutionary Approach*, 2d ed. Blackwell Scientific Publications, Oxford, U.K.

Cox, F. E. G. 1982. "Immunology." In F. E. G. Cox (ed.), *Modern Parasitology*, pp. 173–203. Blackwell Scientific Publications, Oxford, U.K.

Cox, G. W., and B. J. Le Boeuf. 1977. Female initiation of male competition: a mechanism of mate selection. *American Naturalist* 111: 317–35.

Coyne, J. A. 1954. Correlation between heterozygosity and rate of chromosomal evolution in animals. *American Naturalist* 123: 725–29.

Coyne, J. A. 1976. Lack of genic similarity between two sibling species of *Drosophila* as revealed by varied techniques. *Genetics* 84: 593–607.

Coyne, J. A., A. A. Felton, and R. C. Lewontin. 1978. Extent of genetic variation at a highly polymorphic esterase locus in *Drosophila pseudoobscura*. Proceedings of the National Academy of Sciences of the United States of America 75: 5090–93.

Cracraft, J. 1983. Species concepts and speciation analysis. pp. 159–87 in R. F. Johnson (ed.). *Current Ornithology* Vol. 1, Plenum Press, New York.

Crawley, M. J. 1983. Herbivory, the Dynamics of Animal-Plant Interactions. *Studies in Ecology*, Vol. 10. Blackwell Scientific Publications, Oxford, U.K.

Crawley, M. J. 1985. Reduction of oak fecundity by low-density herbivore populations. *Nature* 314: 163–64.

Crawley, M. J. 1987. Benevolent herbivores? *Trends in Ecology and Evolution* 2: 167–69.

Creel, S. R., and P. M. Waser. 1991. Failures of reproductive suppression in dwarf mongooses (*Helogale parvula*): Accident or adaptation? *Behavioral Ecology* 2: 7–15.

Cristoffer, C., and J. Eisenberg. 1985. *On the Captive Breeding and Reintroduction of the Florida Panther in Suitable Habitats*. Task #1, Report #2, Florida Game and Fresh Water Fish Commission and Panther Technical Advisory Committee, Tallahassee.

Crocker, R. L., and J. Major. 1955. Soil development in relation to vegetation and surface age at Glacier Bay, Alaska. *Journal of Ecology* 43: 427–48.

Cronin, E. W., and P. W. Sherman. 1977. A resource-based mating system: the orange rumped honey guide. *Living Bird* 15: 5–32.

Crosby, A. W. 1986. *Ecological Imperialism: The Biological Expansion of Europe 900–1900*. Cambridge University Press, Cambridge, U.K.

Crow, J. F., and M. Kimura. 1970. *An Introduction to Population Genetics Theory*. Harper and Row, New York.

Csada, R. D., P. C. James, and R. H. M. Espie. 1996. The "file drawer problem" of non-signficant results: Does it apply to biological research? *Oikos* 76: 591–93.

Cummins, K. W. 1974. Structure and function of stream ecosystems. *Bioscience* 24: 631–41. Springer Verlag, Berlin.

Currie, D. J. 1991. Energy and large-scale patterns of animal- and plant-species richness. *American Naturalist* 137: 27–49.

Currie, D. J., and V. Paquin. 1987. Large-scale biogeographical patterns of species richness of trees. *Nature* 329: 326–27.

Cyr, H., and M. L. Pace. 1993. Magnitude and patterns of herbivory in aquatic and terrestrial ecosystems. *Nature* 361: 148–50.

Daday, H. 1954. Gene frequencies in wild populations of *Trifolium repens* L. I. Distribution by latitude. *Heredity* 8: 61–78.

Darlington, P. J., Jr. 1959. Area, climate and evolution. *Evolution* 13: 488–510.

Darwin, C. 1859. *On the Origin of Species by Means of Natural Selection*. John Murray, London.

Davidson, D. W. 1993. The effects of herbivory and granivory on terrestrial plant succession. *Oikos* 68: 23–35.

Davidson, J. 1944. On the relationship between temperature and rate of development of insects at constant temperatures. *Journal of Animal Ecology* 13: 26–38.

Davidson, J., and H. G. Andrewartha. 1948b. The influence of rainfall evaporation, and atmospheric temperature on fluctuations in the size of a natural population of *Thrips imaginis*. *Journal of Animal Ecology* 17: 200–22.

Davidson, J., and H. G. Andrewartha 1948a. Annual trends in a natural population of *Thrips imaginis*. (Thysanoptera). *Journal of Animal Ecology* 17: 193–99.

Davis, M. B. 1983. Holocene vegetational history of the eastern United States. In H. E. Wright Jr. (ed.) *Late-Quaternay Environments of the*

United States, Vol. II, pp. 166–81. The Holocene, University of Minnesota Press, Minnesota.

Davis, M. B. 1986. Climatic instability, time lags, and community disequilibrium. In J. Diamond and T. J. Case (eds.). *Community Ecology*, pp. 269–84. Harper and Row, New York.

Davis, M. G., and Zabinski. 1992. Changes in geographical range resulting from greenhouse warming: Effects on biodiversity in forests. In *Global warming and biodiversity*, pp. 297–308. R. Peters and T. Lovejoy. Yale University Press, New Haven, Connecticut.

Davies, N. B. 1978. Territorial defense in the speckled wood butterfly (*Pararge aegeria*): The resident always wins. *Animal Behaviour* 26: 138–47.

Davies, N. B., B. J. Hatchwell, T. Robson, and T. Burke. 1992. Paternity and parental effort in dunnocks *Prunella modularis*: How good are male chick-feeding rules. *Animal Behavior* 43: 729–45.

Dawkins, R. 1989. *The Selfish Gene*. 2d ed. Oxford University Press, Oxford, U.K.

Dawkins, R., and J. R. Krebs. 1979. Arms races between and within species. *Proceedings of the Royal Society of London Series B* 205: 489–511.

Dayan, T., and D. Simberloff. 1994a. Character displacement, sexual dimorphism and morphological variation among British and Irish mustelids. *Ecology* 75: 1063–73.

Dayan, T., and D. Simberloff. 1994b. Morphological relationships among coexisting heteromyids: an incisive dental character. *The American Naturalist* 143: 462–77.

Dayton, P. K., and M. J. Tegner. 1984. The importance of scale in community ecology: a kelp forest example with terrestrial analogs. In P. W. Price, C. N. Slobodchikoff, and W. S. Gaud (eds.), *A New Ecology: Novel Approaches to Interactive Systems*, pp. 457–81. Wiley, New York.

Dean, T. A., and L. E. Hurd. 1980. Development in an estuarine fouling community: The influence of early colonists on later arrivals. *Oecologia* 46: 295–301.

DeAngelis, D. L. 1975. Stability and connectance in food web models. *Ecology* 56: 238–43.

DeAngelis, D. L. 1992. *Dynamics of nutrient cycling and food webs*. Chapman and Hall, New York.

DeAngelis, D. L., and J. C. Waterhouse. 1987. Equilibrium and non-equilibrium concepts in ecological models. *Ecological Monographys* 57: 1–21.

Debach, P. S., and R. A. Sundby. 1963. Competitive displacement between ecological homologues. *Hilgardia* 43: 105–66.

Demaynadier, P., and M. L. Hunter. 1994. Keystone support. *Bioscience* 44: 2.

den Boer, P. J. 1968. Spreading of risk and stabilization of animal numbers. *Acta Biotheoretica* 18: 165–94.

den Boer, P. J. 1981. On the survival of populations in a heterogeneous and variable environment. *Oecologia* 50: 39–53.

Denniston, C. 1978. Small population size and genetic diversity: implications for endangered species. In S. A. Temple (ed.), Endangered Birds: management techniques for preserving threatened species, pp. 281–89. University of Wisconsin Press, Madison.

Denno, R. F., M. S. McClure, and J. R. Ott. 1995. Interspecific interactions in phytophagous insects: competition re-examined and resurrected. *Annual Review of Entomology* 40: 297–331.

Desowitz, R. S. 1981. *New Guinea tapeworms and Jewish grandmothers: Tales of parasites and people*. W. W. Norton, New York.

De Vos, A., R. H. Manville, and G. Van Gelder. 1956. Introduced mammals and their influence on native biota. *Zoologica* 41: 163–94.

Dial, K.P., and J. M. Marzluff. 1988. Are the smallest organisms the most diverse? *Ecology* 69: 1620–24.

Diamond, J. 1986a. The environmentalist myth. *Nature* 324: 19–20.

Diamond, J. 1986b. Overview: laboratory experiments, field experiments, and natural experiments. In J. Diamond and T. J. Case (eds.), *Community Ecology*, pp. 3–22. Harper and Row, New York.

Diamond, J. M. 1975. Assembly of species communities. In M. L. Cody and J. M. Diamond (eds.) *Ecology and evolution of communities*, pp. 342–44. Harvard Univ. Press, Harvard. MA.

Diamond, J. M. 1978. Niche shifts and the rediscovery of competition: Why did field biologists so long overlook the widespread evidence for interspecific competition that had already impressed Darwin? *American Scientist* 66: 322–31.

Diamond, J. M., and R. M. May. 1977. Species turnover rates on islands: Dependence on census interval. *Science* 197: 266–70.

Dirzo, R., and J. L. Harper. 1982. Experimental studies on slug-plant interactions. IV. The performance of cyanogenic and acyanogenic morphs of *Trifolium repens* in the field. *Journal of Ecology* 70: 119–38.

Di Silvestro, R. L. 1988. U.S. demand for carved ivory hastens African elephant's end. *Audubon* 90(2): 14.

Diver, C. 1929. Fossil records of Mendelian mutants. *Nature* 124: 183.

Dixon, A. F. G., P., Kindlmann, J. Leps and J. Holman. 1987. Why are there so few species of aphids, especially in the tropics. *American Naturalist* 129: 580–92.

Dobson, A., and M. Crawley. 1994. Pathogens and the structure of plant communities. *Trends in Ecology and Evolution* 9: 393–98.

Dobson, A. P., J. P. Rodriguez, W. M. Roberts, and D. S. Wilcove. 1997. Geographic distribution of endangered species in the United States. *Science* 275: 550–53.

Dobzhansky, T. 1936. Studies on hybrid sterility. II. Localization of sterility factors in *Drosophila pseudoobscura* hybrids. *Genetics* 21: 113–35.

Dobzhansky, T. 1950. Evolution in the tropics. *American Scientist* 38: 209–21.

Dobzhansky, T. 1970. *Genetics of the Evolutionary Process*. Columbia University Press, New York.

Donoghue, M. J. 1985. A critique of the biological species concept and recommendation for a phylogenetic alternative. *The Bryologist* 88: 172–81.

Downhower, J. F., and K. B. Armitage. 1971. The yellow-bellied marmot and the evolution of polygamy. *American Naturalist* 105: 355–70.

Drake, J. A. 1990. Communities as assembled structures: do rules govern pattern? *Trends in Ecology and Evolution* 5: 159–64.

Drake, J. A. 1991. Community-assembly mechanics and the structure of an experimental species ensemble. *American Naturalist* 137: 1–26.

Drake, J. A., T. E. Flum, G. J. Witteman, T. Voskuil, A. M. Holyman, C. Creson, D. A. Kenny, G. R. Huxel, C. S. Laurie and J. R. Ducan 1993. The construction and assembly of an ecological landscape. *Journal of Animal Ecology* 62: 117–730.

Drake, J. A., G. R. Huxel, and C. I. Hewitt. 1996. Microcosms as models for generating and testing community theory. *Ecology* 77: 670–77.

Drollette, D. 1997. Wide use of rabbit virus is good news for native species. *Science* 275: 154.

Dunbar, M. J. 1980. The blunting of Occam's Razor, or to hell with parsimony. *Canadian Journal of Zoology* 58: 123–28.

Duncan, P., and N. Vigne. 1979. The effect of group size in horses on the rate of attacks by blood-sucking flies. *Animal Behavior* 27: 623–25.

Dyson, F. J. 1988. *Infinite in All Directions*. Harper and Row, New York.

Eadie, J. MA., L. Broekhoven, and P. Colgan. 1987. Size ratios and artificats: Hutchinson's rule revisiteda. *American Naturalist* 129: 1–17.

Ebenhard, T. 1988. Introduced birds and mammals and their ecological effects. *Swedish Wildlife Research* 13: 1–107.

Ebert, D., and W. D. Hamilton. 1996. Sex against virulence: The coevolution of parasitic diseases. *Trends in Ecology and Evolution* 11: 79–82.

Eckholm, E. P. 1976. *Losing Ground*. Norton, New York.

Edmondson, W. T. 1944. Ecological studies of sessile Rotatoria I. factors limiting distribution. *Ecological Monographs* 14: 31–66.

Edmunds, M. 1974. *Defence in Animals*. Harlow, Essex, U.K.

Edwards, C. A., and G. W. Heath. 1963. The role of soil animals in breakdown of leaf material. In D. Doiksen and J. van der Pritt (eds.), *Soil Organisms*, pp. 76–84. North-Holland, Amsterdam.

Edwards, P. J., and M. P. Gillman. 1987. Herbivores and plant succession. In A. J. Gray, M. J. Crawley, and P. J. Edwards (eds.), *Colonization, succession and stability*, pp. 295–314. Blackwell Scientific Publishers, Oxford.

Edwards, P. J., and S. P. Wratten. 1985. Induced plant defenses against insect grazing: fact or artifact? *Oikos* 44: 70–74.

Edwards, R. W., and M. P. Brooker 1982. *The ecology of the Wye*. Junk, The Hague.

Edwards-Jones, G., and V. K. Brown. 1993. Successional trends in insect herbivore population densities: A field test of a hypothesis. *Oikos* 66: 463–71.

Efron, B. 1982. *The Jackknife, the Bootstrap, and other resampling plans*. Society of Industrial and Applied Mathematics, Philadelphia.

Egler, F. E. 1954. Vegetation science concepts: initial floristic composition—a factor in old-field development. *Vegetatio* 4: 412–417.

Ehler, L. E. 1979. Assessing competitive interactions in parasite guilds prior to introduction. *Environmental Entomology* 8: 558–560.

Ehler, L. E., and R. W. Hall. 1982. Evidence for competitive exclusion of introduced natural enemies in biological control. *Environmental Entomology* 1: 1–4.

Ehleringer, J. R., and R. K. Manson. 1993. Evolutionary and ecological aspects of photosynthetic pathway variation. *Annual Review of Ecology and Systematics* 24: 411–39.

Ehrlich, P. R. 1975. The population biology of coral reef fishes. *Annual Review of Ecology and Systematics* 6: 213–47.

Ehrlich, P. R., and A. H. Ehrlich. 1981. *Extinction: The causes and consequences of the disappearance of species*. Random House, New York.

Ehrlich, P. R., A. Ehrlich, and J. P. Holden. 1977. *Ecoscience: Population, Resources, Environment*. Freeman, San Francisco.

Ehrlich, P. R., and P. H. Raven. 1964. Butterflies and plants: a study in coevolution. *Evolution* 18: 586–608.

Ehrlich, P. R., and E. O. Wilson. 1991. Biodiversity studies: science and policy. *Science* 253: 758–62.

Ehrlich, P. R., and J. Roughgarden. 1987. *The Science of Ecology*. Macmillan, New York.

Ehrlich, P. R., and E. O. Wilson. 1991. Biodiversity studies: science and policy. *Science* 253: 758–62.

Einstein, A., and L. Infeld. 1938. *The evolution of physics: from early concepts to relativity and quanta*. Simon and Schuster, New York.

Eisner, T., and D. J. Aneshansley. 1982. Spray aiming in bombardier beetles: jet deflection by the Coanda effect. *Science* 215: 83–85.

Eisner, T., and J. Meinwald. 1966. Defensive secretions of arthropods. *Science* 153: 1341–1350.

Elner, R. W., and R. N. Hughes. 1978. Energy maximization in the diet of the shore crab, *Carcinus maenas*. *Journal of Animal Ecology* 47: 103–16.

Elton, C. 1927. *Animal Ecology*. Sidgwick and Jackson, London.

Elton, C. 1942. *Voles, mice and lemmings Problems in population dynamics*. Clarendon Press, Oxford.

Elton, C. 1958. *The Ecology of Invasions by Animals and Plants*. Methuen, London.

Elton, C., and M. Nicholson. 1942. The ten-year cycle in numbers of the lynx in Canada. *Journal of Animal Ecology* 11: 215–244.

Enright, J. T. 1976. Climate and population regulation. *Oecologia* 24: 295–310.

Erickson, E., and S. L. Buchmann. 1983. Electrostatics and pollination. In C. E. Jones and R. J. Little (eds.), *Handbook of Experimental Pollination Biology*, pp. 173–184. Van Nostrand Reinhold, New York.

Erlinge, S., G. Göransson, G. Högstedt, G. Jansson, O. Liberg, J. Loman, I. N. Nilsson, T. von Schantz, and M. Sylvén. 1984. Can vertebrate predators regulate their prey? *American Naturalist* 123: 125–33.

Erwin, P. H., J. W. Valentine, and J. J. Sepkoski. 1987. A comparative study of diversification events: The early Paloeozoic versus the Mesozoic. *Evolution* 41: 1177–86

Erwin, T. L. 1982. Tropical forests: Their richness in Coleoptera and other arthropod species. *Coleopterists Bulletin* 36: 74–75.

Erwin, T. L. 1983. Beetles and other insects of tropical forest canopies at Manaus, Brazil, sampled by insecticidal fogging. In S. L. Sutton, T. C. Whitmore, and A. C. Chadwick (eds.), *Tropical Rain Forest: Ecology and Management*, pp. 59–75. Blackwell Scientific Publications, Oxford, U.K.

Erwin, T. L. 1988. The tropical forest canopy: The heart of biotic diversity. In E.O. Wilson (ed.), *Biodiversity*, pp. 123–29. National Academy Press, Washington, D.C.

Estes, J. A., and D. O. Duggins. 1995. Sea otters and kelp forests in Alaska: Generality and variation in a community ecological paradigm. *Ecological Monographs* 65: 75–100.

Estes, J. A., R. J. Jameson, and E. B. Rhode. 1982. Activity and prey selection in the sea otter: influence of population status on community structure. *American Naturalist* 120: 242–58.

Everitt, B. 1980. *Cluster Analysis*. 2d ed. Halsted Press, New York.

Ewekm J. J., M. M. Mazzarino, and C. W. Berish. 1991. Tropical soil fertility changes under monocultures and successional communities of different structure. *Ecological applications*. 1: 289–302.

Facelli, J. M., and E. Facelli. 1993. Interactions after death: Plant litter controls priority effects in a successional plant community. *Oecologia* 95: 277–82.

Facelli, J. M., and P. A. Pickett. 1990. Markovian chains and the role of history in succession. *Trends in Ecology and Evolution* 8: 27–30.

Facelli, J. M., and E. Facelli 1993. Interactions after death: Plant litter controls priority effects in a successional plant community. *Oecologia* 95: 277–82.

Faeth, S. H. 1985. Quantitative defense theory and patterns of feeding by oak insects. *Oecologia* 68: 34–40.

Faeth, S. H. 1988. Plant-mediated interactions between seasonal herbivores: enough for evolution or coevolution? In K. C. Spencer (ed.), *Chemical Mediation of Coevolution*, pp. 391–414. Academic Press, New York.

Faeth, S. H., and K. E. Hammon. 1997a. Fungal endophytes in oak trees: Long–term patterns of abundance and associations with leafminers. *Ecology* 78: 810–19.

Faeth, S. H., and K. E. Hammon. 1997b. Fungal endophytes in oak trees: Experimental analyses of interactions with leafminers. *Ecology* 78: 820–27.

Fagerström, T. 1987. On theory, data, and mathematics in ecology. *Oikos* 50: 258–61.

Fahim, H. M. 1981. *Dams, People and Development: The Aswan High Dam Case*. Pergamon, New York.

Fairweather, P. 1990. Is predation capable of interacting with other community processes on rocky reefs? *Australian Journal of Ecology* 13: 453–64.

Falkowski, P. G., Z. Dublinsky, L. Muscatine, and L. McCloskey. 1993. Population control in symbiotic corals. *BioScience* 43: 606–11.

Fastie, C. L. 1995. Causes and ecosystem consequences of multiple pathways of primary succession at Glacier Bay, Alaska. *Ecology* 76: 1899–912.

Fauth, J. E., J. Bernardo, M. Camara, W. J. Resetarits Jr., J. Van Buskirk, and S. A. McCollum. 1996. Simplifying the jargon of community ecology: a conceptual approach. *The American Naturalist* 147: 282–86.

Fay, P. A., and D. C. Hartnett. 1991. Constraints on growth and allocation patterns of *Silphium integrifolium* (Asteraceae) caused by a cynipid gall wasp. *Oecologia* 88: 243–50.

Feeny, P. 1970. Seasonal changes in the oak leaf tannins and nutrients as a cause of spring feeding by winter moth caterpillars. *Ecology* 51: 565–81.

Feeny, P. 1976. Plant apparency and chemical defense. *Recent Advances in Phytochemistry* 10: 1–40.

Fenchel, T. 1974. Intrinsic rate of natural increase: The relationship with body size. *Oecologia* 14: 317–26.

Fenner, F., and F. Ratcliffe. 1965. *Myxamatosis.* Cambridge University Press, Cambridge.

Fenner, M. 1987. Seed characteristics in relation to succession. In A. J. Gray, M. J. Crawley, and P. J. Edwards (eds.), *Colonization, succession and stability*, pp. 103–14. Blackwell Scientific Publishers, Oxford.

Fergus, C. 1991. The Florida panther verges on extinction. *Science* 251: 1178–80.

Ferguson, K. I., and P. Stiling. 1996. Nonadditive effects of multiple natural enemies on aphid populations. *Oecologia* 108: 375–79.

Ferson, S., and P. Downey, P. Klerks, M. Weissburg, I Kroot, S. Stewart, G. Jacquez, J. Ssemakula, R. Malenky, and K. Anderson. 1986. Competing reviews, or why do Connell and Schoener diagree? *The American Naturalist* 127: 517–76.

Firey, W. J. 1960. *Man, Mind and Land: Theory of Resource Use.* Free Press, Glencoe, III, and Greenwood Press, London.

Fischer, A. G. 1960. Latitudinal variation in organic diversity. *Evolution* 14: 64–81.

Fischer, J., N. Simon, and J. Vincent. 1969. *The Red Book—Wildlife in Danger*. Collins, London.

Fisher, R. A. 1930. *The Genetical Theory of Natural Selection*. Clarendon Press, Oxford

Fitzgibbon, C. D. 1989. A cost to individuals with reduced vigilance in groups of Thompson's gazelles hunted by cheetahs. *Animal Behaviour* 37: 508–10.

Fitzsimmons, A. K. 1996. Stop the parade. *BioScience* 46: 78–79.

Fleiss, J. L. 1981. *Statistical Methods for Rates and Proportions.* John Wiley, New York.

Flessa, K. W., and D. Jablonski. 1985. Declining Phanerozoic background extinction rates: effects of taxonomic structure? *Nature* 313: 216–18.

Floyd, T. 1996. Top-down impact on creosotebush herbivores in a spatially and temporally complex environment. *Ecology* 77: 1544–55.

Food and Agriculture Organization. 1986. *Production Yearbook.* FAO, Rome.

Forbes, S. A. 1880. The food of fishes. *Bulletin Illinois State Laboratory of Natural History* 1: 18–32.

Forbes, S. A. 1883. The food relations of the Carabidae and Coccindellidae. *Bulletin Illinois State Laboratory of Natural History* 1: 33–64.

Forbes, S. A. 1887. The lake as a microcosm. *Bulletin Science Association of Peoria, Illinois 1887*, 77–87.

Ford, R. G., and F. A. Pitelka. 1984. Resource limitation in the California vole. *Ecology* 65: 122–36.

Foster, G. M., and B. G. Anderson. 1979. *Medical Anthropology.* Wiley, New York.

Foster, M. S. 1990. Organization of macroalgal assemblages in the Northeast Pacific. The assumption of homogeneity and the illusion of generality. *Hydrobiologia* 192: 21–23.

Foster, M. S. 1991. Rammed by the Exxon Valdez: A reply to Paine. *Oikos* 62: 93–96.

Foulds, W., and J. P. Grime. 1972a. The influence of soil mositure on the frequency of cyanogenic plans in populations of *Trifolium repens* and *Lotus corniculatus. Heredity* 28: 143–46.

Foulds, W., and J. P. Grime. 1972b. The response of cyanogenic and acyanogenic pheotypes of *Trifolium repens* to soil moisture supply. *Heredity* 28: 181–87.

Fowler, S. V., and J. H. Lawton. 1985. Rapidly induced defenses and talking trees: The devil's advocate position. *American Naturalist* 126: 181–95.

Fowler, S. V., and M. MacGarvin. 1985. The impact of hairy wood ants, *Formica lugubris*, on the guild structure of herbivorous insects on birch, *Beteula pubescens. Journal of Animal Ecology* 54: 847–55.

Fox, B. J. 1983. Mammal species diversity in Australian heathlands: The importance of pyric succession and habitat diversity. In F. J. Kruger, D. T. Mitchell and J. U. M. Jervis (eds.) *Mediterranean-type ecosystems: The role of nutrients*, pp. 473–89. Springer-Verlag, Berlin.

Fox, B. J., and J. H. Brown. 1993. Assembly rules for functional groups in North American desert roden communities. *Oikos* 67: 358–70.

Fox, B.J., and J. H. Brown. 1995. Reaffirming the validity of the assembly rule for functional groups or guilds: A reply to Wilson. *Oikos* 73: 125–32.

Fox, D. R., and J. Ridsill-Smith. 1995. Tests for density-dependence revisited. *Oecologia* 103: 435–43.

Fox, L. R., and P. A. Morrow. 1992. Eucalypt responses to fertilization and reduced herbivory. *Oecologia* 89: 214–22.

France, R. 1992. The North Atlantic latitudinal gradient in species richness and geographical range of freshwater crayfish and amphipods. *American Naturalist* 139: 342–54.

Frankel, O. H., and M. E. Soulé. 1981. *Conservation and evolution*. Cambridge University Press, Cambridge, U.K.

Frankham, R. 1995. Conservation genetics. *Annual Review of Genetics* 29: 305–27.

Frankham, R. 1996. Relationship of genetic variation to population size in wildlife. *Conservation Biology* 10: 1500–1508.

Franks, F., S. F. Mathias and R. H. M. Hatley. 1990. Water, temperature and life. *Philosophical Transactions of the Royal Society*, Series B, 326: 517–83.

Franklin, I. R. 1980. Evolutionary change in small populations. In M. E. Soulé and B. A. Wilcox (eds.). *Conservation Biology: An Evolutionary-Ecological Perspective*. pp. 135–39. Sinauer Associates Sunderland, Massachusetts.

Fretwell, S. D., and H. L. Lucas. 1970. On territorial behaviour and other factors influencing habitat distribution in birds. *Acta Biotheoretica* 19: 16-36.

Fryer, G., and T. D. Iles. 1972. *The Cichlid Fishes of the Great Lakes of Africa*. T. F. H. Publications, Neptune City, N.J.

Fuller, C. A., and A. R. Blaustein. 1996. Effects of the parasite *Eimeria arizonensis* on survival of deer mice (*Peromyosus mariculatus*). *Ecology* 77: 2196–202.

Futuyma, D. J. 1983. Evolutionary interactions among herbivorous and plants. In D. J. Futuyma and M. Slatkin (eds.), *Coevolution*, pp. 207–31. Sinauer Associates, Sunderland, Mass.

Futuyma, D. J. 1986. *Evolutionary Biology*, 2nd ed. Sinauer Associates, Sunderland, Massachusetts.

Futuyma, D. 1995. Science on trial: the case for evolution. Sinauer Associates, Sunderland, Mass, 2d ed.

Futuyma, D. J., and S. C. Peterson. 1985. Genetic variation in the use of resources by insects. *Annual Review of Entomology* 30: 217–38.

Futuyma, D. J., and S. S. Wasserman, 1980. Resource concentration and herbivory in oak forests. *Science* 210: 920–22.

Gange, A. C. 1995. Positive effects of endophyte infections on sycamore aphids. *Oikos* 75: 500–510.

GAO. 1994. Ecosystem management: Additional actions needed to adequately test a promising approach. AGO/RCED-94-111. U.S. General Accounting Office, Washington, D.C.

Gaston, K. J. 1991. The magnitude of global insect species richness. *Conservation Biology* 5: 283–96.

Gauch, H. G., R. A. Whittaker, and S. B. Singer. 1981. A comparative study of non-metric ordinations. *Journal of Ecology* 69: 135–52.

Gause, G. F. 1932. Experimental studies on the struggle for existence. I. Mixed population of two species of yeast. *Journal of Experimental Biology* 9: 389–402.

Gause, G. F. 1934. *The struggle for existence.* Macmillan (Hafner Press), New York (reprinted 1964).

Gehring, C. A., N. S. Cobb, and T. G. Whitham. 1997. Three-way interactions among ectomycorrhizal mutualists, scale insects and resistant and susceptible pinyon pines. *The American Naturalist* 149: 824–41.

Gentle, W., F. R. Humphreys, and M. J. Lambert, 1965. An examination of *Pinus radiata* phosphate fertilizer trial fifteen years after treatment. *Forest Science* 11: 315–24.

Gentry, A. H. 1988. Tree species of upper Amazonian forests. *Proceedings of the National Academy of Science of the United States of America* 85: 156.

George, C. J. 1972. The role of the Aswan High Dam in changing the fisheries of the southeastern Mediterranean. In M. Taghi Farvar and J. P. Milton (eds.), *The careless technology: Ecology and international development*, pp. 159–78. Natural History Press, New York.

Gerrish, G., D. Mueller-Dombois, and K. W. Bridges. 1988. Nutrient limitation and *Metrosideros* forest dieback in Hawaii. *Ecology* 69: 723–27.

Gessel, S. P. 1962. Progress and problems in mineral nutrition of forest trees. In T. T. Kozlowski (ed.), *Tree Growth*, pp. 221–35. Ronald Press, New York.

Gibbens, R. P., K. M. Havstad, D. D. Billheimer, and C. H. Herbel. 1993. Creosotebush vegetation after 50 years of lagomorph exclusion. *Oecologia* 94: 210–17.

Gilbert, F., and J. Owen, 1990. Size, shape, competition and community structure in hoverflies (Diptera: Syrphidae). *Journal of Animal Ecology* 59: 21–39.

Gilbert, F. S. 1980. The equilibrium theory of island biogeography: Fact or fiction. *Journal of Biogeography* 7: 209–35.

Gill, D. E. 1974. Intrinsic rate of increase, saturation density, and competitive ability. II. The evolution of competitive ability. *American Naturalist* 108: 103–16.

Gilman, A. P., D. P. Peakall, D. J. Hallett, G. A. Fox, and R. J. Norstrom. 1979. *Animals as Monitors of Environmental Pollutants.* National Academy Press, Washington, D.C.

Gilpin, M. E. 1987. Spatial structure and population vulnerability. In M. E. Soule (ed), *Viable Populations for Conservation*, pp. 125–40. Cambridge University Press, Cambridge.

Gilpin, M. E., and J. M. Diamond. 1984. Are species co-occurrences on islands non-random, and are null hypotheses useful in community ecology? In L. G. Abele, D. S. Simberloff, D. R. Strong, and A. B. Thistle, 1984. *Community ecology: conceptual issues and the evidence*, pp. 297–315 and 332–343. Princeton University Press, Princeton, New Jersey, USA.

Gleason, H. A. 1926. The individualistic concept of the plant association. *Torrey Botanical Club Bulletin* 53: 7–26.

Glenn-Lewin, D. C., R.K. Peet, and T. T. Vebler (eds.). 1992. *Plant succession: Theory and predictions.* Chapman & Hall, London.

Godfrey, L. R. 1983. *Scientists Confront Creationism.* W. W. Norton, New York.

Goldberg, D., and A. Novoplansky. 1997. On the relative importance of competition in unproductive environments. *Journal of Ecology* 85: 409–18.

Goldsmith, F. B. 1973. The vegetation of exposed sea cliffs at South Stack, Anglesey I. The multivariate approach. *Journal of Ecology* 61: 787–818.

Goldsmith, F. B. 1983. Evaluating nature. In A. Warren and F. B. Goldsmith (eds.), *Conservation in Perspective*, pp. 233–46. John Wiley, Chichester, U.K.

Goldwasser, L., and J. Roughgarden. 1997. Sampling effects and the estimation of food-web properties. *Ecology* 78: 41–54.

Golley, F. B. 1993. *A history of the ecosystem concept.* Yale University Press, New Haven CT.

Goodall, D. W. 1954. Objective methods for the comparison of vegetation III: An essay in the use of factor analysis. *Australian Journal of Botany* 1: 39–63.

Goodland, R. J. 1975. The tropical origin of ecology: Eugen Warming's jubilee. *Oikos* 26: 240–45.

Goodman, D. 1975. The theory of diversity-stability relationships in ecology. *Quarterly Review of Biology* 50: 237–66.

Gordon, H. S. 1954. The economic theory of a common property resource: the fishery. *Journal of Political Economics* 62: 124–42.

Gordon, M. S. 1968. *Animal function: Principles and applications.* Macmillan, New York.

Gorlick, D. L., P. D. Atkins, and G. S. Losey. 1978. Cleaning stations as water holes, garbage dumps, and sites for the evolution of reciprocal altruism. *American Naturalist* 112: 341–53.

Gotelli, N. 1995. *A Primer of Ecology.* Sinauer Associates, Sunderland, Massachusetts.

Gotelli, N. J., and G. R. Graves. 1996. *Null models in ecology.* Smithsonian Institution press, Washington, D.C.

Götmark, F., M. AnAhlund, and M. O. G. Eriksson. 1986. Are indices reliable for assessing conservation value of natural areas? An avian case study. *Biological Conservation* 38: 55–73.

Gould, S. J., and R. C. Lewontin. 1979. The spandrels of San Marco and the Panglossian paradigm: A critique of the adaptationist programme. *Proceedings of the Royal Society of London* B 205: 581–98.

Goulding, M. 1980. *The fishes and the forest: Explorations in amazonian natural history.* University of California Press, Berkeley and Los Angeles.

Grant, B., and R. J. Howlett. 1988. Background selection by the peppered moth (*Biston betularia* Linn): Individual differences. *Biological Journal of the Linnean Society* 33: 217–32.

Grant, B. S., D. F. Owen, and C. A. Clarke. 1995. Parallel rise and fall of melanic peppered moths in America and Britain. *Journal of Heredity* 87: 351–357.

Grant, K. A., and V. Grant. 1964. Mechanical isolation of *Salvia apiana* and *Salvia mellifera* (Labiatae). *Evolution* 18: 196–212.

Grant, P. R., and B. R. Grant. 1987. The extraordinary El Niño event of 1982–1983: Effects on Darwin's finches on Isla Genovesa, Galapagos. *Oikos* 49: 55–66.

Grant, V. 1977. *Organismic evolution.* Freeman, San Francisco.

Grant, V. 1981. *Plant speciation*, 2d ed. Columbia University Press, New York.

Grant, V. 1985. *The evolutionary process: A critical review of evolutionary theory.* Columbia University Press, New York.

Predation: Is nature red in tooth and claw? The color patterns and behaviors of many animals suggest predation is of vital importance.

▲ Aposematic or warning coloration. *Dendrobates histronicus*, a poison dart frog. (Dr. E. R. Degginger, 248.)

▲ If predation were not important, why would this katydid mimic leaves? (Laval, Animals Anials, 4G447S-2.)

◀The monarch butterfly, Batesian model or Mullerian mimic? (Degginger, Animals Animals, 520202-1.)

▶ Batesian mimicry: a harmless hoverfly, *Epistrophe balteata*, imitating a harmful wasp. (Vock/Okapia, Photo Researchers, Inc., 6Z0719.)

Predation continued.

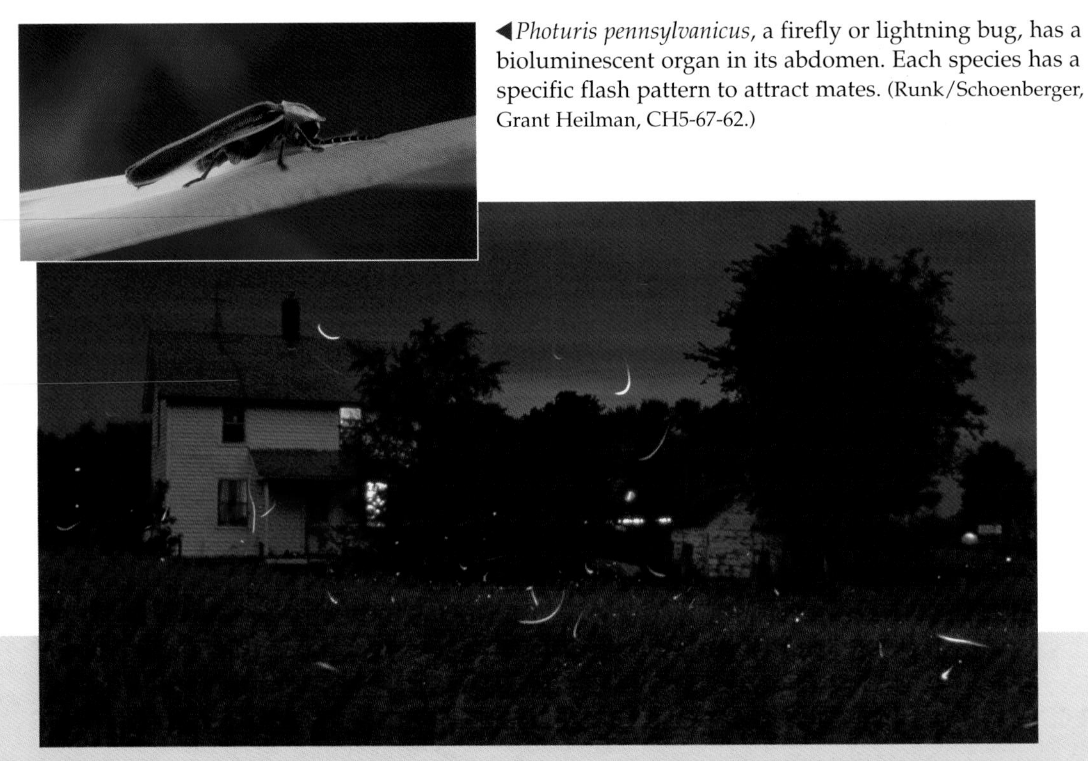

◀*Photuris pennsylvanicus*, a firefly or lightning bug, has a bioluminescent organ in its abdomen. Each species has a specific flash pattern to attract mates. (Runk/Schoenberger, Grant Heilman, CH5-67-62.)

▲Fireflies gathered over this farmyard in Iowa are visible as faint streaks of green. Some females mimic the signals of other species so that they can feed on the "mates" they attract. This could be thought of as aggressive mimicry, allowing a predator better access to its prey. (Kent, Photo Researchers, Inc., 7R2777.)

▲Bluff. This frilled lizard, *Chlamydosaurus kingii*, from Australia, tries to surprise potential predators by extending the frill around its neck to appear larger. (Uhlenhunt, Animals Animals, 544996-R.)

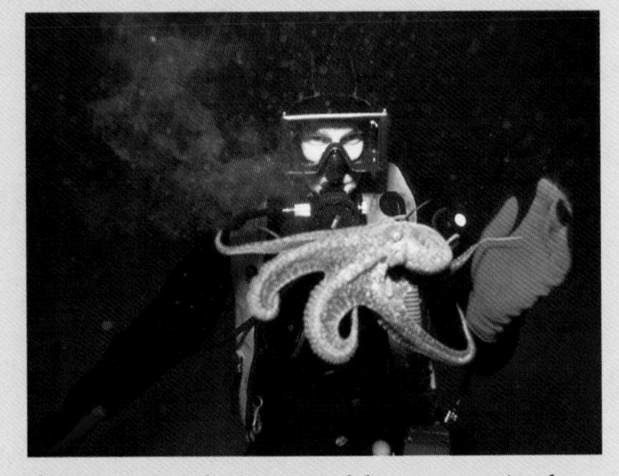

▲Chemical defense is used by some animals like the bombardier beetle, the skunk, or in this case, the octopus, which releases a cloud of ink. (Wu, Peter Arnold, AN-BE-15.)

Many features of plants suggest that herbivory is important in nature.

▲As these spiny oak-worm caterpillars feed on water-oak leaves, their digestion will become less efficient as tannins bind to substances in their guts and prevent digestion. (Peter Stiling.)

◀Bittersweet, *Solanum dulcamara* (also sometimes called deadly nightshade) defends itself against herbivores by being mildly poisonous. An infusion of juice from the European deadly nightshade, *Atropa belladonna*, was formerly dropped in women's eyes causing dilation of the pupils to produce a "wide-eyed" look—hence bella-donna, beautiful lady. (Dr. E. R. Degginger, 260.)

▲Spines on the desert cactus *Echinocereus mohavensis* in Joshua Tree National Monument, California, protect it from herbivory. (Gerlach, Animals Animals, 632235.)

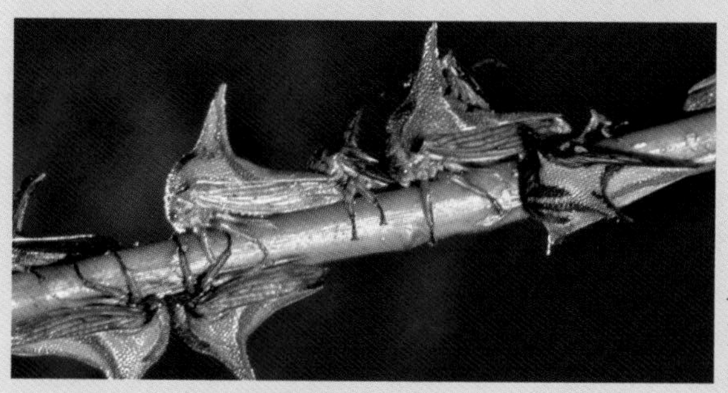

◀Some "thorns" are actually insects mimicking thorns, like these membracid treehoppers in South Florida. (Kolar, Animals Animals I-6078.)

Herbivory continued.

▲ Thorns on *Acacia collinsii*, in Costa Rica, provide homes for *Pseudomyrmex ferruginea* ants, whose aggresive behavior further deters any herbivores. (Dimijian, Photo Researchers, 7Q0700.)

▲ Inside this frothy mass on a stinging nettle leaf lives a cercopid, an insect which feeds on xylem fluid. The cercopid kicks up the excreted xylem fluid into a frothy mass, deterring predators. (Peter Stiling.)

▲ Leaf abscission kills endophytic insects within leaves, like these leafminers. (Peter Stiling.)

▲ The array of spines on the Io caterpillar are testament to strong pressure on it by natural enemies. If the plant can delay insect development long enough, a parasite or predator may kill the caterpillar, thus doing the plant a favor. Tri-trophic interactions involve the influence of plants on natural enemies of their herbivores. (Peter Stiling.)

Gray, J. S. 1981. *The ecology of marine sediments*. Cambridge Univ. Press, Cambridge, U.K.

Gray, J. S. 1987. Species-abundance patterns. In J. H. R. Gee and P. S. Giller (eds.). *Organization of communites: Past and present*, pp. 53–68. Blackwell Scientific Publications, Oxford.

Greene, J. C. 1959. *The death of Adam: Evolution and its impact on Western thought*. Iowa State University Press, Ames, Ia.

Greenslade, P. J. M. 1983. Adversity selection and the habitat templet. *American Naturalist* 122: 352–65.

Greenwood, P. J., and P. H. Harvey, and M. Slatkin (eds.), *Evolution, Essays in Honour of John Maynard Smith*. Cambridge University Press, Cambridge.

Greenwood, P. J., and P. Wheeler. 1985. The evolution of sexual size dimorphism in birds and mammals: a "hot-blooded hypothesis", pp. 287–99.

Greenwood, P. J. 1980. Mating systems, philopatry, and dispersal in birds and mammals. *Animal Behaviour* 28: 1140–62.

Grime, J. P. 1977. Evidence for the existence of three primary strategies in plants and its relevance to ecological and evolutionary theory. *American Naturalist* 111: 1169–94.

Grime, J. P. 1979. *Plant strategies and vegetation process*. John Wiley, New York.

Grime, J. P. 1993. Ecology sans frontieres, *Oikos* 68: 385–92.

Grime, J. P., J. H. C. Cornelissen, K. Thompson, and J. G. Hodgson. 1996. Evidence of a causal connection between anti-herbivore defense and the decomposition rate of leaves. *Oikos* 77: 489–94.

Grimm, V., and C. Wissel. 1997. Babel, or the ecological stability discussions: An inventory and analysis of terminology and a guide for avoiding confusion. *Oecologia* 109: 323–34.

Grinnell, J. 1918. The niche relationships of the California thrasher. *Auk* 34: 427–33.

Gronemeyer, P. A., B. J. Dilger, J. L. Bouzat, and K. N. Paige. 1997. The effects of herbivory on paternal fitness in scarlet gilia: Better moms also make better pops. *American Naturalist* 150: 592–602.

Guiler, E. R. 1961. *Australian Journal of Science* 23: 207–10.

Gulland, J. 1988. The end of whaling? *New Scientist* 120(1636): 42–47.

Gunn, A. S. 1990. Preserving rare species. In *Earthbound: Introductory essays in environmental ethics*, T. Reagan (ed.), pp. 289–308. Waveland Press, Inc. Prospect Heights, Il.

Gupta, A. P., and R. C. Lewontin. 1982. A study of reaction norms in natural populations of *Drosophila pseudoobscura*. *Evolution* 36: 934–48.

Gurrevitch, J., L. L. Morrow, A. Wallace, and J.S. Walsh. 1992. A meta-analysis of field experiments on competition. *American Naturalist* 140: 539–72.

Hacker, S. D., and M. D. Bertness. 1996. Trophic consequences of a positive plant interaction. *The American Naturalist* 148: 559–75.

Hacker, S. D., and S. D. Gaines. 1997. Some implications of direct positive interactions for community species diversity. *Ecology* 78: 1990–2003.

Hairston, J. G., Sr. 1989. *Ecological experiments: purpose, design, and execution*. Cambridge University Press, Cambridge, U.K.

Hairston, N. G. 1969. On the relative abundance of species. *Ecology* 50: 1091–94.

Hairston, N. G. 1991. The literature glut: Causes and consequences: reflections of a dinosaur. *Bulletin of the Ecological Society of America* 72: 171–74.

Hairston, N. G., J. D. Allen, R. K. Colwell, D. J. Futuyma, J. Howell, M. D. Lubin, J. Mathias, and J. H. Vandermeer. 1968. The relationship between species diversity and stability: an experimental approach with protozoa and bacteria. *Ecology* 49: 1091–1101.

Hairston, N. G., and N. G. Hairston. 1993. Cause-effect relationships in energy flow, tropic structure, and interspecific interactions. *American Naturalist* 142: 379–411.

Hairston, N.G. Jr., and N. G. Hairston, Sr. 1997. Does food wed complexity eliminate trophic-level dynamics. *The American Naturalist* 149: 1001–1007.

Hairston, N. G., F. E. Smith, and L. B. Slobodkin. 1960. Community structure, population control, and competition. *American Naturalist* 44: 421–25.

Haldane, J. B. S. 1953. Animal populations and their regulation. *Penguin Modern Biology* 15: 9–24.

Haldane, J. B. S. 1963. The acceptance of a scientific idea. Reprinted in R. L. Weber 1982, *More Random Walks in Science*, (ed.) Institue of Physics, Bristol,

Hall, B. G. 1991. Adaptive evolution that requires multiple spontaneous mutations: mutations involving base substitutions. *Proceedings of the National Academy of Sciences* 88: 5882–86.

Hall, R. W., L. E. Ehler, and B. Bisabri-Ershadi. 1980. Rates of success in classical biological control of arthropods. *Bulletin of the Entomological Society of America* 26: 111–14.

Hall, S. J., and D. G. Raffaelli. 1991. Food web patterns: Lessons from a species rich web. *Journal of Animal Ecology* 60: 823–41.

Hall, S. J., D. G. Raffaelli, and S. Thrush. 1994. Patchiness and disturbance in shallow water benthic assemblages. In P. Giller, A. Hildrew, and D. G. Raffaelli (eds.), *Aquatic ecology: Scale, pattern and process*, pp. 333–75. Blackwell Scientific, Oxford, U.K.

Hamburg, S. P., and C. V. Cogbill. 1988. Historical decline of red spruce populations and climatic warming. *Nature* 331: 428–31.

Hamilton, W. D. 1964. The genetical evolution of social behaviour. I, II. *Journal of Theoretical Biology* 7: 1–52.

Hamilton, W. D. 1967. Extraordinary sex ratios. *Science* 156: 477–488.

Hamilton, W. D. 1971. Geometry for the selfish herd. *Journal of Theoretical Biology* 31: 295–311.

Hamrick, J. L., and M. J. W. Godt. 1990. Allozyme diversity in plant species. In A. H. D. Brown, M. T. Clegg, A. L. Kehles, and B. S. Weir (eds.), *Plant population genetics, breeding, and genetic resources*, pp. 43–63. Sinauer Associates, Sunderland, Mass.

Hansen, A. J. 1986. Fighting behavior in bald eagles: a test of game theory. *Ecology* 67: 787–97.

Hanski, I. 1989. Metapopulation dynamics: does it help to have more of the same? *Trends in Ecology and Evolution* 4: 113–114.

Hanski I. 1991. Single-species metapopulation dynamics: concepts, models, and observations. *Biological Journal of the Linnean Society* 42: 17–38.

Hanski, I., T. Pakkala, M. Kuussarri, and G. Lei. 1995. Metapopulation persistence of an endangered butterfly in a fragmented environment. *Oikos* 72: 21–28.

Harcourt, A. H., P. H. Harvey, S. G. Larson, and R. V. Short. 1981. Testis weight, body weight, and breeding system in primates. *Nature* 293: 55–57.

Hardin, G. 1957. The threat of clarity. *American Journal of Psychiatry* 114: 392–96.

Hardin, G. 1960. The competitive exclusion principle. *Science* 162: 1243–48.

Hardin, G. 1968. The tragedy of the commons. *Science* 162: 1243–48.

Harley, J. L., and S. E. Smith. 1983. *Mycorrhizal Symbiosis*. Academic Press, London.

Harper, J. L. 1977. *The population biology of plants*. Academic Press, London.

Harris, H. 1966. Enzyme polymorphisms in man. *Proceedings of the Royal Society of London Series B* 164: 298–310.

Harris, P. M. 1996. Competitive equivalence in a community of lichens on rock. *Oecologia* 108: 663–68.

Harrison, R. G. 1980. Dispersal polymorphism in insects. *Annual Review of Ecology and Systematics* 11: 95–118.

Harrison, S. 1991. Local extinction in a metapopulation context: An empirical evaluation. *Biological Journal of the Linnean Society* 42: 73–88.

Harrison, S., and N. Cappuccino. 1995. Using density-manipulation experiments to study population regulation. In N. Cappuccino and P. W. Price (eds.), *Population dynamics: New approaches and synthesis*, pp. 131–48. Academic Press, San Diego.

Harrison, S., D. D. Murphy, and P. R. Ehrlich. 1988. Distribution of the bay checkerspot butterfly, *Euphydryas editha bayensis*: Evidence for a metapopulation model. *The American Naturalist* 132: 360–82.

Hart, D. D., and R. J. Horwitz. 1991. Habitat diversity and the species-area relationship: Alternative models and tests. In S. S. Bell, E. D. McCoy, and H. Mushinsky (eds.), *Habitat Structure: The physical arrangement of objects in space*, pp. 47–68. Chapman and Hall, London.

Harvey, P. H., J. J. Bull, and R. J. Paxton. 1983. Looks pretty nasty. *New Scientist* 97: 26–27.

Harvey, P. H., M. Kavanagh, and T. H. Clutton-Brock. 1978. Sexual dimorphism in primate teeth. *Journal of Zoology* 186: 475–86.

Harvey, P. H., and H. C. J. Godfray. 1987. How species divide resources. *American Naturalist* 129: 318–20.

Harvey, P. H., and A. H. Harcourt. 1982. Sperm competition, testes size and breeding systems in primates. In R. C. Smith (ed.), *Sperm Competition and the Evolution of Animal Mating Systems*, pp. 589–600. Academic Press, London.

Hassell, M. P., J. Latto, and R. M. May. 1989. Seeing the wood for the trees: Detecting density dependence from existing life-table studies. *Journal of Animal Ecology* 58: 883–92.

Hastings, A., and S. Harrison. 1994. Metapopulation dynamics and genetics. *Annual Review of Ecology and Systematics* 25: 167–88.

Hastings, J. R., and R. M. Turner. 1965. *The Changing Mile*. University of Arizona Press, Tucson.

Hattersley, P. W. 1983. The distribution of C_3 and C_4 grasses in Australia in relation to climate. *Oecologia* 57: 113–28.

Haukioja, E. 1980. On the role of plant defenses in the fluctuation of herbivore populations. *Oikos* 35: 202–213.

Hawkins, B. A. 1992. Parasitoid-host food webs and donor control. *Oikos* 65: 159–62.

Hawkins, B. A. 1994. *Pattern and process in host-parasitoid interactions*. Cambridge University Press, Cambridge, U.K.

Hawkins, B. A., and N. J. Mills. 1996. Variability in parasitoid community structure. *Journal of Animal Ecology* 65: 501–16.

Hawkins, C. P., and J. A. MacMahon. 1989. Guilds: The multiple meanings of a concept. *Annual Review of Entomology* 34: 423–51.

Haury, L. R., J. A. McGowan, and P. H. Wiebe. 1978. Patterns and processes in the time-space scales of plankton distributions. In J. H. Steele (ed.), *Spatial patterns in plankton communites*, pp. 277–327. Plenum, New York, New York.

Havens, K. E. 1997. Unique structural peroperties of pelagic food webs. *Oikos* 78: 75–80.

Hay, M. 1981. The functional morphology of turf-forming seaweeds: Persistence in stressful marine habitats. *Ecology* 62: 739–50.

Heads, S. P. Lawler, D. Magda, C. D. Thomas, L. J. Thompson, and S. Young. 1993. The Ecotron: A controlled environmental facility for the investigation of populations and ecosystem processes. *Philosophical Transactions of the Royal Society of London*, B. 341: 181–94.

Heal, O. W., and S. F. Maclean. 1975. Comparative productivity in ecosystems —secondary productivity. In W. H. van Dobben and R. H. Lowwe-McConnell (eds.), *Unifying Concepts in Ecology*, pp. 89–108. Dr. W. Junk. The Hague.

Heaney, L. R. 1986. Biogeography of mammals in S.E. Asia: estimates of rates of colonization, extinction, and speciation. *Biological Journal of the Linnaean Society* 28: 127–65.

Heatwole, H., and R. Levins. 1972. Trophic structure, stability and faunal change during recolonisation. *Ecology* 53: 531–34.

Hedrick, P. W., and P. S. Miller. 1992. Conservation genetics: Techniques and fundamentals. *Ecological applications* 2: 30–46.

Heinrich, B. 1976. Flowering phenologies: Bog, woodland, and disturbed habitats. *Ecology* 57: 890–99.

Heinrich, B. 1979. *Bumblebee Economics*. Harvard University Press, Cambridge, Mass.

Heltshe, J. F., and N. E. Forrestor 1983. Estimating species richness using the jackknife procedure. *Biometrics* 39: 1–11.

Hengeveld, R. 1990. *Dynamic biogeography*, Cambridge University Press, Cambridge, U.K.

Heron, A. C. 1972. Population ecology of a colonizing species: The pelagic tunicate *Thalia democratica*. *Oecologia* 10: 269–93, 294–312.

Hershkovitz, P. 1977. *Living New World Monkeys (Platyrhini)*, Vol. 1. University of Chicago Press, Chicago.

Heske, E. J., J. H. Brown, and S. Mistry. 1994. Long-term experimental study of chihuahuan desert rodent community: 13 years of competition. *Ecology* 75: 438–45.

Heslop-Harrison, J. 1964. Forty years of genecology. *Advances in Ecological Research* 2: 159–247.

Hessler, R., P. Lonsdale, and J. Hawkins. 1988. Patterns on the ocean floor. *New Scientist* 117(1605): 47–51.

Heywood, V. H. 1989. Patterns, extents, and modes of invasions by terrestrial plants. In J. A. Drake, H. A. Mooney, F. di Castri, R. H. Groves, F. Kruger, M. Rejmanek, and M. Williamson (eds.), *Biological Invasion: A Global Perspective*, pp. 31–60. Wiley, Chicester, England.

Heywood, V. H. (ed.) 1995. *Global biodiversity assessment*. United Nations Environment Programme, Cambridge University Press, Cambridge, U.K.

Hibbard, C. W. 1960. *An interpretation of Pliocene climates in North America*. Annual Report of the Michigan Academy of Arts and Sciences Letters 62: 5–30.

Hibler, C. P., R. E. Lange, and C. Metzger. 1972. Transplacental transmission of *Protostrongylus* spp. in big-horn sheep. *Journal of Wildlife Diseases* 8: 389.

Hildrew, A. G., C. R. Townsend, and A. Hasham. 1985. The predatory chironomidae of an iron-rich stream: Feeding ecology and food web structure. *Ecological Entomology* 10: 403–413.

Hill, D., R. Wright, and M. Street. 1987. Survival of Mallard ducklings *Anas platyrhynchos* and competition with fish for invertebrates on a flooded gravel quarry in England. *Ibis* 129: 159–67.

Hill, K. D. 1993. The endangered species act: What do we mean by species? *Environmental Affairs* 20: 239–64.

Hiura, T. 1995. Gap formation and species diversity in Japanese beech forests: A test of the intermediate disturbance hypothesis on a geographic scale. *Oecologia* 104: 265–71.

Hixon, M. A., and W. N. Brostoff. 1996. Succession and herbivory: effects of differential fish grazing on Hawaiian coral-reef algae. *Ecological Monographs* 66: 67–90.

Hjalten, J., and P. W. Price. 1997. Can plants gain protection from herbivory by association with unpalatable neighbours: A field experiment in a willow-sawfly system. *Oikos* 78: 317–22.

Hjermann, D. O., and R. Ankerims. 1996. Landscape ecology of the wart-biter *Decticus verrucivorus* in a patchy landscape. *Journal of Animal Ecology* 65: 768–80.

Hokkanen, H., and D. Pimentel. 1984. New approach for selecting biological control agents. *Canadian Entomologist* 116: 1109–21.

Holdgate, M. W. 1960. The fauna of the mid-Atlantic islands. *Proceedings of the Royal Society of London B* 152: 550–67.

Holdridge, L.R. 1967. *Life Zone Ecology*. Tropical Science Center, San José.

Holling, C. S. 1973. Resilience and stability of ecological systems. *Annual Review of Ecology and Systematics* 4: 1–23.

Holling, C. S. (ed.). 1978. *Adaptive Environmental Assessment and Management*. John Wiley and Sons, Chichester, U.K.

Holm, L. G., L. W. Weldon, and R. D. Blackburn. 1969. Aquatic weeds. *Science* 166: 699–709.

Holman, J. A. 1985. New evidence on the status of Ladds quarry. *National Geographic Research* 1: 569–70.

Holt, R. D. 1977. Predation, apparent competition and the structure of prey communities. *Theoretical Population Biology* 12: 197–229.

Holt, R. D. 1997. Personal communication.

Holyoak, K., and P. H. Crowley. 1993. Avoiding eroneously high levels of detection in combinations of semi-independent tests. *Oecologia* 95: 103–14.

Holyoak, M., and S. P. Lawler, 1996. The role of dispersal in predator-prey metapopulation dynamics. *Journal of Animal Ecology* 65: 640–52.

Hooper, D. U., P. M., and Vitousek. 1997. The effects of plant composition and diversity on ecosystem processes. *Science* 277: 1302–05.

Horn, H. S. 1966. Measurement of "overlap" in comparative ecological studies. *American Naturalist* 100: 419–24.

Horn, H. S. 1971. *Adaptive Geometry of Trees*. Princeton University Press, Princeton, N.J.

Horn, H. S. 1975. Markovian processes of forest succession. In M. L. Cody and J. Diamond (eds.), *Ecology and Evolution of Communities*, pp. 196–213. Harvard University Press, Cambridge, Mass.

Horn, H. S. 1976. Succession. In R. M. May (ed.), *Theoretical Ecology: Principles and Applications*, pp. 187–204. Sanders, Philadelphia.

Horn, H. S., and R. M. May. 1977. Limits to similarity among coexisting competitors. *Nature* 270: 660–61.

Houston, A. I., and N. B. Davies. 1985. The evolution of cooperation and life history in the dunnock, *Prunella modularis*. In R. M. Sibly and R. H. Smith (eds.), *Behavioural Ecology, Ecological Consequences of Adaptive Behaviour*, pp. 471–87. Blackwell Scientific Publications, Oxford.

Howard, R. D. 1978. The evolution of mating strategies in bullfrogs, *Rana catesbeiana*. *Evolution* 32: 850–71.

Howard, W. E. 1949. Dispersal, amount of inbreeding, and longevity in a local population of prairie deer mice on the George Reverse, southern Michigan. *Contributions from the Laboratory of Vertebrate Biology of the University of Michigan* 43: 1–50.

Howarth, F. G. 1983. Classical biological control: Panacea or Pandora's box? *Proceedings of the Hawaii Entomological Society* 24: 239–44.

Hubbard., A. L., S. McOrist, T. W. Jones, R. Boid, R. Scott, and M. Easterbee. 1992. Is survival of European wildcats *Felis silvestris* in Britain threatened by interbreeding with domestic cats? *Biological Conservation* 61: 203–8.

Hubbell, S. P., and R. B. Foster. 1986. Biology, chance, and history and the structure of tropical forest tree comunities. In J. Diamond and T. J. Case (eds.). *Community Ecology*, pp. 314–30. Harper and Row, New York.

Hubbell, S. P., and R. B. Foster. 1987. The spatial context of regeneration in a neotropical forest. In A. J. Gray, M. J. Crawley and P. J. Edwards (eds.), *Colonization, succession and stability*, pp. 395–412. Blackwell Scientific Publishers, Oxford.

Hudson, P. J., A. P. Dobson, and D. Newborn. 1992. Do parasites make prey vulnerable to predation? Red grouse and parasites. *Journal of Animal Ecology* 61: 681–92.

Hudson, P. J., D. Newborn and A. P. Dobson. 1992. Regulation and stability of a free-living host-parasite system: *Trichostrongylus tenuis* in the red grouse I. Monitoring and parasite reduction experiments. *Journal of Animal Ecology* 61: 477–486.

Huffaker, C. B. 1958. Experimental studies on predation: Dispersion factors and predator-prey oscillation. *Hilgardia* 27: 343–83.

Huffaker, C. B., and C. E. Kennett. 1959. A ten-year study of vegetational changes associated with biological control of Klamath weed. *Journal of Range Management* 12: 69–82.

Huffaker, C. B., and C. E. Kennett. 1969. Some aspects of assessing efficiency of natural enemies. *Canadian Entomologist* 101: 425–40.

Huffaker, C. B., K. P. Shea and S. G. Herman. 1963. Experimental studies on predation: Complex dispersion and levels of food in an acarine predator-prey interaction. *Hilgardia* 34: 305–330.

Hulme, P. E. 1996. Herbivores and the performance of grassland plants: A comparison of anthropod, mollusc and rodent herbivores. *Journal of Ecology* 84: 43–51.

Hungate, R. E. 1975. The rumen microbial system. *Annual Review of Ecology and Systematics* 6: 39–66.

Hunter, M. D., and P. W. Price 1992. Playing chutes and ladders: Heterogeneity and the relative roles of bottom-up and top-down forces in natural communites. *Ecology* 73: 724–32.

Huntly, N. 1991. Herbivores and the dynamics of communities and ecosystems. *Annual review of ecology and systematics* 22: 477–503.

Hurlbert, S. H. 1971. The nonconcept of species diversity: A critique and alternative parameters. *Ecology* 52: 577–86.

Hutchinson, G. E. 1958. Concluding remarks. *Cold Spring Harbor Symposia on Quantitative Biology* 22: 415–27.

Hutchinson, G. E. 1959. Homage to Santa Rosalia, or why are there so many kinds of animals? *American Naturalist* 93: 145–59.

Hutchinson, J. B. 1965. Crop-plant evolution: A general discussion. In J. B. Hutchinson (ed.), *Essays on Crop Plant Evolution*, pp. 166–81. Cambridge University Press, New York.

Hurlbert, S. H., W. Loayza, and T. Moreno. 1986. Fish-flamingo-plankton interactions in the Peruvian Andes. *Limnology and Oceanography* 31: 457–68.

Huston, M. A. 1979. A general hypothesis of species diversity. *American Naturalist* 113: 81–101.

Huston, M. A. 1994. *Biological diversity: The coexistence of species on changing landscapes*. Cambridge University Press, Cambridge, U.K.

Huston, M. A. 1997. Hidden treatments in ecological experiments: re-evaluating the ecosystem function of biodiversity. *Oecologia* 110: 449–60.

Hutcheson, K. 1970. A test for comparing diversities based on the Shannon formula. *Journal of Theoretical Biology* 29: 151–54.

Hutchinson, G. E., and R. H. MacArthur. 1959. A theoretical ecological model of size distributions among species and animals. *American Naturalist* 93: 117–25.

Hutchinson, J. B. 1965. Crop-plant evolution: A general discussion. In J. B. Hutchinson (ed.), *Essays on Crop Plant Evolution*, pp 166–81. Cambridge University Press, New York.

Iason, G. R., C. D. Duck, and T. H. Clutton-Brock. 1986. Grazing and reproductive success of red deer: The effect of local enrichment by gull colonies. *Journal of Animal Ecology* 55: 507–15.

Inouye, R. S., N. J. Huntly, D. Tilman, J. R. Tester, M. Stillwell, and K. C. Zinnel. 1987. Old-field succession on a Minnesota sand plain. *Ecology* 68: 12–26.

International Union for Conservation of Nature and Natural Resources. 1980. *World Conservation Strategy*. International Union for Conservation of Nature and Natural Resources, United National Environmental Program, World Wildlife Fund. Gland, Switzerland.

International Union for Conservation of Nature and Natural Resources. 1985. *The United Nations List of National Parks and Protected Areas*. Gland, Switzerland.

Itioka, T., and T. Lowe. 1996. Density-dependent ant attendance and its effects on the parasitism of a honeydew-producing scale insect, Ceroplates rubens. *Oecologia* 106: 448–54.

Jablonski, D. 1986a. Evolutionary consequences of mass extinctions. In D. M. Raup and D. Jablonski (eds.), *Patterns and Processes in the History of Life*, pp. 313–30. Springer-Verlag, Berlin.

Jablonski, D. 1993. The tropics as a source of evolutionary novelty through geological time. *Nature* 364: 142–44.

Jaccard, P. 1912. The distribution of the flora of the alpine zone. *New Phytologist* 11: 37–50.

Jackson, M. V., and T. E. Williams. 1979. Response of grass swards to fertilizer N under cutting or grazing. *Journal of Agricultural Science*, Cambridge, 92: 549–62.

Jackson, W., and J. Piper. 1989. The necessary marriage between ecology and agriculture. *Ecology* 70: 1591–93.

Jaksic, F. M., and R. G. Medel. 1990. Objective recognition of guilds: testing for statistically signifcant species clusters. *Oecologia* 82: 87–92.

Jaksic, G. M. 1981. Abuse and misuse of the term "guild" in ecological studies. *Oikos* 37: 397–400.

Jaksic, F. M., and M. Delibes. 1987. A comparative analysis of food-niche relationships and trophic guild structure on how assemblages of vertebrate predators differ in species richness: Causes, correlations, and consequences. *Oecologia* 71: 461–72.

Jaksic, F. M., and R. G. Medel. 1990. Objective recognition of guilds: testing for statistically signifcant species clusters. *Oecologia* 82: 87–92.

Janzen, D. H. 1966. Coevolution of mutualism between ants and acacias in Central America. *Evolution* 20: 249–75.

Janzen D. H. 1968. Host plants as islands in evolutionary and comtemporary time. *American Naturalist* 102: 592–95.

Janzen, D. H. 1970. Herbivores and the number of tree species in tropical forests. *American Naturalist* 104: 501–28.

Janzen, D. H. 1973. Host plants as islands II. Competition in evolutionary and contemporary time. *American Naturalist* 107: 786–90.

Janzen, D. H. 1979a. How to be a fig. *Annual Review of Ecology and Systematics* 10: 13–51.

Janzen, D. H. 1979b. New horizons in the biology of plant defenses. In G. A. Rosenthal and D. H. Janzen (eds.), *Herbivores: Their Interaction with Secondary Plant Metabolites*, pp. 331–50. Academic Press, New York.

Janzen, D. H. 1979c. Why fruit rots. *Natural History Magazine* 88(6): 60–64.

Janzen, D. H., and P. S. Martin. 1982. Neotropical anachronisms: the fruits the gomphotheres ate. *Science* 215: 19-27.

Jarvis, J. U. M., M. J. O'Riain, N. C. Bennett, and P. W. Sherman 1994. Mammalian eusociality: A family affair. *Trends in Ecology and Evolution* 9: 47–51.

Jarvis, J. V. M. 1981. Eusociality in a mammal: Co-operative breeding in naked mole rat colonies. *Science* 212: 241–50.

Jarvis, J. V. M., and J. B. Sale. 1971. Burrowing and burrow patterns of East African mole rats—*Tachyoryctes, Heliophobius*, and *Heterocephalus*. *Journal of Zoology* 163: 451–79.

Jedrzerjewski, W., B. Jedrzerjewski, H. Okarma, and A. L. Ruprecht. 1992. Wolf predation and snow cover as mortality factors in the ungulate community of the Bialowieza National Park, Poland. *Oecologia* 90: 27–36.

Jenkins, B., R. L. Kitching, and S. L. Pimm. 1992. Productivity, disturbance, and food web structure at a local spatial scale in experimental container habitats. *Oikos* 65: 249–55.

Jenny, J. 1980. *The Soil Resource: Origin and Behavior*. Springer-Verlag, New York.

Jimenez, J. A., K. A. Hughes, G. Alaks, L. Graham, and R. C. Lacy. 1994. An experimental study of inbreeding depression in a natural habitat. *Science* 266: 271–73.

Johnson, C. G. 1969. *Migration and Dispersal of Insects by Flight*. Methuen, London.

Johnson, D. M., and P. Stiling. 1996. Host specificity of *Cactoblastis cactorum* Berg, on exotic *Opuntia*—feeding moth, in Florida. *Environmental Entomology* 25: 745–48.

Johnson, E. 1990. Treating the dirt: Environmental ethics and moral theory. In T. Reagan (ed.), *Earthbound: Introductory essays in environmental ethics*, pp. 336–58. Waveland Press, Prospect Heights, Ill.

Johnson, K . H., K. A. Vogt, H. J. Clark, O. J. Schmitz, and D. J. Vogt. 1996. Biodiversty and the productivity and stability of ecosystems. *Trends in Ecological and Evolution* 11: 372–77.

Johnson, M. S. 1978. The botanical significance of derelict industrial sites in Britain. *Environmental Conservation* 5: 223–38.

Johnson, M. S., P. D. Putwain, and R. J. Holliday. 1978. Wildlife conservation values of derelict metalliferous mine workings in Wales. *Biological Conservation* 14: 131–48.

Johnston, C. A., J. Pastor, and R. J. Naiman. 1992. Effects of beaver and moose on boreal forest landscapes. In R. Haines-Young, D. R. Green, and S. Cousin (eds.), *Landscape Ecology and Geographic Information Systems*, pp. 237–254. Taylor and Francis Ltd., London.

Johnston, R. J., D. Gregory, and D. M. Smith. 1994. *The dictionary of human geography*. Basil Blackwell, Oxford, U.K.

Jones, D. A. 1973. Co-evolution and cyanogenesis. In V. H. Heywood (ed.), *Taxonomy and Ecology*, pp. 213–42. Academic Press, New York.

Jones, D. F. 1924. The attainment of homozygosity in inbred strains of maize. *Genetics* 9: 405–18.

Jones, J. S., B. H. Leith, and P. Rawlings. 1977. Polymorphism in *Cepaea*: a problem with too many solutions. *Annual Review of Ecology and Systematics* 8: 109–43.

Jongman, R. H. G., C. J. F. Ter Braak, and O. F. R. Van Tongeren (eds.). 1995. *Data analysis in community and landscape ecology*, Cambridge University Press, Cambridge, U.K.

Jordan, C. F., and J. R. Kline. 1972. Mineral cycling: Some basic concepts and their application in a tropical rain forest. *Annual Review of Ecology and Systematics* 3: 33–49.

Jukes, T. H. 1983. Evolution of the amino acid code. In M. Nei and R. K. Koehn (eds.), *Evolution of Genes and Proteins*, pp. 191–207. Sinauer Associates, Sunderland, Mass.

Karban, R. 1987. Effects of clonal variation of the host plant, interspecific competition, and climate on the population size of a folivorous thrips. *Oecologia* 74: 298–303.

Karban, R. 1989a. Community organization of *Erigeron glaucus* folivores: Effects of competition, predation, and host plant. *Ecology* 70: 1028–39.

Karban, R. 1989b. Fine-scale adaptation of herbivorous thrips to individual host plants. *Nature* 340: 60–61.

Karban, R. 1992. Plant variation: Its effect on populations of herbivorous insects. In R. S. Fritz and E. L. Simms (eds.), *Plant resistance to herbivores and pathogens: ecology, evolution and genetics*, pp. 193–215. University of Chicago Press, Chicago.

Karban, R. 1997. Evolution of prolonged development: A life table analysis for periodical cicadas. *American Naturalist* 150: 446–61.

Karban, R., and J. R. Carey. 1984. Induced resistance of cotton seedlings to mites. *Science* 225: 53–54.

Karban, R., and R. E. Ricklefs. 1984. Leaf traits and species richness and abundance of lepidopteran larvae on deciduous trees in southern Ontario. *Oikos* 43: 165–70.

Karr, J. R. 1991. Avian survival rates and the extinction process on Barro Colorado Island, Panama. *Conservation Biology* 4: 391–97.

Karran, J. D. 1987. A comparison of levels of genetic polymorphism and self-compatibility in geographically restricted and widespread plant congeners. *Evolutionary Ecology* 1: 47–58.

Keddy, P. 1990. Is mutualism really irrelevant to ecology? *Bulletin of the Ecological Society of America* 71: 101–102.

Keith, L. B. 1983. Role of food in hare population cycles. *Oikos* 40: 385–95.

Keller, M. A. 1984. Reassessing evidence for competitive exclusion of introduced natural enemies. *Environmental Entomology* 13: 192–95.

Kellert, G.R., and E. O. Wilson (eds.). 1993. *Biophilia*. Island Press.

Kennedy, G. G., and J. D. Barbour. 1992. Resistance variation in natural and managed systems. In R. S. Fritz and E. L. Simms (eds.), *Plant resistance to herbivores and pathogens: Ecology, evolution and genetics*, pp. 13–41. The University of Chicago Press, Chicago.

Kent, M., and P. Coker. 1992. *Vegetation description and analysis: A practical approach*. CRC Press, Boca Raton, Florida.

Kenward, R. E. 1978. Hawks and doves: Factors affecting success and selection in goshawk attacks on wood-pigeons. *Journal of Animal Ecology* 47: 449–60.

Kerfoot, W. C., and A. Sih. (eds.). 1987. *Predation: Direct and indirect impacts on aquatic communities*. University of New England Press, Hanover, N.H.

Kerr, J. T., and D. J. Currie. 1995. Effects of human activity on global extinction risk. *Conservation Biology* 9: 1528–38.

Kerr, R. A. 1988. Whom to blame for the Great Storm? *Science* 239: 1238–39.

Kerr, R. A. 1997. Greenhouse forecasting still cloudy. *Science* 276: 1040–42.

Kethley, J. B., and D. E. Johnston. 1975. Resource tracking patterns in bird and mammal ectoparasites. *Miscellaneous Publications of the Entomological Society of America* 9: 231–36.

Kettlewell, H. B. D. 1955. Selection experiments on industrial melanism in the Lepidoptera. *Heredity* 10: 287–301.

Kettlewell, H. B. D. 1973. *The Evolution of Melanism*. Clarendon Press, Oxford.

Kilburn, P. D. 1966. Analysis of the species-area relation. *Ecology* 47: 831–43.

Kimura, M. 1983a. *The Neutral Theory of Molecular Evolution*. Cambridge University Press, Cambridge, U.K.

Kimura, M. 1983b. The neutral theory of molecular evolution. In M. Nei and R. K. Koehn (eds.), *Evolution of Genes and Proteins*, pp. 208–233. Sinauer Associates, Sunderland, Mass.

King, C. E. 1964. Relative abundance of species and MacArthur's Model. *Ecology* 45: 716–27.

King, D. A. 1986. Tree form, height growth, and susceptibility to wind damage in *Acer saccharum*. *Ecology* 67: 980–90.

Kishk, M. A. 1986. Land degradation in the Nile Valley. *Ambio* 15: 226–30.

Klein, D. R. 1968. The introduction, increase, and crash of reindeer on St. Matthew Island. *Journal of Wildlife Management* 32: 350–67.

Knoll, A. H. 1986. Patterns of change in plant communities through geological fire. In J. Diamond and T. J. Case, *Community Ecology*, pp. 126–42. Harper and Row, New York.

Knowlton, N. 1979. Reproductive synchrony, parental investment and the evolutionary dynamics of sexual selection. *Animal Behaviour* 27: 1022–83.

Koblentz-Mishke, I. J., V. V. Volkounsky, and R. J. B. Kabanova. 1970. Plankton primary production of the world ocean. In W. S. Wooster (ed.), *Scientific Exploration of the South Pacific*, pp. 183–193. National Academy of Sciences, Washington, D.C.

Koehn, R., W. J. Diehl, and T. M. Scott. 1988. The differential contribution by individual enzymes of glycolysis and protein catabolism to the relationship between heterozygosity and growth rate in the coot clam *Mulinia lateralis*. *Genetics* 118: 121–30.

Koehn, R. K., A. J. Zera, and J. G. Hall. 1983. Enzyme polymorphism and natural selection. In M. Nei and R. K. Koehn (eds.), *Evolution of Genes and Proteins*, pp. 115–36. Sinauer Associates, Sunderland, Mass.

Kozhov, M. 1963. Lake Baikal and its life. *Monographs in Biology* 11: 1–344.

Krebs, C. J. 1985a. *Ecology: the Experimental Analysis of Distribution and Abundance*, 3d ed. Harper and Row, New York.

Krebs, C. J. 1985b. Do changes in spacing behaviour drive population cycles in small mammals? In R. M. Sibly and R. H. Smith (eds.), *Behavioural Ecology: Ecological Consequences of Adaptive Behaviour*, pp. 295–312. Blackwell Scientific Publications, Oxford.

Krebs, C. J. 1988. The experimental approach to rodent population dynamics. *Oikos* 52: 143–49.

Krebs, C. J. 1989. *Ecological Methodology*. Harper and Row, New York.

Krebs, C. J., S. Boutin, R. Boonstra, A. R. E. Sinclair, J. N. M. Smith, M. R. T. Dale, and R. Turkington. 1995. Impact of food and predators on the snowshoe hare cycle. *Science* 269: 112–15.

Krebs, J. R. 1971. Territory and breeding density in the great tit, *Parus major* L. *Ecology* 52: 2–22.

Krebs, J. R., and N. B. Davies. 1993. *An introduction to behavioural ecology*, 3d ed. Blackwell Scientific Publications, Oxford.

Krementz, D. G., M. J. Conroy, J. E. Hines, and H. F. Percival. 1988. The effects of hunting on survival rates of American black ducks. *Journal of Wildlife Management* 52: 214–26.

Kruuk, H. 1964. Predators and anti-predator behaviour of the black headed gull, *Larus ridibundus*. *Behaviour Supplement* 11: 1–129.

Kruuk, H., and M. Turner. 1967. Comparative notes on predation by lion, leopard, cheetah and wild dog in the Serengeti area, East Africa. *Mammalia* 31: 1–27.

Kucera, C. L. 1981. Grasslands and fire. In H. A. Mooney, T. M. Bonnicksen, M. L. Christensen, J. E. Lotan, and W. A. Reiners (eds.), *Fire Regimes and Ecosystem Properties*, pp. 9–111. U.S.D.A. Forest Service General Technical Report WO-26, Washington, D.C.

Kuris, A. M., A. R. Blaustein, and J. J. Alio. 1980. Hosts as islands. *American Naturalist* 116: 570–86.

Kurtén, B. 1963. Return of a lost structure in the evolution of the felid dentition. *Societas Scientiarum Fennica Arsbok-Vuosikirja* 26: 3–11.

LaBreque, M. 1992. To model the otherwise unmodelable. *Mosaic* 23: 12–23.

Lacey, R. C. 1987. Loss of genetic diversity from unmanaged populations: Interacting effects of drift, mutation, immigration, selection, and population subdivision. *Conservation Biology* 1: 143–58.

Lack, D. 1968. *Ecological Adaptations for Breeding in Birds*. Methuen, London.

Lack, D. 1971. *Ecological isolation in birds*. Blackwell Scientific Publications, Oxford, U.K.

Lamarck, J. B. P. de. 1809. *Philosophie Zoologique*. Paris.

Lancaster, J., and A. L. Robertson. 1995. Microcrustacean prey and macroinvertebrate predators in a stream food web. *Freshwater Biology* 34: 123–34.

Lande, R. 1976. The maintenance of genetic variability by mutation in a polygenic character with linked loci. *Genetic Research* 26: 221–35.

Lande, R. 1995. Mutation and Conservation. *Conservation Biology* 9: 782–91.

Lank, D. B., L. W. Oring, and S. J. Maxson. 1985. Mate and nutrient limitation of egg-laying in a polyandrous shorebird. *Ecology* 66: 1513–24.

Larcher, W. 1980. *Physiological Plant Ecology*. Springer-Verlag, New York.

Latham, R. E., and R. E. Ricklefs. 1993. Continental comparisons of temperate-zone tree species diversity. In R. E. Ricklefs and P. Schluter (eds.), *Species diversity in ecological communities*, pp. 294–314. University of Chicago Press, Chicago.

Laurie, H., P. J. Mustart and R. M. Lowling. 1997. A shared niche? The case of the species pair *Protea obtusifolia - Leucadendron meridianum*. *Oikos* 79: 127–36.

Lawler, A. 1997. Changing the game. *Science* 276: 1787.

Lawler, S. P. 1993. Species richness, species composition, and population dynamics of protists in experimental microcosms. *Journal of Animal Ecology* 64: 711–19.

Lawler, S. P., and P. J. Morin. 1993. Food web architecture and population dynamics in laboratory microcosms of protists. *American Naturalist* 141: 675–86.

Lawlasisr, L. R. 1978. A comment on randomly constructed ecosystem models. *American Naturalist*. 112: 443–47.

Lawrence, W. F. 1991. Ecological correlates of extinction proneness in Australian tropical rain forest mammals. *Conservation Biology* 5: 79–89.

Lawton, J. H. 1984. Non-competitive populations, non-convergent communities, and vacant niches: The herbivores of bracken. In D. R. Strong, D. Simberloff, L. G. Abele, and A. B. Thistle (eds.), *Ecological Communities: Conceptual Issues and the Evidence*, pp. 67–100. Princeton University Press, Princeton.

Lawton, J. H. 1987. Are there assembly rules for successional communities? In A. J. Gray, M. J. Crawley, and P. J. Edwards (eds.) *Colonization, succession and stability*, pp. 225–44. Blackwell, Oxford, England.

Lawton, J. H. 1991a. Species richness and population dynamics of animal assemblages. Patterns in body size: Abundance space. *Philosophical Transactions of the Royal Society*, B. 330: 283–91.

Lawton, J. H. 1991b. Are species useful? *Oikos* 62: 3–4.

Lawton, J. H. 1994. What do species do in ecosystems. *Oikos* 71: 367–74.

Lawton, J. H., and C. G. Jones. 1995. Linking species and ecosystems: organisms as ecosystem engineers. In C. G. Jones and J. H. Lawton (eds.), *Linking species and ecosystems*, pp. 141–50. Chapman and Hill, New York, New York.

Lawton, J. H., and R. M. May. 1984. The birds of Selborne. *Nature* 306: 732–33.

Lawton, J. H., and R. M. May. 1995. *Extinction Rates*. Oxford University Press, Oxford, U.K.

Lawton, J. H., and S. McNeill. 1979. Between the devil and the deep blue sea: On the problem of being a herbivore. In R. M. Anderson, B. D. Taylor, and L. R. Taylor (eds.), *Population Dynamics*, pp. 223–44. Blackwell Scientific Publications, Oxford.

Lawton. J. H., S. Naeem, R. M. Woodfin, V. K. Brown, A. Gange, H. C.J. Godfray, P. A. Heads, S. Lawler, D. Magda, C.D. Thomas, L.J. Thompson and S. Young. 1993. The ecotron: a controlled environmental facility for the investigation of population and ecosystem processes. *Philosophical Transactions of the Royal Society of London*. B341:181–194.

Lawton, J. H., and D. R. Strong. 1981. Community patterns and competition in folivorous insects. *American Naturalist* 118: 317–38.

Lawton, J. H., T. M. Lewinsohn and S. G. Compton. 1993. Patterns of diversity for the insect herbivores on bracken. In R. E. Ricklefs and D. Schluter (eds.), *Species Diversity in Ecological Communities*, pp. 178–84. University of Chicago Press, Chicago.

Leader-Williams, N., and S. D. Albon. 1988. Allocation of resources for conservation. *Nature* 336: 533–35.

Leader-Williams N., S. Dalban, and P. S. M. Berry. 1990. Illegal exploitation of black rhinoceros and elephant populations: Patterns of decline, law enforcement and patrol effort in Luangwa valley, Sambia. *Journal of Applied Ecology* 27: 1055–87.

Le Boeuf, B. J. 1974. Male-male competition and reproductive success in elephant seals. *American Zoologist* 14: 163–76.

Le Boeuf, B. J., and S. Kaza (eds.). 1981. *The Natural History of Año Nuevo*. Boxwood Press, Pacific Grove, Calif.

Lee, J. A., and B. Greenwood. 1976. The colonization by plants of calcareous wastes from the salt and alkali industry in Cheshire. *Biological Conservation* 1: 131–49.

Lees, D. R. 1981. Industrial melanism: Genetic adaptation of animals to air pollution. In J. A. Bishop and L. M. Cook (eds.), *Genetic Consequences of Man-made Change*, pp. 129–176. Academic Press, London.

Lees, D. R., and E. R. Creed. 1975. Industrial melanism in *Biston betularia*: the role of selective predation. *Journal of Animal Ecology* 44: 67–83.

Lehman, T. 1993. Ectoparasites: Direct impact on host fitness. *Parasitology Today* 9: 8–13.

Leibold, M. A. 1989. Resource edibility and the effects of predators and productivity on the outcome of trophic interactions. *American Naturalist* 134: 922–49.

Lenski, R. E., and J. E. Mittler. 1993. The directed mutation controversy and Neo-Darwinism. *Science* 259: 188–94.

Lerner, I. M. 1954. *Genetic Homeostasis*. Oliver and Boyd, Edinburgh.

Lessels, C. M. 1991. The evolution of life histories. In J. R. Krebs and N. B. Davies (eds.), *Behavioral Ecology*, pp. 32–68. 3d ed. Blackwell Scientific Publications, Oxford.

Letcher, A. J., and P. H. Harvey. 1994. Variations in geographical range size among mammals of the Palearctic. *American Naturalist*. 144: 30–42.

Leverich, W. J., and P. A. Levin. 1979. Age-specific survivorship and reproduction in *Phlox drummondi*. *American Naturalist* 113: 881–903.

Levin, D. A. 1979. The nature of plant species. *Science* 204: 381–84.

Levin, D. A. 1981. Dispersal versus gene flow in plants. *Annals of the Missouri Botanical Garden* 68: 233–53.

Levin, D. A. 1983. Polyploidy and novelty in flowering plants. *American Naturalist* 122: 1–25.

Levin, S. A. 1970. Community equilibria and stability, and on extension of the competitive exclusion principle. *American Naturalist* 104: 413–23.

Levin, S. A. 1974. Dispersion and population interactions. *American Naturalist* 108: 207–28.

Levin, S. A. 1976. Competitive interactions in ecosystems. *American Naturalist*. 110: 903–10.

Levin, S. A., 1992. The problem of pattern and scale in ecology. *Ecology* 73: 1943–67.

Levine, S. 1976. Competitive interactions in ecosystems. *American Naturalist*. 110: 903–10.

Levins, R. 1968. *Evolution in Changing Environments*. Princeton University Press, Princeton, N.J.

Levins, R., and B. B. Schultz. 1996. Effects of density dependence, feedback and environmental sensitivity on correlations among predators, prey and plant resources: Models and practical implications. *Journal of Animal Ecology* 66: 802–12.

Lewin, R. 1983. Santa Rosalia was a goat. *Science* 221: 636–39.

Lewin, R. 1986. In ecology, change brings stability. *Science* 234: 1071–73.

Lewontin, R. C. 1974. The analysis of variance and the analysis of causes. *American Journal of Human Genetics* 26: 400–411.

Lewontin, R. C., and J. L. Hubby. 1966. A molecular approach to the study of genic heterozygosity in natural populations. II. Amount of variation and degree of heterozygosity in natural populations of *Drosophila pseudoobscura*. *Genetics* 54: 595–609.

Lieberei, R., B. Biehl, A. Giesemann, and N. T. V. Junqueira. 1989. Cyanogenesis inhibits active defense reactions in plants. *Plant physiology* 90: 33–36.

Liebert, T. G., and P. M. Brakefield. 1987. Behavioral studies on the peppered moth *Biston betularia* and a discussion of the role of pollution and epiphytes in industrial melanism. *Biological Journal of the Linnen Society* 31: 129–50.

Lieth, H. 1975. Primary productivity in ecosystems: Comparative analysis of global patterns. In W. H. van Dobben and R. H. Lowe-McConnell (eds.), *Unifying Concepts in Ecology*, pp. 67–88. Dr. W. Junk, The Hague.

Likens, F. H., N. Bormann, N. M. Johnson, D. W. Fisher, and R. S. Pierce. 1970. The effects of forest cutting and herbivore treatment on nutrient budgets in the Hubbard Brook Watershed-Ecosystem. *Ecological Monograph* 40: 23–47.

Likens, G. E., F. H. Bormann, and H. M. Johnson. 1981. Interactions between major biogeochemical cycles in terrestrial ecosystems. In G. E. Likens (ed.) *Some perspectives of the major biogeochemical cycles*, pp. 93–120. Wiley, New York.

Lindeman, R. L. 1942. The trophic-dynamic aspect of ecology. *Ecology* 23: 399–18.

Lindow, S. E. 1985. Ecology of *Pseudomonas syringae* relevant to the field use of ice-deletion mutants constructed in vitro for plant frost control. In H. O. Halvorson, D. Pruner, and M. Rogul (eds.). *Engineered Organisms in the Environment: Scientific issues*, pp. 23–35. American Society for Microbiology, Washington, D.C.

Little, L. R., and M. A. Maun. 1996. The *"Ammophila"* problem revisited: A role for mycorthizal fungi. *Journal of Ecology* 84: 1–7

Lloyd, J. E. 1975. Aggressive mimicry in *Photuris* fireflies: Signal repertoires by femme fatales. *Science* 187: 452–53.

Lloyd, P. H., and O. A. E. Rasa. 1984. Status, reproductive success and fitness in the cape mountain zebra (*Equus zebra zebra*). *Behavioral Ecology and Sociobiology* 25: 441–20.

Loader, C., and H. Damman. 1991. Nitrogen content of food plants and vulnerability of *Pieris rapae* to natural enemies. *Ecology* 72: 1586–90.

Loehle, C. 1987. Tree life history strategies: The role of defenses. *Canadian Journal of Forest Research* 18: 209–22.

Loehle, C. 1988. Problems with the triangular model for representing plant strategies. *Ecology* 59: 284–86.

Loik, M. E., and S. P. Nobel. 1993. Freezing tolerance and water relations of *Opuntia fragilis* from Canada and the United States. *Ecology* 74: 1722–32.

Loman, J., and T. von Schantz. 1991. Birds in a farmland: More species in small than in large habitat islands. *Conservation Biology* 5: 176–188.

Losey, J. E., A. R. Ives, J. Harman, F. Ballantyre, and C. Brown. 1997. A polymorphism maintained by opposite patterns of parasitism and predation. *Nature* 388: 269–72.

Losos, J. B., S. Naeem, and R. K. Colwell. 1989. Hutchinsonian ratios and statistical power. *Evolution* 43: 1820–26.

Lotka, A. J. 1922. The stability of the normal age distribution. *Proceedings of the National Academy of Science* 8: 339–45.

Lotka, A. J. 1925. *Elements of Physical Biology*. (Reprinted, Dover Publications, New York, 1956).

Lovelock, J. 1991. *Healing Gaia: Practical medicine for the planet*. Harmony Books, New York.

Lovelock, J. E. 1979. *Gaia: A New Look at Life on Earth*. Oxford University Press, Oxford.

Loye, J., and M. Zuk (eds.). 1991. *Bird-parasite interactions*. Oxford University Press, Oxford.

Lucas, G., and H. Synge. 1978. *Red Data Book*. International Union for the Conservation of Nature and Natural Resources, Marges, Switzerland.

Luck, R. F., and H. Podoler. 1985. Competitive exclusions of *Aphytis lignanensis* by *A. melinus*: potential role of host size. *Ecology* 66: 904–13.

Luckinbill, L. S. 1973. Coexistence in laboratory populations of *Paramecium aurelia* and its predator *Didinium nasutum*. *Ecology* 59: 1320–27.

Luckinbill, L. S. 1974. The effects of space and enrichment on a predator-prey system. *Ecology* 55: 1142–47.

Ludwig, J. A., and J. F. Reynolds. 1988. *Statistical ecology: A primer on method and computing*. John Wiley, New York.

Luria, S. E., and M. Delbruck. 1943. Of bacteria, from virus sensitivity to virus resistance. *Genetics* 28: 491–511.

Lutz, W., W. Sanderson, and S. Scherbov. 1997. Doubling of world population unlikely. *Nature* 387: 803–5.

Lyell, C. 1969. *Principles of Geology*. 3 volumes. Reprint of 1830–1833 ed. Introduction by M. J. S. Rudwick. Johnson Reprint Collection, New York.

Lynch, J. M. 1990. *The rhizosphere*. John Wiley, New York.

Mabey, R. 1980. *The Common Ground: A Place for Nature in Britain's Future?* Hutchinson, London.

MacArthur, R. H. 1955. Fluctuations of animal populations, and a measure of community stability. *Ecology* 36: 538–536.

MacArthur, R. H., and J. W. MacArthur. 1961. On bird species diversity. *Ecology* 42: 594–98.

MacArthur, R. H., and E. O. Wilson. 1967. *The Theory of Island Biogeography*. Princeton University Press, Princeton, N.J.

MacArthur, R. H. 1957. On the relative abundance of bird species. *Proceedings of the National Academy of Science of the United States of America*.. 43: 293–5.

MacArthur, R. H. 1960. On the relative abundance of species. *American Naturalist* 94: 25–36.

MacArthur, R. H. 1969. Patterns of communities in the tropics. *Biological Journal of the Linnean Society* 1: 19–30.

MacArthur, R. H. 1972. *Geographical Ecology*. Harper and Row. New York.

MacArthur, R. H., and E. O. Wilson. 1963. An equilibrium theory of insular biogeography. *Evolution* 17: 373–87.

MacDonald, P.W. 1996. Dangerous liaisons and disease. *Nature* 379: 400–401.

Mace, G. M., and R. Lande. 1991. Assessing Extinction Threats: Towards a reevaluation of IUCN threatened species categories. *Conservation Biology* 5: 148–57.

MacEwen, A., and M. MacEwen. 1983. National parks: A cosmetic conservation system. In A. Warren and F. B. Goldsmith (eds.), *Conservation in Perspective*, pp. 391–409. Wiley, Chichester, England.

MacPherson, E., and C. M. Duarte. 1994. Patterns in species richness, site, and latitudinal range of east Atlantic fishes. *Ecography* 17: 242–48.

Maddux, G. D., and N. Cappuccino. 1986. Genetic determination of plant susceptibility to an herbivorous insect depends on environmental context. *Evolution* 40: 863–66.

Maehr, P. S., and G. B. Caddick. 1995. Demographics and genetic introgression in the Florida panther. *Conservation Biology* 9: 1295–98.

Magurran, A. E. 1988. *Ecological diversity and its management*. Princeton University Press, Princeton.

Mahoney, M. J. 1977. Publication prejudices: An experimental study of confirmatory bias in the peer review system. *Cognitive Therapy and Research* 1: 161–75.

Maiorana, V. C. 1978. An explanation of ecological and developmental constants. *Nature* 273: 375–77.

Mallet, J., J. T. Longino, D. Murawski, A. Murawski, and A. Simpson de Gamboa. 1987. Handling effects in *Heliconius*: Where do all the butterflies go? *Journal of Animal Ecology* 56: 377–86.

Malthus, T. R. 1798. *An Essay on the Principle of Population, as It Affects the Future Improvement of Society, with Remarks on the Speculations of Mr. Godwin, M. Condorcet and Other Writers*. J. Johnson, London.

Mandelbrot, B. B. 1983. *The Fractal Geometry of Nature*. Freeman, San Francisco.

Margalef, R. 1969. Diversity and stability: A practical proposal and a model of interdependence. *Brookhaven Symposium of Biology* 22: 25–37.

Martin, T.E. 1993. Nest predation among vegetation layers and habitat types: Revising the dogmas. *American Naturalist* 141: 897–13.

Martinat, P. J. 1987. The role of climatic variation and weather in forest insect outbreaks. In P. Barbosa and J. C. Schultz (eds.), *Insect Outbreaks*, pp. 241–268. Academic Press, San Diego.

Martinez, N. D. 1992. Constant connectance in community food webs. *American Naturalist* 139: 1208–18.

Martinez, N. D., and J. H. Lawton. 1995. Scale and food-web structure from local to global. *Oikos* 73: 148–54.

Marquis, R. J., and C. J. Whelan. 1994. Insectivorous birds increase growth of white oak through consumption of leaf-chewing insects. *Ecology* 75: 2007–14.

Marquis, R.J., and C. J. Whelan. 1996. Plant morphology and recruitment of the third tropic level: Subtle and little-recognized defenses? *Oikos* 75: 330–34.

Marvier, M. A. 1996. Parasitic plant-host interactions: Plant performance and indirect effects on parasite-feeding herbivores. *Ecology* 77: 1398–1409.

Massey, A. B. 1925. Antagonism of the walnuts (*Juglans nigra* L. and *J. cinerea* L.) in certain plant associations. *Phytopathology* 15: 773–84.

Matson, P. A., W. J. Parton, A. G. Power, and M J. Swift. 1997. Agricultural intensification and ecosystem properties. *Science* 277: 504–9.

Mattson, W. J. 1980. Herbivory in relation to plant nitrogen content. *Annual Review of Ecology and Systematics* 11: 119–161.

Mattson, W. J., and N. D. Addy. 1975. Phytophagous insects as regulators of forest primary production. *Science* 190: 515–22.

Mauffette, Y., and W. C. Oechel. 1989. Seasonal variation in leaf chemistry of the coast live oak *Quercus agrifolia* and implications for the California oak moth *Phrygaridia californica*. *Oecologia* 79: 439–45.

May, R. M. 1973. *Stability and Complexity in Model Ecosystems*. Princeton University Press, Princeton, N.J.

May, R. M. 1974. General Introduction. In M. B. Usher and M. H. Williamson (eds.) *Ecological Stability*, pp. 1–14. Chapman and Hall, London.

May, R. M. 1975. Patterns of species abundance and diversity. In M. L. Cody and J. Diamond (ed.), *Ecology and Evolution of Communities*, pp. 81–120. Harvard University Press, Cambridge, Mass.

May, R. M. 1976. Models for two interacting populations. In R. M. May (ed.), *Theoretical Ecology: Principles and Applications*, pp. 49–70. Sanders, Philadelphia.

May, R. M. 1977. Food lost to pests. *Nature* 267: 669–70.

May, R. M. 1978. The dynamics and diversity of insect faunas. In L. A. Mound and N. Waloff (eds.), *Diversity of Insect Faunas*, pp. 188–204. Blackwell Scientific Publications, Oxford.

May, R. M. 1979. Fluctuations in abundance of tropical insects. *Nature* 278: 505–7.

May, R. M. 1981. Population biology of parasitic infections. In K. S. Warren and E. F. Purcell (eds.), *The Current Status and Future of Parasitology*, pp. 208–235. Josiah Macy Jr. Foundation, New York.

May, R. M. 1983. Parasitic infections as regulators of animal populations. *American Scientist* 71: 36–45.

May, R. M. 1986. How many species are there? *Nature* 324: 514–15.

May, R. M. 1988. How many species are there on earth? *Science* 241: 1441–49.

May, R. M. 1990. Taxonomy as destiny. *Nature* 347: 129–30.

May, R. M., and R. M. Anderson. 1979. Population biology of infectious diseases. *Nature* 280: 455–61.

May, R. M., and R. H. MacArthur. 1972. Niche overlap as a function of environmental viability. *Proceedings of the National Academy of Sciences of the United States of America* 69: 1109–13.

Maynard Smith, J. 1968. *Mathematical Ideas in Biology*. Cambridge University Press, New York.

Maynard Smith, J. 1976a. Group selection. *Quarterly Review of Biology* 51: 277–83.

Maynard Smith, J. 1976b. Evolution and the theory of games. *American Scientist* 64: 41–45.

Maynard Smith, J. 1978. The ecology of sex. In J. R. Krebs and N. B. Davies (eds.), *Behavioural Ecology, an Evolutionary Approach*, pp. 159–179. Sinauer Associates, Sunderland, Mass.

Maynard Smith, J. 1979. Game theory and the evolution of behavior. *Proceedings of the Royal Society of London* B 205: 475–88.

Maynard Smith, J. 1982. *Evolution and the Theory of Games*. Cambridge University Press, Cambridge.

Mayr, E. 1942. *Systematics and the Origin of Species*. Columbia University Press, New York.

Mayr, E. 1963. *Animal Species and Evolution*. Harvard University Press, Cambridge, Mass.

Mayr, E. 1964. The nature of colonization in birds. In H. G. Baker and G. L. Stebbins (eds.) *The genetics of colonizing species*, pp. 29–43. Academic Press, New York.

McCann, T. S. 1981. Aggression and sexual activity of male southern elephant seals. *Journal of the Zoological Society of London* 195: 295–310.

McCollum, H., and A. Dobson. 1995. Detecting disease and parasite threats to endangered species and ecosystems. *Trends in Ecology and Evolution* 10: 190–94.

McCoy, E. D., and J. R. Rey. 1987. Terrestrial arthropods on northwest Florida saltmarshes: Hymenoptera (Insecta). *Florida Entomologist*, 70: 90–97.

McCune, B. N., and T. F. H. Allen. 1985. Will similar forests develop on similar sites? *Canadian Journal of Botany* 63: 367–76.

McFalls, J. A., Jr. 1991. Population: A lively introduction. *Population Bulletin* 46(2): 1–43.

McGarrahan, E. 1997. Much-studied butterfly winks out on Stanford Reserve. *Science* 275: 479–80.

McGraw, J. B., and F. S. Chapin. 1989. Competitive ability and adaptation to fertile and infertile soils in two *Eriophorum* species. *Ecology* 70: 736–49.

McIntosh, R. P. 1967. The continuum concept of vegetation. *Botanical Review* 33: 130–87.

McIntosh, R. P. 1985. *The background of ecology: concept and theory*. Cambridge University Press, Cambridge, U.K.

McIntosh, R. P. 1987. Pluralism in ecology. *Annual Review of Ecology and Systematics* 18: 321–41.

McKaye, K. R. 1983. Ecology and breeding behavior of a cichlid fish, *Cyrtocara eucinostomus*, in a large lake in Lake Malawi, Africa. *Environmental Biology of Fish* 8: 81–96.

McKaye, K. R., S. M. Louda, and J. R. Stauffer. 1990. Bower size and male reproductive success in a chichlid fish lek. *American Naturalist* 135: 597–613.

McKey, D. 1979. The distribution of secondary compounds within plants. In G. A. Rosenthal and D. H. Janzen (eds.), *Herbivores: Their Interaction with Secondary Plant Metabolites*, pp. 55–133. Academic Press, New York.

McKitrick, M. C., and R. M. Zink. 1988. Species concepts in ornithology. *Condor* 90: 1–14.

McLaren, B. E., and R. C. Peterson. 1994. Wolves, moose, and tree rings on Isle Royale. *Science* 266: 1555–58.

McNab, B. K. 1973. Energetics and the distribution of vampires. *Journal of Mammology* 54: 131–44.

McNaughton, S. J. 1986. On plants and herbivores. *American Naturalist* 128: 765–70.

McNaughton, S. J. 1988. Mineral nutrition and spatial concentrations of African ungulates. *Nature* 334: 343–45.

McNaughton, S. J., M. Oesterheld, D. A. Frank, and K. J. Williams. 1989. Ecosystem-level patterns of primary productivity and herbivory in terrestrial habitats. *Nature* 341: 142–44.

McNaughton, S. J., M. Oesterheld, D. A. Frank, and K. J. Williams. 1991. Primary and secondary production in terrestrial ecosystems. In J. Cole, G. Lovett and S. Findlay (eds.), *Comparative analyses of ecosystems: patterns, mechanisms, and theories*. pp. 120–39. Springer-Verlag, New York.

McNeely, J. A., K. R. Miller, W. V. Reid., R. A. Muttermeir and T. B. Werner. 1990. *Conserving the World's Biological Diversity*. IUCN, Gland, Switzerland.

McQueen, D. J., M. R. S. Johannes, J. R. Post, T. J. Stewart, and D. R. S. Lean 1989. Bottom-up and top-down impacts on freshwater pelagic community structure. *Ecological Monographs* 59: 289–309.

Meffe, G. K., and C. R. Carroll. 1994. *Principles of Conservation Biology*. Sinauer Associates, Inc., Sunderland, Mass.

Mellanby, K. 1967. *Pesticides and Pollution*. Fontana, London.

Menge, B. A. 1995. Indirect effects in marine rocky intertidal interaction webs: patterns and importance. *Ecological Monographs* 65: 21–74.

Menge, B.A. 1997. Detection of direct versus indirect effects: Were experiments long enough? *The American Naturalist* 149: 801–23.

Menge, B. A., B. Daley, and P.A. Wheeler. 1995. Control of interaction strength in marine benthic communities. In G. A. Polis and K. O. Winemiller (eds.), *Food webs: Integration of patterns and dynamics*, pp. 258–274. Chapman and Hall, New York.

Menge B. A., and T. M. Farrell. 1989. Community structure and interaction webs in shallow marine hard-bottom communities: Test of an environmental stress model. *Advances in Ecological Research* 19: 189–262.

Menge, B. A., and J. P. Sutherland. 1976. Species diversity gradients: Synthesis of the roles of predation, competition, and temporal heterogeneity. *American Naturalist* 110: 351–69.

Menge, B. A., and J. Sutherland. 1987. Community regulation: Variation in disturbance, competition, and predation in relation to environmental stress and recruitment. *American Naturalist*. 130: 730–757.

Menkinick, E. F. 1964. A comparison of some species-individuals diversity indices applied to samples of field insects. *Ecology* 45: 859–61.

Menzel, D. W., and J. H. Ryther. 1961. Nutrients limiting the production of phytoplankton in the Sargasso Sea, with special reference to iron. *Deep-Sea Research* 7: 276–81.

Meriggi, A., and S. Lovari. 1996. A review of wolf predation in Southern Europe: Does the wolf prefer wild prey to livestock? *Journal of Applied Ecology* 33: 1561–1671.

Meyer, G. A. 1993. A comparison of the impacts of leaf-and sap-feeding insects on growth and allocation of goldenrod. *Ecology* 74: 1101–16.

Milinski, M. 1979. An evolutionarily stable feeding strategy in sticklebacks. *Zeitschrift für Tierpsychologie* 51: 36–40.

Mills, K. H., S. M. Chalanchuk, L. C. Mohr, and I. J. Davies. 1987. Responses of fish populations in Lake 223 to 8 years of experimental artification. *Canadian Journal of Fisheries and Aquatic Sciences*, Series 44 (Supplement 4): 114–25.

Mills, L. S., M. E. Soule, and D. F. Doak. 1993. The Keystone-species concept in ecology and conservation. *Bioscience* 43: 219–24.

Mills, N. 1992. Parasitoid guilds, life-styles, and host ranges in the parasitoid complexes of tortricoid hosts (Lepidoptera: Tortricoidea). *Environmental Entomology* 21: 230–39.

Mills, S. 1982. On the edge of a precipice. *Birds* 9: 57–60.

Milne, A. 1961. Definition of competition among animals. *Society for Experimental Biology (Symposia)* 15: 40–61.

Milner-Gulland, E. J., and N. Leader-Williams. 1992. A model of incentives for the illegal exploitation of black rhinos and elephants: Poaching pays in Luangwa Valley, Zambia. *Journal of Applied Ecology* 29: 388–401.

Minorsky, P. V. 1985. An heuristic hypothesis of chilling injury in plants: A role for calcium as the primary physiological transducer of injury. *Plant Cell and Environment* 18: 75–94.

Mitchell, G. C. 1986. Vampire bat control in Latin America. In *Ecological Knowledge and Environmental Problem-Solving Concepts and Case Studies*, pp. 151–64. National Academy Press, Washington, D.C.

Mitchell, J. G. 1994. Uncle Sam's undeclared war against wildlife. *Wildlife Conservation* 97: 20–31.

Mittermeier, R. A. 1988. Primate diversity and the tropical forest: Case studies from Brazil and Madagascar and the importance of the megadiversity countries. In E. O. Wilson and F. M. Peter (eds.) *Biodiversity*, pp. 145–154. National Academic Press, Washington, D.C.

Mittermeier, R. A., and T. B. Werner. 1990. Wealth of plants and animals unites "megadiversity" countries. *Tropicus* 4: 1, 4–5.

Mitter, C., and D. R. Brooks. 1983. Phylogenetic aspects of coevolution. In D. J. Futuyma and M. Slatkin (eds.), *Coevolution*, pp. 65–98. Sinauer Associates, Sunderland, Mass.

Mittlebach, G. G., C. W. Osenberg, and M. A. Leibold. 1988. Trophic relations and ontogenic niche shifts in aquatic ecosystems. In B. Ebenman and L. Person (eds.). *Size Structured Populations*. pp. 219–35. Springer-Verlag, Berlin.

Mittlebach, G. W. 1988. Competition among refuging sunfishes and effects of fish density on littoral zone invertebrates. *Ecology* 61: 614–23.

Mittlebach, G. G., A. M. Turner, D. J. Hall, J. E. Rettig, and C. W. Osenberg. 1995. Pertubation and resilience: A long-term whole-lake study of predator extinction and reintroduction. *Ecology* 76: 2347–60.

Mlot, C. 1993. Predators, prey, and natural disasters attract ecologists. *Science* 261: 1115.

Mobius, K. 1877. *Die Auster- und die Austernwirtschaft*. Berlin: Verlag Von Wiegandt, Hempel and Pary. Translated by H. J. Rice, pp. 683–751, in *Report of the Commissioner for 1880*, Part VIII, U.S. Commission of Fish and Fisheries.

Moen, J., and L. Oksanen. 1991. Ecosystem trends. *Nature* 355: 510.

Moffat, A. S. 1994. Theoretical ecology: Winning its spurs in the real world. *Science* 263: 1090–92.

Moller, A. P. 1993. Ectoparasites enhance the cost of reproduction in their hosts. *Journal of Animal Ecology* 62: 304–22.

Moore, J. 1995. The behavior of parasitized animals. *BioScience* 45: 89–96.

Moore, J. A. 1961. A cellular basis for genetic isolation. In W. F. Blair (ed.), *Vertebrate Speciation*, pp. 62–68. University of Texas Press, Austin.

Moore, J. A. 1985. Science as a way of knowing—human ecology. *American Zoologist* 25: 483–637.

Mopper, S., M. Beck, D. Simberloff, and P. Stiling. 1995. Local adaptation and agents of selection in a mobile insect. *Evolution* 49: 810–15.

Mopper, S., and S. Strauss. 1997. *Genetic variation and local adaptation in natural insect populations: Effects of ecology, life history, and behavior*. Chapman and Hall, London.

Moran, M. D., T. P. Rooney, and L.E. Hurd. 1996. Top-down cascade from a bitrophic predator in an old-field community. *Ecology* 77: 2219–27.

Moran, N. A., and T. G. Whitham. 1990. Interspecific competition between root-feeding and leaf-galling aphids mediated by host-plant resistance. *Ecology* 71: 1050–58.

Moran, R. J., and W. L. Palmer. 1963. Ruffed grouse introductions and population trends on Michigan islands. *Journal of Wildlife Management* 27: 606–14.

Moran, V. C., and T. R. E. Southwood. 1982. The guild composition of arthropod communities in trees. *Journal of Animal Ecology* 51: 289–306.

Morin, P. J., and S. P. Lawler. 1995. Effects of food chain length and ominivory on population dynamics in experimental food webs. In G. A. Polis and K. O. Winemiller (eds.). *Food webs: Integration of patterns and dynamics*, pp. 218–230. Chapman and Hall, N.Y.

Morin, P. J., S. P. Lawler, and E. A. Johnson. 1988. Competition between aquatic insects and vertebrates: Interaction strength and higher order interactions. *Ecology* 69: 1401–09.

Morisita, M. 1959. Measuring of interspecific association and similarity between communities. *Memoirs of the faculty of Science of Kyushu University*. Series E (Biology) 3: 65–80.

Morrell, V. 1993. Australian pest control by virus causes concern. *Science* 261: 683–84.

Morrell, V. 1996. Genes vs. teams: Weighing group tactics in evolution. *Science* 273: 739–40.

Morrell. V. 1997. How the malaria parasite manipulates its hosts. *Science* 278: 223.

Morrin, D. J., M. M. Holland, and D. M. Lawrence. 1993. Profiles of ecologists: Results of a survey of the membership of the Ecological Society of America. Part III. Environmental science capabilities and funding. *ESA Bulletin* 74: 237–49.

Morris, R. F. 1959. Single-factor analysis in population dynamics. *Ecology* 40: 580–88.

Morris, R. F. 1963. The dynamics of epidemic spruce budworm populations. *Memoirs of the Entomological Society of Canada* 31: 1–332.

Morris, R. F., and C. A. Miller. 1954. The development of life tables for the spruce budworm. *Canadian Journal of Zoology* 32: 283–301.

Morris, W. F. 1992. The effects of natural enemies, competition, and host plant water availability on an aphid population. *Oecologia* 90: 359–65.

Morris, W. F. 1996. Mutualism denied? Nectar-robbing bumble bees do not reduce female or male success of bluebells. *Ecology* 77: 1451–62.

Morrison, C. W. 1997. The insular biogeography of small Bahamian cays. *Journal of Ecology* 85: 441–54.

Morrison, D. A., R. T. Buchney, B. J. Bewick, and G. J. Cary. 1996. Conservation conflicts over burning bush in south-eastern Australia. *Biological Conservation* 76: 167–75.

Morrison, G., M Auerbach, and E. D. McCoy. 1979. Anomalous diversity of tropical parasitoids: a general phenomenon? *The American Naturalist* 114: 303–7.

Morse, D. R., N. E. Stork, and J. H. Lawton. 1988. Species number, species abundance and body length relationships of arboreal beetles in Bornean lowland rain forest trees. *Ecological Entomology* 13: 25–37.

Morton, N. E., J. F. Crow, and H. J. Muller. 1956. An estimate of the mutational damage in man from data on consanguineous marriages. *Proceedings of the National Academy of Sciences of the United States of America* 42: 855–863.

Moss, B. 1989. *Ecology of freshwater: Man and medium*. 2d ed. Blackwell Scientific Publishers, Oxford.

Muller, C. H. 1966. The role of chemical inhibition (allelopathy) in vegetational composition. *Bulletin of the Torrey Botanical Club* 93: 332–51.

Muller, C. H. 1970. Phytotoxins as plant habitat variables. *Recent Advances in Phytochemistry* 3: 105–21.

Muller, F. 1879. *Ituna* and *Thyridis*, a remarkable case of mimicry in butterflies, translated from the German by R. Meldola. *Proceedings of the Entomological Society of London* 27: 20–29.

Mumme, R. L. 1992. Do helpers increase reproductive success? An experimental analysis in the Florida scrub jay. *Behavioral Ecology and Sociobiology* 31: 319–28.

Munroe, E. G. 1948. The geographical distribution of butterflies in the West Indies. Ph. D. Diss., Cornell University, Ithaca, N.Y.

Murdoch, W. W. 1975. Diversity, complexity, stability and past control. *Journal of Applied Ecology* 12: 795–807.

Murdoch, W. W., C. J. Briggs, and R. M. Nisbet. 1996. Competitive displacement and biological control in parasitoids: A model. *American Naturalist*. 148: 807–26.

Murie, A. 1944. *Wolves of Mount McKinley*, Fauna of National Parks, U.S. Fauna Series Number 5, Washington, D.C.

Murray, B. G., Jr. 1986. The structure of theory, and the role of competition in community dynamics. *Oikos* 46: 145–58.

Murray, B. G., Jr. 1992. Research methods in physics and biology. *Oikos* 64: 594–96.

Myers, J. H., and K. S. Williams. 1984. Does tent caterpillar attack reduce the food quality of red alder foliage? *Oecologia* 62: 74–79.

Myers, J. H., and K. S. Williams. 1987. Lack of short-or long-term inducible defenses in the red alder-western test caterpillar system. *Oikos* 48: 73–78.

Myers, J. P., P. G. Connors, and F. A. Pitelka. 1981. Optimal territory size and the sanderling: Compromises in a variable environment. In A. C. Kamil and T. D. Sargent (eds.), *Foraging Behavior, Ecological, Ethological and Psychological Approaches*, pp. 135–158. Garland STPM Press, New York.

Myers, N. 1988. Threatened biotas: `Hot spots' in tropical forests. *The Environmentalist* 8(3): 187–208.

Myers, N. 1990. The biodiversity challenges: Expanded hot-spots analysis. *The Environmentalist* 10: 243–56.

Naeem, S., K. Håkansson, J. H. Lawton, M. J. Crawley, and L. J. Thompson. 1996. Biodiversity and plant productivity in a model assemblage of plant species. *Oikos* 76: 259–64.

Naeem, S., L. J. Tompson, S. P. Lawler, J. H. Lawton, and R. M. Woodfin. 1994. Declining biodiversity can alter the performance of ecosystems. *Nature* 368: 734–37.

Naeem, S., L. J. Thompson, S. P. Lawler, J. H. Lawton, and R. M. Woodfin. 1995. Biodiversity and ecosystem functioning: Empirical evidence from experimental microcosms. *Endeavour* 19: 58–63.

Naiman, R. J., J. M. Melillo, and J. M. Hobbie. 1986. Ecosystem alteration of border forest streams by beaver (*Castor canadiensis*). *Ecology* 67: 1254–69.

National Academy of Sciences. 1984. *Science and Creationism: A View from the National Academy of Sciences.* National Academy Press, Washington, D.C.

Navarrette, S. A., and B. A. Menge. 1996. Keystone predation and interaction strength: Interactive effects of predators on their main prey. *Ecological Monographs* 66: 409–29.

Nee, S., P. H. Harvey, and R. M. May. 1991. Lifting the veil on abundance patterns. *Proceedings of the Royal Society of London*, Ser. B. 243: 161–63.

Neel, J. V. 1983. Frequency of spontaneous and induced "point" mutations in higher eukaryotes. *Journal of Heredity* 74: 2–15.

Nei, M. 1975. *Molecular population genetics and evolution.* North-Holland, Amsterdam.

Nei, M. 1983. Genetic polymorphism and the role of mutation in evolution. In M. Nei and R. K. Koehn (eds.). *Evolution of Genes and Proteins*, pp. 165–190. Sinauer Associates, Sunderland Massachusetts.

Nei, M., and D. Graur. 1984. Extent of protein polymorphism and the neutral mutation theory. *Evolutionary Biology* 17: 73–118.

Neil, M. 1983. Genetic polymorphism and the role of mutation in evolution. In M. Nei and R. K. Koehn (eds.), *Evolution of Genes and Proteins*, pp. 165–90. Sinauer Associates, Sunderland, Massachusetts.

Neill, S. R. S. J., and J. M. Cullen. 1974. Experiments on whether schooling by their prey affects the hunting behaviour of cephalopods and fish predators. *Journal of the Zoological Society of London* 172: 549–69.

Nevo, E., A. Bieles, and R. Ben-Shlomo. 1984. The evolutionary significance of genetic diversity: Ecological, demographic and life-history correlates. In G.S . Mani (ed.), *Evolutionary Dynamics of Genetic Diversity*, pp. 193–213. Springer-Velag, Berlin, Germany.

Newsome, A. 1990. The control of vertebrate pests by vertebrate predators. *Trends in Ecology and Evolution* 5: 187–91.

Newsome, A. E., I Parer, and P. C. Catling. 1989. Prolonged prey suppression by carnivores-predator-removal experiments. *Oecologia* 78: 458–67.

Newton, I., and I. Wyllie. 1992. Recovery of a sparrowhawk population in relation to declining pesticide contamination. *Journal of Applied Ecology* 29: 476–84.

Ngugi, A. W. 1988. Cultural aspects of fuelwood shortage in the Kenyan highlands. *Journal of Biogeography* 15: 165–70.

Nieminen, M. 1996. Migration of moth species in a network of small islands. *Oecologia* 108: 643–51.

Niemelä, P., J. Tuomi, R. Mannila, and P. Ojala. 1984. The effect of previous damage on the quality of Scots pine foliage as food for dipronid sawflies. *Zeitschift angewandte Entomologie* 98: 33–43.

Niklas, K. J. 1986. Large-scale changes in animal and plant terrestrial communities. In D. M. Raup and D. Jablonski (eds.), *Patterns and Processes in the History of Life*, pp. 383–405. Springer-Verlag, Berlin.

Niklas, K. J., B. H. Tiffney, and A. H. Knoll. 1980. Apparent changes in the diversity of fossil plants. *Evolutionary Biology* 12: 1–89.

Nilsson, S. G., and U. Wästljung. 1987. Seed predation and cross-pollination in mast-seeding beech (*Fagus sylvatica*) patches. *Ecology* 68: 260–65.

Nobele, I. R., and R. Dirzo. 1997. Forests as human-dominated ecosystems. *Science* 277: 522–25.

Noble, I. R. 1981. Predicting successional change. In *Fire regimes and ecosystem properties*, pp. 278–300. H. A. Mooney (ed.). U.S. Dept. Agriculture, Forest Service, General Technical Report WO-26.

Numbers, R. L. 1982. Creationism in 20th-century America. *Science* 218: 538–44.

Nunney, L., and K. A. Campbell. 1993. Assessing minimum viable population size: Demography seeks population genetics. *Trends in Ecology and Evolution* 5: 234–39.

Nye, P. H., and P. B. Tinker. 1977. *Solute movement in the soil-root system.* Blackwell Scientific Publications, Oxford.

O'Brien, S. J., M. E. Roelke, L. Marker, A. Newmann, C. A. Winkler, O. Metzler, L. Colly, J. F. Evermann, M. Bush, and D. E. Wildt. 1985. Genetic basis for species vulnerability in the cheetah. *Science* 227: 1428–34.

O'Brien, S. J., D. E. Wildt, D. Goldman, C. R. Merril, and M. Bush. 1983. The cheetah is depauperate in genetic variation. *Science* 221: 459–a62.

O'Connor, R. J. 1991. Long-term bird population studies in the U.S. *Ibis* 133, supplement 1: 30–48.

O'Riain, M. J., J. U. M. Jarvis, and C. E. Faulkes. 1996. A dispersive morph in the naked mole-rat. *Nature* 380: 619–21.

Odum, E. P. 1957. The ecosystem approach in the teaching of ecology illustrated with simple class data. *Ecology* 38: 531–35.

Odum, E. P. 1959. *Fundamentals of Ecology*, 2d ed. Saunders, Philadelphia.

Odum, E. P. 1969. The strategy of ecosystem development. *Science* 164: 262–70.

Odum, E. P. 1971. *Fundamentals of Ecology*, 3d ed. Saunders, Philadelphia.

Odum, E. P. 1997. *Ecology: The Bridge between Science and Society*. Sinauer Associates. Sunderland. M.A.

Odum, H. T., J. E. Cantlon, and L. S. Kornicher. 1960. An organization heirarchy postulate for the interpretation of species-individual distributions, species entropy, ecosystem evolution, and the meaning of a species-variety index. *Ecology* 41: 395–99.

Oksanen, L., S. D. Fretwell, J. Arruda, and P. Niemela. 1981. Exploitation ecosystems in gradients of primary productivity. *American Naturalist* 118: 240–61.

Oksanen, L., T. Oksanen, P. Ekerholm, J. Moen, P. Lundberg, M. Schneider, and M. Aunapuu, 1995. Studies and dynamics of Arctic-subarctic grazing webs in relation to primary productivity. In G. A. Polis and K. O. Winemiller (eds.), *Food webs: Integration of patterns and dynamics*, pp. 231–242. Chapman and Hall, New York.

Olff, H., J. Huisman, R. B. F. Van Tooren. 1993. Species dynamics and nutrient accumulation during early primary succession in coastal sand dunes. *Journal of Ecology* 81: 693–706.

Olson, J. S. 1958. Rates of succession and soil changes on southern Lake Michigan sand dunes. *Botanical Gazette* 119: 125–70.

Onuf, C. P. 1978. Nutritive value as a factor in plant-insect interactions with an emphasis on field studies. In G. G. Montgomery (ed.), *The Ecology of Arboreal Folivores*, pp. 85–96. Smithsonian Press, Washington, D.C.

Onuf, C. P., J. M. Teal, and I. Valiela. 1977. Interactions of nutrients, plant growth, and herbivory in a mangrove ecosystem. *Ecology* 58: 514–26.

Oostiny, H. J., and W. D. Billings. 1942. Factors affecting vegetational zonation on costal dunes. *Ecology* 23: 131–42.

Opler, P. A. 1974. Oaks as evolutionary islands for leaf-mining insects. *American Scientist* 62: 67–73.

Orians, G. H. 1969a. The number of bird species in some tropical forests. *Ecology* 50: 783–97.

Orians, G. H. 1969b. On the evolution of mating systems in birds and mammals. *American Naturalist* 103: 589–603.

Orians, G. H. 1975. Diversity, stability, and maturity in natural ecosystems. In W. H. van Dobben and R. H. Lowe-McConnell (eds.), *Unifying concepts in Ecology*, pp. 139–58. Dr. W. Junk, The Hague.

Orians, G. H., and J. F. Whittenberger. 1991. Spatial and temporal scales in habitat selection. *American Naturalist* 137: S29–S49.

Oring, L. W. 1981. Avian-mating systems. In D. S. Farner and J. R. King (eds.), *Avian Biology, Vol. 6*. Academic Press, London.

Osmond, C. H., and J. Monro. 1981. Prickly pear. In D. J. Carr and S. G. M. Carr (eds.), *Plants and Man in Australia*, pp. 194–222. Academic Press, New York.

Oster, G. F., and E. O. Wilson. 1978. *Caste and Ecology in the Social Insects*. Monographs in Population Biology No. 12. Princeton University Press, Princeton, N.J.

Osterhaus, A. D. M. E, J. Groen, P. deVries, F. G. E. M. Uytadehaag, B. Klingeborn, and R. Zarnke. 1988. Canine distemper virus in seals. *Nature* 335: 403–4.

Owen, D. F. 1961. Industrial melanism in North American moths. *American Naturalist* 95: 227–33.

Owen, D. F. 1962. The evolution of melanism in six species of North American geometrid moths. *Annals of the Entomological Society of America* 55: 699–703.

Owen, D. F. 1966. Polymorphism in Pleistocene land snails. *Science* 152: 71–72.

Owen, D. F. 1980. *Camouflage and Mimicry*. Oxford University Press, Oxford.

Owen, D. F., and J. Owen. 1974. Species diversity in temperate and tropical Ichneumonidae. *Nature* 249: 583–84.

Owen, D. F., and D. L. Whiteley. 1986. Reflexive selection: Moment's hypothesis resurrected. *Oikos* 47: 117–20.

Owen, D. F., and R. G. Wiegert. 1987. Leaf eating as mutualism. In P. Barbosa and J. C. Schultz (eds.), *Insect Outbreaks*, pp. 81–95. Academic Press, San Diego.

Owen, O. S. 1975. *Natural Resource Conservation: An Ecological Approach*, 2d ed. Macmillan, New York.

Pacala, S. W., M. P. Hassell, and R. M. May 1990. Host-paraistoid associations in patchy environments. *Nature* 344: 150–53.

Packer, C. 1977. Reciprocal altruism in *Papio anubis*. *Nature* 265: 441–43.

Packer, C., D. A. Gilbert, A. E. Pusey, and S. J. O'Brien. 1991. A molecular genetic analysis of kinship and cooperation in African lions. *Nature* 351: 562–65.

Packer, C. P. 1997. Virus hunter. *Natural History* 106(9): 36–41.

Pagel, M. D., P. H. Harvey, and H. G. J. Godfray. 1991. Species-abundance, biomass, and resource-use distributions. *American Naturalist* 138: 836–50.

Pagel, M.D., R.M. May and A.R. Collie. 1991. Ecological aspects of the geographical distribution and diversity of mammalian species. *American Naturalist* 137: 791–815.

Paige, K. N. 1992. Overcompensation in response to mammalian herbivory: from mutualistic to antagonistic interactions. *Ecology* 73: 2076–85.

Paige, K. N. 1994. Herbivory and *Ipomopsis aggregata*: Differences in response, differences in experimental protocol (a reply to Bergelson and Crawley). *American Naturalist* 143: 739–49.

Paine, R. T. 1966. Food web complexity and species diversity. *American Naturalist* 100: 65–75.

Paine, R. T. 1969. A note on trophic complexity and community stability. *American Naturalist*. 103: 91–93.

Paine, R. T. 1980. Food webs: linkage, interaction strength and community infrastructure. *Journal of Animal Ecology* 49: 667–85.

Paine, R. T. 1988. Food webs: road maps of interactions or grist for theoretical development? *Ecology* 69: 1648–54.

Paine, R. T. 1991. Between Scylla and Charybdis: Do some kinds of criticism merit a response? *Oikos* 62: 90–92.

Paine, R. T. 1992. Food-web analysis through field measurement of per capita interaction strength. *Nature* 355: 73–75.

Paine, R. T., and S. A. Levin. 1981. Intertidal landscapes: Disturbance and the dynamics of pattern. *Ecological Monographs* 51: 145–78.

Paine, R. T., J. L. Reuesnik, A. Sun, E. L. Soulanille, M. J. Worham, C. D. G. Harley, D. R. Brumbaugh, and D. L. Secord. 1996. Trouble on oiled waters: Lessons from the Exxon Valdez oil spill. *Annual Review of Ecology and Systematics* 27: 197–235.

Palmer, W. L. 1962. Ruffed grouse flight capability over water. *Journal of Wildlife Management* 26: 338–39.

Painter, R. H. 1951. *Insect resistance in crop plants*. MacMillan, N.Y.

Park, T. 1948. Experimental studies of interspecies competition. I. Competition between populations of the flour beetles *Tribolium confusum* Duval and *Tribolium castaneum* Herbst. *Ecological Monographs* 18: 265–307.

Park, T. 1954. Experimental studies of interspecies competition. II. Temperature, humidity, and competition in two species of *Tribolium*. *Physiological Zoology* 27: 177–238.

Park, T., P. H. Leslie, and D. B. Metz. 1964. Genetic strains and competition in populations of *Tribolium*. *Physiological Zoology* 37: 97–162.

Parker, C., and C. R. Riches. 1993. *Parasitic weeds of the world: Biology and control*. C.A.B. International, Wallingford, U.K.

Parker, M., and R. B. Root. 1981. Insect herbivores limit habitat distribution of a native composite *Machaeranthera canescens*. *Ecology* 62: 1390–92.

Parry, G. D. 1981. The meanings of *r*- and *K*-selection. *Oecologia* 48: 260–64.

Parsons, K. A., and A. A. de la Cruz. 1980. Energy flow and grazing behavior of canocephaline grasshoppers in a *Juncus roemerianus* marsh. *Ecology* 61: 1045–50.

Pauly, D., and V. Christensen. 1995. Primary production required to sustain global fisheries. *Nature* 374: 255–57.

Payne, I. 1987. A lake perched on piscine peril. *New Scientist* 115(1575): 50–54.

Peabody, R. R. 1931. *The common sense of drinking*. Little Brown, Boston.

Peakall, R., A. J. Beattie, and S. J. James. 1987. Pseudocopulation of an orchid by male ants: a test of two hypotheses accounting for the rarity of ant pollination. *Oecologia* 78: 522–24.

Pearl, R. 1928. *The Rate of Living*. Knopf, New York.

Pearl, R., and L. J. Reed. 1920. On the rate of growth of the population of the United States since 1790 and its mathematical representation. *Proceedings of the National Academy of Sciences of the United States of America* 6: 275–88.

Pearsall, W. H. 1954. Biology and land use in East Africa. *New Biology* 17: 9–26.

Peet, R. K. 1974. The measurement of species diversity. *Annual Review of Ecology and Systematics* 5: 285–307.

Peet, R. K., and O. L. Loucks. 1977. A gradient analysis of southern Wisconsin forests. *Ecology* 58: 485–99.

Pellmyr, O., and C. J. Huth. 1994. Evolutionary stability of mutualism between yuccas and yucca moths. *Nature* 372: 257–60.

Pellmyr, O., L. Leebens-Mack, and C. J. Huth. 1996. Non-mutualistic yucca moths and their evolutionary consequences. *Nature* 380: 155–56.

Pennings, S. C., and R. M. Callaway. 1996. Impact of a parasitic plant on the structure and dynamics of salt marsh vegetation. *Ecology* 77: 1410–19.

Persson, L. 1985. Asymetrical competition: Are larger animals competitively superior? *American Naturalist* 126: 261–66.

Petts, G. E. 1985. *Impounded Rivers: Perspectives for Ecological Management*. Wiley, Chichester, U.K.

Pianka, E. R. 1980. Guild structure in desert lizards. *Oikos* 35: 194–201.

Pianka, E. R. 1970. On *r*- and *K*-selection. *American Naturalist* 104: 592–97.

Pianka, E. R. 1976. Competition and niche theory. In R. M. May (ed.), *Theoretical Ecology: Principles and Applications*, pp. 114–141. Blackwell Scientific Publications, Oxford.

Pianka, E. R. 1986. *Ecology and natural history of desert lizards*. Princeton Press. Princeton, N.J.

Pianka, E. R. 1988. *Evolutionary Ecology*, 4th ed. Harper and Row, NY.

Pielou, E. C. 1966. The measurement of diversity in different types of biological collections. *Journal of Theoretical Biology* 13: 131–44.

Pilgram, T., and D. Western. 1986. Inferring hunting patterns on African elephants from tusks in the international ivory trade. *Journal of Applied Ecology* 23: 503–14.

Pimentel, D. 1986. Biological invasions of plants and animals in agriculture and forestry. In H. A. Mooney and J. A. Drake (eds.), *Ecology of Biological Invasions of North America and Hawaii*, pp. 149–62. Springer-Verlag, New York.

Pimentel, D. 1988. Herbivore population feeding pressure on plant hosts: Feedback evolution and host conservation. *Oikos* 53: 289–302.

Pimentel, D., D. Andow, R. Dyson-Hudson, D. Gallahan, S. Jacobson, M. Irish, G. Kroop, A. Moss, I. Schreiner, M. Shepard, T. Thompson, and B. Vinzant. 1980. Environmental and social costs of pesticides: a preliminary assessment. *Oikos* 34: 126–40.

Pimentel, D., S. A. Levins, and A. B. Soans. 1975. On the evolution of energy balance in some exploiter-victim systems. *Ecology* 56: 381–90.

Pimm, S. L. 1979. The structure of food webs. *Theoretical Population Ecology* 16: 144–58.

Pimm, S. L. 1980. Food web design and the effect of species deletion. *Oikos* 35: 139–47.

Pimm, S. L. 1982. *Food Webs*. Chapman and Hall, London.

Pimm, S. L. 1984. Food chains and return times. In D. R. Strong, D. Simberloff, L. G. Abele, and A. B. Thistle (eds.), *Ecological Communities: Conceptual Issues and the Evidence*, pp. 397–412. Princeton University Press, Princeton, N.J.

Pimm, S. L., H. L. Jones, and J. Diamond. 1988. On the risk of extinction. *American Naturalist* 132: 757–85.

Pimm, S. L., and R. L. Kitching. 1987. The determinants of food chain lengths. *Oikos* 50: 302–7.

Pimm, S. L., J. H. Lawton, and J. E. Cohen. 1991. Food web patterns and their consequences. *Nature* 350: 669–74.

Pimm, S. L., and A. Redfearn. 1988. The variability of population densities. *Nature* 334: 613–14.

Pirie, N. W. 1969. *Food Resources: Conventional and Novel*. Penguin, Harmondsworth, U.K.

Pitelka, F. A. 1964. The nutrient-recovery hypothesis for Arctic microtine cycles. I. Introduction. In D. J. Crisp (ed.), *Grazing in Terrestrial and Marine Environments*, pp. 55–56. Blackwell Scientific Publications, Oxford.

Platt, A. W. 1941. The influence of some environmental factors on the expressions of the solid stem character in certain wheat varieties. *Scientific agriculture* 22: 216–23.

Pleszczynska, W. K. 1978. Microgeographic prediction of polygyny in the lark bunting. *Species* 201: 935–37.

Podoler, H., and D. Rogers. 1975. A new method for the identification of key factors from life-table data. *Journal of Animal Ecology* 44: 85–115.

Polis, G. A. 1991. Complex trophic interactions in deserts: an empirical critique of food-web theory. *American Naturalist* 138: 123–55.

Polis, G. A., and D. R. Strong. 1996. Food web complexity and community dynamics. *American Naturalist* 147: 813–46.

Polis, G. A., and K. O. Winemiller, 1995. *Food webs: Integration of patterns and dynamics*. Chapman and Hall, N.Y.

Pollard, A. J. 1992. The importance of deterrence: Responses of grazing animals to plant variation. In *Plant resistance to herbivores and pathogens: ecology, evolution, and genetics*, pp. 216–239. R. S. Fritz and E. L. Simms (eds.). The University of Chicago Press, Chicago.

Popper, K. R. 1972a. *The Logic of Scientific Discovery*, 3d ed. Hutchinson, London.

Popper, K. R. 1972b. *Objective Knowledge: An Evolutionary Approach*. Clarendon Press, Oxford.

Porter, K. 1995. Integrating the microbial loop and the classic food chain into a realistic planktonic food web. In G. A. Polis and K. O. Winemiller (eds.), *Food webs: Integration of patterns and dynamics*, pp. 51–54. Chapman and Hall, New York

Portwood, D. 1978. *Common-sense suicide: The final right*. Dodd, Mead, New York.

Potter, D. A., and T. W. Kimmerer. 1988. Do holly leaf spines really deter herbivory? *Oecologia* 75: 216–21.

Potts, G. R., and M. J. Aebischer. 1995. Population dynamics of the grey partridge (*Perdix perdix*) 1793–1993, *Monitoring Modelling and Management* 137: 529–37.

Pound, R., and F. E. Clements. 1899. *The phytogeography of Nebraska*. Reprinted, Arno Press, New York, 1977.

Power, M. E. 1984. Depth distributions of armored catfish: Predator-induced resource avoidance? *Ecology* 65: 523–28.

Power, M. E. 1992. Top-down and bottom-up forces in food webs: Do plants have primacy? *Ecology* 73: 733–46.

Power, M. E., and L. S. Mills. 1995. The Keystone cops meet in Hilo, *Trends in Ecology and Evolution* 10: 182–84.

Power, M. E., M. S. Parker and J. T. Wooton. 1995. Disturbance and food chain length in rivers. In G. A. Polis and K. O. Winemiller (eds.), *Food webs: Integration of patterns and dynamics*, pp. 286–297. Chapman and Hall, New York.

Prescott-Allen, C., and R. Prescott-Allen. 1986. *The First Resource: Wild Species in the North American Economy*. Yale University Press, New Haven, Conn.

Prestidge, R. A., and S. McNeill. 1983. The role of nitrogen in the ecology of grassland Auchenorrhyncha. In J. A. Lee, S. McNeil, and I. H. Rorison (eds.), *Nitrogen as an Ecological Factor: 22nd Symposium of the British Ecological Society*, pp. 257–83. Blackwell Scientific Publications, Oxford.

Preston, F. W. 1948. The commonness and rarity of species. *Ecology* 29: 254–83.

Preston, F. W. 1960. Time and space and the variation of species. *Ecology* 41: 611–27.

Preston, F. W. 1962. The canonical distribution of commonness and rarity. *Ecology* 43: 185–215, 410–32.

Price, P. W. 1970. Characteristics permitting coexistence among parasitoids of a sawfly in Quebec. *Ecology* 51: 445–54.

Price, P. W. 1975. *Insect Ecology*. John Wiley, New York.

Price, P. W. 1980. *Evolutionary Biology of Parasites*. Princeton University Press, Princeton, New Jersey.

Price, P. W., C. E. Bouton, P. Gross, B. A., McPherson, J. M. Thompson, and A. E. Weiss. 1980. Interactions among three trophic levels: Influence of plants on interactions between insect herbivores and natural enemies. *Annual Review of Ecology and Systematics* 11: 41–65.

Prins, H. H. T., and F. J. Weyerhaeuser. 1987. Epidemics in populations of wild ruminants: Anthrax and impala, rinderpest, and buffalo in Lake Manyara National Park, Tanzania. *Oikos* 49: 28–38.

Prins, H. H. T., and I. Douglas-Hamilton. 1990. Stability in a multi-species assemblage of large herbivores in East Africa. *Oecologia* 83: 392–400.

Proctor, J., and S. Proctor. 1978. *Color in Plants and Flowers*. Everest, New York.

Pugnaire, F., P Haase, and J. Puigdefabregas. 1996. Facilitation between higher plant species in a semiarid environment. *Ecology* 77: 1420–26.

Pulliam, H. R., and N. M. Haddad. 1994. Human population growth and the carrying capacity concept. *Bulletin of the Ecological Society of America* 75: 141–57.

Pullin, A. S., and J. E. Gilbert. 1989. The stinging nettle, *Urtica dioica*, increases trichome density after herbivore and mechanical damage. *Oikos* 54: 275–80.

Pusey, A. E. 1987. Sex-biased dispersal and inbreeding avoidance in birds and mammals. *Trends in Ecology and Evolution* 2: 295–300.

Putman, R. J. 1994. *Community ecology*. Chapman & Hall, London.

Quinn, J. F. 1983. Mass extinctions in the fossil record. *Science* 219: 1239–40.

Quinn, J. F., and A. E. Dunham. 1983. On hypothesis testing in ecology and evolution. *American Naturalist* 122: 602–17.

Quinn, J. F., and S. P. Harrison. 1988. Effects of habitat fragmentation and isolation on species richness: Evidence from biogeographic patterns. *Oecologia* 75: 132–40.

Rabinowitz, D. 1981. Seven forms of rarity. In H. Synge (ed.), *The Biological Aspects of Rare Plant Conservation*, pp. 205–17. John Wiley, London.

Raffaelli, D. G., and S. J. Hall. 1995. Assessing the relative importance of trophic links on food webs. In G. A. Polis and K. O. Winemiller (eds.). *Food webs: Integration of patterns and dynamics*, pp. 185–91. Chapman and Hall, New York.

Raffaelli, D. G., and S. J. Hall. 1992. Compartments and predation in an estuarine food web. *Journal of Animal Ecology* 61: 551–60.

Raich, J. W., A. E. Russell, and P. M. Vitusek. 1997. Primary productivity and ecosystem development along an elevational gradient on Mauna-Loa, Hawaii. *Ecology* 78: 707–21.

Ralls, K., and J. Ballou. 1983. Extinction: Lessons from zoos. In C. M. Schonewald-Cox, S. M. Chambers, B. MacBryde, and L. Thomas (eds.), *Genetics and Conservation: A reference for managing wild animal and plant populations*, pp. 164–184. Benjamin/Cummings, Menlo Park, Cal.

Ranta, E., J. Lindstrom, V. Kaitala, H. Kokko, H. Linden, and E. Helle. 1997. Solar activity and hare dynamics: A cross-continental comparison. *American Naturalist* 149: 765–75.

Rapoport, E. 1982. *Areography: Geographical strategies of species*. Pergamon Press, NY.

Ratcliffe, D. A. 1980. *The peregrine falcon*. Buteo books, Vermillion, S.D.

Raup, D. M. 1962. Computer as aid in describing form in gastropod shells. *Science* 138: 150–52.

Raup, D. M. 1966. Geometric analysis of shell coiling: General problems. *Journal of Paleontology* 40: 1178–90.

Raup, D. M. 1978. Cohort analysis of genetic survivorship. *Paleobiology* 4: 1–15.

Raup, D. M. 1979. Biases in the fossil record of species and genera. *Bulletin of the Carnegie Museum of Natural History* 13: 85–91.

Raup, D. M. 1984. Evolutionary radiations and extinctions. In H. D. Holl and A. F. Trendall (eds.), *Patterns of Change in Earth Evolution*, pp. 5–14. Springer-Verlag, Berlin.

Raup, D. M. 1991. A kill curve for Phanerozoic marine species. *Paleobiology* 17: 37–48.

Raup, D. M., S. J. Gould, T. J. M. Schopf, and D. Simberloff. 1973. Stochastic models of phylogeny and the evolution of diversity. *Journal of Geology* 81: 525–42.

Raup, D. M., and J. J. Sepkoski Jr. 1984. Periodicities of extinctions in the geologic past. *Proceedings of the National Academy of Sciences of the United States of America* 81: 801–5.

Rausher, M. D., K. Iwao, E. L. Simms, N. Ohsaki, and D. Hall. 1993. Induced resistance in *Ipomoea purpurea*. *Ecology* 74: 20–29.

Ray, C., and A. Hastings. 1996. Density dependence: are we searching at the wrong spatial scale? *Journal of Animal Ecology* 65: 556–66.

Raynaud, D., J. Jouzel, J. M. Barnola, J. Chappellaz, R. J. Delmas, and C. Lorius. 1993. The ice-core record of greenhouse gases. *Science* 259: 926–34.

Reddinguis, J. 1996. Tests for density dependence. *Oecologia* 108: 640–42.

Reed, J. M., P. D. Doerr, and J. R. Walters. 1988. Minimum viable population size of the red-cockaded woodpecker. *Journal of Wildlife Management* 52: 385–91.

Reeve, J. D. 1988. Environmental variability, migration, and persistence in host-parasitoid systems. *The American Naturalist* 132: 810–36.

Reice, S. R. 1994. Nonequilibrium determinants of biological community structure. *American Scientist* 82: 424–35.

Reid, G. K. 1961. *Ecology of inland waters and estuaries*. Reinhold, New York.

Reiners, W. A. 1986. Complementary models for ecosystems. *American Naturalist* 127: 59–73.

Rejmaiek, M., and J.M. Randall. 1994. Invasive alien plants in California: 1993 summary and comparison with other areas in North America. *Madroño* 41: 161–77.

Rennie, P. J. 1957. The uptake of nutrients by timber forest and its importance to timber production in Britain. *Quarterly Journal of Forestry* 51: 101–15.

Resnick, D. N., F. H. Shaw, F. H. Rodd, and R. G. Shaw. 1977. Evaluation of the rate of evolution in natural populations of guppies (*Poecilia reticulata*). *Science* 275: 1934–37.

Revelle, P., and C. Revelle. 1984. *The Environment: Issues, and Choices for Society*, 2d ed. Willard Grant Press, Boston.

Rey, J. R. 1981. Ecological biogeography of arthropods on *Spartina* islands in northwest Florida. *Ecological Monographs* 51: 237–65.

Rey, J. R. 1983. Insular Ecology of salt marsh arthropods: Species-level patterns. *Journal of Biogeography* 12: 96–107.

Rey, J. R., E. D. McCoy, and D. R. Strong Jr. 1981. Herbivore pests, habitat islands, and the species-area relation. *American Naturalist* 117: 611–22.

Reynolds, S. G. 1988. Some factors of importance in the integration of pastures and cattle with coconuts (*Cocos nucifera*). *Journal of Biogeography* 15: 31–39.

Rhoades, D. F. 1979. Evolution of plant chemical defense against herbivores. In G. A. Rosenthal and D. H. Janzen (eds.), *Herbivores: Their Interaction with Secondary Plant Metabolites*, pp. 3–54. Academic Press, New York.

Rhoades, D. F., and R. G. Cates. 1976. Toward a general theory of plant antiherbivore chemistry. *Recent Advances in Phytochemistry* 10: 168–213.

Rhymer, J. M., and D. Simberloff. 1996. Extinction by hybridization and introgression. *Annual Review of Ecology and Systematics* 27: 83–109.

Ricklefs, R. E., and D. Schluter. 1993. Species diversity: Regional and historical influences. In R. E. Ricklefs and D. Schluler (eds.), *Species diversity in ecological communities*, pp. 350–64. University of Chicago Press, Chicago.

Ringwood, A. E., S. E. Kesson, N. G. Ware, W. Hibberson, and A. Major. 1979. Immobilisation of high level nuclear reactor wastes in SYNROC. *Nature* 278: 219–23.

Ritland, D. B. 1994. Variation in palatability of queen butterflies (*Danaus gilippus*) and implications regarding mimicry. *Ecology* 75: 732–46.

Ritland, D. B., and L. P. Brower. 1991. The viceroy butterfly is not a batesian mimic. *Nature* 350: 497–98.

Roberts, L. 1988. Is there life after climate change? *Science* 242: 1010–12.

Roelke-Parker, M. E., L. Munson, C. Packer, R. Kock, S. Cleaveland, M. Carpenter, S. J. O'Brien, A. Pospischil, R. Hofmann-Lehmann, H. Lutz, G. L. M. Mwamengole, M. N. Mgasa, G. A. Machange, B. A.

Summers, and M. J. G. Appel. 1996. A canine distemper virus epidemic in Serengeti lions *(Panthera leo)*, *Nature* 379: 441–45.

Rohde, K. 1982. *Ecology of marine parasites*. University of Queensland Press, St. Lucia, Queensland, Australia.

Rohde, K. 1992. Latitudinal gradients in species diversity: The search for the primary cause. *Oikos* 65: 514–27.

Rohde, K. 1997. The larger area of tropics does not explain latitudinal gradients in species diversity. *Oikos* 79: 169–72.

Rohde, K., M. Heap, and D. Heap. 1993. Rapoport's rule does not apply to marine teleosts and cannot explain latitudinal gradients in species richness. *American Naturalist* 142: 1–16.

Rojas, M. 1992. The species problem and conservation: What are we protecting? *Conservation Biology* 6: 170–78.

Roland, J. 1988. Decline of winter moth populations in North America: Direct versus indirect effect of introduced parasites. *Journal of Animal Ecology* 57: 523–31.

Rolstad, J. 1991. Consequences of forest fragmentation for the dynamics of bird populations: Conceptual tissues and the evidence. *Biological Journal of the Linnean Society* 42: 149–63.

Romeo, J. T., J. A. Saunders and P. Barbosa (eds.). 1996. Phytochemical diversity and redundancy in ecological interactions. *Recent Advances in Phytochemistry* vol. 30.

Rood, J. P. 1978. Dwarf mongoose helpers at the den. *Zeitschrift für Tierpsychologie* 48: 277–87.

Rood, J. P. 1990. Group size, survival, reproduction, and routes to breeding in dwarf mongooses. *Animal behavior* 39: 566–72.

Room, P. M., K. L. S. Harley, I. W. Forno, and P. P. D. Sands. 1981. Successful biological control of the floating weed *Salvinia*. *Nature* 294: 78–80.

Root, R. 1967. The niche exploitation pattern of the blue-gray gnatcatcher. *Ecological Monographs* 37: 317–50.

Root, R. 1973. Organization of a plant-arthropod association in simple and diverse habitats: The fauna of collards (*Brassica oleracea*). *Ecological Monographs* 43: 95–124.

Root, R. B. 1996. Herbivore pressure on goldenrods (*Solidago altissima*): Its variation and cumulative effects. *Ecology* 77: 1074–87.

Root, T. 1988. Energy constraints in avian distributions and abundances. *Ecology* 69: 330–39.

Root, T. L. 1993. Effects of global climate change on north American birds and their communities. In P. M. Kareiva, J. G. Kingsolver, and R. B. Huey (eds.). *Biotic interactions and global change*, pp. 280–92. Sinauer Associates, Sunderland, Mass.

Rosenheim, J. A., H. Kaya, L. E. Ehler, J. J. Marios and B. A. Jaffee. 1995. Intraguild predators among biological control agents: Theory and evidence. *Biological Control* 5: 303–35.

Rosenzweig, M. C. 1995. *Species diversity in space and time*. Cambridge University Press, Cambridge, U.K.

Rosenzweig, M. L. 1968. Net primary productivity of terrestrial communities: Prediction from climatological data. *American Naturalist* 102: 67–74.

Rosenzweig, M. L., and R. H. MacArthur. 1963. Graphical representation and stability conditions of predator-prey interactions. *American Naturalist* 97: 209–23.

Rosenzweig, M. L., and E. A. Sandlin. 1997. Species diversity and latitudes: Listening to area's signal. *Oikos* 80: 172–76.

Rotheray, G. E. 1981. Host searching and oviposition behavior of some parasitoids of aphidophagous Syrphidae. *Ecological Entomology* 6: 79–87.

Rotheray, G. E. 1986. Colour, shape, and defense in aphidophagous syrphid larvae (*Diptera*). *Zoological Journal of the Linnaean Society* 88: 201–16.

Rotheray, G. E., and P. Barbosa. 1984. Host-related factors affecting oviposition behavior in *Brachymeria intermedia*. *Entomologia Experimentalis et Applicata* 35: 141–45.

Roughgarden, J. 1983. Competition and theory in community ecology. *American Naturalist* 122: 583–601.

Roughgarden, J. 1989. The Structure and assembly of communities. In J. Roughgarden, R. M. May and S. A. Levin (eds.). *Perspectives in Ecological Theory*, pp. 203–226. Princeton University Press, Princeton, N.J.

Routledge, R. D. 1979. Diversity indices: Which ones are admissible. *Journal of Theoretical Biology* 76: 503–15.

Rowe, J. S., and B. V. Barnes. 1994. Geo-ecosystems and bio-ecosystems. *ESA Bulletin* 75: 40–41.

Roy, K., D. Jablonski, and J. W. Valentine. 1994. Eastern Pacific molluscan provinces and latitudinal diversity gradient: No evidence for "Rapoport's rule." *Proceedings of the National Academy of Sciences of the United States of America* 91: 8871–74.

Royal Society Study Group. 1983. *The nitrogen cycle of the United Kingdom*. The Royal Society, London.

Royama, T. 1996. A fundamental problem in key factor analysis. *Ecology* 77: 87–93.

Rubenstein, D. I., and R. W. Wrangham (eds.). 1986. *Ecological Aspects of Social Evolution: Birds and Mammals*. Princeton University Press, Princeton, N.J.

Ruohomaki, K., F. S. Chapin III, E. Haukioja, S. Neuvonen, and J. Suomela. 1996. Delayed inducible resistance in mountain birch in response to fertilization and shade. *Ecology* 77: 2302–11.

Ryan, F. J. 1955. Spontaneous mutation in non-dividing bacteria. *Genetics* 40: 726–38.

Ryther, J. H. 1963. Geographic variation in productivity. In M. N. Hill (ed.), *The Sea*, vol. 2, 2d ed., pp. 347–80. Wiley-Interscience, New York.

Saffo, M. B. 1992. Coming to terms with a field: words and concepts in symbiosis. *Symbiosis* 14: 17–31.

Sailer, R. I. 1983. History of insect introductions. In C. Graham and C. Wilson (eds.), *Exotic Plant Pests and North American Agriculture*, pp. 15–38. Academic Press, New York.

Sala, O. S., W. J. Parton, L. A. Joyce, and W. K. Lauenroth. 1988. Primary production of the central grassland region of the United States. *Ecology* 69: 40–45.

Salt, G. 1970. *The cellular defense reactions of insects*. Cambridge University Press, New York.

Salt, G. W. 1967. Predation in an experimental protozoan population (*Woodruffia - Paramecium*). *Ecological Monographs* 37: 113–44.

Sanders, H. L. 1968. Marine benthic diversity: A comparative study. *American Naturalist* 102: 243–82.

Sanderson, I. T. 1945. *Living Treasure*. Viking Press, New York.

Sargent, T. D. 1968. Cryptic moths: Effects on background selection of painting the circumocular scales. *Science* 159: 100–101.

Schaffer, M. L., and F. B. H. Samson. 1985. Population size and extinction: a note on determining critical population sizes. *American Naturalist* 125: 144–52.

Scheffer, V. B. 1951. The rise and fall of a reindeer herd. *Scientific Monthly* 73: 356–62.

Scheiner, S. M. 1993. Theories, hypotheses, and statistics. In S. M. Scheiner and J. Gurevitch (eds.), *Design and analysis of ecological experiments*, pp. 1–13. Chapman and Hall, New York.

Scheiner, S. M., and J. Gurevitch. 1993. *Design and analysis of ecological experiments*. Chapman and Hall, New York.

Schemske, D. W. 1980. The evolutionary significance of extrafloral nectar production by *Costus woodsonii* (*Zingiberaceae*): An experimental analysis of ant protection. *Journal of Ecology* 68: 959–67.

Schemske, D. W. 1983. Limits to specialization and coevolution in plant-animal mutualisms. In M. H. Nitecki (ed.), *Coevolution*, pp. 67–109. University of Chicago Press, Chicago.

Schierenbeck, K. A., R. N. Mack, and R. R. Sharitz. 1994. Effects of herbivory on growth and biomass allocation in nature and introduced species of *Lonicera*. *Ecology* 75: 1661–72.

Schindler, D. W. 1974. Eutrophication and recovery in experimental lakes: Implications for lake management. *Science* 184: 397–99.

Schindler, D. W. 1977. Evolution of phosphorus limitation in lakes. *Science* 195: 260–62.

Schlesinger, W. H. 1991. *Biogeochemistry: An analysis of global change*. Academic Press, New York.

Schluter, D. 1986. Character displacement between distantly related taxa? Finches and bees in the Galapogos. *American Naturalist* 127: 95–102.

Schluter, D., and R. E. Ricklefs. 1993. Convergence and the regional component of species diversity. In R. Ricklefs and D. Schluter (eds.) *Species diversity in ecological communities*, pp. 230–40. University of Chicago Press, Chicago.

Schmida, A., and M. V. Wilson. 1985. Biological determinants of species diversity. *Journal of Biogeography* 12: 1–20.

Schmitz, O. J., and T. D. Nudds. 1994. Parasite-mediated competition in deer and moose: How strong is the effect of meningeal worm on moose? *Ecological Applications* 4: 91–103.

Schoener, A. 1974. Experimental zoogeography: Colonization of marine mini-islands. *American Naturalist* 108: 715–37.

Schoener, A., E. R. Long, and J. R. DePalma. 1978. Geographic variation in artificial island colonisation curves. *Ecology* 59: 367–82.

Schoener, T. W. 1974. Resource partitioning in ecological communities. *Science* 185: 27–39.

Schoener, T. W. 1976. The species-area relation with archipelagoes: models and evidence from island land birds. *Proceedings of the XVI International Onthological Congress*: 629–42.

Schoener, T. W. 1983. Field experiments on interspecific competition. *American Naturalist* 122: 240–85.

Schoener, T. W. 1985. Some comments on Connell's and my reviews of field experiments in interspecific competition. *American Naturalist* 125: 730–40.

Schoener, T. W. 1986. Overview: kinds of ecological communities—ecology becomes pluralistic. In J. Diamond and T. J. Case (eds.), *Community Ecology*, pp. 467–479. Harper and Row, New York.

Schoener, T. W. 1989. Food webs from the small to the large. *Ecology* 70: 1559–89.

Schoener, T. W. 1993. On the relative importance of direct versus indirect effects in ecological communities. In H. Kawanabe, J. E. Cohen and K. Iwasaki (eds.), *Mutualism and Community Organization: Behavioural, Theoretical and Food-web Approaches*, pp. 365–411. Oxford University Press, New York.

Schoener, T. W., and D. A. Spiller. 1987. Effect of lizards on spider populations: manipulative reconstruction of a natural experiment. *Science* 236: 949–52.

Schoenly, K., R. A. Beaver, and T. A. Heumier. 1991. On the trophic relations of insects: a food-web approach. *The American Naturalist* 137: 597–638.

Schopf, Y. J. M. 1974. Permo-Triassic extinctions: a relation to sea-floor spreading. *Journal of Geology* 82: 129–43.

Schrader-Frechette, K. S., and E. D. McCoy. 1993. Method in ecology: Strategies for conservation. Cambridge Press, Cambridge, U.K.

Schultz, A. M. 1964. The nutrient-recovery hypothesis for Arctic microtine cycles. II. Ecosystem variables in relation to Arctic microtine cycles. In D. J. Crisp (ed.), *Grazing in Terrestrial and Marine Environments*, pp. 57–68. Blackwell Scientific Publications, Oxford.

Schultz, J. C., and I. T. Baldwin. 1982. Oak leaf quality declines in response to defoliation by gypsy moth larvae. *Science* 217: 149–51.

Schulze, E. D., and H. A. Mooney (eds.). 1993. *Biodiversity and Ecosystem Function*. Springer Verlag, New York.

Schupp, E. W. 1986. *Azteca* protection of *Cecropia*: Ant occupation benefits juvenile trees. *Oecologia* 70: 379–85.

Schutt, D. A. 1976. The effect of plant oestrogens on animal reproduction. *Endeavour* 35: 110–13.

Scott, G. R. 1970. Rinderpest. In J. W. Davis, L. H. Karstad, and D. O. Trainer (eds.), *Infectious Diseases of Wild Mammals*, pp. 20–35. Iowa State University Press, Ames, Ia.

Scriber, J. M., and F. Slansky. 1981. The nutritional ecology of immature insects. *Annual Review of Entomology* 26: 183–211.

Seghers, B. H. 1974. Schooling behaviour in the guppy *Poecilia reticulata*: An evolutionary response to predation. *Evolution* 28: 486–89.

Selander, R. K. 1976. Genic variation in natural populations. In F. J. Ayala (ed.), *Molecular Evolution*, pp. 21–45. Sinauer Associates, Sunderland, Mass.

Sepkoski, J. J., Jr. 1978. A kinetic model of Phanerozoic taxonomic diversity, I. Analysis of marine orders. *Paleobiology* 4: 223–51.

Sepkoski, J. J., Jr. 1979. A kinetic model of Phanerozoic taxonomic diversity. II. Early Phanerozoic families and multiple equilibria. *Paleobiology* 5: 222–51.

Sepkoski, J. J., Jr. 1984. A kinetic model of Phanerozoic taxonomic diversity. III. Post-Paleozoic families and mass extinctions. *Paleobiology* 10: 246–67.

Sepkoski J. J. Jr. 1992. Phylogenetic and ecologic patterns in the Phanerozoic history of marine biodiversity. In M. Eldridge (ed.). *Systematics, Ecology and the Biodiversity Crisis*, pp. 77–100. Columbia University Press, New York.

Shachak, M., C. G. Jones and Y. Granot. 1987. Herbivory in rocks and the weathering of a desert. *Science* 236: 1098–99.

Shannon, C. E., and W. Weaver. 1949. *The mathematical theory of communication*. University of Illinois Press, Urbana.

Shaver, G. R., and J. M. Melillo. 1984. Nutrient budgets of marsh plants: efficiency concepts and relation to availability. *Ecology* 68: 1491–1510.

Sherman, P. W. 1977. Nepotism and the evolution of alarm cells. *Science* 197: 1246–53.

Sibly, R. M. 1983. Optimal group size is unstable. *Animal Behavior* 31: 947–48.

Sih, A. 1982. Foraging strategies and the avoidance of predation by an aquatic insect, *Notoneeta hoffmani*. *Ecology* 68: 786–96.

Sih, A. 1991. Reflections on the power of a grand paradigm. *Bulletin of the Ecological Society of America* 72: 174–78.

Sih, A., P. Crawley, M. McPeek, J. Petranka, and K. Strohmeir. 1985. Predation, competition, and prey communities: A review of field experiments. *Annual Review of Ecology and Systematics* 16: 269–311.

Silvertown, J. W. 1980. The evolutionary ecology of mast seeding in reeds. *Biological Journal of the Linnaean Society* 14: 235–50.

Silvertown, J. W., M. Franco, and K. McConway. 1992. A demographic interpretation of Grime's Triangle. *Functional Ecology* 6: 130–36.

Silvertown, J. W., M. Franco, I. Pisanty, and A. Mendoza. 1993. Comparative planned demography: relative importance of life-cycle components to the finite rate of increase in woody and herbaceous perennials. *Journal of Ecology* 81: 465–76.

Simberloff, D. S. 1972. Properties of rarefaction diversity measures. *American Naturalist* 106: 414–15.

Simberloff, D. S. 1974. Equilibrium theory of island biogeography and ecology. *Annual Review of Ecology and Systematics* 5: 161–79.

Simberloff, D. S. 1976a. Experimental zoogeography of islands: Effects of island size. *Ecology* 57: 629–48.

Simberloff, D. S. 1976b. Species turnover and equilibrium island biogeography. *Science* 194: 572–78.

Simberloff, D. 1978a. Colonization of islands by insects: Immigration, extinction, and diversity. In L. A. Mound and N. Waloff (eds.), *Diversity of Insect Faunas*, pp. 139–53. Blackwell Scientific Publications, Oxford.

Simberloff, D. S. 1978b. Using island biographic distributions to determine if colonization is stochastic. *American Naturalist* 112: 713–26.

Simberloff, D. 1983. Competition theory, hypothesis-testing, and other community ecological buzzwords. *American Naturalist* 122: 626–35.

Simberloff, D. 1986a. Design of natural reserves. In M. B. Usher (ed.), *Wildlife Conservation Evaluation*, pp. 316–37 . Chapman and Hall, London.

Simberloff, D. 1986b. The proximate causes of extinction. In D. M. Raup and D. Jablonski (eds.), *Patterns and Processes in the History of Life*, pp. 259–76. Springer-Verlag, Berlin.

Simberloff, D. 1988. The contribution of population and community biology to conservation science. *Annual Review of Ecology and Systematics* 19: 473–512.

Simberloff, D. S., and W. J. Boecklen. 1981. Santa Rosalia reconsidered: size ratios and competition. *Evolution* 35: 1206–28.

Simberloff, D. S., B. J. Brown, and S. Lowrie. 1978. Isopod and insect root borers may benefit Florida mangroves. *Science* 201: 630–32.

Simberloff, D., and T. Dyan. 1991. The guild concept and the structure of ecological communities. *Annual Review of Ecology and Systematics* 22: 115–43.

Simberloff, D. S., J. A. Farr, J. Cox, and D. W. Mehlman. 1993. Movement corridors: conservation bargains or poor investments? *Conservation Biology* 6: 493–504.

Simberloff, D., and P. Stiling. 1996. How risky is biological control? *Ecology* 77: 1965–74.

Simberloff, D., and E. O. Wilson. 1969. Experimental zoogeography of islands: The colonization of empty islands. *Ecology* 50: 278–96.

Simberloff, D. S., and E. O. Wilson. 1970. Experimental zoogeography of islands: A two year record of recolonization. *Ecology* 51: 934–37.

Simmons, I. G. 1981. *The Ecology of Natural Resources*, 2d ed. Edward Arnold, London.

Simon, J. L. 1986. Disappearing species, deforestation, and data. *New Scientist* 110(1508): 60–63.

Simpson, G. G. 1969. Species density of North American recent mammals. *Systematic Zoology* 13: 57–73.

Simpson, G. H. 1949. Measurement of diversity. *Nature* 163: 688.

Sinclair, A. R. E. 1986. Testing multi-factor causes of population limitation: an illustration using snowshoe hares. *Oikos* 47: 360–64.

Sinclair, A. R. E. 1989. Population regulation in animals. In J. M. Cherret (ed.), *Ecological Concepts*, pp. 197–242. Blackwell Scientific Publications, Oxford.

Sinclair, A. R. E., J. M. Gosline, G. Holdsworth, C. J. Krebs, S. Boutin, J. N. M. Smith, R. Boonstra, and M. Dale. 1993. Can the solar cycle and climate synchronize the snowshoe hare cycle in Canada? Evidence from tree rings and ice cones. *American Naturalist* 141: 173–98.

Sinclair, A. R. E., and M. Norton-Griffiths. 1979. *Serengeti. Dynamics of an ecosystem*. University of Chicago Press, Chicago, Illinois.

Skinner, G. J., and J. B. Whittaker. 1981. An experimental investigation of interrelationships between the wood-ant (*Formica rufa*) and some tree-canopy herbivores. *Journal of Animal Ecology* 50: 313–26.

Sláma, K. 1969. Plants as a source of materials with insect hormone activity. *Entomologia Experimentalis et Applicata* 12: 721–28.

Slobodkin, L. B. 1986a. The role of minimalization in art and science. *The American Naturalist* 127: 257–65.

Slobodkin. L. B. 1986b. Natural philosophy rampant. *Paleobiology* 12: 111–18.

Slobodkin, L. B. 1988. Intellectual problems of applied ecology. *BioScience* 38: 337–42.

Slobodkin, L. C., F. E. Smith, and N. G. Hairston. 1967. Regulation in terrestrial ecosystems in gradients of primary productivity. *American Naturalist* 118: 240–61.

Smallwood, P. D., and W. D. Peters. 1986. Grey squirrel food preferences: The effects of tannin and fat concentration. *Ecology* 67: 168–74.

Smiley, J. 1986. Ant constancy at *Passiflora* extrafloral nectaries: Effects on caterpillar survival. *Ecology* 67: 516–21.

Smith, F. D. M., R. M. May, and P. H. Harvey. 1994. Geographical ranges of Australian mammals. *Journal of Animal Ecology* 63: 441–50.

Smith, J. N. M. 1962. Detoxification mechanisms. *Annual Review of Entomology* 7: 465–80.

Smith, J. N. M., C. J. Krebs, A. R. E. Sinclair, and R. Boonstra. 1988. Population biology of snowshoe hares. II. Interactions with winter food plants. *Journal of Animal Ecology* 57: 269–86.

Smith, K. L., Jr. 1985. Deep-sea hydrothermal vent mussels: nutritional state and distribution at the Galapagos rift. *Ecology* 66: 1067–80.

Smith, R. A. H., and A. D. Bradshaw. 1979. The use of heavy metal tolerant plant populations for the reclamation of metalliferous wastes. *Journal of Applied Ecology* 16: 595–612.

Smith, R. L. 1974. *Ecology and Field Biology*, 2d ed. Harper and Row, New York.

Sober, E., and R. C. Lewontin. 1982. Artifact, cause and genic selection. *Philosophy of Science* 49: 157–80.

Soderstrom, T. R., and C. E. Calderon. 1971. Insect pollination in tropical rain forest grasses. *Biotropica* 3: 1–16.

Solow, A. R., and J. H. Steele. 1990. On sample size, statistical power, and the detection of density dependence. *Journal of Animal Ecology* 59: 1073–76.

Sommer, U. 1990. Phytoplankton nutrient competition: From laboratory to lake. In J. B. Grace and D. Tilman (eds.), *Perspectives on plant competition*, pp. 193–213. Academic Press, New York.

Sorensen, T. 1948. A method of establishing groups of equal amplitude in plant sociology based on similarity of species content. *Det. Kong. Danske Vidensk. Selsk. Biol. Skr* (Copenhagen) 5(4): 1–34.

Soulé, M. E. 1976. Allozyme variation: Its determinants in space and time. In F. J. Ayala (ed.), *Molecular Evolution*, pp. 60–77. Sinauer Associates, Sunderland, Mass.

Soulé, M. E. 1980. Thresholds for survival: maintaining fitness and evolutionary potential. In M. E. Soulé and B. A. Wilcox (eds.), *Conservation Biology: An Evolutionary-Ecological Perspective*, pp. 151–70. Sinauer Associates, Sunderland, Massachusetts.

Soulé, M. E., and D. Simberloff. 1986. What do genetics and ecology tell us about the design of nature reserves? *Biological Conservation* 35: 19–40.

Soulé, M. E., and B. A. Wilcox (eds.). 1980. *Conservation Biology*, Sinauer Associates, Sunderland, Mass.

Sousa, W. P. 1979. Disturbance in marine intertidal boulder fields: the nonequilibrium maintenance of species diversity. *Ecology* 60: 1225–39.

Southwick, C. H. (ed.). 1985. *Global Ecology*. Sinauer Associates, Sunderland, Mass.

Southwick, E. E. 1984 Photosynthate allocation to floral nectar: A neglected energy investment. *Ecology* 65: 1775–79.

Southwood, T. R. E. 1961. The number of species of insect associated with various trees. *Journal of Animal Ecology* 30: 1–8.

Southwood, T. R. E. 1976. Bionomic strategies and population parameters. In *Theoretical Ecology: Principles and Applications*, pp. 26–48. Saunders, Philadelphia.

Southwood, T. R. E. 1977. The relevance of population dynamic theory to pest status. In J. M. Cherrett and G. R. Sagor (eds.), *Origins of Pest, Parasite, Disease and Weed Problems*, pp. 35–54. Blackwell Scientific Publications, Oxford.

Southwood, T. R. E. 1978. *Ecological Methods with Particular Reference to the Study of Insect Populations*, 2d ed. Methuen, London.

Southwood, T. R. E. 1987. The concept and nature of the community. In J. H. R. Gee and P. S. Giller (eds.), *Organization of communities: Past and present*, pp. 3–27. Blackwell Scientific Publications.

Sparks, J. 1982. *Discovering Animal Behaviour*. BBC Publications, London.

Spear, L. 1988. A naturalist at large: The halloween mask episode. *Natural History* 97(6): 40–8.

Spencer, M., and P. H. Warren. 1996. The effects of energy input, immigration, and habitat size on food web structure: A microcosm experiment. *Oecologia* 108: 764–70.

Spencer, W. P. 1957. Genetic studies on *Drosophila mulleri*. I. Genetic analysis of a population. *Texas University Publication* 5721: 186–205.

Springer, A. 1992. Walleye Pollock: How much difference do they really make? *Fisheries Oceanography* 1: 80–96.

Stangel, P. W., M. R. Lennartz, and M. H. Smith. 1992. Genetic variation and the population structure of red-cockaded woodpeckers. *Conservation Biology* 6: 283–92.

Stanley, S. M. 1975. A theory of evolution above the species level. *Proceedings of the National Academy of Sciences of the United States of America* 72: 646–50.

Stanley, S. M. 1979. *Macroevolution: Pattern and Process*. Freeman, San Francisco.

Stark, R. W. 1964. Recent trends in forest entomology. *Annual Review of Entomology* 10: 303–24.

Steadman, D. W. 1995. Prehistoric extinctions of Pacific Island birds: Biodiversity meets zooarchaeology. *Science* 267: 1123–30.

Stearns, S. C. 1976. Life-history tactics: A review of the ideas. *Quarterly Review of Biology* 51: 3–47.

Stearns, S. C. 1977. The evolution of life history traits. *Annual Review of Ecology and Systematics* 8: 145–71.

Stearns, S. C. 1992. *The evolution of life histories*. Oxford University Press, Oxford.

Stebbins, G. L. 1950. *Variation and Evolution in Plants*. Columbia University Press, New York.

Stebbins, G. L. 1974. *Flowering Plants: Evolution Above the Species Level*. Harvard University Press, Cambridge, Mass.

Steele, J. H. 1974. *The Structure of Marine Ecosystems*. Blackwell Scientific Publications, Oxford.

Steele, J. H., and E. W. Henderson. 1984. Modeling long-term fluctuations in fish stocks. *Science* 224: 985–87.

Stenseth, N. C. 1983. Causes and consequences of dispersal in small mammals. In I. R. Swingland and P. J. Greenwood (eds.), *The Ecology of Animal Movement*, pp. 63–101. Oxford University Press, Oxford.

Stephens, D. W., and J. R. Krebs. 1986. *Foraging Theory*. Princeton University Press, Princeton.

Stern, C. 1973. *Principles of Human Genetics*, 3d ed. Freeman, San Francisco.

Sternberg, J. G., G. P. Waldbauer, and M. R. Jeffords. 1977. Batesian mimicry: Selective advantage of color pattern. *Science* 195: 681–83.

Stevens, G. C. 1989. The latitudinal gradient in geographical range: how so many species coexist in the tropics. *American Naturalist* 133: 240–56.

Stevens, G. C., and J. F. Fox. 1991. The causes of treeline. *Annual Review of Ecology and Systematics* 22: 177–91.

Stevenson-Hamilton, J. 1957. Tsetse fly and the rinderpest epidemic of 1896. *South African Journal of Science* 58: 216.

Stewart, A. J. A. 1986a. Nymphal colour/pattern polymorphism in the leafhoppers *Eupteryx urticae* (F.) and *E. cyclops* Matsumara (Hemiptera: Auchenorrhyncha): Spatial and temporal variation in morph frequencies. *Biological Journal of the Linnaean Society* 27: 79–101.

Stewart, A. J. A. 1986b. The inheritance of nymphal colour/pattern polymorphism in the leafhoppers *Eupteryx urticae* (F.) and *E. cyclops* Matsumara (Hemiptera: Auchenorrhyncha). *Biological Journal of the Linnaean Society* 27: 57–77.

Stiling, P. D. 1980. Colour polymorphism in some nymphs of the genus *Eupteryx*. *Ecological Entomology* 5: 175–78.

Stiling, P. D. 1987. The frequency of density dependence in insect-host-parasitoid systems. *Ecology* 68: 844–56.

Stiling, P. D. 1988a. Eating a thin line. *Natural History* 97(2): 62–67.

Stiling, P. D. 1988b. Density-dependent processes and key factors in insect populations. *Journal of Animal Ecology* 57: 581–93.

Stiling, P. D. 1990. Calculating the establishment rates of parasitoids in classical biological control. *Bulletin of the Entomological Society of America* 36: 225–30.

Stiling, P. 1993. Why do natural enemies fail in classical biological control programs? *American Entomologist* 39: 31–37.

Stiling, P. D. 1994. What do ecologists do? *Bulletin of the Ecological Society of America* 75: 116–21.

Stiling, P. D., B. V. Brodbeck, and D. R. Strong. 1982. Foliar nitrogen and larval parasitism as determinants of leafminer distribution patterns on *Spartina alterniflora*. *Ecological Entomology* 7: 447–52.

Stiling, P., and A. M. Rossi. 1995. Coastal insect herbivore communities are affected more by local environmental conditions than by plant genotype. *Ecological Entomology* 20: 184–90.

Stiling, P., and A. M. Rossi. 1996. Complex effects of genotype and environment on insect herbivores and their natural enemies. *Ecology* 77: 2212–18.

Stiling, P., and A. M. Rossi. 1997a. Experimental manipulations of top-down and bottom-up factors in an in-trophic system. *Ecology* 78: 1602–06.

Stiling, P., and A. M. Rossi. 1997b. Deme formation in a dispersive gall-forming midge. In S. Mopper and S. Strauss (eds.), *Genetic Structure in Natural Inspect Populations: Effects of Ecology, Life History, and Behavior*, pp. 22–36. Chapman and Hall).

Stiling, P., A. M. Rossi, B. Hungate, P. Dijkstra, C.R. Hinkle, W.N. Knott III, and B. Drake. Decreased survivorship of leaf-mining insects in elevated CO_2: reduced forage quality and an indirect trophic cascade. *Ecological Applications* (in press).

Stiling, P. D., and D. Simberloff. 1989. Leaf abscission: Induced defense against pests or response to damage? *Oikos* 55: 43–49.

Stiling, P. D., and D. R. Strong. 1982. Egg density and the intensity of parasitism in *Prokelisia marginata (Homogotera: Delphacidae)*. *Ecology* 63: 1630–35.

Stiling, P. D., and D. R. Strong. 1983. Weak competition among *Spartina* stem borers by means of murder. *Ecology* 64: 770–78.

Stiling, P. D., A. Throckmorton, J. Silvanima, and D. Strong. 1991. Does scale affect the incidence of density dependence? A field test with insect parasitoids. *Ecology* 72: 2143–54

Stone, L., T. Dayan, and D. Simberloff. 1996. Community-wide assembly patterns unmasked: The importance of species' differing geographical ranges. *The American Naturalist* 148: 997–1015.

Stork, M. S., and C. J. C. Lyal. 1993. Extinction or "co-extinction" rates. *Nature* 366: 307.

Stork, N. E. 1987. Guild structure of arthropods from Bornean rain forest trees. *Ecological Entomology* 12: 69–80.

Stork, N. E. 1988. Insect diversity: facts, fiction and speculation. *Biological Journal of the Linnean Society* 35: 321–37.

Strassman, J. E. 1989. Altruism and relatedness at colony foundation in social insects. *Trends in Ecology and Evolution* 4: 371–74.

Strathmann, R. R. 1978. Progressive vacating of adaptive types during the Phanerozoic. *Evolution* 32: 907–14.

Strauss, S. Y. 1987. Direct and indirect effects of host-plant fertilization on an insect community. *Ecology* 68: 1670–78.

Strauss, S. Y. 1988. Determining the effects of herbivory using damaged plants. *Ecology* 69: 1628–30.

Strauss, S. Y. 1991. Indirect effects in community ecology: Their definition, study, and importance. *Trends in Ecology and Evolution* 6: 206–10.

Strickberger, M. W. 1986. *Genetics*, 3d ed. Macmillan, New York.

Strong, D. R. 1974a. Nonasymptotic species richness models and the insects of British trees. *Proceedings of the National Academy of Sciences of the United States of America* 71: 2766–69.

Strong, D. R. 1974b. The insects of British trees: Community equilibrium in ecological time. *Annals of the Missouri Botanical Garden* 61: 692–701.

Strong, D. R. 1979. Biogeographic dynamics of insect-host plant communities. *Annual Review of Entomology* 24: 89–119.

Strong, D. R. 1980. Null hypotheses in ecology. *Synthése* 43: 271–85.

Strong, D. R., Jr. 1983. Natural variability and the manifold mechanisms of ecological communities. *American Naturalist* 122: 636–60.

Strong, D. R. 1986. Density-vague population change. *Trends in Ecology and Evolution* 1: 39–42.

Strong, D. R. 1988. Insect host range (special feature). *Ecology* 69: 885.

Strong, D. R. 1992. Are trophic cascades all wet? Differentiation and donor control in speciose ecosystems. *Ecology* 73: 747–754.

Strong, D. R., J. H. Lawton, and T. R. E. Southwood. 1984. *Insects on Plants: Community Patterns and Mechanisms*. Blackwell Scientific Publications, Oxford.

Strong, D. R., E. D. McCoy, and J. R. Rey. 1977. Time and the number of herbivore species: the pests of sugarcane. *Ecology* 58: 167–75.

Strong, D. R., and P. Stiling. 1983. Wing dimorphism changed by experimental density manipulation in a planthopper (*Prokelisia marginata, Homoptera Delphacidae*). *Ecology* 64: 206–9.

Strong, D. R., L. Szyska, and D. Simberloff. 1979. Tests of community-wide character displacement against null hypotheses. *Evolution* 33: 897–913.

Struhsaker, T. T. 1967. Social structure among velvet monkeys (*Cercopithecus aethiops*). *Behaviour* 29: 83–121.

Stubbs, M. 1977. Density dependence in the life-cycles of animals and its importance in K- and r-strategies. *Journal of Animal Ecology* 46: 677–88.

Sugden, A. M. 1994. 100 issues of TREE. *Trends in Ecology and evolution* 9: 353–54.

Sugihara, G. 1980. Minimal Community Structure: An explanation of species abundance patterns. *American Naturalist* 116: 770–87.

Sugihara, G. 1981. $S = CA^2$, $z = 1/4$: A reply to Connor and McCoy. *American Naturalist* 117: 790–93.

Sugihara, G. 1995. From out of the blue. *Nature* 378: 559–60.

Sugihara, G., K. Schoenley, and A. Trombla. 1989. Scale invariance in food web structure. *Science* 245: 48–52.

Sunquist, F. 1988. Zeroing in on keystone species. *International Wildlife* 18(5): 18–23.

Suppe, F. 1977. *The Structure of Scientific Theories*, 2d ed. University of Illinois Press, Urbana.

Sutcliffe, J. 1977. *Plants and Temperature*. Edward Arnold, London.

Sutherland, J. P. 1974. Multiple stable points in natural communities. *American Naturalist* 108: 859–73.

Sutherland, J. P. 1981. The fouling community at Beaufort, North Carolina: A study in stability. *American Naturalist* 118: 499–579.

Sutherland, J. P., and R. H. Karlson. 1977. Development and stability of the fouling community at Beaufort, North Carolina. *Ecological Monographs* 47: 425–46.

Swift, M. J., O. W. Heal, and J. M. Anderson. 1979. *Decomposition in Terrestrial Ecosystems*. Blackwell Scientific Publications, Oxford.

Tansley, A. G. 1935. The use and abuse of vegetational concepts and terms. *Ecology* 16: 284–307.

Tapper, S.C., G.R. Potts, and M. H. Brockless. 1996. The effect of an experimental reduction in predator pressure on the breeding series and population density of grey partridges *Perdix perdix*. *Journal of Applied Ecology* 33: 965–78.

Taylor, L. R. 1978. Bates, Williams, Hutchinson: A variety of diversities. In L. A. Mount and N. Waloff (eds.), *Diversity of Insect faunas*, 9th Symposium of the Royal Entomological Society, pp. 1–18. Blackwell, Oxford.

Taylor, M. 1997. Traveling the Australian dog fence. *National Geographic* 191(4): 181–237.

Teal, J. M. 1962. Energy flow in the salt marsh ecosystem of Georgia. *Ecology* 43: 614–24.

Teeri, J. A., and L. G. Stowe. 1976. Climatic patterns and the distribution of C_4 grasses in North America. *Oecologia* 28: 1–12.

Temple, S. A. 1977. Plant-animal mutualism: Co-evolution with dodo leads to near extinction of plant. *Science* 197: 885–86.

Templeton, A. R., and B. Read. 1983. The elimination of inbreeding depression in a captive herd of Speke's gazelle. In C. M. Schonewald-Cox, S. M. Chambers, B. MacBryde, and L. Thomas (eds.), *Genetics and Conservation*, pp. 241–61. Benjamin/Cummings, Menlo Park, CA.

Terborgh, J. 1973. On the notion of favorableness in plant ecology. *American Naturalist* 107: 481–501.

Terborgh, J. 1986. Keystone plant resources in the tropical forest. In M. E. Soulé (ed). *Conservation biology: The science of scarcity and diversity*, pp. 330–44. Sinauer, Sunderland, MA.

Thomas, C. D. 1990. Fewer species. *Nature* 347: 237.

Thomas, C. D. 1995. Local extinctions, colonizations, and distributions: Habitat tracking by British butterflies. In S. Leather, A. Watt, and M. Mills (eds.), *Individuals, Populations, and Patterns in Ecology*, pp. 120–42. Blackwell, Oxford.

Thomas, C. D., D. Ng, C. M. Singer, J. L. B. Mallet, C. Parmesan, and H. L. Billington. 1987. Incorporation of a European weed into the diet of a North American herbivore. *Evolution* 41: 892–901.

Thompson, D. W. 1942. *On Growth and Form*, 2d ed. Cambridge University Press, Cambridge.

Thompson, J. N. 1988. Variation in interspecific interactions. *Annual Review of Ecology and Systematics* 19: 65–87.

Thompson, R. 1988. What's going wrong with the weather? *New Scientist* 117(1605): 65.

Thorne, E. T., and E. S. Williams. 1988. Disease and endangered species: The black-footed ferret as a recent example. *Conservation Biology* 2: 66–74.

Thornhill, N. W. (ed.). 1993. *The natural history of inbreeding and outbreeding: Theoretical and empirical perspectives*. University of Chicago Press, Chicago.

Thorpe, K. W., and P. Barbosa. 1986. Effects of consumption of high and low nicotine tobacco by *Manduca sexta* (Lepidoptera: Sphingidae) on survival of gregarious endoparasitoid *Cotesia congregata* (Hymenoptera: Braconidae). *Journal of Chemical Ecology* 12: 1329–37.

Thurston, J. M. 1969. The effect of liming and fertilizers on the botanical composition of permanent grassland and on the yield of hay. In

I. H. Rorison (ed.) *Ecological Aspects of Mineral Nutrition of Plank*, pp. 3–10. Blackwell, Oxford.

Tilman, D. 1982. *Resource competition and community structure*. Princeton University Press, Princeton, N. J.

Tilman, D. 1985. The resource ratio hypothesis of plant succession. *American Naturalist* 125: 827–52.

Tilman, D. 1987. The importance of mechanisms of interspecific competition. *American Naturalist* 129: 769–74.

Tilman, D. 1988. *Plant strategies and the dynamics and structure of plant communities*. Princeton University Press, Princeton, N.J.

Tilman, D. 1996. Biodiversity: Population versus ecosystem stability. *Ecology* 77: 350–63.

Tilman, D., and J. A. Downing. 1994. Biodiversity and stability in grasslands. *Nature* 367: 363–65.

Tilman, D., J. Knops, D. Wedin, P. Reich, M. Ritchie, and E. Siemann. 1997. The influence of functional diversity and composition on ecosystem processes. *Science* 277: 1300–02.

Tilman, D. , D. Wedin, and J. Knops. 1996. Productivity and sustainability influenced by biodiversity in grassland ecosystems. *Nature* 379: 718–20.

Tinbergen, L. 1960. The natural control of insects in pine woods. *I. Nérlandaise de Zoologie* 13: 265–343.

Tokeshi, M. 1990. Niche apportionment or random assortment: species abundance patterns revisited. *Journal of Animal Ecology* 59: 1129–46.

Toksöz, M. N. 1975. The subduction of the lithosphere. *Scientific American* 233(5): 88–98.

Tonn, W. M., and J. J. Magnuson. 1982. Patterns in the species composition and richness of fish assemblages in northern Wisconsin lakes. *Ecology* 63: 1149–60.

Toone, W. D., and M. P. Wallace. 1994. The extinction in the wild and reintroduction of the California condor (*Gymnogyps californianus*). In P. J. S. Olney, G. M. Mace, and A. T. C. Keistner (eds.), *Creative Conservation: Interactive management of wild and captive animals*, pp 411–19. Chapman & Hall, London.

Tracy, C. R., and T. L. George. 1992. On the determinants of extinction. *American Naturalist* 139: 102–22.

Travis, J. 1989. Results of the survey of the membership of the Ecological Society of America: 1987–1988. *Bulletin of the Ecological Survey of America* 70: 78–88.

Trenberth, K. E. 1997. The use and abuse of climate models. *Nature* 386: 131–33.

Trewartha, G. T. 1969. *A Geography of Population: World Patterns*. John Wiley, New York.

Trivers, R. L. 1971. The evolution of reciprocal altruism. *Quarterly Review of Biology* 46: 35–37.

Trivers, R. L., and H. Hare. 1976. Haploidiploidy and the evolution of social insects. *Science* 191: 249–63.

Trostel, K., A. R. E. Sinclair, C. J. Walters, and C. J. Krebs. 1987. Can predation cause the 10-year hare cycle? *Oecologia* 74: 185–92.

Tscharntke, T. 1989. Changes in shoot growth of *Phragmites australis* caused by the gall maker *Giraudiella inclusa* (Diptera: Cecidomyiidae). *Oikos* 54: 370–77.

Tscharntke, T. 1992. Coexistence, tritrophic interactions, and density dependence in a species-rich parasitoid community. *Journal of Animal Ecology* 61: 59–67.

Tschinkel, W. R. 1990. Personal communication with author.

Tukey, J. 1958. Bias and confidence in not quite large samples. *Annals of Mathematics and Science* 29: 614.

Turchin, P. 1990. Rarity of density dependence or population regulation with lags? *Nature* 344: 660–63.

Turesson, G. 1922. The species and variety as ecological units. *Hereditas* 3: 100–13.

Turesson, G. 1930. The selective effect of climate upon the plant species. *Hereditas* 14: 99–152.

Turner, J. R. G., C. M. Gratehouse, and C. A. Carey. 1987. Does solar energy control organic diversity? Butterflies, moths, and the British climate. *Oikos* 48: 195–205.

Turner, J. R. G., J. J. Lennon, and J. A. Lawrenson. 1988. British bird species distributions and the energy theory. *Nature* 335: 539–41.

Turner, T. 1983. Facilitation as a successional mechanism in a rocky intertidal community. *American Naturalist* 121: 729–38.

Udvardy, M. D. F. 1975. A classification of the biogeographical provinces of the world. *IUCN Occasional Paper No. 18*. IUCN, Gland, Switzerland.

Uhazy, L. S., J. C. Holmes, and J. G. Stelfox. 1973. Lungworms in the rocky mountain bighorn sheep of western Canada. *Canadian Journal of Zoology* 51: 817–24.

Underwood, A. J. 1986. What is a community? Patterns and processes. In D. M. Rays and D. Jablonski (eds.), *The history of life*, pp. 351–67. Springer-Verlag, Berlin.

U.S. Congress Office of Technology Assessment. 1987. *Technologies to Maintain Biological Diversity, OTA-F-330*. U.S. Government Printing Office, Washington, D.C.

U.S. Department of the Interior and U.S. Department of Commerce, U.S. Fish and Wildlife Service, and U.S. Bureau of the Census. 1982. *1980 National Survey of Fishing, Hunting, and Wildlife-Associated Recreation*. Washington, D.C.

Uyeda, S. 1978. *The New View of the Earth: Moving Continents and Moving Oceans*. Freeman, San Francisco.

Valentine, J. F., K. L. Heck, J. Busby, and D. Webb. 1997. Experimental evidence that herbivory increases shoot density and productivity in a subtropical turtlegrass (*Thalassia testudinum*) meadow. *Oecologia* 112: 193–200.

Van Andel, J., J. P. Bakker, and A. P. Grootjans. 1993. Mechanisms of vegetation succession: Review of concepts and perspectives. *Acta Botanica Neerlandica* 42: 413–33.

Van Blaricom, G. R., and J. A. Estes. 1988. *The community ecology of sea otters*. Springer-Verlag, New York.

van der Schalie, H. 1972. WHO project Egypt 10: A case history of a schistosomiasis control project. In M. Taghi Farvar and J. P. Milton (eds.), *The careless technology: Ecology and international development*, pp. 116–36. Natural History Press, New York.

van Hylckama, T. E. A. 1975. Water resources. In W. W. Murdoch (ed.), *Environment*, 2d ed., pp. 147–65. Sinauer Associates, Sunderland, Mass.

van Lenteren, J. C. 1980. Evaluation of control capabilities of natural enemies: does art have to become science? *Netherlands Journal of Zoology* 30: 369–81.

Van der Putten, C. Van Dijk, and B. A. M. Peters. 1993. Plant-specific soil-borne diseases contribute to succession in foredune vegetation. *Nature* 362: 53–56.

Vandermeer, J. 1980. Indirect mutualism: Variations on a theme by Stephen Levine. *American Naturalist* 114: 441–48.

Vane-Wright, R. I., C. J. Humphries, and P. H. Williams. 1991. What to protect?: Systematics and the agony of choice. *Biological Conservation* 55: 235–54.

Vanni, M. J. 1995. Nutrient transport and recycling by consumers in lake food webs: Implications for algal communities. In G.A. Polis and K. Winemiller (eds.), *Food webs: Integration of patterns and dynamics*, pp. 81–95. Chapman and Hall, NY.

Vanni, M. J., C. Luecke, J. Kitchell, Y. Allen, J. Pente, and J. Magnuson. 1990. Effects on lower trophic levels of massive fish kill mortality. *Nature* 344: 333–35.

Van Scoy, K., and K. Coale. 1994. Dumping iron in the Pacific. *New Scientist*, December 4, 1994, pp. 32–35.

Van Valen, L. M. 1973. A new evolutionary law. *Evolutionary Theory* 1: 1–30.

Van Valen, L. M. 1975. Time to ecological equilibrium. *Nature* 253: 684.

Varley, G. C. 1970. The concept of energy flow applied to a woodland community. In A. Watson (ed.), *Animal Populations in Relation to Their Food Resources*, pp. 389–405. Blackwell Scientific Publications, Oxford.

Varley, G. C. 1971. The effects of natural predators and parasites on winter moth populations in England. In *Proceedings, Tall Timbers Conference on Ecological Animal Control by Habitat Management No. 2*, pp. 103–16. Tall Timbers Research Station, Tallahassee, Florida.

Varley, G. C., G. R. Gradwell, and M. P. Hassell. 1973. *Insect Population Ecology: An Analytical Approach*. Blackwell Scientific Publications, Oxford.

Verhulst, P. F. 1838. Notice sur la loi que la population suit dans son accroissement. *Correspondence in Mathematics and Physics* 10: 113–121.

Vickery, W. L., and T. D. Nudds. 1991. Testing for density-dependent effects in sequential censuses. *Oecologia* 85: 419–23.

Vince, S. W., I. Valiela, and J. M. Teal. 1981. An experimental study of the structure of herbivorous insect communities in a salt marsh. *Ecology* 62: 1662–78.

Vitousek, P. M. J. D. Aber, R. W. Hawarth, G. E. Likens, P. A. Mattson, D. W. Schindler, W. H. Schlesinger, and D. G. Tilman. 1997a. Human alteration of the global nitrogen cycle: Sources and consequences. *Ecological Applications* 7: 737–50.

Vitousek, P. M., C.M. D'Antonio, L. L. Loope, and R. Westbrooks. 1996. Biological invasions as global environmental change. *American Scientist* 84: 468–78.

Vitousek, P. M., H. A. Mooney, J. Lubchenco, and J. M. Mellilo. 1997b. Human domination of the earth's ecosystems. *Science* 277: 494–99.

Volterra, V. 1926. Fluctuations in the abundance of a species considered mathematically. *Nature* 118: 558–60.

Vrijenhoek R. C. 1994. Genetic diversity and fitness in small populations. In V. Loeschcke, J. Tomiuk, and S. K. Jain (eds.), *Conservation Genetics*, pp. 37–53. Birkhauser, Verlag, Basel.

Vuilleumier, F. 1970. Insular biogeography in continental regions. I. The northern Andes of South America. *American Naturalist* 104: 373–88.

Waage, J. K. 1982. Sib-mating and sex ratio strategies in scelionid wasps. *Ecological Entomology* 7: 103–12.

Wade, D., J. Ewel, and R. Hotstetler. 1980. *Fire In South Florida Ecosystems*. USDA Forest Service General Technical Report SE-17, Asheville, North Carolina.

Waldbauer, G. P. 1988. Aposematism and Batesian mimicry. *Evolutionary Biology* 22: 261–86.

Walde, S. J., and W. W. Murdoch. 1988. Spatial density dependence in parasitoids. *Annual Review of Entomology* 33: 441–66.

Walker, B. H. 1992. Biodiversity and ecological redundancy. *Conservation Biology* 6: 18–23.

Walker, L. R., and F. S. Chapin III. 1987. Interactions among processes controlling successional change. *Oikos* 50: 131–35.

Walker, T. J. 1982. Sound traps for sampling male crickets (*Orthoptera: Gryllotalpidae: Scapteriscus*). *Florida Entomologist* 65: 13–25.

Wall, G., and C. Wright. 1977. *The Environmental Impact of Outdoor Recreation*. Waterloo University, Ontario. Department of Geography Publication Series No. 11.

Wallace, A. R. 1876. *The geographical distribution of animals*, vols. 1 and 2. Reprint, Hafner, New York 1962.

Wallace, J. B., and J. O'Hop. 1985. Life on a fast pad: Waterlily leaf beetle impact on water lilies. *Ecology* 66: 1534–44.

Waloff, N., and O. W. Richards. 1977. The effect of insect fauna on growth, mortality, and natality of broom, *Sorothamnus scoparius*. *Journal of Applied Ecology* 14: 787–98.

Walsh, J. 1973. The wake of the *Torrey Canyon*. In W. Jackson (ed.), *Man and the Environment*, 2d ed., pp. 60–62. William C. Brown Company, Dubuque, Ia.

Walton, S. 1981. Aswan revisited: U.S.-Egypt Nile project studies high dam's effects. *BioScience* 31: 9–13.

Ward, P., and A. Zahavi. 1973. The importance of certain assemblages of birds as "information-centres" for food finding. *Ibis* 115: 517–34.

Wardle, D. A., K. I. Bonner, and K. S. Nicholson. 1996. Biodiversity and plant litter: Experimental evidence which does not support the view that enhanced species richness improves ecosystem function. *Oikos* 79: 247–58.

Wardle, P. A., O. Zackrisson, G. Hornberg, and C. Gallet. 1997. The influence of island area on ecosystem properties. *Science* 277: 1296–99.

Waring, G. L., and N. S. Cobb. 1992. The impact of plant stress on herbivore population dynamics. In E. Bernays (ed.), *Insect-plant interactions*, pp. 168–226. vol. 4, CRC Press, Boca Raton, FL.

Waring, R. H. 1989. Ecosystems: Fluxes of matter and energy. In J. M. Cherrett (ed.), *Ecological Concepts*, pp. 17–42. Blackwell Scientific Publishers, Oxford.

Warming, J.G.B. 1895. *Oecology of plants: An introduction to the study of plant communities*, English version (modified), Clarendon Press, Oxford (1909).

Warren, A., and F. B. Goldsmith. 1983. An introduction to nature conservation. In A. Warren and F. B. Goldsmith (eds.), *Conservation in Perspective*, pp. 1–15. Wiley, Chichester, U.K.

Warren, M. S. 1992. The conservation of British butterflies.In R. L. H. Dennis (ed.), *The ecology of butterflies in Britain*, pp. 246–74. Oxford Science Publishers, Oxford.

Warren, S. D., H. L. Black, D. A. Eastmond, and W. H. Whaley. 1988. Structure function of buttresses of *Tachigalia versicolor*. *Ecology* 69: 532–36.

Waser, N. M., L. Chittka, M. V. Price, N. M. Williams, and J. Ollerton. 1996. Generalization in pollination systems, and why it matters. *Ecology* 77: 1043–60.

Wasserman, A. O. 1957. Factors affecting interbreeding in sympatric species of spadefoots (genus *Scaphiopus*). *Evolution* 11: 320–38.

Watson, A. 1967. Territory and population regulation in the red grouse. *Nature* 215: 1274–75.

Watson, A., R. Moss, and R. Parr. 1984. Effects of food enrichment on numbers and spacing behavior of red grouse. *Journal of Animal Ecology* 53: 663–78.

Webb, S. D. 1987. Community patterns in extinct terrestrial invertebrates. In J. H. R. Gee and P. S. Giller (eds.), *Organization of Communities: Past and Present*, pp. 439–48. Blackwell Scientific Publications, Oxford.

Weck, J., and C. Wiebecke. 1961. *Weltwirtschaft and Deutschlands Forst- und Holzwirtschaft*. Bayerischer Landwirtschaftsverlag, Munich.

Weiher, E., and P. A. Keddy. 1995. Assembly rules, null models, and trait dispersion: New questions from old patterns. *Oikos* 74: 159–64.

Weismann, A. 1893. *The Germ-Plasm: A Theory of Heredity*, English (translation). Walter Scott, London.

Werner, E. E., and J. F. Gilliam. 1984. The ontogenetic niche and species interactions in size-structured populations. *Annual Review of Ecology and Systematics* 15: 393–426.

Werren, J. H. 1983. Sex ratio evolution under local mate competition in a parasitic wasp. *Evolution* 37: 116–24.

West, C. 1985. Factors underlying the late seasonal appearance of the lepidopterous leaf-mining guild on oak. *Ecological Entomology* 10: 111–20.

West Eberhard, M. J. 1975. The evolution of social behaviour by kin selection. *Quarterly Review of Biology* 50: 1–33.
Westoby, M. 1997. What does "ecology" mean? *Trends in Ecology and Evolution* 12: 166.
Whicker, F. W., and V. Schultz. 1982. *Radioecology: Nuclear Energy and the Environment*, vols. 1 and 2. CRC Press, Boca Raton, FL.
White, G. 1789. *The natural history and antiquities of Selborne in the county of Southampton*. Reprinted, Macmillan, London, 1900.
White, G. G. 1981. Current status of prickly pear control by *Cactoblastis cactorum* in Queensland. In *Proceedings of the Fifth International Symposium for the Biological Control of Weeds, Brisbane, 1980*, pp. 609–616.
White, M. J. D. 1978. *Modes of Speciation*. Freeman, San Francisco.
White, T. C. R. 1978. The importance of relative shortage of food in animal ecology. *Oecologia* 33: 71–86.
White, T. C. R. 1984. The abundance of invertebrate herbivores in relation to the availability of nitrogen in stressed food plants. *Oecologia* 63: 90–105.
Whitehead, D. R., and C. E. Jones. 1969. Small islands and the equilibrium theory of insular biogeography. *Evolution* 23: 171–79.
Whitham, T. G. 1978. Habitat selection by *Pemphigus* aphids in response to resource limitation and competition. *Ecology* 59: 1164–76.
Whitham, T. G. 1979. Territorial behaviour of *Pemphigus* gall aphids. *Nature* 279: 324–25.
Whitham, T. G. 1980. The theory of habitat selection examined and extended using *Pemphigus* aphids. *American Naturalist* 115: 449–66.
Whitham, T. G., J. Maschiniski, K. C. Larson, and K. M. Paige. 1991. In P. W. Price, T. M. Lewinsohn, G. W. Fernandes, and W. W. Benson (eds.), *Plant-animal interaction: Evolutionary ecology in tropical and temperate regions*, pp. 227–56. John Wiley, New York.
Whittaker, R. H. 1953. A consideration of climax theory: The climax as a population and pattern. *Ecological Monographs* 23: 41–78.
Whittaker, R. H. 1954. The ecology of serpentine soils. I. introduction. *Ecology* 35: 258–59.
Whittaker, R. H. 1961. Experiments with radiophosphorus tracer in aquarium microcosms. *Ecological Monographs* 31: 157–88.
Whittaker, R. H. 1967. Gradient analysis of vegetation. *Biological Reviews* 42: 207–64.
Whittaker, R. H. 1970. *Communities and ecosystems*. MacMillan, London.
Whittaker, R. H. 1972. Evolution and measurement of species diversity. *Taxon* 21: 213–51.
Whittaker, R. H. 1975. *Communities and Ecosystems*, 2d ed. Macmillan, New York.
Whittaker, R. H., and G. E. Likens. 1975. The biosphere and man. In H. Lieth and R. H. Whittaker (eds.), *Primary Productivity of the Biosphere*, pp. 305–28. Springer-Verlag Ecological Studies, vol. 14, Berlin.
Whittaker, R. J., M. B. Bush, and F. K. Richards, 1989. Plant recolonization and vegetation succession on the Krakatau Islands, Indonesia. *Ecological Monographs* 59: 59–123.
Wickler, W. 1968. *Mimicry in Plants and Animals*. McGraw-Hill, New York.
Wiens, J. A. 1977. On competition and variable environments. *American Scientist* 65: 590–97.
Wiens, J. A. 1983. Avian community ecology: An iconoclastic view. In A. H. Brush and G. A. Clark (eds.), *Perspectives in ornithology*, pp. 355–403. Cambridge University Press, Cambridge.
Wiens, J. A. 1984. On understanding a non-equilibrium world: Myth and reality in community patterns and process. In D. R. Strong, D. Simberloff, L. G. Abele, and A. B. Thistle (eds.), *Ecological Communities*, pp. 439–57. Princeton Univ. Press, Princeton, N. J.
Wiens, J. A. 1985. Habitat selection in variable environments: Shrub-steppe birds. In M. Cody (ed.), *Habitat selection in birds*, pp. 227–51. Academic Press, Orlando, FL.
Wiens, J. A. 1986. Spatial scale and temporal variation in studies of shrub-steppe birds. In J. Diamond and T. J. Case (eds.), *Community Ecology*, pp. 154–72. Harper and Row, New York.
Wiens, J. A. 1991. Ecological similarity of shrub-desert avifaunas of Australia and North America. *Ecology* 72: 479–95.
Wiens, J. A., J. F. Addicott, T. J. Case, and J. Diamond. 1986. Overview: The importance of spatial and temporal scale in ecological investigations. In J. Diamond and T. J. Case (eds.), *Community Ecology*, pp. 145–53. Harper and Row, New York.
Wiens, J. A., and J. R. Rotenberry. 1980. Patterns of morphology and ecology in grassland and shrub steppe bird populations. *Ecological Monographs* 50: 287–308.
Wigley, T. M. L., P. D. Jones, and P. M. Kelly. 1980. Scenarios for a warm high-CO_2 world. *Nature* 238: 17–21.
Wigley, T. M. L., and S. C. B. Raper. 1991. Detection of the enhanced greenhouse effect on climate. In J. Jager and H. L. Gerguson (eds.), *Climate change: Science, impacts and policy*, pp. 231–42. Cambridge University Press, Cambridge.
Wilbur, H. M., and R. A. Alford. 1985. Priority effects in experimental pond communities. *Ecology* 66: 1106–14.
Wilcove, D. S., McMillan, and K. C. Winston. 1993. What exactly is an endangered species?: An analysis of the U.S. Endangered species list: 1985–1991. *Conservation Biology* 7: 87–93.
Wilcox, B. A. 1986. Extinction models and conservation. *Trends in Ecology and Evolution* 1: 46–48.
Wiley, R. H. 1973. Territoriality and non-random mating in the sage grouse, *Centrocerus urophasianus*. *Animal Behaviour Monographs* 6: 87–169.
Wilkinson, P. F., and C. C. Shank. 1977. Rutting-fight mortality among musk oxen-on Banks Island, Northwest Territories, Canada. *Animal Behaviour* 24: 756–58.
Williams, A. G., and T. G. Whitham. 1986. Premature leaf abscission: An induced plant defense against gall aphids. *Ecology* 67: 1619–27.
Williams, C. B. 1964. *Patterns in the Balance of Nature and Related Problems in Quantitative Ecology*. Academic Press, New York.
Williams, D. W., and A. M. Leibhold. 1995. Detection of delayed density dependence: Effects of autocorrelation in an exogenous factor. *Ecology* 76: 1005–8.
Williams, D. W., and A. M. Leibhold. 1997. Detection of delayed density dependence: Reply. *Ecology* 78: 320–22.
Williams, E. D. 1978. *Botanical Composition of the Park Grass Plots at Rothamstead 1856–1976. Rothamstead Experimental Station*, Harpenden, UK.
Williams, G. C. 1966. *Adaptation and Natural Selection*. Princeton University Press, Princeton, NJ.
Williams, K.S., K. G. Smith and F.M. Stephen. 1993. Emergence of 13-year periodical cicadas (*Cicadidae: Magicicada*): Phenology, mortality, and predator satiation. *Ecology* 74: 1143–82.
Williams, N. H. 1983. Floral fragrances as cues in animal behavior. In C. E. Jones and R. J. Little (eds.), *Handbook of Experimental Pollination Biology*, pp. 50–72. Van Nostrand Reinhold, New York.
Williams, P. H., K. J. Gaston, and C. J. Humphries. 1994. Do conservationists and molecular biologists value differences between organisms in the same way. *Biodiversity Letters* 2: 67–78.
Williams, P., D. Gibbons, C. Margules, A. Rebelo, C. Humphries, and R. Presey. 1996. A comparison of richness hotspots, rarity hotspots, and complementary areas for conserving diversity of British birds. *Conservation biology* 10: 155–74.
Williams, P. H., and C. J. Humphries. 1994. Biodiversity, taxonomic

relatedness, and endemism in conservation. In P. L. Forey, C. J. Humphries and R. I. Vare-Wright (eds.), *Systematics and Conservation Evaluation*, pp. 269–88. Clarendon Press, Oxford.

Williamson, M. H. 1983. The land-bird community of Stockholm: Ordination and turnovers. *Oikos* 41: 378–84.

Williamson M. 1987. Are communities ever stable? In A. J. Gray, M. J. Crawley, and P. J. Edwards (eds.), *Colonization, succession and stability*, pp. 353–71. Blackwell Scientific Publications, Oxford.

Williamson, M. 1989. The MacArthur and Wilson theory today: True but trivial. *Journal of Biogeography* 16: 3–4.

Williamson, M. H., and K. C. Brown. 1986. The analysis and modelling of British invasions. *Proceedings and Transactions of the Royal Society of London B* 314: 502–22.

Willis, E. O. 1972. Do birds flock in Hawaii, a land without predators? *California Birds* 3: 1–8.

Wilson, D. 1995. Endophyte: The evolution of a term and clarification of its use and definition. *Oikos* 73: 274–76.

Wilson, D. B. (ed.). 1983. *Did the Devil Make Darwin Do It?* Iowa State University Press, Ames, IA.

Wilson, D. S. 1997. Introduction: multilevel selection theory comes of age. *American Naturalist* 150: S1–S4.

Wilson, E. O. 1980. Caste and division of labor in leaf-cutting ants (*Hymenoptera: Formicidae: Atta*). I. The overall pattern in *A. sexdens. Behavioral Ecology and Sociobiology* 7: 143–56.

Wilson, E. O. 1985. The biological diversity crisis. *BioScience* 35: 700–6.

Wilson, E. O. 1988. The current state of biological diversity. In E. O. Wilson and F. M. Peter (eds.), *Biodiversity*, pp. 3–18. National Academy Press, Washington, D.C.

Wilson, E. O. 1992. *The diversity of life*. Belknap Press, Cambridge Mass.

Wilson, E. O., and W. L. Brown. 1953. The subspecies concept and its taxonomic applications. *Systematic Zoology* 2: 97–111.

Wilson, E. O., and E. O. Willis. 1975. Applied biogeography. In M. L. Cody and J. Diamond (eds.), *Ecology and Evolution of Communities*, pp. 522–34. Harvard University Press, Cambridge, Mass.

Wilson, J. B. 1993. Would we recognize a broken-stick community if we found one? *Oikos* 67: 181–83.

Wilson, J. B. 1995. Null models for assembly rules: The Jack Horner effect is more insidious than the Narcissus effect. *Oikos* 72: 139–44.

Wilson, J. B. 1996. The myth of constant predator: prey ratios. *Oecologia* 106: 272–76.

Wilson, J. B., and S. H. Roxbough. 1994. A demonstration of guild-based assembly rules for a plant community and determination of intrinsic guilds. *Oikos* 69: 267–76.

Wilson, J. B., I. Ullmann, and P. Bannister. 1996. Do species assemblages every recur? *Journal of Ecology* 84: 471–74.

Winemiller, K. O. 1990. Spatial and temporal variation in tropical fish trophic networks. *Ecological Monographs* 60: 331–67.

Wint, G. R. W. 1983. The effect of foliar nutrients upon the growth and feeding of a lepidopteran larva. In J. A. Lee, S. McNeill, and I. H. Rorison (eds.), *Nitrogen as an Ecological Factor*. 22nd Symposium of the British Ecological Society, pp. 301–20. Blackwell Scientific Publications, Oxford.

Winterbourn, M. J., and C. R. Townsend. 1991. Streams and rivers: One-way flow systems. In R. S. K. Barnes and K. H. Mann (eds.), *Fundamentals of aquatic ecology*, pp. 230–44. Blackwell Scientific Publishers, Oxford.

Wise, D. H. 1981. A removal experiment with darkling beetles: Lack of evidence for interspecific competition. *Ecology* 62: 727–38.

Wise, M. J., and C. F. Sacchi. 1996. Impact of two specialist insect herbivores on reproduction of horse nettle *Solanum carolinense*. *Oecologia* 108: 328–37.

Witmer, M. C. 1991. The dodo and the tambaldcoque tree: An obligate mutualism reconsidered. *Oikos* 61: 133–37.

Witz, B. W. 1989. Antipredator mechanisms in arthropods: A twenty-year literature survey. *Florida Entomologist* 73: 71–99.

Woinarski, J. C. Z., O. Price, and D. P. Faith. 1996. Application of a taxon priority system for conservation planning by selecting areas which are most distinct from environments already reserved. *Biological Conservation* 76: 147–59.

Woiwood, I. P., and I. Hanski. 1992. Patterns of density dependence in moths and aphids. *Journal of Animal Ecology* 61: 619–29.

Wolda, H. 1981. Similarity indices, sample size, and diversity. *Oecologia* 30: 296–302.

Wolda, H. 1983. "Long-term" stability of tropical insect populations. *Researches on Population Ecology*, Tokyo, Supplement No. 3: 112–26.

Wolda, H. 1986. Spatial and temporal variation in abundance in tropical animals. In S. L. Sutton, T. C. Whitmore, and A. C. Chadwick (eds.), *Tropical Rain Forest: Ecology and Management*, pp. 93–105. Blackwell Scientific Publications, Oxford.

Wolda, H., and E. Broadhead. 1985. The seasonality of Psocoptera in two tropical forests in Panama. *Journal of Animal Ecology* 54: 519–30.

Wolda, H., and B. Dennis. 1993. Density dependence tests: Are they? *Oecologia* 95: 581–91.

Wolfenden, G. E. 1975. Florida scrub jay helpers at the nest. *Auk* 92: 1–15.

Wolfenden, G. E., and J. W. Fitzpatrick. 1984. *The Florida Scrub Jay: Demography of a Cooperative Breeding Bird*. Princeton University Press, Princeton, NJ.

Wood, D. M., and R. del Moral. 1987. Mechanisms of early primary succession in sabal pine habitats on Mount St. Helens. *Ecology* 8: 780–90.

Wood, T. K., and S. I. Guttman. 1983. *Enchenopa binotata* complex: Sympatric speciation? *Science* 220: 310–12.

Woodley, J. D., E. A. Chornesky, P. A. Clifford, J. B. C. Jackson, L. S. Kaufman, N. Knowlton, J. C. Lang, M. P. Pearson, J. W. Porter, M. C. Rooney, K. W. Rylaarsdam, V. J. Tunnicliffe, C. M. Wahle, J. L. Wulft, A. S. G. Curtis, M. D. Dallmeyer, B. P. Jupp, M. A. R. Koehl, J. Neigel, and E. M. Sides. 1981. Hurricane Allen's impact on Jamaican coral reefs. *Science* 214: 749–55.

Woodwell, G. M., C. F. Warster, and P. A. Isaacson. 1967. DDT residues in an east coast estuary. *Science* 156: 821–24.

Wooten, M. C., and M. H. Smith. 1985. Large mammals are genetically less variable. *Evolution* 34: 210–12.

Wootton, J. T. 1997. Estimates and tests of per capita interction strength: Diet, abundance, and impact of intertidally foraging birds. *Ecological Monographs* 67: 45–64.

Wootton, J. T., and M. E. Power. 1993. Productivity, consumers, and the structure of a river food chain. *Proceedings of the National Academy of Science*. 90: 1384–87.

World Conservation Monitoring Centre. 1992. *Global biodiversity: status of the earth's living resources*. Chapman and Hall, London.

Worthen, W. B., S. Mayrose, and R. G. Wilson. 1994. Complex interactions between biotic and abiotic factors: Effects on mycophagous fly communities. *Oikos* 69: 277–86.

Wright, D. A. 1983. Species-energy theory: An extension of species-area theory. *Oikos* 41: 496–506.

Wright, M. 1988. Mixed blessings of the flooding in Sudan. *New Scientist* 119(1631) 44–47.

Wright, S. J. 1981. Extinction-mediated competition: The *Anolis* lizards and insectivorous birds of the West Indies. *American Naturalist* 117: 181–92.

Wulff, J. L. 1997. Mutualisms among species of coral reef sponges. *Ecology* 78: 146–59.

Wynne-Edwards, V. C. 1962. *Animal Dispersion in Relation to Social Behaviour*. Oliver and Boyd, Edinburgh.

Wynne-Edwards, V. C. 1977. Intrinsic population control and introduction. In F. J. Ebling and D. M. Stoddart (eds.), *Population Control by Social Behavior*, pp. 1–22. Institute of Biology, London.

Young, A. 1988. Agroforestry and its potential to contribute to land development in the tropics. *Journal of Biogeography* 15: 19–30.

Young, T. P. 1987. Increased thorn length in *Acacia depranolobium*—An induced response to browsing. *Oecologia* 71: 436–38.

Yule, G. U. 1949. Measurement of diversity. *Nature* 163: 688.

Zangerl, A. R., and F. A. Bazzaz. 1992. Theory and pattern in plant defense allocation. In R. S. Fritz and E. L. Simms (eds.), *Plant resistance to herbivores and pathogens: Ecology, evolution, and genetics*, pp. 383–91. Univ. of Chicago Press, Chicago.

Zar, J. H. 1984. *Biostatistical Analysis*. 2nd ed. Prentice-Hall, Englewood Cliffs, New Jersey.

Zaret, T. M., and R. T. Paine. 1973. Species introduction in a tropical lake. *Science* 182: 449–55.

Zelitch. I. 1971. *Photosynthesis, photorespiration, and plant productivity*. Academic Press, New York.

Zimmerman, B. L., and R. O. Bierregaard. 1986. Relevance of the equilibrium theory of island biogeography and species-area relations to conservation with a case from Amazonia. *Journal of Biogeography* 13: 133–43.

Zucker, W. V. 1983. Tannins: Does structure determine function? An ecological perspective. *American Naturalist* 121: 335–65.

Index